Arithmetic Operations:

$$ab + ac = a(b+c)$$

$$\frac{a}{b} + \frac{c}{d} = \frac{ad+bc}{bd}$$

$$\frac{a+b}{c} = \frac{a}{c} + \frac{b}{c}$$

$$\frac{\left(\frac{a}{b}\right)}{\left(\frac{c}{d}\right)} = \frac{ad}{bc}$$

$$a\left(\frac{b}{c}\right) = \frac{ab}{c}$$

$$\frac{a-b}{c-d} = \frac{b-a}{d-c}$$

$$\frac{ab+ac}{a} = b+c$$

$$\frac{\left(\frac{a}{b}\right)}{c} = \frac{a}{bc}$$

$$\frac{a}{\left(\frac{b}{c}\right)} = \frac{ac}{b}$$

Exponents and Radicals:

$$a^0 = 1 \ (a \neq 0)$$

$$\frac{a^x}{a^y} = a^{x-y}$$

$$\left(\frac{a}{b}\right)^x = \frac{a^x}{b^x}$$

$$\sqrt[n]{a^m} = a^{m/n} = (\sqrt[n]{a})^m$$

$$a^{-x} = \frac{1}{a^x}$$

$$(a^x)^y = a^{xy}$$

$$\sqrt{a} = a^{1/2}$$

$$\sqrt[n]{ab} = \sqrt[n]{a}\,\sqrt[n]{b}$$

$$a^x a^y = a^{x+y}$$

$$(ab)^x = a^x b^x$$

$$\sqrt[n]{a} = a^{1/n}$$

$$\sqrt[n]{\frac{a}{b}} = \frac{\sqrt[n]{a}}{\sqrt[n]{b}}$$

Algebraic Errors to Avoid:

$\dfrac{a}{x+b} \neq \dfrac{a}{x} + \dfrac{a}{b}$ (To see this error, let $a = b = x = 1$.)

$\sqrt{x^2+a^2} \neq x + a$ (To see this error, let $x = 3$ and $a = 4$.)

$a - b(x-1) \neq a - bx - b$ (Remember to distribute negative signs. The equation should be $a - b(x-1) = a - bx + b$.)

$\dfrac{\left(\frac{x}{a}\right)}{b} \neq \dfrac{bx}{a}$ (To divide fractions, invert and multiply. The equation should be

$$\frac{\frac{x}{a}}{b} = \frac{\frac{x}{a}}{\frac{b}{1}} = \left(\frac{x}{a}\right)\left(\frac{1}{b}\right) = \frac{x}{ab}.)$$

$\sqrt{-x^2+a^2} \neq -\sqrt{x^2-a^2}$ (We can't factor a negative sign outside of the square root.)

$\dfrac{a+bx}{a} \neq 1+bx$ (This is one of many examples of incorrect cancellation. The equation should be $\dfrac{a+bx}{a} = \dfrac{a}{a} + \dfrac{bx}{a} = 1 + \dfrac{bx}{a}$.)

$\dfrac{1}{x^{1/2}-x^{1/3}} \neq x^{-1/2} - x^{-1/3}$ (This error is a sophisticated version of the first error.)

$(x^2)^3 \neq x^5$ (The equation should be $(x^2)^3 = x^2 x^2 x^2 = x^6$.)

Conversion Table:

1 centimeter = 0.394 inches	1 joule = 0.738 foot-pounds	1 mile = 1.609 kilometers
1 meter = 39.370 inches	1 gram = 0.035 ounces	1 gallon = 3.785 liters
= 3.281 feet	1 kilogram = 2.205 pounds	1 pound = 4.448 newtons
1 kilometer = 0.621 miles	1 inch = 2.540 centimeters	1 foot-lb = 1.356 joules
1 liter = 0.264 gallons	1 foot = 30.480 centimeters	1 ounce = 28.350 grams
1 newton = 0.225 pounds	= 0.305 meters	1 pound = 0.454 kilograms

PRECALCULUS

PRECALCULUS

Roland E. Larson
Robert P. Hostetler

THE PENNSYLVANIA STATE UNIVERSITY
THE BEHREND COLLEGE

With the assistance of
David E. Heyd

THE PENNSYLVANIA STATE UNIVERSITY
THE BEHREND COLLEGE

D. C. Heath and Company
Lexington, Massachusetts / Toronto

PREFACE

Success in college level mathematics courses for those interested in any one of a variety of disciplines such as computer science, engineering, management, statistics, or one of the natural sciences begins with a firm understanding of algebraic and trigonometric concepts. The goal of our textbook is to further the preparation of students, who have completed two years of high school algebra, in such important areas as graphical techniques, algebraic and transcendental functions, and analytic geometry. These are some of the fundamental elements used in the calculus and other mathematical endeavors that many students pursue.

The features of our book have been designed to create a comprehensive teaching instrument that employs effective pedagogical techniques.

• *Order of Topics.*   Chapter 1 provides a quick review of the basic concepts of algebra. The algebra of polynomial and rational functions (Chapters 1 through 3) is brought to a logical conclusion with a discussion of the Fundamental Theorem of Algebra. Then, in Chapters 4 through 7 coverage is given to the transcendental functions and their applications. Chapters 8 through 10 include additional topics in algebra: systems of linear equations, matrices, sequences, and series. Analytic geometry is discussed in Chapters 11 and 12.

• *Algebra of Calculus.*   Special emphasis has been given to the *algebra of calculus.* Many examples and exercises consist of algebra problems that arise in the study of calculus. These examples are clearly identified.

• *Examples.*   The text contains over 550 examples, each carefully chosen to illustrate a particular concept or problem-solving technique. Each example is titled for quick reference and many examples include side comments (set in color) to justify or explain the steps in the solution.

• *Exercises.*   Over 3800 exercises are included that are designed to build competence, skill, and understanding. Each exercise set is graded in difficulty to allow students to gain confidence as they progress. To help prepare students for calculus, we stress a graphical approach in many sections and have included numerous graphs in the exercises.

• *Graphics.* The ability to visualize a problem is a critical part of a student's ability to solve a problem. This text includes over 900 figures.

• *Applications.* Throughout the textual material we have included numerous word problems that give students concrete ideas about the usefulness of the topics included.

• *Calculators.* Although we do not require the use of calculators in any section, techniques for calculator use are provided at appropriate places throughout the text. In addition, calculators have allowed us to include many realistic applications that are often excluded from other texts because of lengthy or tedious computations. Exercises meant to be solved with the help of a calculator are clearly indicated.

• *Supplements.* An *Instructor's Guide* by Meredythe M. Burrows is available and it includes answers to the even-numbered exercises as well as sample tests for each chapter.

We would like to thank the many people who have helped us at various stages of this project. Their encouragement, criticisms, and suggestions have been invaluable to us. The following reviewers offered many excellent ideas: Ben P. Bockstage, Broward Community College; Daniel D. Bonar, Denison University; H. Eugene Hall, DeKalb Community College; William B. Jones, University of Colorado; Jimmie D. Lawson, Louisiana State University; Jerome L. Paul, University of Cincinnati; and Shirley C. Sorensen, University of Maryland.

The mathematicians listed below completed a survey conducted by D. C. Heath and Company in 1983 which helped outline our topical coverage: Stan Adamski, University of Toledo; Daniel D. Anderson, University of Iowa; James E. Arnold, University of Wisconsin; Prem N. Bajaj, Wichita State University; Imogene C. Beckemeyer, Southern Illinois University; Bruce Blake, Clemson University; Dale E. Boye, Schoolcraft College; Sarah W. Bradsher, Danville Community College; John H. Brevit, Western Kentucky University; Milo F. Bryn, South Dakota State University; Gary G. Carlson, Brigham Young University; Louis J. Chatterley, Brigham Young University; Mary Clarke, Cerritos College; Lee G. Corbin, College of the Canyons; Milton D. Cox, Miami University; Robert G. Cromie, St. Lawrence University; Bettyann Daley, University of Delaware; Clinton O. Davis, Brevard Community College; Karen R. Dougan, University of Florida; Richard B. Duncan, Tidewater Community College; Don Duttenhoeffer, Brevard Community College; Bruce R. Ebanks, Texas Technological University; Susan L. Ehlers, St. Louis Community College; Delvis A. Fernandez, Chabot College; Leslee Francis, Brigham Young University; August J. Garver, University of Missouri; Douglas W. Hall, Michigan State University; James E. Hall, University of Wisconsin; Nancy Harbour, Brevard Community College; Ferdinand Haring, North Dakota State University; Cecilia Holt, Calhoun Community College; James Howard, Ferris State College; Don Jefferies, Orange Coast College; William B. Jones, University of Colorado; Eugene F. Krause, University of Michigan; Richard Langlie, North Hennerin Community College; Jimmie D. Lawson, Louisiana State University; John Linnen, Ferris State College; Joseph T. Mathis, William Jewell College; Walter M. Potter, University of Wisconsin; Sandra M. Powers, College of Charleston; Nancy J. Poxon, California State University; George C. Ragland, St. Louis Community College; Ralph J. Redman, University of Southern Colorado; Emilio O. Roxin, University of Rhode Island; Charles I. Sherrill, University of Colorado; Donald R. Snow, Brigham Young University; William F. Stearns, University of Maine; Don L. Stevens, Eastern Oregon State College; Warren Strickland, Del Mar College; Billy J. Taylor, Gainesville Junior College; Henry Tjoelker, California State University; Richard G. Vinson, University of Southern Alabama; Glorya Welch, Cerritos College; Dennis Weltman, North Harris County College; Larry G. Williams, Schoolcraft College; and Rey Ysais, Cerritos College.

We would also like to give special thanks to our publisher, D. C. Heath and Company, and in particular, the following people: Tom Flaherty, editorial director; Mary Lu Walsh, mathematics editor; Mary LeQuesne, developmental editor; Cathy Cantin, production editor; Nancy Blodget, designer; Carolyn Johnson, editorial assistant; and Mike O'Dea, manufacturing supervisor.

Several of our colleagues also worked on this project with us. David E. Heyd assisted us with the text and Meredythe Burrows wrote the *Instructor's Guide*. Three students helped with the computer graphics and accuracy checks: Wendy Hafenmaier, Timothy Larson, and A. David Salvia. Linda Matta spent many hours carefully typing the instructor's manual and proofreading the galleys and pageproofs. Deanna Larson had the enormous job of typing the entire manuscript.

On a personal level, we are grateful to our children for their interest and support during the three years the book was being written and produced, and to our wives, Deanna Larson and Eloise Hostetler, for their love, patience, and understanding.

If you have suggestions for improving this text, please feel free to write us. Over the past several years we have received many useful comments from both instructors and students and we value this very much.

Roland E. Larson
Robert P. Hostetler

THE LARSON AND HOSTETLER PRECALCULUS SERIES

To accommodate the different methods of teaching college algebra, trigonometry, and analytic geometry, we have prepared four volumes. These separate titles are described below.

COLLEGE ALGEBRA

A text designed for a one-term course covering standard topics such as algebraic functions, exponential and logarithmic functions, matrices, determinants, probability, sequences, and series.

TRIGONOMETRY

This text is used in a one-term course covering the trigonometric functions, exponential and logarithmic functions, and analytical geometry.

ALGEBRA AND TRIGONOMETRY

This title combines the content of the two texts mentioned above (with the exception of analytic geometry). It is comprehensive enough for two terms of courses or may be covered, with careful selection, in one term.

PRECALCULUS

With this book, students cover the algebraic and trigonometric functions, and analytic geometry in preparation for a course in calculus. This may be used in a one- or two-term course.

CONTENTS

INTRODUCTION TO CALCULATORS

This text includes some examples and exercises that make use of a scientific calculator. A calculator can assist you in both learning and applying mathematics. Moreover, a calculator can significantly extend the range of practical applications. Instructions in the use of a calculator will be given as we encounter new functions and applications. Of necessity, the instructions that we provide are somewhat general and may not agree precisely with the steps required by your calculator.

One of the basic differences in calculators is their internal hierarchy (priority) of operations. For use with this text, we recommend a calculator with the following features.

1. (At least) 8-digit display with scientific notation

2. Four arithmetic operations: $\boxed{+}$, $\boxed{-}$, $\boxed{\times}$, $\boxed{\div}$

3. Exponential keys: $\boxed{y^x}$ or $\boxed{a^x}$, $\boxed{e^x}$ or $\boxed{\text{INV}}$ $\boxed{\ln x}$

4. Natural logarithm: $\boxed{\ln x}$

5. Pi: $\boxed{\pi}$

6. Inverse, reciprocal, square root: $\boxed{\text{INV}}$, $\boxed{1/x}$, $\boxed{\sqrt{}}$

7. Trigonometric functions: $\boxed{\sin}$, $\boxed{\cos}$, $\boxed{\tan}$

8. Memory: $\boxed{M}$ or $\boxed{\text{STO}}$

9. Parentheses: $\boxed{(}$, $\boxed{)}$

10. Change sign key: $\boxed{+/-}$ (Note that this is not the subtraction key. It is used to enter negative numbers into the calculator.)

In this text, all calculator steps will be given with *algebraic logic*, that is, the calculator logic using the normal algebraic order of operations. For example, the calculation 4.69[5 + 2(6.87 − 3.042)] can be performed with the following sequence of steps:

4.69 $\boxed{\times}$ $\boxed{(}$ 5 $\boxed{+}$ 2 $\boxed{\times}$ $\boxed{(}$ 6.87 $\boxed{-}$ 3.042 $\boxed{)}$ $\boxed{)}$ $\boxed{=}$

which should yield the value 59.35664. Without parentheses, we would work from the inside out with the following sequence to obtain the same result:

6.87 $\boxed{-}$ 3.042 $\boxed{=}$ $\boxed{\times}$ 2 $\boxed{+}$ 5 $\boxed{=}$ $\boxed{\times}$ 4.69 $\boxed{=}$

When rounding off decimals, we use the following rules:

1. Determine the number of positions you wish to keep. The digit in the last position you keep is called the rounding digit, and the digit in the first position you discard is called the decision digit.

2. If the decision digit is 5 or greater (≥ 5), round up by adding 1 to the rounding digit.

3. If the decision digit is 4 or less (≤ 4), round down by leaving the rounding digit unchanged.

4. Keep the decimal point in the same place.

We cannot control the internal round-off that occurs in calculators. What does your calculator display when you compute 2 ÷ 3? Some calculators simply truncate (drop) the digits that exceed their display range (of eight digits) and display .66666666. Others have an internal round-off subroutine and display .66666667. Although the second display is more accurate, *both* of these decimal representations of $\frac{2}{3}$ contain a round-off error. One of the best ways to minimize error due to round-off is to *leave numbers in your calculator* until your calculations are complete. If you want to save a number for future use, store it in your calculator memory.

1 | REVIEW: REAL NUMBERS, THE PLANE, AND ALGEBRA

The Real Number System 1.1

In this text we will assume that you are familiar with basic algebra and will begin our study of precalculus with a look at the **real number system.** We use real numbers every day to describe quantities like age, miles per gallon, container size, population, and so on. To represent real numbers we use symbols such as

$$9, \quad -5, \quad \sqrt{2}, \quad \pi, \quad \tfrac{4}{3}, \quad 0.6666\ldots,$$
$$28.21, \quad 0, \quad \text{and} \quad \sqrt[3]{-32}$$

The set of real numbers is made up of the following five subsets:

Natural Numbers	$\{1, 2, 3, 4, \ldots\}$
Whole Numbers	$\{0, 1, 2, 3, \ldots\}$
Integers	$\{\ldots, -3, -2, -1, 0, 1, 2, 3, \ldots\}$
Rational Numbers	{all numbers of the form p/q}* or {all terminating or repeating decimals}
Irrational Numbers	{all nonrepeating, nonterminating decimals}

The Real Number Line

The model we use to represent the real number system is called the **real number line.** It consists of a horizontal line with an arbitrary point (the

---

*Rational numbers can be expressed as the ratio of two integers; that is, they can be written in the form p/q, where p and q are integers with $q \neq 0$.

origin) labeled 0. Positive units are measured to the right from the origin and negative units are measured to the left, as shown in Figure 1.1.

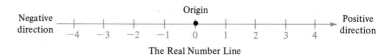

The Real Number Line

FIGURE 1.1

The importance of the real number line lies in the fact that *each point on the line corresponds to one and only one real number* and *each real number corresponds to one and only one point on the real line*. This type of relationship is called a **one-to-one correspondence.** (See Figure 1.2.)

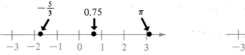
Every real number corresponds to a point on the real line.

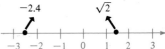
Every point on the real line corresponds to a real number.

One–to–One Correspondence

FIGURE 1.2

The number associated with a point on the real line is called the **coordinate** of the point. For example, in Figure 1.2, $-\frac{5}{3}$ is the coordinate of the left-most point, and $\sqrt{2}$ is the coordinate of the right-most point.

Ordering the Real Numbers

The real number line is useful in demonstrating the order property of real numbers. We say that the real number a **is less than** the real number b if a lies to the left of b on the real number line. Symbolically, we denote this relationship by

$a < b$ *(a is less than b)*

or equivalently

$b > a$ *(b is greater than a)*

Arithmetically, we have

$a < b$ if and only if $b - a > 0$ *(b − a is positive)*
$a > b$ if and only if $b - a < 0$ *(b − a is negative)*

Remark: In mathematics we use the phrase "if and only if" as a means of stating two implications in one sentence. For instance, the statement $a < b$ if and only if $b - a > 0$ means

if $a < b$ then $b - a > 0$ *and*   if $b - a > 0$ then $a < b$

The symbols $<$ and $>$ are referred to as **inequality signs.** They are sometimes combined with an equal sign as follows:

$$a \le b \qquad\qquad \textit{(a is less than or equal to b)}$$

$$b \ge a \qquad\qquad \textit{(b is greater than or equal to a)}$$

Inequalities are useful in denoting subsets of the real numbers. For instance,

Inequality		**Subset of Reals**
$x \le 2$	denotes	All real numbers less than or equal to 2.
$-2 \le x < 3$	denotes	All real numbers between -2 and 3, including -2 (but not including 3).
$x > -5$	denotes	All real numbers greater than -5.

Subsets of real numbers are sometimes expressed in the **interval** forms shown in Table 1.1. (We use the symbols ∞ and $-\infty$ to denote positive and negative infinity.)

TABLE 1.1
Intervals on the Real Line

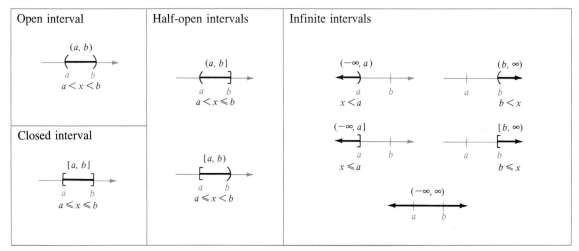

It should be clear from our discussion of order that for any two real numbers a and b *exactly one* of the following is true:

$$a = b, \qquad a < b, \qquad \text{or} \qquad a > b$$

This is referred to as the **law of trichotomy.**

We have compared two numbers using the order relations $<$ and $>$. Two numbers can also be compared using their **absolute value.** By the absolute value of a number, we mean its *magnitude (its value disregarding* *its sign)*. We denote the absolute value of *a* by $|a|$. Thus, the absolute value of -5 is

$$|-5| = 5$$

DEFINITION OF ABSOLUTE VALUE

For any real number *a*:

$$|a| = \begin{cases} a, & \text{if } a > 0 \\ -a, & \text{if } a < 0 \end{cases}$$

The absolute value of a number can never be negative, for if *a* is negative ($a < 0$), then $-a$ is actually positive. For instance, let $a = -5$, then

$$|a| = |-5| = -a = -(-5) = 5$$

Absolute value is useful in finding the distance between two numbers on the real number line. To see how this is done, consider the numbers -3 and 4, shown in Figure 1.3.

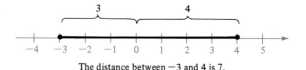

The distance between -3 and 4 is 7.

FIGURE 1.3

To find the distance between these two points, we subtract *either* number from the other and then take the absolute value of the resulting difference. Thus, using absolute value, we have

$$\text{distance} = |4 - (-3)| = |4 + 3| = |7| = 7$$

or

$$\text{distance} = |-3 - 4| = |-7| = 7$$

DISTANCE BETWEEN TWO POINTS ON THE REAL LINE

For any real numbers *a* and *b*:

the distance between *a* and *b* is

$$\text{distance} = d(a, b) = |b - a| = |a - b|$$

EXAMPLE 1
Distance and Absolute Value

Use absolute value to denote each of the following.

(a) The distance between $\sqrt{7}$ and 4

(b) The distance between *c* and -2 is at least 7

(c) The distance between x and 2.3 is less than 1

(d) x is closer to 0 than to -4

Solution:

(a) $d(\sqrt{7}, 4) = |4 - \sqrt{7}| = 4 - \sqrt{7} \approx 4 - 2.646 \approx 1.354$

(b) Since $d(c, -2) = |c + 2|$, we have

$$|c + 2| \geq 7$$

(c) Since $d(x, 2.3) = |x - 2.3|$, we have

$$|x - 2.3| < 1$$

(d) Since $d(x, 0) = |x - 0| = |x|$ and $d(x, -4) = |x - (-4)| = |x + 4|$, we have

$$|x| < |x + 4|$$

Recall that *expressions* are the main building blocks of algebra. Some examples are

$$5t - 7, \qquad \frac{x}{x + 2}, \qquad \sqrt{x - 1}, \qquad \text{and} \qquad s^2 - 3s$$

where s, t, and x are *variables*. Unless specified otherwise, variables in this text will represent real numbers. On occasion it is necessary to restrict the numbers represented by a variable. For example, in the expression $x/(x + 2)$ we cannot use $x = -2$ since this would require division by zero, which we know is undefined. For the expression $\sqrt{x - 1}$, we restrict the numbers to $x \geq 1$ to avoid square roots of negative numbers, which yield nonreal results. The set of all *permissible* (or usable) values for a variable is called the **domain** of the variable.

Equations

An **equation** is a statement that two expressions are equal. Some examples of equations in one variable x are

$$3x - 5 = 7, \qquad x^2 - x - 6 = 0, \qquad x^2 - 9 = (x + 3)(x - 3)$$

$$\sqrt{2x} = 4, \qquad \frac{x}{x + 2} = \frac{2}{3}$$

To **solve** an equation means to find all values of the unknown (the variable) for which the equation is a true statement. Those values for which an equation is true are called **solutions** or **roots** of the equation. For instance, $x = 4$ is a solution of the equation $3x - 5 = 7$, because $3(4) - 5 = 7$ is a true statement. Similarly, $x = 8$ is a solution to $\sqrt{2x} = 4$, because $\sqrt{2(8)} = \sqrt{16} = 4$.

Equations that are true for *every* real number for which all terms of the equation are defined are called **identities.** For instance,

$$x^2 - 9 = (x + 3)(x - 3)$$

is an identity because every real number is a solution to this equation.

Most equations have values in their domains that are not solutions. For example, $x = 1$ is not a solution to $x^2 - 4 = 0$. We call such equations **conditional equations.**

The algebraic process of *solving an equation* usually generates a chain of intermediate equations, each with the same solution(s) as the original. Such equations are called **equivalent equations.**

GENERATING EQUIVALENT EQUATIONS

A given equation is transformed into an *equivalent equation* by:

1. Adding or subtracting the same quantity from both sides.

2. Multiplying or dividing both sides by the same nonzero* quantity.

*When multiplying or dividing by a *variable*, check to see that the resulting equation has the same solutions as the given one.

EXAMPLE 2
Solving an Equation

Solve the following equation for x:

$$6(x - 1) + 4 = 3(7x + 3)$$

Solution:
We have

$$6(x - 1) + 4 = 3(7x + 3)$$
$$6x - 6 + 4 = 21x + 9$$
$$6x - 2 = 21x + 9$$
$$-15x = 11$$
$$x = -\tfrac{11}{15}$$

These five equations are all equivalent, and we say they form the *steps* of the solution.

The solution method in Example 2 works well for equations that reduce to the *linear* form $ax = b$ (or $ax - b = 0$). *Quadratic* (second degree) equations in the standard form $ax^2 + bx + c = 0$ $(a \neq 0)$ can be solved either by factoring or by the *quadratic formula*.

THE QUADRATIC FORMULA

The solutions of a quadratic equation in standard form

$$ax^2 + bx + c = 0, \qquad a \neq 0$$

are obtained from the formula

$$x = \frac{-b \pm \sqrt{b^2 - 4ac}}{2a}$$

Since the square root of a negative number is not a real number, some concern should exist about the quantity under the radical sign, $b^2 - 4ac$. This quantity is called the **discriminant,** and it determines the nature of the roots of a quadratic equation.

ROOTS OF A QUADRATIC EQUATION

Given $ax^2 + bx + c = 0,$ $a \neq 0$:

1. If $b^2 - 4ac > 0$, the equation has **two distinct real roots**

$$x = \frac{-b \pm \sqrt{b^2 - 4ac}}{2a}$$

2. If $b^2 - 4ac = 0$, the equation has a **double root**

$$x = -\frac{b}{2a}$$

3. If $b^2 - 4ac < 0$, the equation has **no real roots.**

Remark: We will discuss the case in which the discriminant is negative in Chapter 3.

EXAMPLE 3
Using the Quadratic Formula

Solve by factoring or the quadratic formula:

(a) $x^2 + 3x = 9$ (b) $3x^2 - 5x - 12 = 0$

Solution:

(a) In standard form, the equation is

$$x^2 + 3x - 9 = 0$$

where $a = 1, b = 3,$ and $c = -9$. By the quadratic formula, we have

$$x = \frac{-b \pm \sqrt{b^2 - 4ac}}{2a} = \frac{-3 \pm \sqrt{(3)^2 - 4(1)(-9)}}{2(1)}$$

$$= \frac{-3 \pm \sqrt{45}}{2} = \frac{-3 \pm 3\sqrt{5}}{2}$$

(b) For

$$3x^2 - 5x - 12 = 0$$

$a = 3, b = -5,$ and $c = -12$, and by the quadratic formula we have

$$x = \frac{-b \pm \sqrt{b^2 - 4ac}}{2a} = \frac{-(-5) \pm \sqrt{(-5)^2 - 4(3)(-12)}}{2(3)}$$

$$= \frac{5 \pm \sqrt{169}}{6}$$

$$x = \frac{5 + 13}{6} = 3 \quad \text{or} \quad x = \frac{5 - 13}{6} = -\frac{4}{3}$$

Note that when the discriminant turns out to be a perfect square (169 in this case), it means we could have factored the quadratic. This particular quadratic factors as $3x^2 - 5x - 12 = (3x + 4)(x - 3)$.

Section Exercises 1.1

In Exercises 1–10, locate each pair of real numbers on the real number line and place the appropriate inequality sign ($<$ or $>$) between them.

1. $\frac{3}{2}$, 7
2. -3.5, 1
3. π, -6
4. -8, $-\frac{25}{2}$
5. -4, -8
6. 1, $-\frac{16}{3}$
7. $\frac{5}{6}$, $\frac{2}{3}$
8. $-\frac{8}{7}$, $-\frac{3}{7}$
9. -1.75, -2.5
10. $-\frac{5}{4}$, 6

In Exercises 11–18, complete the two missing descriptions of the given interval.

Interval Notation	Inequality Notation	Graph
11. _____	_____	(graph: open at -2, open at 0, over $-3, -2, -1, 0$)
12. $(-\infty, -4]$	_____	_____
13. _____	$3 \le x \le \frac{11}{2}$	_____
14. $(-1, 7)$	_____	_____
15. _____	_____	(graph: over 98, 99, 100, 101, 102)
16. _____	$10 < x < \infty$	_____
17. $(\sqrt{2}, 8]$	_____	_____
18. _____	$\frac{1}{3} < x \le \frac{22}{7}$	_____

In Exercises 19–24, use inequality notation to rewrite the given expression.

19. x is negative.
20. y is greater than 5 and less than or equal to 12.
21. Burt's age, A, is at least 30.
22. The yield, Y, is estimated to be no more than 45 bushels per acre.
23. The monthly rate of inflation, R, is expected to range from $R = \frac{1}{3}\%$ to $R = 1\%$.
24. The price, P, of unleaded gasoline is not expected to go above $1.50 per gallon during the coming year.

In Exercises 25–32, find the distance between a and b.

25.
(graph: $a = -1$, $b = 3$, over $-2, -1, 0, 1, 2, 3, 4, 5$) x

26.
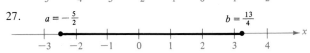
(graph: $a = -4$, $b = -\frac{3}{2}$, over $-5, -4, -3, -2, -1, 0, 1, 2$) x

27.
(graph: $a = -\frac{5}{2}$, $b = \frac{13}{4}$, over $-3, -2, -1, 0, 1, 2, 3, 4$) x

28.
(graph: $a = \frac{1}{4}$, $b = \frac{11}{4}$, over $-2, -1, 0, 1, 2, 3, 4, 5$) x

29. $a = 126, b = 75$
30. $a = -126, b = -75$
31. $a = 9.34, b = -5.65$
32. $a = \frac{16}{5}, b = \frac{112}{75}$

In Exercises 33–38, use absolute value notation to describe the given expression.

33. The distance between x and 5 is no more than 3.
34. The distance between x and -10 is at least 6.
35. The distance between z and $\frac{3}{2}$ is greater than 1.
36. The distance between z and 0 is less than 8.
37. y is closer to 0 than to 8.
38. y is at most 2 units from a.

In Exercises 39–48, determine if the given value of the unknown is a solution of the equation.

39. $5x - 3 = 3x + 5$
   (a) $x = 0$ (b) $x = -5$ (c) $x = 4$ (d) $x = 10$
40. $7 - 3x = 5x - 17$
   (a) $x = -3$ (b) $x = 0$ (c) $x = 8$ (d) $x = 3$
41. $3x^2 + 2x - 5 = 2x^2 - 2$
   (a) $x = -3$ (b) $x = 1$ (c) $x = 4$ (d) $x = -5$
42. $5x^3 + 2x - 3 = 4x^3 + 2x - 11$
   (a) $x = 2$ (b) $x = -2$ (c) $x = 0$ (d) $x = 10$
43. $\dfrac{5}{2x} - \dfrac{4}{x} = 3$
   (a) $x = -\frac{1}{2}$ (b) $x = 4$ (c) $x = 0$ (d) $x = \frac{1}{4}$
44. $3 + \dfrac{1}{x + 2} = 4$
   (a) $x = -1$ (b) $x = -2$ (c) $x = 0$ (d) $x = 5$
45. $(x + 5)(x - 3) = 20$
   (a) $x = 3$ (b) $x = -5$ (c) $x = 5$ (d) $x = -7$

46. $\sqrt[3]{x - 8} = 3$
   (a) $x = 2$ (b) $x = -2$ (c) $x = 35$ (d) $x = 8$

47. $x + \frac{1}{2}\sqrt{x - 3} = 0$
   (a) $x = -4$ (b) $x = 4$ (c) $x = \frac{1}{2}$ (d) $x = \frac{9}{4}$

48. $x^4 - 3x^3 = 4x^2 + 12x$
   (a) $x = 0$ (b) $x = -2$ (c) $x = 2$ (d) $x = 3$

In Exercises 49–58, determine if the equation is conditional or an identity.

49. $3x - 10 = 4x$
50. $x^2 + 2(3x - 2) = x^2 + 8(x + 2) - 2x - 20$
51. $x^2 - 8x + 5 = (x - 4)^2 - 11$
52. $4[(x + \frac{1}{2})^2 - 6] = 4x^2 - 4x - 23$
53. $\frac{x}{3} + \frac{x}{4} = 1$
54. $\frac{5}{x} + \frac{3}{x} = 24$
55. $3 + \frac{1}{x + 1} = \frac{4x}{x + 1}$
56. $2x(x^2 - 7x + 12) = 2x(x^2 - 4x) - 6(x^2 - 4x)$
57. $\frac{1}{t}(t + 2)(t - 1) = \frac{t(t + 2) - t - 2}{t}$
58. $(x + 2)^2(x + 1)^2 = 0$

In Exercises 59–88, solve the given equation and check your answer.

59. $8x - 5 = 3x + 10$ 60. $7x + 3 = 3x - 13$
61. $\frac{x}{2} - 6 = 4x - \frac{3}{4}$ 62. $\frac{x}{5} - \frac{x}{2} = 3$
63. $2(x + 5) - 7 = 3(x - 2)$
64. $4(13t - 15) + 3(2t - 38) = 0$
65. $6[x - (2x + 3)] = 8 - 5x$
66. $8(x + 2) - 3(2x + 1) = 2(x + 5)$
67. $\frac{3}{2}(z + 5) - \frac{1}{4}(z + 24) = 0$
68. $0.6z + 1.1 = 0.3z - 4$
69. $0.25x + 0.75(10 - x) = 3$
70. $0.60x + 0.40(100 - x) = 50$
71. $\frac{100 - 4u}{3} = \frac{5u + 6}{4} + 6$
72. $\frac{16 + y}{y} + \frac{32 + y}{y} = 100$
73. $\frac{5x - 4}{5x + 4} = \frac{2}{3}$ 74. $\frac{10x + 3}{5x + 6} = \frac{1}{2}$
75. $10 - \frac{13}{x} = 4 + \frac{5}{x}$ 76. $\frac{15}{x} - 4 = \frac{6}{x} + 3$
77. $\frac{7}{2x + 1} - \frac{8x}{2x - 1} = -4$
78. $\frac{4}{u - 1} + \frac{6}{3u + 1} = \frac{15}{3u + 1}$
79. $\frac{1}{x - 3} + \frac{1}{x + 3} = \frac{10}{x^2 - 9}$

80. $\frac{1}{x - 2} + \frac{3}{x + 3} = \frac{4}{x^2 + x - 6}$
81. $(x + 2)^2 + 5 = (x + 3)^2$
82. $(x + 1)^2 + 2(x - 2) = (x + 1)(x - 2)$
83. $(x + 2)^2 - x^2 = 4(x + 1)$
84. $4(x + 1) - 3x = x + 5$
85. $(2x + 1)^2 = 4(x^2 + x + 1)$
86. $6x + ax = 2x + 5$
87. $4 - 2(x - 2b) = ax + 3$
88. $5 + ax = 12 - bx$

In Exercises 89–100, use the quadratic formula to find all real roots.

89. $16x^2 + 8x - 3 = 0$ 90. $25x^2 - 20x + 3 = 0$
91. $x^2 - 2x - 2 = 0$ 92. $x^2 - 10x + 22 = 0$
93. $x^2 + 14x + 44 = 0$ 94. $x^2 + 6x - 4 = 0$
95. $x^2 + 8x - 4 = 0$ 96. $4x^2 - 4x - 4 = 0$
97. $9x^2 - 12x - 3 = 0$ 98. $16x^2 - 40x + 22 = 0$
99. $36x^2 + 24x - 7 = 0$ 100. $x^2 + 3x - 1 = 0$

Calculator Exercises

101. (a) Use a calculator to order the following real numbers starting with the smallest:

$$\frac{7071}{5000}, \quad \frac{584}{413}, \quad \sqrt{2}, \quad \frac{47}{33}, \quad \frac{127}{90}$$

    (b) Which of the rational numbers in part (a) is closest to $\sqrt{2}$?

102. Use a calculator to order the following real numbers starting with the smallest:

$$\frac{26}{15}, \quad \sqrt{3}, \quad 1.73\overline{20}, \quad \frac{381}{220}, \quad \sqrt{10} - \sqrt{2}$$

In Exercises 103–110, use a calculator to solve the equations. (Round your answers to three decimal places.)

103. $0.275x + 0.725(500 - x) = 300$
104. $2.763 - 4.5(2.1x - 5.1432) = 6.32x + 5$
105. $\frac{x}{0.6321} + \frac{x}{0.0692} = 1000$
106. $\frac{2}{7.398} - \frac{4.405}{x} = \frac{1}{x}$
107. $5.1x^2 - 1.7x - 3.2 = 0$
108. $10.4x^2 + 8.6x + 1.2 = 0$
109. $7.06x^2 - 14.85x + 3.92 = 0$
110. $-0.052x^2 + 0.101x + 0.193 = 0$

Exponents and Radicals

1.2

Expressions like a^2 or x^3 are said to be in **exponential form.** In general, if n is a positive integer, we have

$$a^n = \underbrace{a \cdot a \cdot a \cdots a}_{n \text{ factors}} \qquad (n \text{ is a positive integer})$$

where n is called the **exponent** (or power) and a is called the **base.** We read a^n as "a to the nth power" or simply "a to the n."

Remark: It is important to recognize the difference between exponential forms like $(-2)^4$ and -2^4. In $(-2)^4$, the parentheses indicate that the power applies to the negative sign as well as the 2, but in $-2^4 = -(2^4)$, the power applies only to the 2. Similarly, in $(5x)^3$, the parentheses indicate that the power applies to the 5 as well as to the x, whereas in $5x^3 = 5(x^3)$, the power applies only to the x.

Multiplying Exponentials with Like Bases

To see what happens when we multiply two exponential expressions having the same (nonzero) base, note the following:

$$2^2 \cdot 2^5 = \underbrace{(2 \cdot 2)}_{2 \text{ factors}} \cdot \underbrace{(2 \cdot 2 \cdot 2 \cdot 2 \cdot 2)}_{5 \text{ factors}} = \underbrace{2 \cdot 2 \cdot 2 \cdot 2 \cdot 2 \cdot 2 \cdot 2}_{7 \text{ factors}}$$

that is,

$$2^2 \cdot 2^5 = 2^{2+5} = 2^7$$

This suggests that when multiplying exponential expressions with the same base, we *add* powers. In general, we have

$$a^m \cdot a^n = a^{m+n}$$

Dividing Exponentials with Like Bases

If we divide 2^5 by 2^2, we get

$$\frac{2^5}{2^2} = \frac{2 \cdot 2 \cdot 2 \cdot 2 \cdot 2}{2 \cdot 2} = 2 \cdot 2 \cdot 2 = 2^3 = 2^{5-2}$$

which suggests that when dividing exponential expressions, we *subtract* powers. That is, for $m > n$, we have

$$\frac{a^m}{a^n} = a^{m-n}, \qquad a \neq 0$$

Two special cases arise from this rule for dividing exponential expressions. First, if $m = n$, then

$$\frac{a^m}{a^n} = a^{m-n} = a^0 = 1, \qquad a \neq 0$$

and we say that *any nonzero number raised to the zero power is 1*. Second, if the power of the denominator is greater than the power of the numerator, then we have

$$\frac{2^2}{2^5} = \frac{2 \cdot 2}{2 \cdot 2 \cdot 2 \cdot 2 \cdot 2} = \frac{1}{2^3} = 2^{2-5} = 2^{-3}$$

This suggests that if n is a positive integer, then

$$\frac{1}{a^n} = a^{-n}, \qquad a \neq 0$$

These relationships can be summarized as follows. A formal proof of each rule can be given using the Principle of Mathematical Induction (see Section 10.4).

PROPERTIES OF EXPONENTS

For real numbers a, b, x, and y and for integers m and n (assuming that all bases and denominators are nonzero):

Property	Example
1. $a^m a^n = a^{m+n}$	$3^2 \cdot 3^4 = 3^{2+4} = 3^6$
2. $\dfrac{a^m}{a^n} = a^{m-n}$	$\dfrac{x^7}{x^4} = x^{7-4} = x^3$
3. $\dfrac{1}{a^n} = a^{-n}$	$\dfrac{1}{y^4} = y^{-4}$
4. $a^0 = 1$	$(x^2 + 1)^0 = 1$
5. $(ab)^m = a^m b^m$	$(5x)^4 = 5^4 x^4$
6. $(a^m)^n = a^{mn}$	$(y^3)^{-4} = y^{3(-4)} = y^{-12} = \dfrac{1}{y^{12}}$
7. $\left(\dfrac{a}{b}\right)^m = \dfrac{a^m}{b^m}$	$\left(\dfrac{2}{x}\right)^3 = \dfrac{2^3}{x^3}$

The preceding rules hold for *all* integers m and n (not just positive ones). For instance, by Property 2, we have

$$\frac{3^4}{3^{-5}} = 3^{4-(-5)} = 3^{4+5} = 3^9$$

As another example, we can use Properties 6 and 7 followed by Property 3 to obtain

$$\left(\frac{4x^{-2}y^3}{5}\right)^2 = \frac{4^2 x^{-4} y^6}{5^2} = \frac{16y^6}{25x^4}$$

When evaluating expressions having negative exponents, you can reduce the likelihood of errors by converting to positive exponents before evaluating.

EXAMPLE 1
Simplifying with
Negative Exponents

Rewrite each of the following with positive exponents and simplify.

(a) $(-3b^4)(5b^2)\left(\dfrac{1}{9}b^{-7}\right)$ (b) $\dfrac{(2x^3)^{-1}(7x^{-4})^0}{4x^{-2}}$ (c) $\left(\dfrac{y^{-1}}{3x^{-2}}\right)\left(\dfrac{3x^2}{y}\right)^{-2}$

Solution:

(a) $(-3b^4)(5b^2)\left(\dfrac{1}{9}b^{-7}\right) = \left(-\dfrac{15}{9}\right)(b^4)(b^2)(b^{-7})$ *Multiply coefficients*

$\qquad\qquad\qquad\qquad = -\dfrac{15}{9}b^{4+2-7}$ *Property 1*

$\qquad\qquad\qquad\qquad = -\dfrac{15}{9}b^{-1}$

$\qquad\qquad\qquad\qquad = -\dfrac{15}{9b} = -\dfrac{5}{3b}$ *Property 3*

(b) Note how negative exponents can be converted to positive exponents by simply shifting the *factors* with negative exponents from numerator to denominator or vice versa.

$\qquad \dfrac{(2x^3)^{-1}(7x^{-4})^0}{4x^{-2}} = \dfrac{(2x^3)^{-1}(1)}{4x^{-2}} = \dfrac{x^2}{4(2x^3)}$ *Shift factors*

$\qquad\qquad\qquad\qquad = \dfrac{x^2}{8x^3} = \dfrac{1}{8x}$ *Subtract exponents*

(c) $\left(\dfrac{y^{-1}}{3x^{-2}}\right)\left(\dfrac{3x^2}{y}\right)^{-2} = \left(\dfrac{x^2}{3y}\right)\left(\dfrac{y}{3x^2}\right)^2 = \left(\dfrac{x^2}{3y}\right)\left(\dfrac{y^2}{9x^4}\right)$

$\qquad\qquad\qquad\qquad = \dfrac{x^2y^2}{27x^4y} = \dfrac{y}{27x^2}$

Remark: Notice that in part (c) of Example 1 *fractions* raised to negative powers were simplified by inverting the fraction and changing the sign of the exponent.

Scientific Notation

Exponents provide an efficient way of writing and computing with the very large (or very small) numbers used in science. It is convenient to write such numbers in **scientific notation.** This notation has the form

$\qquad c \times 10^n$

where $1 \le c < 10$ and n is an integer.

EXAMPLE 2 | Convert the following numbers to the indicated form.
Scientific Notation

Decimal Form	**Scientific Form**
(a) ⬜	1.345×10^2
(b) 0.0000782	⬜
(c) 836,100,000	⬜

Solution:

(a) Since the exponent is positive, we move the decimal point to the right to get

$$1.345 \times 10^2 = 134.5$$

2 places

(b) The number 0.0000782 is very small. Hence the exponent is negative and we write

$$0.0000782 = 7.82 \times 10^{-5}$$

5 places

(c) In this case, $c = 8.361$ and we write

$$836,100,000 = 8.361 \times 10^8$$

8 places

Exponents and the Calculator

Most scientific calculators automatically switch their display to scientific notation when computing with large (or small) numbers that exceed the display range. Try multiplying 98,900,000 × 500. If your calculator follows standard conventions, its display should be

4.945	10

This means that $c = 4.945$ and the exponent is $n = 10$.

To allow you to *enter* numbers in scientific notation, your calculator should have an exponential entry key labeled $\boxed{\text{EE}}$ or $\boxed{\text{EXP}}$. If you wanted to perform the preceding multiplication using scientific notation, you could begin by writing

$$98,900,000 \times 500 = (9.89 \times 10^7)(5.0 \times 10^2)$$

and then enter

9.89 $\boxed{\text{EE}}$ 7 $\boxed{\times}$ 5 $\boxed{\text{EE}}$ 2 $\boxed{=}$

Remark: A number such as 10^{15} is entered

1 $\boxed{\text{EE}}$ 15   and *not* just $\boxed{\text{EE}}$ 15

This latter sequence would be interpreted as 0×10^{15}, which is zero.

The Exponential Key

Scientific calculators are capable of evaluating exponential expressions via the keys $\boxed{y^x}$ and $\boxed{e^x}$. The second of these two keys will be discussed in Chapter 4. To use the first key, remember that y is the *base* and x is the *exponent*. Thus, to calculate 3^6, we can enter

	Display
3 $\boxed{y^x}$ 6 $\boxed{=}$	729

Negative exponents are entered into a calculator by pressing the change-sign key $\boxed{+/-}$ immediately after entering the exponent. For instance, to enter 27^{-6} we use the following sequence:

	Display
27 $\boxed{y^x}$ 6 $\boxed{+/-}$ $\boxed{=}$	2.5812 -09

Radicals

Solutions to equations such as $r^2 = 49$ or $s^3 = 125$ involve finding the **roots** of numbers. For example, we say:

If $x^2 = 49$,   then x is the **square root** of 49.

If $x^3 = 125$, then x is the **cube root** of 125.

In general, we have the following definition of an nth root of a number.

DEFINITION OF nth ROOT

> For nonnegative real numbers a and b and positive integer n:
>
> a is an nth root of b if $a^n = b$

Remark: In the special case where n is an *odd* positive integer, we extend the definition of nth root to include negative values for a and b.

In denoting roots of numbers, we use the **radical** form  except for $n = 2$, which is simply written as $\sqrt{}$.

DEFINITION OF

> For a and b nonnegative and n a positive integer *or* for a and b negative and n an odd positive integer:
>
> $a = \sqrt[n]{b}$ if and only if $a^n = b$
>
> The radical, $\sqrt[n]{b}$, is called the **principal nth root of b,** n is called the **index** of the radical, and b is called the **radicand.**

Remark: It is important that you understand the significance of the restrictions placed on the numbers a, b, and n in this definition. For instance, if $b = -9$ $(b < 0)$ and $n = 2$, then there is *no real* number a such that $a^2 = -9$, since the square of every real number is nonnegative. [In Section 3.4 we show how even roots of negative numbers fit into the set of *complex* numbers.]

Odd roots of negative numbers are real numbers. For example,

$$\sqrt[3]{-64} = -4 \quad \text{because} \quad (-4)^3 = (-4)(-4)(-4) = -64$$

For even roots of positive numbers, we obtain more than one answer. For example, since $5^2 = 25$ and $(-5)^2 = (-5)(-5) = 25$, we conclude that *both* $5 = \sqrt{25}$ and $-5 = -\sqrt{25}$ are *square roots* of 25. In general, if n is even and b positive, then there are *two* real nth roots of b, namely

$$a = \sqrt[n]{b} \qquad\qquad \textit{(principal nth root of b)}$$

and

$$-a = -\sqrt[n]{b} \qquad\qquad \textit{(negative nth root of b)}$$

In summary, we have the following situation with respect to nth roots of a number.

REAL ROOTS OF $a^n = b$

Condition	Roots	Examples
n even, $b > 0$	$a = \sqrt[n]{b},\ -a = -\sqrt[n]{b}$	$a^2 = 5 \implies a = \sqrt{5},\ -a = -\sqrt{5}$
n even, $b < 0$	no real roots	$a^4 = -10 \implies a$ is not real
n odd, $b > 0$	$a = \sqrt[n]{b} > 0$	$a^3 = 27 \implies a = \sqrt[3]{27} = 3$
n odd, $b < 0$	$a = \sqrt[n]{b} < 0$	$a^3 = -27 \implies a = \sqrt[3]{-27} = -3$

Remark: It is *incorrect* to write

$$\sqrt{81} = \pm 9 \qquad\qquad \textit{Common mistake}$$

The radical $\sqrt{81}$ denotes only the principal (positive) square root of 81. We use $-\sqrt{81}$ to denote the negative square root of 81.

EXAMPLE 3
Evaluating Radicals

Evaluate the following radicals.

(a) $\sqrt[5]{-32}$ (b) $\sqrt[3]{\dfrac{125}{64}}$ (c) $-\sqrt{\dfrac{1}{121}}$

Solution:

(a) $\sqrt[5]{-32} = -2$ because $(-2)^5 = -32$.

(b) $\sqrt[3]{\dfrac{125}{64}} = \dfrac{5}{4}$ because $\left(\dfrac{5}{4}\right)^3 = \dfrac{5^3}{4^3} = \dfrac{125}{64}$.

(c) $-\sqrt{\dfrac{1}{121}} = -\dfrac{1}{11}$ because $-\left(\dfrac{1}{11}\right)^2 = -\left(\dfrac{1}{121}\right)$.

Rational Exponents

Some roots of numbers raised to powers can be evaluated more easily by writing the radicals in exponential form using *rational exponents*.

DEFINITION OF RATIONAL EXPONENTS

For integer m, natural number n, and real number b such that $\sqrt[n]{b}$ exists:

$$b^{m/n} = (\sqrt[n]{b})^m = \sqrt[n]{b^m}$$

Remark: Note in this definition that the denominator is the *index* for the corresponding radical form and the numerator is the *power* of the radical (or radicand).

$$b^{m/n} = (\sqrt[n]{b})^m = \sqrt[n]{b^m}$$

with "power" labeling the exponent m and "index" labeling n.

Two important special cases arise from the definition of rational exponents.

1. If $m = 1$, then

$$b^{1/n} = \sqrt[n]{b} = \text{principal } n\text{th root of } b$$

2. If $m = n$, then

$$b^{n/n} = \sqrt[n]{b^n} = b \qquad\qquad (n \text{ is odd})$$
$$b^{n/n} = \sqrt[n]{b^n} = |b| \qquad\qquad (n \text{ is even})$$

Note the difference between the parts of the second case, particularly the occurrence of $|b|$ when n is even. For instance,

$$\sqrt[4]{(-6)^4} = |-6| = 6 \qquad \text{and} \qquad \sqrt{x^2} = |x|$$

EXAMPLE 4
Changing Radicals to Exponential Form

Write each of the following in exponential form.

(a) $\sqrt{(3xy)^5}$ (b) $2x\sqrt[4]{x^3}$

Solution:

(a) In this case the base is $(3xy)$, the power is 5, and the index is 2. Thus

$$\sqrt{(3xy)^5} = \sqrt[2]{(3xy)^5} = (3xy)^{5/2}$$

(b) $2x\sqrt[4]{x^3} = (2x)(x^{3/4}) = 2x^{1+(3/4)} = 2x^{7/4}$

EXAMPLE 5
Changing from Exponential
to Radical Form

Write each of the following in radical form.

(a) $(x^2 + y^2)^{3/2}$ (b) $2y^{3/4}z^{1/4}$

(c) $a^{-3/2}$ (d) $(4x^2y^{-3})^{2/3}$

Solution:

(a) $(x^2 + y^2)^{3/2} = (\sqrt{x^2 + y^2})^3 = \sqrt{(x^2 + y^2)^3}$

(b) $2y^{3/4}z^{1/4} = 2(y^3z)^{1/4} = 2\sqrt[4]{y^3z}$

(c) $a^{-3/2} = \dfrac{1}{a^{3/2}} = \dfrac{1}{\sqrt{a^3}}$

(d) $(4x^2y^{-3})^{2/3} = \sqrt[3]{(4x^2y^{-3})^2} = \sqrt[3]{4^2x^4y^{-6}} = \sqrt[3]{\dfrac{16x^4}{y^6}}$

Since radicals can be written in exponential form using rational exponents, it follows that radicals possess properties similar to those of exponents. Compare the following list of properties with those given in the previous section.

PROPERTIES OF RADICALS

For integer m, natural number n, and real numbers a and b such that $\sqrt[n]{a}$ and $\sqrt[n]{b}$ are real:

Property	**Example**
1. $\sqrt[n]{a^m} = (\sqrt[n]{a})^m = a^{m/n}$	$\sqrt[3]{8^2} = (\sqrt[3]{8})^2 = 8^{2/3}$
2. $\sqrt[n]{a} \cdot \sqrt[n]{b} = \sqrt[n]{ab}$	$\sqrt{5} \cdot \sqrt{7} = \sqrt{5 \cdot 7} = \sqrt{35}$
3. $\dfrac{\sqrt[n]{a}}{\sqrt[n]{b}} = \sqrt[n]{\dfrac{a}{b}}$ $(b \neq 0)$	$\dfrac{\sqrt[3]{27}}{\sqrt[4]{9}} = \sqrt[4]{\dfrac{27}{9}} = \sqrt[4]{3}$
4. $\sqrt[n]{\sqrt[m]{a}} = \sqrt[mn]{a}$	$\sqrt[3]{\sqrt{10}} = \sqrt[6]{10}$
5. $(\sqrt[n]{a})^n = a$	$(\sqrt[6]{15})^6 = 15$
6. For n even, $\sqrt[n]{a^n} = |a|$	$\sqrt{(-12)^2} = |-12| = 12$
For n odd, $\sqrt[n]{a^n} = a$	$\sqrt[3]{(-12)^3} = -12$

Simplifying Radicals

An expression involving radicals is in **simplest form** when the following conditions are satisfied:

1. All possible factors have been removed from under the radical sign.

2. The index for the radical has been reduced as far as possible.

3. All fractions have radical-free denominators (accomplished by a process called *rationalizing the denominator*).

EXAMPLE 6
Simplifying Radicals

Simplify the following radical expressions, reducing the index when possible.

(a) $\sqrt{75x^3}$

(b) $\sqrt[3]{24a^4c^8}$

(c) $\sqrt[3]{\sqrt{125}}$

(d) $\sqrt[4]{27xy^3}\,\sqrt[4]{\dfrac{3y^3}{x^3}}$

Solution:

(a) $\sqrt{75x^3} = \sqrt{25(x^2)(3x)}$ *Find largest square factors*

$\qquad = \sqrt{25}\,\sqrt{x^2}\,\sqrt{3x}$ *Property 2 of radicals*

$\qquad = 5x\,\sqrt{3x}$ *Find roots of perfect squares*

(b) $\sqrt[3]{24a^4c^8} = \sqrt[3]{(8)(a^3)(c^6)(3ac^2)}$ *Find largest cube factors*

$\qquad = \sqrt[3]{8}\,\sqrt[3]{a^3}\,\sqrt[3]{c^6}\,\sqrt[3]{3ac^2}$ *Property 2 of radicals*

$\qquad = 2ac^2\,\sqrt[3]{3ac^2}$ *Find roots of perfect cubes*

(c) $\sqrt[3]{\sqrt{125}} = \sqrt[6]{125} = \sqrt[6]{5^3} = 5^{3/6} = 5^{1/2} = \sqrt{5}$

(d) $\sqrt[4]{27xy^3}\,\sqrt[4]{\dfrac{3y^3}{x^3}} = \sqrt[4]{\dfrac{81xy^6}{x^3}}$ *Multiply radicals*

$\qquad = \sqrt[4]{(3^4y^4)\dfrac{y^2}{x^2}}$ *Find largest 4th powers*

$\qquad = 3y\,\sqrt[4]{\dfrac{y^2}{x^2}}$ *Find 4th root*

$\qquad = (3y)\dfrac{y^{1/2}}{x^{1/2}}$ *Write with fractional exponents*

$\qquad = 3y\,\sqrt{\dfrac{y}{x}}$ *Rewrite in radical form*

Rationalizing Radicals

Our third simplification technique involves a *rationalizing* procedure that removes radicals from either the numerator or the denominator of a fraction. In algebra, we usually emphasize rationalizing the denominator. However, in calculus it is helpful to be able to rationalize the numerator also. In both instances, we make use of the form $a + b\sqrt{m}$ and its **conjugate** $a - b\sqrt{m}$. Note that the product of this conjugate pair has no radical:

$$(a + b\sqrt{m})(a - b\sqrt{m}) = a^2 - b^2m$$

EXAMPLE 7
Rationalizing the Denominator

Eliminate the radical in the denominator of each of the following.

(a) $\dfrac{5}{2\sqrt{3}}$

(b) $\dfrac{\sqrt{3} + \sqrt{x}}{\sqrt{3} - \sqrt{x}}$

Solution:

(a) $\dfrac{5}{2\sqrt{3}} = \dfrac{5}{2\sqrt{3}} \cdot \dfrac{\sqrt{3}}{\sqrt{3}}$ *Multiply by $\sqrt{3}/\sqrt{3}$*

$= \dfrac{5\sqrt{3}}{2(3)} = \dfrac{5\sqrt{3}}{6}$

(b) $\dfrac{\sqrt{3} + \sqrt{x}}{\sqrt{3} - \sqrt{x}} = \dfrac{\sqrt{3} + \sqrt{x}}{\sqrt{3} - \sqrt{x}} \cdot \dfrac{\sqrt{3} + \sqrt{x}}{\sqrt{3} + \sqrt{x}}$ *Multiply by conjugate*

$= \dfrac{(\sqrt{3})^2 + 2\sqrt{3}\sqrt{x} + (\sqrt{x})^2}{(\sqrt{3})^2 - (\sqrt{x})^2}$

$= \dfrac{3 + 2\sqrt{3x} + x}{3 - x}$

Although the sum or difference of two radicals can often be simplified by multiplying by its conjugate, it is not so easy to simplify a sum or difference that is part of a single radicand. In particular, note that

$$\sqrt{a + b} \qquad \textbf{DOES NOT EQUAL} \qquad \sqrt{a} + \sqrt{b}$$

For example,

$$\sqrt{16 + 9} = \sqrt{25} = 5$$

whereas

$$\sqrt{16} + \sqrt{9} = 4 + 3 = 7$$

Watch out for the following version of this common error:

$$\sqrt{x^2 + y^2} \qquad \textbf{DOES NOT EQUAL} \qquad x + y$$

EXAMPLE 8
Combining Radicals

Simplify and combine terms of

$$\sqrt{12} - 3\sqrt{27} + 2\sqrt{48}$$

Solution:
We have

$\sqrt{12} - 3\sqrt{27} + 2\sqrt{48}$

$= \sqrt{4 \cdot 3} - 3\sqrt{9 \cdot 3} + 2\sqrt{16 \cdot 3}$ *Find square factors*

$= 2\sqrt{3} - 9\sqrt{3} + 8\sqrt{3}$ *Find square roots*

$= (2 - 9 + 8)\sqrt{3} = \sqrt{3}$ *Combine like terms*

Radicals and Calculators

There are two methods of evaluating radicals on most calculators. For square roots, you use the *square root key* $\boxed{\sqrt{}}$. For other roots, you should first convert the radical to exponential form and then use the *exponential key* $\boxed{y^x}$.

EXAMPLE 9
Evaluating Radicals
with a Calculator

Use a calculator to evaluate each of the following radicals.

(a) $\sqrt[3]{56}$ (b) $\sqrt[3]{-4}$ (c) $(1.2)^{-1/6}$

Solution:

(a) First, we write in exponential form $\sqrt[3]{56} = 56^{1/3}$. Then there are several options:

Calculator Steps

1 ÷ 3 = STO 56 y^x RCL =   *Use memory key*

56 y^x (1 ÷ 3) =   *Use parentheses*

56 y^x 3 1/x =   *Use reciprocal key*

For each of these three keystroke sequences, the answer is

$$\sqrt[3]{56} \approx 3.8258624$$

(b) Since

$$\sqrt[3]{-4} = \sqrt[3]{(-1)(4)} = \sqrt[3]{-1} \cdot \sqrt[3]{4} = -\sqrt[3]{4}$$

we can attach the negative sign of the radicand at the end of the keystroke sequence as follows:

Calculator Steps   **Display**

4 y^x (1 ÷ 3) = +/−   −1.5874011

(c) **Calculator Steps**   **Display**

1.2 y^x (1 ÷ 6 +/−) =   .97007012

Section Exercises 1.2

In Exercises 1–6, evaluate each expression for the given value of x.

1. $-3x^3$ $(x = 2)$
2. $\dfrac{x^2}{2}$ $(x = 6)$
3. $4x^{-3}$ $(x = 2)$
4. $7x^{-2}$ $(x = 4)$
5. $6x^0 - (6x)^0$ $(x = 10)$
6. $5(-x)^3$ $(x = 3)$

In Exercises 7–30, simplify the given expression.

7. $5x^4(x^2)$
8. $(8x^4)(2x^3)$
9. $6y^2(2y^4)^2$
10. $z^{-3}(3z^4)$
11. $10(x^2)^2$
12. $(4x^3)^2$
13. $\dfrac{7x^2}{x^{-3}}$
14. $\dfrac{r^4}{r^6}$
15. $\dfrac{12(x + y)^3}{9(x + y)}$
16. $(2x^2yz^5)^0$

17. $(-2x^2)^3(4x^3)^{-1}$
18. $\dfrac{(4y^{-2})(8y^4)}{6y^2}$
19. $\left(\dfrac{3z^2}{x}\right)^{-2}$
20. $\left(\dfrac{x^{-3}y^4}{5}\right)^{-3}$
21. $(4a^{-2}b^3)^{-3}$
22. $(5x^2y^4z^6)^3(5x^2y^4z^6)^{-3}$
23. $(2x^2 + y^2)^4(2x^2 + y^2)^{-4}$
24. $[(x^2y^{-2})^{-1}]^{-1}$
25. $\left(\left(\dfrac{y^2}{x^2}\right)^{-1}\right)^2$
26. $\left(\dfrac{4x^{-2}}{3}\right)^{-3}$
27. $\left(\dfrac{a^{-2}}{b^{-2}}\right)\left(\dfrac{b}{a}\right)^3$
28. $(-3x^2)(4x^{-3})(\tfrac{1}{6}x)$
29. $\left(\dfrac{10x^{-2}y^3}{x^4}\right)^0$
30. $(-2r^2s^3u^{-1})^{-3}$

31. Change each of the following from decimal form to scientific notation.
   (a) 93,000,000   (b) 900,000,000   (c) 0.00000435

32. Change each of the following from decimal form to scientific notation.
   (a) 0.000087   (b) 6.87   (c) 0.004392

33. Change each of the following from scientific notation to decimal form.
   (a) 1.91×10^6 (b) 2.345×10^{11}
   (c) 6.21×10^0 (d) 8.52×10^{-3}
   (e) 7.021×10^{-5} (f) 3.798×10^{-8}

34. Change each of the following from scientific notation to decimal form.
   (a) 2.65×10^7 (b) 9.4675×10^4
   (c) 3.0025×10^8 (d) 1.0909×10^{-4}
   (e) 3.2×10^{-7} (f) 4.6666×10^{-5}

In Exercises 35–46, complete the two missing descriptions.

$\sqrt[n]{b^m} = a$	$b^{m/n} = a$	$a^{n/m} = b$
35. $\sqrt{9} = 3$		
36. $\sqrt[3]{64} = 4$		
37.	$32^{1/5} = 2$	
38.	$-(144^{1/2}) = -12$	
39.		$14^2 = 196$
40.		$8.5^3 = 614.125$
41. $\sqrt[3]{-216} = -6$		
42.		$-3^5 = -243$
43.	$27^{2/3} = 9$	
44. $(\sqrt[4]{81})^3 = 27$		
45. $\sqrt[4]{81^3} = 27$		
46.		$16^{5/4} = 32$

In Exercises 47–54, evaluate each radical without using a calculator.

47. (a) $\sqrt{\frac{9}{4}}$ 48. (a) $-\sqrt{25}$
   (b) $\sqrt[3]{\frac{27}{8}}$ (b) $-\sqrt[3]{-27}$
49. (a) $16^{3/2}$ 50. (a) $(\sqrt[4]{16})^3$
   (b) $8^{2/3}$ (b) $(\sqrt[3]{-125})^3$
51. (a) $(\sqrt[6]{326})^6$ 52. (a) $121^{-1/2}$
   (b) $\sqrt[4]{562^4}$ (b) $32^{-3/5}$
53. (a) $64^{-2/3}$ 54. (a) $(-\frac{27}{8})^{-1/3}$
   (b) $(\frac{9}{4})^{-1/2}$ (b) $-1/(144^{-1/2})$

In Exercises 55–60, simplify by removing all possible factors from under the radical. (Assume $x > 0$, $y > 0$.)

55. (a) $\sqrt{8}$ 56. (a) $\sqrt[3]{\frac{16}{27}}$
   (b) $\sqrt{18}$ (b) $\sqrt[3]{\frac{24}{125}}$

57. (a) $\sqrt[3]{16x^5}$ 58. (a) $\sqrt[4]{(3x^2y^3)^4}$
   (b) $\sqrt[4]{32x^4z^5}$ (b) $\sqrt[3]{54x^7}$
59. (a) $\sqrt{75x^2y^{-4}}$ 60. (a) $\sqrt[3]{96b^6c^3}$
   (b) $\sqrt{5(x-y)^3}$ (b) $\sqrt{72(x+1)^4}$

In Exercises 61–64, simplify by reducing the index of the radical as far as possible. (Assume $x > 0$, $y > 0$.)

61. (a) $\sqrt[4]{x^2}$ 62. (a) $\sqrt[6]{(x+y)^3}$
   (b) $\sqrt[6]{x^2}$ (b) $\sqrt[6]{(x+y)^4}$
63. (a) $\sqrt[8]{(3x^2y^3)^2}$ 64. (a) $\sqrt[4]{9x^2y^2}$
   (b) $\sqrt[10]{(6x^2y^4)^5}$ (b) $\sqrt[6]{16(x-y)^4}$

In Exercises 65–72, simplify each radical by rationalizing the denominator and reducing the resulting fraction to lowest terms.

65. (a) $\frac{1}{\sqrt{3}}$ 66. (a) $\frac{5}{\sqrt{10}}$
   (b) $\frac{3}{\sqrt{21}}$ (b) $\frac{21}{\sqrt{7}}$
67. (a) $\frac{8}{\sqrt[3]{2}}$ 68. (a) $\frac{5}{\sqrt[3]{(5x)^2}}$
   (b) $\frac{1}{\sqrt[3]{12}}$ (b) $\frac{3}{\sqrt[5]{(3x)^3}}$
69. (a) $\frac{2x}{5-\sqrt{3}}$ 70. (a) $\frac{5}{\sqrt{15}-2}$
   (b) $\frac{16}{6+\sqrt{10}}$ (b) $\frac{x}{\sqrt{2}+\sqrt{3}}$
71. (a) $\frac{3}{\sqrt{x}+\sqrt{y}}$ 72. (a) $\frac{5}{3\sqrt{x}-5}$
   (b) $\frac{8}{\sqrt{2}-2\sqrt{3}}$ (b) $\frac{34}{5\sqrt{2}-4}$

In Exercises 73–76, simplify each radical by rationalizing the numerator and reducing the resulting fraction to lowest terms.

73. (a) $\frac{\sqrt{13}}{2}$ 74. (a) $\frac{\sqrt{26}}{2}$
   (b) $\frac{\sqrt{2}}{3}$ (b) $\frac{\sqrt{y}}{6y}$
75. (a) $\frac{\sqrt{3}-\sqrt{2}}{x}$ 76. (a) $\frac{\sqrt{x+2}-\sqrt{x}}{2}$
   (b) $\frac{\sqrt{15}+3}{12}$ (b) $\frac{\sqrt{a-b}+\sqrt{a}}{b}$

In Exercises 77–84, combine and/or simplify the given radicals.

77. (a) $5\sqrt{x}-3\sqrt{x}$ 78. (a) $4\sqrt{27}-\sqrt{75}$
   (b) $6\sqrt{2}+7\sqrt{2}$ (b) $5\sqrt[3]{2}+\sqrt[3]{54}$
79. (a) $2\sqrt{4xy}-2\sqrt{9xy}+10\sqrt{xy}$
   (b) $3\sqrt{ab/2}+5\sqrt{2ab}$

80. (a) $2\sqrt{80} + \sqrt{125} - \sqrt{500}$

(b) $\sqrt{\dfrac{25x}{y}} + 3\sqrt{\dfrac{x}{36y}} - \sqrt{\dfrac{32x}{50y}}$

81. (a) $\sqrt{5x^2y}\ \sqrt{3y}$

(b) $\dfrac{\sqrt{54a^2}}{\sqrt{2a^4}}$

82. (a) $\sqrt[3]{\dfrac{4z^2}{y^5}}\ \sqrt[3]{\dfrac{2z}{y}}$

(b) $\dfrac{\sqrt[4]{16b}}{\sqrt[4]{2b^5}}$

83. (a) $\sqrt{\sqrt{\sqrt{32}}}$

(b) $\sqrt{\sqrt[3]{\sqrt{10a^7b}}}$

84. (a) $\sqrt{50}\ \sqrt[3]{2}$

(b) $\sqrt{50}\ \sqrt[3]{10}$

Calculator Exercises

In Exercises 85–90, use a calculator to perform the indicated calculations.

85. (a) $2400(1 + 0.06)^{20}$

(b) $750\left(1 + \dfrac{0.11}{365}\right)^{800}$

(c) $\dfrac{(2.414 \times 10^4)^6}{(1.68 \times 10^5)^5}$

(d) $(9.3 \times 10^6)^3(6.1 \times 10^{-4})^4$

86. (a) $\dfrac{3000}{[1 + (0.05/4)]^4}$

(b) $\dfrac{4 - 1.25^6}{1 - 0.625^4}$

(c) $\dfrac{(3.28 \times 10^{-6})^{10}}{(5.34 \times 10^{-3})^{25}}$

(d) $(2.52 \times 10^4)^5(1.63 \times 10^{-3})^7$

87. (a) $(0.000345)(8,980,000,000)$

(b) $\dfrac{67,000,000 + 93,000,000}{0.0052}$

88. (a) $\dfrac{848,000,000}{1,624,000}$

(b) $\dfrac{0.0000928 - 0.0000021}{0.0061}$

89. The speed of light is 11,160,000 miles per minute. The distance from the sun to the earth is 93,000,000 miles. Find the time it takes for light to travel from the sun to the earth.

90. The *per capita public debt* is defined as the gross debt divided by the population. Find the per capita debt of the United States in 1980 if the gross debt was 839.2 billion dollars and the population was 220 million.

91. Use a calculator to approximate each of the following to four decimal places.

(a) $\sqrt{57}$

(b) $\sqrt[5]{562}$

(c) $\sqrt[3]{45^2}$

(d) $(-10)^{4/5}$

(e) $(15.25)^{-1.4}$

(f) $(9.42 \times 10^5)^{2/3}$

92. Use a calculator to approximate each of the following to four decimal places.

(a) $\sqrt[6]{125}$

(b) $\sqrt[5]{-65}$

(c) $\sqrt{75 + 3\sqrt{8}}$

(d) $225^{-2/3}$

(e) $(2.65 \times 10^{-4})^{1/3}$

(f) $(9.3 \times 10^7)^2$

93. The amount A after t years in a savings account earning r percent interest compounded n times per year is

$$A = P\left(1 + \dfrac{r}{n}\right)^{nt}$$

where P is the (original) principal. Complete the following table for $500 deposited in an account earning 12% compounded daily. (Note that 12% interest implies that $r = 0.12$.)

t	5	10	20	30	40	50
A						

(The key sequence for programming a Texas Instruments programmable calculator is

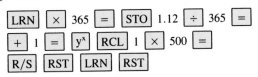

If you enter the time in years and press the run/start key R/S, the calculator will display the value for A.)

94. The time, t, it takes for a funnel to empty when filled with water to a height of h is given by

$$t = 0.03[12^{5/2} - (12 - h)^{5/2}], \qquad 0 \le h \le 12$$

Find t (to two decimal places) for $h = 7$ centimeters.

95. To calculate uniform depreciation by the declining balance method, we use the formula

$$R = N\left[1 - \left(\dfrac{S}{C}\right)^{1/N}\right]$$

where R is the percentage depreciation each year, N is the useful life of the item, C is the original cost, and S is the salvage value. Calculate R (to two decimal places) for each of the following.

(a) $N = 8$, $C = \$10,400$, $S = \$1,500$

(b) $N = 4$, $C = \$11,200$, $S = \$3,200$

Polynomials and Factoring

1.3

An **algebraic expression** is a collection of variables and real numbers (called **constants**) organized in some manner through use of additions, subtractions, multiplications, divisions, or radicals. A **term** of an algebraic expression is any product of a constant and one or more variables raised to powers. One of the simplest kinds of algebraic expressions is the **polynomial.** Some examples are

$$2x + 5, \qquad 3x^4 - 7x^2 + 2x + 4, \qquad 5x^2y^2 - xy + 3$$

The first two are *polynomials in x,* and the last one is a *polynomial in x and y.* The terms of a polynomial in x have the form cx^k, where c is called the **coefficient** and k the **degree** of the term.

DEFINITION OF A POLYNOMIAL IN x

> For real numbers $a_0, a_1, a_2, \ldots, a_n$ and nonnegative integer n:
> A **polynomial in x** is an expression of the form
>
> $$a_n x^n + a_{n-1}x^{n-1} + \cdots + a_1 x + a_0$$
>
> where $a_n \neq 0$. The polynomial is of **degree** n. The numbers $a_0, a_1, a_2, \ldots, a_n$ are called the **coefficients,** and a_n is called the **leading coefficient.**

Remark: Polynomials with one, two, or three terms are called **monomials, binomials,** or **trinomials,** respectively.

Note in the above definition that the polynomial is written in *order* with decreasing powers of x. This is referred to as **standard form.** Note also that the polynomial is written as a *sum.* Consequently, the coefficients take on the sign between terms. For instance, the polynomial

$$2x^3 - 5x^2 - 3x + 1 = 2x^3 + (-5)x^2 + (-3)x + 1$$

has coefficients 2, -5, -3, and 1. Following are some other examples of polynomials rewritten in standard form, together with terms, degrees, and coefficients.

Polynomial	Standard Form	Degree	Leading Coefficient	Terms	Coefficients
$4x^2 - 5x^7 - 2 + 3x$	$-5x^7 + 4x^2 + 3x - 2$	7	-5	$-5x^7, 4x^2, 3x, -2$	$-5, 4, 3, -2$
$4 - 9x^2$	$-9x^2 + 4$	2	-9	$-9x^2, 4$	$-9, 4$
8	8	0 ($8 = 8x^0$)	8	8	8

For polynomials in more than one variable, the degree of a *term* is the sum of the powers of the variables in the term. The degree of the *polynomial* is the highest degree of all its terms. For instance, the polynomial

$$5x^3y - x^2y^2 + 2xy - 5$$

has two terms of degree 4, one term of degree 2, and one term of degree 0. The degree of the polynomial is 4.

Operations with Polynomials

We can add and subtract polynomials in much the same way that we add and subtract real numbers. We simply add or subtract the coefficients of *like terms*, which are terms that have the same variables to the same powers.

EXAMPLE 1
Sums and Differences of Polynomials

(a) Add $5x^3 - 7x^2 - 3$ and $x^3 + 2x^2 - x + 8$.
(b) Subtract $3x^4 - 4x^2 + 3x$ from $7x^4 - x^2 - 4x + 2$.

Solution:

(a) $(5x^3 - 7x^2 - 3) + (x^3 + 2x^2 - x + 8)$

$\qquad = (5x^3 + x^3) + (2x^2 - 7x^2) - x + (8 - 3)$ *Group like terms*

$\qquad = (5 + 1)x^3 + (2 - 7)x^2 - x + 5$ *Distributive Law*

$\qquad = 6x^3 - 5x^2 - x + 5$ *Combine like terms*

(b) $(7x^4 - x^2 - 4x + 2) - (3x^4 - 4x^2 + 3x)$

$\qquad = 7x^4 - x^2 - 4x + 2 - 3x^4 + 4x^2 - 3x$

$\qquad = (7x^4 - 3x^4) + (4x^2 - x^2) + (-3x - 4x) + 2$

$\qquad = 4x^4 + 3x^2 - 7x + 2$

Remark: A very common mistake is to fail to change the sign of *each* term inside parentheses preceded by a minus sign.

$$-(3x^2 - 7xy + 3) = -3x^2 + 7xy - 3$$
$$\neq -3x^2 - 7xy + 3 \qquad \textit{Common mistake}$$

To find the product of two polynomials, the Distributive Laws

$$a(b + c) = ab + ac \qquad \text{and} \qquad (a + b)c = ac + bc$$

are useful. For example, if we treat $(3x - 2)$ as a *single* quantity in the Distributive Law, we have

$(3x - 2)(5x + 7)$

$\qquad = (3x - 2)(5x) + (3x - 2)(7)$ *Left Distributive Law*

$\qquad = (3x)(5x) - 2(5x) + (3x)(7) - 2(7)$ *Right Distributive Law*

$\qquad = 15x^2 - 10x + 21x - 14$ *Law of Exponents*

$\qquad = 15x^2 + 11x - 14$

When multiplying two polynomials, you must take care to see that each term of one polynomial is multiplied by each term of the other one. The following vertical arrangement works well for such multiplications.

$$
\begin{array}{r}
2x^2 - 3x + 5 \\
x - 4 \\
\hline
2x^3 - 3x^2 + 5x \\
- 8x^2 + 12x - 20 \\
\hline
2x^3 - 11x^2 + 17x - 20
\end{array}
$$

$x(2x^2 - 3x + 5)$

$-4(2x^2 - 3x + 5)$

Combine like terms

We list next some special binomial products that should be memorized.

SPECIAL BINOMIAL PRODUCTS

Form	Product	Example
(Binomial)(Binomial)	$(ax + b)(cx + d)$ $= acx^2 + adx + bcx + bd$ $= acx^2 + (ad + bc)x + bd$ $= mx^2 + nx + r$	$(3x - 2)(x + 5) = 3x^2 + 15x - 2x - 10$ $= 3x^2 + 13x - 10$
(Binomial)^2	$(u + v)^2 = u^2 + 2uv + v^2$ $(u - v)^2 = u^2 - 2uv + v^2$	$(3x - 2)^2 = (3x)^2 + 2(3x)(-2) + (-2)^2$ $= 9x^2 - 12x + 4$
(Sum)(Difference)	$(u + v)(u - v) = u^2 - v^2$	$(7x + 4)(7x - 4) = (7x)^2 - (4)^2$ $= 49x^2 - 16$
(Binomial)^3	$(u + v)^3 = u^3 + 3u^2v + 3uv^2 + v^3$	$(x + 2)^3 = x^3 + 3x^2(2) + 3x(2)^2 + (2)^3$ $= x^3 + 6x^2 + 12x + 8$
	$(u - v)^3 = u^3 - 3u^2v + 3uv^2 - v^3$	$(2x - 1)^3 = (2x)^3 - 3(2x)^2(1) + 3(2x)(1)^2 - (1)^3$ $= 8x^3 - 12x^2 + 6x - 1$

These special products can be generalized so that the u and v terms are replaced by quantities enclosed in parentheses. For instance,

$$[(a + 2) + (b - 1)]^2 = (a + 2)^2 + 2(a + 2)(b - 1) + (b - 1)^2$$
$$[u + v]^2 = u^2 + 2uv + v^2$$

where each of the resulting terms can be expanded according to the same rules.

Evaluating Algebraic Expressions

We frequently need to **evaluate** an algebraic expression. This means that a specific number is assigned to each variable in the expression. When evaluating, we need to be aware of what numbers are *permissible* to use in place of the variables. (In this context we consider permissible numbers as those that yield *real*-valued answers.) The set of permissible values that can be assigned to the variable in an algebraic expression is commonly referred to as the **domain** of the variable. Thus, the domain of x in the expression $\sqrt{x - 2}$ is all $x \geq 2$. Similarly, the domain of x in the expression $3/(x^2 - 4)$ is all $x \neq \pm 2$.

EXAMPLE 2
Evaluating Algebraic Expressions

Evaluate the following expression for $x = -1$.

$$2x^3 - 5x^2 + 3x$$

Solution:
For $x = -1$,

$$
\begin{aligned}
2x^3 - 5x^2 + 3x &= 2(-1)^3 - 5(-1)^2 + 3(-1) \quad &\textit{Replace x with } -1 \\
&= 2(-1) - 5(1) - 3 \quad &\textit{Raise to powers} \\
&= -2 - 5 - 3 = -10 \quad &\textit{Simplify}
\end{aligned}
$$

We have seen how to multiply polynomials to get new polynomials. Now we show how to find the factors whose product will yield a given polynomial. This process of writing a polynomial as a product is called **factoring.** Factoring is an important tool for reducing fractional expressions and for solving equations and inequalities. Unless noted otherwise, we will limit our discussion of factoring to polynomials whose factors have integer coefficients.

Factoring Out a Monomial

We start with polynomials that can be written as the product of a monomial and another polynomial. The technique here is to use the Distributive Laws

$$a(b + c) = ab + ac \qquad \text{and} \qquad (a + b)c = ac + bc$$

from *right to left* rather than from left to right. We look for a monomial that is common to each term of the polynomial. For instance, the polynomial $9x^3 - 27x^4y + 18x^2$ has the monomial $9x^2$ as a factor in each term. Hence, we write

$$
\begin{aligned}
9x^3 - 27x^4y + 18x^2 &= (9x^2)(x) - (9x^2)(3x^2y) + (9x^2)(2) \\
&= 9x^2(x - 3x^2y + 2)
\end{aligned}
$$

EXAMPLE 3
Factoring Out the Greatest Common Factor

Factor out the greatest common factor in each of the following.

(a) $6y^3z - 4yz^2 + 2y^2z^3$ (b) $(x + 2)(a + b) + (x + 2)(a - b)$

Solution:

(a) $2yz$ is common to all three terms, so we write

$$
\begin{aligned}
6y^3z - 4yz^2 + 2y^2z^3 &= (2yz)(3y^2) - (2yz)(2z) + (2yz)(yz^2) \\
&= 2yz(3y^2 - 2z + yz^2)
\end{aligned}
$$

(b) In this case, the binomial factor $(x + 2)$ is common to both terms, so we write

$$(x + 2)(a + b) + (x + 2)(a - b)$$
$$= (x + 2)[(a + b) + (a - b)] \qquad \textit{Distributive Law}$$
$$= (x + 2)[a + b + a - b] \qquad \textit{Remove parentheses}$$
$$= (x + 2)(2a) \qquad \textit{Combine terms}$$

Factoring polynomials is more complicated than finding products of polynomials. In fact, it can be quite difficult to factor polynomials of degree greater than 2.

FACTORING SPECIAL POLYNOMIAL FORMS

Polynomial	Factored Form	Example
Difference of Two Squares	$u^2 - v^2 = (u + v)(u - v)$	$9x^2 - 4 = (3x)^2 - (2)^2$ $= (3x + 2)(3x - 2)$
Perfect Square Trinomial	$u^2 + 2uv + v^2 = (u + v)^2$	$x^2 + 6x + 9 = x^2 + 2(x)(3) + (3)^2$ $= (x + 3)^2$
Trinomial with Binomial Factors	$mx^2 + nx + r = (ax + b)(cx + d)$ $[m = ac, r = bd, n = ad + bc]$	$3x^2 - 2x - 5 = (3x - 5)(x + 1)$
Sum and Difference of Two Cubes	$u^3 + v^3 = (u + v)(u^2 - uv + v^2)$ $u^3 - v^3 = (u - v)(u^2 + uv + v^2)$	$x^3 + 8 = (x)^3 + (2)^3$ $= (x + 2)(x^2 - 2x + 4)$ $27x^3 - 1 = (3x)^3 - (1)^3$ $= (3x - 1)(9x^2 + 3x + 1)$

Difference of Two Squares

One of the easiest special polynomial forms to recognize and to factor is the difference of two squares. Think of this form as

$$u^2 - v^2 = (u + v)(u - v)$$

difference   opposite signs

To recognize perfect square terms, look for terms whose coefficients are squares of integers and whose variables have *even* powers.

EXAMPLE 4
Factoring the Difference
of Two Squares

Factor the following.

(a) $9 - 25x^2$ 　　　　　　　　　(b) $16x^4 - y^4$

Solution:

(a) Using the difference of two squares formula with $u = 3$ and $v = 5x$, we get

$$9 - 25x^2 = (3)^2 - (5x)^2 = (3 + 5x)(3 - 5x)$$

(b) Applying the formula twice, we get

$$16x^4 - y^4 = (4x^2)^2 - (y^2)^2 \qquad \text{\textit{1st application}}$$
$$= (4x^2 + y^2)(4x^2 - y^2)$$
$$= (4x^2 + y^2)[(2x)^2 - (y)^2]$$
$$= (4x^2 + y^2)(2x + y)(2x - y) \quad \text{\textit{2nd application}}$$

Perfect Square Trinomials

A perfect square trinomial is a trinomial that is the square of a binomial. It has the form

$$u^2 + 2uv + v^2 = (u + v)^2 \qquad \text{or} \qquad u^2 - 2uv + v^2 = (u - v)^2$$

same sign   same sign

Note the following characteristics of a perfect square trinomial.

1. The first term is a square, u^2.
2. The last term is a square, v^2.
3. The middle term is twice the product of the terms u and v.
4. The sign of the middle term determines the sign in the binomial.

EXAMPLE 5
Factoring Perfect
Square Trinomials

Factor $16x^2 + 8x + 1$.

Solution:

$$u^2 = 16x^2 = (4x)^2$$
$$v^2 = (1)^2$$
$$2uv = 2(4x)(1) = 8x$$

Thus,

$$16x^2 + 8x + 1 = (4x + 1)^2$$

Remark: When attempting to factor using one of the four special polynomial forms, you should first check for any common monomial factors and remove them before proceeding further.

Trinomials with Binomial Factors

Some trinomials that are not perfect squares can be factored into the product of two binomials according to the formula

$$mx^2 + nx + r = (ax + b)(cx + d)$$

This is simply the **FOIL** method in reverse. The goal is to find a combination of factors of m and r so that the outside and inside (**O** and **I**) products yield the middle term nx. This means that in the following scheme we want $nx = $ **O** + **I**.

$$mx^2 + nx + r = (?x + ?)(?x + ?)$$

$$nx = \mathbf{O} + \mathbf{I} \qquad \mathbf{O}$$

Remark: It is impossible to factor some trinomials $mx^2 + nx + r$ into the product of two binomials with rational coefficients. Such trinomials are called **irreducible quadratic polynomials.**

EXAMPLE 6
Trinomials with Binomial Factors

Factor the following.

(a) $x^2 - 7x + 12$ (b) $8x^2 + 22x + 9$

Solution:

(a) Consider

$$mx^2 + nx + r = x^2 - 7x + 12$$

Since $r = +12$, its factors have like signs, and since $n = -7$, both signs will be negative. Let's try $r = 12 = (-2)(-6)$.

$$x^2 - 7x + 12 \stackrel{?}{=} (x - 2)(x - 6) \qquad \textit{Test possible factors}$$
$$\mathbf{O} + \mathbf{I} = -8x \neq -7x \qquad \textit{Fails } \mathbf{O} + \mathbf{I} \textit{ Test}$$

A quick $\mathbf{O} + \mathbf{I}$ Test (in your head) of the factors $(x - 1)$ and $(x - 12)$ will show that they don't work either, so we are left with the following (correct) factorization.

$$x^2 - 7x + 12 \stackrel{?}{=} (x - 3)(x - 4) \qquad \textit{Test possible factors}$$
$$\mathbf{O} + \mathbf{I} = -7x \qquad \textit{Satisfies } \mathbf{O} + \mathbf{I} \textit{ Test}$$

(b) To factor $8x^2 + 22x + 9$, we choose combinations of factors from among

Factors of 8	**Factors of 9**
1, 2, 4, 8	1, 3, 9

The following tabular solution method is one of the most efficient ways to find and test *all* combinations of factors of m and r.

TABULAR METHOD FOR FACTORING TRINOMIALS

$$8x^2 + 22x + 9 = (?x + ?)(?x + ?)$$

Factors of $m = 8$	**Factors of** $r = 9$	**Factor Combinations** m	r	$\mathbf{O} + \mathbf{I} \stackrel{?}{=} 22$
1, 2, 4, 8	1, 3, 9	(1)(8)	(1)(9)	$\mathbf{O} + \mathbf{I} = 17$
		(1)(8)	(3)(3)	$\mathbf{O} + \mathbf{I} = 27$
		(2)(4)	(1)(9)	$\mathbf{O} + \mathbf{I} = 22$
		(2)(4)	(3)(3)	$\mathbf{O} + \mathbf{I} = 18$

Correct Factorization: $8x^2 + 22x + 9 = (2x + 1)(4x + 9)$

Remark: You will find that much of this work can be done in your head, and you can save even more time finding the correct factorization of trinomials. Note also that if $O + I$ yields the right number but the wrong sign, simply change the sign of *both* factors of *r*.

Sum or Difference of Cubes

The next two formulas show that sums and differences of cubes factor quite simply. Pay special attention to the signs of the terms.

$$u^3 + v^3 = (u + v)(u^2 - uv + v^2) \quad u^3 - v^3 = (u - v)(u^2 + uv + v^2)$$

with "like signs" labeling the $+$ in $(u+v)$ and the $-$ in $(u-v)$, and "unlike signs" labeling the $-uv$ and $+uv$ terms.

EXAMPLE 7
Sum or Difference of Cubes

Factor $y^3 - 27x^3$.

Solution:
Consider $u = y$ and $v = (3x)$. Then

$$y^3 - 27x^3 = (y)^3 - (3x)^3$$
$$= (y - 3x)[y^2 + y(3x) + (3x)^2]$$
$$= (y - 3x)(y^2 + 3xy + 9x^2)$$

Factoring by Grouping

Sometimes polynomials with more than three terms can be factored by a method called **factoring by grouping.** It is not always obvious which terms to group, and sometimes several different groupings will work. The goal is to find groupings that lead to the special factorizations discussed in this section.

EXAMPLE 8
Factoring by Grouping

Factor completely $4x^2 - 4x - y^2 + 1$.

Solution:
Consider a grouping that leads to a special trinomial.

$$4x^2 - 4x - y^2 + 1$$
$$= (4x^2 - 4x + 1) - y^2 \qquad \textit{Perfect square trinomial}$$
$$= (2x - 1)^2 - y^2 \qquad \textit{Difference of squares}$$
$$= [(2x - 1) + y][(2x - 1) - y] \qquad \textit{Factor}$$
$$= (2x - 1 + y)(2x - 1 - y) \qquad \textit{Simplify}$$

A general guideline to follow in factoring polynomials is to (1) factor out any common monomial factor, (2) factor according to one of the special polynomial forms, or (3) factor by grouping.

Section Exercises 1.3

In Exercises 1–20, perform the indicated operations and express each result as a polynomial.

1. Add: $-3x^2 - 13x + 7$ and $14x - 15$
2. Add: $3x^4 - 19x^2 + 1$ and $17x^3 + 4x$
3. Subtract: $-8x^3 - 14x^2 - 17$ from $15x^2 - 6$
4. Subtract: $13x^4 - 5x + 15$ from $15x^4 - 18x - 19$
5. $(9x^5 + 10x^3 + 8x) + (-7x^3 + 19x^2 - 7x - 9) + (5x^4 + 16x^3)$
6. $(-14x^4 - 5x^2 - 6) + (-2x^3 - 9x + 5) + (-8x^4 + 7x)$
7. $(18x^7 + 16x^3 - 7) - (-7x^7 + 19)$
8. $(-4x^2 - 2x - 8) - (9x^2 - 10x) + (-7x - 4)$
9. $(-17x^3 - 16xy - 9y^2) + (18x^3 + 4xy)$
10. $(-13x^2 - 19y^2 + 6xy) - (9xy + 2x^2 - 16y^2)$
11. Multiply: $(x^3 - 2x + 1)$ by $(x - 5)$
12. Multiply: $(x - y + 1)$ by $(x + y - 1)$
13. $(x^2 + 9)(x^2 - x - 4)$
14. $(x - 2)(x^2 + 2x + 4)$
15. $(x + 3)(x^2 - 3x + 9)$
16. $(2x^2 + 3y^3)(4x^4 - 6x^2y^3 + 9y^6)$
17. $(x^2 + 1)(x + 1)(x - 1)$
18. $(x^2 + x - 2)(x^2 - x + 2)$
19. $(x + \sqrt{5})(x - \sqrt{5})(x + 4)$
20. $(2x - y)(x + 3y) + 3(2x - y)$

In Exercises 21–46, use the special products given in this section to perform the indicated multiplication of binomials.

21. $(x + 3)(x + 4)$
22. $(x - 5)(x + 10)$
23. $(3x - 5)(2x + 1)$
24. $(7x - 2)(4x - 3)$
25. $(x + 6)^2$
26. $(3x - 2)^2$
27. $(2x - 5y)^2$
28. $[(x + 1) - y]^2$
29. $[(x - 3) + y]^2$
30. $(x + 3)(x - 3)$
31. $(x + 2y)(x - 2y)$
32. $(2x + 3y)(2x - 3y)$
33. $(2r^2 - 5)(2r^2 + 5)$
34. $(3a^3 - 4b^2)(3a^3 + 4b^2)$
35. $(x + 1)^3$
36. $(x - 2)^3$
37. $(2x - y)^3$
38. $(3x + 2y)^3$
39. $[(x + 1) - y]^3$
40. $[(z - 2) + y]^3$
41. $(\sqrt{x} + \sqrt{y})(\sqrt{x} - \sqrt{y})$
42. $(5 + 2\sqrt{s})(5 - 2\sqrt{s})$
43. $(4r^3 - 3s^2)^2$
44. $(8y + \sqrt{z})^2$
45. $(10m - 3n)(4n + 3m)$
46. $(3x + y)^3$

In Exercises 47–54, factor out the common factor.

47. $3x + 6$
48. $5y - 30$
49. $xy - xz$
50. $-2cx - 10cy$
51. $9a^2b - 12ab^3$
52. $16x^2y^2 + 8xy^2 - 20xy$
53. $(x + y)z^2 - (x + y)$
54. $(x - 1)(y + z) + (x - 1)(y - z)$

In Exercises 55–62, factor each difference of squares.

55. $x^2 - 36$
56. $z^2 - 100$
57. $16y^2 - 9$
58. $49 - 9y^2$
59. $(x - 1)^2 - 4$
60. $25 - (z + 5)^2$
61. $81 - y^2$
62. $x^4 - 16$

In Exercises 63–70, factor each perfect square trinomial.

63. $x^2 - 4x + 4$
64. $x^2 + 10x + 25$
65. $4x^2 + 4x + 1$
66. $9x^2 - 12x + 4$
67. $x^2 - 4xy + 4y^2$
68. $25y^2 + 40yz + 16z^2$
69. $a^2b^2 - 2abc + c^2$
70. $x^4 - 2x^2z + z^2$

In Exercises 71–78, factor each trinomial with distinct binomial factors.

71. $x^2 + x - 2$
72. $x^2 + 5x + 6$
73. $x^2 - 5x + 6$
74. $2x^2 - x - 1$
75. $3x^2 - 5x + 2$
76. $x^2 - xy - 2y^2$
77. $2y^2 + 21yz - 36z^2$
78. $60x^2 - 61xy + 15y^2$

In Exercises 79–86, factor each sum or difference of cubes.

79. $x^3 - 8$
80. $x^3 - 27$
81. $y^3 + 64$
82. $z^3 + 125$
83. $x^3 - 27y^3$
84. $64x^3 + 125y^3$
85. $x^6 + 64$
86. $(x - a)^3 + b^3$

In Exercises 87–92, factor by grouping.

87. $xy - y + xz - z$
88. $xy - xz + 5y - 5z$
89. $5xy - 3y + 10x - 6$
90. $2ab + 6a - 7b - 21$
91. $7 - 10x - 7y + 10xy$
92. $ar - as - 8r + 8s$

In Exercises 93–106, completely factor each expression.

93. $2x^3 - 2x^2 - 4x$
94. $2ay^3 - 7ay^2 - 15ay$
95. $63rs^2 - 7r^3$
96. $80 - 5z^2$
97. $4x^3y - 4x^2y^2 + xy^3$
98. $150x + 120x^2 + 24x^3$
99. $6x^4 - 48xy^3$
100. $(x^2 + y^2)^2 - 4x^2y^2$
101. $(x^2 + 2y^2)^2 - 9x^2y^2$
102. $9x(3x - 9)^2 + (3x - 9)^3$
103. $4x^2(2x - 1) + 2x(2x - 1)^2$
104. $27x^2y - x^2y^4$
105. $2x^2z^2 + 2x^2z + 4xyz^2 + 4xyz$
106. $2r^2u + 10ru - 6r^2v - 30rv$
107. What is the degree of the product of two polynomials? (Assume one is of degree m and the other of degree n.)

Calculator Exercises

108. Use a calculator to evaluate $6x^3 - 4.2x^2 + 2.7$ when
    (a) $x = 5$
    (b) $x = 1.5$
    (c) $x = -0.43$
    (d) $x = \frac{2}{3}$

109. Use a calculator to evaluate $5/(x^2 + 2x)$ when
    (a) $x = 3.2$
    (b) $x = -1.5$
    (c) $x = -2$
    (d) $x = \frac{4}{3}$

110. Use the formula

$$(x + a)^5 = x^5 + 5x^4a + 10x^3a^2 + 10x^2a^3 + 5xa^4 + a^5$$

to factor

$$x^5 - 10x^4 + 40x^3 - 80x^2 + 80x - 32$$

111. Use the formula

$$(x + a)^3 = x^3 + 3x^2a + 3xa^2 + a^3$$

to factor

$$27x^3 + 27x^2y + 9xy^2 + y^3$$

Solving Inequalities   1.4

Inequalities were introduced in Section 1.1 in the context of *order* on the real number line. Here we look at inequalities involving a variable. To *solve* such inequalities means to find the set of all real numbers for which the statement is true. In most cases, these solution sets consist of intervals on the real line and we call them **solution intervals** of the inequality.

EXAMPLE 1
Solving Linear Inequalities

Solve the following inequalities:

(a) $5x - 7 < 3x + 9$ (b) $1 - \dfrac{3x}{2} \geq x - 4$ (c) $-3 \leq 6x - 1 < 3$

Solution:

(a) $5x - 7 < 3x + 9$ *Given*

      $5x - 3x < 9 + 7$ *Subtract 3x and add 7*

           $2x < 16$ *Combine terms*

            $x < 8$ *Divide by 2*

Thus, the solution interval is $x < 8$.

(b) $1 - \dfrac{3x}{2} \geq x - 4$ *Given*

     $2 - 3x \geq 2x - 8$ *Multiply by LCD*

        $-5x \geq -10$ *Subtract 2x, subtract 2*

                               Divide by -5 and reverse the

          $x \leq 2$ *inequality*

Thus, the solution interval is $x \leq 2$.

(c) In this case, we have two inequalities which we solve simultaneously.

      $-3 \leq 6x - 1 < 3$ *Given*

       $-2 \leq 6x < 4$ *Add 1*

      $-\dfrac{2}{6} \leq x < \dfrac{4}{6}$ *Divide by 6*

      $-\dfrac{1}{3} \leq x < \dfrac{2}{3}$ *Reduce fractions*

Many important uses of inequalities involve absolute values. In Section 1.1 we used absolute value to denote the distance between two points on the number line. That is,

$$d(a, b) = |a - b| = \text{distance between points } a \text{ and } b$$

This means that

$$d(x, 0) = |x| = \text{distance between } x \text{ and } 0$$

As a consequence, we give the following interpretation of absolute value inequalities.

ABSOLUTE VALUE INEQUALITIES

For $a > 0$:

Inequality	**Interpretation**	**Sketch**
$|x| < a$	All points x whose distance from 0 is *less* than a	
$|x| > a$	All points x whose distance from 0 is *greater* than a	

In summary:

$|x| < a$ if and only if $-a < x < a$

$|x| > a$ if and only if $x < -a$ or $x > a$

Remark: Informally, it may help to think of *less than* as denoting ''inside-ness'' and *greater than* as denoting ''outsideness.'' That is,

$|x| < a$ has solutions *inside* the interval $(-a, a)$

$|x| > a$ has solutions *outside* the interval $(-a, a)$

EXAMPLE 2
Solving Absolute
Value Inequalities

Solve the inequalities and sketch the solution intervals.

(a) $|x - 5| < 2$ (b) $|x + 3| \geq 7$

Solution:

(a) We seek all points x whose distance from 5 is less than 2, as shown in Figure 1.4.

$$|x - 5| < 2 \qquad \text{\textit{Given}}$$
$$-2 < x - 5 < 2 \qquad \text{\textit{Interpret absolute value}}$$
$$3 < x < 7 \qquad \text{\textit{Add 5}}$$

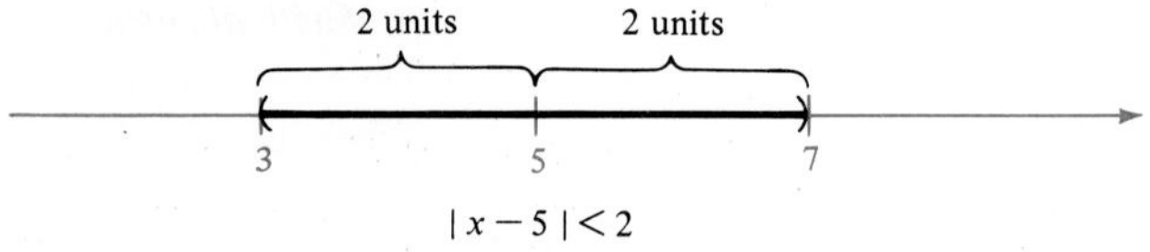

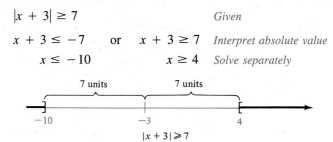

FIGURE 1.4

(b) We seek all points x whose distance from -3 is greater than or equal to 7, as shown in Figure 1.5.

$$|x + 3| \geq 7 \qquad \qquad \textit{Given}$$

$$x + 3 \leq -7 \quad \text{or} \quad x + 3 \geq 7 \qquad \textit{Interpret absolute value}$$

$$x \leq -10 \qquad \qquad x \geq 4 \qquad \textit{Solve separately}$$

FIGURE 1.5

Solving Inequalities by Factoring

Many nonlinear inequalities lend themselves to two solution methods used for quadratic equations—*factoring* and *completing the square*. With the factoring method, we use the principle that a polynomial can change signs *only* at its zeros (the values that make the polynomial zero). Between two consecutive zeros a polynomial must be entirely positive or entirely negative. This means that when the real zeros of a polynomial are put in order, they divide the real line into intervals in which the polynomial has no sign changes. For example, the polynomial

$$x^2 - x + 6 = (x - 3)(x + 2)$$

can change signs only at $x = -2$ and $x = 3$. In the context of polynomial inequalities, we call these values the **critical numbers** and the resulting intervals on the real line the **test intervals** of the inequality. We need to test only *one value* from each interval to solve a polynomial inequality.

EXAMPLE 3
Solving Inequalities by Factoring

Find the solution intervals for each of the following:

(a) $x^2 < x + 6$ \qquad\qquad (b) $2x^3 + 5x^2 \geq 12x$

Solution:

(a) $\qquad\qquad x^2 < x + 6 \qquad\qquad$ *Given*

$\qquad x^2 - x - 6 < 0 \qquad$ *Standard form*

$\qquad (x - 3)(x + 2) < 0 \qquad$ *Factor*

Critical numbers:

$$x = -2, \qquad x = 3$$

Test intervals:

$$x < -2, \qquad -2 < x < 3, \qquad x > 3$$

To test an interval, we choose a representative number in the interval and compute the sign of each factor. For example, for any $x < -2$, both of the factors $(x - 3)$ and $(x + 2)$ are negative. Consequently, the product (of two negatives) is positive and the inequality is *not* satisfied in the interval $x < -2$. We suggest the testing format shown in Figure 1.6.

Test: Is $(x - 3)(x + 2) < 0$?

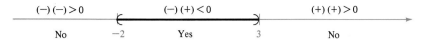

FIGURE 1.6

Since the inequality is satisfied only by the center test interval, we conclude that the solution interval is

$$-2 < x < 3$$

(b) $\qquad 2x^3 + 5x^2 \geq 12x$

$2x^3 + 5x^2 - 12x \geq 0$

$x(2x - 3)(x + 4) \geq 0$

Critical numbers:

$$x = -4, \qquad x = 0, \qquad x = \tfrac{3}{2}$$

Test intervals:

$$x < -4, \qquad -4 < x < 0, \qquad 0 < x < \tfrac{3}{2}, \qquad x > \tfrac{3}{2}$$

Test: Is $x(2x - 3)(x + 4) \geq 0$?

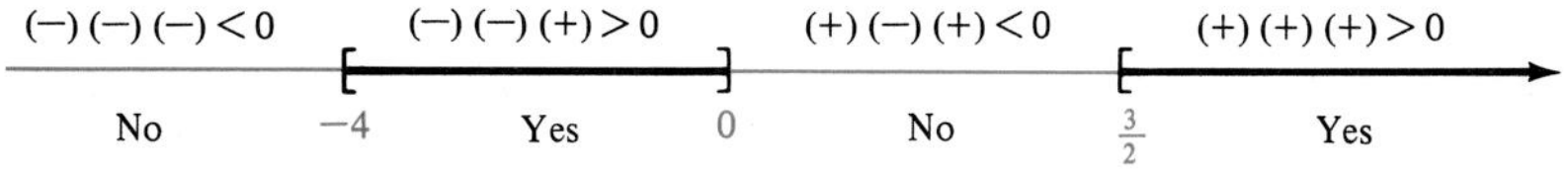

$(-)(-)(-) < 0$	$(-)(-)(+) > 0$	$(+)(-)(+) < 0$	$(+)(+)(+) > 0$
No	Yes	No	Yes
-4	0	$\tfrac{3}{2}$	

FIGURE 1.7

Solution intervals:

$$-4 \leq x \leq 0 \qquad \text{or} \qquad x \geq \tfrac{3}{2}$$

The concept of critical numbers can be extended to inequalities involving fractional expressions. Specifically, an expression that is the ratio of two polynomials can change signs only at its zeros (the values that make the numerator zero) *and* at its undefined values (the values that make the denominator zero). For example, the expression

$$\frac{x - 1}{(x - 2)(x + 3)}$$

can change signs only at $x = 1$, $x = 2$, and $x = -3$. Thus, these three values are the critical numbers of the inequality

$$\frac{x - 1}{(x - 2)(x + 3)} < 0$$

EXAMPLE 4
Inequality Involving
a Fractional Expression

Solve

$$\frac{2x - 7}{x - 5} \leq 3$$

Solution:
We should rewrite the inequality with 0 alone on the right.

$$\frac{2x - 7}{x - 5} \leq 3 \qquad \qquad \textit{Given}$$

$$\frac{2x - 7}{x - 5} - 3 \leq 0 \qquad \qquad \textit{Subtract 3}$$

$$\frac{2x - 7 - 3x + 15}{x - 5} \leq 0 \qquad \qquad \textit{Combine terms}$$

$$\frac{-x + 8}{x - 5} \leq 0$$

Critical numbers:

$$x = 5, \qquad x = 8$$

Test intervals:

$$x < 5, \qquad 5 < x < 8, \qquad x > 8$$

Test: Is $(8 - x)/(x - 5) \leq 0$?

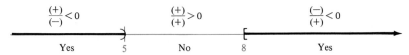

FIGURE 1.8

Solution intervals:

$$x < 5 \qquad \text{or} \qquad x \geq 8$$

Note that we allow the critical number 8 in the solution intervals, but we do not allow $x = 5$ since it yields a zero denominator in the original expression.

Before looking at our next example, note these two properties of absolute value:

$$|ab| = |a| \cdot |b| \quad \text{and} \quad \left|\frac{a}{b}\right| = \frac{|a|}{|b|}, \quad b \neq 0$$

In Exercises 69 and 70 you are asked to prove these properties.

EXAMPLE 5
An Inequality Involving
Absolute Value

Solve

$$|x - 2| \leq |4x + 1|$$

Solution:

$$|x - 2| \leq |4x + 1| \qquad \textit{Given}$$

$$\frac{|x - 2|}{|4x + 1|} \leq 1 \qquad \textit{Divide by } |4x + 1|$$

$$\left|\frac{x - 2}{4x + 1}\right| \leq 1 \qquad \textit{Property of absolute value}$$

$$-1 \leq \frac{x - 2}{4x + 1} \leq 1 \qquad \textit{Interpret absolute value}$$

$$\frac{x - 2}{4x + 1} \geq -1 \qquad\qquad \frac{x - 2}{4x + 1} \leq 1$$

$$\frac{x - 2}{4x + 1} + 1 \geq 0 \qquad\qquad \frac{x - 2}{4x + 1} - 1 \leq 0$$

$$\frac{5x - 1}{4x + 1} \geq 0 \qquad\qquad \frac{-3x - 3}{4x + 1} \leq 0$$

Critical numbers: *Critical numbers:*

$$x = -\tfrac{1}{4}, \quad x = \tfrac{1}{5} \qquad\qquad x = -1, \quad x = -\tfrac{1}{4}$$

Now, combining these two sets of critical numbers, we have the following test intervals:

Test intervals:

$$x < -1, \quad -1 < x < -\tfrac{1}{4}, \quad -\tfrac{1}{4} < x < \tfrac{1}{5}, \quad x > \tfrac{1}{5}$$

To test these intervals it is convenient to choose x-values and substitute back into the original equation.

Interval	x-value	$\|x - 2\|$	$\|4x + 1\|$	$\|x - 2\| \leq \|4x + 1\|$?
$x < -1$	-2	4	7	yes
$-1 < x < -\frac{1}{4}$	$-\frac{1}{2}$	$\frac{5}{2}$	1	no
$-\frac{1}{4} < x < \frac{1}{5}$	0	2	1	no
$x > \frac{1}{5}$	1	1	5	yes

Since $x = -1$ and $x = \frac{1}{5}$ both satisfy the original inequality, the solution intervals are as follows.

Solution intervals: $x \leq -1$ or $x \geq \frac{1}{5}$

For inequalities involving an irreducible quadratic, the factoring method is inappropriate, and we use a different method for finding the solution intervals. This method involves *completing the square* and is based on the following relationship:

PERFECT SQUARE INEQUALITIES

For $a > 0$:

$x^2 < a$ if and only if $-\sqrt{a} < x < \sqrt{a}$

$x^2 > a$ if and only if $x < -\sqrt{a}$ or $x > \sqrt{a}$

EXAMPLE 6
Solving Inequalities by
Completing the Square

Find the solution interval(s) for

$$3x^2 - 6x \leq 8$$

Solution:

$3x^2 - 6x \leq 8$	*Given*
$3x^2 - 6x - 8 \leq 0$	*Standard form*

Since this quadratic is not factorable, we complete the square.

$x^2 - 2x \leq \frac{8}{3}$	*Divide by coefficient of x^2*
$x^2 - 2x + 1 \leq \frac{8}{3} + 1$	*Add $(\frac{1}{2} \cdot 2)^2$*
$(x - 1)^2 \leq \frac{11}{3}$	*Perfect square inequality*
$\|x - 1\| < \sqrt{\frac{11}{3}}$	*Absolute value form*
$-\sqrt{\frac{11}{3}} < x - 1 < \sqrt{\frac{11}{3}}$	*Interpret absolute value*
$1 - \sqrt{\frac{11}{3}} < x < 1 + \sqrt{\frac{11}{3}}$	*Solution interval*
$-0.91 < x < 2.9$	*Decimal approximation*

EXAMPLE 7
Applications of Inequalities

(a) A projectile is fired straight upward from ground level with an initial velocity of 384 feet per second. During what time period will its height exceed 2048 feet?

(b) A subcompact car can be rented from Company A for $180 per week with no charge for mileage. A similar car can be rented from Company B for $100 per week, plus 20 cents per mile driven. For what weekly mileage m does it cost less to rent from Company A?

Solution:

(a) Recall that the position of an object moving in a vertical path is given by $s = -16t^2 + v_0t + s_0$. In this case, $s_0 = 0$ and $v_0 = 384$. Thus, we seek the solution interval for the inequality

$$-16t^2 + 384t > 2048 \qquad \text{\textit{Divide by} } -16 \text{ \textit{and reverse}}$$
$$t^2 - 24t < -128 \qquad \text{\textit{inequality}}$$
$$t^2 - 24t + 128 < 0 \qquad \text{\textit{Standard form}}$$
$$(t - 8)(t - 16) < 0 \qquad \text{\textit{Factored form}}$$

Critical numbers: $t = 8, \qquad t = 16$

Test intervals: $t < 8, \qquad 8 < t < 16, \qquad t > 16$

Solution interval: $8 < t < 16$

(b) We need to solve the inequality

(weekly cost from B) > (weekly cost from A)
$$100 + 0.20m > 180$$
$$0.20m > 80$$

Solution interval: $m > 400$ miles

Section Exercises 1.4

In Exercises 1–4, determine whether or not the given value of x satisfies the inequality.

1. $5x - 12 > 0$
   (a) $x = 3$ (b) $x = -3$ (c) $x = \frac{5}{2}$ (d) $x = \frac{3}{2}$

2. $x + 1 < \dfrac{2x}{3}$
   (a) $x = 0$ (b) $x = 4$ (c) $x = -4$ (d) $x = -3$

3. $0 < \dfrac{x - 2}{4} < 2$
   (a) $x = 4$ (b) $x = 10$ (c) $x = 0$ (d) $x = \frac{7}{2}$

4. $-1 < \dfrac{3 - x}{2} \le 1$
   (a) $x = 0$ (b) $x = \sqrt{5}$ (c) $x = 1$ (d) $x = 5$

In Exercises 5–40, solve the inequality and graph the solution on the real number line.

5. $x - 5 \ge 7$
6. $2x > 3$
7. $4x + 1 < 2x$
8. $2x + 7 < 3$
9. $2x - 1 \ge 0$
10. $3x + 1 \ge 2$
11. $4 - 2x < 3$
12. $x - 4 \le 2$
13. $-4 < \dfrac{2x - 3}{3} < 4$
14. $0 \le \dfrac{x + 3}{2} < 5$
15. $\frac{3}{4} > x + 1 > \frac{1}{4}$
16. $-1 < -\dfrac{x}{3} < 1$
17. $|x| < 5$
18. $|2x| < 6$
19. $\left|\dfrac{x}{2}\right| > 3$
20. $|5x| > 10$

21. $|x + 2| < 5$

22. $|3x + 1| \geq 4$

23. $\left|\dfrac{x - 3}{2}\right| \geq 5$

24. $|2x + 1| < 5$

25. $|9 - 2x| < 1$

26. $\left|1 - \dfrac{2x}{3}\right| < 1$

27. $x^2 \leq 9$

28. $x^2 < 5$

29. $x^2 > 4$

30. $(x - 3)^2 \geq 1$

31. $(x + 2)^2 < 25$

32. $(x + 6)^2 \leq 8$

33. $x^2 + 4x + 4 \geq \frac{9}{4}$

34. $x^2 - 6x + 9 < \frac{16}{9}$

35. $x^2 + x - 6 < 0$

36. $4x^3 - 12x^2 > 0$

37. $3(x - 1)(x + 1) > 0$

38. $6(x + 2)(x - 1) < 0$

39. $4x^3 - 6x^2 < 0$

40. $x^2 + 2x - 3 > 0$

In Exercises 41–48, find the interval(s) in which the given expression is defined.

41. $\sqrt{x^2 - 7x + 12}$

42. $\sqrt{x^2 - 4}$

43. $\sqrt[4]{4 - x^2}$

44. $\sqrt{144 - 9x^2}$

45. $\sqrt{12 - x - x^2}$

46. $\sqrt{x^2 + 4}$

47. $\sqrt{x^2 - 3x + 3}$

48. $\sqrt[4]{-x^2 + 2x - 2}$

49. Use absolute values to define each interval (or pair of intervals) on the real line.

(a)

(b)

50. Use absolute values to define each interval (or pair of intervals) on the real line.

(a) All real numbers within 10 units of 12.

(b)

In Exercises 51–60, solve each inequality.

51. $\dfrac{x + 6}{x + 1} < 2$

52. $\dfrac{x + 12}{x + 2} \geq 3$

53. $\dfrac{3x - 5}{x - 5} > 4$

54. $\dfrac{5 - 2x}{1 + 3x} < 6$

55. $\dfrac{4}{x + 5} > \dfrac{1}{2x + 3}$

56. $\dfrac{5}{x - 6} > \dfrac{3}{x + 2}$

57. $\dfrac{4}{5x - 6} \leq \dfrac{6}{7x + 9}$

58. $|x + 6| > |2x - 5|$

59. $|1 - x| \leq |x + 2|$

60. $|2x + 3| < |3x + 4|$

61. A projectile is fired straight upward from ground level with an initial velocity of 160 feet per second.

(a) At what instant will it be back at ground level?

(b) During what time period will its height exceed 384 feet?

62. A rectangle with a perimeter of 100 meters is to have an area of at least 500 square meters. Within what bounds must the length of the rectangle lie?

63. P dollars invested at simple interest rate r for t years grows to an amount

$$A = P + Prt$$

If an investment of $1000 is to grow to an amount greater than $1250 in two years, then the interest rate must be greater than what percentage?

64. A family establishes a business selling mini-donuts at a shopping mall. The cost of making a dozen donuts is $1.45, and the donuts sell for $2.95 per dozen. In addition to the cost of the ingredients, the business must pay $25 per day for rent and utilities. If the daily profit varies between $50 and $200, between what levels (in dozens) do the daily sales vary?

65. In the manufacture and sale of a certain product, the revenue for selling x units is

$$R = 115.95x$$

and the cost of producing x units is

$$C = 95x + 750$$

In order for a profit to be realized, it is necessary that R be greater than C. For what values of x will this product return a profit?

66. A utility company has a fleet of vans for which the annual operating cost per van is estimated to be

$$C = 0.32m + 2300$$

where C is measured in dollars and m is measured in miles. If the company wants the annual operating cost to be less than $10,000, then m must be less than what value?

67. The heights, h, of two-thirds of the members of a certain population satisfy the inequality

$$\left|\dfrac{h - 68.5}{2.7}\right| \leq 1$$

where h is measured in inches. Determine the interval on the real line in which these heights lie.

68. The estimated daily production, p, at a refinery is given by

$$|p - 2{,}250{,}000| < 125{,}000$$

where p is measured in barrels of oil. Determine the high and low production levels.

69. Given any two real numbers a and b, prove that

$$|ab| = |a| \cdot |b|$$

70. Given any two real numbers a and b, prove that

$$\left|\dfrac{a}{b}\right| = \dfrac{|a|}{|b|}$$

The Cartesian Plane and the Distance Formula

1.5

Just as we can represent the real numbers geometrically by points on the real line, we can represent ordered pairs of real numbers by points in a plane. An **ordered pair** (x, y) of real numbers has x as its *first* member and y as its *second* member. The model for representing ordered pairs is called the **rectangular coordinate system,** or the **Cartesian plane.** It is developed by considering two real lines intersecting at right angles (Figure 1.9).

The horizontal real line is traditionally called the **x-axis,** and the vertical real line is called the **y-axis.** Their point of intersection is called the **origin,** and the lines divide the plane into four parts called **quadrants** (Figure 1.10).

We identify each point in the plane by an ordered pair (x, y) of real numbers x and y, called the **coordinates** of the point. The number x represents the directed distance from the y-axis to the point, and y represents the directed distance from the x-axis to the point (Figure 1.11). For the point (x, y), the first coordinate is referred to as the x-coordinate or **abscissa,** and the second or y-coordinate is referred to as the **ordinate.**

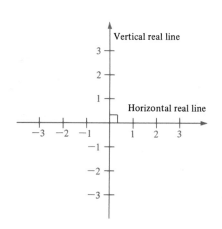

FIGURE 1.9

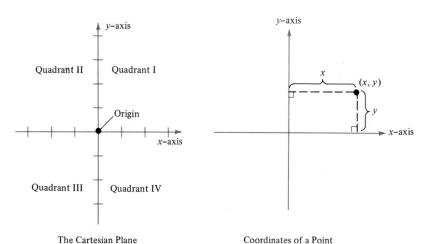

The Cartesian Plane

FIGURE 1.10

Coordinates of a Point

FIGURE 1.11

Remark: We use the notation (x, y) to denote both a point in the plane and an open interval on the real line. Generally this should cause no confusion because the nature of a specific problem will show which we are talking about.

EXAMPLE 1
Plotting Points in the Cartesian Plane

Locate the points $(-1, 2)$, $(3, 4)$, $(0, 0)$, $(3, 0)$, and $(-2, -3)$ in the Cartesian plane.

Solution:
The solution is shown in Figure 1.12.

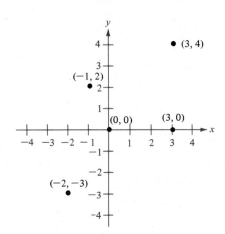

FIGURE 1.12

Development of the Distance Formula

In Section 1.1 we determined the distance between two points x_1 and x_2 on the real line. We will now find the distance between two points in the plane. Recall from the Pythagorean Theorem that, for a right triangle with hypotenuse c and sides a and b, we have the relationship $a^2 + b^2 = c^2$. Conversely, if $a^2 + b^2 = c^2$, then the triangle is a right triangle (Figure 1.13).

Suppose we wish to determine the distance d between two points (x_1, y_1) and (x_2, y_2) in the plane. Using these two points, a right triangle can be formed, as shown in Figure 1.14. We see that the length of the vertical side of the triangle is $|y_2 - y_1|$. Similarly, the length of the horizontal side of the triangle is $|x_2 - x_1|$. By the Pythagorean Theorem, we then have

$$d^2 = |x_2 - x_1|^2 + |y_2 - y_1|^2 \quad \text{or} \quad d = \sqrt{|x_2 - x_1|^2 + |y_2 - y_1|^2}$$

Replacing $|x_2 - x_1|^2$ and $|y_2 - y_1|^2$ by the equivalent expressions $(x_2 - x_1)^2$ and $(y_2 - y_1)^2$, we can write

$$d = \sqrt{(x_2 - x_1)^2 + (y_2 - y_1)^2}$$

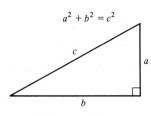

Pythagorean Theorem

FIGURE 1.13

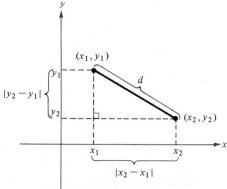

FIGURE 1.14   Distance Between Two Points

We choose the positive square root for d because the distance between two points is not a directed distance. We have therefore established the following rule:

DISTANCE FORMULA

> The distance d between two points (x_1, y_1) and (x_2, y_2) in the plane is given by
>
> $$d = \sqrt{(x_2 - x_1)^2 + (y_2 - y_1)^2}$$

EXAMPLE 2
Finding the Distance
Between Two Points

Find the distance between the points $(-2, 1)$ and $(3, 4)$.

Solution:
Applying the Distance Formula, we have

$$
\begin{aligned}
d &= \sqrt{[3 - (-2)]^2 + (4 - 1)^2} \\
&= \sqrt{(5)^2 + (3)^2} = \sqrt{25 + 9} \\
&= \sqrt{34} \approx 5.83
\end{aligned}
$$

EXAMPLE 3
An Application of the
Distance Formula

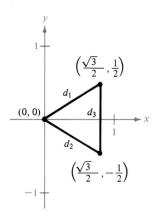

FIGURE 1.15

Use the Distance Formula to show that the triangle formed by $(0, 0)$, $(\sqrt{3}/2, 1/2)$, $(\sqrt{3}/2, -1/2)$ is equilateral.

Solution:
Refer to Figure 1.15. The three sides have lengths

$$d_1 = \sqrt{\left(\frac{\sqrt{3}}{2} - 0\right)^2 + \left(\frac{1}{2} - 0\right)^2} = \sqrt{\frac{3}{4} + \frac{1}{4}} = \sqrt{1} = 1$$

$$d_2 = \sqrt{\left(\frac{\sqrt{3}}{2} - 0\right)^2 + \left(-\frac{1}{2} - 0\right)^2} = \sqrt{\frac{3}{4} + \frac{1}{4}} = \sqrt{1} = 1$$

$$d_3 = \sqrt{\left(\frac{\sqrt{3}}{2} - \frac{\sqrt{3}}{2}\right)^2 + \left[\frac{1}{2} - \left(-\frac{1}{2}\right)\right]^2} = \sqrt{0 + 1} = \sqrt{1} = 1$$

Since $d_1 = d_2 = d_3$, we can conclude that the triangle is equilateral.

Remark: In Example 3, the figure provided was not really essential to the solution of the problem. *Nevertheless,* we strongly recommend that you get in the habit of including sketches with your problem solutions even if they are not specifically required. Throughout our many years of teaching mathematics, we have found that students who have a good grasp of the visual aspects of mathematics are very often the same students who do well with the technical aspects of the subject.

Next we introduce a rule for finding the coordinates of the midpoint of the line segment joining two points in the plane.

MIDPOINT RULE

The midpoint of the line segment joining points (x_1, y_1) and (x_2, y_2) is

$$\left(\frac{x_1 + x_2}{2}, \frac{y_1 + y_2}{2}\right)$$

Proof:
To find the midpoint of a line segment, we merely find the "average" values of the respective coordinates of the two endpoints. To prove this we refer to Figure 1.16, and show that

$$d_1 = d_2 \quad \text{and} \quad d_1 + d_2 = d_3$$

Using the Distance Formula, we obtain

$$d_1 = \sqrt{\left(\frac{x_1 + x_2}{2} - x_1\right)^2 + \left(\frac{y_1 + y_2}{2} - y_1\right)^2}$$

$$= \frac{1}{2}\sqrt{(x_2 - x_1)^2 + (y_2 - y_1)^2}$$

$$d_2 = \sqrt{\left(x_2 - \frac{x_1 + x_2}{2}\right)^2 + \left(y_2 - \frac{y_1 + y_2}{2}\right)^2}$$

$$= \frac{1}{2}\sqrt{(x_2 - x_1)^2 + (y_2 - y_1)^2}$$

$$d_3 = \sqrt{(x_2 - x_1)^2 + (y_2 - y_1)^2}$$

Thus, it follows that

$$d_1 = d_2 \quad \text{and} \quad d_1 + d_2 = d_3$$

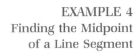

Midpoint Rule

FIGURE 1.16

EXAMPLE 4
Finding the Midpoint
of a Line Segment

Find the midpoint of the line segment joining the points $(-3, -5)$ and $(3, 9)$.

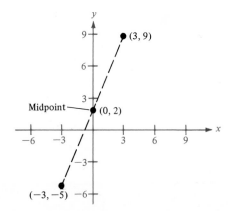

FIGURE 1.17

Solution:

Figure 1.17 shows the two given points together with their midpoint. By the Midpoint Rule, we have

$$\left(\frac{-3 + 3}{2}, \frac{-5 + 9}{2}\right) = (0, 2)$$

EXAMPLE 5

Finding Points at a Specified Distance from a Given Point

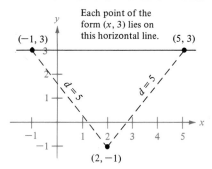

Each point of the form $(x, 3)$ lies on this horizontal line.

FIGURE 1.18

Find x so that the distance between $(x, 3)$ and $(2, -1)$ is 5.

Solution:

As we begin this problem we don't know how many values of x satisfy the given requirements. Even so, we can use the Distance Formula to write the following equation:

$$d = 5 = \sqrt{(x - 2)^2 + (3 + 1)^2}$$
$$25 = (x^2 - 4x + 4) + 16$$
$$0 = x^2 - 4x - 5$$
$$0 = (x - 5)(x + 1)$$
$$x = 5 \text{ or } -1$$

Now we see that there are two solutions and conclude that the points $(5, 3)$ and $(-1, 3)$ each lie 5 units from the point $(2, -1)$. (See Figure 1.18.)

Section Exercises 1.5

In Exercises 1–8, plot the points, find the distance between the points, and find the midpoint of the line segment joining the points.

1. $(2, 1)$, $(4, 5)$
2. $(-3, 2)$, $(3, -2)$
3. $(\frac{1}{2}, 1)$, $(-\frac{3}{2}, -5)$
4. $(\frac{2}{3}, -\frac{1}{3})$, $(\frac{5}{6}, 1)$
5. $(2, 2)$, $(4, 14)$
6. $(-3, 7)$, $(1, -1)$
7. $(1, \sqrt{3})$, $(-1, 1)$
8. $(-2, 0)$, $(0, \sqrt{2})$
9. Show that the points $(4, 0)$, $(2, 1)$, $(-1, -5)$ are vertices of a right triangle.
10. Show that the points $(1, -3)$, $(3, 2)$, $(-2, 4)$ are vertices of an isosceles triangle.
11. Show that the points $(0, 0)$, $(1, 2)$, $(2, 1)$, $(3, 3)$ are vertices of a rhombus. (A rhombus is a four-sided figure whose sides are all of the same length.)
12. Show that the points $(0, 1)$, $(3, 7)$, $(4, 4)$, $(1, -2)$ are vertices of a parallelogram.
13. Use the Distance Formula to determine if the points $(0, -4)$, $(2, 0)$, and $(3, 2)$ lie on a straight line.
14. Use the Distance Formula to determine if the points $(0, 4)$, $(7, -6)$, and $(-5, 11)$ lie on a straight line.
15. Use the Distance Formula to determine if the points $(-2, 1)$, $(-1, 0)$, and $(2, -2)$ lie on a straight line.

16. Find y so that the distance from the origin to the point $(3, y)$ is 5.
17. Find x so that the distance from the origin to the point $(x, -4)$ is 5.
18. Find x so that the distance from $(2, -1)$ to the point $(x, 2)$ is 5.
19. Find the relationship between x and y so that the point (x, y) is equidistant from $(4, -1)$ and $(-2, 3)$.
20. Find the relationship between x and y so that the point (x, y) is equidistant from $(3, \frac{5}{2})$ and $(-7, -1)$.
21. Use the Midpoint Rule successively to find the three points that divide the line segment joining (x_1, y_1) and (x_2, y_2) into four equal parts.
22. Use the result of Exercise 21 to find the points that divide into four equal parts the line segment joining these points:
    (a) $(1, -2)$, $(4, -1)$ (b) $(-2, -3)$, $(0, 0)$
23. Prove that

$$\left(\frac{2x_1 + x_2}{3}, \frac{2y_1 + y_2}{3}\right)$$

is one of the points of trisection of the line segment joining

(x_1, y_1) and (x_2, y_2). Also, find the midpoint of the line segment joining

$$\left(\frac{2x_1 + x_2}{3}, \frac{2y_1 + y_2}{3}\right) \quad \text{and} \quad (x_2, y_2)$$

to find the second point of trisection of the line segment joining (x_1, y_1) and (x_2, y_2).

24. Use the results of Exercise 23 to find the points of trisection of the line segment joining these points:
    (a) $(1, -2)$, $(4, 1)$ (b) $(-2, -3)$, $(0, 0)$

In Exercises 25 and 26, use Figure 1.19 showing average rates for home mortgages between January, 1980 and May, 1981.

25. Approximate the average mortgage rate for
    (a) May, 1980   (b) December, 1980
    (c) July, 1980   (d) May, 1981
26. Approximate the *increase* in average mortgage rates from
    (a) July, 1980 to May, 1981
    (b) January, 1981 to May, 1981

In Exercises 27–30, use Figure 1.20 showing the Dow-Jones Industrial Average (DJIA) from 1929 to 1980.

27. Approximate the DJIA for
    (a) June, 1949   (b) January, 1970
    (c) December, 1953   (d) March, 1963

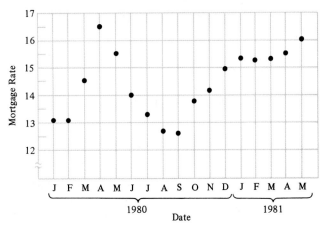

FIGURE 1.19   Home Mortgage Rates

28. Approximate the *increase* (or *decrease*) in the DJIA from
    (a) the high of 1929 to the low of 1932
    (b) the low of 1974 to the high of 1976
29. In which years did the DJIA go above 1000?
30. Approximate the *percentage increase* (or *decrease*) in the DJIA from
    (a) January, 1940 to January, 1950
    (b) January, 1973 to January, 1975

FIGURE 1.20   Dow–Jones Average of Industrial Stock Prices

Calculator Exercises

In Exercises 31–36, use a calculator to (a) find the midpoint of the line segment joining the two points, and (b) find the distance between the points.

31. $(6.2, 5.4)$, $(-3.7, 1.8)$

32. $(-16.8, 12.3)$, $(5.6, 4.9)$
33. $(-36, -18)$, $(48, -72)$
34. $(1.451, 3.051)$, $(5.906, 11.360)$
35. $(0.721, -1.106)$, $(-0.345, -0.093)$
36. $(-8.62, 18.25)$, $(-8.62, 4.67)$

Graphs of Equations   1.6

News magazines frequently show graphs that compare the rate of inflation, the gross national product, wholesale prices, or the unemployment rate to the time of year. Industrial firms and businesses use graphs to report their monthly production and sales statistics. Such graphs provide a simple geometrical picture of the way one quantity changes with respect to another.

Frequently, the relationship between two quantities is expressed in the form of an equation. In this section, we introduce the basic procedure for determining the geometric picture associated with an algebraic equation. Consider the equation

$$3x + y = 7$$

If $x = 2$ and $y = 1$, the equation is satisfied, and we call the point $(2, 1)$ a **solution point** of the equation. Of course, there are other solution points. To make up a table of solution points, we choose arbitrary values for x and determine the corresponding values for y. To determine the values for y, it is convenient to replace the equation by the equivalent form

$$y = 7 - 3x$$

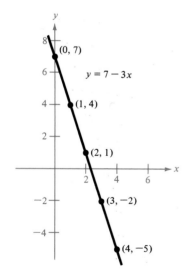

x	0	1	2	3	4
y	7	4	1	-2	-5

Thus, $(0, 7)$, $(1, 4)$, $(2, 1)$, $(3, -2)$, and $(4, -5)$ are all solution points of the equation $3x + y = 7$. We could continue this process indefinitely and obtain infinitely many solution points for the equation $3x + y = 7$. We call the collection of all such solution points the **graph** of the equation $3x + y = 7$, as shown in Figure 1.21.

FIGURE 1.21

DEFINITION OF THE GRAPH OF AN EQUATION

> The **graph of an equation** involving two variables x and y is the collection of all points in the plane that are solution points to the equation.

EXAMPLE 1
Sketching the Graph
of an Equation

Sketch the graph of the equation $y = x^2 - 2$.

Solution:
First, we make a table of values (solution points) by choosing several convenient values of x and calculating the corresponding values of y.

x	-2	-1	0	1	2	3
$y = x^2 - 2$	2	-1	-2	-1	2	7

Next, we plot these points in the plane, as in Figure 1.22. Finally, we connect the points by a smooth curve, as in Figure 1.23.

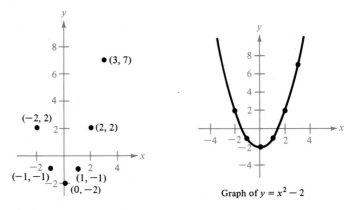

Graph of $y = x^2 - 2$

FIGURE 1.22   **FIGURE 1.23**

We call this method of sketching a graph the **point-plotting method.** It has three basic steps.

THE POINT-PLOTTING METHOD OF GRAPHING

> 1. Make up a table of several solution points of the equation.
> 2. Plot these points in the plane.
> 3. Connect the points with a smooth curve.

Steps 1 and 2 of the point-plotting method can usually be accomplished with ease. However, Step 3 can be the source of some major difficulties. For instance, how would you connect the four points in Figure 1.24? Without additional points or further information about the equation, any one of the three graphs in Figure 1.25 would be reasonable.

Obviously, with too few solution points, we could badly misrepresent the graph of a given equation. Just how many points should be plotted? For

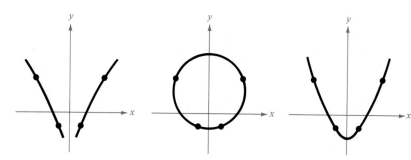

FIGURE 1.25

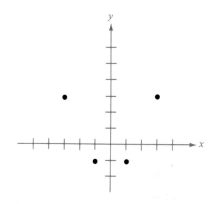

FIGURE 1.24

a straight-line graph, two points are sufficient. For more complicated graphs we need many more points. More sophisticated techniques will be discussed in later sections, but for now plot enough points so as to reveal the essential behavior of the graph. A programmable calculator is useful for determining the many solution points needed for an accurate graph.

We suggest that in choosing points to plot you start with those points that are easiest to calculate. Two such points are those having zero as either their *x*- or *y*-coordinate. These points are called **intercepts,** because they are points at which the graph intersects the *x*- or *y*-axis.

DEFINITION OF INTERCEPTS

> The point $(a, 0)$ is called an ***x*-intercept** of the graph of an equation if it is a solution point of the equation. To find the *x*-intercepts, let *y* be zero and solve the equation for *x*.
>
> The point $(0, b)$ is called a ***y*-intercept** of the graph of an equation if it is a solution point of the equation. To find the *y*-intercepts, let *x* be zero and solve the equation for *y*.

Remark: Some texts denote the *x*-intercept as the *x*-coordinate of the point $(a, 0)$ rather than the point itself. Unless it is necessary to make a distinction, we will use "intercept" to mean either the point or the coordinate.

Of course, it is possible that a particular graph will have no intercepts, or it may have several. For instance, consider the four graphs in Figure 1.26.

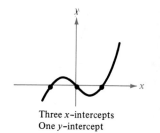

Three *x*-intercepts
One *y*-intercept

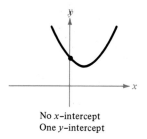

No *x*-intercept
One *y*-intercept

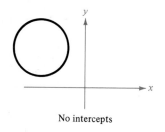

No intercepts

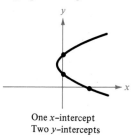

One *x*-intercept
Two *y*-intercepts

FIGURE 1.26

EXAMPLE 2
Finding x- and y-Intercepts

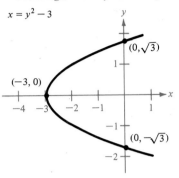

$x = y^2 - 3$

FIGURE 1.27

EXAMPLE 2

Finding x- and y-Intercepts

Find the x- and y-intercepts for the graph of

$$y^2 - 3 = x$$

Solution:

Let $y = 0$. Then $-3 = x$.

    x-intercept: $(-3, 0)$

Let $x = 0$. Then $y^2 - 3 = 0$ has solutions $y = \pm\sqrt{3}$.

    y-intercepts: $(0, \sqrt{3})$, $(0, -\sqrt{3})$

(See Figure 1.27.)

    The graph shown in Figure 1.27 is said to be "symmetric" with respect to the x-axis. This means that if the Cartesian plane were folded along the x-axis, the portion of the graph above the x-axis would then coincide with the portion below the x-axis. Symmetry with respect to the y-axis can be described in a similar manner.

    Knowing the symmetry of a graph *before* attempting to sketch it is beneficial, for then we need only half as many solution points as we otherwise would. We define three basic types of symmetry as follows (see Figure 1.28).

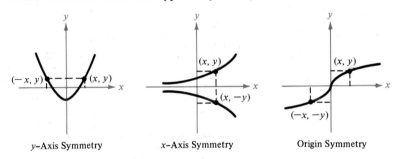

y–Axis Symmetry   x–Axis Symmetry   Origin Symmetry

FIGURE 1.28

DEFINITION OF SYMMETRY

A graph is said to be **symmetric with respect to the y-axis** if, whenever (x, y) is on the graph, $(-x, y)$ is also on the graph.

A graph is said to be **symmetric with respect to the x-axis** if, whenever (x, y) is on the graph, $(x, -y)$ is also on the graph.

A graph is said to be **symmetric with respect to the origin** if, whenever (x, y) is on the graph, $(-x, -y)$ is also on the graph.

    Suppose we apply this definition of symmetry to the graph of the equation $y = x^2 - 2$. By replacing x by $-x$, we obtain

$$y = (-x)^2 - 2 \quad \text{or} \quad y = x^2 - 2$$

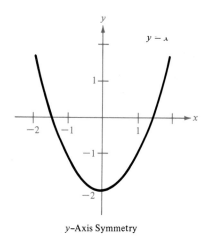

y-Axis Symmetry

FIGURE 1.29

Since this substitution does not change the equation, it follows that if (x, y) is a solution point of the equation, then $(-x, y)$ must also be a solution point. Therefore, the graph of $y = x^2 - 2$ is symmetric with respect to the *y*-axis. (See Figure 1.29.)

A similar test can be made for symmetry with respect to the *x*-axis or to the origin. These three tests are summarized as follows.

TESTS FOR SYMMETRY

> The graph of an equation is symmetric with respect to:
> 1. the *y*-axis if replacing x by $-x$ yields an equivalent equation
> 2. the *x*-axis if replacing y by $-y$ yields an equivalent equation
> 3. the origin if replacing x by $-x$ *and* y by $-y$ yields an equivalent equation

EXAMPLE 3
Testing for Symmetry

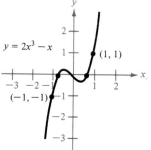

Origin Symmetry

FIGURE 1.30

Show that the graph of $y = 2x^3 - x$ is symmetric with respect to the origin.

Solution:
By replacing x by $-x$ and y by $-y$, we have

$$-y = 2(-x)^3 - (-x)$$
$$-y = -2x^3 + x$$

Now by multiplying both sides of the equation by -1, we have

$$y = 2x^3 - x$$

which is the original equation. Therefore, the graph of $y = 2x^3 - x$ is symmetric with respect to the origin. See Figure 1.30.

EXAMPLE 4
Using Symmetry as an Aid to Graphing

First, plot the points above the *x*-axis, then use symmetry to complete the graph.

FIGURE 1.31

Sketch the graph of $x - y^2 = 1$.

Solution:
The graph is symmetric with respect to the *x*-axis, since replacing y by $-y$ yields

$$x - (-y)^2 = 1$$
$$x - y^2 = 1$$

This means that the graph below the *x*-axis is a mirror-image of the graph above the *x*-axis. Hence we can first sketch the portion above the *x*-axis and then reflect it to obtain the entire graph (Figure 1.31).

$x = y^2 + 1$	1	2	5
y	0	1	2

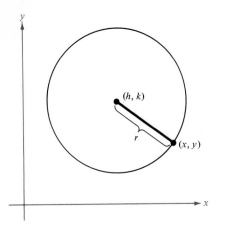

FIGURE 1.32

So far in this section we have studied the Point-Plotting Method of graphing and two additional concepts (intercepts and symmetry) that can be used to streamline this graphing procedure. A third graphing aid is that of *equation recognition,* the ability to recognize the general shape of a graph simply by looking at its equation. A **circle** is one type of graph whose equation is easily recognized.

Figure 1.32 shows a circle of radius r with center at (h, k). The point (x, y) is on this circle if and only if its distance from the center (h, k) is r. This means that a circle consists of the set of all points (x, y) that are at a given positive distance r from a fixed point (h, k). Expressing this relationship in terms of the Distance Formula, we have

$$\sqrt{(x - h)^2 + (y - k)^2} = r$$

By squaring both sides of this equation, we obtain the **standard form of the equation of a circle.**

STANDARD FORM OF THE EQUATION OF A CIRCLE

> The point (x, y) lies on the circle of radius r and center (h, k) if and only if
>
> $$(x - h)^2 + (y - k)^2 = r^2$$

As a special case, *the equation of a circle with its center at the origin* is simply

$$x^2 + y^2 = r^2$$

The Unit Circle

In Chapter 5 we will use the circle given by

$$x^2 + y^2 = 1$$

to introduce the trigonometric functions. We call the graph of this equation the **unit circle.** It has a radius of 1, and its center is at the origin. There are several points on the unit circle that play a special role in trigonometry. We discuss some of the properties of these special points in the next two examples. Study both examples carefully.

EXAMPLE 5
Points on the Unit Circle

Use the Distance Formula to show that the points shown in Figure 1.33 divide the unit circle into 8 equal parts.

Solution:
Referring to Figure 1.33, we use the Distance Formula as follows:

$$d_1 = \sqrt{\left(\frac{\sqrt{2}}{2} - 1\right)^2 + \left(\frac{\sqrt{2}}{2} - 0\right)^2}$$

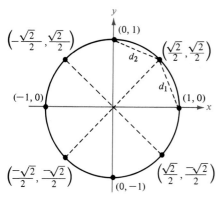

FIGURE 1.33

EXAMPLE 6
Points on the Unit Circle

$$= \sqrt{\frac{2 - 4\sqrt{2} + 4}{4} + \frac{2}{4}} = \sqrt{2 - \sqrt{2}}$$

$$d_2 = \sqrt{\left(\frac{\sqrt{2}}{2} - 0\right)^2 + \left(\frac{\sqrt{2}}{2} - 1\right)^2}$$

$$= \sqrt{\frac{2}{4} + \frac{2 - 4\sqrt{2} + 4}{4}} = \sqrt{2 - \sqrt{2}}$$

In a similar way, we could show that the length of the chord connecting any two adjacent points is $\sqrt{2 - \sqrt{2}}$. From this we can conclude that these 8 points divide the unit circle into 8 equal parts.

Use the Distance Formula to show that the points shown in Figure 1.34 divide the unit circle into 12 equal parts.

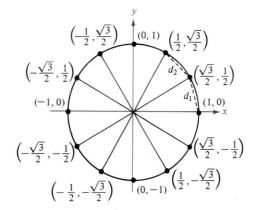

FIGURE 1.34

Solution:
Referring to Figure 1.34, we use the Distance Formula as follows:

$$d_1 = \sqrt{\left(\frac{\sqrt{3}}{2} - 1\right)^2 + \left(\frac{1}{2} - 0\right)^2}$$

$$= \sqrt{\frac{3 - 4\sqrt{3} + 4}{4} + \frac{1}{4}} = \sqrt{2 - \sqrt{3}}$$

$$d_2 = \sqrt{\left(\frac{1}{2} - \frac{\sqrt{3}}{2}\right)^2 + \left(\frac{\sqrt{3}}{2} - \frac{1}{2}\right)^2}$$

$$= \sqrt{\frac{1 - 2\sqrt{3} + 3}{4} + \frac{3 - 2\sqrt{3} + 1}{4}} = \sqrt{2 - \sqrt{3}}$$

In a similar way, we could show that the length of the chord connecting any two adjacent points is $\sqrt{2 - \sqrt{3}}$. From this we can conclude that these 12 points divide the unit circle into 12 equal parts.

Section Exercises 1.6

In Exercises 1–6, match the given equation with its graph. [The graphs are labeled (a), (b), (c), (d), (e), and (f).]

1. $y = x - 2$
2. $y = -\frac{1}{2}x + 2$
3. $y = x^2 + 2x$
4. $y = \sqrt{9 - x^2}$
5. $y = |x| - 2$
6. $y = x^3 - x$

(a)

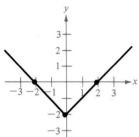

(b)

(c)

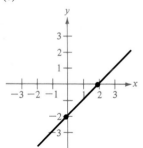

(d)

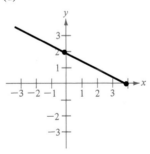

(e)

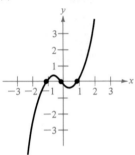

(f)

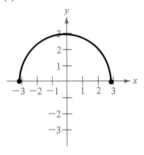

In Exercises 7–16, find the intercepts.

7. $y = 2x - 3$
8. $y = (x - 1)(x - 3)$
9. $y = x^2 + x - 2$
10. $y^2 = x^3 - 4x$
11. $y = x^2\sqrt{9 - x^2}$
12. $xy = 4$

13. $y = \dfrac{x - 1}{x - 2}$
14. $y = \dfrac{x^2 + 3x}{(3x + 1)^2}$
15. $x^2y - x^2 + 4y = 0$
16. $y = 2x - \sqrt{x^2 + 1}$

In Exercises 17–26, check for symmetry about both axes and the origin.

17. $y = x^2 - 2$
18. $y = x^4 - x^2 + 3$
19. $x^2y - x^2 + 4y = 0$
20. $(xy)^2 - x^2 - 4y^2 = 0$
21. $y^2 = x^3 - 4x$
22. $xy^2 = -10$
23. $y = x^3 + x$
24. $xy = 1$
25. $y = \dfrac{x}{x^2 + 1}$
26. $y = x^3 + x - 3$

In Exercises 27–39, use the methods of this section to sketch the graph of each equation. Identify the intercepts and test for symmetry.

27. $y = x$
28. $y = x + 3$
29. $y = -3x + 2$
30. $y = 2x - 3$
31. $y = 1 - x^2$
32. $y = x^2 + 3$
33. $y = x^3 + 2$
34. $y = x^3 - 1$
35. $y = (x + 2)^2$
36. $y = \dfrac{1}{x^2 + 1}$
37. $y = 2x^4$
38. $y = |x - 2|$
39. $y = \sqrt{x - 3}$
40. (a) Sketch the graph of $y = 3x^4 - 4x^3$ by completing the accompanying table and plotting the resulting points.

x	-1	0	1	2
y				

(b) Find additional points satisfying $y = 3x^4 - 4x^3$ by completing the accompanying table. Now refine the graph of part (a).

x	-0.75	-0.50	-0.25	0.25	0.5	1.33
y						

In Exercises 41–44, determine whether the given points lie on the graph of the given equation.

41. Equation: $2x - y - 3 = 0$
    (a) $(1, 2)$ (b) $(1, -1)$ (c) $(4, 5)$
42. Equation: $x^2 + y^2 = 4$
    (a) $(1, -\sqrt{3})$ (b) $(\frac{1}{2}, -1)$ (c) $(\frac{3}{2}, \frac{7}{2})$

43. Equation: $x^2y - x^2 + 4y = 0$
    (a) $(1, \frac{1}{5})$ (b) $(2, \frac{1}{2})$ (c) $(-1, -2)$
44. Equation: $x^2 - xy + 4y = 3$
    (a) $(0, 2)$ (b) $(-2, -\frac{1}{6})$ (c) $(3, -6)$
45. The Consumer Price Index for the 1970s is given in the following table.

Year	1970	1971	1972	1973	1974
CPI	116.3	121.3	125.3	133.1	147.7

Year	1975	1976	1977	1978	1979
CPI	161.2	170.5	181.5	195.3	211.1

A mathematical model for the CPI during this 10-year period is

$$y = 0.55t^2 + 5.85t + 114.41$$

where y represents the CPI and t represents the year, with $t = 0$ corresponding to 1970.
    (a) Graphically compare the actual CPI during the 10-year period with the one predicted by the model.
    (b) Use the model to predict the CPI for 1985.
46. From the model in Exercise 45, we obtain the model

$$V = \frac{100}{0.55t^2 + 5.85t + 114.41}$$

where V represents the purchasing power of the dollar (in terms of constant 1967 dollars) and t represents the year, with $t = 0$ corresponding to 1970. Use this model to complete the following table, and then graph your results.

t	0	2	4	6	8	10	12
V							

47. The farm population in the United States as a percentage of the total population is given in the following table:

Year	1950	1955	1960	1965	1970	1975	1979
%	15.3	11.6	8.7	6.4	4.8	4.2	3.4

A mathematical model for these data is given by

$$y = \frac{100}{4.90 + 0.79t}$$

where y represents the percentage and t represents the year, with $t = 0$ corresponding to 1950.
    (a) Graphically compare the actual percentage with that given by the model.
    (b) Use the model to predict the farm population as a percentage of the total population in 1990.
48. The average number of acres per farm in the United States is given in the following table:

Year	1950	1960	1965	1970	1975	1978
Number of Acres	213	297	340	374	391	401

A mathematical model for these data is given by

$$y = -0.13t^2 + 10.43t + 211.3$$

where y represents the acreage and t represents the year, with $t = 0$ corresponding to 1950.
    (a) Graphically compare the actual number of acres per farm with that given by the model.
    (b) Use the model to predict the average number of acres per farm in the United States in 1985.
49. Show that the points given in Example 5 lie on the graph of $x^2 + y^2 = 1$.
50. Show that the points given in Example 6 lie on the graph of $x^2 + y^2 = 1$.

In Exercises 51–54, show that any point (x, y) on the unit circle satisfies the given equation. (Assume $x > 0$, $y > 0$.)

51. $\left(\dfrac{1}{x} - \dfrac{1}{y}\right)^2 = \dfrac{1 - 2xy}{x^2y^2}$

52. $\dfrac{1 + y}{x} + \dfrac{x}{1 + y} = \dfrac{2}{x}$

53. $\dfrac{y^2 + 4y + 3}{x^2} = \dfrac{3 + y}{1 - y}$

54. $\dfrac{y/x}{y - 2(y/x)} = \dfrac{1}{x - 2}$

55. Use the following steps to find the points that divide the unit circle into 16 equal parts. (Refer to Example 5 of this section.)
    (a) Show that $y = \dfrac{\sqrt{2}}{2 + \sqrt{2}} x$ is the equation of the line through the origin and the midpoint of the chord connecting the points $(1, 0)$ and $\left(\dfrac{\sqrt{2}}{2}, \dfrac{\sqrt{2}}{2}\right)$.
    (b) Find the point of intersection of the unit circle and the equation of the line of part (a).
    (c) Use symmetry to find the other required points on the unit circle.

Algebraic Errors and the Algebra of Calculus

1.7

Before we wrap up our review of the fundamental concepts of algebra, we want to look at some common algebraic errors. Many of these errors are made because they are the *easiest* things to do. This is a strong temptation, especially when the error makes the remainder of the problem much simpler (a big temptation in calculus). Regardless of *why* they are made, we feel that it is helpful to review a list of errors to avoid.

ALGEBRAIC ERRORS TO AVOID

	Error	**Correct Form**	**Comment**
1. Parentheses	$a - (x - b) \neq a - x - b$	$a - (x - b) = a - x + b$	Change all signs when distributing negative through parentheses.
	$(a + b)^2 \neq a^2 + b^2$	$(a + b)^2 = a^2 + 2ab + b^2$	Don't forget middle term when squaring binomials.
	$\left(\frac{1}{2} a\right)\left(\frac{1}{2} b\right) \neq \frac{1}{2}(ab)$	$\left(\frac{1}{2} a\right)\left(\frac{1}{2} b\right) = \frac{1}{4}(ab) = \frac{ab}{4}$	1/2 occurs twice as a factor.
2. Fractions	$\dfrac{a}{x + b} \neq \dfrac{a}{x} + \dfrac{a}{b}$	Leave as $\dfrac{a}{x + b}$	Don't add denominators when adding fractions.
	$\dfrac{\left(\dfrac{x}{a}\right)}{b} \neq \dfrac{bx}{a}$	$\dfrac{\left(\dfrac{x}{a}\right)}{b} = \left(\dfrac{x}{a}\right)\left(\dfrac{1}{b}\right) = \dfrac{x}{ab}$	Multiply by reciprocal when dividing.
	$\dfrac{1}{a} + \dfrac{1}{b} \neq \dfrac{1}{a + b}$	$\dfrac{1}{a} + \dfrac{1}{b} = \dfrac{a + b}{ab}$	Use definition for adding fractions.
	$\dfrac{1}{3x} \neq \dfrac{1}{3} x$	$\dfrac{1}{3x} = \dfrac{1}{3} \cdot \dfrac{1}{x}$	Use definition for multiplying fractions.
	$\dfrac{1}{3} x \neq \dfrac{1}{3x}$	$\dfrac{1}{3} x = \dfrac{1}{3} \cdot x = \dfrac{x}{3}$	Be careful when using a slash to denote division.
	$\dfrac{1}{x} + 2 \neq \dfrac{1}{(x + 2)}$	$\dfrac{1}{x} + 2 = \dfrac{1}{x} + 2$ $= \dfrac{1 + 2x}{x}$	Be careful when using a slash to denote division.

3. Exponents and Radicals

$(x^2)^3 \neq x^5$	$(x^2)^3 = x^{2\cdot3} = x^6$	Multiply exponents when an exponential form is raised to a power.
$x^2 \cdot x^3 \neq x^6$	$x^2 \cdot x^3 = x^{2+3} = x^5$	Add exponents when multiplying exponentials with like bases.
$2x^3 \neq (2x)^3$	$2x^3 = 2(x^3)$	Exponents have priority over coefficients.
$\dfrac{1}{x^{1/2} - x^{1/3}} \neq x^{-1/2} - x^{-1/3}$	Leave as $\dfrac{1}{x^{1/2} - x^{1/3}}$	Don't shift term-by-term from denominator to numerator.
$\sqrt{5x} \neq 5\sqrt{x}$	$\sqrt{5x} = \sqrt{5}\sqrt{x}$	Radicals apply to every factor inside radical.
$\sqrt{x^2 + a^2} \neq x + a$	Leave as $\sqrt{x^2 + a^2}$	Don't apply radicals term-by-term.
$\sqrt{-x^2 + a^2} \neq -\sqrt{x^2 - a^2}$	Leave as $\sqrt{-x^2 + a^2}$ or write as $\sqrt{a^2 - x^2}$	Don't factor negatives out of square roots.

4. Cancellations

$\dfrac{a + bx}{a} \neq 1 + bx$	$\dfrac{a + bx}{a} = \dfrac{a}{a} + \dfrac{bx}{a}$ $= 1 + \dfrac{b}{a}x$	Cancel common factors, *not* common terms.
$\dfrac{a + ax}{a} \neq a + x$	$\dfrac{a + ax}{a} = \dfrac{a(1 + x)}{a}$ $= 1 + x$	Factor *before* canceling.
$1 + \dfrac{x}{2x} \neq 1 + \dfrac{1}{x}$	$1 + \dfrac{x}{2x} = 1 + \dfrac{1}{2} = \dfrac{3}{2}$	When canceling factors with like bases, subtract exponents, *not* coefficients.

The Algebra of Calculus

In our review of algebra to this point we have required that most answers be expressed in simplest form. In calculus you will often have to reverse this procedure. At times we need *unsimplified* algebraic forms in order to perform the operations of calculus. Consequently, it is often necessary to take a simplified algebraic expression and **unsimplify** it. The following list, taken from a standard calculus text, shows some of the "backwards" algebra needed in calculus.

SOME ALGEBRA OF CALCULUS

	Required Simplest Algebraic Form	**Desired Unsimplified Form for Calculus**	**Comment**
1. Unusual Factoring	$\dfrac{5x^4}{8}$	$\dfrac{5}{8}(x^4)$	Factor out fractional coefficient.
	$\dfrac{x^2 + 3x}{-6}$	$-\dfrac{1}{6}(x^2 + 3x)$	Factor out fractional coefficient.
	$2x^2 - x - 3$	$2\left(x^2 - \dfrac{x}{2} - \dfrac{3}{2}\right)$	Factor out leading coefficient.
	$\dfrac{x}{2}(x + 1)^{-1/2} + (x + 1)^{1/2}$	$\dfrac{(x + 1)^{-1/2}}{2}[x + 2(x + 1)]$	Remove factor with negative exponent.
2. Inserting Required Factors	$(2x - 1)^3$	$\dfrac{1}{2}(2x - 1)^3(2)$	Multiply and divide by desired factor.
	$7x^2(4x^3 - 5)^{1/2}$	$\dfrac{7}{12}(4x^3 - 5)^{1/2}(12x^2)$	Multiply and divide by desired factor.
	$\dfrac{4x^2}{9} - 4y^2 = 1$	$\dfrac{x^2}{9/4} - \dfrac{y^2}{1/4} = 1$	Invert and *divide*.
3. Rewriting with Negative Exponents	$\dfrac{9}{-5x^3}$	$\dfrac{9}{-5}(x^{-3})$	Move factor to numerator and change sign of exponent.
	$\dfrac{7}{\sqrt{2x - 3}}$	$7(2x - 3)^{-1/2}$	Move factor to numerator and change sign of exponent.
4. Writing a Fraction as a Sum of Terms	$\dfrac{x + 2x^2 + 1}{\sqrt{x}}$	$x^{1/2} + 2x^{3/2} + x^{-1/2}$	Divide each term by $x^{1/2}$.
	$\dfrac{1 + x}{x^2 + 1}$	$\dfrac{1}{x^2 + 1} + \dfrac{x}{x^2 + 1}$	Rewrite fraction as sum of fractions.
	$\dfrac{2x}{x^2 + 2x + 1}$	$\dfrac{2x + 2 - 2}{x^2 + 2x + 1} =$	Add and subtract terms in numerator.
		$\dfrac{2x + 2}{x^2 + 2x + 1} - \dfrac{2}{(x + 1)^2}$	Rewrite fraction as difference of fractions.
	$\dfrac{x^2 - 2}{x + 1}$	$x - 1 - \dfrac{1}{x + 1}$	Long division.
	$\dfrac{x + 7}{x^2 - x - 6}$	$\dfrac{2}{x - 3} - \dfrac{1}{x + 2}$	Use method of *partial fractions*.

Remark: (a) Notice that factorization is *not* limited to integer factors. Non-integer factoring simply boils down to *multiplying and dividing* by the desired factor.

(b) When we multiply like factors we add exponents. When we factor we are undoing multiplication and so we *subtract* exponents. Hence

$$x(x + 1)^{-1/2} + (x + 1)^{1/2}$$

factors as

$$(x + 1)^{-1/2}[x + (x + 1)]$$

(c) *Long division* is needed to express

$$\frac{x^2 + 2}{x + 1} \quad \text{as} \quad x - 1 - \frac{1}{x + 1}$$

and we will cover that in Section 3.2.

(d) *Partial fractions* are needed to express

$$\frac{x + 7}{x^2 - x - 6} \quad \text{as} \quad \frac{2}{x - 3} - \frac{1}{x + 2}$$

and we will cover that in Section 3.7.

The next two examples fill in some details for many of the steps given in the table.

EXAMPLE 1 Inserting Factors and Rewriting with Negative Exponents	(a) Explain the following. $$\frac{4x^2}{9} - 4y^2 = \frac{x^2}{9/4} - \frac{y^2}{1/4}$$ (b) Rewrite the expression so that the denominator is free of terms involving *x*. $$\frac{2}{x^3} - \frac{1}{\sqrt{x}} + \frac{3}{16x^2}$$

Solution:

(a) To get rid of fractions in a denominator we invert and multiply. Hence to *put* fractions in the denominator we do the opposite: Invert and *divide*. Thus, we have

$$\frac{4x^2}{9} - 4y^2 = \frac{4}{9}(x^2) - 4(y^2) = \frac{x^2}{9/4} - \frac{y^2}{1/4}$$

(b) $\dfrac{2}{x^3} - \dfrac{1}{\sqrt{x}} + \dfrac{3}{16x^2} = \dfrac{2}{x^3} - \dfrac{1}{x^{1/2}} + \dfrac{3}{(4x)^2}$

$$= 2x^{-3} - x^{-1/2} + 3(4x)^{-2}$$

EXAMPLE 2
Writing a Fraction
as a Sum of Terms

Rewrite each fraction as the sum of two or more terms.

(a) $\dfrac{x + 2x^2 + 1}{\sqrt{x}}$ $\qquad$ (b) $\dfrac{1 + x}{x^2 + 1}$ $\qquad$ (c) $\dfrac{2x}{x^2 + 2x + 1}$

Solution:

(a) If the denominator is a monomial, we make one fraction for each term in the numerator, then reduce each. Thus, we have

$$\frac{x + 2x^2 + 1}{\sqrt{x}} = \frac{x}{x^{1/2}} + \frac{2x^2}{x^{1/2}} + \frac{1}{x^{1/2}} \qquad \text{\textit{Separate into 3 fractions}}$$

$$= x^{1/2} + 2x^{3/2} + x^{-1/2} \qquad \text{\textit{Subtract exponents}}$$

(b) Again, we can make a fraction for each term in the numerator, but reductions are not possible. We have

$$\frac{1 + x}{x^2 + 1} = \frac{1}{x^2 + 1} + \frac{x}{x^2 + 1} \qquad \text{\textit{Separate into 2 fractions}}$$

(c) We can always add and subtract the same terms in an expression. So we can write

$$\frac{2x}{x^2 + 2x + 1} = \frac{2x + 2 - 2}{x^2 + 2x + 1} \qquad \text{\textit{Add and subtract 2}}$$

$$= \frac{2x + 2}{x^2 + 2x + 1} - \frac{2}{x^2 + 2x + 1} \qquad \text{\textit{Separate into 2 fractions}}$$

$$= \frac{2x + 2}{x^2 + 2x + 1} - \frac{2}{(x + 1)^2}$$

Section Exercises 1.7

In Exercises 1–24, find and correct any errors.

1. $2x - (3y + 4) = 2x - 3y + 4$

2. $\dfrac{4}{16x - (2x + 1)} = \dfrac{4}{14x + 1}$

3. $5z + 3(x - 2) = 5z + 3x - 2$

4. $x(yz) = (xy)(xz)$

5. $-\dfrac{x - 3}{x - 1} = \dfrac{3 - x}{1 - x}$

6. $\dfrac{(x - 1)(x + 3)}{(5 - x)(-x)} = \dfrac{(1 - x)(x + 3)}{x(5 - x)}$

7. $a\left(\dfrac{x}{y}\right) = \dfrac{ax}{ay}$

8. $(5z)(6z) = 30z$

9. $(4x)^2 = 4x^2$

10. $\left(\dfrac{x}{y}\right)^3 = \dfrac{x^3}{y}$

11. $\sqrt{x + 9} = \sqrt{x} + 3$

12. $\dfrac{x + 5}{y + 5} = \dfrac{x}{y} + 1$

13. $\dfrac{6x + y}{6x - y} = \dfrac{x + y}{x - y}$

14. $(-2)^6 = -2^6$

15. $\dfrac{1}{x + y^{-1}} = \dfrac{y}{x + 1}$

16. $\dfrac{1}{a^{-1} + b^{-1}} = \left(\dfrac{1}{a + b}\right)^{-1}$

17. $\dfrac{4 + x}{xy^{-1}} = \dfrac{4 + xy}{x}$

18. $x(x + 5)^{1/2} = (x^2 + 5x)^{1/2}$

19. $\sqrt[3]{x^3 + 7x^2} = x^2\sqrt[3]{x + 7}$

20. $\dfrac{1}{x^{1/2} + y^{1/2}} = (x + y)^{-1/2}$

21. $\dfrac{3}{x} + \dfrac{4}{y} = \dfrac{7}{x + y}$

22. $\dfrac{7x - 5(x + 3)}{x(x + 3)} = 12$

23. $\dfrac{1}{2y} = \dfrac{1}{2}y$

24. $4y + \dfrac{1}{2y} = 4.5y$

In Exercises 25–39, factor each expression so that at least one factor is a polynomial with integer coefficients.

25. $\frac{2}{3}x^2 + \frac{1}{3}x + 5$

26. $\frac{3}{4}x + \frac{1}{2}$

27. $\sqrt{x} + (\sqrt{x})^3$

28. $x^{1/3} - 5x^{4/3}$

29. $(2x)x^{1/2} + \frac{1}{2}x^{-1/2}(x^2 + 1)$

30. $2x^{1/3} + \frac{1}{3}x^{-2/3}(2x - 1)$

31. $-\frac{2}{3}x(1 - 2x)^{-2/3} + (1 - 2x)^{1/3}$

32. $\dfrac{\frac{1}{2}x^{-1/2}(x + 2)^3 - 3x^{1/2}(x + 2)^2}{(x + 2)^6}$

33. $\dfrac{\dfrac{x^2}{\sqrt{x^2 + 1}} - \sqrt{x^2 + 1}}{x^2}$

34. $\dfrac{1}{2\sqrt{x}} + 5x^{3/2} - 10x^{5/2}$

35. $\frac{1}{10}(2x + 1)^{5/2} - \frac{1}{6}(2x + 1)^{3/2}$

36. $\frac{4}{3}(x + 3)^{3/2} - 7(x + 3)^{1/2}$

37. $\frac{3}{7}(t + 1)^{7/3} - \frac{3}{4}(t + 1)^{4/3}$

38. $\frac{2}{3}(1 - x)^{3/2} - \frac{4}{5}(1 - x)^{5/2} + \frac{2}{7}(1 - x)^{7/2}$

39. $\frac{7}{20}(2x - 1)^{5/2} + \frac{1}{6}(2x - 1)^{3/2} - \frac{3}{4}(2x - 1)^{1/2}$

In Exercises 40–49, insert the required factor in the parentheses.

40. $x^2(x^3 - 1)^4 = (\qquad)(x^3 - 1)^4(3x^2)$

41. $x(1 - 2x^2)^3 = (\qquad)(1 - 2x^2)^3(-4x)$

42. $5x\sqrt[3]{1 + x^2} = (\qquad)\sqrt[3]{1 + x^2}(2x)$

43. $\dfrac{1}{\sqrt{x}(1 + \sqrt{x})^2} = (\qquad)\dfrac{1}{(1 + \sqrt{x})^2}\left(\dfrac{1}{2\sqrt{x}}\right)$

44. $\dfrac{4x + 6}{(x^2 + 3x + 7)^3} = (\qquad)\dfrac{1}{(x^2 + 3x + 7)^3}(2x + 3)$

45. $\dfrac{x + 1}{(x^2 + 2x - 3)^2} = (\qquad)\dfrac{1}{(x^2 + 2x - 3)^2}(2x + 2)$

46. $\dfrac{1}{(x - 1)\sqrt{(x - 1)^4 - 4}} = \dfrac{(\qquad)}{(x - 1)^2\sqrt{(x - 1)^4 - 4}}$

47. $\dfrac{9x^2}{25} + \dfrac{16y^2}{49} = \dfrac{x^2}{(\quad)} + \dfrac{y^2}{(\quad)}$

48. $\dfrac{36(x - 1)^2}{169} + (y + 5)^2 = \dfrac{(x - 1)^2}{(\quad)} + (y + 5)^2$

49. $\dfrac{3}{x} + \dfrac{5}{2x^2} - \dfrac{3}{2}x = (\qquad)(6x + 5 - 3x^3)$

In Exercises 50–55, write each fraction as a sum of terms, as in Example 2.

50. $\dfrac{x^3 - 5x^2 + 4}{x^2}$

51. $\dfrac{16 - 5x - x^2}{x}$

52. $\dfrac{2x^5 - 3x^3 + 5x - 1}{x^{3/2}}$

53. $\dfrac{4x^3 - 7x^2 + 1}{x^{1/3}}$

54. $\dfrac{3x^2 - 5}{x^3 + 1}$

55. $\dfrac{x^2 + 4x + 8}{x^4 + 1}$

Review Exercises / Chapter 1

In Exercises 1–24, describe the *error* and then make the necessary correction.

1. $\frac{7}{16} + \frac{3}{16} = \frac{10}{32}$

2. $\frac{15}{32} - \frac{21}{32} = \frac{-6}{0}$

3. $10(4 \cdot 7) = 40 \cdot 70$

4. $(\frac{1}{3}x)(\frac{1}{3}y) = \frac{1}{3}xy$

5. $4(\frac{3}{7}) = \frac{12}{28}$

6. $\frac{2}{9} \times \frac{4}{9} = \frac{8}{9}$

7. $\frac{15}{16} \div \frac{2}{3} = \frac{5}{8}$

8. $15 \div 2 + 3 = 15 \div 5 = 3$

9. $12 + 8 \times 6 = 20 \times 6 = 120$

10. $\frac{-3}{4} = -\frac{3}{4}$

11. $2[5 - (3 - 2)] = 2[5 - 3 - 2]$

12. $-3(-x + y) = 3x + 3y$

13. $(2x)^4 = 2x^4$

14. $\left(\dfrac{y}{8}\right)^2 = \dfrac{y^2}{8}$

15. $(5 + 8)^2 = 5^2 + 8^2$

16. $(-x)^6 = -x^6$

17. $(3^4)^4 = 3^8$

18. $6^{-2} = -6^2$

19. $\sqrt{3^2 + 4^2} = 3 + 4$

20. $\sqrt{10x} = 10\sqrt{x}$

21. $\sqrt{7x}\sqrt[3]{2} = \sqrt{14x}$

22. $\sqrt[4]{\sqrt[4]{2}} = \sqrt[8]{2}$

23. Since $-5 < -3$, it follows that $-2(-5) < -2(-3)$

24. Since $4 < 7$, it follows that $\frac{1}{4} < \frac{1}{7}$

In Exercises 25–40, perform the indicated operations.

25. $-10(7 - 5)$

26. $-10(7)(-5)$

27. $-|16 - 5|$

28. $|5 - 16|$

29. $|-3| + 4(-2) - 6$

30. $16 - 8 \div 4$

31. $\sqrt{5}\sqrt{125}$

32. $\dfrac{\sqrt{72}}{\sqrt{2}}$

33. $6[4 - 2(6 + 8)]$

34. $-4[16 - 3(7 - 10)]$

35. $\left(\dfrac{3^2}{5^2}\right)^{-3}$

36. $6^{-4}(-3)^5$

37. $2(-5)^2$

38. $\left(\dfrac{25}{16}\right)^{-1/2}$

39. $(3 \times 10^4)^2$

40. $\dfrac{1}{(4 \times 10^{-2})^3}$

In Exercises 41–44, use absolute value notation to describe each distance.

41. The distance between x and 7 is at least 4.
42. The distance between x and 25 is no more than 10.
43. The distance between y and -30 is less than 5.
44. The distance between y and $1/2$ is more than 2.

In Exercises 45–48, graph each interval on the real number line.

45. $|x - 2| < 1$

46. $|x| \le 4$

47. $|x - \frac{3}{2}| \ge \frac{3}{2}$

48. $|x + 3| > 4$

In Exercises 49–60, find (a) the distance between the two points and (b) the coordinates of the midpoint of the line segment between the two points.

49. $(0, 0), (6, 0)$

50. $(0, 0), (0, 10)$

51. $(-2, -1), (2, 2)$

52. $(-1, 4), (2, 0)$

53. $(2, 1) \ (14, 6)$

54. $(-2, 2), (3, -10)$

55. $(5, -2), (5, 6)$

56. $(-3, -4), (-3, 5)$

57. $(-1, 0), (6, 2)$

58. $(1, 6), (4, 2)$

59. $(\frac{1}{3}, \frac{4}{3}), (\frac{2}{5}, \frac{4}{5})$

60. $(0.64, 0.45), (1.32, 4.68)$

In Exercises 61–84, perform the indicated operations and simplify.

61. $(x^2 - 2x + 1)(x^3 - 1)$

62. $(x^3 - 3x)(2x^2 + 3x + 5)$

63. $\left(x^2 - \dfrac{1}{x}\right)(x^2 + 1)$

64. $(t^5 - 3t)\left(\dfrac{1}{t^2} + t\right)$

65. $(y^2 - y)(y^2 + 1)(y^2 + y + 1)$

66. $(3z^3 + 4z)(z - 5)(z + 1)$

67. $\dfrac{x}{x^3 - 1} \cdot \dfrac{x - 1}{x^3}$

68. $\dfrac{4x^2 - 1}{(2x)(x^2 + 2x + 1)} \cdot \dfrac{x + 1}{4x^2 + 4x + 1}$

69. $\dfrac{x^2}{x^4 - 2x^2 - 8} \cdot \dfrac{x^2 + 4x + 2}{x^2 + 2}$

70. $\dfrac{x^2 - 1}{x^3 + x} \cdot \dfrac{x^4 - 1}{(x + 1)^2}$

71. $\dfrac{1}{x - 1} - \dfrac{1}{x + 2}$

72. $\dfrac{2}{x} - \dfrac{3}{x - 1} + \dfrac{4}{x + 1}$

73. $x - 1 + \dfrac{1}{x + 2} + \dfrac{1}{x - 1}$

74. $2x + \dfrac{3}{2(x - 4)} - \dfrac{1}{2(x + 2)}$

75. $x + 3 + \dfrac{6}{x - 1} + \dfrac{4}{(x - 1)^2} + \dfrac{1}{(x - 1)^3}$

76. $\dfrac{1}{x - 1} + \dfrac{1 - x}{x^2 + x + 1}$

77. $\dfrac{1}{x} - \dfrac{x - 1}{x^2 + 1}$

78. $\dfrac{1}{6(x - 2)} - \dfrac{1}{6(x + 2)} + \dfrac{1}{3(x^2 + 2)}$

79. $\dfrac{1}{x - 2} + \dfrac{1}{(x - 2)^2} + \dfrac{1}{x + 2} - \dfrac{1}{(x + 2)^2}$

80. $\dfrac{1}{L}\left(\dfrac{1}{y} - \dfrac{1}{L - y}\right)$ where L is a constant

81. $\dfrac{x^3}{x^3 - 1} \div \dfrac{x^2}{x^2 - 1}$

82. $\dfrac{4x - 6}{(x - 1)^2} \div \dfrac{2x^2 - 3x}{x^2 + 2x - 3}$

83. $\dfrac{\dfrac{x^2(5x - 6)}{2x + 3}}{\dfrac{5x}{2x + 3}}$

84. $\dfrac{\dfrac{x^3 + y^3}{x^2 + y^2}}{x^2 - xy + y^2}$

In Exercises 85–90, simplify each expression.

85. $\dfrac{\dfrac{1}{x} - \dfrac{1}{y}}{x^2 - y^2}$

86. $\dfrac{\dfrac{1}{x} - \dfrac{1}{y}}{\dfrac{1}{x} + \dfrac{1}{y}}$

87. $\dfrac{\left(\dfrac{3a}{a^2} - 1\right)}{\dfrac{a}{x} - 1}$

88. $\dfrac{\dfrac{1}{2x - 3} - \dfrac{1}{2x + 3}}{\dfrac{1}{2x} - \dfrac{1}{2x + 3}}$

89. $\dfrac{\dfrac{3}{2(x - 4)} - \dfrac{1}{2x}}{x^2 - 3x - 10}$

90. $\dfrac{1 - \dfrac{1}{1 + 1/a}}{\dfrac{1}{a + 1} - \dfrac{1}{a - 1}}$

In Exercises 91–100, insert the missing factor or factors.

91. $x^3 - 1 = (x - 1)(\qquad)$

92. $x^6 - y^6 = (x - y)(x + y)(\qquad)(\qquad)$

93. $x^4 - 2x^2y^2 + y^4 = (x + y)^2(\quad)^2$

94. $a^6 + 2a^3b^3 + b^6 = (a + b)^2(\quad)^2$

95. $\frac{3}{4}x^2 - \frac{5}{8}x + 4 = \frac{1}{12}(\quad)$

96. $\frac{2}{3}x^4 - \frac{3}{8}x^3 + \frac{5}{6}x^2 = \frac{x^2}{24}(\quad)$

97. $\dfrac{(x + 1)^{1/2} - \dfrac{x}{2(x + 1)^{1/2}}}{x + 1} = \left[\dfrac{1}{(\quad)}\right](x + 2)$

98. $\dfrac{x^3}{(x^2 - 1)^{1/2}} + 2x(x^2 - 1)^{1/2} = \left[\dfrac{x}{(x^2 - 1)^{1/2}}\right](\quad)$

99. $z^4 - 5z^3 + 8z - 40 = (z - 5)(\quad)(\quad)$

100. $x^2 - 7xy + 10y^2 + x^3 - 2x^2y = (x - 2y)(\quad)$

In Exercises 101–114, sketch the graph of the equations.

101. $y - 2x - 3 = 0$
102. $3x + 2y + 6 = 0$
103. $x - 5 = 0$
104. $y + 4 = 0$
105. $y = |x - 3|$
106. $y = 8 - |x|$
107. $y = \sqrt{5 - x}$
108. $y = x^2 - 4x$
109. $y = x^3 - x$
110. $y + 2x^2 = 0$
111. $y^2 - x + 3 = 0$
112. $x^2 + y^2 = 0$
113. $y = \sqrt{25 - x^2}$
114. $y = \sqrt{x + 2}$

Calculator Exercises

115. Calculate 15^4 in two ways. First use the exponential key $\boxed{y^x}$. Second, enter 15 and then press the square key $\boxed{x^2}$ twice. Why do these two methods give the same result?

116. Calculate $\sqrt[5]{107}\ \sqrt[5]{1145}$ in two ways. First use the keystroke sequence

$$107\ \boxed{y^x}\ .2\ \boxed{\times}\ 1145\ \boxed{y^x}\ .2\ \boxed{=}$$

Second, use the sequence

$$107\ \boxed{\times}\ 1145\ \boxed{=}\ \boxed{y^x}\ .2\ \boxed{=}$$

Why do these two methods give the same result?

117. Enter any number between 0 and 1 in a calculator and press the square key $\boxed{x^2}$ repeatedly. What number does the calculator display seem to be approaching?

118. Enter any positive number other than 1 in a calculator. What number is approached as the square root key $\boxed{\sqrt{x}}$ is pressed repeatedly?

119. Complete the following table.

n	1	10	10^2	10^4	10^6	10^{10}
$\dfrac{5}{n}$						

What number is $5/n$ approaching as n increases without bound?

2 FUNCTIONS AND THEIR GRAPHS

Functions 2.1

Many common relationships involve two variables in such a way that the value of one of the variables depends on the value of the other. For example, the sales tax on an item depends on its selling price. The distance an object moves in a given time depends on its speed.

Consider the relationship between the area of a circle and its radius. This relationship can be expressed by the equation $A = \pi r^2$. We have within the set of positive numbers a free choice for the value of r. The value of A then depends on our choice of r. Thus, we refer to A as the **dependent variable** and r as the **independent variable.**

Of particular interest are relationships such that to every value of the independent variable there corresponds *one and only one* value of the dependent variable. We call this type of correspondence a **function.**

DEFINITION OF A FUNCTION

A **function** is a relationship between two variables such that to each value of the independent variable there corresponds exactly one value of the dependent variable.

The collection of all values assumed by the independent variable is called the **domain** of the function, and the collection of all values assumed by the dependent variable is called the **range** of the function.

Functions and Formulas

We often specify functions by formulas. In such cases we *exclude* from the domain of the function all real numbers for which the formula is undefined.

Two common instances in which the domain of a function is restricted by its formula are:

1. *Even Roots of Negative Numbers.* For example,

   Domain:

   $$y = \sqrt{x + 2} \qquad\qquad x \geq -2$$

   has negative values inside the square root if $x < -2$. Thus, the domain is restricted to all real numbers greater than or equal to -2.

2. *Division by Zero.* For example,

   Domain:

   $$y = \frac{4}{x^2 - 9} \qquad\qquad \text{all real } x \neq \pm 3$$

   has a zero denominator when $x = \pm 3$. Hence the domain is restricted to all real numbers except $x = \pm 3$.

EXAMPLE 1 Determining Functional Relationships from Equations	Which of the following equations define functional relationships between the variables x and y? (a) $x + y = 1$ (b) $x^2 + y^2 = 1$ (c) $x^2 + y = 1$ (d) $x + y^2 = 1$ **Solution:** To determine if an equation defines a functional relationship between its variables, we isolate the dependent variable on the left-hand side, as shown in the second and fourth columns of Table 2.1. Note that those equations that assign two values ($\pm$) to the dependent variable for each assigned value of the independent variable do not define functions.

TABLE 2.1

Original equation	*x as the dependent variable*	*Is x a function of y?*	*y as the dependent variable*	*Is y a function of x?*
(a) $x + y = 1$	$x = 1 - y$	yes	$y = 1 - x$	yes
(b) $x^2 + y^2 = 1$	$x = \pm\sqrt{1 - y^2}$	no (two values of x for some values of y)	$y = \pm\sqrt{1 - x^2}$	no (two values of y for some values of x)
(c) $x^2 + y = 1$	$x = \pm\sqrt{1 - y}$	no	$y = 1 - x^2$	yes
(d) $x + y^2 = 1$	$x = 1 - y^2$	yes	$y = \pm\sqrt{1 - x}$	no

EXAMPLE 2
Finding the Domain
and Range of a Function

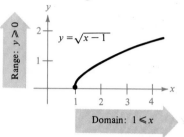

$y = \sqrt{x-1}$

Range: $y \geq 0$

Domain: $1 \leq x$

FIGURE 2.1

Determine the domain and range for the function of x defined by

$$y = \sqrt{x-1}$$

Solution:

Since $\sqrt{x-1}$ is not defined for $x - 1 < 0$ (that is, for $x < 1$), we must have $x \geq 1$. Therefore, the domain of the function is $x \geq 1$.

To find the range, we observe that $y = \sqrt{x-1}$ is never negative. Moreover, as x takes on the various values in the domain, y takes on all nonnegative values, and we find the range to be $y \geq 0$.

The graph of the function lends further support to our conclusions (Figure 2.1).

EXAMPLE 3
A Function Defined
by More Than One Equation

If $x < 1$,
$y = 1 - x$

If $x \geq 1$,
$y = \sqrt{x-1}$

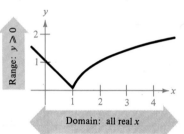

Range: $y \geq 0$

Domain: all real x

FIGURE 2.2

Determine the domain and range for the function of x given by

$$y = \begin{cases} \sqrt{x-1}, & \text{if } x \geq 1 \\ 1 - x & \text{if } x < 1 \end{cases}$$

Solution:

Since $x \geq 1$ or $x < 1$, the domain of the function is the entire set of real numbers.

On the portion of the domain for which $x \geq 1$, the function behaves as in Example 2. For $x < 1$, $1 - x$ is positive, and therefore the range of the function is the interval $[0, \infty)$.

Again, a graph of the function helps to verify our conclusions (Figure 2.2).

Remark: Sometimes the domain is *implied*, as in Example 2 ($\sqrt{x-1}$ is defined only if $x \geq 1$). On other occasions, the physical nature of the problem may restrict the domain to a certain subset of the real numbers. For instance, in the function for the area of a circle, $A = \pi r^2$, we list the domain as all $r > 0$.

Functional Notation

When using an equation to define a function, we generally isolate the dependent variable on the left side of the equation. For instance, writing the equation $x + 2y = 1$ in the form

$$y = \frac{1-x}{2}$$

indicates that y is the dependent variable. In **functional notation** this equation has the form

$$f(x) = \frac{1-x}{2}$$

This notation has the advantage of clearly identifying the dependent variable as $f(x)$ while at the same time providing a name "f" for the function. [The symbol $f(x)$ is read "f of x."]

To denote the value of the dependent variable when $x = 3$, we use the symbol $f(3)$ as follows:

$$f(3) = \frac{1 - (3)}{2} = \frac{-2}{2} = -1$$

Similarly,

$$f(0) = \frac{1 - (0)}{2} = \frac{1}{2}$$

$$f(-2) = \frac{1 - (-2)}{2} = \frac{3}{2}$$

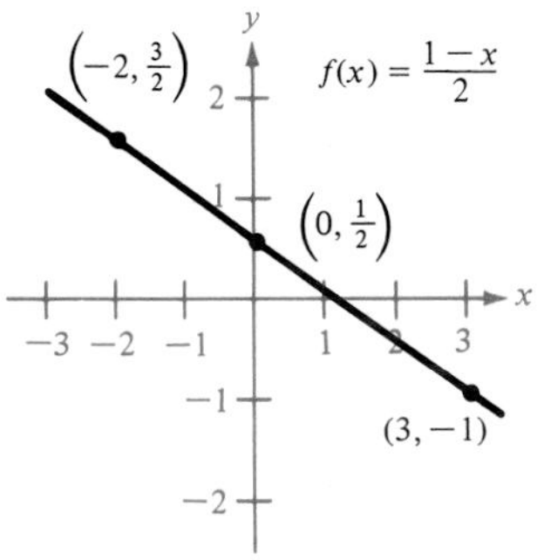

The values $f(3)$, $f(0)$, and $f(-2)$ are called **functional values,** and they lie in the range of f. This means that the values $f(3)$, $f(0)$, and $f(-2)$ are y-values, and thus the points $(3, f(3))$, $(0, f(0))$, and $(-2, f(-2))$ lie on the graph of f. (See Figure 2.3.)

The role of the variable x, in an equation that defines a function, is simply that of a "placeholder." For instance, the function

$$f(x) = 2x^2 - 4x + 1$$

can be properly described by the form

$$f(\ ) = 2(\ )^2 - 4(\ ) + 1$$

where parentheses are used instead of x. Therefore, to evaluate $f(-2)$, we simply place -2 in each set of parentheses:

$$f(-2) = 2(-2)^2 - 4(-2) + 1 = 2(4) + 8 + 1 = 17$$

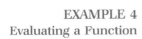

$\left(-2, \frac{3}{2}\right)$ $f(x) = \frac{1-x}{2}$

$\left(0, \frac{1}{2}\right)$

$-3\ -2\ -1\quad 1\qquad 3$

$(3, -1)$

FIGURE 2.3

EXAMPLE 4

Evaluating a Function

For the function f defined by $f(x) = x^2 - 4x + 7$, evaluate

(a) $f(3x)$ (b) $f(x - 1)$ *(c) $\dfrac{f(x + \Delta x) - f(x)}{\Delta x}$

Solution:
We begin by writing the equation for f in the form

$$f(\ ) = (\ )^2 - 4(\ ) + 7$$

(a) $f(3x) = (3x)^2 - 4(3x) + 7$ *Replace x with 3x*
$\qquad\quad = 9x^2 - 12x + 7$ *Expand terms*

(b) $f(x - 1) = (x - 1)^2 - 4(x - 1) + 7$ *Replace x with (x − 1)*
$\qquad\qquad\quad = x^2 - 2x + 1 - 4x + 4 + 7$ *Expand terms*
$\qquad\qquad\quad = x^2 - 6x + 12$ *Collect like terms*

*Δ is the uppercase Greek letter "delta." The symbol Δx is read as "delta-x" and is commonly used in calculus to denote a small change in x.

(c) $\dfrac{f(x + \Delta x) - f(x)}{\Delta x}$

$$= \frac{[(x + \Delta x)^2 - 4(x + \Delta x) + 7] - [x^2 - 4x + 7]}{\Delta x}$$

$$= \frac{x^2 + 2x\Delta x + (\Delta x)^2 - 4x - 4\Delta x + 7 - x^2 + 4x - 7}{\Delta x}$$

$$= \frac{2x\Delta x + (\Delta x)^2 - 4\Delta x}{\Delta x}$$

$$= \frac{\Delta x(2x + \Delta x - 4)}{\Delta x}$$

$$= 2x + \Delta x - 4$$

A function whose range consists of a single value is called a **constant function.** For example, $f(x) = -4$ is a constant function and no matter what value we choose for x, the resulting functional value is -4. That is, $f(-2) = -4$, $f(0) = -4$, $f(10) = -4$, and so on.

Although we generally use f as a convenient function name and x as the independent variable, we can use other symbols. For instance, the following equations all define the same function:

$$f(x) = x^2 - 4x + 7$$
$$f(t) = t^2 - 4t + 7$$
$$g(s) = s^2 - 4s + 7$$

Functions and Calculators

The basic steps in evaluating a function by means of a programmable calculator are

1. Program the function into the calculator.
2. Evaluate the function at specified values.

For example, for the function

$$C(x) = \frac{5000 + 0.56x}{x}$$

the calculator steps for a programmable (Texas Instruments) calculator are

| LRN | STO | 1 | × | .56 | + | 5000 | = | ÷ | RCL | 1 | = |
| R/S | RST | LRN | RST |

Once this program has been entered, the function can be evaluated at several points simply by entering various x-values as follows:

Calculator Steps	Display
10 R/S	500.56
100 R/S	50.56
1000 R/S	5.56
2000 R/S	3.06
5000 R/S	1.56
10000 R/S	1.06

We conclude this section with a glossary of terms related to the concept of a function.

GLOSSARY OF THE TERMINOLOGY OF FUNCTIONS

Function: A relationship between two variables such that to each value of the independent variable there corresponds exactly one value of the dependent variable.

Function Notation: $y = f(x)$

f is the **name** of the function
y is the **dependent variable**
x is the **independent variable**
$f(x)$ is the **value of the function at x**

Domain: The collection of all values of the independent variable for which the function is defined. If x is in the domain of f, we say that f is **defined** at x. If x is not in the domain of f, we say that f is **undefined** at x.

Range: The collection of all values assumed by the dependent variable.

Implied Domain: If f is defined by an equation and the domain is not specified, then we assume the domain to consist of all real numbers for which the equation is defined. [For example, the implied domain of the function $f(x) = \sqrt{x}$ is the collection of nonnegative real numbers.]

Constant Function: A function whose range consists of a single number.

Section Exercises 2.1

1. Given $f(x) = 2x - 3$, find
    (a) $f(1)$ (b) $f(0)$
    (c) $f(-3)$ (d) $f(b)$
    (e) $f(x - 1)$ (f) $f(\frac{1}{4})$

2. Given $f(x) = x^2 - 2x + 2$, find
    (a) $f(\frac{1}{2})$ (b) $f(3)$
    (c) $f(-1)$ (d) $f(c)$
    (e) $f(x + \Delta x)$ (f) $f(2)$

3. Given $f(x) = \sqrt{x + 3}$, find
     (a) $f(-3)$ (b) $f(-2)$
     (c) $f(0)$ (d) $f(6)$
     (e) $f(x + \Delta x)$ (f) $f(c)$

4. Given $f(x) = 1/\sqrt{x}$, find
     (a) $f(1)$ (b) $f(4)$
     (c) $f(2)$ (d) $f(\frac{1}{4})$
     (e) $f(x + \Delta x)$ (f) $f(x + \Delta x) - f(x)$

5. Given $f(x) = |x|/x$, find
     (a) $f(2)$ (b) $f(-2)$
     (c) $f(-100)$ (d) $f(100)$
     (e) $f(x^2)$ (f) $f(x - 1)$

6. Given $f(x) = |x| + 4$, find
     (a) $f(2)$ (b) $f(-2)$
     (c) $f(3)$ (d) $f(x^2)$
     (e) $f(x + \Delta x)$ (f) $f(x + \Delta x) - f(x)$

7. Given $f(x) = x^2 - x + 1$, find

$$\frac{f(2 + \Delta x) - f(2)}{\Delta x}$$

8. Given $f(x) = 1/x$, find

$$\frac{f(1 + \Delta x) - f(1)}{\Delta x}$$

9. Given $f(x) = x^3$, find

$$\frac{f(x + \Delta x) - f(x)}{\Delta x}$$

10. Given $f(x) = 3x - 1$, find

$$\frac{f(x) - f(1)}{x - 1}$$

11. Given $f(x) = 1/\sqrt{x - 1}$, find

$$\frac{f(x) - f(2)}{x - 2}$$

12. Given $f(x) = x^3 - x$, find

$$\frac{f(x) - f(1)}{x - 1}$$

13. Let $f(x) = x^2 - 1$. Find all real numbers x such that $f(x) = 8$.

14. Let $f(x) = x^3 - x$. Find all real numbers x such that $f(x) = 0$.

15. Let

$$f(x) = \frac{3}{x - 1} + \frac{4}{x - 2}$$

Find all real numbers x such that $f(x) = 0$.

16. Let

$$f(x) = a + \frac{b}{x}$$

Find all real numbers x such that $f(x) = 0$.

In Exercises 17–26, find the domain and range of the given function.

17. $f(x) = \sqrt{x - 1}$

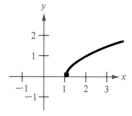

18. $f(x) = \sqrt{1 - x}$

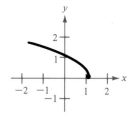

19. $f(x) = x^2$

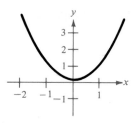

20. $f(x) = 4 - x^2$

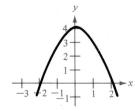

21. $f(x) = \sqrt{9 - x^2}$

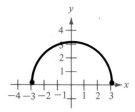

22. $f(x) = \sqrt{25 - x^2}$

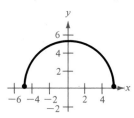

23. $f(x) = 1/|x|$

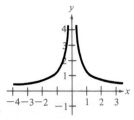

24. $f(x) = |x - 2|$

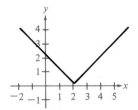

25. $f(x) = |x|/x$

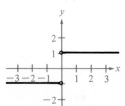

26. $f(x) = \sqrt{x^2 - 4}$

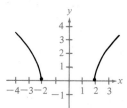

In Exercises 27–36, identify the equations that determine y as a function of x.

27. $x^2 + y^2 = 4$
28. $x = y^2$
29. $x^2 + y = 4$
30. $x + y^2 = 4$
31. $2x + 3y = 4$
32. $x^2 + y^2 - 2x - 4y + 1 = 0$
33. $y^2 = x^2 - 1$
34. $y = \pm\sqrt{x}$
35. $x^2y - x^2 + 4y = 0$
36. $xy - y - x - 2 = 0$

In Exercises 37 and 38, find a formula for the given function $V = f(x)$ and give its domain.

37. The value V of a farm having $500,000 worth of buildings, livestock, and equipment in terms of the number of acres on the farm. (Each acre is valued at $1750.)
38. The value V of wheat at $4.45 per bushel as a function of the number of bushels.
39. A company produces a product for which the variable cost is $12.30 per unit and the fixed costs are $98,000. The product sells for $17.98. Let x be the number of units produced.
    (a) Write the total cost C as a function of the number of units produced.
    (b) Write the revenue R as a function of the number of units produced.
    (c) Write the profit P as a function of the number of units produced. (Note: $P = R - C$.)
40. The inventor of a new game believes that the variable cost for producing the game is $0.95 per unit and the fixed costs are $6,000. He plans to wholesale the game for $1.69. Let x be the number of games sold.
    (a) Write the total cost C as a function of the number of games sold.
    (b) Write the average cost per unit $\overline{C} = C/x$ as a function of x.

41. The demand function for a particular commodity is given by

$$p = \frac{14.75}{1 + 0.01x}, \qquad 0 \le x$$

where p is the price per unit and x is the number of units sold.
    (a) Find x as a function of p.
    (b) Use the result of part (a) to find the number of units sold when the price is $10.00.
42. A power station is on one side of a river that is one-half mile wide. A factory lies downstream three miles on the other side of the river. It costs $10 per foot to run the power lines on land and $15 per foot to run them underwater. Write the cost C of running the line from the power station to the factory as a function of x. (See Figure 2.4.)

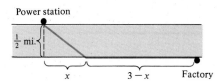

FIGURE 2.4

43. A radio manufacturer charges $90 per unit for units that cost $60 to produce. To encourage large orders from distributors, the manufacturer will reduce the price by $0.01 per unit for each unit in excess of 100. (For example, an order of 200 units would have a price of $89 per unit.) This price reduction is discontinued when the price per unit drops to $75.
    (a) Write the price per unit p as a function of the order size x.
    (b) Write the profit P as a function of the order size x. (Note: $P = R - C = px - 60x$.)
    (c) Find the total profit for an order of 1000 units.
44. Assume that the amount of money deposited in a bank is proportional to the square of the interest rate the bank pays on the money. That is, $d = kr^2$, where d is the total deposit, r is the interest rate, and k is the proportionality constant. Assuming the bank can reinvest the money for a return of 18%, write the bank's profit P as a function of the interest rate r.

Graphs of Functions   2.2

In Section 2.1 we discussed functions from an algebraic (analytic) point of view. Here we study functions from a geometric (graphic) perspective. In this setting it is convenient to view a function f as a correspondence between two sets of real numbers.

ALTERNATIVE DEFINITION OF A FUNCTION

A **function** f from a set X to a set Y is a correspondence that assigns to each element x in X exactly one element y in Y. We call y the **image** of x under f and denote it by $f(x)$. The **domain** of f is the set X, while the **range** consists of all images of elements in X.

This section is divided into three parts, each related to one aspect of the graph of a function:

1. The **shape** of the graph.
2. The **position** of the graph relative to the x- and y-axes.
3. **Transformations** of the graph.

As you study this section, remember that the graph of the function $y = f(x)$ consists of all points $(x, f(x))$ where

$$x = \text{the directed distance from the } y\text{-axis}$$
$$f(x) = \text{the directed distance from the } x\text{-axis}$$

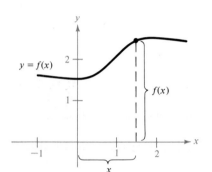

FIGURE 2.5

as shown in Figure 2.5.

Since (by the definition of a function) there corresponds exactly one y-value for each x-value, it follows that a vertical line can intersect the graph of a function at most once. This observation provides us with a convenient visual test for functions.

THE VERTICAL LINE TEST FOR FUNCTIONS

An equation defines y as a function of x if no vertical line in the plane intersects the graph of the equation at more than one point.

EXAMPLE 1
Vertical Line Test for Functions

Which of the graphs in Figure 2.6 represent functions of x?

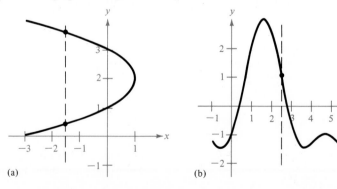

(a)   (b)

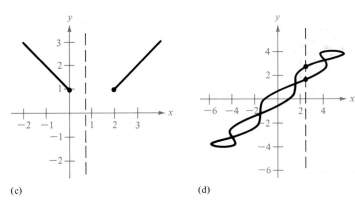

(c)   (d)

FIGURE 2.6

Solution:

(a) This is *not* the graph of a function, since we can find a vertical line that intersects the graph twice.

(b) This *is* the graph of a function, since every vertical line intersects this graph at most once.

(c) This *is* the graph of a function. (Note that if a vertical line does not intersect the graph, it simply means that the function is undefined for this particular value of *x*.)

(d) This is *not* the graph of a function, since we can find a vertical line that intersects the graph twice.

In Section 2.1 we mentioned that a function is called **constant** if its range consists of a single *y*-value. For example,

$$h(x) = 2 \qquad \text{and} \qquad g(x) = -1$$

are constant functions. Since the *y*-value of a constant function is the same for all *x*, it follows that the graph of a constant function consists of a horizontal line. (See Figure 2.7.)

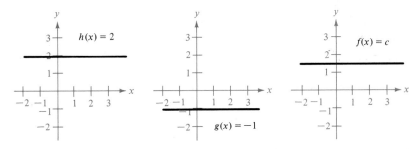

The graph of a constant function is a horizontal line.

FIGURE 2.7

GRAPH OF A CONSTANT FUNCTION

> The graph of $f(x) = c$ is a horizontal line with a y-intercept at $(0, c)$.

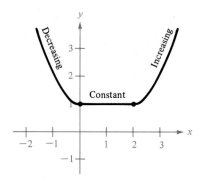

FIGURE 2.8

The more we know about the graph of a function, the more we know about the function itself. Consider the graph shown in Figure 2.8. As we move from left to right, this graph falls for negative values of x, is constant from $x = 0$ to $x = 2$, and then rises for x greater than 2. Correspondingly, we say that the function is:

decreasing on the interval $(-\infty, 0)$,

constant on the interval $(0, 2)$, and

increasing on the interval $(2, \infty)$.

DEFINITION OF INCREASING AND DECREASING FUNCTIONS

> A function f is said to be **increasing** on an interval if, for any two numbers x_1 and x_2 in the interval,
>
> $$x_1 < x_2 \quad \text{implies} \quad f(x_1) < f(x_2)$$
>
> A function f is said to be **decreasing** on an interval if, for any two numbers x_1 and x_2 in the interval,
>
> $$x_1 < x_2 \quad \text{implies} \quad f(x_1) > f(x_2)$$
>
> A function f is said to be **constant** on an interval if, for any two numbers x_1 and x_2 in the interval,
>
> $$f(x_1) = f(x_2)$$

EXAMPLE 2
Increasing and
Decreasing Functions

In Figure 2.9 determine the intervals over which each function is increasing, decreasing, or constant.

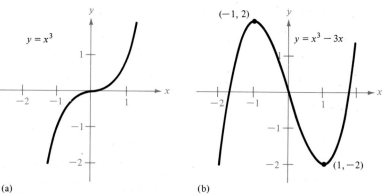

(a)   (b)

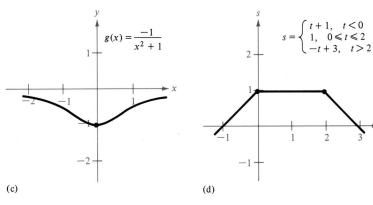

(c)   (d)

FIGURE 2.9

Solution:

(a) Although it might appear that there is an interval about zero over which this function is constant, we see that if $x_1 < x_2$, then $x_1^3 < x_2^3$, and we conclude that the function is increasing for all x.

(b) This function is

increasing on the interval $(-\infty, -1)$,

decreasing on the interval $(-1, 1)$, and

increasing on the interval $(1, \infty)$.

(c) This function is

decreasing on the interval $(-\infty, 0)$ and

increasing on the interval $(0, \infty)$.

(d) This function is

increasing on the interval $(-\infty, 0)$,

constant on the interval $(0, 2)$, and

decreasing on the interval $(2, \infty)$.

Remark: The points at which a function changes its increasing, decreasing, or constant behavior are especially important in producing an accurate graph of the function. These points often identify maximum or minimum values of the function. Techniques for finding the exact location of these special points are developed in calculus.

Position of a Function's Graph Relative to the x- and y-Axes

In Section 1.6 we defined an x-intercept to be a point $(x, 0)$ at which a graph crosses the x-axis. When the graph of a function crosses the x-axis, we say that the function has a **zero** at that point. For example, the function $f(x) = x - 4$ has a zero at $(4, f(4))$, since $f(4) = 4 - 4 = 0$.

In Section 1.6, we also discussed different types of symmetry. In the terminology of functions, we say that a function is **even** if its graph is symmetric with respect to the y-axis; a function is **odd** if its graph is symmetric with respect to the origin. Thus, our symmetry tests in Section 1.6 yield the following test for even and odd functions.

TEST FOR EVEN AND ODD FUNCTIONS

> The function $y = f(x)$ is **even** if
>
> $$f(-x) = f(x)$$
>
> The function $y = f(x)$ is **odd** if
>
> $$f(-x) = -f(x)$$

EXAMPLE 3
Even and Odd Functions

Determine whether the following functions are even, odd, or neither.

(a) $g(x) = x^3 - x$ (b) $h(x) = x^2 + 1$ (c) $f(x) = x^3 - 1$

Solution:

(a) This function is odd, since

$$g(-x) = (-x)^3 - (-x) = -x^3 + x = -(x^3 - x) = -g(x)$$

(b) This function is even, since

$$h(-x) = (-x)^2 + 1 = x^2 + 1 = h(x)$$

(c) By substituting $-x$ for x, we have

$$f(-x) = (-x)^3 - 1 = -x^3 - 1$$

Now, since

$$f(x) = x^3 - 1 \quad \text{and} \quad -f(x) = -x^3 + 1$$

we conclude that

$$f(-x) \neq f(x) \quad \text{and} \quad f(-x) \neq -f(x)$$

which implies that the function is neither even nor odd.

Transformations of a Function's Graph

Some families of graphs all have the same basic shape. Consider the graph of $y = x^2$, as shown in Figure 2.10. Now compare this graph to those shown in Figure 2.11.

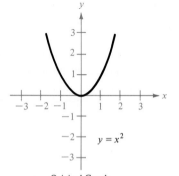

Original Graph

FIGURE 2.10

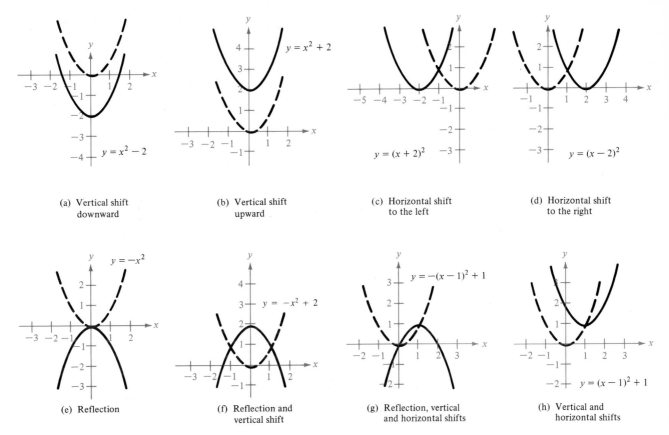

(a) Vertical shift
    downward

(b) Vertical shift
    upward

(c) Horizontal shift
    to the left

(d) Horizontal shift
    to the right

(e) Reflection

(f) Reflection and
    vertical shift

(g) Reflection, vertical
    and horizontal shifts

(h) Vertical and
    horizontal shifts

FIGURE 2.11   *Transformations of the Graph of* $y = x^2$

Each of the graphs in Figure 2.11 is a **transformation** of the graph of $y = x^2$. The three basic types of transformations involved in these eight graphs are (1) horizontal shifts, (2) vertical shifts, and (3) reflections.

BASIC TYPES OF TRANSFORMATIONS $(c > 0)$

Original Graph: $y = f(x)$
Horizontal Shift c units to the **right:** $y = f(x - c)$
Horizontal Shift c units to the **left:** $y = f(x + c)$
Vertical Shift c units **downward:** $y = f(x) - c$
Vertical Shift c units **upward:** $y = f(x) + c$
Reflection (about the x-axis): $y = -f(x)$
Reflection (about the y-axis): $y = f(-x)$

EXAMPLE 4
Transformations of the
Graph of a Function

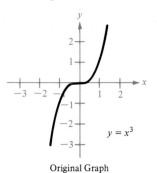

Original Graph

FIGURE 2.12

Use the graph of $y = x^3$ shown in Figure 2.12 to sketch the graph of each of the following functions:

(a) $y = x^3 + 1$ (b) $y = (x - 1)^3$

(c) $y = -x^3$ (d) $y = (x + 2)^3 + 1$

Solution:

The graphs are shown in Figure 2.13.

(a)

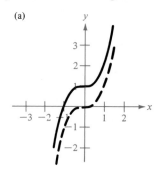

Vertical shift: 1 up
$y = f(x) + 1 = x^3 + 1$

(b)

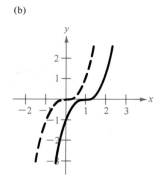

Horizontal shift: 1 right
$y = f(x - 1) = (x - 1)^3$

(c)

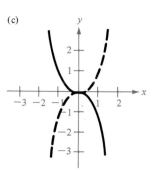

Reflection
$y = -f(x) = -x^3$

(d)

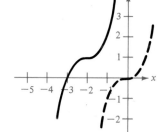

Vertical shift: 1 up
Horizontal shift: 2 left
$y = f(x + 2) + 1 = (x + 2)^3 + 1$

FIGURE 2.13

    Horizontal shifts, vertical shifts, and reflections are called **rigid** transformations because the basic shape of the graph is unchanged. As illustrated in Figures 2.11 and 2.13, the only change caused by a rigid transformation is the *position* of the graph in the *xy*-plane.

    Nonrigid transformations are those which cause a *distortion* of the original graph. For example, the graph of

$$y = cf(x)$$

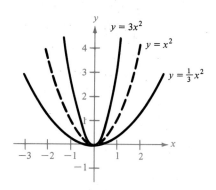

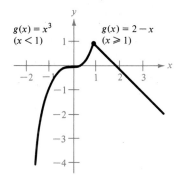

FIGURE 2.14

is a nonrigid vertical transformation of the graph of $y = f(x)$. If $c > 1$, we say the transformation **stretches,** and if $0 < c < 1$ we say the transformation **shrinks** the graph of $y = f(x)$. Figure 2.14 shows the effect of the coefficients $c = 3$ and $c = \frac{1}{3}$ on the graph of $f(x) = x^2$.

On occasion we encounter a function that is defined differently on distinct subsets of its domain. Such is the case with

$$g(x) = \begin{cases} x^3, & \text{if } x < 1 \\ 2 - x, & \text{if } x \geq 1 \end{cases}$$

whose graph is shown in Figure 2.15. One commonly used function that fits into this category is the **greatest integer function**

$$f(x) = [x] = \text{the greatest integer less than or equal to } x$$

More explicitly, f is described by

$$f(x) = \begin{cases} \quad \vdots & \quad \vdots \\ -1, & \text{if } -1 \leq x < 0 \\ 0, & \text{if } \quad 0 \leq x < 1 \\ 1, & \text{if } \quad 1 \leq x < 2 \\ \quad \vdots & \quad \vdots \end{cases}$$

and its graph is shown in Figure 2.16.

FIGURE 2.15

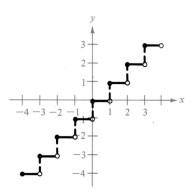

FIGURE 2.16   *Greatest Integer Function*

Section Exercises 2.2

In Exercises 1–10, use the vertical line test to determine if y is a function of x.

1. $y = x^2$

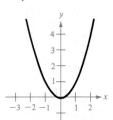

2. $y = x^3 - 1$

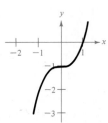

3. $x - y^2 = 0$

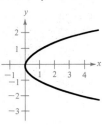

4. $x^2 + y^2 = 9$

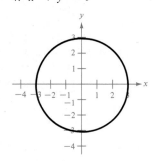

5. $\sqrt{x^2 - 4} - y = 0$

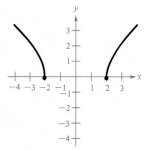

6. $x - xy + y + 1 = 0$

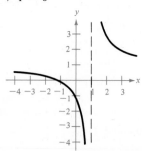

7. $x^2 = xy - 1$

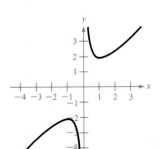

8. $x = |y|$

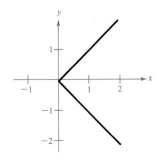

9. $x^2 - 4y^2 + 4 = 0$

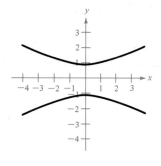

10. $y = |x - 3| + |x - 1|$

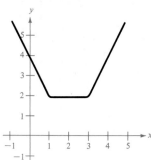

In Exercises 11–20, (a) determine the intervals over which the function is increasing, decreasing, or constant, and (b) determine if the function is even, odd, or neither.

11. $f(x) = 2x$

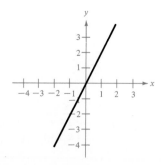

12. $f(x) = x^2 - 2x$

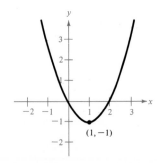

(1, −1)

13. $f(x) = x^3 - 3x^2$

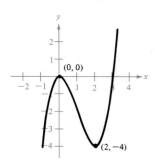

14. $f(x) = \sqrt{x^2 - 4}$

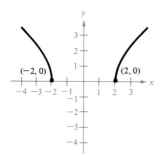

15. $f(x) = 3x^4 - 6x^2$

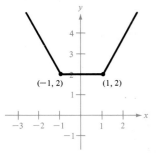

16. $f(x) = x^{2/3}$

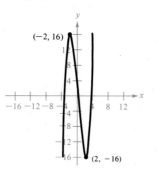

17. $f(x) = |x + 1| + |x - 1|$

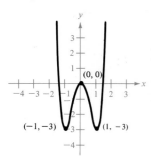

18. $f(x) = x^3 - 12x$

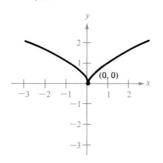

19. $f(x) = x\sqrt{x + 3}$

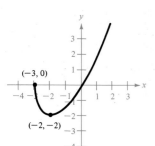

20. $f(x) = |2x - 3|$

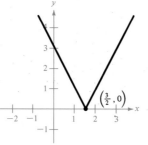

In Exercises 21–30, sketch the graph of the function.

21. $f(x) = 5 - 3x$

22. $f(x) = 2x - 3$

23. $f(x) = x^2 - 4$

24. $f(x) = -x^2 + 2x$

25. $g(t) = \sqrt[3]{t - 1}$

26. $h(t) = \dfrac{1}{t^2 + 1}$

27. $f(x) = |x + 2|$

28. $f(x) = |x| + |x + 2|$

29. $f(x) = \begin{cases} x + 3, & \text{if } x \le 0 \\ 3, & \text{if } 0 < x \le 2 \\ 2x - 1, & \text{if } x > 2 \end{cases}$

30. $f(x) = \begin{cases} 2x + 1, & \text{if } x \le -1 \\ x^2 - 2, & \text{if } x > -1 \end{cases}$

31. Use the accompanying graph of $f(x) = \sqrt{x}$ to sketch the graph of each of the following.
   (a) $y = f(x) + 2 = \sqrt{x} + 2$
   (b) $y = -f(x) = -\sqrt{x}$
   (c) $y = f(x - 2) = \sqrt{x - 2}$
   (d) $y = f(x + 3) = \sqrt{x + 3}$
   (e) $y = 2 - f(x - 4) = 2 - \sqrt{x - 4}$
   (f) $y = f(2x) = \sqrt{2x}$
   (g) $y = 2f(x) = 2\sqrt{x}$

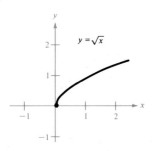

32. Use the accompanying graph of $f(x) = \sqrt[3]{x}$ to sketch the graph of each of the following.
   (a) $y = f(x) - 1 = \sqrt[3]{x} - 1$
   (b) $y = f(x + 1) = \sqrt[3]{x + 1}$
   (c) $y = f(x - 1) = \sqrt[3]{x - 1}$
   (d) $y = -f(x - 2) = -\sqrt[3]{x - 2}$
   (e) $y = f(x + 1) - 1 = \sqrt[3]{x + 1} - 1$
   (f) $y = f(x/2) = \sqrt[3]{x/2}$
   (g) $y = \frac{1}{2}f(x) = \frac{1}{2}\sqrt[3]{x}$

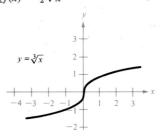

33. Use the accompanying graph of $f(x) = x\sqrt{x + 3}$ to write formulas for the functions whose graphs are shown in parts (a) through (d).

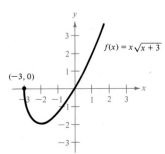

(a)

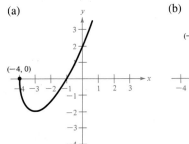

(b)

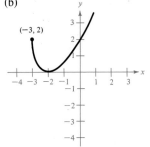

(c)

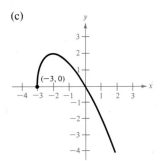

(d)

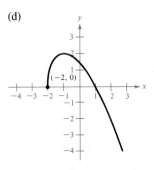

34. Use the accompanying graph of $f(x) = 1/(x^2 + 1)$ to write formulas for the functions whose graphs are shown in parts (a) through (d).

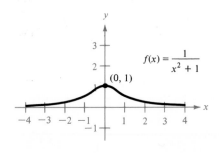

(a)

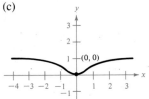

(b)

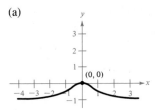

(c)

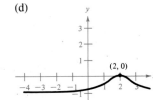

(d)

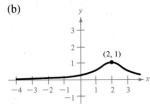

35. Show that a function of the form

$$f(x) = a_{2n+1}x^{2n+1} + \cdots + a_3x^3 + a_1x$$

is odd.

36. Show that a function of the form

$$f(x) = a_{2n}x^{2n} + a_{2n-2}x^{2n-2} + \cdots + a_2x^2 + a_0$$

is even.

37. Show that the product of two even functions is an even function.

38. Show that the product of two odd functions is an even function.

39. Show that the product of an odd function and an even function is odd.

Linear Functions and Slope

2.3

The simplest (and perhaps the most useful) type of function is a **linear function**

$$f(x) = ax + b, \qquad a \neq 0$$

As we shall see in this section, the graph of a linear function is a (nonvertical) line. In this text, we follow the convention of using the term **line** to mean a *straight* line.

The Slope of a Line

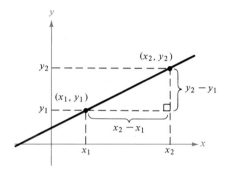

FIGURE 2.17

By the **slope** of a (nonvertical) line, we mean the number of units a line rises (or falls) vertically for each unit of horizontal change from left to right. For instance, consider the two points (x_1, y_1) and (x_2, y_2) on the line in Figure 2.17. As we move from left to right along this line, a change of $(y_2 - y_1)$ units in the vertical direction corresponds to a change of $(x_2 - x_1)$ units in the horizontal direction:

$$y_2 - y_1 = \text{the change in } y$$

and

$$x_2 - x_1 = \text{the change in } x$$

Thus, the slope of the line in Figure 2.17 is given by the ratio of these changes.

DEFINITION OF THE SLOPE OF A LINE

> The **slope** of the line passing through the points (x_1, y_1) and (x_2, y_2) is
>
> $$m = \frac{y_2 - y_1}{x_2 - x_1}$$
>
> where $x_1 \neq x_2$.

Remark: Note that

$$\frac{y_2 - y_1}{x_2 - x_1} = \frac{-(y_1 - y_2)}{-(x_1 - x_2)} = \frac{y_1 - y_2}{x_1 - x_2}$$

Hence it does not matter which pair of coordinates we subtract to find the slope *as long as* we are consistent and both "subtracted coordinates" come from the same point.

In our first example (Figure 2.18), you will see that as we move from left to right:

1. A line with positive slope ($m > 0$) *rises*.

2. A line with negative slope ($m < 0$) *falls*.

3. A line with zero slope ($m = 0$) is *horizontal*.

EXAMPLE 1
Finding the Slope of a Line
Passing Through Two Points

Find the slope of the line containing each of the following pairs of points:

(a) $(-2, 0)$ and $(3, 1)$ (b) $(-1, 2)$ and $(2, 2)$ (c) $(0, 4)$ and $(1, -1)$

Solution:

(a) The slope is

$$m = \frac{1 - 0}{3 - (-2)} = \frac{1}{3 + 2} = \frac{1}{5}$$

(b) The slope is

$$m = \frac{2 - 2}{2 - (-1)} = \frac{0}{3} = 0$$

(c) The slope is

$$m = \frac{-1 - 4}{1 - 0} = \frac{-5}{1} = -5$$

See Figure 2.18.

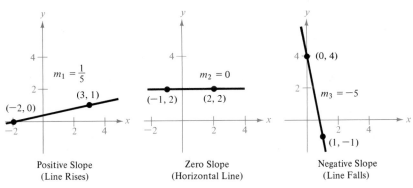

Positive Slope
(Line Rises)

Zero Slope
(Horizontal Line)

Negative Slope
(Line Falls)

FIGURE 2.18

Remark: In Example 1, note that we did not consider the slope of a vertical line. This was not an oversight! The definition of slope does not apply to vertical lines. Informally, we say that the slope of a vertical line is *undefined*. Consider the points $(3, 4)$ and $(3, 1)$ on the line shown in Figure 2.19. In attempting to find the slope of this line, we obtain

$$m = \frac{4 - 1}{3 - 3} \qquad \text{\textit{Undefined division by zero}}$$

Since division by zero is undefined, it follows that the slope of a vertical line must also be undefined.

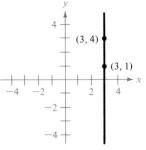

Slope is undefined.
(vertical line)

FIGURE 2.19

It is important to realize that the slope of a line is independent of the particular points used to calculate the slope. *Any* two points on the line can be used. This can be verified from the similar triangles shown in Figure 2.20. (Recall that the ratios of corresponding sides of similar triangles are equal.)

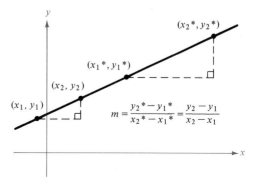

$$m = \frac{y_2{}^* - y_1{}^*}{x_2{}^* - x_1{}^*} = \frac{y_2 - y_1}{x_2 - x_1}$$

Any two points on a line can be used
to determine its slope.

FIGURE 2.20

Equations of Lines

If we know the slope of a line and one point on the line, how can we determine the equation of the line? Figure 2.21 leads us to the answer to this question. For, if (x_1, y_1) is a point lying on a line of slope m, and (x, y) is any *other* point on the line, then

$$\frac{y - y_1}{x - x_1} = m$$

This equation, involving the two variables x and y, can be rewritten in the form

$$y - y_1 = m(x - x_1)$$

which is called the **point-slope equation** of a line.

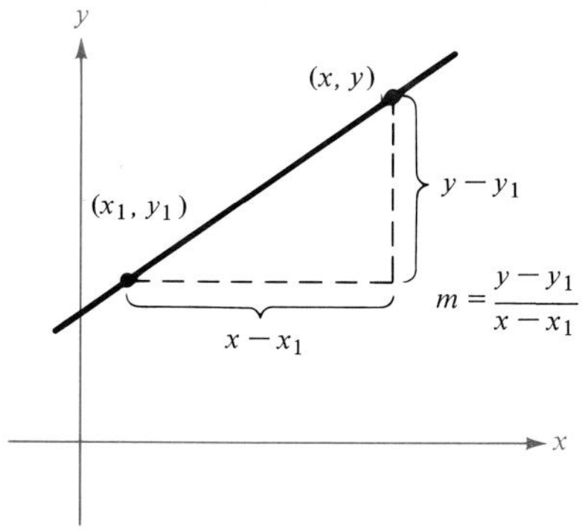

FIGURE 2.21

POINT-SLOPE EQUATION OF A LINE

> The equation of the line with slope m passing through the point (x_1, y_1) is given by
>
> $$y - y_1 = m(x - x_1)$$

EXAMPLE 2
The Point-Slope Equation
of a Line

Find the equation of the line with slope of 3 and passing through the point $(1, -2)$.

Solution:
Using the point-slope form,

$$y - y_1 = m(x - x_1)$$

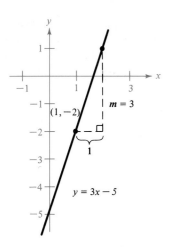

FIGURE 2.22

we have

$$y - (-2) = 3(x - 1)$$
$$y + 2 = 3x - 3$$

or

$$y = 3x - 5$$

See Figure 2.22.

We can combine the definition of the slope of a line with the point-slope equation to obtain the following **two-point equation** of a line.

TWO-POINT EQUATION OF A LINE

The equation of the line passing through the points (x_1, y_1) and (x_2, y_2) is given by

$$y - y_1 = \frac{y_2 - y_1}{x_2 - x_1} (x - x_1)$$

where $x_1 \neq x_2$.

EXAMPLE 3
The Two-Point Equation of a Line

The total U.S. sales (including inventories) during the first two quarters of 1978 were 539.9 and 560.2 billion dollars, respectively. Assuming a *linear growth pattern*, predict the total sales during the fourth quarter of 1978.

Solution:
In Figure 2.23 we let x represent the quarter and y represent the sales in billions of dollars, and we let $(1, 539.9)$ and $(2, 560.2)$ be two points on the line representing the total U.S. sales. Using the two-point equation of a line, we have

$$y - y_1 = \frac{y_2 - y_1}{x_2 - x_1} (x - x_1)$$

$$y - 539.9 = \frac{560.2 - 539.9}{2 - 1} (x - 1)$$

$$y = 20.3(x - 1) + 539.9$$

$$y = 20.3x + 519.6$$

Now, we estimate the fourth quarter sales ($x = 4$) to be

$$y = (20.3)(4) + 519.6 = 600.8 \text{ billion dollars}$$

(In this particular case, the estimate proves to be quite good. The actual fourth quarter sales in 1978 were 600.5 billion dollars.)

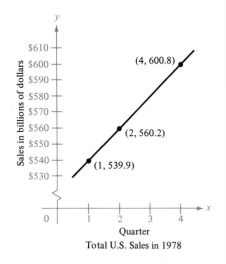

Total U.S. Sales in 1978

FIGURE 2.23

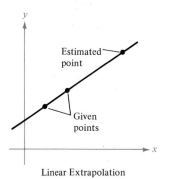

Linear Extrapolation

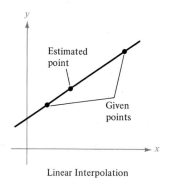

Linear Interpolation

FIGURE 2.24

The prediction method illustrated in Example 3 is called **linear extrapolation.** Note that the extrapolated point does not lie between the given points. (See Figure 2.24.) When the estimated point lies *between* two given points, we call the procedure **linear interpolation.**

Graphing Lines in the Plane

Many problems in analytic geometry can be classified in two basic categories:

1. Given a graph, what is its equation?
2. Given an equation, what is its graph?

We can use the point-slope and two-point equations for the first category to find an equation of a line given its geometric description. However, these two forms *are not* particularly useful for the second category. The form that is best suited to graphing lines is called the **slope-intercept** form for the equation of a line.

THE SLOPE-INTERCEPT FORM FOR THE EQUATION OF A LINE

The graph of the equation

$$y = mx + b$$

is a line having a *slope* of m and a *y-intercept* at $(0, b)$.

Remark: Note that the linear function

$$f(x) = ax + b$$

is in slope-intercept form. Thus, the graph of a linear function is a line having a slope of a and a y-intercept at $(0, b)$.

EXAMPLE 4
Graphing Linear Equations

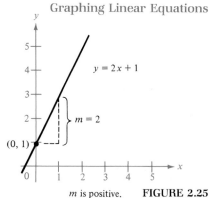

$y = 2x + 1$

$m = 2$

$(0, 1)$

m is positive.   **FIGURE 2.25**

Sketch the graphs of the following linear equations:

(a) $y = 2x + 1$

(b) $y = 2$

(c) $3y + x - 6 = 0$

Solution:

(a) The y-intercept occurs at $b = 1$, and since the slope is 2 we know that this line rises 2 units for each unit the line moves to the right. (See Figure 2.25.)

(b) The y-intercept occurs at $b = 2$, and since the slope is zero we know that the line is horizontal. That is, it doesn't rise or fall. (See Figure 2.26.)

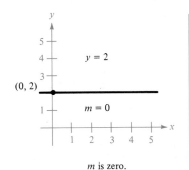

m is zero.

FIGURE 2.26

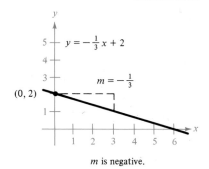

m is negative.

FIGURE 2.27

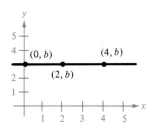

Horizontal Line: $y = b$

FIGURE 2.28

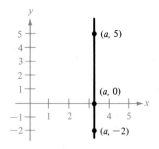

Vertical Line: $x = a$

FIGURE 2.29

(c) We begin by writing the equation in slope-intercept form:

$$3y + x - 6 = 0$$
$$3y = -x + 6$$
$$y = -\frac{1}{3}x + 2$$

Now we see that the y-intercept occurs at $b = 2$ and the slope is $m = -\frac{1}{3}$. Thus, this line falls 1 unit for every 3 units the line moves to the right. (See Figure 2.27.)

From the slope-intercept equation of a line, we can see that a horizontal line ($m = 0$) has an equation of the form

$$y = (0)x + b \qquad \text{or} \qquad y = b \qquad \textit{Horizontal line}$$

This is consistent with the fact that each point on a horizontal line through $(0, b)$ has a y-coordinate of b. (See Figure 2.28.)

Similarly, each point on a vertical line through $(a, 0)$ has an x-coordinate of a. (See Figure 2.29.) Hence a vertical line has an equation of the form

$$x = a \qquad \qquad \textit{Vertical line}$$

This equation cannot be written in the slope-intercept form, since the slope of a vertical line is undefined. However, *every* line has an equation that can be written in the **general form**

$$Ax + By + C = 0 \qquad \qquad \textit{General form}$$

where A and B are not *both* zero. If $A = 0$ (and $B \neq 0$), the equation can be reduced to the form $y = b$, which represents a horizontal line. If $B = 0$ (and $A \neq 0$), the general equation can be reduced to the form $x = a$, which represents a vertical line.

We now have identified the following six forms of equations of lines.

EQUATIONS OF LINES

General Equation:	$Ax + By + C = 0$
Vertical Line:	$x = a$
Horizontal Line:	$y = b$
Slope-Intercept Equation:	$y = mx + b$
Point-Slope Equation:	$y - y_1 = m(x - x_1)$
Two-Point Equation:	$y - y_1 = \dfrac{y_2 - y_1}{x_2 - x_1}(x - x_1)$

Section Exercises 2.3

In Exercises 1–6, estimate the slope of the given line from its graph.

1.

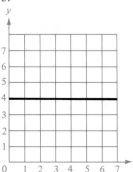

2.

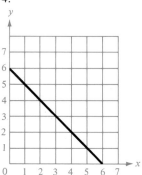

3.

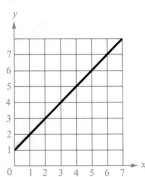

4.

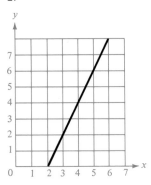

5.

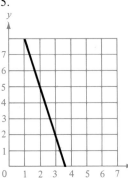

6.

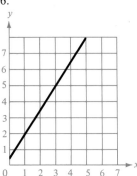

In Exercises 7–14, plot the points and find the slope of the line passing through each pair of points.

7. $(3, -4)$, $(5, 2)$

8. $(-2, 1)$, $(4, -3)$

9. $(\frac{1}{2}, 2)$, $(6, 2)$

10. $(-\frac{3}{2}, -5)$, $(2, -1)$

11. $(-6, -1)$, $(-6, 4)$

12. $(2, 1)$, $(2, 5)$

13. $(1, 2)$, $(-2, -2)$

14. $(\frac{7}{8}, \frac{3}{4})$, $(\frac{5}{4}, -\frac{1}{4})$

In Exercises 15–33, find an equation for the indicated line and sketch its graph.

15. through $(2, 1)$ and $(0, -3)$
16. through $(-3, -4)$ and $(1, 4)$
17. through $(0, 0)$ and $(-1, 3)$
18. through $(-3, 6)$ and $(1, 2)$
19. through $(2, 3)$ and $(2, -2)$
20. through $(6, 1)$ and $(10, 1)$
21. through $(1, -2)$ and $(3, -2)$
22. through $(\frac{7}{8}, \frac{3}{4})$ and $(\frac{5}{4}, -\frac{1}{4})$
23. through $(0, 3)$: $m = \frac{3}{4}$
24. through $(-1, 2)$: m is undefined
25. through $(0, 0)$: $m = \frac{2}{3}$
26. through $(-1, -4)$: $m = \frac{1}{4}$
27. through $(0, 5)$: $m = -2$
28. through $(-2, 4)$: $m = -\frac{3}{5}$
29. y-intercept at 2: $m = 4$
30. y-intercept at $-\frac{2}{3}$: $m = \frac{1}{6}$
31. y-intercept at $\frac{2}{3}$: $m = \frac{3}{4}$
32. y-intercept at 4: $m = 0$
33. vertical line with x-intercept at 3
34. Show that the line with intercepts $(a, 0)$ and $(0, b)$ has the following equation:

$$\frac{x}{a} + \frac{y}{b} = 1, \qquad a \neq 0, b \neq 0$$

In Exercises 35–42, use the result of Exercise 34 to write an equation of the indicated line.

35. x-intercept $(2, 0)$; y-intercept $(0, 3)$
36. x-intercept $(-3, 0)$; y-intercept $(0, 4)$
37. x-intercept $(-\frac{1}{6}, 0)$; y-intercept $(0, -\frac{2}{3})$
38. x-intercept $(-\frac{2}{3}, 0)$; y-intercept $(0, -2)$

39. Point on line $(1, 2)$
    x-intercept $(a, 0)$
    y-intercept $(0, a)$
    $a \neq 0$

40. Point on line $(-3, 4)$
    x-intercept $(a, 0)$
    y-intercept $(0, a)$
    $a \neq 0$

41. Point on line $(\frac{3}{2}, \frac{1}{2})$
    x-intercept $(2a, 0)$
    y-intercept $(0, a)$
    $a \neq 0$

42. Point on line $(-3, 1)$
    x-intercept $(a, 0)$
    y-intercept $(0, -a)$
    $a \neq 0$

43. Find the equation of the line giving the relationship between the temperature in degrees Celsius, C, and degrees Fahrenheit, F. Use the fact that water freezes at 0° Celsius (32° Fahrenheit) and boils at 100° Celsius (212° Fahrenheit).

44. Use the result of Exercise 43 to complete the table.

C		−10°	10°			177°
F	0°			68°	90°	

45. A manufacturer pays its assembly line workers $4.50 per hour, *plus* an additional piecework rate of $0.75 per unit produced. Write a linear equation for the hourly wages W in terms of the number of units produced per hour x.

46. A small business purchases a piece of equipment for $875. After five years the equipment will be outdated and have no value. Write a linear equation giving the value V of the equipment during the five years it will be used.

47. A company constructs a warehouse for $825,000. It has an estimated useful life of 25 years, after which its value is expected to be $75,000. If straight-line depreciation is used, write a linear equation giving the value V of the warehouse during its 25 years of useful life.

48. A real estate office handles an apartment complex with 50 units. When the rent is $280 per month, all 50 units are occupied. However, when the rent is $325 per month, the average number of occupied units drops to 47. Assume that the relationship between the monthly rent p and the demand x is linear.
   (a) Write the equation of the line giving the demand x in terms of the rent p.
   (b) (Linear Extrapolation) Use this equation to predict the number of units occupied if the rent is raised to $355.
   (c) (Linear Interpolation) Predict the number of units occupied if the rent is lowered to $295.

49. The amount (in billions of dollars) spent by the United States for energy imports between 1975 and 1978 is given in the following table.

Year	1975	1976	1977	1978
t	0	1	2	3
Imports: y	$ 96	$121	$148	$172

(a) Assuming an approximately linear relation between y and t, write an equation for the line passing through (0, 96) and (3, 172).
(b) (Linear Interpolation) Use this equation to estimate the amount spent in 1976 and 1977. Compare the estimate with the actual amount.
(c) (Linear Extrapolation) Predict the amount spent on energy imports in 1980.
(d) What information is given by the slope of the line in part (a)?

50. A particular brand of woodstove sells for $739.40, and a cord of wood sells for $105.00.
   (a) Write an equation giving the total cost, C, in terms of the number, x, of cords of wood purchased.
   (b) Find the total cost of burning 6 cords of wood in this stove.

51. A contractor purchases a piece of equipment for $26,500. The equipment's operator is paid $9.50 per hour, and it uses an average of $5.25 per hour for fuel and maintenance.
   (a) Write a linear equation giving the total cost, C, of operating this equipment t hours.
   (b) If customers are charged $25 per hour of machine use, write an equation for the revenue, R, derived from t hours of use.
   (c) Use the formula for profit ($P = R - C$) to write an equation for the profit derived from t hours of use.
   (d) (Break-Even Point) Use the result of part (c) to find the number of hours this equipment must be used to yield a profit of 0 dollars.

52. A sales representative uses her own car as she travels for her company. The cost to the company is $75 per day for lodging and meals, plus $0.22 per mile driven. Write a linear equation giving the daily cost, C, to the company in terms of x, the number of miles driven.

Quadratic Functions and Their Graphs   2.4

Polynomial functions are the most widely used functions in algebra, and you would be wise to become familiar with their equations and graphs.

DEFINITION OF A POLYNOMIAL FUNCTION

> The function
>
> $$f(x) = a_n x^n + a_{n-1} x^{n-1} + \cdots + a_2 x^2 + a_1 x + a_0$$
>
> is called a **polynomial function of degree n.** The numbers a_i are called **coefficients,** with a_n the **leading coefficient** and a_0 the **constant term** of the polynomial function (n is a nonnegative integer).

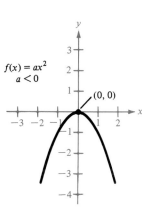

$f(x) = ax^2$
$a < 0$

(0, 0)

In Section 2.3 we studied first degree polynomial functions of the form $f(x) = ax + b$ (linear functions), and we saw that their graphs are straight lines. In this section we will look at second degree polynomial functions of the form $f(x) = ax^2 + bx + c$.

DEFINITION OF A QUADRATIC FUNCTION

> The function
>
> $$f(x) = ax^2 + bx + c, \qquad a \neq 0$$
>
> is called a **quadratic function,** and its graph is called a **parabola.**

Remark: It is common practice to use the subscript notation for coefficients of general polynomial functions, but for specific polynomial functions (of low degree) we generally use the simpler a, b, c, etc., form for the coefficients.

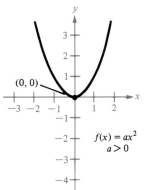

(0, 0)

$f(x) = ax^2$
$a > 0$

FIGURE 2.30

0th Degree: $f(x) = a$	*Constant function*
1st Degree: $f(x) = ax + b$	*Linear function*
2nd Degree: $f(x) = ax^2 + bx + c$	*Quadratic function*
3rd Degree: $f(x) = ax^3 + bx^2 + cx + d$	*Cubic function*

The simplest type of quadratic function is $f(x) = ax^2$, in which b and c are both zero. We call this a **monomial** quadratic function because it has only one term. The graph of $f(x) = ax^2$ is a parabola that opens upward if $a > 0$ and downward if $a < 0$ (see Figure 2.30).

On the graph of $f(x) = ax^2$, we call the point $(0, 0)$ the **vertex** of the parabola. If $a > 0$, the vertex is a minimum point. If $a < 0$, the vertex represents a maximum point on the parabola.

EXAMPLE 1
Graphing Monomial
Quadratic Functions

Sketch the graph of

(a) $f(x) = x^2$ 　　　　(b) $f(x) = \tfrac{1}{2}x^2$ 　　　　(c) $f(x) = 2x^2$

Solution:

(a) We begin by plotting a few points on the graph.

x	-2	-1	0	1	2
$y = x^2$	4	1	0	1	4

Then we connect these points by a smooth curve, as shown in Figure 2.31.

(b) See Figure 2.32.   (c) See Figure 2.33.

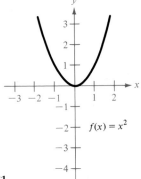

FIGURE 2.31

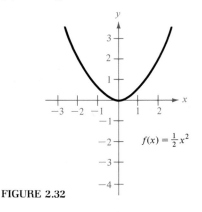

FIGURE 2.32

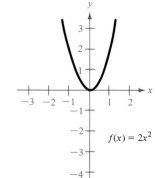

FIGURE 2.33

Remark: Note that in Example 1 the graph of a monomial quadratic function ($y = ax^2$) has *symmetry with respect to the y-axis*. Also note that the coefficient a determines how widely the parabola opens. If $|a|$ is small, the parabola opens more widely than if $|a|$ is large.

We can use our knowledge of the basic shape of the graph of $f(x) = ax^2$ to graph any quadratic function by using the techniques developed in Section 2.2. You may wish to review Example 4 in Section 2.2.

EXAMPLE 2
Graphing General
Quadratic Functions

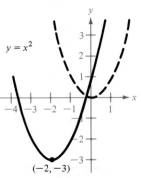

FIGURE 2.34 $f(x) = (x + 2)^2 - 3$

Sketch the graph of

$$f(x) = (x + 2)^2 - 3$$

Solution:
The basic graph we use to sketch this quadratic function is $y = x^2$. The graph of $f(x) = (x + 2)^2 - 3$ represents a horizontal shift of 2 units to the left and a vertical shift of 3 units down, as shown in Figure 2.34.

Remark: Note that the graph in Example 2 is symmetric with respect to the vertical line passing through the vertex of the parabola. We call this line the **axis** of the parabola.

The equation in Example 2 is written in **standard form** $f(x) = a(x - h)^2 + k$, which is especially convenient for sketching.

STANDARD FORM OF THE EQUATION OF A PARABOLA
(Vertical Axis)

The quadratic function

$$f(x) = a(x - h)^2 + k, \qquad a \neq 0$$

is said to be in **standard form.** The axis is given by the vertical line $x = h$, and the vertex occurs at the point (h, k). If $a > 0$, the parabola opens upward, and if $a < 0$, the parabola opens downward.

To write a quadratic function in standard form, we use a procedure called *completing the square,* demonstrated in the next example.

EXAMPLE 3
Writing a Quadratic Function
in Standard Form

Sketch the graph of the following quadratics by first writing each one in standard form.

(a) $f(x) = 2x^2 + 8x + 7$ (b) $f(x) = -x^2 + 6x - 8$

Solution:

(a) Since the coefficient of x^2 is different from 1, it is convenient to group the x terms and factor a 2 out of each x term before completing the square.

$$\begin{aligned}
f(x) &= 2x^2 + 8x + 7 && \textit{Given} \\
&= 2(x^2 + 4x) + 7 && \textit{Factor 2 out of x terms} \\
&= 2(x^2 + 4x + 4 - 4) + 7 && \textit{Add and subtract (half of 4)}^2 \\
& \qquad\quad (2)^2 \\
&= 2(x^2 + 4x + 4) - 2(4) + 7 && \textit{Regroup terms} \\
&= 2(x^2 + 4x + 4) - 8 + 7 && \textit{Simplify} \\
&= 2(x + 2)^2 - 1 && \textit{Standard form}
\end{aligned}$$

Now we see that this graph can be obtained from the graph of $y = 2x^2$ by shifting 2 units to the left and 1 unit down (Figure 2.35).

(b) In this case, we factor -1 out of the x terms.

$$\begin{aligned}
f(x) &= -x^2 + 6x - 8 && \textit{Given} \\
&= -(x^2 - 6x) - 8 && \textit{Factor} -1 \textit{ out of x terms} \\
&= -(x^2 - 6x + 9 - 9) - 8 && \textit{Add and subtract (half of 6)}^2 \\
& \qquad\quad (3)^2 \\
&= -(x^2 - 6x + 9) - (-9) - 8 && \textit{Regroup terms} \\
&= -(x^2 - 6x + 9) + 9 - 8 && \textit{Simplify} \\
&= -(x - 3)^2 + 1 && \textit{Standard form}
\end{aligned}$$

Therefore, the graph can be obtained from the graph of $y = -x^2$ by shifting 3 units to the right and 1 unit up, as shown in Figure 2.36.

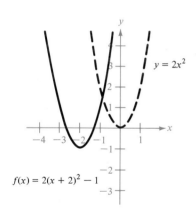

$y = 2x^2$

$f(x) = 2(x + 2)^2 - 1$

FIGURE 2.35

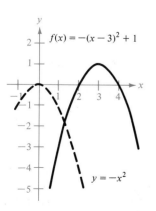

$f(x) = -(x - 3)^2 + 1$

$y = -x^2$

FIGURE 2.36

In addition to finding the vertex of the graph of a quadratic function, it is often useful to find the *x*- and *y*-intercepts. The graph of

$$y = ax^2 + bx + c, \qquad a \neq 0$$

has a *y*-intercept at $(0, c)$. The *x*-intercepts can be determined by solving the quadratic equation $ax^2 + bx + c = 0$.

EXAMPLE 4
Finding Intercepts of a Quadratic Function

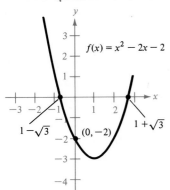

FIGURE 2.37

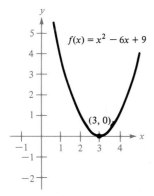

FIGURE 2.38

Sketch the graphs of

(a) $f(x) = x^2 - 2x - 2$ (b) $f(x) = x^2 - 6x + 9$
(c) $f(x) = x^2 + 2x + 2$

and label the *x*- and *y*-intercepts.

Solution:

(a) The *y*-intercept occurs when $x = 0$:

$$y = f(0) = 0^2 - 2(0) - 2 = -2$$

The *x*-intercepts (if any) occur when $y = 0$:

$$y = x^2 - 2x - 2 = 0$$

By the quadratic formula, we have

$$x = \frac{-(-2) \pm \sqrt{(-2)^2 - 4(1)(-2)}}{2(1)}$$

$$= \frac{2 \pm \sqrt{12}}{2} = \frac{2 \pm 2\sqrt{3}}{2}$$

$$= 1 \pm \sqrt{3}$$

See Figure 2.37.

(b) The *y*-intercept occurs at $(0, 9)$ and the *x*-intercept occurs when

$$x^2 - 6x + 9 = 0$$

or

$$(x - 3)^2 = 0$$

which implies that there is only one *x*-intercept at $(3, 0)$, as shown in Figure 2.38.

(c) The *y*-intercept occurs at $(0, 2)$. When we try to find the *x*-intercepts of this function, we obtain

$$x^2 + 2x + 2 = 0$$

and since the discriminant

$$b^2 - 4ac = 2^2 - 4(1)(2) = 4 - 8 = -4$$

we see that there are no *x*-intercepts, as shown in Figure 2.39.

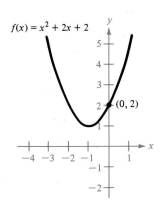

$f(x) = x^2 + 2x + 2$

(0, 2)

FIGURE 2.39

The graphs in Example 4 illustrate the three possible cases for the occurrence of x-intercepts for quadratic functions.

INTERCEPTS OF A PARABOLA

The graph of $f(x) = ax^2 + bx + c$ has

1. *two* x-intercepts if $b^2 - 4ac > 0$
2. *one* x-intercept if $b^2 - 4ac = 0$
3. *no* x-intercept if $b^2 - 4ac < 0$

The y-intercept is the point $(0, c)$.

Vertex and Axis of a Quadratic Function

If a parabola has two x-intercepts, then its axis is a vertical line halfway between its x-intercepts. Thus, by averaging the x-intercepts

$$x = \frac{-b + \sqrt{b^2 - 4ac}}{2a} \quad \text{and} \quad x = \frac{-b - \sqrt{b^2 - 4ac}}{2a}$$

we can obtain the midpoint

$$\frac{\dfrac{-b + \sqrt{b^2 - 4ac}}{2a} + \dfrac{-b - \sqrt{b^2 - 4ac}}{2a}}{2} = \frac{-(2b/2a)}{2} = -\frac{b}{2a}$$

Consequently, the graph of $f(x) = ax^2 + bx + c$ has a vertical axis at

$$x = -\frac{b}{2a}$$

and a vertex at

$$\left(-\frac{b}{2a},\ f\left(-\frac{b}{2a}\right)\right)$$

This provides us with a way to find the vertex of the graph of $f(x) = ax^2 + bx + c$ without writing the equation in standard form. The formula is also valid for one or no x-intercept.

EXAMPLE 5
Finding the Vertex and Extreme Values for a Quadratic Function

Find the vertex for the graph of

$$f(x) = 20 - 8x + 3x^2$$

and determine the minimum (or maximum) value of $f(x)$.

Solution:
By rewriting the function as

$$f(x) = 3x^2 - 8x + 20$$

we see that $a = 3$, $b = -8$, and $c = 20$. Thus, the *x*-coordinate of the vertex is

$$x = -\frac{b}{2a} = \frac{-(-8)}{2(3)} = \frac{4}{3}$$

and the *y*-coordinate is

$$f\left(\frac{4}{3}\right) = 3\left(\frac{16}{9}\right) - 8\left(\frac{4}{3}\right) + 20 = \frac{16}{3} - \frac{32}{3} + \frac{60}{3} = \frac{44}{3}$$

Therefore, the vertex is the point

$$\left(\frac{4}{3}, \frac{44}{3}\right)$$

Since $a = 3$ $(a > 0)$, the graph opens upward and it follows that

$$f\left(\frac{4}{3}\right) = \frac{44}{3}$$

is the *minimum* value of $f(x)$.

EXAMPLE 6
An Application: Finding the
Maximum Value and Intercept

An object is propelled straight up into the air with an initial velocity of 32 feet per second (beginning at a height of 6 feet). The height at any time t is given by

$$s(t) = -16t^2 + 32t + 6$$

where $s(t)$ is measured in feet and t is measured in seconds.

(a) Find the maximum height attained by the object before it begins falling back to the ground.
(b) When does the object hit the ground?

Solution:

(a) The maximum height of the object occurs at the vertex of the graph of s. We note that $a = -16$, $b = 32$, and $c = 6$. Thus, the value of t at the vertex is

$$t = -\frac{b}{2a} = \frac{-32}{2(-16)} = 1$$

and the value of $s(t)$ is

$$s(1) = -16 + 32 + 6 = 22$$

Now since the vertex occurs at the point $(1, 22)$, we conclude that the maximum height is 22 feet.

(b) Note in part (a) that the object reached its maximum height after one second. To find the time that it hits the ground, we set $s(t) = 0$ and solve for t:

$$s(t) = -16t^2 + 32t + 6 = 0$$

Using the quadratic formula, we have

$$t = \frac{-(32) \pm \sqrt{(32)^2 - 4(-16)(6)}}{2(-16)}$$

$$= \frac{-32 \pm \sqrt{1408}}{-32}$$

$$= \frac{-32 \pm 8\sqrt{22}}{-32} = \frac{4 \pm \sqrt{22}}{4}$$

Finally, choosing the positive value of t, we find that the object hits the ground when

$$t = \frac{4 + \sqrt{22}}{4} \approx 2.17 \text{ seconds}$$

Section Exercises 2.4

In Exercises 1–10, use the basic graphs of $y = ax^2$, as given in Figure 2.30, and the concept of horizontal and vertical translations to match the quadratic function with the correct graph.

1. $f(x) = (x - 3)^2$
2. $f(x) = (x + 5)^2$
3. $f(x) = x^2 - 4$
4. $f(x) = x^2 + 1$
5. $f(x) = 5 - x^2$
6. $f(x) = 4 - (x - 1)^2$
7. $f(x) = (x + 2)^2 - 2$
8. $f(x) = 3x^2 - 5$
9. $f(x) = 9 - 2x^2$
10. $f(x) = 2(x - 1)^2 + 1$

(e)

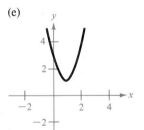

(f)

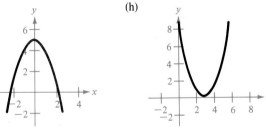

(a)

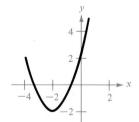

(b)

(g)

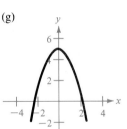

(h)

(c)

(d)

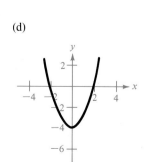

(i)

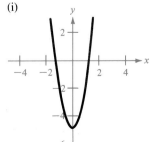

(j)

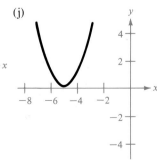

In Exercises 11–16, use the given graph to write (a) the standard form, and (b) the general form of the equation of the parabola.

11.

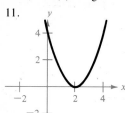

12.

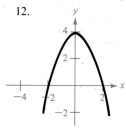

13.

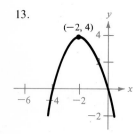

14.

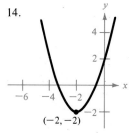

15.

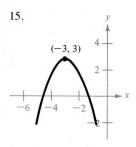

16.

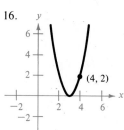

In Exercises 17–36, sketch the graph of the quadratic function and identify the vertex and the intercepts.

17. $f(x) = x^2 - 5$

18. $f(x) = x^2 + 4$

19. $f(x) = 3x^2 + 1$

20. $f(x) = \frac{1}{2}x^2 - 4$

21. $f(x) = (x + 5)^2 - 6$

22. $f(x) = (x - 6)^2 + 3$

23. $f(x) = 16 - x^2$

24. $f(x) = 25 - x^2$

25. $f(x) = x^2 - 8x + 16$

26. $f(x) = x^2 + 2x + 1$

27. $f(x) = x^2 + 2x - 3$

28. $f(x) = x^2 - 4x - 1$

29. $f(x) = x^2 - 16x + 54$

30. $f(x) = x^2 + 8x + 11$

31. $f(x) = x^2 - x + \frac{5}{4}$

32. $f(x) = x^2 + 3x + \frac{1}{4}$

33. $f(x) = -x^2 + 2x + 5$

34. $f(x) = -x^2 - 4x + 1$

35. $f(x) = 4x^2 - 4x + 21$

36. $f(x) = 2x^2 - x + 1$

37. A rectangle with a perimeter of 100 inches has a width and length of x and $50 - x$ inches, respectively. Find x so that the area of the rectangle is maximum $[A = x(50 - x)]$.

38. Find the number of units x that produce a maximum revenue R if $R = 900x - 0.1x^2$.

39. A manufacturer of lighting fixtures has daily production costs of $C = 800 - 10x + 0.25x^2$. How many fixtures x should be produced each day to minimize costs?

40. Let x be the amount (in hundreds of dollars) a company spends on advertising, and let P be the profit. If $P = 230 + 20x - 0.5x^2$, what amount of advertising gives the maximum profit?

41. A manufacturer of radios charges $90 per unit when the average production cost per unit is $60. However, to encourage large orders from distributors, the manufacturer will reduce the charge by $0.10 per unit for each unit ordered in excess of 100 (for example, there would be a charge of $88 per radio for an order of 120 radios). Find the largest order size the manufacturer should allow so as to realize maximum profit.

42. Find the quadratic function that has a maximum at $(3, 4)$ and passes through the point $(1, 2)$.

43. Find the quadratic function that has a minimum at $(5, 12)$ and passes through the point $(7, 15)$.

44. Find two quadratic functions (one opening upward and the other downward) that have x-intercepts $(-1, 0)$ and $(3, 0)$.

45. Find two quadratic functions (one opening upward and the other downward) that have x-intercepts $(-2.5, 0)$ and $(2, 0)$.

Composite and Inverse Functions   2.5

Two functions can be combined in various ways to create new functions. For example, if

$$f(x) = 2x - 3$$

and

$$g(x) = x^2 - 1$$

we can form the functions

$$f(x) + g(x) = (2x - 3) + (x^2 - 1) = x^2 + 2x - 4 \qquad (Sum)$$

$$f(x) - g(x) = (2x - 3) - (x^2 - 1) = -x^2 + 2x - 2 \qquad (Difference)$$

$$f(x)g(x) = (2x - 3)(x^2 - 1) = 2x^3 - 3x^2 - 2x + 3 \qquad (Product)$$

$$\frac{f(x)}{g(x)} = \frac{2x - 3}{x^2 - 1}, \qquad x \neq \pm 1 \qquad (Quotient)$$

These *combination* functions may have domains that differ from those of the original functions. For instance, the domain of $f(x)/g(x)$ has the restriction $x \neq \pm 1$, which does not occur in the original functions.

We can combine two functions in yet another way, called the **composition** of two functions.

DEFINITION OF COMPOSITE FUNCTION

Let f and g be functions such that the range of g is in the domain of f. Then the function whose values are given by $f(g(x))$ is called the **composite** of f with g.

Remark: Some texts use the notation $(f \circ g)(x)$ to denote the composite function $f(g(x))$.

It is important to realize that the composite of f with g may not be equal to the composite of g with f. This is illustrated in the following example.

EXAMPLE 1
Forming Composite Functions

Given $f(x) = 2x - 3$ and $g(x) = x^2 - 1$, find

(a) $f(g(x))$ (b) $g(f(x))$

Solution:

(a) Since $f(x) = 2(x) - 3$

we have

$$f(g(x)) = 2(g(x)) - 3$$
$$= 2(x^2 - 1) - 3 = 2x^2 - 5$$

(b) Since $g(x) = (x)^2 - 1$

we have

$$g(f(x)) = (f(x))^2 - 1 = (2x - 3)^2 - 1$$
$$= 4x^2 - 12x + 8$$

Although the compositions formed in Example 1 look fairly straightforward, this procedure cannot be performed haphazardly. That is, you must take care to see that the domains and ranges of f and g have a proper fit with each other.

TABLE 2.2
Forming Composite Functions

$$g(x) = -x^2 \qquad f(x) = \sqrt{x + 1} \qquad f(g(x)) = \sqrt{-x^2 + 1}$$

Domain	$[-1, 1]$	$\nearrow [-1, \infty)$	$[-1, 1]$
Range	$[-1, 0]$	$[0, \infty)$	$[0, 1]$

Remark: The domain of g is restricted in order to make the range of g lie within the domain of f.

$$g(x) = \frac{1}{x^2 + 1} \qquad f(x) = -\sqrt{x} \qquad f(g(x)) = -\sqrt{\frac{1}{x^2 + 1}}$$

Domain	$(-\infty, \infty)$	$\nearrow [0, \infty)$	$(-\infty, \infty)$
Range	$(0, 1]$	$(-\infty, 0]$	$[-1, 0)$

Remark: The range of g lies within the domain of f.

Note in Example 1 that $f(g(x)) \neq g(f(x))$ and, in general, these two composite functions are *not* equal. An important case in which they are equal occurs when

$$f(g(x)) = g(f(x)) = x$$

We call such functions **inverses** of each other, as stated in the following definition.

DEFINITION OF INVERSE FUNCTIONS

Two functions f and g are **inverses** of each other if

$$f(g(x)) = x \qquad \text{for each } x \text{ in the domain of } g$$

and

$$g(f(x)) = x \qquad \text{for each } x \text{ in the domain of } f$$

We denote g by f^{-1} (read "f inverse").

Remark: For inverse functions f and g, the range of g must be equal to the domain of f, and vice versa. Also note that whenever we write $f^{-1}(x)$, we will *always* be referring to the inverse of the function f and *not* to the reciprocal of f.

EXAMPLE 2
Demonstrating Inverse Functions

Show that the following functions are inverses of each other:

$$f(x) = 2x^3 - 1 \qquad \text{and} \qquad g(x) = \sqrt[3]{\frac{x + 1}{2}}$$

Solution:

First, note that both composite functions exist, since the domain and range of both f and g consist of the set of all real numbers. The composite of f with g is

$$f(g(x)) = 2\left(\sqrt[3]{\frac{x+1}{2}} \right)^3 - 1$$

$$= 2\left(\frac{x+1}{2}\right) - 1 = x + 1 - 1 = x$$

The composite of g with f is

$$g(f(x)) = \sqrt[3]{\frac{(2x^3 - 1) + 1}{2}}$$

$$= \sqrt[3]{\frac{2x^3}{2}} = \sqrt[3]{x^3} = x$$

Since $f(g(x)) = g(f(x)) = x$, f and g are inverses of each other.

The following rule suggests a geometrical interpretation of inverse functions.

GRAPHS OF INVERSE FUNCTIONS

> The graph of f contains the point (a, b) if and only if the graph of f^{-1} contains the point (b, a).

The rule can be interpreted geometrically to mean that the graph of f^{-1} can be obtained by reflecting the graph of f in the line $y = x$ (Figure 2.40). This geometric property suggests an *algebraic* procedure for finding the inverse of a function. Since (a, b) lies on the graph of f if and only if (b, a) lies on the graph of f^{-1}, we can find the inverse of a function by algebraically interchanging the roles of x and y.

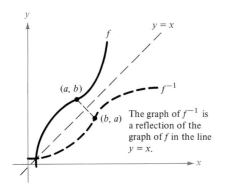

The graph of f^{-1} is a reflection of the graph of f in the line $y = x$.

FIGURE 2.40

EXAMPLE 3
Finding the Inverse of a Function

Find the inverse of the function given by $f(x) = \sqrt{2x - 3}$.

Solution:
Substituting y for $f(x)$, we have $y = \sqrt{2x - 3}$. Now to find the inverse function, we simply solve for x in terms of y. Since y is nonnegative, squaring both sides gives an equivalent equation:

$$\sqrt{2x - 3} = y \qquad \textit{Given y as a function of x}$$

$$2x - 3 = y^2$$

$$2x = y^2 + 3$$

$$x = \frac{y^2 + 3}{2} \qquad \textit{Solve for x as a function of y}$$

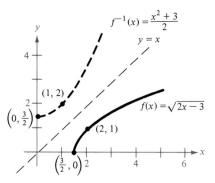

FIGURE 2.41

$$f^{-1}(y) = \frac{(y)^2 + 3}{2} \qquad \textit{Write in functional notation}$$

$$f^{-1}(x) = \frac{x^2 + 3}{2} \qquad \textit{Use x as independent variable}$$

Now that we have a formula for f^{-1}, we find its domain by finding the range of f.

Range of f: $[0, \infty)$ $\Rightarrow$ Domain of f^{-1}: $[0, \infty)$

Finally, we conclude that the inverse function is

$$f^{-1}(x) = \frac{x^2 + 3}{2}, \qquad x \geq 0$$

Note the reflective property of these two graphs in Figure 2.41.

Remark: To avoid being confused by the interplay of variables in Example 3, remember that the independent variable is really a "dummy variable" which serves merely as a placeholder. Thus,

$$f^{-1}(y) = \frac{y^2 + 3}{2}, \qquad f^{-1}(a) = \frac{a^2 + 3}{2}, \quad \text{and} \quad f^{-1}(x) = \frac{x^2 + 3}{2}$$

all represent the same function.

The following summary should help you remember the basic steps for finding the inverse of a function.

FINDING THE INVERSE OF A FUNCTION

To find the inverse of f:
1. Write the function in the form $y = f(x)$
2. If possible, solve for x in terms of y: $x = g(y)$
3. Interchange x and y: $y = g(x)$
4. Check to see that   Domain of f = Range of g
   and   Range of f = Domain of g
5. The inverse of f is $f^{-1}(x) = g(x)$

Not all functions possess an inverse. *For a function f to have an inverse it is necessary that f be one-to-one.* A function is **one-to-one** if no two elements in its domain correspond to the same element in its range.

Horizontal Line Test

Graphically, we can see that a function $y = f(x)$ is not one-to-one if a horizontal line intersects the graph of the function more than once. For instance, in Figure 2.42, the line $y = 3$ intersects the graph of the function twice, and hence the function is not one-to-one.

EXAMPLE 4

A Function That Has No Inverse

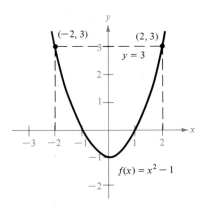

f is not one-to-one
and has no inverse.

FIGURE 2.42

Find the inverse (if it exists) of the function given by

$$f(x) = x^2 - 1$$

(Assume the domain of f is the set of all real numbers.)

Solution:
We note that

$$f(2) = (2)^2 - 1 = 3$$

and $\quad f(-2) = (-2)^2 - 1 = 3$

Thus, f is not one-to-one, and it has no inverse. This same conclusion can be obtained by substituting y for $f(x)$ and solving for x as follows:

$$x^2 - 1 = y$$
$$x^2 = y + 1$$
$$x = \pm\sqrt{y + 1}$$

This last equation does not define x as a function of y, and thus f has no inverse. (See Figure 2.42.)

As you work through the exercises for this section, develop the habit of checking the range and domain of each function. The problem of finding an inverse function can be difficult (or even impossible) for two reasons. First, given $y = f(x)$, it may be algebraically difficult to solve for x in terms of y. Second, if f is not one-to-one, then f^{-1} does not exist.

In later chapters we will study two important classes of inverse functions: logarithmic functions and inverse trigonometric functions.

Section Exercises 2.5

In Exercises 1–8, find (a) $f(x) + g(x)$, (b) $f(x) - g(x)$, (c) $f(x) \cdot g(x)$, (d) $f(x)/g(x)$, (e) $f(g(x))$, if defined, and (f) $g(f(x))$, if defined.

1. $f(x) = x + 1,$ $\quad g(x) = x - 1$
2. $f(x) = 2x - 5,$ $\quad g(x) = 1 - x$
3. $f(x) = x^2,$ $\quad g(x) = 1 - x$
4. $f(x) = 2x - 5,$ $\quad g(x) = 5$
5. $f(x) = x^2 + 5,$ $\quad g(x) = \sqrt{1 - x}$
6. $f(x) = \sqrt{x^2 - 4},$ $\quad g(x) = x^2/(x^2 + 1)$
7. $f(x) = 1/x,$ $\quad g(x) = 1/x^2$
8. $f(x) = x/(x + 1),$ $\quad g(x) = x^3$

In Exercises 9–16, (a) show that f and g are inverse functions by showing that $f(g(x)) = x$ and $g(f(x)) = x$, and (b) graph f and g on the same set of coordinate axes.

9. $f(x) = x^3,$ $\quad g(x) = \sqrt[3]{x}$
10. $f(x) = 1/x,$ $\quad g(x) = 1/x$

11. $f(x) = 5x + 1,$ $\quad g(x) = (x - 1)/5$
12. $f(x) = 3 - 4x,$ $\quad g(x) = (3 - x)/4$
13. $f(x) = \sqrt{x - 4},$ $\quad g(x) = x^2 + 4, x \geq 0$
14. $f(x) = 9 - x^2, x \geq 0,$ $\quad g(x) = \sqrt{9 - x}, x \leq 9$
15. $f(x) = 1 - x^3,$ $\quad g(x) = \sqrt[3]{1 - x}$
16. $f(x) = 1/(1 + x^2), x \geq 0,$ $\quad g(x) = \sqrt{(1 - x)/x},$
$$0 < x \leq 1$$

In Exercises 17–28, find the inverse of f. Then graph both f and f^{-1}.

17. $f(x) = 2x - 3$
18. $f(x) = 3x$
19. $f(x) = x^5$
20. $f(x) = x^3 + 1$
21. $f(x) = \sqrt{x}$
22. $f(x) = x^2, x \geq 0$
23. $f(x) = \sqrt{4 - x^2}, 0 \leq x \leq 2$
24. $f(x) = \sqrt{x^2 - 4}, x \geq 2$
25. $f(x) = \sqrt[3]{x - 1}$
26. $f(x) = 3\sqrt[5]{2x - 1}$
27. $f(x) = x^{2/3}, x \geq 0$
28. $f(x) = x^{3/5}$

In Exercises 29–40, determine if the given function is one-to-one and if so find its inverse.

29. $f(x) = ax + b, a \neq 0$
30. $f(x) = \sqrt{x - 2}$
31. $f(x) = x^2$
32. $f(x) = x^4$

33. $f(x) = x/\sqrt{x^2 + 5}$
34. $f(x) = |x - 2|$
35. $f(x) = 3$
36. $f(x) = x\sqrt{2 - x}$
37. $f(x) = 1/x$
38. $f(x) = (3x + 4)/5$
39. $f(x) = \sqrt{2x + 3}$
40. $f(x) = x^2/(x^2 + 1)$

Review Exercises / Chapter 2

In Exercises 1–10, find the general form of the equation of the line through the two points.

1. $(0, 0), (6, 0)$
2. $(0, 0), (0, 10)$
3. $(-2, -1), (2, 2)$
4. $(-1, 4), (2, 0)$
5. $(2, 1), (14, 6)$
6. $(-2, 2), (3, -10)$
7. $(5, -2), (5, 6)$
8. $(-3, -4), (-3, 5)$
9. $(-1, 0), (6, 2)$
10. $(1, 6), (4, 2)$

11. Determine the value of t so that the points $(-2, 5), (0, t)$, and $(1, 1)$ are on the same line.
12. Determine the value of t so that the points $(-6, 1), (1, t)$, and $(10, 5)$ are on the same line.
13. Show that the points $(1, 1), (8, 2), (9, 5)$, and $(2, 4)$ are vertices of a parallelogram.
14. Show that the points $(4, 5), (-2, 4)$, and $(3, -1)$ are the vertices of an isosceles triangle.
15. Determine the domain and range of the function $f(x) = \sqrt{25 - x^2}$.
16. Determine the domain and range of the function $h(t) = |t + 1|$.
17. If $f(x) = x^2 + 1$, find
    (a) $f(2)$ (b) $f(-4)$ (c) $f(t^2)$
    (d) $-f(x)$ (e) $f(x + \Delta x)$
18. If $f(x) = \sqrt[4]{x + 5}$, find
    (a) $f(11)$ (b) $f(-2)$ (c) $f(s - 2)$
    (d) $f(-x)$ (e) $f(x + \Delta x)$
19. Given the functions $f(x) = 1/x$ and $g(x) = x^2 + 1$, find
    (a) $f[g(x)]$ (b) $g[f(x)]$
20. Given the functions $f(x) = 1/(2x - 4)$ and $g(x) = 1/x$, find
    (a) $f[g(x)]$ (b) $g[f(x)]$

In Exercises 21–30, analyze the function and sketch its graph.

21. $f(x) = (x + \frac{3}{2})^2 + 1$
22. $f(x) = (x - 4)^2 - 3$
23. $f(x) = -(x - 1)^3$
24. $f(x) = (x + 1)^3$

25. $f(x) = x^4 - x^3 - 2x^2$
26. $f(x) = -2x^3 - x^2 + x$
27. $f(x) = x^3 - 3x$
28. $f(x) = -x^3 + 3x - 2$
29. $f(x) = x(x + 3)^2$
30. $f(x) = x^4 - 4x^2$

In Exercises 31–35, use completion of the square to find the maximum or minimum of the quadratic function.

31. $f(x) = x^2 - 2x$
32. $f(x) = x^2 + 8x + 10$
33. $f(x) = 6x - x^2$
34. $f(x) = 3 + 4x - x^2$
35. $f(x) = x^2 + 3x - 8$
36. Let x be the amount (in hundreds of dollars) a company spends on advertising and let P be the profit. If

$$P = 230 + 20x - \tfrac{1}{2}x^2$$

what amount of advertising gives the maximum profit?
37. Find the number of units x that produces a maximum revenue R if $R = 900x - 0.1x^2$.
38. A real estate office handles 50 apartment units. When the rent is $360 per month, all units are occupied. However, on the average, for each $20 increase in rent, 1 unit becomes vacant. Each occupied unit requires an average of $12 per month for service and repairs. What should be charged to realize the most profit?

In Exercises 39–44, (a) find the inverse, f^{-1}, of the given function, (b) sketch the graphs of $f(x)$ and $f^{-1}(x)$ on the same axes, and (c) verify that $f^{-1}[f(x)] = f[f^{-1}(x)] = x$.

39. $f(x) = \frac{1}{2}x - 3$
40. $f(x) = 5x - 7$
41. $f(x) = \sqrt{x + 1}$
42. $f(x) = x^3 + 2$
43. $f(x) = x^2 - 5, x \geq 0$
44. $f(x) = \sqrt[3]{x + 1}$
45. The function $f(x) = 2(x - 4)^2$ does not have an inverse. Give a restriction on the domain of f so that the restricted function has an inverse and then find the inverse.
46. Repeat Exercise 45 for the function $f(x) = |x - 2|$.

3 POLYNOMIAL AND RATIONAL FUNCTIONS

Graphs of Polynomial Functions

3.1

At this point you should be able to sketch an accurate graph of polynomial functions of degree 0, 1, or 2.

Degree	Function	Graph
0th Degree	$f(x) = a$	Horizontal line
1st Degree	$f(x) = ax + b$	Line of slope a
2nd Degree	$f(x) = ax^2 + bx + c$	Parabola

The graphs of polynomial functions of degree higher than 2 are more difficult to classify and to sketch. However, in this section you will begin to recognize some of the basic characteristics of these polynomial graphs. We continue to use the standard graphing techniques of point-plotting, finding x- and y-intercepts, and testing for symmetry about the y-axis or the origin.

Continuity

A polynomial function is continuous.* This means that the graph of a polynomial function has no breaks or gaps. (See Figure 3.1.)

Smooth Curve

The graph of a polynomial function has only smooth rounded turns, as indicated in Figure 3.2.

*Continuous functions are dealt with much more completely in calculus.

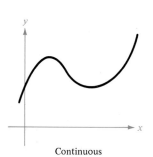

Continuous

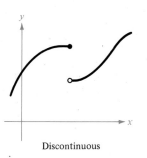

Discontinuous

FIGURE 3.1

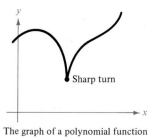

Polynomial functions have
smooth, rounded graphs.

The graph of a polynomial function
cannot have a sharp pointed turn.

FIGURE 3.2

The Leading Coefficient Test

Although the graph of a polynomial function can have several turns, eventually the graph will rise or fall without bound as x moves to the right or left. Symbolically, we write

$$f(x) \rightarrow \infty \qquad \text{as} \qquad x \rightarrow \infty$$

to mean that $f(x)$ increases without bound as x moves to the right without bound. (∞ is the mathematical symbol for infinity.) Whether the graph of

$$f(x) = a_n x^n + a_{n-1}x^{n-1} + \cdots + a_2 x^2 + a_1 x + a_0$$

eventually rises or falls can be determined by the function's degree (odd or even) and by the leading coefficient, as indicated in Figures 3.3 and 3.4.

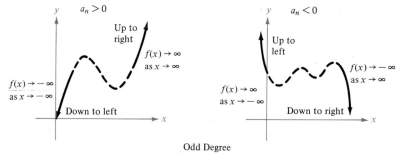

Odd Degree

FIGURE 3.3

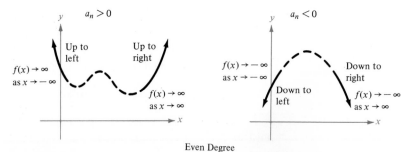

Even Degree

FIGURE 3.4

Note that the dashed portions of the graphs in Figures 3.3 and 3.4 indicate that the leading coefficient test determines *only* the right and left behavior of the graph.

EXAMPLE 1
Determining the Right and Left Behavior of a Polynomial Graph

Use the Leading Coefficient Test to determine the right and left behavior of the graphs of the following polynomial functions:

(a) $f(x) = 3x^3 - x^2 + 4$ (b) $f(x) = -x^4 + 2x - 1$
(c) $f(x) = -2x^5 + 4x^4 + 2$

Solution:

(a) Since the degree is odd and the leading coefficient is positive, the graph moves up to the right and down to the left, as shown in Figure 3.5.
(b) Since the degree is even and the leading coefficient is negative, the graph moves down to the right and left, as shown in Figure 3.6.
(c) Since the degree is odd and the leading coefficient is negative, the graph moves down to the right and up to the left, as shown in Figure 3.7.

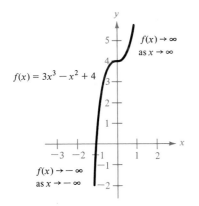

FIGURE 3.5

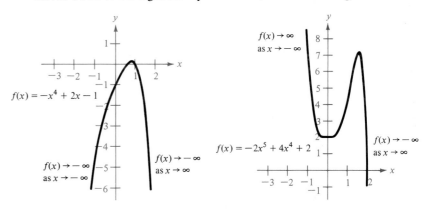

FIGURE 3.6   **FIGURE 3.7**

One of the most important problems in algebra is determining the *zeros* of polynomial functions. There is a strong interplay between graphical and algebraic approaches to this problem. Sometimes we use information about the graph of a function to help us find its zeros, and in other problems we use information about the zeros of a function to help us sketch its graph.

REAL ZEROS OF POLYNOMIAL FUNCTIONS

If f is a polynomial function and a is a real number, then the following statements are equivalent:

1. $x = a$ is a *zero* of the function f.
2. $x = a$ is a *root* of the polynomial equation $f(x) = 0$.
3. $(x - a)$ is a *factor* of the polynomial $f(x)$.
4. $(a, 0)$ is an *x-intercept* of the graph of the function.

Graphing a polynomial function is often easier if we are able to factor the polynomial.

EXAMPLE 2

Finding Zeros of Polynomial
Functions by Factoring

Find all real zeros of the following polynomial functions:

(a) $f(x) = x^3 - x^2 - 2x$ (b) $f(x) = -2x^4 + 2x^2$
(c) $f(x) = (-x^5 + 3x^3 + 4x)/4$

Solution:

(a) $f(x) = x^3 - x^2 - 2x = x(x^2 - x - 2)$
$$= x(x - 2)(x + 1)$$

Thus, the real zeros are $x = 0$, $x = 2$, and $x = -1$. From Figure 3.8, we see that $(0, 0)$, $(2, 0)$, and $(-1, 0)$ are the x-intercepts of the graph of this function.

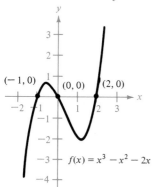

FIGURE 3.8

(b) $f(x) = -2x^4 + 2x^2 = -2x^2(x^2 - 1)$
$$= -2x^2(x - 1)(x + 1)$$

Thus, the real zeros are $x = 0$, $x = 1$, and $x = -1$. From Figure 3.9, we see that $(0, 0)$, $(1, 0)$, and $(-1, 0)$ are the x-intercepts of the graph of this function.

(c) $f(x) = \frac{1}{4}(-x^5 + 3x^3 + 4x) = -\frac{1}{4}x(x^4 - 3x^2 - 4)$
$$= -\frac{1}{4}x(x^2 - 4)(x^2 + 1)$$
$$= -\frac{1}{4}x(x - 2)(x + 2)(x^2 + 1)$$

Thus, the real zeros are $x = 0$, $x = 2$, and $x = -2$. From Figure 3.10, we see that $(0, 0)$, $(2, 0)$, and $(-2, 0)$ are the x-intercepts of the graph of this function. Note that the quadratic factor $(x^2 + 1)$ has no real roots and consequently produces no real zeros of the function.

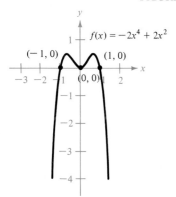

FIGURE 3.9

Remark: Note that in Example 2 each of the graphs happened to have three x-intercepts. An important result in algebra states that *the number of real zeros of a polynomial function cannot exceed its degree*. This result is related to the Fundamental Theorem of Algebra, to be discussed in Section 3.5. For now, we will use this important result and postpone its proof.

NUMBER OF REAL ZEROS OF A POLYNOMIAL FUNCTION

A polynomial function of degree n has at most n real zeros.

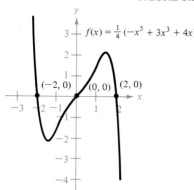

FIGURE 3.10

Note that this theorem tells us nothing about how we might *find* the zeros, or even if any real zeros exist. In fact, many polynomial functions have no real zeros. For example, the function $f(x) = x^2 + 1$ has no real zeros.

The next theorem gives us information about the existence of real zeros of a polynomial function. It is called the Intermediate Value Theorem for Polynomial Functions, and it follows from the fact that polynomial functions are continuous. The theorem indicates that if $(a, f(a))$ and $(b, f(b))$ are two points on the graph of a polynomial such that $f(a) \neq f(b)$, then for any number d between $f(a)$ and $f(b)$ there must be a number c between a and b such that $f(c) = d$. See Figure 3.11.

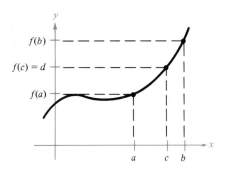

If d lies between $f(a)$ and $f(b)$ then there exists c between a and b such that $f(c) = d$.

FIGURE 3.11   *Intermediate Value Theorem*

INTERMEDIATE VALUE THEOREM

> If f is a polynomial function such that $a < b$ and $f(a) \neq f(b)$, then f takes on every value between $f(a)$ and $f(b)$ in the interval $[a, b]$.

This theorem helps us locate the real zeros of polynomial functions. Specifically, if we can find a value $x = a$ where a polynomial function is positive, and another value $x = b$ where it is negative, then we can conclude that the function has at least one real zero between these two values. For example, the function

$$f(x) = x^3 + x^2 + 1$$

is negative when $x = -2$ and positive when $x = -1$. It follows from the Intermediate Value Theorem that f must have a real zero somewhere between -2 and -1, as shown in Figure 3.12.

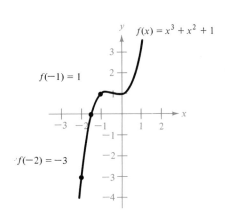

f has a zero between -2 and -1

FIGURE 3.12

EXAMPLE 3
Graphically Estimating Zeros
of Polynomial Functions

Use the Intermediate Value Theorem to estimate the real zero of

$$f(x) = x^3 - x^2 + 1$$

to the nearest tenth.

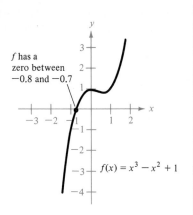

f has a
zero between
−0.8 and −0.7

$f(x) = x^3 - x^2 + 1$

FIGURE 3.13

Solution:
We begin by computing a few functional values as follows:

x	-2	-1	0	1
$f(x)$	-11	-1	1	1

Now since $f(-1)$ is negative and $f(0)$ is positive, we may conclude that the function has a zero between -1 and 0. To pinpoint this zero even further, we divide the interval $[-1, 0]$ into tenths and evaluate the function at each subdivision. Now we see that one real zero lies between -0.8 and -0.7. This is a tedious procedure, but we could continue it to estimate this zero to any desired accuracy. The graph of this function is shown in Figure 3.13.

x	-1	-0.9	-0.8	-0.7	-0.6	-0.5	-0.4	-0.3	-0.2	-0.1	0
$f(x)$	-1	-0.539	-0.152	0.167	0.424	0.625	0.776	0.883	0.952	0.989	1

In the last two examples in this section, we will demonstrate the use of the various sketching aids we have developed up to this point.

EXAMPLE 4
Sketching the Graph
of a Polynomial Function

Sketch the graph of $f(x) = x^4 + 1$.

Solution:
Since the leading coefficient is positive and the degree is even, we know that the graph moves up to the right and left. Also, by setting $f(x)$ equal to zero we see that there are no x-intercepts, since $x^4 = -1$ has no real solutions. By plotting the y-intercept $(0, 1)$, we have the rough sketch shown in Figure 3.14.

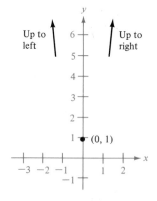

Up to
left

Up to
right

$(0, 1)$

FIGURE 3.14

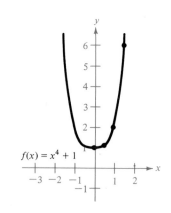

$f(x) = x^4 + 1$

FIGURE 3.15

To complete the sketch, we observe that the graph has symmetry with respect to the *y*-axis, and plot additional points, as shown in Figure 3.15.

x	0.5	1	1.5
f(x)	1.0625	2	6.0625

EXAMPLE 5
Sketching the Graph
of a Polynomial Function

Sketch the graph of

$$f(x) = 3x^4 - 4x^3$$

Solution:
Since the leading coefficient is positive and the degree is even, we know that the graph moves up to the right and left. Also, by setting *f(x)* equal to zero, we have

$$f(x) = 3x^4 - 4x^3 = 0$$
$$x^3(3x - 4) = 0$$

Thus, the *x*-intercepts occur at $(0, 0)$ and $(\frac{4}{3}, 0)$, as shown in Figure 3.16.

To complete the sketch, we plot a few additional points, as shown in Figure 3.17.*

x	−1	0.5	1	1.5
f(x)	7	−0.3125	−1	1.6875

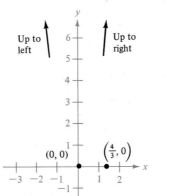

FIGURE 3.16

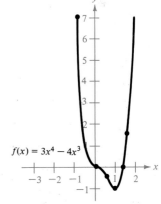

$f(x) = 3x^4 - 4x^3$

FIGURE 3.17

*In calculus you will learn how to find and then verify that a point such as $(1, -1)$ is a minimum point.

Section Exercises 3.1

In Exercises 1–10, use the Leading Coefficient Test or the number of real zeros of a polynomial function to match the polynomial functions with the correct graph.

1. $f(x) = -3x + 5$
2. $f(x) = 2x - 3$
3. $f(x) = x^2 - 2x$
4. $f(x) = -2x^2 - 8x - 9$
5. $f(x) = 2x^3 - 3x^2 - 12x + 8$
5. $f(x) = -\frac{1}{3}x^3 + x - \frac{2}{3}$
7. $f(x) = -\frac{1}{4}x^4 + 2x^2$
8. $f(x) = 3x^4 + 4x^3$
9. $f(x) = x^5 - 5x^3 + 4x$
10. $f(x) = x^3 - 1$

(a)

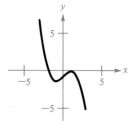

(b)

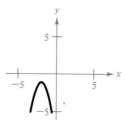

(c)

(d)

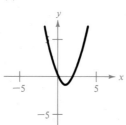

(e)

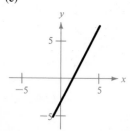

(f)

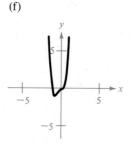

(g)

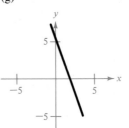

(h)

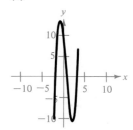

(i)

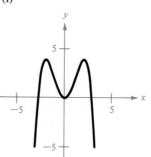

(j)

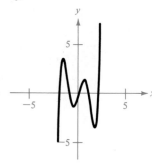

In Exercises 11–20, use the Leading Coefficient Test to determine the right- and left-hand behavior of the graph of the polynomial function.

11. $f(x) = 2x^2 - 3x + 1$
12. $f(x) = \frac{1}{3}x^3 + 5x$
13. $f(x) = 5 - \frac{7}{2}x - 3x^2$
14. $f(x) = -2.1x^5 + 4x^3 - 2$
15. $f(x) = 2x^5 - 5x + 7.5$
16. $f(x) = 1 - x^6$
17. $f(x) = 6 - 2x + 4x^2 - 5x^3$
18. $f(x) = \dfrac{3x^4 - 2x + 5}{4}$
19. $f(x) = -\frac{2}{3}(x^2 - 5x + 3)$
20. $f(x) = -\frac{7}{8}(x^3 + 5x^2 - 7x + 1)$

In Exercises 21–35, find all the real zeros of the polynomial function.

21. $f(x) = x^2 - 25$
22. $f(x) = 49 - x^2$
23. $f(x) = x^2 - 6x + 9$
24. $f(x) = x^2 + 10x + 25$
25. $f(x) = x^2 + x - 2$
26. $f(x) = \frac{1}{2}(x^2 + 5x - 3)$
27. $f(x) = 3(x^2 - 4x + 1)$
28. $f(x) = 5(x^2 - 2x - 1)$
29. $f(x) = x^3 - 4x^2 + 4x$
30. $f(x) = x^4 - x^3 - 20x^2$

31. $f(x) = \frac{1}{2}(x^4 - 1)$ 32. $f(x) = x^5 + x^3 - 6x$
33. $f(x) = 2(x^4 - x^2 - 20)$ 34. $f(x) = x^5 - 6x^3 + 9x$
35. $f(x) = 5(x^4 + 3x^2 + 2)$

In Exercises 36–44, find a polynomial function that has the given zeros.

36. $0, -3$ 37. $0, 10$ 38. $-4, 5$
39. $2, -6$ 40. $0, 2, 5$ 41. $0, -2, -3$
42. $1, -\sqrt{2}, \sqrt{2}$ 43. $4, -3, 3, 0$ 44. $-2, -1, 0, 1, 2$
45. (a) Sketch the graph of $f(x) = x^3$.
    (b) Using the graph of part (a) as a model, sketch the graph of
        (i) $f(x) = (x - 2)^3$ (ii) $f(x) = x^3 - 2$
        (iii) $f(x) = (x - 2)^3 - 2$ (iv) $f(x) = -\frac{1}{2}x^3$
46. (a) Sketch the graph of $f(x) = x^4$.
    (b) Using the graph of part (a) as a model, sketch the graph of
        (i) $f(x) = (x + 3)^4$ (ii) $f(x) = x^4 - 3$

(iii) $f(x) = 4 - x^4$ (iv) $f(x) = \frac{1}{4}(x - 1)^4$

In Exercises 47–56, sketch the graph of the polynomial function using the sketching aids of this section.

47. $f(x) = x^3 - 3x^2$ 48. $f(x) = 1 - x^5$
49. $f(x) = x^3 - 4x$ 50. $f(x) = -(x^4 - 2x^3)$
51. $f(x) = x^4 - 2x^3 + x^2$ 52. $f(x) = \frac{1}{4}x^4 - 2x^2$
53. $f(x) = 12x - x^3$ 54. $f(x) = (x + 1)^5 - 2$
55. $f(x) = x^7 - 2$ 56. $f(x) = 1 - x^6$

In Exercises 57–60, follow the procedure given in Example 3 to estimate the zero of $f(x)$ in the given interval $[a, b]$. (List your approximation to the nearest tenth.)

57. $f(x) = x^3 + x - 1, [0, 1]$
58. $f(x) = x^5 + x + 1, [-1, 0]$
59. $f(x) = x^4 - 10x^2 - 11, [3, 4]$
60. $f(x) = -x^3 + 3x^2 + 9x - 2, [4, 5]$

Polynomial Division and Synthetic Division   3.2

Up to this point in our study we have added, subtracted, and multiplied polynomials. In this section, we look at a procedure for dividing polynomials. This procedure has many important applications and is especially valuable in factoring and finding the zeros of polynomial functions.

Consider the graph of

$$f(x) = 6x^3 - 19x^2 + 16x - 4$$

As shown in Figure 3.18, a zero occurs near $x = 2$. You could verify this by evaluating the function at $x = 2$.

$$f(2) = 6(2^3) - 19(2^2) + 16(2) - 4$$
$$= 48 - 76 + 32 - 4 = 0$$

Since $x = 2$ is a zero of the polynomial function, we know that $(x - 2)$ is a factor of $f(x)$. This means that there exists a second degree polynomial $q(x)$ such that

$$f(x) = (x - 2) \cdot q(x)$$

To find $q(x)$, we can make use of a process called **long division of polynomials,** which is patterned after the long division for real numbers. As with real numbers, the algorithm makes repeated use of the pattern

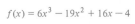

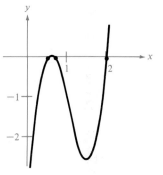

FIGURE 3.18

partial quotient $\Rightarrow$ multiply $\Rightarrow$ subtract

Study the following example to see how this pattern is used.

EXAMPLE 1
Long Division of Polynomials

Divide the polynomial $f(x) = 6x^3 - 19x^2 + 16x - 4$ by $(x - 2)$, and use the result to factor $f(x)$ completely.

Solution:

Partial quotients

$$
\begin{array}{r}
6x^2 - 7x + 2 \\
x - 2 \overline{\smash{)}6x^3 - 19x^2 + 16x - 4} \\
\underline{6x^3 - 12x^2} \\
-7x^2 + 16x \\
\underline{-7x^2 + 14x} \\
2x - 4 \\
\underline{2x - 4} \\
0
\end{array}
$$

Multiply $6x^2(x - 2)$
Subtract
Multiply $-7x(x - 2)$
Subtract
Multiply $2(x - 2)$
Subtract

Now we see that

$$6x^3 - 19x^2 + 16x - 4 = (x - 2)(6x^2 - 7x + 2)$$

and by factoring the quadratic, we have

$$6x^3 - 19x^2 + 16x - 4 = (x - 2)(2x - 1)(3x - 2)$$

Note that this factorization agrees with the graph of f (Figure 3.18) in that the three x-intercepts occur at

$$x = 2, \qquad x = \frac{1}{2}, \qquad \text{and} \qquad x = \frac{2}{3}$$

The process of long division is (theoretically) summarized in a well known mathematical theorem called the **Division Algorithm.**

THE DIVISION ALGORITHM

If $f(x)$ and $d(x)$ are polynomials such that $d(x) \neq 0$, and the degree of $d(x)$ is less than (or equal to) the degree of $f(x)$, then there exist unique polynomials $q(x)$ and $r(x)$ such that

$$f(x) = d(x)q(x) + r(x)$$

Dividend   Divisor   Quotient   Remainder

where either $r(x) = 0$ or the degree of $r(x)$ is less than the degree of $d(x)$.

Remark: We can write a polynomial division in either

Division Tableau Form   or   *Fractional Form*

$$\frac{q(x)}{d(x)\overline{)f(x)}}$$

$$\frac{f(x)}{d(x)} = q(x) + \frac{r(x)}{d(x)}$$

$$\begin{array}{c} \cdot \\ \cdot \\ \cdot \\ \hline r(x) \end{array}$$

where the rational expression $f(x)/d(x)$ is an **improper fraction** if the degree of the numerator is greater than (or equal to) the degree of the denominator. Similarly, we say that the rational expression $r(x)/d(x)$ is **proper** if the degree of the numerator is less than the degree of the denominator.

EXAMPLE 2
Long Division of Polynomials

Perform the indicated divisions.

(a) $\dfrac{3x - 1}{x + 2}$

(b) $(x^3 - 1) \div (x - 1)$

(c) $x^2 + 2x - 3\overline{)2x^4 + 4x^3 - 5x^2 + 2x - 2}$

Solution:

(a) This fraction is improper because the numerator and denominator are of the same degree. Since

$$\begin{array}{r} 3 \\ x + 2\overline{)3x - 1} \\ \underline{3x + 6} \\ -7 \end{array}$$

we have

$$\frac{3x - 1}{x + 2} = 3 - \frac{7}{x + 2}$$

(b) Because there is no x^2-term or x-term in the dividend, we line up the subtraction by using zero coefficients (or leaving a space) for these missing terms.

$$\begin{array}{r} x^2 + x + 1 \\ x - 1\overline{)x^3 + 0x^2 + 0x - 1} \\ \underline{x^3 - x^2} \\ x^2 \\ \underline{x^2 - x} \\ x - 1 \\ \underline{x - 1} \\ 0 \end{array}$$

(c)
$$
\begin{array}{r}
2x^2 \qquad\quad + 1 \\
x^2 + 2x - 3\overline{)2x^4 + 4x^3 - 5x^2 + 2x - 2} \\
\underline{2x^4 + 4x^3 - 6x^2} \\
x^2 + 2x - 2 \\
\underline{x^2 + 2x - 3} \\
1
\end{array}
$$

Note that the first subtraction eliminated two terms from the dividend. When this happens, the quotient will contain a missing term. Thus, we write the quotient as

$$
\frac{2x^4 + 4x^3 - 5x^2 + 2x - 2}{x^2 + 2x - 3} = 2x^2 + 1 + \frac{1}{x^2 + 2x - 3}
$$

Synthetic Division

If the divisor is of the form $(x - a)$, there is a nice shortcut for this special case of long division by linear factors. The shortcut is called **synthetic division.** To see how synthetic division works, we take another look at Example 1.

$$
\begin{array}{r}
6x^2 - 7x + 2 \\
x - 2\overline{)6x^3 - 19x^2 + 16x - 4} \\
\underline{6x^3 - 12x^2} \\
-7x^2 + 16x \\
\underline{-7x^2 + 14x} \\
2x - 4 \\
\underline{2x - 4} \\
0
\end{array}
$$

We can retain the essential steps in this method by using only the coefficients, as follows:

$$
\begin{array}{r}
6 \quad -7 \quad\ \ 2 \\
-2\overline{)6 \quad -19 \quad 16 \quad -4} \\
\underline{6 \quad -12} \\
-7 \quad 16 \\
\underline{-7 \quad 14} \\
2 \quad -4 \\
\underline{2 \quad -4} \\
0
\end{array}
$$

Since the coefficients shown in color are duplicates of those in the quotient or the dividend, we can omit them and then condense vertically to the form

$$
\begin{array}{r}
6 \quad -7 \quad\ \ 2 \\
-2\overline{)6 \quad -19 \quad 16 \quad -4} \\
\underline{-12 \quad 14 \quad -4} \\
0
\end{array}
$$

Moving the quotient down to the bottom row, we have the alternative form

$$
\begin{array}{r}
-2\overline{)\,6 \quad -19 \quad\;\; 16 \quad -4} \\
-12 \quad\;\; 14 \quad -4 \\
\hline
6 \quad\; -7 \quad\;\;\; 2 \quad\;\;\; 0
\end{array}
$$

Now we can change from subtraction to addition (and reduce the likelihood of errors) by changing the sign of the divisor and of row two. Thus, we obtain the following synthetic division array for Example 1:

$$
\begin{array}{r|rrrr}
2 & 6 & -19 & 16 & -4 \\
& & 12 & -14 & 4 \\
\hline
& 6 & -7 & 2 & 0
\end{array}
$$

Don't worry if you have a hard time following the equivalence of the tableaus for long division and synthetic division. The important thing is that you be able to *use* the synthetic division algorithm. We list the pattern for a cubic polynomial. The pattern for higher degree polynomials is similar.

SYNTHETIC DIVISION FOR A CUBIC POLYNOMIAL

To divide $ax^3 + bx^2 + cx + d$ by $(x - k)$, we use the following pattern:

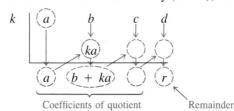

Coefficients of quotient   Remainder

Vertical Pattern: Add terms.
Diagonal Pattern: Multiply by k.

EXAMPLE 3
Using Synthetic Division

Use synthetic division to perform the following division:

$$
x + 3\overline{)\,x^4 \quad\quad -10x^2 - 2x + 4}
$$

Solution:
We can set up the array as follows:

Dividend $\longrightarrow$ $x^4 \quad\quad -10x^2 - 2x + 4$

Divisor $\longrightarrow -3$ | $1 \quad\quad 0 \quad -10 \quad -2 \quad\quad 4$
$(x + 3)$ $-3 \quad\quad 9 \quad\quad 3 \quad -3$

$\qquad\qquad\qquad\qquad 1 \quad -3 \quad\; -1 \quad\quad 1 \quad\quad 1 \longleftarrow$ Remainder

Quotient $\longrightarrow \quad x^3 - 3x^2 - \quad x + 1$

Thus, we have

$$\frac{x^4 - 10x^2 - 2x + 4}{x + 3} = x^3 - 3x^2 - x + 1 + \frac{1}{x + 3}$$

The remainder obtained in the synthetic division process has an important interpretation which is described by a theorem called the Remainder Theorem.

THE REMAINDER THEOREM

If a polynomial $f(x)$ is divided by $(x - k)$, then the remainder is

$$r = f(k)$$

Proof:
From the Division Algorithm, we have

$$f(x) = (x - k)q(x) + r(x)$$

and since the degree of $r(x)$ must be less than the degree of $(x - k)$, we know that $r(x)$ must be a constant

$$r(x) = r$$

Now, by evaluating $f(x)$ at $x = k$, we have

$$\begin{aligned}
f(k) &= (k - k)q(k) + r \\
&= (0)q(k) + r \\
&= r
\end{aligned}$$

EXAMPLE 4
Evaluating a Polynomial
by the Remainder Theorem

Use the Remainder Theorem to evaluate the following function at $x = -2$:

$$f(x) = 3x^3 + 8x^2 + 5x - 7$$

Solution:
Using synthetic division, we have

$$\begin{array}{r|rrrr}
-2 & 3 & 8 & 5 & -7 \\
   &   & -6 & -4 & -2 \\
\hline
   & 3 & 2 & 1 & -9
\end{array}$$

Since the remainder is $r = -9$, we conclude that

$$f(-2) = -9$$

To check this result, we can use the standard procedure for evaluating functions.

$$\begin{aligned}
f(-2) &= 3(-2)^3 + 8(-2)^2 + 5(-2) - 7 \\
&= -24 + 32 - 10 - 7 = -9
\end{aligned}$$

We have already made considerable use of the equivalence of factors of polynomials and zeros of polynomial functions. Using the Remainder Theorem we can now prove this important result. It is known as the Factor Theorem.

FACTOR THEOREM

A polynomial $f(x)$ has a factor $(x - k)$ if and only if $f(k) = 0$.

Proof:
Using the division algorithm with the factor $(x - k)$, we have

$$f(x) = (x - k)q(x) + r(x)$$

By the Remainder Theorem, $r(x) = r = f(k)$, and we have

$$f(x) = (x - k)q(x) + f(k)$$

where $q(x)$ is a polynomial of lesser degree than $f(x)$. Now if $f(k) = 0$, then

$$f(x) = (x - k)q(x)$$

and we see that $(x - k)$ is a factor of $f(x)$. Conversely, if $(x - k)$ is a factor of $f(x)$, then division of $f(x)$ by $(x - k)$ yields a remainder of 0. Hence, by the Remainder Theorem, we have

$$f(k) = 0$$

EXAMPLE 5
Using Synthetic Division to Find Factors of a Polynomial

Show that $(x - 2)$ and $(x + 3)$ are factors of the polynomial

$$f(x) = 2x^4 + 7x^3 - 4x^2 - 27x - 18$$

Then find the remaining factors of $f(x)$.

Solution:
Using synthetic division with 2 and -3 successively, we get

$$
\begin{array}{r|rrrrr}
2 & 2 & 7 & -4 & -27 & -18 \\
  &   & 4 & 22 & 36 & 18 \\
\hline
  & 2 & 11 & 18 & 9 & 0
\end{array}
$$

$\Rightarrow$ *0 remainder*
                (x − 2) is a factor

$$
\begin{array}{r|rrrr}
-3 & 2 & 11 & 18 & 9 \\
   &   & -6 & -15 & -9 \\
\hline
   & 2 & 5 & 3 & 0
\end{array}
$$

$\Rightarrow$ *0 remainder*
                (x + 3) is a factor

The resulting quadratic factors as

$$2x^2 + 5x + 3 = (2x + 3)(x + 1)$$

Thus, the complete factorization of $f(x)$ is

$$f(x) = 2x^4 + 7x^3 - 4x^2 - 27x - 18$$
$$= (x - 2)(x + 3)(2x + 3)(x + 1)$$

Horner's Method

Synthetic division provides us with a method for evaluating polynomials that is especially useful on a calculator. Let's reconsider the polynomial function given in Example 4,

$$f(x) = 3x^3 + 8x^2 + 5x - 7$$

To evaluate $f(k)$, synthetic division yields

k	3	8	5	-7
		$3k$	$(3k + 8)k$	$[(3k + 8)k + 5]k$
	3	$3k + 8$	$(3k + 8)k + 5$	$[(3k + 8)k + 5]k - 7$

Now by the Remainder Theorem we know that

$$f(k) = [(3k + 8)k + 5]k - 7$$

In terms of x, we then have

$$3x^3 + 8x^2 + 5x - 7 = [(3x + 8)x + 5]x - 7$$

We call this **Horner's Method** of writing a polynomial. It can be applied to any polynomial by successively factoring out x from each non-constant term, as demonstrated in the following example.

EXAMPLE 6
Horner's Method

Use Horner's Method to rewrite the polynomial function

$$f(x) = 5x^4 - 3x^3 + x^2 - 8x + 7$$

Solution:

$$f(x) = 5x^4 - 3x^3 + x^2 - 8x + 7$$
$$= (5x^3 - 3x^2 + x - 8)x + 7 \qquad \textit{Factor x from first 4 terms}$$
$$= ([5x^2 - 3x + 1]x - 8)x + 7 \qquad \textit{Factor x from first 3 terms}$$
$$= ([(5x - 3)x + 1]x - 8)x + 7 \qquad \textit{Factor x from first 2 terms}$$

Now, notice how easily we can evaluate $f(k)$ for the polynomial function in Example 6 by entering the following calculator steps:

$$5 \boxed{\times} k \boxed{-} 3 \boxed{=} \boxed{\times} k \boxed{+} 1$$
$$\boxed{=} \boxed{\times} k \boxed{-} 8 \boxed{=} \boxed{\times} k \boxed{+} 7$$

If k is a large number or a decimal value, we could save time by storing k and then using the $\boxed{\text{RCL}}$ key each time k is needed.

EVALUATING A POLYNOMIAL WITH A CALCULATOR

To evaluate the polynomial

$$f(x) = a_nx^n + a_{n-1}x^{n-1} + \cdots + a_1x + a_0$$

on a calculator with algebraic logic, use the key sequence

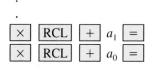

If $a_i = 0$, then omit $\boxed{+}$ a_i, and if $a_i < 0$, use $\boxed{-}$ instead of $\boxed{+}$.

Remark: Note how the calculator routine *repeats* the five-stroke sequence

$$\boxed{\times}\ \boxed{\text{RCL}}\ \boxed{+}\ a_i\ \boxed{=}$$

Section Exercises 3.2

In Exercises 1–20, divide the first polynomial by the second using long division.

1. $4x^3 - 7x^2 - 11x + 5, 4x + 5$
2. $6x^3 - 16x^2 + 17x - 6, 3x - 2$
3. $x^4 + 5x^3 + 6x^2 - x - 2, x + 2$
4. $6x^4 + 9x^3 + 10x^2 + x - 21, 2x + 3$
5. $3x^4 + 3x^3 - 10x^2 - 6x + 8, x^2 - 2$
6. $4x^5 - 8x^4 + 4x^3 + 7x^2 - 14x + 7, 4x^3 + 7$
7. $7x + 3, x + 2$
8. $8x - 5, 2x + 1$
9. $x^3, x^2 - 5$
10. $6x^3 + 10x^2 + x + 8, 2x^2 + 1$
11. $x^3 - 9, x^2 + 1$
12. $x^4 + 3x^2 + 1, x^2 - 2x + 3$
13. $x^5 + 7, x^3 - 1$
14. $5x^4, x^2 - 5x + 1$
15. $4x^4 + 5, 2x^2 + 6x - 3$
16. $11x^4 - 9x^3 + 33x^2 - 27x, 11x^2 - 9x$
17. $x^3 - 21x, 5 + 4x - x^2$
18. $x^3 - x + 3, x^2 + x - 2$
19. $2x^3 - 4x^2 - 15x + 5, x^2 - 2x - 8$
20. $x^4, (x - 1)^3$

In Exercises 21–40, use synthetic division to divide the first polynomial by the second.

21. $3x^3 - 17x^2 + 15x - 25, x - 5$
22. $5x^3 + 18x^2 + 7x - 6, x + 3$
23. $4x^3 - 9x + 8x^2 - 18, x + 2$
24. $9x^3 - 16x - 18x^2 + 32, x - 2$
25. $-x^3 + 75x - 250, x + 10$
26. $3x^3 - 16x^2 - 72, x - 6$
27. $5x^3 - 6x^2 + 8, x - 4$
28. $5x^3 + 6x + 8, x + 2$
29. $10x^4 - 50x^3 - 800, x - 6$
30. $x^5 - 13x^4 - 120x + 80, x + 3$
31. $x^5 - 32, x - 2$
32. $x^4 - 625, x - 5$
33. $x^3 + 512, x + 8$
34. $5x^3, x + 3$
35. $-3x^4, x - 2$
36. $-3x^4, x + 2$
37. $5 - 3x + 2x^2 - x^3, x + 1$
38. $180x - x^4, x - 6$
39. $4x^3 + 16x^2 - 23x - 15, x + \frac{1}{2}$
40. $3x^3 - 4x^2 + 5, x - \frac{3}{2}$

41. Use synthetic division to find (a) $f(1)$, (b) $f(-2)$, (c) $f(\frac{1}{2})$ and (d) $f(8)$ when

$$f(x) = 4x^3 - 13x + 10$$

42. Use synthetic division to find (a) $f(1)$, (b) $f(-2)$, (c) $f(5)$ and (d) $f(-10)$ when

$$f(x) = 0.4x^4 - 1.6x^3 + 0.7x^2 - 2$$

In Exercises 43–50, use synthetic division to show that r is a root of the third degree polynomial equation, and use the result to factor the polynomial completely.

43. $x^3 - 7x + 6 = 0, r = 2$
44. $x^3 - 28x - 48 = 0, r = -4$

45. $2x^3 - 15x^2 + 27x - 10 = 0, r = \frac{1}{2}$
46. $48x^3 - 80x^2 + 41x - 6 = 0, r = \frac{2}{3}$
47. $x^3 - 1.9x^2 + 1.1x - 0.2 = 0, r = 0.4$
48. $x^3 + 1.5x^2 - 7.21x - 0.735 = 0, r = -3.5$
49. $x^3 - 3x^2 + 2 = 0, r = 1 + \sqrt{3}$
50. $x^3 - x^2 - 13x - 3 = 0, r = 2 - \sqrt{5}$

In Exercises 51–54, use Horner's method to evaluate the polynomial at the indicated value of x.

51. $f(x) = x^3 - 6x^2 + 12x - 8$ at 5 and -4.5
52. $f(x) = 3x^4 + 6x^3 - 10x^2 - 7x + 2$ at -4.8 and 0.02
53. $f(x) = -5x^4 + 8.5x^3 + 10x - 3$ at 1.08 and -5.4
54. $f(x) = -2x^5 + 4x^3 - 6x^2 + 10$ at 4 and -3.7

Rational Zeros of Polynomial Functions   3.3

In Section 3.1 we saw that an *n*th degree polynomial function can have at most *n* real zeros. Although we do not have a general procedure for finding these real zeros, there is a procedure for finding the *rational* zeros of a polynomial function with rational coefficients. This procedure is called the **Rational Zero Test.** Before looking at this test, let's look at the zeros of a few simple polynomial functions.

Function	Real Zeros	Comment
$f(x) = 3x + 4$	$x = -\frac{4}{3}$	One rational zero
$f(x) = x - \sqrt{2}$	$x = \sqrt{2}$	One irrational zero
$f(x) = 2x^2 - 8$	$x = -2, 2$	Two rational zeros
$f(x) = x^2 - 3$	$x = -\sqrt{3}, \sqrt{3}$	Two irrational zeros
$f(x) = x^2 + 1$	None	No real zeros

Several valuable observations can be made concerning the nature of the zeros:

1. Only four of these polynomial functions have rational coefficients. The Rational Zero Test does not apply to functions with irrational coefficients such as $\sqrt{2}$.

2. A polynomial function can have rational coefficients but no rational zeros.

3. The leading coefficient and constant term of a polynomial function are significant in determining the rational zeros.

THE RATIONAL ZERO TEST

> If the polynomial function
>
> $$f(x) = a_n x^n + a_{n-1} x^{n-1} + \cdots + a_2 x^2 + a_1 x + a_0$$
>
> has *integer* coefficients, then every rational zero of f has the form
>
> $$\text{rational zero} = \frac{p}{q}$$
>
> where
>
> $p = $ a factor of the constant term a_0
>
> $q = $ a factor of the leading coefficient a_n

Proof:

We begin by assuming that p/q is a rational zero of f and that p/q is written in reduced form (that is, p and q have no common factors other than 1). Since $f(p/q) = 0$, we have

$$a_n \left(\frac{p}{q}\right)^n + a_{n-1} \left(\frac{p}{q}\right)^{n-1} + \cdots + a_2 \left(\frac{p}{q}\right)^2 + a_1 \left(\frac{p}{q}\right) + a_0 = 0$$

Multiplying by q^n, we have

$$a_n p^n + a_{n-1} p^{n-1} q + \cdots + a_2 p^2 q^{n-2} + a_1 p q^{n-1} + a_0 q^n = 0$$

Now we rewrite this equation in two convenient forms, one having p as a factor and one having q as a factor.

$$p(a_n p^{n-1} + a_{n-1} p^{n-2} q + \cdots + a_2 p q^{n-2} + a_1 q^{n-1}) = -a_0 q^n$$
$$q(a_{n-1} p^{n-1} + \cdots + a_2 p^2 q^{n-3} + a_1 p q^{n-2} + a_0 q^{n-1}) = -a_n p^n$$

From the first equation we see that p is a factor of $a_0 q^n$. However, since p and q have no factors in common, it follows that p must be a factor of a_0. Similarly, from the second equation we see that q is a factor of $a_n p^n$, and again, since p and q have no factors in common, it follows that q must be a factor of a_n.

To use the Rational Zero Test to find zeros of a polynomial function, we list all rational numbers whose numerators are factors of the constant term and whose denominators are factors of the leading coefficient. Once the list of *possible rational zeros* is formed, we use trial and error to determine which, if any, are actual zeros of the polynomial.

If the leading coefficient is 1, then the possible rational zeros are simply the factors of a_0. This situation is illustrated in Example 1.

EXAMPLE 1
Leading Coefficient is 1

Use the Rational Zero Test for each of the following functions:

(a) $f(x) = x^3 + x + 1$ (b) $f(x) = x^4 - x^3 + x^2 - 3x - 6$

Solution:

(a) For

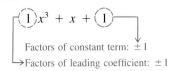

Factors of constant term: ± 1

Factors of leading coefficient: ± 1

the factors of the leading coefficient are ± 1, hence the possible rational zeros are simply the factors of the constant term.

Possible rational zeros $= \pm 1$

By testing these potential zeros, we see that neither works.

$$f(1) = (1)^3 + 1 + 1 = 3$$
$$f(-1) = (-1)^3 + (-1) + 1 = -1$$

Thus, we conclude that this polynomial function has *no* rational zeros.

(b) For

$$x^4 - x^3 + x^2 - 3x - 6$$

the leading coefficient is 1, hence the possible rational zeros are the factors of the constant term.

Possible rational zeros $= \pm 1, \pm 2, \pm 3, \pm 6$

By testing these potential zeros, you should conclude that $x = -1$ and $x = 2$ are the only two that work. Check the others to be sure.

The process of checking several possible rational roots can be tedious, and you can shorten the process in several different ways. First, a programmable calculator is useful. Second, the graph of the function can be used to estimate the location of the real zeros. A third shortcut is to find one zero, then divide the polynomial by the corresponding linear factor to obtain a polynomial of lower degree that may be more easily factorable. In particular, if you succeed in reducing the degree to two, then you can apply the quadratic formula.

Synthetic division makes this third shortcut even more convenient. For example, in part (b) of Example 1, after finding that $x = -1$ and $x = 2$ are zeros of

$$f(x) = x^4 - x^3 + x^2 - 3x - 6$$

we could save quite a bit of work by dividing both $(x + 1)$ and $(x - 2)$ into

$f(x)$ as follows:

$$
\begin{array}{r|rrrrr}
-1 & 1 & -1 & 1 & -3 & -6 \\
   &   & -1 & 2 & -3 & 6 \\
\hline
   & 1 & -2 & 3 & -6 & 0 \\
\end{array}
$$

$$
\begin{array}{r|rrrr}
2 & 1 & -2 & 3 & -6 \\
  &   & 2 & 0 & 6 \\
\hline
  & 1 & 0 & 3 & 0 \\
\end{array}
$$

Thus, we have

$$f(x) = (x + 1)(x - 2)(x^2 + 3)$$

and since the factor $(x^2 + 3)$ has no real roots, we see that $x = -1$ and $x = 2$ are the only real zeros of f.

   If the leading coefficient of the polynomial is something other than 1, the list of possible rational zeros can increase dramatically. In such cases, the shortcuts to the Rational Zero Test can be especially valuable.

EXAMPLE 2
Leading Coefficient Other Than 1

Find the rational roots of the following polynomial functions:

(a) $f(x) = 2x^3 + 3x^2 - 8x + 3$
(b) $f(x) = 10x^3 - 15x^2 - 16x + 12$

Solution:

(a) For $2x^3 + 3x^2 - 8x + 3$,

$$\text{Possible rational zeros} = \frac{\text{factors of 3}}{\text{factors of 2}}$$

$$= \frac{\pm 1, \ \pm 3}{\pm 1, \ \pm 2} = \pm 1, \ \pm 3, \ \pm \frac{1}{2}, \ \pm \frac{3}{2}$$

By synthetic division, we can determine that $x = 1$ is a zero *and* we can obtain the factored version of the polynomial.

$$
\begin{array}{r|rrrr}
1 & 2 & 3 & -8 & 3 \\
  &   & 2 & 5 & -3 \\
\hline
  & 2 & 5 & -3 & 0 \\
\end{array}
$$

Now we see that

$$
\begin{aligned}
f(x) &= (x - 1)(2x^2 + 5x - 3) \\
     &= (x - 1)(2x - 1)(x + 3) \qquad \textit{Factor quadratic}
\end{aligned}
$$

which implies that the zeros of f are $x = 1, \frac{1}{2},$ and -3.

$f(x) = 10x^3 - 15x^2 - 16x + 12$

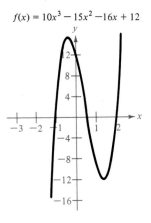

FIGURE 3.19

(b) For $10x^3 - 15x^2 - 16x + 12$,

$$\text{Possible rational zeros} = \frac{\text{factors of } 12}{\text{factors of } 10}$$

$$= \frac{\pm 1, \ \pm 2, \ \pm 3, \ \pm 4, \ \pm 6, \ \pm 12}{\pm 1, \ \pm 2, \ \pm 5, \ \pm 10}$$

With so many possibilities for zeros, it will be worth our time to stop and make a sketch of this function. From Figure 3.19, it looks as though three reasonable choices for zeros would be $x = -1$, $x = 0.5$ and $x = 2$. To test these choices, we use synthetic division as follows:

Test $x = -1$:

$$
\begin{array}{r|rrrr}
-1 & 10 & -15 & -16 & 12 \\
   &   & -10 & 25 & -9 \\
\hline
   & 10 & -25 &   9 & \boxed{3}
\end{array}
$$ ⟹ *$x = -1$ is not a zero of f*

Test $x = 0.5$:

$$
\begin{array}{r|rrrr}
0.5 & 10 & -15 & -16 & 12 \\
    &   &   5 & -5 & -10.5 \\
\hline
    & 10 & -10 & -21 & \boxed{1.5}
\end{array}
$$ ⟹ *$x = 0.5$ is not a zero of f*

Test $x = 2$:

$$
\begin{array}{r|rrrr}
2 & 10 & -15 & -16 & 12 \\
  &   & 20 & 10 & -12 \\
\hline
  & 10 &   5 & -6 & \boxed{0}
\end{array}
$$ ⟹ *$x = 2$ is a zero of f*

Thus, we have

$$f(x) = (x - 2)(10x^2 + 5x - 6)$$

and by the quadratic formula we find the two additional zeros to be

$$x = \frac{-5 + \sqrt{265}}{20} \approx 0.5639$$

and

$$x = \frac{-5 - \sqrt{265}}{20} \approx -1.0639$$

Remark: Remember that the last entry in the bottom row (of synthetic division) must be zero in order to conclude that the test value is a zero of f.

Upper and Lower Bounds for Rational Zeros

Occasionally, we gain additional information from synthetic division by recognizing certain patterns in the last row. In particular, if the test value is positive and each entry in the last row is positive (or zero), then we can conclude that the test value is an **upper bound** for the zeros of f. Similarly, if the test value is negative and the entries in the last row are alternately positive and negative (zero entries count as positive or negative in this test), then we can conclude that the test value is a **lower bound** for the zeros of f.

For example, in part (b) of Example 2, a test of $x = 3$ yields

$$
\begin{array}{r|rrrr}
\textit{Test } x = 3: \quad 3 & 10 & -15 & -16 & 12 \\
 & & 30 & 45 & 87 \\
\hline
 & 10 & 15 & 29 & 99
\end{array}
$$

Since the test value is positive and all entries in the last row are positive, we can conclude that $x = 3$ is an upper bound for the zeros of f. A test of $x = -2$ yields

$$
\begin{array}{r|rrrr}
\textit{Test } x = -2: \quad -2 & 10 & -15 & -16 & 12 \\
 & & -20 & 70 & -108 \\
\hline
 & 10 & -35 & 54 & -96
\end{array}
$$

Since the test value is negative and the entries in the last row are alternately positive and negative, we can conclude that $x = -2$ is a lower bound for the zeros of f. This upper and lower bound information is useful because we can then restrict our test values to possible rational zeros that lie between -2 and 3.

The next example shows how to extend the Rational Zero Test to cover polynomial functions with rational (but not all integer) coefficients. The basic procedure consists of finding the lowest common denominator (LCD) of the coefficients and then factoring 1/LCD out of the polynomial.

EXAMPLE 3
Polynomial Functions with
Rational Coefficients

Find the rational roots of the polynomial function

$$
f(x) = x^3 - \frac{1}{2}x^2 + \frac{1}{3}x - \frac{1}{6}
$$

Solution:
Since the LCD of the coefficients is 6, we rewrite the function as follows:

$$
\begin{aligned}
f(x) &= \frac{6}{6}x^3 - \frac{3}{6}x^2 + \frac{2}{6}x - \frac{1}{6} \\
&= \frac{1}{6}(6x^3 - 3x^2 + 2x - 1)
\end{aligned}
$$

Now, the zeros of f must coincide with the roots of

$$
6x^3 - 3x^2 + 2x - 1 = 0
$$

$$\text{Possible rational zeros} = \frac{\text{factors of } 1}{\text{factors of } 6}$$

$$= \frac{\pm 1}{\pm 1, \ \pm 2, \ \pm 3, \ \pm 6} = \pm 1, \ \pm\frac{1}{2}, \ \pm\frac{1}{3}, \ \pm\frac{1}{6}$$

By testing these potential zeros, we see that $x = \frac{1}{2}$ works.

$$
\begin{array}{c|cccc}
0.5 & 6 & -3 & 2 & -1 \\
    &   & 3 & 0 & 1 \\
\hline
    & 6 & 0 & 2 & 0
\end{array}
$$

and we have

$$6x^3 - 3x^2 + 2x - 1 = \left(x - \frac{1}{2}\right)(6x^2 + 2) = (2x - 1)(3x^2 + 1)$$

Finally, since the factor $(3x^2 + 1)$ has no real roots, we conclude that $x = \frac{1}{2}$ is the only rational zero of f.

Section Exercises 3.3

In Exercises 1–30, use the Rational Zero Test and synthetic division as aids to finding the real zeros of the function.

1. $f(x) = x^3 - 6x^2 + 11x - 6$
2. $f(x) = x^3 - 7x - 6$
3. $f(x) = x^3 - 4x^2 - x + 4$
4. $f(x) = x^3 - 9x^2 + 20x - 12$
5. $f(x) = x^3 + 12x^2 + 21x + 10$
6. $f(x) = x^3 + 6x^2 + 12x + 8$
7. $f(x) = x^3 - 6x^2 + 12x - 8$
8. $f(x) = x^3 + 2x^2 - 7x + 4$
9. $f(x) = x^3 - 4x^2 + 5x - 2$
10. $f(x) = x^3 - 9x^2 + 27x - 27$
11. $f(x) = 2x^3 + 3x^2 - 1$
12. $f(x) = 3x^3 - 19x^2 + 33x - 9$
13. $f(x) = 4x^3 - 3x - 1$
14. $f(x) = 12x^3 - 4x^2 - 27x + 9$
15. $f(x) = 8x^3 - 6x^2 - 5x + 3$
16. $f(x) = 2x^3 - x^2 + 8x - 4$
17. $f(x) = 4x^3 + 3x^2 + 8x + 6$
18. $f(x) = 3x^3 - 2x^2 + 15x - 10$
19. $f(x) = x^4 - 3x^2 + 2$
20. $f(x) = x^4 - 7x^2 + 12$
21. $f(x) = x^4 - x^3 - 2x - 4$
22. $f(x) = x^4 - x^3 - 29x^2 - x - 30$

23. $f(x) = x^4 - 5x^2 + 4$
24. $f(x) = x^4 - 6x^3 + 11x^2 - 6x$
25. $f(x) = x^4 - 13x^2 - 12x$
26. $f(x) = 2x^4 + 7x^3 - 26x^2 + 23x - 6$
27. $f(x) = 2x^4 - 11x^3 - 6x^2 + 64x + 32$
28. $f(x) = x^5 - x^4 - 3x^3 + 5x^2 - 2x$
29. $f(x) = x^5 - 7x^4 + 10x^3 + 14x^2 - 24x$
30. $f(x) = 6x^4 - 11x^3 - 51x^2 + 99x - 27$

In Exercises 31–35, (a) list the possible rational zeros of the function, (b) sketch the graph of the function so that some of the possible zeros of part (a) can be disregarded, and (c) determine the real zeros of the polynomial function.

31. $f(x) = 32x^3 - 52x^2 + 17x + 3$
32. $f(x) = 6x^3 - x^2 - 13x + 8$
33. $f(x) = 4x^3 + 7x^2 - 11x - 18$
34. $f(x) = 6x^3 - 31x^2 - 7x + 60$
35. $f(x) = 4x^3 - 21x - 10$

In Exercises 36–40, find the rational zeros of the polynomial function.

36. $f(x) = x^3 - \frac{3}{2}x^2 - \frac{23}{2}x + 6$
37. $f(x) = x^3 - \frac{1}{4}x^2 - x + \frac{1}{4}$
38. $f(x) = x^3 + \frac{11}{6}x^2 - \frac{1}{2}x - \frac{1}{3}$
39. $f(x) = x^3 - \frac{2}{3}x^2 - \frac{5}{9}x + \frac{2}{9}$

40. $f(x) = x^3 + \frac{19}{4}x^2 + \frac{15}{8}x - \frac{9}{2}$

In Exercises 41–44, use synthetic division to determine if each value of x is an upper bound, lower bound, or neither of the zeros of f.

41. $f(x) = x^4 - 4x^3 + 15$
    (a) $x = 4$ (b) $x = -1$ (c) $x = 3$

42. $f(x) = 2x^3 - 3x^2 - 12x + 8$
    (a) $x = 2$ (b) $x = 4$ (c) $x = -1$

43. $f(x) = x^4 - 4x^3 + 16x - 16$
    (a) $x = -1$ (b) $x = -3$ (c) $x = 5$

44. $f(x) = 2x^4 - 8x + 3$
    (a) $x = 1$ (b) $x = 3$ (c) $x = -4$

Complex Numbers   3.4

So far, we have dealt only with the set of real numbers. Although the real number system is sufficient for most applications of algebra and trigonometry, it does have a serious inadequacy related to even roots of negative numbers. For instance, the square of a real number is always nonnegative. Hence, roots like

$$\sqrt{-1} \quad \text{and} \quad \sqrt{-4}$$

cannot be real numbers since their respective squares are negative:

$$[(-1)^{1/2}]^2 = (-1)^1 = -1 \quad \text{and} \quad [(-4)^{1/2}]^2 = (-4)^1 = -4$$

To resolve this problem, mathematicians created an expanded system of numbers based upon the **imaginary** (nonreal) **number i,** whose square is defined to be -1.

DEFINITION OF THE IMAGINARY NUMBER i

> The **imaginary number i** is defined by
>
> $$i = \sqrt{-1}$$
>
> where $i^2 = -1$.

With this single addition to the real number system, we can develop the system of **complex numbers.**

DEFINITION OF A COMPLEX NUMBER

> If a and b are real numbers, then the number
>
> $$a + bi$$
>
> is called a **complex number,** where a is called the **real part** and b the **imaginary part** of the number.

Remark: The form $a + bi$ is called the **standard form** of a complex number.

We can see that the set of real numbers is a subset of the set of complex numbers, since every real number *a* can be written as a complex number using $b = 0$. That is, for every real number *a*,

$$a = a + 0i$$

Powers of *i*

Using the rules for powers of real numbers, we see that

$$i = \sqrt{-1}$$
$$i^2 = -1$$
$$i^3 = i^2 \cdot i = -i$$
$$i^4 = i^2 \cdot i^2 = (-1)(-1) = 1$$
$$i^5 = i^4 \cdot i = i$$

The pattern begins to repeat after the 4th power. Therefore, to compute the value of i^n for any natural number *n*, we simply factor out the multiples of 4 in the exponent and compute the remaining portion. For instance,

$$i^{38} = i^{36} \cdot i^2 = (i^4)^9 \cdot i^2 = (1)^9(-1) = -1$$

Operations with Complex Numbers

Since a complex number consists of a real part plus a multiple of *i*, we define operations with complex numbers in a manner consistent with the rules for real numbers and the imaginary unit.

EQUALITY OF TWO COMPLEX NUMBERS

> If $a + bi$ and $c + di$ are two complex numbers written in standard form, then
>
> $$a + bi = c + di$$
>
> if and only if $a = c$ and $b = d$.

To add (or subtract) two complex numbers, we add (or subtract) the real and imaginary parts of the numbers separately.

ADDITION AND SUBTRACTION OF COMPLEX NUMBERS

> If $a + bi$ and $c + di$ are two complex numbers written in standard form, then their sum and difference are
>
> **Sum:** $(a + bi) + (c + di) = (a + c) + (b + d)i$
>
> **Difference:** $(a + bi) - (c + di) = (a - c) + (b - d)i$

The **additive identity** in the complex number system is zero (the same as in the real number system). Furthermore, the **additive inverse** of the complex number $a + bi$ is given by

Given Number	Additive Inverse
$a + bi$	$-(a + bi) = -a - bi$

Thus, we have

$$(a + bi) + (-a - bi) = 0 + 0i = 0$$

EXAMPLE 1
Adding and Subtracting Complex Numbers

Write the following sums and differences in standard form.

(a) $(3 - i) + (2 + 3i)$ (b) $2i + (-4 - 2i)$
(c) $7 - (5 - i) + (-2 + 3i)$

Solution:

(a) $(3 - i) + (2 + 3i) = 3 - i + 2 + 3i$
$$= 3 + 2 - i + 3i$$
$$= (3 + 2) + (-1 + 3)i$$
$$= 5 + 2i$$

(b) $2i + (-4 - 2i) = 0 + 2i - 4 - 2i$
$$= -4 + 2i - 2i$$
$$= -4$$

(c) $7 - (5 - i) + (-2 + 3i) = 7 + 0i - 5 + i - 2 + 3i$
$$= 7 - 5 - 2 + i + 3i$$
$$= 4i$$

Remark: Note in part (b) of Example 1 that the sum of two complex numbers can be a real number.

Many of the properties of real numbers are valid for complex numbers as well. Some such properties are

Associative Property of Addition and Multiplication

Commutative Property of Addition and Multiplication

Distributive Property of Multiplication over Addition

Notice how these properties come into play when two complex numbers are multiplied.

$(a + bi)(c + di) = a(c + di) + bi(c + di)$ *Distributive Law*
$\qquad\qquad = ac + (ad)i + (bc)i + (bd)i^2$ *Distributive Law*
$\qquad\qquad = ac + (ad)i + (bc)i + (bd)(-1)$ *Definition of i*
$\qquad\qquad = ac - bd + (ad)i + (bc)i$ *Commutative Law*
$\qquad\qquad = (ac - bd) + (ad + bc)i$ *Associative Law*

We summarize this result as follows.

MULTIPLICATION OF COMPLEX NUMBERS

If $a + bi$ and $c + di$ are two complex numbers written in standard form, then their product is

$$(a + bi)(c + di) = (ac - bd) + (ad + bc)i$$

Remark: Rather than trying to memorize this rule, you may just want to remember how the Distributive Law is used to derive it.

EXAMPLE 2 Multiplying Complex Numbers	Find the following products. (a) $(i)(-3i)$ (b) $(2 - i)(4 + 3i)$ (c) $(3 + 2i)(3 - 2i)$ **Solution:** (a) $(i)(-3i) = -3i^2 = -3(-1) = 3$ (b) $(2 - i)(4 + 3i) = 8 + 6i - 4i - 3i^2$ $\qquad\qquad\qquad\quad = 8 + 6i - 4i - 3(-1)$ $\qquad\qquad\qquad\quad = 8 + 3 + 6i - 4i$ $\qquad\qquad\qquad\quad = 11 + 2i$ (c) $(3 + 2i)(3 - 2i) = 9 - 6i + 6i - 4i^2$ $\qquad\qquad\qquad\qquad = 9 - 4(-1) = 9 + 4$ $\qquad\qquad\qquad\qquad = 13$

Complex Conjugates

Pairs of complex numbers of the forms

$$(a + bi) \quad \text{and} \quad (a - bi)$$

are called **complex conjugates.** That is, the conjugate of $(3 + 2i)$ is $(3 - 2i)$. As illustrated in part (c) of Example 2, the product of two complex conjugates is a real number:

$$(a + bi)(a - bi) = a^2 - abi + abi - b^2 i^2$$
$$= a^2 - b^2(-1)$$
$$= a^2 + b^2$$

We can use complex conjugates to perform division in the complex number system. That is, if we want to find the quotient

$$\frac{a + bi}{c + di}$$

we simply multiply numerator and denominator by the conjugate of the denominator as follows:

$$\frac{a + bi}{c + di} = \frac{a + bi}{c + di}\left(\frac{c - di}{c - di}\right)$$

$$= \frac{(a + bi)(c - di)}{c^2 + d^2}$$

EXAMPLE 3
Dividing Complex Numbers

Write the following in standard form.

(a) $\dfrac{1}{1 + i}$ (b) $\dfrac{2 + 3i}{4 - 2i}$

Solution:

(a) $\dfrac{1}{1 + i} = \dfrac{1}{1 + i}\left(\dfrac{1 - i}{1 - i}\right)$

$$= \frac{1 - i}{1 - (-1)} = \frac{1 - i}{2} = \frac{1}{2} - \frac{1}{2}i$$

(b) $\dfrac{2 + 3i}{4 - 2i} = \dfrac{2 + 3i}{4 - 2i}\left(\dfrac{4 + 2i}{4 + 2i}\right)$

$$= \frac{(2 + 3i)(4 + 2i)}{16 + (-2)^2} = \frac{1}{20}(8 + 12i + 4i + 6i^2)$$

$$= \frac{1}{20}(8 - 6 + 12i + 4i) = \frac{1}{20}(2 + 16i) = \frac{1}{10} + \frac{4}{5}i$$

In algebra it is common to obtain a result such as $\sqrt{-3}$. In standard form this complex number can be written as

$$\sqrt{-3} = \sqrt{(3)(-1)} = \sqrt{3}\sqrt{-1} = \sqrt{3}\,i$$

In general, we have the following rule for writing the square root of a negative number in standard form.

STANDARD FORM OF SQUARE ROOT OF A NEGATIVE NUMBER

If $a > 0$, then
$$\sqrt{-a} = \sqrt{a}\,i$$

Thus,

$$\sqrt{-16} = \sqrt{16}\,i = 4i$$

and

$$\frac{-2 - \sqrt{-2}}{2} = \frac{-2}{2} - \frac{\sqrt{2}\,i}{2} = -1 - \frac{\sqrt{2}}{2}i$$

Remark: When working with square roots of negative numbers, be sure to convert to standard form *before* multiplying. For instance, consider the following:

$$\textit{Correct:}\quad \sqrt{-1}\sqrt{-1} = i \cdot i = i^2 = -1$$

$$\textit{Incorrect:}\quad \sqrt{-1}\sqrt{-1} = \sqrt{(-1)(-1)} = \sqrt{1} = 1$$

EXAMPLE 4
Manipulating Expressions
with Radicals

Expand the following expression.

$$(-1 + \sqrt{-3})^2 + 2(-1 + \sqrt{-3}) + 4$$

Solution:

$$(-1 + \sqrt{-3})^2 + 2(-1 + \sqrt{-3}) + 4$$
$$= (-1 + \sqrt{3}\, i)^2 + 2(-1 + \sqrt{3}\, i) + 4$$
$$= (-1)^2 - 2\sqrt{3}\, i + (\sqrt{3})^2(i^2) - 2 + 2\sqrt{3}\, i + 4$$
$$= 1 + 3(-1) - 2 + 4 - 2\sqrt{3}\, i + 2\sqrt{3}\, i$$
$$= (1 - 3 - 2 + 4) + (-2\sqrt{3} + 2\sqrt{3})i$$
$$= 0$$

Section Exercises 3.4

1. Write out the first 20 positive integer powers of i (i.e., i, i^2, i^3, . . . , i^{20}), and express each in the simplest form of i, $-i$, 1, and -1. What is the pattern of the simplest form?

In Exercises 2–20, write the complex number in standard form and give the complex conjugate.

2. $3 + \sqrt{-16}$
3. $4 + \sqrt{-9}$
4. $-10 - \sqrt{-15}$
5. $-7 - \sqrt{-64}$
6. $\sqrt{-4} - 5$
7. $\sqrt{-81} - 7$
8. $1 + \sqrt{-8}$
9. $2 - \sqrt{-27}$
10. 45
11. $\sqrt{-75}$
12. $4i^2 - 2i^3$
13. $-6i + i^2$
14. $(-i)^3$
15. $8 + (-\sqrt{-7})^3$
16. $(\sqrt{-4})^2 - 5$
17. 8
18. $2 + \sqrt{3}$
19. $5 - \sqrt{27}$
20. $-5i^5$

In Exercises 21–60, perform the indicated operation and write the result in standard form.

21. $(5 + i) + (6 - 2i)$
22. $(13 - 2i) + (-5 + 6i)$
23. $(8 - i) - (4 - i)$
24. $(3 + 2i) - (6 + 13i)$
25. $6i - (5 + 3i)$
26. $6 - (6 - 3i)$
27. $(-2 + \sqrt{-8}) + (5 - \sqrt{50})$
28. $(8 + \sqrt{-18}) - (4 + 3\sqrt{2}\, i)$
29. $-(\frac{3}{2} + \frac{5}{2}i) + (\frac{5}{3} + \frac{11}{3}i)$
30. $(1.6 + 3.2i) + (-5.8 + 4.3i)$
31. $(1 + i)(3 - 2i)$
32. $(6 - 2i)(2 - 3i)$
33. $(4 + 5i)(4 - 5i)$
34. $(6 + 7i)(6 - 7i)$
35. $6i(5 - 2i)$
36. $-8i(9 + 4i)$
37. $(15i)(6i)$
38. $4(3 - 2i)$
39. $-8(5 - 2i)$
40. $(\sqrt{5} - \sqrt{3}\, i)(\sqrt{5} + \sqrt{3}\, i)$
41. $(\sqrt{14} + \sqrt{10}\, i)(\sqrt{14} - \sqrt{10}\, i)$
42. $(3 + \sqrt{-5})(7 - \sqrt{-10})$
43. $(4 + 5i)^2$
44. $(2 - 3i)^3$
45. $(4 + 5i)^3$
46. $\dfrac{1}{2 + 3i}$
47. $\dfrac{4}{4 - 5i}$
48. $\dfrac{3}{1 - i}$
49. $\dfrac{2 + i}{2 - i}$
50. $\dfrac{8 - 7i}{1 - 2i}$
51. $\dfrac{6 - 7i}{i}$
52. $\dfrac{8 + 20i}{2i}$
53. $(12 + 7i) \div (2 + i)$
54. $(3 + 2i) \div (3 - 2i)$

55. $\dfrac{1}{(2i)^3}$

56. $\dfrac{1}{(4 - 5i)^2}$

57. $\dfrac{5}{(1 + i)^3}$

58. $\dfrac{(2 - 3i)(5i)}{2 + 3i}$

59. $\dfrac{(21 - 7i)(4 + 3i)}{2 - 5i}$

60. $\dfrac{1}{i^3}$

In Exercises 61–64, find real numbers a and b so that the equation is true.

61. $a + bi = -10 + 6i$

62. $a + bi = 13 + 4i$

63. $(a - 1) + (b + 3)i = 5 + 8i$

64. $(a + 6) + 2bi = 6 - 5i$

65. Show that the sum of a complex number and its conjugate is a real number.

66. Show that the difference of a complex number and its conjugate is an imaginary number.

67. Show that the product of a complex number and its conjugate is a real number.

68. Show that the conjugate of the product of two complex numbers is the product of their conjugates.

69. Show that the conjugate of the sum of two complex numbers is the sum of their conjugates.

Complex Zeros and the Fundamental Theorem of Algebra

3.5

We have already been using the fact that an nth degree polynomial function can have at most n real zeros. In this section we will improve upon that result to see that, in the complex number system, every nth degree polynomial function has *precisely* n zeros. This important result is derived from the **Fundamental Theorem of Algebra.** The proof of this theorem is outside the scope of this text, and we will be content with clearly stating and illustrating the theorem.

THE FUNDAMENTAL THEOREM OF ALGEBRA

> If f is a polynomial function of degree $n > 0$, then f has at least one zero in the complex number system.

Remark: We have stated the Fundamental Theorem in its standard form. Using the equivalency of zeros and factors, we can easily establish the following result.

NUMBER OF ZEROS OF A POLYNOMIAL FUNCTION

> If f is a polynomial function of degree $n > 0$, then f has precisely n zeros in the complex number system.

Proof:

Using the Fundamental Theorem, we know that f must have at least one zero c_1. In terms of factors, we know that $(x - c_1)$ is a factor of $f(x)$, and we have

$$f(x) = (x - c_1)f_1(x)$$

If $f_1(x)$ is of degree greater than zero, we can reapply the Fundamental Theorem to conclude that f must also have a zero c_2, which implies that

$$f(x) = (x - c_1)(x - c_2)f_2(x)$$

It is clear that the degree of $f_1(x)$ is $n - 1$, the degree of $f_2(x)$ is $n - 2$, and that we can repeatedly apply the Fundamental Theorem n times until we obtain

$$f(x) = a(x - c_1)(x - c_2) \cdots (x - c_n)$$

where a is the leading coefficient of the polynomial $f(x)$. Now, since $f(x)$ has precisely n linear factors, we apply the Factor Theorem to conclude that f has precisely the following n zeros:

$$c_1, c_2, \ldots, c_n$$

The Fundamental Theorem of Algebra tells us that a polynomial function of degree n has precisely n zeros in the complex number system. These zeros can be real or complex and may be repeated as illustrated by the following functions.

Function	Degree	Factored Form	Zeros
$f(x) = x - 2$	1	$x - 2 = 0$	$x = 2$
$f(x) = x^2 - 9$	2	$(x - 3)(x + 3) = 0$	$x = -3, 3$
$f(x) = x^2 + 4$	2	$(x - 2i)(x + 2i) = 0$	$x = -2i, 2i$
$f(x) = x^3 - 4x^2 + 5x - 2$	3	$(x - 1)(x - 1)(x - 2) = 0$	$x = 1, 1, 2$
$f(x) = x^4 - 1$	4	$(x - 1)(x + 1)(x - i)(x + i) = 0$	$x = -1, 1, -i, i$

Note that neither the Fundamental Theorem nor its corollary gives us any information on how to find the zeros of a polynomial function. To do that, we rely on the techniques developed in earlier sections.

In Section 1.1 we pointed out that the discriminant $(b^2 - 4ac)$ can be used to classify the two zeros given by the quadratic formula. In particular, if the discriminant is negative, then the quadratic function $f(x) = ax^2 + bx + c$ has two complex zeros given by

$$x = \frac{-b \pm \sqrt{b^2 - 4ac}}{2a}$$

EXAMPLE 1
Using the Discriminant

Without solving the equations, determine how many real zeros each of the following has.

(a) $f(x) = 4x^2 - 20x + 25$ (b) $f(x) = 13x^2 + 7x + 1$
(c) $f(x) = 5x^2 - 8x$

Solution:

(a) By setting $f(x)$ equal to zero, we have

$$4x^2 - 20x + 25 = 0$$

Now applying the quadratic formula, we have $a = 4$, $b = -20$, and $c = 25$, with a discriminant of

$$b^2 - 4ac = 400 - 4(4)(25) = 400 - 400 = 0$$

Since the discriminant is zero, there is one double zero.

(b) In this case, $a = 13$, $b = 7$, and $c = 1$, with

$$b^2 - 4ac = 49 - 4(13)(1) = 49 - 52 = -3 < 0$$

Since the discriminant is negative, there are *no* real zeros.

(c) Setting $f(x)$ equal to zero, we have

$$5x^2 - 8x = 0$$

with $a = 5$, $b = -8$, and $c = 0$. Thus,

$$b^2 - 4ac = 64 - 4(5)(0) = 64 > 0$$

Since the discriminant is positive, there are *two* real zeros.

Remark: Recall the connection between zeros and roots. That is, the zeros of a function f correspond to the roots of the equation $f(x) = 0$. In Example 1 we classified the *zeros* of a quadratic function, whereas in our next example we find the *roots* of a quadratic equation.

EXAMPLE 2
Using the Quadratic Formula to
Find Complex Roots

Solve by using the quadratic formula. If the discriminant is negative, list the complex roots in $a + bi$ form.

$$6x^2 - 2x + 5 = 0$$

Solution:
We have $a = 6$, $b = -2$, $c = 5$, and

$$x = \frac{-b \pm \sqrt{b^2 - 4ac}}{2a} = \frac{-(-2) \pm \sqrt{4 - 4(6)(5)}}{2(6)}$$

$$= \frac{2 \pm \sqrt{-116}}{12} = \frac{2 \pm 2\sqrt{-29}}{12}$$

In $a + bi$ form, the complex roots are

$$x = \frac{1}{6} + \frac{\sqrt{29}}{6}\, i \quad \text{and} \quad x = \frac{1}{6} - \frac{\sqrt{29}}{6}\, i$$

Notice that the two complex zeros in Example 2 are conjugates. That is, they are of the form

$$a + bi \quad \text{and} \quad a - bi$$

This is not a coincidence. In fact, the following result tells us that if a polynomial (with real coefficients) has one complex zero $(a + bi)$, then it *must* also have the conjugate $(a - bi)$ as a zero.

COMPLEX ZEROS OCCUR IN CONJUGATE PAIRS

> If $a + bi$ $(b \neq 0)$ is a complex zero of a polynomial function f, with real coefficients, then the conjugate $a - bi$ is also a complex zero of f.

Proof:
We begin by letting $d(x)$ be defined by

$$d(x) = [x - (a + bi)][x - (a - bi)] = x^2 - 2ax + (a^2 + b^2)$$

By the Division Algorithm, we have

$$f(x) = d(x) \cdot q(x) + r(x)$$

where the degree of $r(x)$ is less than the degree of $d(x)$. This implies that

$$r(x) = cx + d$$

Since $f(x)$ and $d(x)$ have real coefficients, the division process for $f(x)/d(x)$ will yield only real coefficients in the quotient and remainder. In particular, we know that c and d are real numbers. Now, since $f(a + bi) = 0$ and $d(a + bi) = 0$, we have $r(a + bi) = 0$, which implies that

$$c(a + bi) + d = (ca + d) + (cb)i = 0$$

and we see that

$$cb = 0 \qquad \text{and} \qquad ca + d = 0$$

Since $b \neq 0$, it follows that $c = 0$, which in turn implies that $d = 0$. Thus, we have

$$f(x) = d(x) \cdot q(x) = [x - (a + bi)][x - (a - bi)] \cdot q(x)$$

from which it follows that $f(a - bi) = 0$.

The preceding result has the following important consequence relating to factors of a polynomial.

FACTORS OF A POLYNOMIAL

> Every polynomial of degree $n > 0$ with real coefficients can be written as the product of linear and quadratic factors with real coefficients, where the quadratic factors have no real zeros.

Proof:
This proof follows quite easily using complex conjugates. To begin, we know that if $f(x)$ is a polynomial of degree n, then it can be *completely* factored in the form

$$f(x) = A(x - c_1)(x - c_2)(x - c_3) \cdots (x - c_n)$$

If each c_i is real, there is nothing more to prove. If any c_i is complex, say

$$c_i = a + bi, \qquad b \neq 0$$

then because the coefficients of $f(x)$ are real we know that the conjugate must also appear as a zero:

$$c_j = a - bi$$

By multiplying these two factors, we obtain the quadratic form

$$(x - c_i)(x - c_j) = [x - (a + bi)][x - (a - bi)]$$
$$= x^2 - 2ax + (a^2 + b^2)$$

where each coefficient is real. This completes the proof.

Remark: A quadratic factor with no real zeros is said to be **irreducible over the reals**. Be sure you see that this is not the same as being *irreducible over the rationals*. For example, the quadratic $x^2 + 1 = (x - i)(x + i)$ is irreducible over the reals (and the rationals). The quadratic $x^2 - 2 = (x - \sqrt{2})(x + \sqrt{2})$ is irreducible over the rationals, but *not* over the reals.

EXAMPLE 3
Factoring a Polynomial

Write the polynomial

$$f(x) = x^4 - x^2 - 20$$

(a) as the product of factors that are irreducible over the *rationals*
(b) as the product of linear factors and quadratic factors that are irreducible over the *reals*
(c) in completely factored form

Solution:

(a) We begin by factoring this fourth degree polynomial into the product of two quadratics, as follows:

$$x^4 - x^2 - 20 = (x^2 - 5)(x^2 + 4)$$

Both of these factors are irreducible over the rationals.
(b) By factoring over the reals, we have

$$x^4 - x^2 - 20 = (x + \sqrt{5})(x - \sqrt{5})(x^2 + 4)$$

Note that the quadratic factor is irreducible over the reals.
(c) In completely factored form, we have

$$x^4 - x^2 - 20 = (x + \sqrt{5})(x - \sqrt{5})(x + 2i)(x - 2i)$$

The polynomial in Example 3 was simple enough to factor directly over the reals and subsequently over the complex numbers. However, for more complicated polynomials factorization by inspection is difficult. In the next example we use synthetic division to factor polynomials with complex zeros.

EXAMPLE 4
Using Synthetic Division
with Complex Zeros

Find all zeros of

$$f(x) = x^4 - 3x^3 + 6x^2 + 2x - 60$$

given that $1 + 3i$ is a zero.

Solution:
Since complex zeros occur in pairs, we divide by $1 + 3i$ and $1 - 3i$:

$1 + 3i$	1	-3	6	2	-60
		$1 + 3i$	$-11 - 3i$	$4 - 18i$	60
	1	$-2 + 3i$	$-5 - 3i$	$6 - 18i$	0
$1 - 3i$	1	$-2 + 3i$	$-5 - 3i$	$6 - 18i$	
		$1 - 3i$	$-1 + 3i$	$-6 + 18i$	
	1	-1	-6	0	

The resulting quadratic factors as

$$x^2 - x - 6 = (x - 3)(x + 2)$$

Therefore, the four zeros of

$$f(x) = x^4 - 3x^3 + 6x^2 + 2x - 60$$

are $-2, 3, 1 + 3i$, and $1 - 3i$.

Occasionally, we want to create polynomials with certain known zeros. For example, to create a polynomial that has $1, -2$, and 5 as zeros, we can write

$$f(x) = (x - 1)(x + 2)(x - 5) = x^3 - 4x^2 - 7x + 10$$

Since complex zeros occur in conjugate pairs, we can create a polynomial given only one part of a conjugate complex pair. For example, suppose that we wanted to find a fourth degree polynomial with real coefficients that has $3 - i$ and $4i$ as complex zeros. Since the conjugate of each of these complex zeros must also be a zero of the polynomial, we write the following factored form:

$$f(x) = [x - (3 - i)][x - (3 + i)][x - (4i)][x - (-4i)]$$
$$= [x^2 - 6x + (9 + 1)][x^2 + 16]$$

which yields the polynomial

$$f(x) = x^4 - 6x^3 + 26x^2 - 96x + 160$$

Section Exercises 3.5

In Exercises 1–8, use the discriminant to determine how many real roots each quadratic equation has.

1. $4x^2 - 4x + 1 = 0$

2. $2x^2 - x - 1 = 0$

3. $3x^2 + 4x + 1 = 0$

4. $x^2 + 2x + 4 = 0$

5. $2x^2 - 5x + 5 = 0$

6. $3x^2 - 6x + 3 = 0$

7. $\frac{1}{5}x^2 + \frac{6}{5}x - 8 = 0$

8. $\frac{1}{3}x^2 - 5x + 25 = 0$

In Exercises 9–20, solve the equation by using the quadratic formula. If the discriminant is negative, list the complex roots in $a + bi$ form.

9. $x^2 - 5 = 0$

10. $3x^2 - 1 = 0$

11. $(x + 5)^2 - 6 = 0$

12. $16 - x^2 = 0$

13. $x^2 - 8x + 16 = 0$

14. $4x^2 + 4x + 1 = 0$

15. $x^2 + 2x + 5 = 0$

16. $54 + 16x - x^2 = 0$

17. $4x^2 - 4x + 5 = 0$

18. $4x^2 - 4x + 21 = 0$

19. $230 + 20x - 0.5x^2 = 0$

20. $6 - (x - 1)^2 = 0$

In Exercises 21–30, find an equation that has the given roots.

21. $1, 5i, -5i$

22. $4, 3i, -3i$

23. $2, 4 + i, 4 - i$

24. $6, -5 + 2i, -5 - 2i$

25. $i, -i, 6i, -6i$

26. $2, 2, 2, 4i, -4i$

27. $-5, -5, 1 + \sqrt{3}i, 1 - \sqrt{3}i$

28. $\frac{2}{3}, -1, 3 + \sqrt{2}i, 3 - \sqrt{2}i$

29. $\frac{3}{4}, -2, -\frac{1}{2} + i, -\frac{1}{2} - i$

30. $0, 0, 4, 1 + i, 1 - i$

In Exercises 31–60, find all the zeros of the function and write the polynomial as a product of linear factors.

31. $f(x) = x^2 + 18x + 77$

32. $f(x) = x^2 - x + 56$

33. $f(x) = x^2 - 4x + 1$

34. $f(x) = x^2 + 10x + 23$

35. $f(x) = x^2 + 25$

36. $f(x) = x^2 + 144$

37. $f(x) = x^4 - 81$

38. $f(x) = x^4 - 625$

39. $f(x) = x^2 - 2x + 2$

40. $f(x) = x^3 - 3x^2 + 4x - 2$

41. $f(x) = x^3 - 6x^2 + 13x - 10$

42. $f(x) = x^3 - 2x^2 - 11x + 52$

43. $f(x) = x^3 - 3x^2 - 15x + 125$

44. $f(x) = x^3 + 11x^2 + 39x + 29$

45. $f(x) = x^3 + 24x^2 + 214x + 740$

46. $f(x) = 2x^3 - 5x^2 + 12x - 5$

47. $f(x) = 16x^3 - 20x^2 - 4x + 15$

48. $f(x) = 9x^3 - 15x^2 + 11x - 5$

49. $f(x) = x^3 - x + 6$

50. $f(x) = x^3 + 9x^2 + 27x + 27$

51. $f(x) = 5x^3 - 9x^2 + 28x + 6$

52. $f(x) = 3x^3 - 4x^2 + 8x + 8$

53. $f(x) = x^4 - 4x^3 + 8x^2 - 16x + 16$

54. $f(x) = x^4 + 6x^3 + 10x^2 + 6x + 9$

55. $f(x) = x^4 + 10x^2 + 9$

56. $f(x) = x^4 + 29x^2 + 100$

57. $f(x) = x^5 + x^4 - 13x^3 - 13x^2 + 36x + 36$

58. $f(x) = x^5 + 2x^4 + 10x^3 + 20x^2 + 9x + 18$

59. $f(x) = x^4 + 2x^3 - x^2 + 4x + 12$

60. $f(x) = x^5 - 8x^4 + 28x^3 - 56x^2 + 64x - 32$

In Exercises 61–64, write the polynomial (a) as the product of factors that are irreducible over the *rationals*, (b) as the product of linear and quadratic factors that are irreducible over the *reals*, and (c) in completely factored form.

61. $f(x) = x^4 + 6x^2 - 27$

62. $f(x) = x^4 - 2x^3 - 3x^2 + 12x - 18$
    (Hint: One factor is $x^2 - 6$.)

63. $f(x) = x^4 - 4x^3 + 5x^2 - 2x - 6$
    (Hint: One factor is $x^2 - 2x - 2$.)

64. $f(x) = x^4 - 3x^3 - x^2 - 12x - 20$
    (Hint: One factor is $x^2 + 4$.)

Rational Functions and Their Graphs   3.6

Just as a rational number can be written as the quotient of two integers, so can a **rational function** be written as the quotient of two polynomials. Specifically, a function f is **rational** if it has the form

$$f(x) = \frac{P(x)}{Q(x)}, \qquad Q(x) \neq 0$$

where $P(x)$ and $Q(x)$ are polynomials.

Characteristics of Rational Functions

The domain of a rational function is the set of all real numbers other than those that make the denominator equal to zero. For example, the domain of

$$f(x) = \frac{x - 1}{(x - 2)(x + 1)}$$

is the set of all real numbers other than $x = 2$ and $x = -1$.

Note some special information about a rational function

$$f(x) = \frac{P(x)}{Q(x)}$$

1. Unless specified otherwise, we will assume that $P(x)$ and $Q(x)$ have no factors in common. That is, $P(x)/Q(x)$ is in *reduced form*.

2. $f(x) = 0$ if and only if $P(x) = 0$. That is, a rational function has zeros at the zeros of its numerator.

3. The zeros of $Q(x)$ make $f(x)$ undefined.

Though the function $f(x) = 1/x$ is one of the simplest rational functions, we can learn some important general characteristics of rational functions by studying its graph.

Symmetry: The graph has symmetry about the origin. Hence, we need only calculate points for positive x-values.

Domain: Note that $x = 0$ is not in the domain of f. Thus, we must pay special attention to the functional values when x is near zero.

Intercepts: Since $x = 0$ is not in the domain of f, there is no y-intercept. Furthermore, since the numerator of $f(x)$ is never zero, there are no x-intercepts.

Table of Values: Note the behavior of $f(x)$ as x approaches 0 and as x approaches very large values.

x	1	0.5	0.1	0.01	0.001	0.0001	0.00001	$\rightarrow$	0
$f(x)$	1	2	10	100	1000	10,000	100,000	$\rightarrow$	∞

x	1	2	10	100	1000	10,000	100,000	$\rightarrow$	∞
$f(x)$	1	0.5	0.1	0.01	0.001	0.0001	0.00001	$\rightarrow$	0

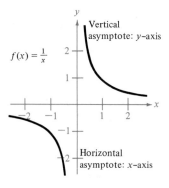

$f(x) = \frac{1}{x}$

Vertical asymptote: y–axis

Horizontal asymptote: x–axis

FIGURE 3.20

Using this table of values and information about symmetry and intercepts, we obtain the graph shown in Figure 3.20.

Note in Figure 3.20 and the accompanying table that the function values *increase without bound* as x approaches 0 from the right. Symbolically, we denote this behavior by

$$\frac{1}{x} \rightarrow \infty \quad \text{as} \quad x \rightarrow 0^+ \qquad \textit{(from the right)}$$

Similarly,

$$\frac{1}{x} \to -\infty \qquad \text{as} \qquad x \to 0^- \qquad\qquad \textit{(from the left)}$$

Under such conditions, the line $x = 0$ (the y-axis) is called a **vertical asymptote** of the graph of f.

When the function values become *arbitrarily close to zero* as x increases (or decreases) without bound, we denote this by

$$\frac{1}{x} \to 0 \qquad \text{as} \qquad x \to \infty \qquad\qquad \textit{(to the right)}$$

$$\frac{1}{x} \to 0 \qquad \text{as} \qquad x \to -\infty \qquad\qquad \textit{(to the left)}$$

Under these conditions, the line $y = 0$ (the x-axis) is called a **horizontal asymptote** of the graph of f.

VERTICAL AND HORIZONTAL ASYMPTOTES

The line $x = a$ is a **vertical asymptote** of the graph of f if

$$f(x) \to \infty \qquad \text{or} \qquad f(x) \to -\infty$$

as $x \to a$, either from the right or from the left.

The line $y = b$ is a **horizontal asymptote** of the graph of f if

$$f(x) \to b$$

as $x \to \infty$ or $x \to -\infty$.

Remark: Note from Figure 3.20 that a vertical asymptote occurs at an x-value for which $f(x)$ is not defined, and in general it is true that the graph of a function will never intersect its *vertical* asymptote. The graph of a function may intersect its horizontal asymptote any number of times. However, eventually (as $x \to \infty$ or $x \to -\infty$) the distance between the horizontal asymptote and the points on the graph must approach zero. Thus, an asymptote can be defined geometrically in the following way.

DEFINITION OF AN ASYMPTOTE

A line is called an **asymptote** to a curve if the distance between the line and a point (x, y) on the curve approaches zero as the distance between the origin and the point increases without bound.

The graph of a rational function may possess one or more asymptotes. We classify asymptotes as either vertical, horizontal, or slant, as shown in Figure 3.21.

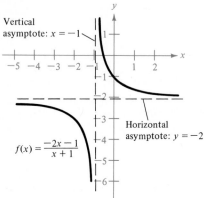

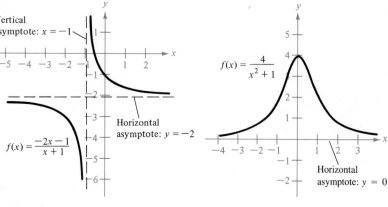

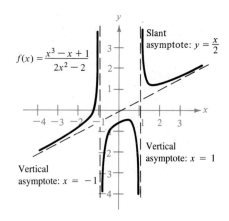

FIGURE 3.21   Asymptotes of Rational Functions

Asymptotes of Hyperbolas

The graph of $f(x) = 1/x$ (see Figure 3.20) is called an **hyperbola.** There are three basic types of rational functions whose graphs are hyperbolas.

$$f(x) = \frac{a}{Ax + B}, \quad g(x) = \frac{ax + b}{Ax + B}, \quad \text{and} \quad h(x) = \frac{ax^2 + bx + c}{Ax + B}$$

where $a \neq 0$ and $A \neq 0$. (We assume that $g(x)$ and $h(x)$ are given in reduced form.)

Each of these graphs has two asymptotes—a vertical asymptote and a horizontal (or slant) asymptote. We can find the vertical asymptote by setting the denominator equal to zero.

  Vertical Asymptote: $Ax + B = 0$

$$x = -\frac{B}{A}$$

To find the horizontal (or slant) asymptote, it is helpful to write the function in the form

$$px + q + \frac{C}{Ax + B}$$

We can see that the line $y = px + q$ is an asymptote of the graph, since

$$\frac{C}{Ax + B} \to 0 \quad \text{as} \quad x \to \pm\infty$$

  Horizontal or Slant Asymptote: $y = px + q$

Remark: The asymptote is horizontal if $p = 0$ and is a slant asymptote if $p \neq 0$.

You will see in Example 1 that if we find and plot the x- and y-intercepts and the asymptotes, we are well on the way to graphing a rational function.

EXAMPLE 1
Sketching the Graphs
of Rational Functions

Sketch the graphs of

(a) $f(x) = \dfrac{3}{x - 2}$ 　　　　　　(b) $f(x) = \dfrac{2x - 1}{x}$

Solution:
For $f(x) = 3/(x - 2)$, we have the following.

(a) *y-intercept:*　Since $f(0) = 3/(0 - 2) = -\frac{3}{2}$, the *y*-intercept occurs at $(0, -\frac{3}{2})$.

　　　　x-intercepts:　None.

　　Vertical Asymptote:　$x - 2 = 0$　*Zero in denominator*
　　　　　　　　　　　　　$x = 2$

　Horizontal Asymptote:　$y = 0$　　　*f(x) → 0 as x → ±∞*

　　Additional Points:

x	-4	1	3	5
$f(x)$	-0.5	-3	3	1

To sketch the graph of f, we begin by plotting the intercepts, asymptotes, and a few additional points, as shown in Figure 3.22. Then we can use these clues to make the final sketch shown in Figure 3.23.

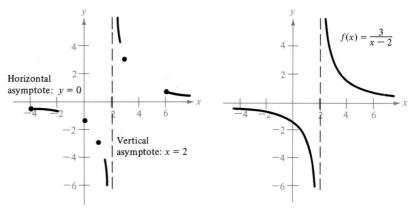

FIGURE 3.22　　　　　　　　　　　FIGURE 2.23

(b) For $f(x) = (2x - 1)/x$, we have the following.

　　　　y-intercept:　None.

　　　　x-intercept:　$2x - 1 = 0$

　　　　　　　　　　　$x = \dfrac{1}{2}$

Vertical Asymptote:　$x = 0$

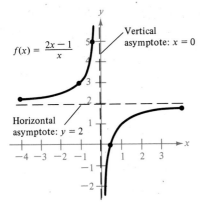

$f(x) = \dfrac{2x - 1}{x}$

Vertical asymptote: $x = 0$

Horizontal asymptote: $y = 2$

FIGURE 3.24

Horizontal Asymptote: By rewriting $f(x)$ as the sum of two fractions, we have

$$f(x) = \frac{2x - 1}{x} = 2 - \frac{1}{x}$$

In this form we see that $y = 2$ is a horizontal asymptote.

Additional Points:

x	-4	-1	4
$f(x)$	2.25	3	1.75

The graph of f is shown in Figure 3.24.

EXAMPLE 2
Graph with Slant Asymptote

Sketch the graph of

$$f(x) = \frac{x^2 - x}{x + 1}$$

Solution:

y-intercept: $(0, 0)$.

x-intercepts: $x^2 - x = 0$
$$x(x - 1) = 0$$
$$x = 0 \text{ or } 1$$

Vertical Asymptote: $x = -1$

Slant Asymptote: We begin by dividing the denominator into the numerator as follows:

$$
\begin{array}{r}
x - 2 \\
x + 1 \overline{)\, x^2 - x } \\
\underline{x^2 + x } \\
-2x \\
\underline{-2x - 2} \\
2
\end{array}
$$

Thus, we have

$$f(x) = \frac{x^2 - x}{x + 1} = x - 2 + \frac{2}{x + 1}$$

Now, we can see that the graph of f has the line $y = x - 2$ as a slant asymptote. The graph of f is shown in Figure 3.25.

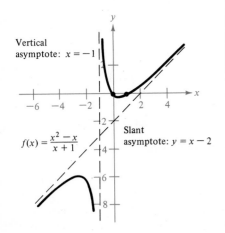

Vertical asymptote: $x = -1$

$f(x) = \dfrac{x^2 - x}{x + 1}$

Slant asymptote: $y = x - 2$

FIGURE 3.25

We can use the same approach for general rational functions that we have been using for functions that graph as hyperbolas. Use of the following guidelines should improve your skill in graphing rational functions of the form

$$f(x) = \frac{P(x)}{Q(x)}$$

For simplicity's sake, we assume that $P(x)$ and $Q(x)$ have no common factors.

GUIDELINES FOR GRAPHING RATIONAL FUNCTIONS

1. *Find and plot y-intercept* by evaluating $f(0)$.

2. *Find and plot x-intercept(s)* by setting the numerator $P(x)$ equal to zero and solving for x. Since $P(x)$ and $Q(x)$ have no common factors, all solutions to $P(x) = 0$ are in the domain of f and are legitimate x-intercepts of the graph.

3. *Find and plot the vertical asymptote(s)* by setting the denominator $Q(x)$ equal to zero and solving for x.

4. *Find and plot the horizontal (or slant) asymptotes* using the following guidelines:
   (a) If the degree of P is *less than* the degree of Q, then the graph has the x-axis ($y = 0$) as a horizontal asymptote.
   (b) If the degree of P is *equal* to the degree of Q, then the graph has a horizontal asymptote given by $y = a_n/b_m$, where a_n is the leading coefficient of P and b_m is the leading coefficient of Q.
   (c) If the degree of P is *one more than* the degree of Q, then the graph has a slant asymptote which can be determined by dividing $Q(x)$ into $P(x)$.
   (d) If the degree of P is *at least two more* than the degree of Q, then the graph has no horizontal or slant asymptotes.

5. *Plot a few additional points* both between and beyond any previously determined x-intercepts and vertical asymptotes. Remember to use any symmetry to the origin or y-axis.

6. *Complete the missing portions of the graph* in a manner that is consistent with the location of intercepts, asymptotes, and additional points.

Remark: The basic idea in applying step 4 is that we can determine the horizontal or slant asymptote of a rational function by comparing the degrees of the numerator and denominator. For instance, in part (a) of Example 1, the degree of the numerator was *less* than the degree of the denominator, thus yielding $y = 0$ as a horizontal asymptote. In part (b), the *equal* degrees yielded $y = 2$ as a horizontal asymptote. Then in Example 2, a slant asymptote was obtained because the degree of the numerator was *one more* than that of the denominator.

EXAMPLE 3
Sketching the Graph
of a Rational Function

Sketch the graph of

$$f(x) = \frac{2}{x^2 + 1}$$

Solution:

y-intercept: Since $f(0) = 2$, the y-intercept occurs at $(0, 2)$.

x-intercept: None.

Vertical Asymptotes: Since $x^2 + 1 = 0$ has no real solutions, there are no vertical asymptotes.

Horizontal Asymptotes: Since the degree of the numerator is less than the degree of the denominator, the x-axis is a horizontal asymptote. [Try to convince yourself that $f(x)$ approaches zero as x increases (or decreases) without bound.]

Additional Points: (Note that the graph is symmetrical with respect to the y-axis.)

x	1	2	3
$f(x)$	1	0.4	0.2

The graph of f is shown in Figure 3.26.

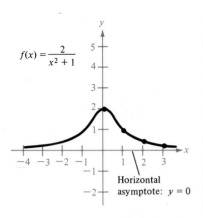

$f(x) = \dfrac{2}{x^2 + 1}$

Horizontal asymptote: $y = 0$

FIGURE 3.26

EXAMPLE 4
Sketching the Graph
of a Rational Function

Sketch the graph of

$$f(x) = \frac{x}{x^2 - x - 2}$$

Solution:

y-intercept: $(0, 0)$.

x-intercept: $(0, 0)$.

Vertical Asymptotes:
$$x^2 - x - 2 = 0$$
$$(x - 2)(x + 1) = 0$$
$$x = -1 \text{ or } 2$$

Horizontal Asymptote: Since the degree of the numerator is less than the degree of the denominator, the x-axis (the line $y = 0$) is a horizontal asymptote.

Additional Points: *Between* and *beyond* the x-intercepts and vertical asymptotes.

x	-3	-0.5	1	3
$f(x)$	-0.3	0.4	-0.5	0.75

The graph of f is shown in Figure 3.27.

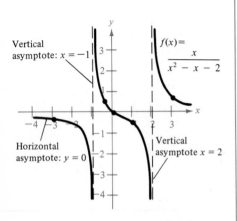

Vertical asymptote: $x = -1$

Horizontal asymptote: $y = 0$

$f(x) = \dfrac{x}{x^2 - x - 2}$

Vertical asymptote $x = 2$

FIGURE 3.27

EXAMPLE 5
Sketching the Graph
of a Rational Function

Sketch the graph of

$$f(x) = \frac{2(x^2 - 9)}{x^2 - 4}$$

Solution: *y-intercept:* $(0, \frac{9}{2})$.

$$\begin{aligned} \text{x-intercepts:} \quad 2(x^2 - 9) &= 0 \\ x^2 - 9 &= 0 \\ (x + 3)(x - 3) &= 0 \\ x &= -3 \text{ or } 3 \end{aligned}$$

Vertical Asymptotes:
$$\begin{aligned} x^2 - 4 &= 0 \\ (x + 2)(x - 2) &= 0 \\ x &= -2 \text{ or } 2 \end{aligned}$$

Horizontal Asymptote: Since the degree of the denominator is equal to the degree of the numerator, the horizontal asymptote is given by the ratio of the leading coefficients. That is, the horizontal asymptote is

$$y = \frac{2}{1} = 2$$

The graph has symmetry with respect to the *y*-axis and is shown in Figure 3.28.

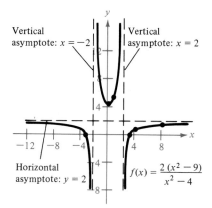

Vertical asymptote: $x = -2$

Vertical asymptote: $x = 2$

Horizontal asymptote: $y = 2$

$$f(x) = \frac{2(x^2 - 9)}{x^2 - 4}$$

FIGURE 3.28

EXAMPLE 6
Sketching the Graph of a
Function with a Slant Asymptote

Sketch the graph of

$$f(x) = \frac{-x^3 + x^2 + 4}{x^2}$$

Solution:

y-intercept: None.

x-intercept: Using the rational zero test, we discover that $x = 2$ is a zero of the numerator. Thus, by factoring $(x - 2)$ out of the numerator, we have

$$-x^3 + x^2 + 4 = 0$$
$$-(x - 2)(x^2 + x + 2) = 0$$

Using the quadratic formula on the quadratic factor, we can determine that the only real zero of the numerator is $x = 2$.

Vertical Asymptote: $x = 0$

Slant Asymptote: Since the degree of the numerator exceeds the degree of the denominator by exactly 1, we divide to obtain

$$f(x) = \frac{-x^3 + x^2 + 4}{x^2} = -x + 1 + \frac{4}{x^2}$$

which implies that the slant asymptote is

$$y = -x + 1$$

The graph of f is shown in Figure 3.29.

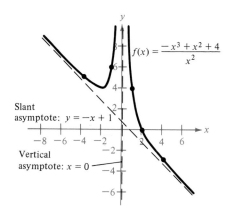

$$f(x) = \frac{-x^3 + x^2 + 4}{x^2}$$

Slant asymptote: $y = -x + 1$

Vertical asymptote: $x = 0$

FIGURE 3.29

If the numerator and denominator of a rational function have a common factor, we must modify our sketching technique as demonstrated in the following example.

EXAMPLE 7
A Rational Function
with Common Factors

Sketch the graph of

$$g(x) = \frac{2x^2 - 7x + 3}{x^2 - 3x}$$

Solution:
We begin by factoring the numerator and denominator as follows:

$$g(x) = \frac{2x^2 - 7x + 3}{x^2 - 3x} = \frac{(2x - 1)(x - 3)}{x(x - 3)}$$

Now, for all points *other than* $x = 3$, the graph of g is the same as the graph of

$$f(x) = \frac{2x - 1}{x}$$

We have already graphed this function in part (b) of Example 1. To sketch the graph of g, we must account for the fact that $x = 3$ is not in the domain of g. Graphically, we indicate that g is undefined when $x = 3$ by an *open dot* at the point $(3, \frac{5}{3})$, as shown in Figure 3.30.

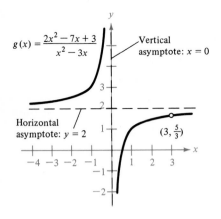

FIGURE 3.30

Remark: In Example 7, the common factor $(x - 3)$ occurred just once as a factor of the numerator and once as a factor of the denominator. By canceling this factor, we formed a new function, since we introduced a new value to the domain. If $(x - 3)$ had occurred as a *multiple* factor of the denominator, then cancellation might not have added a new value to the domain, and in such cases we would not have formed a new function. For example, the following functions both exclude $x = 1$ from their domains and are therefore equal functions:

$$f(x) = \frac{x - 1}{(x - 1)^2} \qquad g(x) = \frac{1}{x - 1}$$

Section Exercises 3.6

In Exercises 1–10, match the rational function with the correct graph.

1. $f(x) = \dfrac{2}{x + 1}$

2. $f(x) = \dfrac{1}{x - 4}$

3. $f(x) = \dfrac{x + 1}{x}$

4. $f(x) = \dfrac{1 - 2x}{x}$

5. $f(x) = \dfrac{2 - x}{x - 1}$

6. $f(x) = \dfrac{x + 2}{x + 1}$

7. $f(x) = \dfrac{x - 2}{x - 1}$

8. $f(x) = -\dfrac{x + 2}{x + 1}$

9. $f(x) = \dfrac{x^2 + 1}{x}$

10. $f(x) = \dfrac{x^2 - 2x}{x - 1}$

(a)

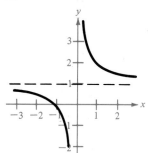

(b)

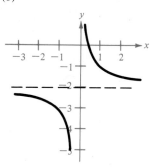

(i)

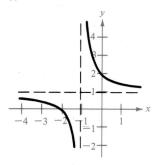

(j)

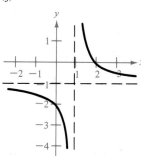

(c)

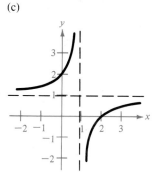

(d)

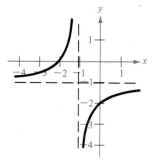

In Exercises 11–30, sketch the graph of the rational function, labeling all intercepts and asymptotes.

11. $f(x) = \dfrac{1}{x + 2}$

12. $f(x) = \dfrac{1}{x - 3}$

13. $f(x) = \dfrac{-1}{x + 2}$

14. $f(x) = \dfrac{1}{3 - x}$

15. $f(x) = \dfrac{1}{x + 2} + 2$

16. $f(x) = \dfrac{1}{x - 3} + 1$

17. $f(x) = \dfrac{x + 1}{x + 2}$

18. $f(x) = \dfrac{x - 2}{x - 3}$

19. $f(x) = \dfrac{2 + x}{1 - x}$

20. $f(x) = \dfrac{3 - x}{2 - x}$

21. $f(x) = \dfrac{3x + 1}{x}$

22. $f(x) = \dfrac{1 - 2x}{x}$

23. $f(x) = \dfrac{5 + 2x}{1 + x}$

24. $f(x) = \dfrac{1 - 3x}{1 - x}$

25. $f(x) = \dfrac{2x^2 + 1}{x}$

26. $f(x) = \dfrac{1 - x^2}{x}$

27. $f(x) = \dfrac{x^2 - x + 1}{x - 1}$

28. $f(x) = \dfrac{x^2}{x - 1}$

29. $f(x) = \dfrac{x^2 + 5x + 8}{x + 3}$

30. $f(x) = \dfrac{2x^2 + x}{x + 1}$

(e)

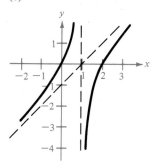

(f)

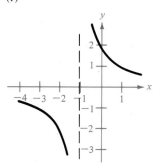

In Exercises 31–40, find all the asymptotes of the rational function.

31. $f(x) = \dfrac{1}{x^2}$

32. $f(x) = \dfrac{4}{(x - 2)^3}$

33. $f(x) = \dfrac{2 + x}{2 - x}$

34. $f(x) = \dfrac{1 - 5x}{1 + 2x}$

35. $f(x) = \dfrac{x^3}{x^2 - 1}$

36. $f(x) = \dfrac{2x^2}{x + 1}$

37. $f(x) = \dfrac{3x^2 + 1}{x^2 + 9}$

38. $f(x) = \dfrac{3x^2 + x - 5}{x^2 + 1}$

39. $f(x) = \dfrac{5x^4 - 2x^2 + 1}{x^2 + 1}$

40. $f(x) = \dfrac{x^5}{2x^3 + 1}$

(g)

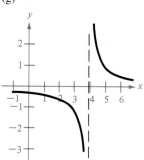

(h)

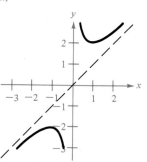

In Exercises 41–60, sketch the graph of the rational function. As a sketching aid, examine the function for intercepts, symmetry, and asymptotes.

41. $f(x) = \dfrac{2 + x}{1 - x}$

42. $f(x) = \dfrac{x - 3}{x - 2}$

43. $f(x) = \dfrac{x^2}{x^2 + 9}$

44. $f(x) = 2 - \dfrac{3}{x^2}$

45. $f(x) = \dfrac{x^2}{x^2 - 9}$

46. $f(x) = \dfrac{x^2 + 1}{x}$

47. $f(x) = \dfrac{x^3}{x^2 - 1}$

48. $f(x) = \dfrac{x^3}{2x^2 - 8}$

49. $f(x) = \dfrac{x}{x^2 + 1}$

50. $f(x) = \dfrac{2x}{x^2 + x - 2}$

51. $f(x) = \dfrac{2x}{1 - x^2}$

52. $f(x) = \dfrac{2x^2 - 5x + 5}{x - 2}$

53. $f(x) = \dfrac{3x}{x^2 - x - 2}$

54. $f(x) = \dfrac{x^2 + x - 2}{x^2 - 2x + 1}$

55. $f(x) = \dfrac{x - 3}{x^2 - 9}$

56. $f(x) = \dfrac{x + 2}{x^2 - 3x - 10}$

57. $f(x) = \dfrac{x^2 - 25}{x + 5}$

58. $f(x) = \dfrac{2x^2 - x - 3}{x + 1}$

59. $f(x) = \dfrac{x^2}{1 + x^4}$

60. $f(x) = \dfrac{4}{1 + x^2}$

61. A right triangle is formed in the first quadrant by the x and y axes and a line segment through the point $(2, 3)$. (See Figure 3.31.)

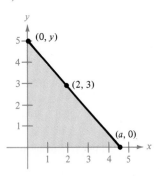

FIGURE 3.31

(a) Show that an equation of the line segment is

$$y = \frac{3(x - a)}{2 - a}$$

(b) Show that the area of the triangle is

$$A = \frac{-3a^2}{2(2 - a)}$$

(c) Sketch the graph of the area function of part (b), and from the graph estimate the value of a ($a > 0$) so that the area is minimum.

62. The cost of producing x units is $C = 0.2x^2 + 10x + 5$, and therefore the average cost per unit is

$$\overline{C} = \frac{C}{x} = 0.2x + 10 + \frac{5}{x}$$

Sketch the graph of the average cost function, and estimate the number of units that should be produced to minimize the average cost.

63. The region shown in Figure 3.32 is bounded by $y = x^2$, $y = 0$, and $x = 2$. It can be shown by techniques of calculus that the area A of this region is approximated by

$$f(n) = \frac{4}{3} \cdot \frac{2n^3 + 3n^2 + n}{n^3}$$

As n increases without bound, $f(n)$ approaches the exact area of the region [i.e., A can be obtained from the horizontal asymptote of $f(n)$]. Find A.

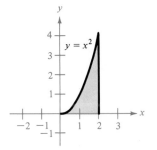

FIGURE 3.32

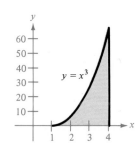

FIGURE 3.33

64. The region shown in Figure 3.33 is bounded by $y = x^3$, $y = 0$, $x = 1$, and $x = 4$. It can be shown by techniques of calculus that the area A of this region is approximated by

$$f(n) = 3 + \frac{27}{2} \cdot \frac{n(n + 1)}{n^2}$$
$$+ \frac{27}{2} \cdot \frac{n(n + 1)(2n + 1)}{n^3}$$
$$+ \frac{81}{4} \cdot \frac{n^2(n + 1)^2}{n^4}$$

As n increases without bound, $f(n)$ approaches the exact area of the region [i.e., A can be obtained from the horizontal asymptote of $f(n)$]. Find A.

Partial Fractions 3.7

In calculus it is often useful to rewrite a rational expression as the sum of simpler rational expressions. For example, the rational expression $(x + 7)/(x^2 - x - 6)$ can be written as the sum of two fractions with linear denominators, as follows:

$$\frac{x + 7}{x^2 - x - 6} = \frac{2}{x - 3} + \frac{-1}{x + 2}$$

The fractions on the right side are called **partial fractions,** and the entire right side is called the **partial fraction decomposition** of the left side.

In Section 3.5 we noted that it is theoretically possible to write any polynomial as the product of linear and irreducible quadratic factors. For instance,

$$x^5 + x^4 - x - 1 = (x - 1)(x + 1)^2(x^2 + 1)$$

where $(x - 1)$ is a linear factor, $(x + 1)^2$ is a repeated linear factor, and $(x^2 + 1)$ is an irreducible quadratic factor.

We can use this factorization to find the partial fraction decomposition of any rational expression having $x^5 + x^4 - x - 1$ as its denominator. Specifically, if $N(x)$ is a polynomial of degree less than five, then the partial fraction decomposition of $N(x)/(x^5 + x^4 - x - 1)$ has the form

$$\frac{N(x)}{x^5 + x^4 - x - 1} = \frac{N(x)}{(x - 1)(x + 1)^2(x^2 + 1)}$$

$$= \frac{A}{x - 1} + \frac{B}{x + 1} + \frac{C}{(x + 1)^2} + \frac{Dx + F}{x^2 + 1}$$

Note that the repeated linear factor $(x + 1)^2$ results in *two* fractions, one for $(x + 1)$ and one for $(x + 1)^2$. If $(x + 1)^3$ were a factor, then we would use three fractions: one for $(x + 1)$, one for $(x + 1)^2$, and one for $(x + 1)^3$. In general, the number of fractions resulting from a repeated linear factor is equal to the number of times the factor is repeated.

DECOMPOSITION OF $N(x)/D(x)$ INTO PARTIAL FRACTIONS

1. *Divide if improper:* If $N(x)/D(x)$ is an improper fraction, then divide the denominator into the numerator to obtain

$$\frac{N(x)}{D(x)} = \text{(a polynomial)} + \frac{N_1(x)}{D(x)}$$

and apply steps 2, 3, and 4 to the proper rational expression $N_1(x)/D(x)$.

2. *Factor denominator:* Completely factor the denominator into factors of the form

$$(px + q)^m \quad \text{and} \quad (ax^2 + bx + c)^n$$

where $(ax^2 + bx + c)$ is irreducible.

3. *Linear factors:* For *each* factor of the form $(px + q)^m$, the partial fraction decomposition must include the following sum of m fractions:

$$\frac{A_1}{(px + q)} + \frac{A_2}{(px + q)^2} + \cdots + \frac{A_m}{(px + q)^m}$$

4. *Quadratic factors:* For *each* factor of the form $(ax^2 + bx + c)^n$, the partial fraction decomposition must include the following sum of n fractions:

$$\frac{B_1x + C_1}{ax^2 + bx + c} + \frac{B_2x + C_2}{(ax^2 + bx + c)^2} + \cdots$$

$$+ \frac{B_nx + C_n}{(ax^2 + bx + c)^n}$$

Algebraic techniques for determining the constants in the numerators are demonstrated in the examples that follow.

EXAMPLE 1
Distinct Linear Factors

Write the partial fraction decomposition for

$$\frac{x + 7}{x^2 - x - 6}$$

Solution:
Since

$$x^2 - x - 6 = (x - 3)(x + 2)$$

we include one partial fraction for each factor and write

$$\frac{x + 7}{x^2 - x - 6} = \frac{A}{x - 3} + \frac{B}{x + 2}$$

Multiplying this equation by the lowest common denominator (LCD), $(x - 3)(x + 2)$, leads to the **basic equation**

$$x + 7 = A(x + 2) + B(x - 3)$$

Since this equation is to be true for all x, we can substitute *convenient* values for x to obtain equations in A and B. These values are the ones that make particular factors zero. To solve for B, we let $x = -2$ and obtain

$$-2 + 7 = A(0) + B(-5)$$
$$5 = -5B$$
$$-1 = B$$

To solve for A, We let $x = 3$ and obtain

$$3 + 7 = A(5) + B(0)$$
$$10 = 5A$$
$$2 = A$$

Therefore, the decomposition is

$$\frac{x + 7}{x^2 - x - 6} = \frac{2}{x - 3} - \frac{1}{x + 2}$$

EXAMPLE 2
Repeated Linear Factors

Write the partial fraction decomposition for

$$\frac{5x^2 + 20x + 6}{x^3 + 2x^2 + x}$$

Solution:
Since

$$x^3 + 2x^2 + x = x(x^2 + 2x + 1) = x(x + 1)^2$$

we include one fraction for each power of x and $(x + 1)$ and write

$$\frac{5x^2 + 20x + 6}{x(x + 1)^2} = \frac{A}{x} + \frac{B}{x + 1} + \frac{C}{(x + 1)^2}$$

Multiplying by the LCD, $x(x + 1)^2$, leads to the *basic* equation

$$5x^2 + 20x + 6 = A(x + 1)^2 + Bx(x + 1) + Cx$$

Let $x = -1$ to eliminate the A and B terms:

$$5 - 20 + 6 = 0 + 0 - C$$
$$C = 9$$

Let $x = 0$ to eliminate the B and C terms:

$$6 = A(1) + 0 + 0$$
$$6 = A$$

We have exhausted the most convenient choices for x, so to find the value of B, we use *any other value* for x along with the calculated values of A and C. Thus, using $x = 1$, $A = 6$, and $C = 9$, we have

$$5 + 20 + 6 = A(4) + B(2) + C$$
$$31 = 6(4) + 2B + 9$$
$$-2 = 2B$$
$$-1 = B$$

Therefore,

$$\frac{5x^2 + 20x + 6}{x(x + 1)^2} = \frac{6}{x} - \frac{1}{x + 1} + \frac{9}{(x + 1)^2}$$

Remark: You need to make as many substitutions for x as there are unknowns $(A, B, C, \ldots)$.

EXAMPLE 3
Distinct Linear and
Quadratic Factors

Write the partial fraction decomposition for

$$\frac{2x^3 - 4x - 8}{(x^2 - x)(x^2 + 4)}$$

Solution:

Since

$$(x^2 - x)(x^2 + 4) = x(x - 1)(x^2 + 4)$$

we include one partial fraction for each factor and write

$$\frac{2x^3 - 4x - 8}{x(x - 1)(x^2 + 4)} = \frac{A}{x} + \frac{B}{x - 1} + \frac{Cx + D}{x^2 + 4}$$

Multiplying by the LCD, $[x(x - 1)(x^2 + 4)]$, yields the *basic* equation

$$2x^3 - 4x - 8 = A(x - 1)(x^2 + 4)$$
$$+ Bx(x^2 + 4) + (Cx + D)(x)(x - 1)$$

If $x = 1$, then

$$-10 = 0 + B(5) + 0$$
$$-2 = B$$

If $x = 0$, then

$$-8 = A(-1)(4) + 0 + 0$$
$$2 = A$$

At this point C and D are yet to be determined. We can find these remaining constants by comparing coefficients in the basic equation. We proceed as follows:

1. Expand the basic equation:

$$2x^3 - 4x - 8 = A(x - 1)(x^2 + 4) + Bx(x^2 + 4)$$
$$+ (Cx + D)(x)(x - 1)$$
$$= Ax^3 - Ax^2 + 4Ax - 4A + Bx^3$$
$$+ 4Bx + Cx^3 + Dx^2 - Cx^2 - Dx$$

2. Collect like terms:

$$2x^3 - 4x - 8 = (A + B + C)x^3 + (-A + D - C)x^2$$
$$+ (4A + 4B - D)x - 4A$$

3. Equate coefficients of like powers on opposite sides:

$$2x^3 + 0x^2 - 4x - 8 = (A + B + C)x^3 + (-A + D - C)x^2$$
$$+ (4A + 4B - D)x - 4A$$

Thus, we have

$$2 = A + B + C, \qquad 0 = -A + D - C,$$
$$-4 = 4A + 4B - D, \qquad -8 = -4A$$

Substituting the known values of $A = 2$ and $B = -2$, we have

$$2 = 2 - 2 + C \qquad -4 = 4(2) + 4(-2) - D$$
$$2 = C \qquad\qquad\qquad 4 = D$$

Finally, we have

$$\frac{2x^3 - 4x - 8}{x(x-1)(x^2+4)} = \frac{A}{x} + \frac{B}{x-1} + \frac{Cx + D}{x^2 + 4}$$

$$= \frac{2}{x} - \frac{2}{x-1} + \frac{2x}{x^2+4} + \frac{4}{x^2+4}$$

In each of the first three examples, we began the solution of the basic equation by substituting values of x that made the linear factors zero. This method works well when the partial fraction decomposition involves *only* linear factors. However, if the decomposition involves a quadratic factor, then the alternative procedure (comparing coefficients in the basic equation) proves to be quite efficient. We suggest the latter procedure for partial fractions involving quadratic denominators. Both methods are outlined in the following summary.

GUIDELINES FOR SOLVING THE BASIC EQUATION

Linear Factors:

1. Substitute the *roots* of the distinct linear factors into the basic equation.

2. For repeated linear factors, use the coefficients determined in part 1 to rewrite the basic equation. Then substitute *other* convenient values for x and solve for the remaining coefficients.

Quadratic Factors:

1. Expand the basic equation.

2. Collect terms according to powers of x.

3. Equate the coefficients of like powers to obtain equations involving A, B, C, etc.

4. Use substitution to solve for $A, B, C, \ldots$.

The second procedure for solving the basic equation is demonstrated in the next example.

EXAMPLE 4
Repeated Quadratic Factors

Write the partial fraction decomposition for

$$\frac{8x^3 + 13x}{(x^2 + 2)^2}$$

Solution:

We include one partial fraction for each power of $(x^2 + 2)$,

$$\frac{8x^3 + 13x}{(x^2 + 2)^2} = \frac{Ax + B}{x^2 + 2} + \frac{Cx + D}{(x^2 + 2)^2}$$

Multiplying by the LCD, $(x^2 + 2)^2$, yields the *basic* equation

$$
\begin{aligned}
8x^3 + 13x &= (Ax + B)(x^2 + 2) + Cx + D \\
&= Ax^3 + 2Ax + Bx^2 + 2B + Cx + D \\
&= Ax^3 + Bx^2 + (2A + C)x + (2B + D)
\end{aligned}
$$

Equating coefficients of like terms,

$$8x^3 + 0x^2 + 13x + 0 = Ax^3 + Bx^2 + (2A + C)x + (2B + D)$$

we have

$$8 = A, \quad 0 = B, \quad 13 = 2A + C, \quad 0 = 2B + D$$

Using the known values $A = 8$ and $B = 0$, we have

$$13 = 2A + C = 2(8) + C \qquad 0 = 2B + D = 2(0) + D$$
$$-3 = C \qquad\qquad\qquad\qquad 0 = D$$

Finally, we conclude that

$$\frac{8x^3 + 13x}{(x^2 + 2)^2} = \frac{8x}{x^2 + 2} + \frac{-3x}{(x^2 + 2)^2}$$

When working the exercises, you should keep in mind that for *improper* rational expressions like

$$\frac{N(x)}{D(x)} = \frac{2x^3 + x^2 - 7x + 7}{x^2 + x - 2}$$

you must first divide to obtain the form

$$\frac{N(x)}{D(x)} = (\text{polynomial}) + \frac{N_1(x)}{D(x)}$$

The proper rational expression $N_1(x)/D(x)$ is then decomposed into its partial fractions by the usual methods.

Section Exercises 3.7

In Exercises 1–30, write the partial fraction decomposition for the rational expression.

1. $\dfrac{1}{x^2 - 1}$

2. $\dfrac{1}{4x^2 - 9}$

3. $\dfrac{3}{x^2 + x - 2}$

4. $\dfrac{x + 1}{x^2 + 4x + 3}$

5. $\dfrac{5 - x}{2x^2 + x - 1}$

6. $\dfrac{3x^2 - 7x - 2}{x^3 - x}$

7. $\dfrac{x^2 + 12x + 12}{x^3 - 4x}$

8. $\dfrac{x^3 - x + 3}{x^2 + x - 2}$

9. $\dfrac{2x^3 - 4x^2 - 15x + 5}{x^2 - 2x - 8}$

10. $\dfrac{x + 2}{x(x - 4)}$

11. $\dfrac{4x^2 + 2x - 1}{x^2(x + 1)}$

12. $\dfrac{2x - 3}{(x - 1)^2}$

13. $\dfrac{x^4}{(x - 1)^3}$

14. $\dfrac{4x^2 - 1}{2x(x + 1)^2}$

15. $\dfrac{3x}{(x - 3)^2}$

16. $\dfrac{6x^2 + 1}{x^2(x - 1)^3}$

17. $\dfrac{x^2 - 1}{x(x^2 + 1)}$

18. $\dfrac{x}{(x - 1)(x^2 + x + 1)}$

19. $\dfrac{x^2}{x^4 - 2x^2 - 8}$

20. $\dfrac{2x^2 + x + 8}{(x^2 + 4)^2}$

21. $\dfrac{x}{16x^4 - 1}$

22. $\dfrac{x^2 - 4x + 7}{(x + 1)(x^2 - 2x + 3)}$

23. $\dfrac{x^2 + x + 2}{(x^2 + 2)^2}$

24. $\dfrac{x^3}{(x + 2)^2(x - 2)^2}$

25. $\dfrac{x^2 + 5}{(x + 1)(x^2 - 2x + 3)}$

26. $\dfrac{x - 1}{x^3 + x^2}$

27. $\dfrac{x + 1}{x^3 + x}$

28. $\dfrac{x^2 - x}{x^2 + x + 1}$

29. $\dfrac{1}{y(L - y)}$, L is a constant

30. $\dfrac{1}{(x + 1)(n - x)}$, n is a positive integer

Review Exercises / Chapter 3

In Exercises 1–16, analyze the equation and sketch its graph.

1. $y = \dfrac{x}{x^2 + 1}$

2. $y = \dfrac{2x}{x^2 + 4}$

3. $y = \dfrac{x^2}{x^2 + 1}$

4. $y = \dfrac{2x^2}{x^2 + 4}$

5. $y = \dfrac{x}{x^2 - 1}$

6. $y = \dfrac{2x}{x^2 - 4}$

7. $y = \dfrac{2x^2}{x^2 - 4}$

8. $y = \dfrac{x^2 + 1}{x + 1}$

9. $y = \dfrac{2x^3}{x^2 + 1}$

10. $xy - 4 = 0$

11. $5y^2 - 4x^2 = 20$

12. $y^2 - 12y - 8x + 20 = 0$

13. $x^2 - 6x + 2y + 9 = 0$

14. $4x^2 + y^2 - 16x + 15 = 0$

15. $x^2 + y^2 - 2x - 4y + 5 = 0$

16. $16x^2 + 16y^2 - 16x + 24y - 3 = 0$

In Exercises 17–22, write the partial fraction decomposition for the rational expression.

17. $\dfrac{x^2}{x^2 + 2x - 15}$

18. $\dfrac{9}{x^2 - 9}$

19. $\dfrac{x^2 + 2x}{x^3 - x^2 + x - 1}$

20. $\dfrac{4x - 2}{3(x - 1)^2}$

21. $\dfrac{3x^3 + 4x}{(x^2 + 1)^2}$

22. $\dfrac{4x^2}{(x - 1)(x^2 + 1)}$

In Exercises 23–34, perform the indicated operations and write the result in standard form.

23. $(7 + 5i) + (-4 + 2i)$

24. $-(6 - 2i) + (-8 + 3i)$

25. $\left(\dfrac{\sqrt{2}}{2} - \dfrac{\sqrt{2}}{2}i\right) - \left(\dfrac{\sqrt{2}}{2} + \dfrac{\sqrt{2}}{2}i\right)$

26. $(13 - 8i) - 5i$

27. $5i(13 - 8i)$

28. $(1 + 6i)(5 - 2i)$

29. $(10 - 8i)(2 - 3i)$

30. $i(6 + i)(3 - 2i)$

31. $\dfrac{6 + i}{i}$

32. $\dfrac{3 + 2i}{5 + i}$

33. $\dfrac{4}{(-3i)}$

34. $\dfrac{1}{(2 + i)^4}$

In Exercises 35–40, divide the first polynomial by the second by using long division.

35. $x^4 - 3x^2 + 2, x^2 - 1$

36. $3x^4, x^2 - 1$

37. $3x^5 - 4x^3 + x^2 + 5, x^2 - 2x + 4$
38. $4x + 7, 3x - 2$
39. $3x^3 + 2x - 1, 2x^3 + x$
40. $-2x^4 - 4x^3 + 3x^2 - 8, x^2 + 4x + 4$

In Exercises 41–46, use synthetic division to divide the first polynomial by the second.

41. $\frac{1}{4}x^4 - 4x^3, x - 2$
42. $2x^3 + 2x^2 - x + 2, x - \frac{1}{2}$
43. $6x^4 - 4x^3 - 27x^2 + 18x, x - \frac{2}{3}$
44. $0.1x^3 + 0.3x^2 - 0.5, x - 5$
45. $2x^3 - 5x^2 + 12x - 5, x - (1 + 2i)$
46. $9x^3 - 15x^2 + 11x - 5, x - (\frac{1}{3} + \frac{2}{3}i)$

In Exercises 47–52, find all zeros of the function.

47. $f(x) = 4x^3 - 11x^2 + 10x - 3$
48. $f(x) = 10x^3 + 21x^2 - x - 6$
49. $f(x) = 6x^3 - 5x^2 + 24x - 20$
50. $f(x) = x^3 - 1.3x^2 - 1.7x + 0.6$
51. $f(x) = 6x^4 - 25x^3 + 14x^2 + 27x - 18$
52. $f(x) = 5x^4 + 126x^2 + 25$

53. Find a fourth degree polynomial with zeros -1, -1, $\frac{1}{3}$, $-\frac{1}{2}$.

54. Find a fourth degree polynomial with zeros $\frac{2}{3}$, 4, $\sqrt{3}i$, $-\sqrt{3}i$.

55. A spherical tank (see Figure 3.34) of radius 50 feet will be two-thirds full when the depth of the fluid is $y_0 + 50$ feet, and y_0 is the root of the equation

$$3y_0^3 - 22{,}500y_0 + 250{,}000 = 0$$

Use the Bisection Method to approximate y_0 to within 0.01 unit. (The equation was derived by calculus techniques.)

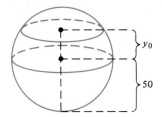

FIGURE 3.34

CHAPTER 4

EXPONENTIAL AND LOGARITHMIC FUNCTIONS

Exponential Functions

4.1

The functions we have dealt with so far have primarily been polynomials and rational functions. In this chapter we will study two new functions: the exponential function and its inverse, the logarithmic function.

Exponential functions are widely used in describing economic and physical phenomena such as compound interest, growth (or decline) of population sizes, and decay of radioactive material. Logarithms have in the past been used extensively to simplify complicated arithmetic calculations. With the advent of electronic computing devices, their computational usefulness has diminished. Nevertheless, properties of logarithms and logarithmic functions still play a critical role in mathematics and its real-world applications.

Consider the familiar functions

$$f(x) = x^3 \qquad \text{and} \qquad g(x) = \sqrt{x} = x^{1/2}$$

which involve a *variable raised to a constant power*. If we interchange the roles of the variable and the constant, we get new functions

$$F(x) = 3^x \qquad \text{and} \qquad G(x) = \left(\frac{1}{2}\right)^x$$

having constant bases and variable exponents. Such functions are called **exponential functions.**

EXPONENTIAL FUNCTION

> The **exponential function with base a** is denoted by
>
> $$f(x) = a^x$$
>
> where $a > 0$, $a \neq 1$, and x is any real number.

From our study of Section 1.2 we know how to interpret real numbers that are raised to integer or rational powers. Now, we need to interpret forms with *irrational* exponents, such as $a^{\sqrt{2}}$ and a^{π}. A technical definition of such forms is beyond the scope of this text. For our purposes it is sufficient to think of

$$a^{\sqrt{2}} \qquad \text{(where } \sqrt{2} \approx 1.414213\text{)}$$

as that value which has the successively closer approximations

$$a^{1.4}, a^{1.41}, a^{1.414}, a^{1.4142}, a^{1.41421}, \ldots$$

Consequently, for exponential functions like

$$f(x) = a^x$$

we assume in this text that a^x exists and that the properties of exponents (Section 1.2) hold for all real numbers x. Hence, the graphs of exponential functions will be continuous, with no holes or jumps.

EXAMPLE 1
Graphs of $y = a^x, a > 1$

On the same coordinate plane, sketch graphs of

(a) $y = 2^x$ (b) $y = (\frac{3}{2})^x$ (c) $y = 4^x$

Solution:
Table 4.1 lists some values for each function, and Figure 4.1 shows their graphs.

TABLE 4.1

x	-2	-1	0	1	2	3
(a) 2^x	1/4	1/2	1	2	4	8
(b) $(3/2)^x$	4/9	2/3	1	3/2	9/4	27/8
(c) 4^x	1/16	1/4	1	4	16	64

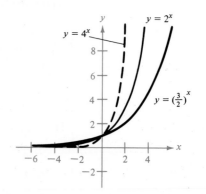

FIGURE 4.1

EXAMPLE 2

Graph of $y = a^{-x} = (1/a)^x$, $a > 1$

On the same coordinate plane, sketch graphs of

(a) $y = (\frac{1}{4})^x$ (b) $y = 2^{-x}$

Solution:

Note in part (b) that we can write

$$2^{-x} = \frac{1}{2^x} = \left(\frac{1}{2}\right)^x$$

Table 4.2 lists some values for each function, and Figure 4.2 shows their graphs.

TABLE 4.2

x	-3	-2	-1	0	1	2
(a) $(1/4)^x$	64	16	4	1	1/4	1/16
(b) 2^{-x}	8	4	2	1	1/2	1/4

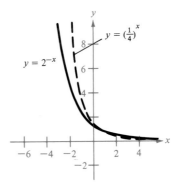

FIGURE 4.2

Exponential Graphs

From Figures 4.1 and 4.2 we can make several important observations about the graphs of $y = a^x$ and $y = a^{-x}$. These are summarized in Figures 4.3 and 4.4.

Graph of $y = a^x$, $a > 1$

1. Domain: all real x
2. Range: all positive reals
3. Intercept: $(0, 1)$
4. a^x *increases* as x increases
5. x-axis is horizontal asymptote to *left*
6. Reflection (in y-axis) of graph of $y = a^{-x}$

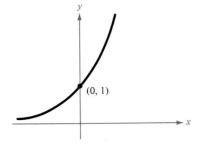

FIGURE 4.3

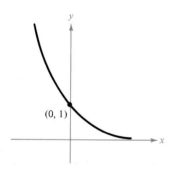

FIGURE 4.4

Graph of $y = a^{-x}$, $a > 1$

1. Domain: all real x
2. Range: all positive reals
3. Intercept: $(0, 1)$
4. a^{-x} *decreases* as x increases
5. x-axis is horizontal asymptote to *right*
6. Reflection (in y-axis) of graph of $y = a^x$

These observations can be used to establish two additional properties of exponential functions not listed back in Section 1.2. These properties are useful in solving equations that involve exponential functions.

SPECIAL PROPERTIES OF EXPONENTIAL FUNCTIONS

For $a > 0$ and $a \neq 1$:

1. If $a^x = a^y$, then $x = y$.
2. If $a^x = b^x$, and $x \neq 0$, then $a = b$.

EXAMPLE 3
Solving Exponential Equations

Solve for x in each of the following:

(a) $(\frac{1}{2})^x = 64$ (b) $2^5 = (x + 3)^5$ (c) $27 = x^{3/2}$

Solution:

(a) Since

$$\left(\frac{1}{2}\right)^x = \frac{1}{2^x} = 2^{-x} \quad \text{and} \quad 64 = 2^6$$

we have

$$2^{-x} = 2^6$$

Thus we conclude that $x = -6$.

(b) Since $a^x = b^x$ implies $a = b$, we have

$$2^5 = (x + 3)^5 \quad \text{implies} \quad 2 = x + 3$$

Thus,

$$-1 = x$$

(c) Rewriting the given equation, we have

$$(27)^{2/3} = (x^{3/2})^{2/3} \qquad \text{*Raise to reciprocal power*}$$
$$(\sqrt[3]{27})^2 = x$$
$$3^2 = 9 = x$$

Applications

Many physical and economic phenomena can be described by exponential functions. One of the most familiar is the exponential growth of an investment on which compound interest is earned. Consider a principal P that is invested at an annual rate r compounded once a year. If the interest is added to the principal at the end of one year, then the new principal P_1 is

$$P_1 = P + Pr = P(1 + r)$$

This pattern of multiplying the previous principal by $1 + r$ is then repeated each successive year. Table 4.3 shows how the results lead to an exponential function.

TABLE 4.3
Compound Interest Formula

Time in Years	Balance After Each Compounding
0	$P = P$
1	$P_1 = P(1 + r)$
2	$P_2 = P_1(1 + r) = P(1 + r)(1 + r) = P(1 + r)^2$
3	$P_3 = P_2(1 + r) = P(1 + r)^2(1 + r) = P(1 + r)^3$
.	.
.	.
.	.
n	$B = P_n = P(1 + r)^n$

EXAMPLE 4
Finding the Balance for Compound Interest

Suppose that \$9000 is invested at an annual rate of 8.5%, compounded annually. Find the balance in the account to the nearest dollar at the end of 3 years.

Solution:
In this case, $P = 9000$, $r = 8.5\% = 0.085$, and $n = 3$. Using the formula

$$B = P(1 + r)^n$$

we have

$$B = 9000(1 + 0.085)^3 = 9000(1.085)^3$$
$$B \approx 9000(1.2773) \approx \$11{,}496$$

EXAMPLE 5
Radioactive Decay

Suppose a 5-gram sample of a radioactive substance decays at a rate such that it loses half of its mass each day.

(a) Verify that after t days its mass is

$$M(t) = \frac{5}{2^t} = 5(2^{-t})$$

(b) Sketch a graph of M.

Solution:

(a) We need to show that $M(0)$ is equal to the original mass 5, and that the mass on a given day is $\frac{1}{2}$ the mass of the previous day. That is, we must show that

$$M(t + 1) = \frac{1}{2} M(t)$$

For $t = 0$, we have

$$M(0) = \frac{5}{2^0} = \frac{5}{1} = 5$$

Furthermore, replacing t by $t + 1$, we get

$$M(t + 1) = \frac{5}{2^{t+1}} = \frac{5}{2 \cdot 2^t} = \frac{1}{2}\left(\frac{5}{2^t}\right) = \frac{1}{2} M(t)$$

(b) Using the values in the following table, we obtain the graph shown in Figure 4.5.

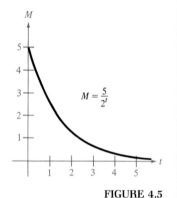

$M = \dfrac{5}{2^t}$

FIGURE 4.5

t	0	1	2	3	4
$M(t)$	5	$\dfrac{5}{2}$	$\dfrac{5}{4}$	$\dfrac{5}{8}$	$\dfrac{5}{16}$

Remark: The radioactive substance in Example 5 is said to have a **half-life** of one day.

Exponential Functions and Calculators

Scientific calculators have an exponential key, $\boxed{y^x}$, that is used to raise a number to a power. Note the key stroke sequences for the following examples:

Exponential	Key Strokes	Display
$(1.085)^3$	1.085 $\boxed{y^x}$ 3 $\boxed{=}$	1.27729
$12^{5/7}$	12 $\boxed{y^x}$ $\boxed{(}$ 5 $\boxed{\div}$ 7 $\boxed{)}$ $\boxed{=}$	5.89989
$29^{-8.6}$	29 $\boxed{y^x}$ 8.6 $\boxed{+/-}$ $\boxed{=}$	2.6508 − 13 (scientific notation)

The Natural Base *e*

Our next example uses an exponential function similar to the one in Example 4 to identify an important irrational number that is the base for the exponential and natural logarithmic functions used in calculus. We denote this number by the letter *e*. To five decimal places, the value of *e* is

$$e \approx 2.71828 \ldots$$

This base arises in many natural phenomena such as the population growth shown in Example 7.

EXAMPLE 6
An Estimation of the Irrational Number *e*

Evaluate the exponential expression

$$\left(1 + \frac{1}{n}\right)^n$$

for several large values of *n* to see that it approaches the value $e \approx 2.71828$ as *n* increases without bound.

Solution:
With a calculator we use the key strokes

$$n \boxed{1/x} \boxed{+} 1 \boxed{=} \boxed{y^x} n \boxed{=}$$

to obtain the values shown in the table that follows.

n	10	100	1000	10,000	100,000	1,000,000
$\left(1 + \frac{1}{n}\right)^n$	2.59374	2.70481	2.71692	2.71815	2.71827	2.71828

From these values it seems reasonable to conclude that

$$\left(1 + \frac{1}{n}\right)^n \to e \quad \text{as} \quad n \to \infty$$

EXAMPLE 7
Population Growth

In Professor Jacob's research, the number of fruit flies in a certain experimental population at time *t* hours is given by the exponential equation

$$Q(t) = 20e^{0.03t}$$

(a) Find the initial number of fruit flies in the population.
(b) How large is the population of fruit flies after 72 hours?
(c) Sketch a graph of *Q*.

Solution:

(a) To find the initial population, we evaluate $Q(t)$ at $t = 0$.

$$Q(0) = 20e^{0.03(0)} = 20e^0 = 20(1) = 20$$

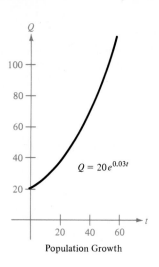

$Q = 20e^{0.03t}$

Population Growth

FIGURE 4.6

(b) After 72 hours, the population size is

$$Q(72) = 20e^{(0.03)(72)} = 20e^{2.16} \approx 173 \qquad \text{(Nearest integer)}$$

(c) The next table shows some values of $Q(t)$ (to the nearest integer), and Figure 4.6 shows the graph of Q.

t	0	5	10	20	40	60
$20e^{0.03t}$	20	23	27	36	66	121

Remark: Many populations in nature have a growth pattern described by the function

$$Q(t) = ce^{kt}$$

where c is the original population, $Q(t)$ is the population at time t, and k is a constant that affects the rate of growth.

We complete this section with another look at compound interest. This time we consider interest that is compounded n times per year. In this case we have the amount of deposit P, the annual rate of interest, r, the number of compoundings per year, n and the number of years, t. The rate per compounding period is therefore r/n, and the balance after t years is

$$B = P\left(1 + \frac{r}{n}\right)^{tn}$$

If we let the number of compoundings, n, increase without bound, we have what is called **continuous compounding.** In the formula for n compoundings per year, let $m = n/r$. Then we have

$$B = P\left(1 + \frac{r}{n}\right)^{nt} = P\left(1 + \frac{1}{m}\right)^{mrt} = P\left[\left(1 + \frac{1}{m}\right)^{m}\right]^{rt}$$

Now as m gets larger and larger, we know from Example 6 that

$$\left[\left(1 + \frac{1}{m}\right)^{m}\right] \rightarrow e$$

Hence, for continuous compounding, it follows that

$$P\left[\left(1 + \frac{1}{m}\right)^{m}\right]^{rt} \rightarrow P[e]^{rt}$$

and we write

$$B = Pe^{rt}$$

FORMULAS FOR COMPOUND INTEREST

The balance in an account with principal P, interest rate r, and time in years t is given by the following formulas:

1. For n compoundings per year:

$$B = P\left(1 + \frac{r}{n}\right)^{nt}$$

2. For continuous compoundings:

$$B = Pe^{rt}$$

EXAMPLE 8
Compounding n Times and Continuously

Suppose $12,000 is invested at an annual rate of 9%. Find the balance after 5 years if it is compounded

(a) quarterly   (b) continuously

Solution:

(a) For quarterly compoundings, we have $n = 4$. Thus, in 5 years at 9% we have a balance of

$$B = P\left(1 + \frac{r}{n}\right)^{nt} = 12,000\left(1 + \frac{0.09}{4}\right)^{4(5)}$$

$$B = 12,000(1.0225)^{20} = 12,000(1.56051) = \$18,726.12$$

(b) Compounded continuously, the balance is

$$B = Pe^{rt} = 12,000e^{0.09(5)} = 12,000e^{0.45}$$
$$= 12,000(1.56831) = \$18,819.72$$

Note that continuous compounding yields

$$\$18,819.72 - \$18,726.12 = \$93.60$$

more than quarterly compounding.

Section Exercises 4.1

In Exercises 1–10, match the given exponential function with one of the graphs to the right or on page 170.

(a)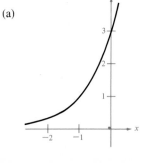

(b)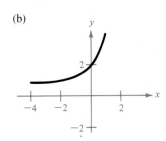

1. $f(x) = 3^x$
2. $f(x) = -3^x$
3. $f(x) = 3^{-x}$
4. $f(x) = 3^{2x}$
5. $f(x) = 3^{x+1}$
6. $f(x) = 3^x + 1$
7. $f(x) = 3^{x/2}$
8. $f(x) = 3^{x-2}$
9. $f(x) = 3^x - 4$
10. $f(x) = 3^x + 3^{-x}$

(c)   *y*

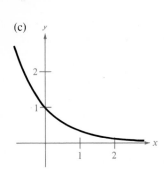

(d)   *y*

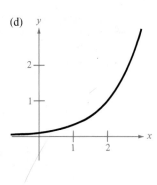

(e)   *y*

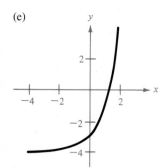

(f)   *y*

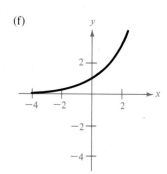

(g)   *y*

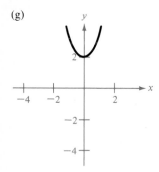

(h)   *y*

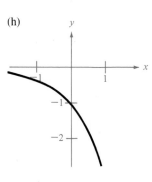

(i)   *y*

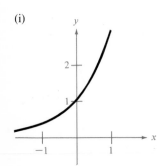

(j)   *y*

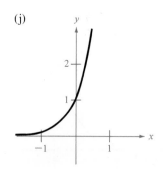

In Exercises 11–20, solve for *x*.

11. $3^x = 81$
12. $5^{x+1} = 125$
13. $(\frac{1}{3})^{x-1} = 27$
14. $(\frac{1}{5})^{2x} = 625$
15. $4^3 = (x + 2)^3$
16. $4^2 = (x + 2)^2$
17. $x^{3/4} = 8$
18. $(x + 3)^{4/3} = 16$
19. $4(2^x) = 1$
20. $e^{-2x} = e^5$

In Exercises 21–37, sketch the graph of the given exponential function.

21. $f(x) = 4^x$
22. $g(x) = 5^x$
23. $h(x) = (\frac{1}{4})^x = 4^{-x}$
24. $f(x) = (\frac{1}{5})^x = 5^{-x}$
25. $f(x) = (\frac{1}{4})^{-x}$
26. $g(x) = (\frac{1}{5})^{-x}$
27. $y = 2^{-x^2}$
28. $y = 3^{-x^2}$
29. $y = 3^{|x|}$
30. $y = 3^{-|x|}$
31. $s(t) = 2^{-t} + 3$
32. $s(t) = \dfrac{3^{-t}}{4}$
33. $f(x) = e^{2x}$
34. $h(x) = e^{x-2}$
35. $A(t) = 500e^{0.15t}$
36. $N(t) = 1000e^{-0.2t}$
37. $g(x) = \dfrac{10}{1 + e^{-x}}$

38. If $1000 is invested at 10% interest, find the amount after 10 years if the interest is compounded
     (a) annually   (b) semiannually   (c) quarterly
     (d) monthly   (e) daily   (f) continuously
39. If $2500 is invested at 12% interest, find the amount after 20 years if the interest is compounded
     (a) annually   (b) semiannually   (c) quarterly
     (d) monthly   (e) daily   (f) continuously
40. The demand equation for a certain product is given by

$$p = 5000\left(1 - \frac{4}{4 + e^{-0.002x}}\right)$$

Find the price of the product if the demand is
     (a) $x = 100$ units   (b) $x = 500$ units
41. The demand equation for a certain product is given by

$$p = 500 - 0.5e^{0.004x}$$

Find the price of the product if the demand is
     (a) $x = 1000$ units   (b) $x = 1500$ units
42. Find the amount of money that should be deposited in an account paying 12% interest compounded continuously to produce a final balance of $100,000 in
     (a) 1 year   (b) 10 years
     (c) 20 years   (d) 50 years
43. A certain type of bacteria increases according to the model

$$P(t) = 100e^{0.2197t}$$

where *t* is time in hours. Find
     (a) $P(0)$ (b) $P(5)$ (c) $P(10)$

44. The population of a town increases according to the model

$$P(t) = 2500e^{0.0293t}$$

where t is time in years, with $t = 0$ corresponding to 1980. Use the model to approximate the population in
   (a) 1985   (b) 1990   (c) 2000
45. The half-life of radioactive radium is approximately 1600 years, and its decay is approximated by the model

$$Q(t) = Q_0e^{-0.0004t}$$

where t is time in years, $Q(t)$ is the quantity at time t, and Q_0 is the initial quantity. What percentage of a given amount remains after 100 years?

If the average time between successive occurrences of some event is λ, then the probability of waiting less than t units of time between successive occurrences can sometimes be approximated by the model

$$F(t) = 1 - e^{-t/\lambda}$$

Use this model in Exercises 46 and 47.

46. Trucks arrive at a terminal at an average of 3 per hour (therefore $\lambda = 20$ minutes). If a truck has just arrived, find the probability that the next arrival will be within
   (a) 10 minutes   (b) 30 minutes   (c) 1 hour
47. The average time between incoming calls at a switchboard is 3 minutes. If a call has just come in, find the probability that the next call will be within
   (a) $\frac{1}{2}$ minute   (b) 2 minutes   (c) 5 minutes
48. A certain automobile gets 28 miles per gallon of gasoline for speeds up to 50 miles per hour. Over 50 miles per hour, the miles per gallon drops at the rate of 12% for each 10 miles per hour. If s is the speed (in miles per hour) and y is the miles per gallon, then

$$y = 28e^{0.6-0.012s}, \qquad s \geq 50$$

Use this function to complete the following table:

Speed	50	55	60	65	70
Miles per gallon					

49. A solution of a certain drug contained 500 units per milliliter when prepared. It was analyzed after 40 days and found to contain 300 units per milliliter. Assuming that the rate of decomposition is proportional to the amount present, the equation giving the amount A after t days is

$$A = 500e^{-0.013t}$$

Use this model to find A when $t = 60$.
50. In calculus it is shown that

$$e^x \approx 1 + x + \frac{x^2}{2} + \frac{x^3}{6} + \frac{x^4}{24}$$

Use this equation to approximate
   (a) e $(x = 1)$ (b) $e^{1/2}$ $(x = \frac{1}{2})$ (c) $e^{-1/2}$ $(x = -\frac{1}{2})$

Logarithms   4.2

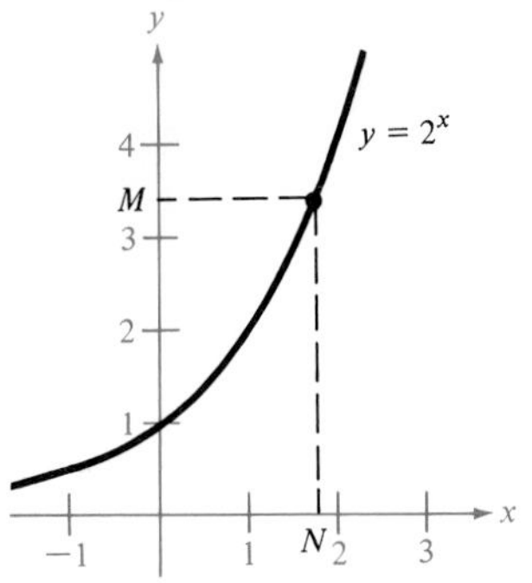

FIGURE 4.7

We noted in Figure 4.1 that the graph of $y = 2^x$ is always increasing and that the range for this exponential function is the set of positive real numbers. This means that to every real number $(M > 0)$ in the range there corresponds a unique real number N in the domain, such that $2^N = M$. For instance, if $M = 3.42$, then from Figure 4.7 we can see that there exists a real number N such that

$$2^N = 3.42$$

If $2^N = 3.42$, we call N the *logarithm of 3.42 to base 2*. Or in equation form, we write

$$N = \log_2 3.42$$

In general, we use the following definition of a logarithm.

DEFINITION OF LOGARITHM TO BASE b

Let $b > 0$ and $b \neq 1$. If M is any positive real number, then there exists a unique real number N such that

$$b^N = M$$

This number N is called the **logarithm of M to base b** and is denoted by

$$N = \log_b M$$

Remark: This definition of logarithms gives us the following equivalence:

$$N = \log_b M \quad \text{if and only if} \quad b^N = M$$

From the definition of logarithms, it should be clear that **A LOGA-RITHM IS AN EXPONENT.** Keep this fact in mind as you work with logarithms. For instance, to evaluate $\log_2 16$ you can ask yourself the question

2 to what power yields 16?

Since $16 = 2^4$, it follows that

$$\log_2 16 = 4$$

In Example 1, we show how to use the equivalent logarithmic and exponential forms

$$\log_b M = N \quad \text{and} \quad b^N = M$$

to evaluate logarithms.

EXAMPLE 1
Evaluating Logarithms

Evaluate the following (if possible).

(a) $\log_4 2$ (b) $\log_3(\frac{1}{81})$ (c) $\log_4 8$ (d) $\log_5(-2)$

Solution:

Logarithmic Form	*Exponential Form*	*Value of N*
(a) Let $N = \log_4 2$	$4^N = 2$	$N = \dfrac{1}{2} = \log_4 2$
(b) Let $N = \log_3\left(\dfrac{1}{81}\right)$	$3^N = \dfrac{1}{81} = \dfrac{1}{3^4}$	$N = -4 = \log_3\left(\dfrac{1}{81}\right)$
(c) Let $N = \log_4 8$	$4^N = 8 = (\sqrt{4})^3$	$N = \dfrac{3}{2} = \log_4 8$
(d) Let $N = \log_5(-2)$	$5^N = -2$	Not possible*

*It is not possible to find a value of N such that $5^N = -2$, since $5^N > 0$ for all N. Thus, $\log_5(-2)$ is not defined.

In each of the following, convert all logarithmic equations to exponential form and exponential equations to logarithmic form. Find the value of x in each case.

(a) $\log_5 x = 2$ (b) $36^x = 6$ (c) $(\frac{1}{27}) = x^{-3}$

(d) $\log_3(x - 1) = 2$ (e) $x = 16^{3/2}$ (f) $\log_8(\frac{1}{2}) = x$

Solution:

Using the equivalent forms

$$\log_b M = N \quad \text{and} \quad b^N = M$$

we can readily make up the following table of solutions. In most cases the value of x is more easily obtained from the exponential form.

Logarithmic Form	Exponential Form	Value of x
(a) $\log_5 x = 2$	$5^2 = x$	$x = 25$
(b) $\log_{36} 6 = x$	$36^x = 6$	$\sqrt{36} = 6,\ x = \dfrac{1}{2}$
(c) $\log_x\left(\dfrac{1}{27}\right) = -3$	$\dfrac{1}{27} = x^{-3}$	$\dfrac{1}{27} = \dfrac{1}{x^3},\ x = 3$
(d) $\log_3(x - 1) = 2$	$3^2 = x - 1$	$x = 1 + 3^2 = 10$
(e) $\log_{16} x = \dfrac{3}{2}$	$16^{3/2} = x$	$x = (\sqrt{16})^3 = 64$
(f) $\log_8\left(\dfrac{1}{2}\right) = x$	$8^x = \dfrac{1}{2}$	$\dfrac{1}{2} = \dfrac{1}{\sqrt[3]{8}} = 8^{-1/3}$
		$x = -\dfrac{1}{3}$

Logarithms are powerful as computational aids primarily because they have fundamental properties that convert complex multiplications, divisions, and exponentiations into simpler additions, subtractions, and multiplications, respectively. The following properties of logarithms can be derived from the properties of exponents that we studied back in Section 1.2.

PROPERTIES OF LOGARITHMS

If $b > 0$, $b \neq 1$, and c is any real number, then the following properties hold:

Property	**Comments**
1. $\log_b b = 1$	1. True because $b^1 = b$.
2. $\log_b 1 = 0$	2. True because $b^0 = 1$.
3. $\log_b(MN) = \log_b M + \log_b N$	3. The log of a product converts to a sum.

4. $\log_b\left(\dfrac{M}{N}\right) = \log_b M - \log_b N$ 4. The log of a quotient converts to a difference.

5. $\log_b(M^c) = c\,\log_b M$ 5. The log of an exponential converts to a product.

6. $\log_b(b^c) = c$ 6. Inverse property.

7. $b^{(\log_b c)} = c$ 7. Inverse property.

8. $\log_b M = \log_b N$ implies $M = N$ 8. One-to-one correspondence.

Proof:

All of these properties follow quite directly from the corresponding properties of exponents. As an example of this correspondence, we list a proof of Property 3. We leave the others up to you.

To prove Property 3, let

$$u = \log_b M \qquad \text{and} \qquad v = \log_b N$$

Using the equivalent forms

$$b^u = M \qquad \text{and} \qquad b^v = N$$

we can multiply to get

$$b^u(b^v) = M(N) \qquad \text{or} \qquad b^{u+v} = MN$$

Or, equivalently, we have

$$\log_b(MN) = u + v = \log_b M + \log_b N$$

Remark: Note that we have no general property that can be used to simplify $\log_b(M \pm N)$.

EXAMPLE 3
Using Properties of Logs

Given $\log_5 2 \approx .4307$, $\log_5 3 \approx .6826$, and $\log_5 7 \approx 1.2091$, use properties of logs to evaluate each of the following:

(a) $\log_5 6$ (b) $\log_5(\frac{7}{27})$ (c) $\log_5 \sqrt[3]{49}$

Solution:

(a) By Property 3,

$$\log_5 6 = \log_5(2 \cdot 3) = \log_5 2 + \log_5 3$$
$$\approx 0.4307 + 0.6826 = 1.1133$$

(b) $\log_5(\frac{7}{27}) = \log_5 7 - \log_5 27$ *Property 4*

$\qquad\qquad = \log_5 7 - \log_5(3^3)$ *Prime factoring*

$\qquad\qquad = \log_5 7 - 3\,\log_5 3$ *Property 5*

$\qquad\qquad \approx 1.2091 - 3(0.6826) = -0.8387$

(c) $\log_5 \sqrt[3]{49} = \log_5(7^2)^{1/3}$ *Rational exponents*

                 $= \log_5(7^{2/3})$ *Property of exponents*

                 $= \frac{2}{3} \log_5 7$ *Property 5*

                 $\approx \frac{2}{3}(1.2091) = 0.80607$

The base of a logarithm can be any positive real number other than 1. For computational purposes, the most widely used base is 10, since our number system uses base 10. For calculus, the most appropriate base is the irrational number e. Logarithms to these bases are given the following special names.

COMMON AND NATURAL LOGARITHMS

> Logarithms to base 10 are called **common logarithms** and are denoted by
>
> $\log_{10} x = \log x$
>
> Logarithms to base e are called **natural logarithms** and are denoted by
>
> $\log_e x = \ln x$

Remark: The eight *properties of logarithms* hold for both common and natural logarithms. On most scientific calculators $\boxed{\text{LOG}}$ and $\boxed{\text{ln}}$ are the keys for common and natural logarithms, respectively.

Sometimes you need to work with logarithms to bases other than 10 or e. In such cases the following *change of base formula* is useful.

CHANGE OF BASE FORMULA

> For $a > 0$, $b > 0$, and $a, b \neq 1$:
>
> $\log_b x = \dfrac{\log_a x}{\log_a b} = \dfrac{1}{\log_a b}(\log_a x)$

Remark: One way of looking at the change of base formula is that logarithms to base b are simply constant multiples of logarithms to base a. The constant multiplier is $1/(\log_a b)$.

EXAMPLE 4
Changing Bases

Change to *common* logarithms, then evaluate using the $\boxed{\text{LOG}}$ key.

(a) $\log_4 30$ (b) $\ln 100$

Change to *natural* logarithms, then evaluate using the $\boxed{\text{ln}}$ key.

(c) $\log e$ (d) $\log_3 x$

Solution:

In all four cases, we apply the change of base formula as indicated.

(a) $\log_4 30 = \dfrac{\log 30}{\log 4} = \dfrac{1.47712}{0.60206} \approx 2.4534$

(b) $\ln 100 = \dfrac{\log 100}{\log e} = \dfrac{2}{\log e} \approx \dfrac{2}{\log 2.71828} \approx 4.6052$

(c) $\log e = \log_{10} e = \dfrac{\ln e}{\ln 10} = \dfrac{1}{\ln 10} \approx 0.4343$

(d) $\log_3 x = \dfrac{\ln x}{\ln 3} \approx \dfrac{\ln x}{1.0986}$

In Example 4, we were asked to find the logarithm of a given number. In science, we are often faced with the reverse problem. For example, we might be asked to solve the following equation for N:

$$\log_b N = u$$

The key to solving this equation lies in Property 7 of logarithms. Using the left and right sides of the equation as exponents for the base b, we have

$$b^{(\log_b N)} = b^u$$
$$N = b^u \qquad\qquad \textit{Property 7}$$

Since we obtain the number N by eliminating a logarithm, we call N the **antilogarithm of u.**

DEFINITION OF ANTILOGARITHM

If

$$\log_b N = u$$

then N is called the **antilogarithm of u for base b.** Specifically, the antilogarithm of u has the value

$$N = b^u$$

EXAMPLE 5
**Finding Antilogarithms
with a Calculator**

Find the antilogarithm for each of the following:

(a) $\log N = 3.718$ (b) $\log N = -2.04$ (c) $\ln N = 0.147$

Solution:

(a) Since $\log N = 3.718$ is equivalent to $10^{3.718} = N$, the calculator sequence 10 $\boxed{y^x}$ 3.718 $\boxed{=}$ yields the antilogarithm value $N = 5223.96$.

Note: Some scientific calculators have an antilogarithm key (base 10) that finds N directly from 3.718.

(b) Again using base 10, we have $\log N = -2.04$ or $10^{-2.04} = N$. Thus, the calculator sequence 10 $\boxed{y^x}$ 2.04 $\boxed{+/-}$ $\boxed{=}$ yields $N = 0.0091201$.

(c) In this case, $\ln N = 0.147$ is equivalent to $e^{0.147} = N$. The calculator sequence .147 $\boxed{e^x}$ $\boxed{=}$ yields $N = 1.15835$.

EXAMPLE 6
Magnitude of Earthquakes

On the Richter Scale, the magnitude R of an earthquake of intensity I is given by

$$R = \log\left(\frac{I}{I_0}\right)$$

where I_0 is the minimum intensity used for comparison. If we let $I_0 = 1$, find the intensity per unit of area for the following earthquakes. (Intensity is a measure of the wave energy of an earthquake.)
(a) San Francisco in 1906, $R = 8.3$
(b) Mexico City in 1978, $R = 7.85$
(c) Predicted in 1985, $R = 6.3$

Solution:

(a) In this case, we have $8.3 = \log I$. We solve for I using the antilogarithm of 8.3 to the base 10.

$$I = 10^{8.3} \approx 199{,}526{,}000$$

(b) For Mexico City, we have $7.85 = \log I$.

$$I = 10^{7.85} \approx 70{,}794{,}600$$

(c) For $R = 6.3$, we have $6.3 = \log I$.

$$I = 10^{6.3} \approx 1{,}995{,}260$$

Note that a drop of 2 units on the Richter Scale (from 8.3 to 6.3) represents an intensity change by a factor of

$$\frac{1}{10^2} = \frac{1}{100}$$

Logarithmic Expressions

In addition to being of assistance in computing with logarithms, the *properties of logarithms* are valuable aids for rewriting logarithmic expressions in forms that simplify the operations of algebra and calculus. Examples 7 and 8 illustrate some cases.

EXAMPLE 7
Rewriting Logarithmic
Expressions

Use the *properties of logarithms* to rewrite each of the following as the sum and/or difference of logarithms:

(a) $\log \dfrac{x^2 y}{5}$ 　　　　　　 (b) $\ln \dfrac{\sqrt{3x - 5}}{7x^3}$ 　　　　　　 (c) $\ln\left(\dfrac{5x^2}{y}\right)^3$

Solution:

(a) $\log \dfrac{x^2 y}{5} = \log x^2 y - \log 5$

$= \log x^2 + \log y - \log 5$

$= 2 \log x + \log y - \log 5$

(b) $\ln \dfrac{\sqrt{3x - 5}}{7x^3} = \ln(3x - 5)^{1/2} - \ln 7x^3$

$= \ln(3x - 5)^{1/2} - \ln 7 - \ln x^3$

$= \dfrac{1}{2}\ln(3x - 5) - \ln 7 - 3\ln x$

(c) $\ln\left(\dfrac{5x^2}{y}\right)^3 = 3\ln \dfrac{5x^2}{y}$

$= 3(\ln 5x^2 - \ln y)$

$= 3(\ln 5 + \ln x^2 - \ln y)$

$= 3(\ln 5 + 2\ln x - \ln y)$

EXAMPLE 8
Rewriting Logarithmic
Expressions

Rewrite as a single logarithm:

(a) $\frac{1}{2}\log x - 3\log(x + 1)$ 　　　　　　 (b) $2\ln(x + 2) - \frac{1}{3}(\ln x + \ln y)$

Solution:

(a) In this case, we use the *properties of logarithms* in the reverse (right to left) direction.

$$\frac{1}{2}\log x - 3\log(x + 1) = \log x^{1/2} - \log(x + 1)^3$$

$$= \log \frac{\sqrt{x}}{(x + 1)^3}$$

(b) Again, in reverse direction we have

$$2\ln(x + 2) - \frac{1}{3}\left(\ln x + \ln y\right) = \ln(x + 2)^2 - (\ln x^{1/3} + \ln y^{1/3})$$

$$= \ln(x + 2)^2 - \ln(xy)^{1/3}$$

$$= \ln \frac{(x + 2)^2}{\sqrt[3]{xy}}$$

Logarithms Using Tables

Although calculators are more efficient than tables in computations with logarithms, it is instructive to see how to work with tables of logarithms. Using base 10, we note first that every positive real number can be written as a product $c \times 10^k$, where $1 \leq c < 10$. For example,

$$1986 = 1.986 \times 10^3, \qquad 5.37 = 5.37 \times 10^0,$$
$$0.0439 = 4.39 \times 10^{-2}$$

Suppose we apply the properties of logarithms to the number 1986:

$$1986 = 1.986 \times 10^3$$
$$\log 1986 = \log(1.986 \times 10^3)$$
$$= \log(1.986) + \log(10^3)$$
$$= \log(1.986) + 3 \log(10)$$
$$= \log(1.986) + 3$$

In general, for any positive real number x (expressible as $x = c \times 10^k$), its logarithm has the **standard form**

$$\log x = \log c + \log 10^k = \log c + k$$

where $1 \leq c < 10$. We call $\log c$ the **mantissa** and k the **characteristic** of $\log x$. Since $\log x$ increases as x increases, it follows, from $1 \leq c < 10$, that

$$\log 1 \leq \log c < \log 10$$
$$0 \leq \log c < 1$$

which means that the *mantissa* always lies between 0 and 1.

The common logarithm table in Appendix C gives four-decimal-place approximations of the *mantissa* for the logarithm of every three-digit number between 1.00 and 9.99. The next example shows how to use these tables to evaluate logarithms.

EXAMPLE 9
Evaluating Logarithms
with Tables

Use the tables in Appendix C to approximate

(a) $\log 85.6$ (b) $\log 0.000329$

Solution:

(a) Since $85.6 = 8.56 \times 10^1$, the characteristic is 1. By the tables, the mantissa is $\log 8.56 = 0.9325$. Therefore,

$$\log 85.6 = (\text{mantissa}) + (\text{characteristic})$$
$$= \log 8.56 + 1$$
$$= 0.9325 + 1 = 1.9325$$

(b) Since $0.000329 = 3.29 \times 10^{-4}$, the characteristic is -4. Using tables for the mantissa, we obtain

$$\log 0.000329 = \log 3.29 + (-4) = 0.5172 - 4 = -3.4828$$

EXAMPLE 10
Finding Antilogarithms
with Tables

Use tables to approximate the value of x if

(a) $\log x = 2.7582$ (b) $\log x = -3.6364$

Solution:

(a) In standard form, we have

$$\log x = 0.7582 + 2 = \log c + k$$

By locating the mantissa 0.7582 in the common logarithm table, we see that it corresponds to log 5.73. Thus, since $k = 2$, we have

$$x = c \times 10^k = 5.73 \times 10^2 = 573$$

(b) To obtain a nonnegative mantissa, we add and subtract 4 to get the standard form

$$\log x = 4 - 3.6364 - 4 = 0.3636 - 4$$

From the logarithm table we find that $0.3636 = \log 2.31$. Thus, since $k = -4$, we have

$$x = c \times 10^k = 2.31 \times 10^{-4} = 0.000231$$

To do numerical computations with logarithms, we combine the properties of logarithms with the procedures shown in Examples 9 and 10.

EXAMPLE 11
Computations Using
Logarithm Tables

Use tables of logarithms to approximate the value of

$$N = \frac{(1.9)^3}{\sqrt{82.7}}$$

Solution:
Using properties of logarithms as in Example 7, we write

$$\log N = 3 \log 1.9 - \frac{1}{2} \log 82.7$$

$$= 3(0.2788) - \frac{1}{2}\left(1.9175\right)$$

$$= 0.8364 - 0.95875$$

$$\approx -0.12235$$

Adding and subtracting 1, we obtain the standard form

$$\log N \approx (1 - 0.12195) - 1 = 0.87805 - 1$$

Using the mantissa value closest to 0.87805 in the table, we have $0.8779 = \log 7.55$. Therefore,

$$\log N \approx 0.8779 - 1$$

$$N = c \times 10^k \approx 7.55 \times 10^{-1} = 0.755$$

Remark: A slightly more accurate approximation could be obtained by using *linear interpolation* (Section 2.3) to find the antilogarithm for 0.87805.

Section Exercises 4.2

In Exercises 1–10, write the equation in exponential form.

1. $4 = \log_2 16$
2. $3 = \log_4 64$
3. $-2 = \log_5(\frac{1}{25})$
4. $-3 = \log_2(\frac{1}{8})$
5. $\frac{1}{2} = \log_{16} 4$
6. $\frac{2}{3} = \log_{27} 9$
7. $0 = \log_7 1$
8. $3 = \log 1000$
9. $2 = \ln e^2$
10. $a = \log_b c$

In Exercises 11–20, write the equation in logarithmic form.

11. $5^3 = 125$
12. $8^2 = 64$
13. $81^{1/4} = 3$
14. $9^{3/2} = 27$
15. $6^{-2} = \frac{1}{36}$
16. $10^{-3} = 0.001$
17. $e^3 = 20.0855 \ldots$
18. $e^0 = 1$
19. $u^v = w$
20. $e^x = 4$

In Exercises 21–34, find the value of the unknown.

21. $\log 1000 = x$
22. $\log 0.1 = x$
23. $\log_4(\frac{1}{64}) = x$
24. $\log_5 25 = x$
25. $\log_3 x = -1$
26. $\log_2 x = -4$
27. $\log_b 27 = 3$
28. $\log_b 2 = \frac{1}{5}$
29. $\ln e^x = 3$
30. $e^{\ln x} = 4$
31. $x^2 - x = \log_5 25$
32. $3x + 5 = \log_2 64$
33. $x - 3 = \log_2 32$
34. $x - x^2 = \log_4(\frac{1}{64})$

In Exercises 35–50, evaluate the logarithm using the properties of logarithms, given $\log_b 2 = 0.3562$, $\log_b 3 = 0.5646$, and $\log_b 5 = 0.8271$.

35. $\log_b 6$
36. $\log_b 15$
37. $\log_b(\frac{3}{2})$
38. $\log_b(\frac{5}{3})$
39. $\log_b 25$
40. $\log_b 18 \ (18 = 2 \cdot 3^2)$
41. $\log_b \sqrt{2}$
42. $\log_b(\frac{9}{2})$
43. $\log_b 40$
44. $\log_b \sqrt[3]{75}$
45. $\log_b(\frac{1}{4})$
46. $\log_b(3b^2)$
47. $\log_b \sqrt{5b}$
48. $\log_b 1$
49. $\log_b\left[\frac{(4.5)^3}{\sqrt{3}}\right]$
50. $\log_b\left(\frac{4}{15^2}\right)$

In Exercises 51–60, use the properties of logarithms to write the expressions as a sum, difference, or multiple of logarithms.

51. $\log_2 xyz$
52. $\ln\left(\frac{xy}{z}\right)$
53. $\ln\sqrt{a - 1}$
54. $\log_3\left(\frac{x^2 - 1}{x^3}\right)^3$
55. $\ln z(z - 1)^2$
56. $\log_4 \sqrt{\frac{x^2}{y^3}}$
57. $\log_b\left(\frac{x^2}{y^2 z^3}\right)$
58. $\log_b\left(\frac{\sqrt{x} y^4}{z^4}\right)$
59. $\ln \sqrt[3]{x}\sqrt{y}$
60. $\ln\left(\frac{x}{\sqrt{x^2 + 1}}\right)$

In Exercises 61–65, write the expression as a single logarithm.

61. $\log_3(x - 2) - \log_3(x + 2)$
62. $3 \ln x + 2 \ln y - 4 \ln z$
63. $\frac{1}{3}[2 \ln(x + 3) + \ln x - \ln(x^2 - 1)]$
64. $2[\ln x - \ln(x + 1) - \ln(x - 1)]$
65. $2 \ln 3 - \frac{1}{2} \ln(x^2 + 1)$

In Exercises 66–73, evaluate the logarithm using the change of base formula and your calculator. Do the problem twice, once with common logarithms and once with natural logarithms.

66. $\log_3 7$
67. $\log_7 4$
68. $\log_{1/2} 4$
69. $\log_4(0.55)$
70. $\log_9(0.4)$
71. $\log_{20} 125$
72. $\log_{15} 1250$
73. $\log_{1/3}(0.015)$

In Exercises 74 and 75, use a calculator to evaluate the logarithms.

74. $\ln\left(\frac{1 + \sqrt{3}}{2}\right)$
75. $\ln(\sqrt{5} - 2)$

76. The time (in hours) necessary for a certain object to cool $10°$ is

$$t = \frac{10 \ln(\frac{1}{2})}{\ln(\frac{3}{4})}$$

Find t.

77. The population of a town will double in

$$t = \frac{10 \ln 2}{\ln 67 - \ln 50}$$

years. Find t.

78. The work (in foot-pounds) done in compressing an initial volume of 9 cubic feet at a pressure of 15 pounds per square inch to a volume of 3 cubic feet is

$$W = 19440(\ln 9 - \ln 3)$$

Find W.

In Exercises 79 and 80, use the Richter Scale (Example 6) for measuring the magnitude of earthquakes.

79. Find the magnitude R of an earthquake of intensity I (let $I_0 = 1$):
   (a) $I = 80,500,000$ (b) $I = 48,275,000$
80. Find the intensity I of an earthquake measuring R on the Richter Scale (let $I_0 = 1$):
   (a) Columbia in 1906, $R = 8.6$
   (b) Los Angeles in 1971, $R = 6.7$

Acidity (pH) is a measure of the hydrogen ion concentration H^+ (measured in moles of hydrogen per liter of solution) of a solution. The formula for pH is

$$pH = -\log[H^+]$$

Use this formula in Exercises 81–84.

81. Compute $[H^+]$ for a solution in which pH $= 5.8$.
82. Compute $[H^+]$ for a solution in which pH $= 3.2$.
83. If $[H^+] = 2.3 \times 10^{-5}$, find the pH.
84. If $[H^+] = 11.3 \times 10^{-6}$, find the pH.

In Exercises 85–87, find the logarithm by using the table in Appendix C.

85. (a) log 417   (b) log 41.7   (c) log 0.0417

86. (a) log 985   (b) log 9.85   (c) log 0.985
87. (a) log 4385   (b) log 43.85   (c) log 0.004385

In Exercises 88–90, find N (antilogarithm) by using the table in Appendix C.

88. (a) $\log N = 4.3979$ (b) $\log N = 1.3979$
   (c) $\log N = -1.6021$
89. (a) $\log N = 2.6294$ (b) $\log N = 0.6294$
   (c) $\log N = -2.3706$
90. (a) $\log N = 6.1335$ (b) $\log N = -3.8665$
   (c) $\log N = 8.1335 - 10$

In Exercises 91–100, use logarithms to approximate the quantity to four-digit accuracy.

91. $\dfrac{(86.4)(8.09)}{38.6}$ 92. $\dfrac{1243}{(42.8)(67.9)}$

93. $(1.05)^{15}$ 94. $(0.2313)^6$

95. $\sqrt[3]{86.5}$ 96. $\sqrt[4]{(4.705)(18.86)}$

97. $\left(\dfrac{18.60}{59.25}\right)^{0.75}$ 98. $(48.8)^{0.75} + (75.4)^{0.45}$

99. $\dfrac{(3165)^{0.19}}{(525)^{0.45}}$ 100. $(5.85 \times 10^4)(16.4 \times 10^5)$

Logarithmic Functions   4.3

Having studied logarithms and their properties, we are now ready to look at a new family of functions known as **logarithmic functions.**

DEFINITION OF A LOGARITHMIC FUNCTION

> The function f defined by
>
> $$f(x) = \log_a x$$
>
> for all positive real numbers x is called the **logarithmic function** to the base a.

Since logarithms are exponents, we would expect logarithmic functions to have a close relationship to the exponential functions studied in Section 4.1. To see the connection between these functions, consider the fact that

$$y = \log_a x \quad \text{if and only if} \quad x = a^y$$

This suggests that for $a > 0$ and $a \neq 1$, the logarithmic function $f(x) = \log_a x$ is the **inverse** of the exponential function $g(x) = a^x$ and vice versa. (See Section 2.5.) Furthermore, this means that their graphs are reflections of each other in the line $y = x$. This is shown in Example 1.

EXAMPLE 1
Graphing a Logarithmic
Function and Its Inverse

On the same set of axes, sketch the graphs of

$$f(x) = 2^x \quad \text{and} \quad g(x) = \log_2 x$$

Solution:
For $f(x) = 2^x$, we make up the following table of values:

x	-2	-1	0	1	2	3
$f(x) = 2^x$	$\dfrac{1}{4}$	$\dfrac{1}{2}$	1	2	4	8

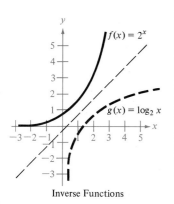

Inverse Functions

FIGURE 4.8

By plotting these points and connecting them with a smooth curve, we have the graph shown in Figure 4.8. Finally, to sketch the graph of $g(x) = \log_2 x$, we use the fact that g is the inverse of f. Notice in Figure 4.8 that the graph of g is obtained by reflecting the graph of f in the line $y = x$.

In Figure 4.8, note how the graph of $g(x) = \log_2 x$ falls sharply as x nears zero from the right. This observation and the fact that g is not defined at $x = 0$ suggest that the y-axis is a *vertical asymptote* for the graph of g. Actually, all logarithmic functions of the form

$$f(x) = \log_b x, \qquad b > 1$$

have the y-axis as a vertical asymptote, as shown in Figure 4.9. Study the general properties of logarithmic functions as we have listed them.

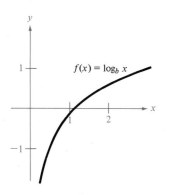

FIGURE 4.9

Graph of $y = \log_b x, b > 1$

1. Domain: All positive reals
2. Range: All reals
3. Intercept: (1, 0)
4. $\log_b x$ increases as x increases
5. y-axis is a vertical asymptote
6. Reflection of the graph of $y = b^x$

EXAMPLE 2
Graphing Logarithmic Functions

On the same set of axes, sketch graphs of

(a) $f(x) = \log x$
(b) $g(x) = \ln x$

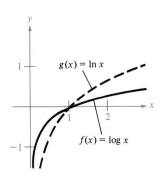

FIGURE 4.10

Solution:

(a) Since

$$f(x) = \log x = \log_{10}x$$

we make up the following table of values. Note that the first three entries in the table can be found without the use of a calculator.

	Without Calculator			With Calculator		
x	$\dfrac{1}{10}$	1	10	2	5	8
$\log_{10}x$	-1	0	1	0.301	0.699	0.903

The graph of f is shown in Figure 4.10.

(b) Since

$$g(x) = \ln x = \log_e x$$

we make up the following table of values. When sketching functions involving the natural logarithm, it is helpful to remember the following approximations:

$$\frac{1}{e} \approx .37, \qquad e \approx 2.72, \qquad e^2 \approx 7.39$$

	Without Calculator			With Calculator		
x	$1/e$	1	e	e^2	2	5	8
$\ln x$	-1	0	1	2	0.693	1.609	2.079

The graph of g is shown in Figure 4.10.

The graphs of logarithmic functions to bases other than 10 or e can be readily obtained from the graph of $f(x) = \log x$ (or $\ln x$) by use of the change of base formula

$$\log_b x = \frac{\log_{10}x}{\log_{10}b} = \left(\frac{1}{\log_{10}b}\right)\log_{10}x$$

This rule shows us that $\log_b x$ is simply a constant multiple of $\log_{10}x$. Hence, a graph of $g(x) = \log_b x$ can be obtained from the graph of $\log_{10}x$ by *multiplying ordinates* (*y*-values) by the factor $1/\log_{10}b$. For instance, the graph of

$$g(x) = \log_4 x$$

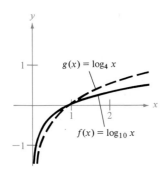

$g(x) = \log_4 x$

$f(x) = \log_{10} x$

FIGURE 4.11

can be obtained from the graph of $f(x) = \log x$ by stretching away from the x-axis by the factor $(1/\log 4) \approx 1.661$, as shown in Figure 4.11. That is, for each x,

$$g(x) = (1.661)f(x)$$

Since logarithms are defined only for positive real numbers, the domain is important in sketching the graphs of logarithmic functions. For instance, Property 5 of logarithms guarantees that

$$\ln(x - 1)^2 = 2 \ln(x - 1)$$

but only for positive values for $(x - 1)$, that is, for $x > 1$. Consequently, the graphs of

$$f(x) = \ln(x - 1)^2 \quad \text{and} \quad g(x) = 2 \ln(x - 1)$$

are not the same because they have different domains, as we will see in the next example.

EXAMPLE 3
Shifts in the Graphs of
Logarithmic Functions

Sketch graphs of

(a) $f(x) = \ln(x - 1)^2$ (b) $g(x) = 2 \ln(x - 1)$

Solution:

As we pointed out in the preceding discussion, these two functions have the same values for $x > 1$. The domains of f and g are as follows:

Function	Domain
$f(x) = \ln(x - 1)^2$ $\qquad$ $g(x) = 2 \ln(x - 1)$	all real $x \neq 1$ $\qquad$ $x > 1$

Using the $\boxed{\text{LN}}$ key on a calculator, we obtain the points shown in the following table. Note the symmetry with respect to the line $x = 1$.

x	-2	-1	0	0.5	1	1.5	2	3	4
$\ln(x - 1)^2$	2.20	1.39	0	-1.39	undef.	-1.39	0	1.39	2.20

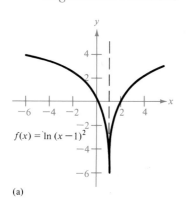

$f(x) = \ln(x - 1)^2$

(a)

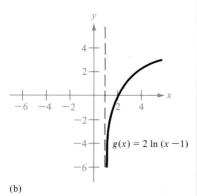

$g(x) = 2 \ln(x - 1)$

(b)

FIGURE 4.12

The graph of f is shown in Figure 4.12(a). To obtain the graph of g, we simply retrace that portion of the graph of f that lies to the right of the line $x = 1$, as shown in Figure 4.12(b). Note that in both cases the x-intercept is $(2, 0)$, a shift of one to the right compared to the graph of $y = \ln x$ in Figure 4.10. Note also that the vertical asymptote falls at $x = 1$, where both functions are undefined.

EXAMPLE 4
Application of a
Logarithmic Function

Students in a precalculus class were given a final exam and then were retested monthly. On equivalent exams given each month thereafter, the average scores decreased according to the *human memory model*

$$F(t) = 75 - 15 \log(t + 1)$$

(a) What was the average score on the original ($t = 0$) exam?
(b) What was the average score at the end of $t = 2$ months? What was the average score at the end of $t = 6$ months?
(c) How long will it take for the average score to drop below 45?
(d) Sketch a graph of F.

Solution:

(a) The original average was

$$F(0) = 75 - 15 \log(0 + 1)$$
$$= 75 - 15 \log 1 = 75 - 15(0)$$
$$= 75$$

(b) After 2 months, the average was

$$F(2) = 75 - 15 \log 3 \approx 75 - 15(.4771)$$
$$\approx 67.8$$

After 6 months, the average was

$$F(6) = 75 - 15 \log 7 \approx 75 - 15(.8451)$$
$$\approx 62.3$$

(c) To find how long it would take for the average score to drop to 45, we set $F(t) = 45$ and solve for t.

$$F(t) = 75 - 15 \log(t + 1) = 45$$
$$-15 \log(t + 1) = -30$$
$$\log(t + 1) = 2$$

Using the antilogarithm to the base 10, we have

$$t + 1 = 10^2$$
$$t = 100 - 1$$
$$= 99 \text{ months}$$

(d) Several points are shown in the following table, and the graph of F is shown in Figure 4.13.

t	0	1	2	6	12
$F(t)$	75	70.5	67.8	62.3	58.3

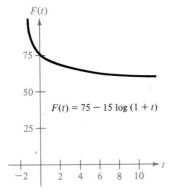

$F(t) = 75 - 15 \log(1 + t)$

FIGURE 4.13

Section Exercises 4.3

In Exercises 1–8, sketch the graph of the function and its inverse on the same set of axes.

1. $f(x) = 3^x$, $g(x) = \log_3 x$
2. $f(x) = 5^x$, $g(x) = \log_5 x$
3. $f(x) = (\frac{1}{2})^x$, $g(x) = \log_{1/2} x$
4. $f(x) = (\frac{1}{3})^x$, $g(x) = \log_{1/3} x$
5. $f(x) = e^x$, $g(x) = \ln x$
6. $f(x) = e^{2x}$, $g(x) = \ln \sqrt{x}$
7. $f(x) = e^{x-1}$, $g(x) = 1 + \ln x$
8. $f(x) = e^x - 1$, $g(x) = \ln(x + 1)$

In Exercises 9–18, use the graph of $y = \ln x$ to match the given function with one of the accompanying graphs.

9. $f(x) = 4 + \ln x$
10. $f(x) = \ln \dfrac{1}{x} = -\ln x$
11. $f(x) = \ln(x - 2)$
12. $f(x) = \ln(x + 5)$
13. $f(x) = \ln x^3 = 3 \ln x$
14. $f(x) = \ln \sqrt{x} = \frac{1}{2} \ln x$
15. $f(x) = -\ln(x + 1)$
16. $f(x) = \ln 2x$
17. $f(x) = \ln x^2$
18. $f(x) = \ln |x - 3|$

(e)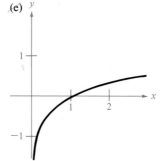

(f)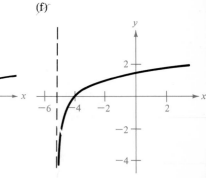

(g)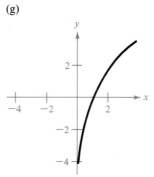

(h)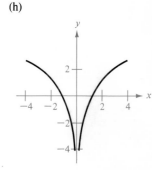

(a)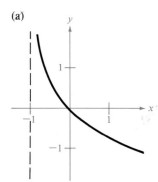

(b)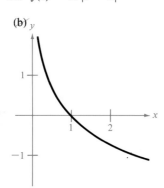

(i)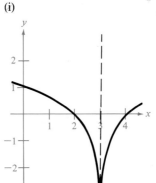

(j)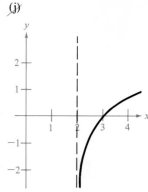

(c)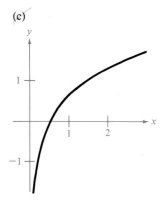

(d)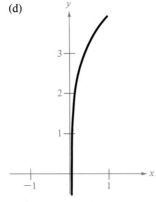

In Exercises 19–26, determine the domain of the function.

19. $f(x) = \log(2x - 1)$
20. $f(x) = \log_4(1 - x)$
21. $g(x) = \ln(-x)$
22. $h(x) = \ln(x - 2) + \ln x$
23. $h(x) = \log_2 \sqrt{4 - x^2}$
24. $g(x) = \log_b(b^{-x})$
25. $f(x) = \log_b |x + 2|$
26. $f(x) = \log_b(x^2 - 4)$

27. The intensity level β, in decibels, of a sound wave is defined by

$$\beta(I) = 10 \log \frac{I}{I_0}$$

where I_0 is an arbitrary intensity of 10^{-16} watts per square centimeter, corresponding roughly to the faintest sound that can be heard. Determine $\beta(I)$ if
(a) $I = 10^{-14}$ watts per square centimeter (whisper)
(b) $I = 10^{-9}$ watts per square centimeter (busy street corner)
(c) $I = 10^{-6.5}$ watts per square centimeter (air hammer)
(d) $I = 10^{-4}$ watts per square centimeter (threshold of pain)
(Intensity is the power transported per unit area of a surface perpendicular to the direction of propagation of the sound wave.)

28. Students in a mathematics class were given an exam and then retested monthly with an equivalent exam. With each passing month the average score for the class decreased according to the human memory model

$$F(t) = 80 - 17 \log(t + 1)$$

(a) What was the average score on the original exam ($t = 0$)?
(b) What was the average score after 4 months?
(c) How long will it take for the average score to drop below 53?

29. A principal P invested at $9\frac{1}{2}\%$, compounded continuously, increases to an amount KP after t years, where t is given by

$$t = \frac{\ln K}{0.095}$$

(a) Complete the following table for the given function:

K	1	2	3	4	6	8	10	12
t								

(b) Use the table of part (a) to graph this function.

30. The time in years for the world population to double if it is increasing at a continuous rate of r is given by

$$t = \frac{\ln 2}{r}$$

Use this function to complete the following table:

r	0.005	0.010	0.015	0.020	0.025	0.030
t						

31. Use a calculator to demonstrate that

$$\frac{\ln x}{\ln y} \neq \ln\left(\frac{x}{y}\right) = \ln x - \ln y$$

by completing the following table:

x	y	$\dfrac{\ln x}{\ln y}$	$\ln \dfrac{x}{y}$	$\ln x - \ln y$
1	2			
3	4			
10	5			
4	0.5			

32. (a) Use a calculator to complete the following table for the function

$$f(x) = \frac{\ln x}{x}$$

x	1	5	10	10^2	10^4	10^6
$f(x)$						

(b) Use the table of part (a) to determine what $f(x)$ approaches as x increases without bound.

Exponential and Logarithmic Equations

4.4

In the first three sections of this chapter, we focused our study mainly on the definitions, properties, graphs, and applications of exponential and logarithmic functions. Here we concentrate on algebraic procedures for solving *equations* involving these functions. In Section 4.1, we solved some simple exponential equations such as $(\frac{1}{2})^x = 64$ by writing each term as a rational power of the base, in this case

$$2^{-x} = 64 = 2^6 \quad \Longrightarrow \quad x = -6$$

However, this method does not work as well for equations like

$$2^x = 7$$

since it is not easy to write 7 as a power of 2. As a general guideline, the following suggestions work well for solving exponential and logarithmic equations.

SUGGESTED RULES FOR SOLVING EXPONENTIAL AND LOGARITHMIC EQUATIONS

> To solve an exponential equation, first take the logarithm of both sides. Then try solving for the unknown.
>
> To solve a logarithmic equation, first rewrite it in exponential form, then try solving for the unknown.

We can use the preceding suggestions to prove the *change of base formula* given in Section 4.2:

$$\log_b x = \frac{\log_a x}{\log_a b}$$

Let $y = \log_b x$, then the equivalent exponential form is $b^y = x$. Therefore,

$$\log_a(b^y) = \log_a x \qquad \qquad \textit{Take } \log_a \textit{ of both sides}$$

$$y \log_a b = \log_a x \qquad \qquad \textit{Property 5}$$

$$y = \frac{\log_a x}{\log_a b} \qquad \qquad \textit{Solve for y}$$

Finally,

$$\log_b x = \frac{\log_a x}{\log_a b} \qquad \qquad \textit{Back substitute for y}$$

EXAMPLE 1
Solving Exponential Equations

Solve for x:

(a) $5^x = 72$ (b) $3^{-x} = 0.026$

Solution:

(a) To solve for x, we take the common logarithm of both sides (natural logarithms would work just as well) to get

$$\log(5^x) = \log 72$$
$$x \log 5 = \log 72 \qquad \text{\textit{Property 5}}$$
$$x = \frac{\log 72}{\log 5} \qquad \text{\textit{Divide by log 5}}$$

The calculator keystroke sequence

72 $\boxed{\text{LOG}}$ $\boxed{\div}$ 5 $\boxed{\text{LOG}}$ $\boxed{=}$

yields $x \approx 2.6572$.

(b) Taking the common logarithm of both sides, we get

$$\log(3^{-x}) = \log 0.026$$
$$-x \log 3 = \log 0.026 \qquad \text{\textit{Property 5}}$$
$$x = -\frac{\log 0.026}{\log 3} \approx 3.3221 \quad \text{\textit{Divide by log 3}}$$

When both members of an equation are exponential functions, a procedure similar to that shown in Example 1 can still be used. However, the algebra is a bit more complicated. Watch how it is done in our next example.

EXAMPLE 2
Solving Exponential Equations

Solve for x:

(a) $4^{2x+3} = 7^{x+2}$ (b) $\dfrac{e^x - e^{-x}}{2} = 5$

Solution:

(a) Taking the common logarithm of both sides, we obtain

$$\log(4^{2x+3}) = \log(7^{x+2})$$
$$(2x + 3) \log 4 = (x + 2) \log 7 \qquad \text{\textit{Property 5}}$$
$$2x \log 4 + 3 \log 4 = x \log 7 + 2 \log 7 \qquad \text{\textit{Distributive Law}}$$
$$2x \log 4 - x \log 7 = 2 \log 7 - 3 \log 4 \qquad \text{\textit{Collect like terms}}$$
$$x(2 \log 4 - \log 7) = 2 \log 7 - 3 \log 4 \qquad \text{\textit{Factor out x}}$$
$$x = \frac{2 \log 7 - 3 \log 4}{2 \log 4 - \log 7} \qquad \text{\textit{Divide}}$$
$$x \approx -0.3231$$

(b) Some preliminary algebra is helpful.

$$\frac{e^x - e^{-x}}{2} = 5 \qquad \text{\textit{Given}}$$

$$e^x - e^{-x} = 10 \qquad \text{\textit{Multiply by 2}}$$

$$e^{2x} - 1 = 10e^x \qquad \text{\textit{Multiply by } } e^x$$

$$(e^x)^2 - 10(e^x) - 1 = 0 \qquad \text{\textit{Quadratic form}}$$

Let $u = e^x$, then we obtain the quadratic equation

$$u^2 - 10u - 1 = 0$$

whose solution, by the Quadratic Formula, is

$$u = \frac{10 \pm \sqrt{100 + 4}}{2} = \frac{10 \pm 2\sqrt{26}}{2} = 5 \pm \sqrt{26}$$

Now, since $u = e^x$, we have

$$u = e^x = 5 \pm \sqrt{26}$$

But e^x is never negative, so the only valid solution is

$$e^x = 5 + \sqrt{26}$$

Taking the *natural* logarithm of both sides, we obtain

$$\ln(e^x) = \ln(5 + \sqrt{26})$$

$$x \ln e = \ln(5 + \sqrt{26}) \qquad \text{\textit{Property 5}}$$

$$x(1) = \ln(5 + \sqrt{26}) \qquad \text{\textit{Property 1}}$$

$$x = \ln(5 + \sqrt{26}) \approx 2.3124$$

In our next example, the *properties of logarithms* play a vital role in the solution procedure.

EXAMPLE 3
Solving Logarithmic Equations

Solve for x:

(a) $\ln(x \mp 2) + \ln(2x - 1) = 2 \ln x$
(b) $\log (5x + 3) - \log(x - 1) = 2$

Solution:

(a) $\ln(x - 2) + \ln(2x - 1) = 2 \ln x \qquad \text{\textit{Given}}$

$$\ln(x - 2)(2x - 1) = \ln x^2 \qquad \text{\textit{Properties 3 and 5}}$$

$$(x - 2)(2x - 1) = x^2 \qquad \text{\textit{Property 8}}$$

$$2x^2 - 5x + 2 = x^2$$

$$x^2 - 5x + 2 = 0 \qquad \text{\textit{Collect like terms}}$$

Using the Quadratic Formula, we obtain

$$x = \frac{5 \pm \sqrt{17}}{2}$$

Since $(x - 2)$ is negative for $x = (5 - \sqrt{17})/2$, the *only valid solution* is

$$x = \frac{5 + \sqrt{17}}{2} \approx 4.5616$$

(b) $\log(5x + 3) - \log(x - 1) = 2$

$$\log\left(\frac{5x + 3}{x - 1}\right) = 2$$

$$\frac{5x + 3}{x - 1} = 10^2$$

$$5x + 3 = 100(x - 1)$$

$$x = \frac{103}{95}$$

We complete this section with two practical applications of exponential equations.

EXAMPLE 4
Compound Interest

An amount of $11,000 is invested in a trust fund at an annual interest rate of 9.5%, compounded continuously.

(a) How long will it take for the initial investment to double in value?
(b) What interest rate is required for the initial investment to double in 10 years' time?

Solution:

(a) For continuous compounding, we have the formula

$$B = Pe^{rt}$$

In this case, $P = 11{,}000$, $r = 0.095$, and $B = 2P = 22{,}000$. Thus, we have

$$22{,}000 = 11{,}000(e^{0.095t})$$

$$2 = e^{0.095t}$$

$$\ln 2 = \ln(e^{0.095t}) \qquad \text{\textit{Take ln of both sides}}$$

$$\ln 2 = 0.095t(\ln e) = 0.095t$$

$$t = \frac{\ln 2}{0.095} \approx 7.3 \text{ years}$$

(b) In this case, $P = 11{,}000$, $B = 22{,}000$, and $t = 10$. Thus, we have

$$22{,}000 = 11{,}000(e^{10r})$$

$$2 = e^{10r}$$

$$\ln 2 = \ln(e^{10r}) = 10r(\ln e) = 10r$$

$$r = \frac{\ln 2}{10} \approx 0.0693 \approx 6.93\%.$$

EXAMPLE 5
Logistic Curve

On a college campus of 5000 students, a single student returned from vacation with a contagious flu virus. The spread of this disease through the student body is given by

$$s(t) = \frac{5{,}000}{1 + 4{,}000\, e^{-0.8t}}$$

where $s(t)$ is the total number infected after t days.

(a) How many are infected after 5 days?
(b) If the college will close if 40% of the students are ill, after how many days will it close?

Solution:

(a) After 5 days, the number infected is

$$s(5) = \frac{5000}{1 + 4000e^{-0.8(5)}}$$

$$= \frac{5000}{1 + 4000e^{-4}} \approx 67 \qquad \textit{(To nearest integer)}$$

One possible calculator key stroke sequence for finding $s(5)$ is

4 $\boxed{+/-}$ $\boxed{e^x}$ $\boxed{\times}$ 4000 $\boxed{+}$ 1 $\boxed{=}$ $\boxed{1/x}$ $\boxed{\times}$ 5000 $\boxed{=}$

(b) In this case, the number infected is

$$(0.40)(5000) = 2000$$

Therefore, we solve for t in the equation

$$2000 = \frac{5000}{1 + 4000e^{-0.8t}}$$

obtaining

$$2000 + 8{,}000{,}000\, e^{-0.8t} = 5000$$

$$e^{-0.8t} = \frac{3000}{8{,}000{,}000} = \frac{3}{8000}$$

$$\ln(e^{-0.8t}) = \ln 3 - \ln 8000$$

$$-0.8t = \ln 3 - \ln 8000$$

$$t = \frac{\ln 3 - \ln 8000}{-0.8} \approx 9.86$$

Hence, in 10 days, at least 40% of the students will be infected and the college will close.

Section Exercises 4.4

In Exercises 1–30, solve for the unknown.

1. $4^x = \frac{1}{16}$
2. $7^x = \frac{1}{49}$
3. $3^x = 243$
4. $8^x = 4$
5. $4^x = 12$ (Use common logarithms)
6. $6^{-x} = 52$ (Use common logarithms)
7. $(8.5)^{-x} = 360$ (Use natural logarithms)
8. $(\frac{3}{4})^x = 0.15$ (Use natural logarithms)
9. $e^{0.09t} = 3$
10. $e^{0.125t} = 8$
11. $\left(1 + \frac{0.10}{12}\right)^{12t} = 2$
12. $\left(1 + \frac{0.065}{365}\right)^{365t} = 4$
13. $\frac{10,000}{1 + 19e^{-t/5}} = 2000$
14. $80e^{-t/2} + 20 = 70$
15. $\left(\frac{1}{1.0775}\right)^N = 0.2247$
16. $3^{2x+1} = 5^{x+2}$
17. $10^{7-x} = 5^{x+1}$
18. $4^{x^2} = 100$
19. $\log_3 x + \log_3(x - 2) = 1$
20. $\log_{10}(x + 3) - \log_{10} x = 1$
21. $x^2 - 2x = \log_5 125$
22. $x - x^2 = \log_4 \frac{1}{16}$
23. $\log_2(x + 5) - \log_2(x - 2) = 3$
24. $\log \sqrt{x + 2} = 1$
25. $\ln x^2 = (\ln x)^2$
26. $\frac{e^x + e^{-x}}{2} = 2$
27. $\frac{e^x + e^{-x}}{e^x - e^{-x}} = 2$
28. $\frac{e^x - e^{-x}}{e^x + e^{-x}} = \frac{1}{4}$
29. $\ln(e^{x^2}) = 4$
30. $\log_b(b^{2x-1}) = 5$

31. $1000 is deposited into a fund at an annual percentage rate of 11%. Find the time for the investment to double if the interest is compounded
   (a) annually   (b) monthly   (c) daily   (d) continuously
32. Repeat Exercise 31, using a percentage rate of $10\frac{1}{2}$% and finding the time it will take for the investment to triple.
33. Complete the following table for the time t necessary for P dollars to triple if interest is compounded continuously at the rate r.

r	2%	4%	6%	8%	10%	12%
t						

34. The demand equation for a certain product is given by
$$P = 500 - 0.5(e^{0.004x})$$
Find the demand x if the price charged is
   (a) $P = $350   (b) $P = $300
35. The demand equation for a certain product is given by
$$P = 5000\left(1 - \frac{4}{4 + e^{-0.002x}}\right)$$
Find the demand x if the price charged is
   (a) $P = $600   (b) $P = $400
36. The population growth of a city is given by $105,300e^{0.015t}$, where t is the time in years with $t = 0$ corresponding to 1985. According to this model, in what year will the city have a population of 150,000?
37. The yield V (in millions of cubic feet per acre) for a forest at age t is given by
$$V = 6.7e^{-48.1/t}$$
Find the time necessary to have a yield of
   (a) 1.3 million cubic feet   (b) 2 million cubic feet
38. In a group project in learning theory, a mathematical model for the proportion P of correct responses after n trials was found to be
$$P = \frac{0.83}{1 + e^{-0.2n}}$$
After how many trials will 60% of the responses be correct?
39. A certain lake is stocked with 500 fish, and their population increases according to the **logistics curve**
$$p(t) = \frac{10,000}{1 + 19e^{-t/5}}$$
where t is measured in months. After how many months will the fish population be 2000?

Newton's Law of Cooling states that the rate of change in the temperature of an object is proportional to the difference between its temperature and the temperature of its environment. If $T(t)$ is the temperature of the object at time t, T_0 is the initial temperature, and T_e is the constant temperature of the environment, then

$$T(t) = T_e + (T_0 - T_e)e^{-kt}$$

Use this model in Exercises 40 and 41.

40. An object in a room at 70° cools from 350° to 150° in 45 minutes. Find

(a) the temperature of the object as a function of time
(b) the temperature after it has cooled for 1 hour
(c) the time necessary for the object to cool to 80°

41. Using Newton's Law of Cooling, determine the reading on a thermometer 5 minutes after it is taken from a room at 72° Fahrenheit to the outdoors where the temperature is 20°, if the reading dropped to 48° after 1 minute.

Review Exercises / Chapter 4

In Exercises 1–15, sketch the graph of the function.

1. $f(x) = 6^x$
2. $f(x) = 0.3^x$
3. $g(x) = 6^{-x}$
4. $g(x) = 0.3^{-x}$
5. $h(x) = e^{-x/2}$
6. $h(x) = 2 - e^{-x/2}$
7. $s(t) = \dfrac{4}{1 + e^{-t}}$
8. $s(t) = 4e^{-2/t},\ t > 0$
9. $f(x) = \ln(x - 3)$
10. $f(x) = \ln|x|$
11. $f(x) = \ln x + 3$
12. $f(x) = \dfrac{\ln x}{4}$
13. $g(x) = \log_2 x$
14. $g(x) = \log_5 x$
15. $h(x) = \ln(e^{x-1})$

In Exercises 16–20, use the properties of logarithms to write the expression as a sum, difference, or multiple of logarithms.

16. $\ln \left| \dfrac{x - 1}{x + 1} \right|$

17. $\ln \left| \dfrac{x^2 + 1}{x} \right|$

18. $\ln \sqrt{\dfrac{x^2 + 1}{x^4}}$

19. $\ln[(x^2 + 1)(x - 1)]$

20. $\ln \sqrt[5]{\dfrac{4x^2 - 1}{4x^2 + 1}}$

In Exercises 21–25, write the expression as a single logarithm.

21. $\frac{1}{2} \ln |2x - 1| - 2 \ln |x + 1|$
22. $5 \ln |x - 2| - \ln |x + 2| - 3 \ln |x|$
23. $2[\ln x + \frac{1}{3} \ln \sqrt{y}]$
24. $\frac{1}{2} \ln (x^2 + 4x) - \ln 2 - \ln x$
25. $\ln 3 + \frac{1}{3} \ln (4 - x^2) - \ln x$

In Exercises 26–30, determine if the statement or equation is true or false.

26. The domain of the function $f(x) = \ln |x|$ is the set of all real numbers.
27. $\ln (x + y) = \ln x + \ln y$
28. $\dfrac{\ln x}{\ln y} = \ln x - \ln y$
29. $\ln \sqrt{x^4 + 2x^2} = \ln(|x|\sqrt{x^2 + 2})$

30. $e^{x-1} = \dfrac{e^x}{e}$

31. The demand equation for a certain product is given by

$$p = 500 - 0.5e^{0.004x}$$

Find the demand x if the price charged is
(a) $p = \$450$
(b) $p = \$400$

32. In a typing class, the average number of words per minute typed after t weeks of lessons was found to be

$$N = \dfrac{157}{1 + 5.4e^{-0.12t}}$$

Find the time necessary to type
(a) 50 words per minute
(b) 75 words per minute

In Exercises 33–36, find the exponential growth function $y = Ce^{kt}$ that passes through the two points.

33.
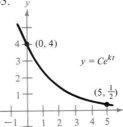

34.
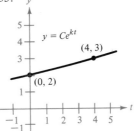

35.
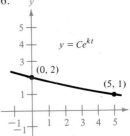

36.
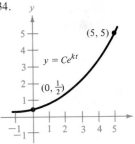

37. A deposit of $750 is made in a savings account for which the interest is compounded continuously. The balance will double in $7\frac{3}{4}$ years.
    (a) What is the annual percentage rate for this account?
    (b) Find the balance in the account after 10 years.

38. A deposit of $10,000 is made in a savings account for which the interest is compounded continuously. The balance will double in 5 years.
    (a) What is the annual percentage rate for this account?
    (b) Find the balance after 1 year.

39. The management at a certain factory has found that the maximum number of units a worker can produce in a day is 30. The learning curve for the number of units N produced per day after a new employee has worked t days is given by

    $$N = 30(1 - e^{kt})$$

    After 20 days on the job, a particular worker produced 19 units.
    (a) Find the learning curve for this worker.
    (b) How many days should pass before this worker is producing 25 units per day?

40. The management in Exercise 39 requires that a new employee be producing at least 20 units per day after 30 days on the job.
    (a) Find the learning curve describing this minimum requirement.

(b) Find the number of days before a minimal achiever is producing 25 units per day.

41. The sales S (in thousands of units) of a new product after it is on the market t years is given by

    $$S = 30(1 - e^{kt})$$

    (a) Find S as a function of t if 5000 units have been sold after 1 year.
    (b) How many units have been sold after 5 years?
    (c) Sketch a graph of this sales function.

42. Complete the following table for the function $f(x) = e^{-x}$:

x	0	1	2	5	10	15
e^{-x}	1	0.36788				

This table demonstrates that for $k > 0$, e^{-kx} approaches 0 as x increases.

43. Use the result of Exercise 42 to determine what each function ($k > 0$) approaches as x increases.
    (a) $f(x) = 30(1 - e^{-0.025x})$
    (b) $f(x) = \dfrac{50}{2 + 3e^{-0.2x}}$
    (c) $f(x) = (100 - a)e^{-kx} + a$
    (d) $f(x) = 5000\left(1 - \dfrac{4}{4 + e^{-0.002x}}\right)$

CHAPTER

5 | TRIGONOMETRY

The Unit Circle and Angle Measurement

5.1

From the Greek, the word **trigonometry** means "measurement of triangles." Initially, trigonometry dealt with the relationships among the sides and angles of triangles. As such it was instrumental in the scientific development of astronomy, navigation, and surveying.

With the advent of calculus in the seventeenth century, and a resulting expansion of knowledge in the physical sciences, a different perspective arose, one that viewed the classic trigonometric relationships as *functions* with the set of real numbers as their domains. Consequently, the applications of trigonometry expanded to include the vast number of physical phenomena that involve rotations or vibrations, including sound waves, light rays, planetary orbits, vibrating strings, pendulums, and orbits of atomic particles.

Our approach to trigonometry will incorporate *both* perspectives. Our goal is to appropriately integrate these two perspectives of trigonometry so that you can readily comprehend this very practical branch of mathematics.

The Unit Circle

The **unit circle** provides a convenient model for discussing some fundamental concepts of trigonometry. Note in Figure 5.1 how a number scale can be placed on the unit circle. Since this circle has a circumference of 2π, we will initially restrict this scale to the interval $0 \le t < 2\pi$. We make two important observations about each real number t in this interval.

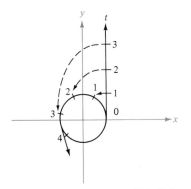

Placing a Number Scale on the Unit Circle

FIGURE 5.1

1. There is a unique point (x, y) on the unit circle that matches t, and conversely. (See Figure 5.2.)

197

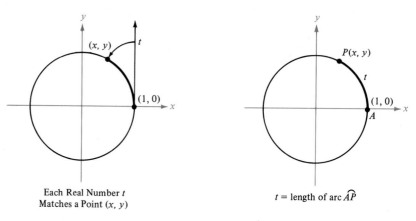

Each Real Number *t*
Matches a Point (*x, y*)

FIGURE 5.2

t = length of arc $\widehat{AP}$

FIGURE 5.3

2. The *length* of the arc measured (counterclockwise) along the unit circle from $A(1, 0)$ to $P(x, y)$ is given by t. (See Figure 5.3.)

This correspondence between real numbers (arc length) and points on the unit circle plays a key role in our study of trigonometry. Recall from Examples 5 and 6 in Section 1.6 that we determined the points that divide the unit circle into 8 and 12 equal arcs (see Figures 5.4 and 5.5). These points give us the arc lengths most commonly used in trigonometry.

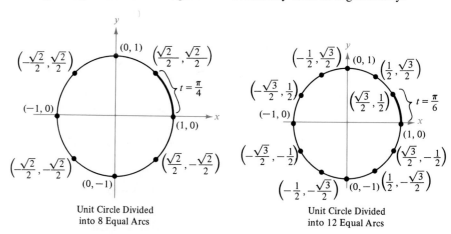

Unit Circle Divided
into 8 Equal Arcs

FIGURE 5.4

Unit Circle Divided
into 12 Equal Arcs

FIGURE 5.5

EXAMPLE 1
Matching Points with Arc Lengths

Find the points on the unit circle that match the arc lengths

(a) $\dfrac{3\pi}{4}$ (b) $\dfrac{7\pi}{6}$ (c) $\dfrac{5\pi}{3}$

Solution:

(a) Refer to Figure 5.4. Since $3\pi/4 = 3(\pi/4)$, we move counterclockwise from $(1, 0)$ through 3 arc segments to the second quadrant point $(-\sqrt{2}/2, \sqrt{2}/2)$.

(b) Refer to Figure 5.5. Since $7\pi/6 = 7(\pi/6)$, we move counterclockwise through 7 arc segments to the third quadrant point $(-\sqrt{3}/2, -1/2)$.

(c) Refer to Figure 5.5. Since $5\pi/3 = 10(\pi/6)$, we move counterclockwise through 10 arc segments to the fourth quadrant point $(1/2, -\sqrt{3}/2)$.

The preceding discussion can be extended so that to *every* real number there corresponds exactly one point on the unit circle. For numbers greater than 2π, we retrace (counterclockwise) the unit circle, making additional revolutions as necessary. For example, since

$$\frac{13\pi}{6} = 2\pi + \frac{\pi}{6} = \text{(one revolution)} + \frac{\pi}{6}$$

the real number $t = 13\pi/6$ matches the same point as $t = \pi/6$, as shown in Figure 5.5. In fact, the numbers $2n\pi + (\pi/6)$, where n is an integer, all correspond to the same point on the unit circle.

For *negative* real numbers, we move *clockwise* around the unit circle to locate corresponding points. For example, in Figure 5.4 we can see that $-\pi$ would match the same point as π, and $-3\pi/2$ would match the same point as $\pi/2$.

Radian Measure

From geometry we know that an arc from point $(1, 0)$ to the point (x, y) on the unit circle (Figure 5.6) subtends a **central angle** θ (the lowercase Greek letter theta). An angle in this position is said to be in **standard position.** [Recall from geometry that an **angle** consists of two rays that originate from a common point (**vertex**) with one ray called the **initial side** and the other the **terminal side.**] To assign a measure to the angle θ in Figure 5.6, it seems natural to use the number t, the length of the intercepted arc on the unit circle. We call t the **radian measure** of angle θ. For example, the angle θ in Figure 5.7 appears to measure about 0.75 radian.

Actually, the radian measure of an angle can be found using a circle of *any* radius, $r > 0$. In Figure 5.8, angle θ subtends an arc of length t on the unit circle and an arc of length s on a circle of radius $r \neq 1$. From geometry, we know that the ratio of the arc lengths is the same as the ratio of the radii. Thus, we have $t/s = 1/r$ or $t = s/r$, and since $\theta = t$, we conclude that θ has radian measure $\theta = s/r$. If $s = r$, then $\theta = 1$; that is, a subtended arc with length equal to the radius has a corresponding angle of 1 radian. In summary, we have the following definition of radian measure.

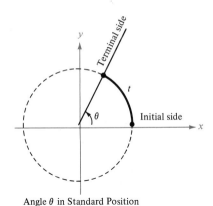

Angle θ in Standard Position

FIGURE 5.6

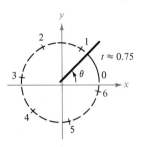

θ has radian measure t.

FIGURE 5.7

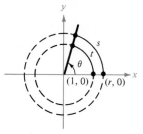

Measure of θ: $\dfrac{t}{1} = \dfrac{s}{r}$

FIGURE 5.8

DEFINITION OF RADIAN MEASURE

> The **radian measure** of a central angle θ of a circle with radius $r > 0$ is given by
>
> $$\theta = \frac{s}{r} \text{ radians}$$
>
> where s is the length of the circular arc subtended by θ (see Figure 5.8).

Remark: Rewriting this formula as $s = r\theta$, we have a way to measure the circular arc length for a given central angle and radius. This is illustrated in Example 3.

EXAMPLE 2
Radian Measure of an Angle

Find the radian measure of a central angle θ subtended by an arc of 28 centimeters on a circle of radius 12 centimeters.

Solution:
By definition,

$$\theta = \frac{s}{r} = \frac{28 \text{ centimeters}}{12 \text{ centimeters}} = \frac{7}{3} \text{ radians} \approx 2.33$$

Remark: Note in Example 2 how the units of measure (centimeters) cancel out, leaving a dimensionless real number. For this reason the word *radian* is often omitted when working with radian measure.

EXAMPLE 3
An Application of Arc Length

A telescope centered in a 44-foot (diameter) hemispherical planetarium requires an opening corresponding to a central angle of 1.1 radians. Find the arc length of the door that exactly covers this opening. (See Figure 5.9.)

Solution:
In this case $r = 22$ feet and $\theta = 1.1$ radians. Thus, we have an arc length of

$$s = r\theta = (22)(1.1) = 24.2 \text{ feet}$$

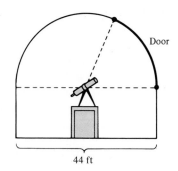

44 ft

FIGURE 5.9

Degree Measure

A second unit for measuring angles is the **degree.** It is determined by constructing a circle centered at the origin, and then dividing its circumference into 360 equal parts. A central angle of 1 degree, denoted by $1°$, subtends an arc equal to $\frac{1}{360}$ of the circumference of the circle. The **degree measure** of an angle θ in standard position is equal to the number of those 360 parts that the terminal side intercepts on the circle (see Figure 5.10). As in radian measure, positive angles are measured counterclockwise and negative angles are measured clockwise.

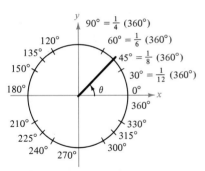

Degree Measure of an Angle

FIGURE 5.10

Parts of Degrees

We can divide degrees into smaller parts in two ways. We can use *decimals,* or we can use *minutes and seconds.* To divide a degree into minutes and seconds, we follow the pattern of time in which each hour is divided into 60 minutes and each minute is divided into 60 seconds. Symbolically, minutes are represented by ′ and seconds by ″. For example 40° 53′ 12″ denotes an angle of 40 degrees, 53 minutes, and 12 seconds. Its *decimal degree* equivalent is calculated as

$$40° \ 53' \ 12'' = \left(40 + \frac{53}{60} + \frac{12}{3600} \right)° \approx 40.887°$$

Remark: Because we use calculators almost exclusively (rather than tables), we will use the decimal degree notation throughout the remainder of this text. A three-place decimal form has accuracy approximately equivalent to the nearest second.

By definition, an angle with an arc of one full circle has degree measure 360°. By the radian measure formula, such an angle has radian measure

$$\theta = \frac{s}{r} = \frac{2\pi r}{r} = 2\pi$$

Consequently, radians and degrees have the relationship

$$360° = 2\pi \text{ radians} \qquad \text{or} \qquad 180° = \pi \text{ radians}$$

CONVERSIONS: DEGREES ↔ RADIANS

1. To convert **degrees,** to **radians,** use

$$1° = \frac{\pi}{180} \approx 0.01745 \text{ radian}$$

2. To convert **radians** to **degrees,** use

$$1 \text{ radian} = \left(\frac{180}{\pi} \right)° \approx 57.3°$$

Some of the *basic angles* used in the study of trigonometry are given in Table 5.1 in both degrees and radians.

TABLE 5.1
Basic Angles in Trigonometry

Radians	$\pi/6$	$\pi/4$	$\pi/3$	$\pi/2$	π	2π
Degrees	30°	45°	60°	90°	180°	360°
Part of Circle	one-twelfth	one-eighth	one-sixth	one-fourth	half	full

EXAMPLE 4
Converting Degrees to Radians

Convert the following angles from degree to radian measure:

(a) 5° (b) −210°

Solution:
From the equation $1° = \pi/180$, we obtain

(a) $5° = 5\left(\dfrac{\pi}{180}\right) = \dfrac{\pi}{36} \approx 0.0873$ radian

(b) $-210° = -210\left(\dfrac{\pi}{180}\right) = \dfrac{-7\pi}{6} \approx -3.6652$ radians

EXAMPLE 5
Converting Radians to Degrees

Convert the following angles from radian to degree measure:

(a) $\dfrac{5\pi}{12}$ (b) −1.7

Solution:
From the equation

$$1 \text{ radian} = \left(\dfrac{180}{\pi}\right)°$$

we obtain

(a) $\dfrac{5\pi}{12}$ radians $= \dfrac{5\pi}{12}\left(\dfrac{180}{\pi}\right) = \dfrac{5(180)}{12} = 5(15) = 75°$

(b) -1.7 radians $= -1.7\left(\dfrac{180}{\pi}\right) \approx -97.40°$

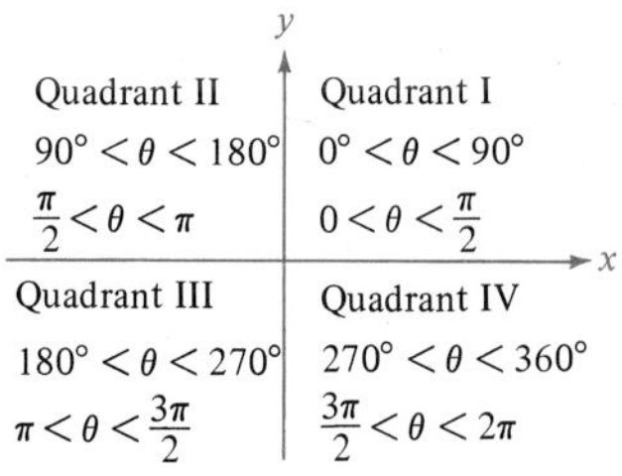

Recall from geometry that a **right angle** has measure 90° and a **straight angle** has measure 180°. Angles with measures between 0° and 90° are called **acute angles,** and their terminal sides lie in Quadrant I. Angles with measures between 90° and 180° are called **obtuse angles,** and their terminal sides lie in Quadrant II. The quadrant locations of all angles with measures between 0° and 360° are shown in Figure 5.11, along with their equivalent radian measures. Angles of 0°, 90°, 270°, and 360° whose terminal sides lie on the coordinate axes are called **quadrant** (or **quadrantal**) **angles.**

	y	
Quadrant II		Quadrant I
$90° < \theta < 180°$		$0° < \theta < 90°$
$\dfrac{\pi}{2} < \theta < \pi$		$0 < \theta < \dfrac{\pi}{2}$
Quadrant III		Quadrant IV
$180° < \theta < 270°$		$270° < \theta < 360°$
$\pi < \theta < \dfrac{3\pi}{2}$		$\dfrac{3\pi}{2} < \theta < 2\pi$

Quadrant Location of Angles

FIGURE 5.11

Sketching Angles

To sketch angles greater than 90° ($\pi/2$ rad), we make use of acute angles called **reference angles.**

DEFINITION OF REFERENCE ANGLES

For any angle θ, its **reference angle** is the acute angle θ' that the terminal side of θ makes with the positive or negative x-axis (see Figure 5.12).

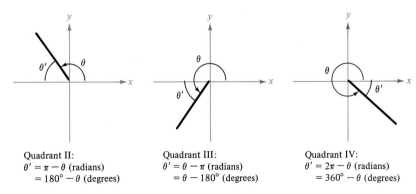

Quadrant II:
$\theta' = \pi - \theta$ (radians)
$\quad = 180° - \theta$ (degrees)

Quadrant III:
$\theta' = \theta - \pi$ (radians)
$\quad = \theta - 180°$ (degrees)

Quadrant IV:
$\theta' = 2\pi - \theta$ (radians)
$\quad = 360° - \theta$ (degrees)

FIGURE 5.12 *Reference Angles*

Remark: The measure of θ' is the *absolute* measure of the angle between the terminal side of θ and the *x*-axis. We ignore direction.

Based on Figure 5.12, we can find θ' for angles with terminal sides in any of the four quadrants, as shown in the following summary. We note that two angles in standard position that have the same terminal side are called **coterminal angles.** For example, in Figure 5.13(c) the angles 225° and $-135°$ are coterminal.

RULES FOR FINDING REFERENCE ANGLES

1. For all θ in $[0, 2\pi]$

Quadrant	θ' *(Radians)*	θ' *(Degrees)*
I	θ	θ
II	$\pi - \theta$	$180° - \theta$
III	$\theta - \pi$	$\theta - 180°$
IV	$2\pi - \theta$	$360° - \theta$

2. For all $\theta > 2\pi$ or $\theta < 0$, find the angle between 0 and 2π that is *coterminal* with θ. Then apply the rules in part (1) to this coterminal angle.

Remark: In the remainder of our work we will round off final answers to four decimal places. For example, we will approximate π as

$$\pi \approx 3.1416$$

EXAMPLE 6
Sketching Angles Using Reference Angles

Make a sketch of each of the following angles in standard position and show the reference angle θ':

(a) $\theta = 300°$ (b) $\theta = 2.3$ (c) $\theta = -135°$

Solution:

(a) Since $\theta = 300°$ lies in Quadrant IV, the angle it makes with the x-axis is

$$\theta' = 360° - 300° = 60°$$

(b) Since $\theta = 2.3$ lies between

$$\frac{\pi}{2} \approx 1.5708 \qquad \text{and} \qquad \pi \approx 3.1416$$

θ is in Quadrant II and thus

$$\theta' = \pi - 2.3 \approx 0.8416$$

(c) First, we determine that $-135°$ is coterminal with

$$-135° + 360° = 225°$$

which lies in Quadrant III. Hence,

$$\theta' = 225° - 180° = 45°$$

Each angle θ and its reference angle θ' are sketched in Figure 5.13.

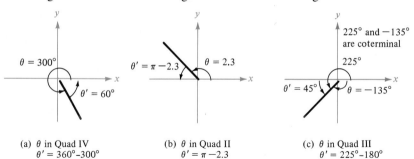

(a) θ in Quad IV $\qquad$ (b) θ in Quad II $\qquad$ (c) θ in Quad III
$\theta' = 360°-300°$ $\qquad\quad$ $\theta' = \pi - 2.3$ $\qquad\quad$ $\theta' = 225°-180°$

FIGURE 5.13 *Reference Angles*

Section Exercises 5.1

In Exercises 1–12, sketch the angles in standard position.

1. $30°$ $\qquad$ 2. $150°$ $\qquad$ 3. $-270°$ $\qquad$ 4. $-120°$

5. $405°$ $\qquad$ 6. $-480°$ $\qquad$ 7. $\dfrac{5\pi}{4}$ $\qquad$ 8. $\dfrac{2\pi}{3}$

9. $-\dfrac{7\pi}{4}$ $\qquad$ 10. $-\dfrac{5\pi}{2}$ $\qquad$ 11. $\dfrac{11\pi}{6}$ $\qquad$ 12. 7π

In Exercises 13–20, determine two coterminal angles (one positive and one negative) for the given angle. Give your answers in degrees.

13.

$\theta = 36°$

14.

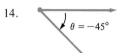

$\theta = -45°$

15.

$\theta = -120°$

16.

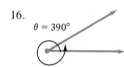

$\theta = 390°$

17. 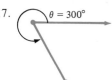 $\theta = 300°$

18. 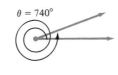 $\theta = 740°$

19. 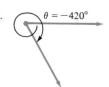 $\theta = -420°$

20. 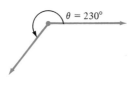 $\theta = 230°$

In Exercises 21–28, determine two coterminal angles (one positive and one negative) for the given angle. Give your answers in radians.

21. $\theta = \dfrac{\pi}{9}$

22. 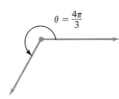 $\theta = \dfrac{4\pi}{3}$

23. 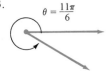 $\theta = \dfrac{11\pi}{6}$

24. 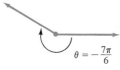 $\theta = -\dfrac{7\pi}{6}$

25. 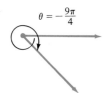 $\theta = -\dfrac{9\pi}{4}$

26. $\theta = -\dfrac{2\pi}{15}$

27. 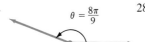 $\theta = \dfrac{8\pi}{9}$

28. $\theta = \dfrac{8\pi}{45}$

In Exercises 29–36, express the given angle in radian measure as a multiple of π.

29. $30°$ 30. $150°$ 31. $315°$ 32. $120°$
33. $-20°$ 34. $-240°$ 35. $-270°$ 36. $144°$

In Exercises 37–44, express the given angle in degree measure.

37. $\dfrac{3\pi}{2}$ 38. $\dfrac{7\pi}{6}$ 39. $-\dfrac{7\pi}{12}$ 40. $\dfrac{\pi}{9}$

41. $\dfrac{7\pi}{3}$ 42. $-\dfrac{11\pi}{30}$ 43. $\dfrac{11\pi}{6}$ 44. $\dfrac{34\pi}{15}$

In Exercises 45–54, convert the angle from degree to radian measure. List your answers to three decimal places.

45. $115°$ 46. $87.4°$ 47. $-216.35°$ 48. $-48.27°$
49. $532°$ 50. $0.54°$ 51. $-0.83°$ 52. $-1425°$
53. $345°$ 54. $741.6°$

In Exercises 55–64, convert the angle from radian to degree measure. List your answers to three decimal places.

55. $\dfrac{\pi}{7}$ 56. $\dfrac{5\pi}{11}$ 57. $\dfrac{15\pi}{8}$ 58. 6.5π

59. -4.2π 60. 4.8 61. -2 62. -0.57
63. 7.5 64. 0.325

In Exercises 65–68, convert the angle measurement to decimal form.

65. (a) $54° \ 45'$ 66. (a) $245° \ 10'$
   (b) $-128° \ 30'$ (b) $2° \ 12'$
67. (a) $85° \ 18' \ 30''$ 68. (a) $-135° \ 36''$
   (b) $330° \ 25''$ (b) $-408° \ 16' \ 25''$

In Exercises 69–84, (a) sketch the given angle, θ, in standard position, (b) show the reference angle, θ', and (c) find the magnitude of the reference angle.

69. $\theta = 203°$ 70. $\theta = 127°$ 71. $\theta = 309°$
72. $\theta = 226°$ 73. $\theta = -245°$ 74. $\theta = -72°$
75. $\theta = -145°$ 76. $\theta = -239°$ 77. $\theta = \dfrac{2\pi}{3}$

78. $\theta = \dfrac{7\pi}{6}$ 79. $\theta = \dfrac{7\pi}{4}$ 80. $\theta = \dfrac{8\pi}{9}$

81. $\theta = 3.5$ 82. $\theta = 5.8$ 83. $\theta = \dfrac{11\pi}{3}$

84. $\theta = -\dfrac{7\pi}{10}$

85. Let r represent the radius of a circle, θ the central angle (measured in radians), and s the length of the arc subtended by the angle, as shown in Figure 5.14. Use the relationship

$$\theta = \frac{s}{r}$$

FIGURE 5.14

to complete the following table:

r	8 ft	15 in.	85 cm		
s	12 ft			96 in.	8642 mi
θ		1.6	$\dfrac{3\pi}{4}$	4	$\dfrac{2\pi}{3}$

86. The minute hand on a clock is 3.5 inches long. (See Figure 5.15.) Through what distance does the tip of the minute hand move in 25 minutes?

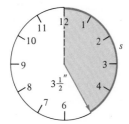

FIGURE 5.15

87. A man bends his elbow through 75°. The distance from his elbow to the top of his index finger is 18.75 inches. (See Figure 5.16.)
   (a) Find the radian measure of this angle.
   (b) Find the distance the tip of the index finger moves.

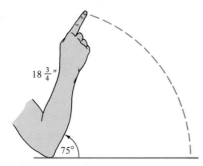

FIGURE 5.16

88. A tractor tire, 5 feet in diameter, is partially filled with a liquid ballast for additional traction. To check the air pressure, the tractor operator rotates the tire until the valve stem is at the top, so that the liquid will not enter the gauge. On a given occasion, the operator notes that the tire must be rotated 80° to have the stem in the proper position. (See Figure 5.17.)
   (a) Find the radian measure of this rotation.
   (b) How far must the tractor be moved to get the valve stem in the proper position?

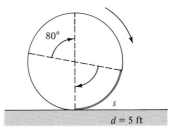

FIGURE 5.17

89. Assuming that the earth is a sphere of radius 4000 miles, what is the difference in latitude of two cities, one of which is 325 miles due north of the other?

90. A car is moving at the rate of 50 miles per hour, and the diameter of its wheels is 2 feet. Find
   (a) the number of revolutions per minute that the wheels are rotating
   (b) the number of radians per minute (*angular speed*) generated by a line segment from the center to the circumference of each wheel

91. Repeat Exercise 90 for the wheels on a truck if they have a diameter of 2.5 feet.

92. A 2-inch-diameter pulley on an electric motor that runs at 1700 revolutions per minute is connected by a belt to a 4-inch-diameter pulley on a saw arbor. Find
   (a) the angular velocity (radians per minute) of each pulley
   (b) the revolutions per minute of the saw

The Trigonometric
Functions   5.2

The unit circle provides a convenient model for introducing the trigonometric functions. As discussed in Section 5.1, we begin by placing a number scale around the unit circle. (See Figure 5.18.) Since each real number t matches up with a single point (x, y) on the unit circle, it follows that the coordinates

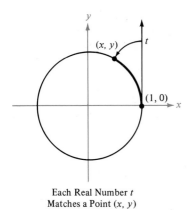

Each Real Number t
Matches a Point (x, y)

FIGURE 5.18

x and y are two distinct functions of the real variable t. We use these two functions to define the six **trigonometric functions** as follows.

DEFINITIONS OF THE SIX TRIGONOMETRIC FUNCTIONS

For any real number t, let (x, y) be the point on the unit circle that corresponds to t.

1. The **sine function** associates each real number t with the y-coordinate of the corresponding point on the unit circle. It is denoted by

$$\sin t = y$$

2. The **cosine function** associates each real number t with the x-coordinate of the corresponding point on the unit circle. It is denoted by

$$\cos t = x$$

3. If $x \neq 0$ then the **tangent** and **secant functions** are defined as

$$\tan t = \frac{\sin t}{\cos t} = \frac{y}{x} \quad \text{and} \quad \sec t = \frac{1}{\cos t} = \frac{1}{x}$$

4. If $y \neq 0$, then the **cotangent** and **cosecant functions** are defined as

$$\cot t = \frac{1}{\tan t} = \frac{x}{y} \quad \text{and} \quad \csc t = \frac{1}{\sin t} = \frac{1}{y}$$

Remark: These functions of real numbers are often called **circular functions** because of their close association with the unit circle.

EXAMPLE 1
Evaluating the Trigonometric Functions Using the Unit Circle

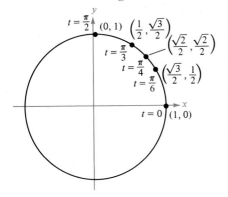

FIGURE 5.19

Find the values of $\sin t$, $\cos t$, and $\tan t$ at

$$t = 0, \quad \frac{\pi}{6}, \quad \frac{\pi}{4}, \quad \frac{\pi}{3}, \quad \text{and} \quad \frac{\pi}{2}$$

Solution:
We begin by associating each value of t with its corresponding coordinates on the unit circle. (See Figure 5.19, which is derived from Figures 5.4 and 5.5.) For $t = 0$, the corresponding point is $(x, y) = (1, 0)$ and we have

$$\sin 0 = y = 0, \quad \cos 0 = x = 1, \quad \text{and} \quad \tan 0 = \frac{y}{x} = \frac{0}{1} = 0$$

For $t = \pi/6$, the corresponding point is $(x, y) = (\sqrt{3}/2, 1/2)$ and we have

$$\sin \frac{\pi}{6} = y = \frac{1}{2}, \quad \cos \frac{\pi}{6} = x = \frac{\sqrt{3}}{2}, \quad \text{and} \quad \tan \frac{\pi}{6} = \frac{y}{x} = \frac{\sqrt{3}}{3}$$

For $t = \pi/4$, $\pi/3$, and $\pi/2$, we use a similar procedure, and we summarize the results as follows:

t	0	$\pi/6$	$\pi/4$	$\pi/3$	$\pi/2$
$\sin t$	0	$1/2$	$\sqrt{2}/2$	$\sqrt{3}/2$	1
$\cos t$	1	$\sqrt{3}/2$	$\sqrt{2}/2$	$1/2$	0
$\tan t$	0	$\sqrt{3}/3$	1	$\sqrt{3}$	undefined

Note from this table that the tangent of $\pi/2$ is *undefined,* since the x-coordinate is zero. Hence, the ratio $\tan t = y/x$ would result in division by zero.

Each of the trigonometric values in Example 1 is positive. This occurs because the coordinate values all lie in Quadrant I. From Figure 5.20 we can determine the signs of the trigonometric functions in each of the four quadrants. For instance, since $\cos t = x$, it follows that $\cos t$ is positive ($x > 0$) in Quadrants I and IV and negative ($x < 0$) in Quadrants II and III. In a similar manner we can verify the rule of signs given next.

RULE OF SIGNS FOR TRIGONOMETRIC FUNCTIONS

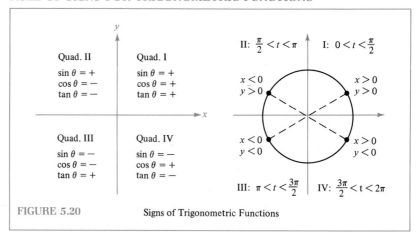

FIGURE 5.20   Signs of Trigonometric Functions

EXAMPLE 2
Evaluating Trigonometric
Functions in Quadrants
II, III, and IV

Find the values of $\sin t$ and $\cos t$ at

$$t = \frac{3\pi}{4}, \quad \frac{5\pi}{4}, \quad \text{and} \quad \frac{7\pi}{4}$$

Solution:
Using the coordinate points shown in Figure 5.21 we have the results shown in the following summary:

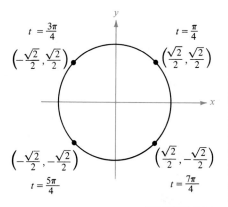

FIGURE 5.21

Quadrant	Correspondences	Function Values
I	$t = \dfrac{\pi}{4} \Rightarrow \left(\dfrac{\sqrt{2}}{2}, \dfrac{\sqrt{2}}{2}\right)$	$\sin \dfrac{\pi}{4} = \dfrac{\sqrt{2}}{2}, \quad \cos \dfrac{\pi}{4} = \dfrac{\sqrt{2}}{2}$
II	$t = \dfrac{3\pi}{4} \Rightarrow \left(\dfrac{-\sqrt{2}}{2}, \dfrac{\sqrt{2}}{2}\right)$	$\sin \dfrac{3\pi}{4} = \dfrac{\sqrt{2}}{2}, \quad \cos \dfrac{3\pi}{4} = \dfrac{-\sqrt{2}}{2}$
III	$t = \dfrac{5\pi}{4} \Rightarrow \left(\dfrac{-\sqrt{2}}{2}, \dfrac{-\sqrt{2}}{2}\right)$	$\sin \dfrac{5\pi}{4} = \dfrac{-\sqrt{2}}{2}, \quad \cos \dfrac{5\pi}{4} = \dfrac{-\sqrt{2}}{2}$
IV	$t = \dfrac{7\pi}{4} \Rightarrow \left(\dfrac{\sqrt{2}}{2}, \dfrac{-\sqrt{2}}{2}\right)$	$\sin \dfrac{7\pi}{4} = \dfrac{-\sqrt{2}}{2}, \quad \cos \dfrac{7\pi}{4} = \dfrac{\sqrt{2}}{2}$

The four points at which the unit circle intersects the x- and y-axes are called **quadrant points.** In the next example we look at the values of the six trigonometric functions at these quadrant points.

EXAMPLE 3
Evaluating Trigonometric Functions at Quadrant Points

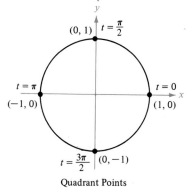

Quadrant Points

FIGURE 5.22

Evaluate the trigonometric functions at

$$t = 0, \quad \frac{\pi}{2}, \quad \pi, \quad \frac{3\pi}{2}, \quad \text{and} \quad 2\pi$$

Solution:
The coordinates (x, y) for these values of t are shown in Figure 5.22. For $t = 0$ *and* $t = 2\pi$, the quadrant point is $(x, y) = (1, 0)$. Thus we have

$$\sin 0 = y = 0 \qquad \csc 0 = \frac{1}{y} \text{ (undefined)}$$

$$\cos 0 = x = 1 \qquad \sec 0 = \frac{1}{x} = 1$$

$$\tan 0 = \frac{y}{x} = 0 \qquad \cot 0 = \frac{x}{y} \text{ (undefined)}$$

A similar analysis at $t = \pi/2$, π, and $3\pi/2$ will yield the following results:

t	Quadrant Point	$\sin t$	$\cos t$	$\tan t$	$\csc t$	$\sec t$	$\cot t$
0	(1, 0)	0	1	0	undefined	1	undefined
$\pi/2$	(0, 1)	1	0	undefined	1	undefined	0
π	(−1, 0)	0	−1	0	undefined	−1	undefined
$3\pi/2$	(0, −1)	−1	0	undefined	−1	undefined	0
2π	(1, 0)	0	1	0	undefined	1	undefined

Remark: For a point (x, y) on the unit circle, note that $-1 \leq x \leq 1$ and $-1 \leq y \leq 1$. Since $\cos t = x$ and $\sin t = y$, we can conclude that the values of $\sin t$ and $\cos t$ also range between -1 and 1.

$$-1 \leq \cos t \leq 1 \qquad \text{and} \qquad -1 \leq \sin t \leq 1$$

Period of Sine and Cosine

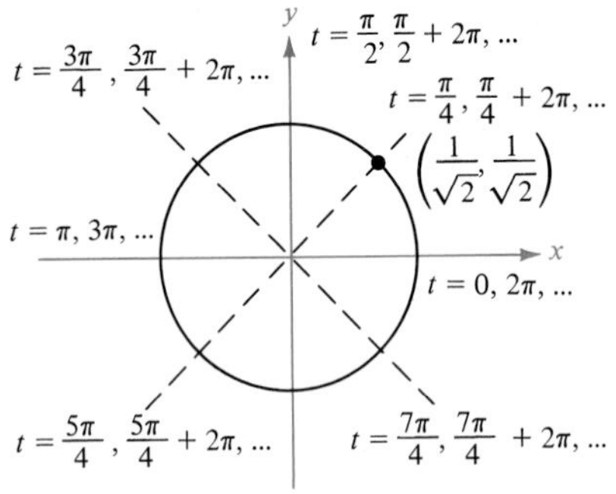

Suppose we add 2π to each value of t in $[0, 2\pi]$, thus completing a *second* revolution around the unit circle, as shown in Figure 5.23. We can see that the values of $\sin(t + 2\pi)$ and $\cos(t + 2\pi)$ correspond to those of $\sin t$ and $\cos t$. For instance,

$$t = \frac{\pi}{4} \qquad \text{and} \qquad t = \left(\frac{\pi}{4} + 2\pi \right)$$

both correspond to the point $(\sqrt{2}/2, \sqrt{2}/2)$, thus yielding

$$\sin \frac{\pi}{4} = \sin \left(\frac{\pi}{4} + 2\pi \right) = \frac{\sqrt{2}}{2}$$

$$\cos \frac{\pi}{4} = \cos \left(\frac{\pi}{4} + 2\pi \right) = \frac{\sqrt{2}}{2}$$

Similar results are obtained for repeated revolutions (positive or negative) on the unit circle. This leads to the general result that

$$\sin(t + 2\pi n) = \sin t \qquad \text{and} \qquad \cos(t + 2\pi n) = \cos t$$

for any integer n and real number t. Functions that behave in such a repetitive (or cyclic) manner are referred to as **periodic functions.**

DEFINITION OF PERIODIC FUNCTION

> A function f is **periodic** if there exists a positive real number c such that
>
> $$f(t + c) = f(t)$$
>
> for all t in the domain of f. The least number c for which f is periodic is called the **period** of f.

Remark: The discussion involving Figure 5.23 indicates that 2π is the period for both the sine and the cosine functions.

$t = \frac{3\pi}{4}, \frac{3\pi}{4} + 2\pi, \ldots$

$t = \frac{\pi}{2}, \frac{\pi}{2} + 2\pi, \ldots$

$t = \frac{\pi}{4}, \frac{\pi}{4} + 2\pi, \ldots$

$\left(\frac{1}{\sqrt{2}}, \frac{1}{\sqrt{2}} \right)$

$t = \pi, 3\pi, \ldots$

$t = 0, 2\pi, \ldots$

$t = \frac{5\pi}{4}, \frac{5\pi}{4} + 2\pi, \ldots$

$t = \frac{7\pi}{4}, \frac{7\pi}{4} + 2\pi, \ldots$

Repeated Revolution of the Unit Circle

FIGURE 5.23

EXAMPLE 4
Using the Period to Evaluate the Sine and Cosine

Evaluate

$$\sin \frac{17\pi}{4} \qquad \text{and} \qquad \cos \frac{7\pi}{3}$$

Solution:
Since $\sin t$ has period 2π, we have

$$\frac{17\pi}{4} = \frac{\pi}{4} + \frac{16\pi}{4} = \frac{\pi}{4} + 4\pi = \frac{\pi}{4} + (2\pi)(2)$$

and it follows that

$$\sin \frac{17\pi}{4} = \sin \left[\frac{\pi}{4} + (2\pi)(2) \right] = \sin \frac{\pi}{4} = \frac{\sqrt{2}}{2}$$

Similarly, since $\cos t$ has a period of 2π, we have

$$\frac{7\pi}{3} = \frac{\pi}{3} + \frac{6\pi}{3} = \frac{\pi}{3} + 2\pi$$

and it follows that

$$\cos \frac{7\pi}{3} = \cos \left[\frac{\pi}{3} + 2\pi \right] = \cos \frac{\pi}{3} = \frac{1}{2}$$

In trigonometry a great deal of time is spent studying relationships between trigonometric functions. You will need to memorize many of these relationships (identities). We begin with some basic identities that are easily established from the definitions of the six trigonometric functions.

FUNDAMENTAL TRIGONOMETRIC IDENTITIES

Reciprocal Identities:

$$\csc t = \frac{1}{\sin t} \qquad \sec t = \frac{1}{\cos t} \qquad \cot t = \frac{1}{\tan t}$$

$$\sin t = \frac{1}{\csc t} \qquad \cos t = \frac{1}{\sec t} \qquad \tan t = \frac{1}{\cot t}$$

$$\tan t = \frac{\sin t}{\cos t} \qquad \cot t = \frac{\cos t}{\sin t}$$

Pythagorean Identities:

$$\sin^2 t + \cos^2 t = 1 \quad 1 + \tan^2 t = \sec^2 t, \quad 1 + \cot^2 t = \csc^2 t$$

Proof:
The reciprocal identities are inherent in the definition of the six trigonometric functions. The Pythagorean identities follow from the fact that the points (x, y) lie on the unit circle.

Remark: We will use the forms

$$\sin^2 t \qquad \text{in place of} \qquad (\sin t)^2$$
$$\tan^3 t \qquad \text{in place of} \qquad (\tan t)^3$$

and so on, for all powers except -1. We will see later that $\sin^{-1} t$, $\cos^{-1} t$, and so on have special meanings not related to powers.

EXAMPLE 5
Using Identities to Evaluate
Trigonometric Functions

If $\sin t = \frac{4}{5}$ and t lies in Quadrant II, find $\cos t$.

Solution:

Given that $\sin t = \frac{4}{5}$, we can use the identity $\sin^2 t + \cos^2 t = 1$ to obtain

$$(\tfrac{4}{5})^2 + \cos^2 t = 1$$
$$\cos^2 t = 1 - \tfrac{16}{25} = \tfrac{9}{25}$$

Since t lies in Quadrant II, we know that $\cos t < 0$, and we take the negative root to obtain

$$\cos t = -\tfrac{3}{5}$$

Matching Real Numbers with Angles

Recall from Section 5.1 that each real number t can be matched with an *angle,* as shown in Figure 5.24. Here a real number t matches the point (x, y), yielding an arc having length t radians. Therefore, on the unit circle each *number t* matches up with a *central angle* of $\theta = t$ radians. This result leads to the following perspective on the trigonometric functions of real numbers:

$$\sin \begin{pmatrix} \text{real} \\ \text{number } t \end{pmatrix} = \sin \begin{pmatrix} \text{angle of} \\ t \text{ radians} \end{pmatrix}$$

$$\sin t = \sin (t \text{ radians})$$

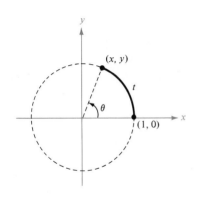

FIGURE 5.24

Similarly, $\cos t = \cos (t \text{ radians})$, $\tan t = \tan (t \text{ radians})$, and so on.

This tie between real numbers and radian measure of angles is *critical,* and from this point on we will not distinguish between trigonometric functions of real numbers and trigonometric functions of angles. When degree measure is desired we will use the degree symbol, as in

$$\cos 60°, \qquad \tan 225°, \qquad \text{and} \qquad \csc 510°$$

If no symbol is attached, as in

$$\sin \frac{\pi}{6}, \qquad \cos 3, \qquad \text{and} \qquad \sec 0.5$$

we will assume that radian measure is being used.

Reference Angles

We are now ready to show how reference angles can be used to find values of trigonometric functions of any angle. Consider the points (x, y) in Figure 5.25 that are determined by the reference angle θ' corresponding to point (x', y') on the unit circle.

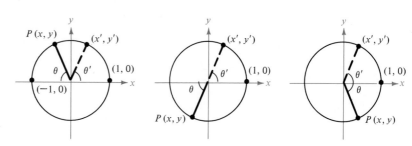

(a) Quadrant II:
$\theta' = \pi - \theta$
$x = -x', y = y'$

(b) Quadrant III:
$\theta' = \theta - \pi$
$x = -x', y = -y'$

(c) Quadrant IV:
$\theta' = 2\pi - \theta$
$x = x', y = -y'$

FIGURE 5.25   *Reference Angles*

In all cases,

$$|x| = x'$$

and

$$|y| = y'$$

Hence, for any angle θ with corresponding point (x, y), we have

$$|\cos \theta| = |x| = x' = \cos \theta'$$
$$|\sin \theta| = |y| = y' = \sin \theta'$$

Similarly,

$$|\tan \theta| = \left|\frac{y}{x}\right| = \frac{|y|}{|x|} = \frac{y'}{x'} = \tan \theta'$$

The same relationships hold for the three reciprocal functions. This means that the value of a trigonometric function of θ agrees with the function value of its corresponding reference angle θ', except possibly in sign.

We can now give the following general rule for evaluating trigonometric functions of any angle.

FINDING VALUES OF THE TRIGONOMETRIC FUNCTIONS

To find the value of a trigonometric function of an angle θ:

1. Determine the function value for the associated reference angle θ'.
2. Determine the quadrant of θ and prefix the appropriate sign.

EXAMPLE 6
Using Reference Angles to
Evaluate Trigonometric Functions

Evaluate

(a) $\cos \dfrac{4\pi}{3}$

(b) $\tan 480°$

Solution:

(a) Since $4\pi/3$ lies in Quadrant III, we know that

$$\cos \frac{4\pi}{3} < 0 \qquad\qquad \textit{Rule of signs}$$

$$\theta' = \frac{4\pi}{3} - \pi = \frac{\pi}{3} \qquad\qquad \textit{Reference angle}$$

Thus,

$$\cos \frac{4\pi}{3} = (-) \cos \frac{\pi}{3} = -\frac{1}{2} \qquad\qquad \textit{Special angle, } \pi/3$$

(b) Since $480°$ is larger than $360°$, we subtract one revolution to obtain $480° - 360° = 120°$. Now since $120°$ lies in Quadrant II, we know that $\tan 120° < 0$ and $\theta' = 180° - 120° = 60°$. Thus,

$$\tan 480° = \tan 120° = -\tan 60° = -\sqrt{3}$$

Section Exercises 5.2

In Exercises 1–8, determine the six trigonometric functions of the given angle *without* using a calculator or tables. (When necessary, use the fact that the trigonometric functions are periodic.)

1. (a) $30°$ (b) $45°$
2. (a) $360°$ (b) $390°$
3. (a) $60°$ (b) $90°$
4. (a) $780°$ (b) $450°$
5. (a) π (b) $\dfrac{3\pi}{2}$
6. (a) 5π (b) $\dfrac{7\pi}{2}$
7. (a) $\dfrac{17\pi}{4}$ (b) $\dfrac{7\pi}{3}$
8. (a) 120π (b) 15π

In Exercises 9–18, determine the quadrant in which θ lies.

9. $\sin \theta < 0$ and $\cos \theta < 0$
10. $\sin \theta > 0$ and $\cos \theta < 0$
11. $\sin \theta > 0$ and $\cos \theta > 0$
12. $\sin \theta < 0$ and $\cos \theta > 0$
13. $\sin \theta > 0$ and $\tan \theta < 0$
14. $\csc \theta > 0$ and $\tan \theta < 0$
15. $\sec \theta > 0$ and $\cot \theta < 0$
16. $\csc \theta < 0$ and $\tan \theta > 0$
17. $\csc \theta < 0$ and $\sec \theta < 0$
18. $\sec \theta < 0$ and $\csc \theta > 0$

In Exercises 19–26, sketch the angle in standard position, determine the reference angle, and evaluate the sine, cosine, and tangent of the given angle *without* using a calculator or tables.

19. (a) $\dfrac{4\pi}{3}$ (b) $\dfrac{2\pi}{3}$
20. (a) $\dfrac{3\pi}{4}$ (b) $\dfrac{5\pi}{4}$
21. (a) $-\dfrac{5\pi}{6}$ (b) $-\dfrac{7\pi}{2}$
22. (a) $-\dfrac{\pi}{2}$ (b) $\dfrac{\pi}{2}$
23. (a) $225°$ (b) $-225°$
24. (a) $300°$ (b) $330°$
25. (a) $750°$ (b) $510°$
26. (a) $\dfrac{10\pi}{3}$ (b) $\dfrac{17\pi}{3}$

In Exercises 27–32, find the value of the trigonometric functions under the given set of conditions.

27. $\sec t = -2$ and $\sin t > 0$
28. $\sin t = 1$
29. $\cot t$ is undefined and $\dfrac{\pi}{2} \le t \le \dfrac{3\pi}{2}$
30. $\sin t = 0$ and $\sec t = -1$
31. $\cos t = 1$
32. $\tan t$ is undefined and $\pi \le t \le 2\pi$

In Exercises 33–38, find two values of θ corresponding to the given functions. List the measure of θ in degrees ($0 \le \theta < 360°$) *and* radians ($0 \le \theta < 2\pi$). Do not use a calculator or tables.

33. (a) $\sin \theta = \dfrac{1}{2}$ (b) $\sin \theta = -\dfrac{1}{2}$
34. (a) $\cos \theta = \dfrac{\sqrt{2}}{2}$ (b) $\cos \theta = -\dfrac{\sqrt{2}}{2}$
35. (a) $\csc \theta = \dfrac{2\sqrt{3}}{3}$ (b) $\cot \theta = -1$
36. (a) $\sec \theta = 2$ (b) $\sec \theta = -2$
37. (a) $\tan \theta = 1$ (b) $\cot \theta = -\sqrt{3}$
38. (a) $\sin \theta = \dfrac{\sqrt{3}}{2}$ (b) $\sin \theta = -\dfrac{\sqrt{3}}{2}$

The Trigonometric Functions and Right Triangles

5.3

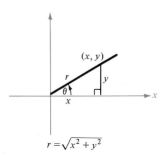

$r = \sqrt{x^2 + y^2}$

FIGURE 5.26

Our second look at the trigonometric *functions* is from a right-triangle perspective. Consider the right triangle shown in Figure 5.26, formed by drawing a line from a point (x, y) on the terminal side of angle θ perpendicular to the *x*-axis.

The sides of the resulting right triangle have lengths x, y, and r, where the distance from the origin $(0, 0)$ to the point (x, y) is $r = \sqrt{x^2 + y^2}$. Note that there are six ratios among the three sides of the triangle:

$$\frac{y}{r}, \quad \frac{x}{r}, \quad \frac{y}{x}, \quad \frac{r}{y}, \quad \frac{r}{x}, \quad \frac{x}{y}$$

For the *similar* right triangles in Figure 5.27, the corresponding ratios are equal. That is,

$$\frac{y'}{r'} = \frac{y}{r} \qquad \frac{x'}{r'} = \frac{x}{r} \qquad \frac{y'}{x'} = \frac{y}{x}$$

$$\frac{r'}{y'} = \frac{r}{y} \qquad \frac{r'}{x'} = \frac{r}{x} \qquad \frac{x'}{y'} = \frac{x}{y}$$

This means that the six ratios among the sides of any right triangle depend only on the angle θ, and not on the relative size of the right triangle.

The preceding six ratios form the basis for our study of triangle trigonometry; in order, they correspond to the **sine, cosine, tangent, cosecant, secant,** and **cotangent of angle θ.**

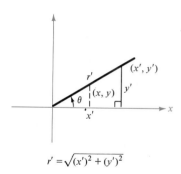

$r' = \sqrt{(x')^2 + (y')^2}$

Similar Right Triangles with Angle θ

FIGURE 5.27

DEFINITION OF THE SIX TRIGONOMETRIC FUNCTIONS

Let θ be an angle in standard position with (x, y) any point (except the origin) on the terminal side of θ, as shown in Figure 5.28. Then the six trigonometric functions of angle θ are defined to be

$$\sin \theta = \frac{y}{r} = \frac{\text{opp.}}{\text{hyp.}} \qquad \csc \theta = \frac{r}{y} = \frac{\text{hyp.}}{\text{opp.}}$$

$$\cos \theta = \frac{x}{r} = \frac{\text{adj.}}{\text{hyp.}} \qquad \sec \theta = \frac{r}{x} = \frac{\text{hyp.}}{\text{adj.}}$$

$$\tan \theta = \frac{y}{x} = \frac{\text{opp.}}{\text{adj.}} \qquad \cot \theta = \frac{x}{y} = \frac{\text{adj.}}{\text{opp.}}$$

Assume that all denominators are nonzero.

opp. = the length of the side *opposite* angle θ

adj. = the length of the side *adjacent* to angle θ

hyp. = the length of the *hypotenuse*

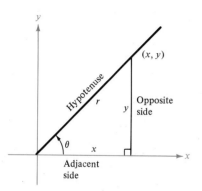

FIGURE 5.28

Remark: To help you memorize these definitions, notice that the functions in the second column are simply the *reciprocals* of the corresponding members of the first column.

EXAMPLE 1
Evaluating Trigonometric Functions

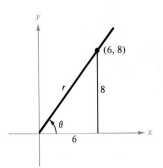

FIGURE 5.29

The terminal side of angle θ (in standard position) goes through the point (6, 8). Find the values of the six trigonometric functions of θ.

Solution:

From the point (6, 8) we drop a perpendicular to the *x*-axis and form a right triangle, as shown in Figure 5.29. By the Pythagorean Theorem, it follows that

$$r = \sqrt{6^2 + 8^2} = \sqrt{100} = 10$$

Thus, we have $x = 6$, $y = 8$, $r = 10$, and consequently, the six trigonometric function values are

$$\sin \theta = \frac{y}{r} = \frac{8}{10} = \frac{4}{5} \qquad \csc \theta = \frac{r}{y} = \frac{5}{4}$$

$$\cos \theta = \frac{x}{r} = \frac{6}{10} = \frac{3}{5} \qquad \sec \theta = \frac{r}{x} = \frac{5}{3}$$

$$\tan \theta = \frac{y}{x} = \frac{8}{6} = \frac{4}{3} \qquad \cot \theta = \frac{x}{y} = \frac{3}{4}$$

EXAMPLE 2
Trigonometric Functions of Angles of a Triangle

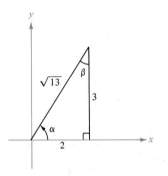

FIGURE 5.30

Use Figure 5.30 to evaluate the following trigonometric functions:

(a) $\sin \alpha$, $\cos \alpha$, $\tan \alpha$ (b) $\sin \beta$, $\cos \beta$, $\cot \beta$

Solution:

(a) *Viewed from angle α:*

$$\text{adj.} = 2, \qquad \text{opp.} = 3, \qquad \text{hyp.} = \sqrt{13}$$

Thus, we have

$$\sin \alpha = \frac{\text{opp.}}{\text{hyp.}} = \frac{3}{\sqrt{13}}$$

$$\cos \alpha = \frac{\text{adj.}}{\text{hyp.}} = \frac{2}{\sqrt{13}}$$

$$\tan \alpha = \frac{\text{opp.}}{\text{adj.}} = \frac{3}{2}$$

(b) *Viewed from angle β:*

$$\text{adj.} = 3, \qquad \text{opp.} = 2, \qquad \text{hyp.} = \sqrt{13}$$

Thus we have

$$\sin \beta = \frac{\text{opp.}}{\text{hyp.}} = \frac{2}{\sqrt{13}}$$

$$\cos \beta = \frac{\text{adj.}}{\text{hyp.}} = \frac{3}{\sqrt{13}}$$

$$\cot \beta = \frac{\text{adj.}}{\text{opp.}} = \frac{3}{2}$$

Remark: Recall from geometry that two acute angles α and β are complementary if $\alpha + \beta = 90°$. Now, note from Example 2 that

cofunctions

$$\sin \alpha = \frac{3}{\sqrt{13}} = \cos \beta$$

complementary angles

and

$$\cos \alpha = \frac{2}{\sqrt{13}} = \sin \beta$$

These results point to some important relationships between trigonometric functions of complementary angles.

COFUNCTIONS OF COMPLEMENTARY ANGLES

For an acute angle θ:

$$\sin(90° - \theta) = \cos \theta \qquad \cos(90° - \theta) = \sin \theta$$

$$\tan(90° - \theta) = \cot \theta \qquad \cot(90° - \theta) = \tan \theta$$

$$\sec(90° - \theta) = \csc \theta \qquad \csc(90° - \theta) = \sec \theta$$

These identities indicate that **co**functions of **co**mplementary angles are equal. For example,

$$\sin 10° = \cos(90° - 10°) = \cos 80° \qquad (10° + 80° = 90°)$$

$$\sec 65° = \csc(90° - 65°) = \csc 25° \qquad (65° + 25° = 90°)$$

In Section 5.2 we obtained trigonometric functional values of three common angles (30°, 45°, and 60°) using the unit circle. These values can also be easily obtained from the right triangles shown in Table 5.2.

TABLE 5.2

Trigonometry of the Special Angles 30°, 45°, and 60°

$\theta = 45° = \pi/4$	Given: Isosceles right triangle with legs equal to 1 and hypotenuse equal to $\sqrt{2}$.	$\sin 45° = \dfrac{1}{\sqrt{2}}$
		$\cos 45° = \dfrac{1}{\sqrt{2}}$
	Note: A 45-45-90 triangle has the sides in the proportion 1 : 1 : $\sqrt{2}$.	$\tan 45° = 1$
$\theta = 30° = \pi/6$ and $\theta = 60° = \pi/3$	Given: An equilateral triangle with sides of length 2 and a height of $\sqrt{3}$.	$\sin 30° = \dfrac{1}{2}$
		$\cos 30° = \dfrac{\sqrt{3}}{2}$
	Note: A 30-60-90 triangle has sides in the proportion 1: $\sqrt{3}$: 2	$\tan 30° = \dfrac{1}{\sqrt{3}}$
		$\sin 60° = \dfrac{\sqrt{3}}{2}$
		$\cos 60° = \dfrac{1}{2}$
		$\tan 60° = \sqrt{3}$

Remark: Because the angles 30°, 45°, and 60° occur frequently in trigonometry, it is a good idea either to memorize their function values or to be able to construct the triangles from which you can determine the function values.

EXAMPLE 3
Using Right Triangles to Evaluate Trigonometric Functions

(a) If θ is an acute angle and $\cos \theta = \frac{5}{13}$, find the values of the remaining five trigonometric functions.

(b) In a right triangle the side opposite angle θ has length 12. If $\cot \theta = \sqrt{3}/3$, find the values of $\tan \theta$, $\cos \theta$, $\sin \theta$, and θ itself.

Solution:

(a) Since $\cos \theta = \frac{5}{13}$, we construct the right triangle shown in Figure 5.31. By the Pythagorean Theorem, it follows that

$$13^2 = y^2 + 5^2$$
$$y^2 = 169 - 25$$
$$y = \sqrt{144} = 12$$

Therefore, we obtain the results

$$\sin \theta = \frac{\text{opp.}}{\text{hyp.}} = \frac{12}{13} \qquad \csc \theta = \frac{13}{12}$$

$$\cos \theta = \frac{\text{adj.}}{\text{hyp.}} = \frac{5}{13} \qquad \sec \theta = \frac{13}{5}$$

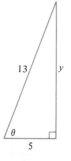

FIGURE 5.31

$$\tan \theta = \frac{\text{opp.}}{\text{adj.}} = \frac{12}{5} \qquad \cot \theta = \frac{5}{12}$$

(b) Since the side opposite θ has length 12, we construct the right triangle shown in Figure 5.32. Using the fact that $\cot \theta = \sqrt{3}/3$, we find that

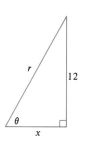

FIGURE 5.32

$$\cot \theta = \frac{\sqrt{3}}{3} = \frac{x}{12}$$

$$x = \frac{12\sqrt{3}}{3} = 4\sqrt{3}$$

Furthermore,

$$r^2 = (4\sqrt{3})^2 + (12)^2 = 48 + 144 = 192$$
$$r = \sqrt{192} = \sqrt{64 \cdot 3} = 8\sqrt{3}$$

Therefore,

$$\tan \theta = \frac{y}{x} = \frac{12}{4\sqrt{3}} = \frac{3}{\sqrt{3}} = \sqrt{3}$$

$$\cos \theta = \frac{x}{r} = \frac{4\sqrt{3}}{8\sqrt{3}} = \frac{1}{2}$$

$$\sin \theta = \frac{y}{r} = \frac{12}{8\sqrt{3}} = \frac{3}{2\sqrt{3}} = \frac{\sqrt{3}}{2}$$

Finally, we recognize θ to be one of the special angles, namely

$$\theta = 60° = \frac{\pi}{3}$$

For angles other than 30°, 45°, and 60°, we use a table or a calculator to find the values of the trigonometric functions. Though our next example shows how to use tables, we will mainly use a calculator to find values of the trigonometric functions.

Most calculators accept angles of any size. However, some older calculators do not accept angles greater than 180°. In such cases, the reference angle approach (in Section 5.2) should be used.

EXAMPLE 4
Using Tables to Evaluate
Trigonometric Functions

Use Appendix D to evaluate each of the following:

(a) cos 208° 10′   (b) tan 53° 40′   (c) sin 15° 24′

Solution:

(a) Since $\theta = 208°\ 10'$ lies in Quadrant III, its reference angle is $\theta = 208°\ 10' - 180° = 28°\ 10'$. Angles of less than 45° are found on the *left* side of the table, and the function names are at the *top*. Thus cos

$28°$ $10'$ $\approx$ 0.8816, and since the cosine is negative in Quadrant III, it follows that

$$\cos 208° 10' = -\cos 28° 10' \approx -0.8816$$

(b) Angles between $45°$ and $90°$ are found on the *right* side, with the function names listed at the *bottom* of the table. Thus, reading up the right side, we find

$$\tan 53° 40' \approx 1.360$$

(c) In this case we use *linear interpolation,* because $15° 24'$ lies *between* the adjacent table values $15° 20'$ and $15° 30'$. We use the scheme

$$10' \left\{ 4' \left\{ \begin{array}{l} \sin 15° 20' = 0.2644 \\ \sin 15° 24' = ? \\ \sin 15° 30' = 0.2672 \end{array} \right. \right\} (0.2672 - 0.2644) = 0.0028 = d$$

$$\sin 15° 24' = \sin 15° 20' + \frac{4}{10}(d)$$

$$= 0.2644 + 0.4(0.0028)$$

$$= 0.2655 \quad \textit{(Rounded to four places)}$$

Remark: If the function *decreases* as the angle increases, then d will have a *negative* value.

Calculators and Trig Functions

Since angles can be measured in either degrees or radians, you must set your calculator to the desired mode before evaluating a trigonometric function. For example, with the mode switch set for *degrees* (dg or deg) we can make the following calculations:

Function	**Keystrokes**	**Display**
tan 78.9°	78.9 $\boxed{\text{tan}}$	5.09704

Switching to *radian* mode, we have

Function	**Keystrokes**	**Display**
$\cos \dfrac{\pi}{12}$	π $\boxed{\div}$ 12 $\boxed{=}$ $\boxed{\cos}$	0.96592 (or 9.6592 − 01)

Reciprocal Functions

Since many calculators do not have keys for the cosecant, secant, and co-tangent, we use their *reciprocal* functions and the $\boxed{1/x}$ key to evaluate such functions.

Function	**Keystrokes**	**Display**
csc 0.58	.58 $\boxed{\text{sin}}$ $\boxed{1/x}$	1.8247376

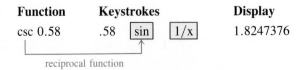
reciprocal function

EXAMPLE 5
Using a Calculator to Evaluate
Trigonometric Functions

Use a calculator to evaluate each of the following:

(a) cot 1.5

(b) sec 5° 40′ 12″

Solution:

(a) Switching to *radian* mode and using the reciprocal key, we have

Function	Keystrokes	Display
cot 1.5	1.5 $\boxed{\tan}$ $\boxed{1/x}$	0.070914 (or 7.0914 − 02)

(b) Converting to decimal form, we have

$$5° \ 40′ \ 12″ = 5° + \left(\frac{40}{60}\right)° + \left(\frac{12}{3600}\right)°$$

$$\approx 5 + 0.66666 + 0.00333 = 5.66999°$$

Hence,

$$\sec 5° \ 40′ \ 12″ \approx \sec 5.67° = \frac{1}{\cos 5.67°} \approx 1.00492$$

The $\boxed{\text{INV}}$ Key

In many applications involving a right triangle, two sides are known and we must find the acute angles of the triangle. The problem we then face is as follows: Given the value of a trigonometric function, find the acute angle associated with that value. For example, if $\sin \theta = \frac{12}{20} = 0.6000$, find θ in degrees.

Since

$$\sin 30° = \frac{1}{2} = 0.5000 \qquad \text{and} \qquad \sin 45° = \frac{1}{\sqrt{2}} \approx 0.7071$$

we conclude that θ lies somewhere between 30° and 45°. With a calculator we can find a more precise value for θ using the **inverse** key $\boxed{\text{INV}}$. In degree mode, we have

Function	Keystrokes	Display
$\sin \theta = \dfrac{12}{20}$	12 $\boxed{\div}$ 20 $\boxed{=}$ $\boxed{\text{INV}}$ $\boxed{\sin}$	36.8699

Thus, we conclude that if $\sin \theta = \frac{12}{20}$, then $\theta \approx 36.87°$. In Section 5.7, we will explain the concepts involved in the use of the $\boxed{\text{INV}}$ key.

EXAMPLE 6
A Right-Triangle Application

If a 40-foot flagpole casts a 30-foot shadow, what is the angle of elevation of the sun (see Figure 5.33)?

Solution:
From Figure 5.33 we see that the *opposite* and *adjacent* sides are known.

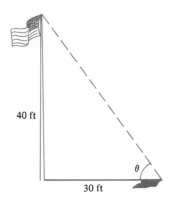

FIGURE 5.33

Thus, we write

$$\tan \theta = \frac{\text{opp.}}{\text{adj.}} = \frac{40}{30}$$

With a calculator in degree mode we use the keystrokes

40 ÷ 30 = INV tan

to obtain $\theta \approx 53.13°$.

Summary

At this point, we have completed our two-pronged discussion of the basic concepts of trigonometry. We have measured angles in both degrees and radians. We have defined the six trigonometric functions as functions of real numbers and as functions of angles of a right triangle.

The *function of real numbers* concept lends itself well to graphing techniques (Sections 5.4, 5.5), to descriptions of inverse trigonometric functions (Section 5.7), and to discussions of trigonometric identities and equations (Chapter 6). Applications of *triangle trigonometry* will occur again in Sections 5.6, 7.1, 7.2, and 7.3.

Perhaps the best advice we can give to make your further study of trigonometry go smoothly is that you put to memory the following basic information:

1. Functions of special angles in Table 5.2.
2. Summary of basic trigonometry inside the book covers.
3. Signs of trigonometric functions in Figure 5.19.
4. Functions of quadrant angles in Example 3 of Section 5.1.

Section Exercises 5.3

In Exercises 1–6, find the indicated trigonometric function from the given one.

1. Given $\sin \theta = \frac{1}{2}$, find $\csc \theta$.

2. Given $\sin \theta = \frac{1}{3}$, find $\tan \theta$.

3. Given $\cos \theta = \frac{4}{5}$, find $\cot \theta$.

4. Given $\sec \theta = \frac{13}{5}$, find $\cot \theta$.

5. Given $\cot \theta = \frac{15}{8}$, find $\sec \theta$.

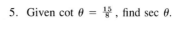

6. Given $\tan \theta = \frac{1}{2}$, find $\sin \theta$.

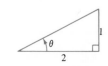

In Exercises 7–10, use the two similar triangles in the accompanying figure. Find (a) the unknown sides of the triangles and (b) the six trigonometric functions of the angles α_1 and α_2.

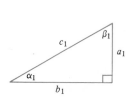 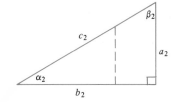

7. $a_1 = 3$, $b_1 = 4$, $a_2 = 9$
8. $b_1 = 12$, $c_1 = 13$, $c_2 = 26$
9. $a_1 = 1$, $c_1 = 2$, $b_2 = 5$
10. $b_1 = 4$, $a_2 = 4$, $b_2 = 10$

In Exercises 11–16, sketch a right triangle corresponding to the trigonometric function of the angle θ, and find the other five trigonometric functions of θ.

11. $\sin \theta = \frac{2}{3}$ 12. $\cot \theta = 5$ 13. $\sec \theta = 2$
14. $\cos \theta = \frac{5}{7}$ 15. $\tan \theta = 3$ 16. $\csc \theta = 4.25$
17. If $\sin 60° = \sqrt{3}/2$, use trigonometric identities to find
  (a) $\cos 60°$ (b) $\tan 60°$
  (c) $\csc 60°$ (d) $\cos 30°$
18. If $\tan 30° = \sqrt{3}/3$, use trigonometric identities to find
  (a) $\sec 30°$ (b) $\cos 30°$
  (c) $\cot 30°$ (d) $\cot 60°$
19. If $\csc \theta = 3$, use trigonometric identities to find
  (a) $\cot \theta$ (b) $\sin \theta$
  (c) $\tan \theta$ (d) $\sec(90° - \theta)$
20. If $\sec \theta = 5$, use trigonometric identities to find
  (a) $\tan \theta$ (b) $\cos \theta$
  (c) $\cot \theta$ (d) $\csc(90° - \theta)$
21. Evaluate the six trigonometric functions of the angle

$$\theta = 45° = \frac{\pi}{4}$$

  by constructing an appropriate triangle.
22. Evaluate the six trigonometric functions of the angle

$$\theta = 30° = \frac{\pi}{6}$$

  by constructing an appropriate triangle.

In Exercises 23–27, use a calculator or table to evaluate each function.

23. (a) $\sin 10°$
    (b) $\cos 80°$
24. (a) $\tan 23.5°$
    (b) $\cot 66.5°$

25. (a) $\sec 42° \, 12'$
    (b) $\csc 48° \, 7'$
26. (a) $\cos 16° \, 18'$
    (b) $\sin 73° \, 56'$
27. (a) $\cot 85° \, 41'$
    (b) $\tan 4° \, 54'$

In Exercises 28–34, use a calculator to evaluate each function.

28. (a) $\cos 4° \, 50' \, 15''$
    (b) $\sec 4° \, 50' \, 15''$
29. (a) $\tan 52° \, 25''$
    (b) $\cot 52° \, 25''$
30. (a) $\sin \frac{\pi}{5}$
    (b) $\csc \frac{\pi}{5}$
31. (a) $\cot \frac{\pi}{16}$
    (b) $\tan \frac{\pi}{16}$
32. (a) $\sec 0.75$
    (b) $\cos 0.75$
33. (a) $\csc 1$
    (b) $\sec \left(\frac{\pi}{2} - 1 \right)$
34. (a) $\tan \frac{1}{2}$
    (b) $\cot \left(\frac{\pi}{2} - \frac{1}{2} \right)$

In Exercises 35–39, find the value of θ in degrees ($0 < \theta < 90°$) and radians ($0 \le \theta \le \pi/2$) without using tables or a calculator.

35. (a) $\sin \theta = \frac{1}{2}$
    (b) $\csc \theta = 2$
36. (a) $\cos \theta = \frac{\sqrt{2}}{2}$
    (b) $\tan \theta = 1$
37. (a) $\sec \theta = 2$
    (b) $\cot \theta = 1$
38. (a) $\tan \theta = \sqrt{3}$
    (b) $\cos \theta = \frac{1}{2}$
39. (a) $\csc \theta = \frac{2\sqrt{3}}{3}$
    (b) $\sin \theta = \frac{\sqrt{2}}{2}$

In Exercises 40–46, find the value of θ in degrees ($0° < \theta < 90°$) and radians ($0 < \theta < \pi/2$) by using the inverse key on a calculator.

40. (a) $\sin \theta = 0.8191$
    (b) $\sin \theta = 0.0175$
41. (a) $\cos \theta = 0.9848$
    (b) $\cos \theta = 0.8746$
42. (a) $\tan \theta = 1.1920$
    (b) $\tan \theta = 0.4663$
43. (a) $\csc \theta = 1.4663$
    (b) $\csc \theta = 1.0098$
44. (a) $\sec \theta = 1.0306$
    (b) $\sec \theta = 2.2812$
45. (a) $\cot \theta = 0.0119$
    (b) $\cot \theta = 0.2679$
46. (a) $\csc \theta = 3.4062$
    (b) $\cot \theta = 2.0405$

47. A 6-foot person standing 12 feet from a streetlight casts an 8-foot shadow. (See Figure 5.34.) What is the height of the streetlight?

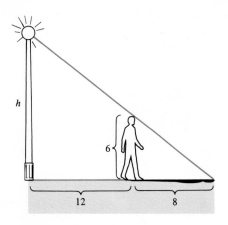

FIGURE 5.34

48. A guy wire is stretched from a broadcasting tower at a point 200 feet above the ground to an anchor 125 feet from the base. (See Figure 5.35.) How long is the wire?

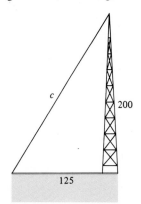

FIGURE 5.35

In Exercises 49–56, solve for x, y, or r, as indicated.

49. Solve for y.

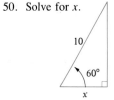

50. Solve for x.

51. Solve for x.

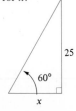

52. Solve for r.

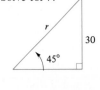

53. Solve for r.

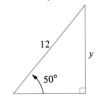

54. Solve for x.

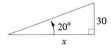

55. Solve for y.

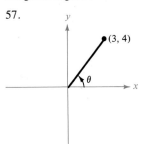

56. Solve for r.

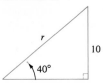

In exercises 57–64, determine all six trigonometric functions for the given angle θ.

57.

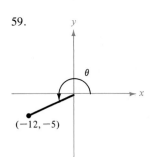

58.

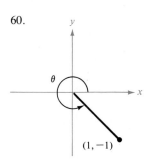

59.

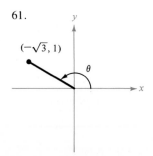

60.

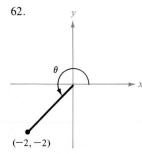

61.

62.

63. The terminal side of θ is in Quadrant III and lies on the line $y = 2x$.

64. The terminal side of θ is in Quadrant IV and lies on the line $4x + 3y = 0$.

65. A 20-foot ladder leaning against the side of a house makes a 75° angle with the ground. (See Figure 5.36.) How far up the side of the house does the ladder reach?

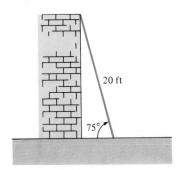

FIGURE 5.36

66. A biologist wants to know the width w of a river in order to properly set instruments to study the pollutants in the water. From point A, she walks downstream 100 feet and sights to point C. From this sighting, she determines that $θ = 50°$. (See Figure 5.37.) How wide is the river?

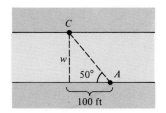

FIGURE 5.37

67. From a 150-foot observation tower on the coast, a Coast Guard officer sights a boat in difficulty. The angle of depression of the boat is 4°. (See Figure 5.38.) How far is the boat from the shoreline?

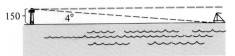

FIGURE 5.38

68. A ramp $17\frac{1}{2}$ feet in length rises to a loading platform that is $3\frac{1}{3}$ feet off the ground. (See Figure 5.39.) Find the angle that the ramp makes with the ground. (Hint: Find the sine of θ and then use the table of trigonometric values in the Appendix to estimate θ.)

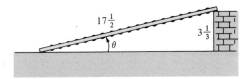

FIGURE 5.39

In Exercises 69–74, determine if the statement is true or false, and give reasons.

69. $\sin 60° \csc 60° = 1$ 70. $\sec 30° = \csc 60°$

71. $\sin 45° + \cos 45° = 1$ 72. $\cot^2 10° - \csc^2 10° = -1$

73. $\dfrac{\sin 60°}{\sin 30°} = \sin 2°$ 74. $\tan (0.8)^2 = \tan^2 (0.8)$

Graphs of Sine and Cosine   5.4

Now that we can determine the values of the trigonometric functions for any angle (or real number), we are ready to tackle the task of graphing these functions. In this section we limit our graphing procedures to the sine and cosine functions. To accommodate graphing on the familiar xy-coordinate system, we will use the variable x in place of θ or t. For example, we will use $y = \sin x$, $y = \cos x$, and so on.

Sine Function

In the previous section we established the following information about the sine function.

Quadrant	Variation in x	Variation in sin x
I	0 to $\pi/2$	0 to 1
II	$\pi/2$ to π	1 to 0
III	π to $3\pi/2$	0 to -1
IV	$3\pi/2$ to 2π	-1 to 0

The domain is all real numbers, and the range is $-1 \le \sin x \le 1$. Since $\sin(x + 2\pi n) = \sin x$ for all integers n, the sine function has a period of 2π.

Using this information and function values from Tables 5.1 and 5.2, we can readily obtain Table 5.3, showing values for $y = \sin x$. Plotting the points from Table 5.3 and connecting them with a smooth curve, we get the solid portion of the graph in Figure 5.40.

TABLE 5.3
Values of sin x

x	0	$\dfrac{\pi}{6}$	$\dfrac{\pi}{4}$	$\dfrac{\pi}{3}$	$\dfrac{\pi}{2}$	$\dfrac{2\pi}{3}$	$\dfrac{3\pi}{4}$	$\dfrac{5\pi}{6}$	π	$\dfrac{7\pi}{6}$	$\dfrac{5\pi}{4}$	$\dfrac{4\pi}{3}$	$\dfrac{3\pi}{2}$	$\dfrac{5\pi}{3}$	$\dfrac{7\pi}{4}$	$\dfrac{11\pi}{6}$	2π
$\sin x$	0	$\dfrac{1}{2}$	$\dfrac{\sqrt{2}}{2}$	$\dfrac{\sqrt{3}}{2}$	1	$\dfrac{\sqrt{3}}{2}$	$\dfrac{\sqrt{2}}{2}$	$\dfrac{1}{2}$	0	$-\dfrac{1}{2}$	$-\dfrac{\sqrt{2}}{2}$	$-\dfrac{\sqrt{3}}{2}$	-1	$-\dfrac{\sqrt{3}}{2}$	$-\dfrac{\sqrt{2}}{2}$	$-\dfrac{1}{2}$	0
	0	0.5	0.71	0.87	1	0.87	0.71	0.5	0	-0.5	-0.71	-0.87	-1	-0.87	-0.71	-0.5	0
		Quadrant I				Quadrant II				Quadrant III				Quadrant IV			

Remark: In Figure 5.40, note how the basic cycle (from $x = 0$ to $x = 2\pi$) is extended to the right and left in a periodic manner. As indicated by the arrows, the graph continues indefinitely to the right and left. The part of the graph corresponding to the interval $0 \le x \le 2\pi$ is called a **sine wave**.

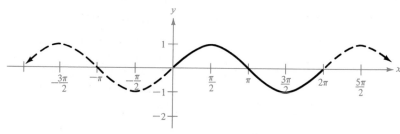

Graph of $y = \sin x$ (period $= 2\pi$)

FIGURE 5.40

Cosine Function

A similar review of the cosine function shows that the domain is all real numbers and the range is $-1 \le \cos x \le 1$. Since $\cos(x + 2\pi n) = \cos x$ for any integer n, the cosine function has a period of 2π. The graph of $y = \cos x$, shown in Figure 5.41, can be sketched by plotting points just as we did for $y = \sin x$.

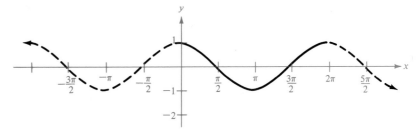

Graph of $y = \cos x$ (period $= 2\pi$)

FIGURE 5.41

TABLE 5.4

Values of $\cos x$

x	0	$\dfrac{\pi}{6}$	$\dfrac{\pi}{4}$	$\dfrac{\pi}{3}$	$\dfrac{\pi}{2}$	$\dfrac{2\pi}{3}$	$\dfrac{3\pi}{4}$	$\dfrac{5\pi}{6}$	π	$\dfrac{7\pi}{6}$	$\dfrac{5\pi}{4}$	$\dfrac{4\pi}{3}$	$\dfrac{3\pi}{2}$	$\dfrac{5\pi}{3}$	$\dfrac{7\pi}{4}$	$\dfrac{11\pi}{6}$	2π
$\cos x$	1	$\dfrac{\sqrt{3}}{2}$	$\dfrac{\sqrt{2}}{2}$	$\dfrac{1}{2}$	0	$-\dfrac{1}{2}$	$-\dfrac{\sqrt{2}}{2}$	$-\dfrac{\sqrt{3}}{2}$	-1	$-\dfrac{\sqrt{3}}{2}$	$-\dfrac{\sqrt{2}}{2}$	$-\dfrac{1}{2}$	0	$\dfrac{1}{2}$	$\dfrac{\sqrt{2}}{2}$	$\dfrac{\sqrt{3}}{2}$	1
	1	0.87	0.71	0.5	0	-0.5	-0.71	-0.87	-1	-0.87	-0.71	-0.5	0	0.5	0.71	0.87	1
		Quadrant I				Quadrant II				Quadrant III				Quadrant IV			

Remark: As with the sine function, the solid part of the graph (called the **cosine wave**) corresponding to the interval $0 \le x \le 2\pi$ is extended indefinitely to the right and left.

Note from Figures 5.40 and 5.41 that the sine graph is symmetric with respect to the *origin*, whereas the cosine graph is symmetric with respect to the *y-axis*. These properties of symmetry follow from the relationship shown in Figure 5.42, where we have

$$\sin(-\theta) = \frac{-y}{r} = -\frac{y}{r} = -(\sin\theta)$$

$$\cos(-\theta) = \frac{x}{r} = \cos\theta$$

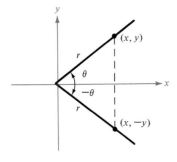

FIGURE 5.42

Key Points on Trig Graphs

To help you memorize the shape of the basic sine and cosine waves, we note *five key points* in each graph: the intercepts, maximum points, and minimum points. For $y = \sin x$ (Figure 5.40), the key points are

$$(0, 0) \qquad \left(\frac{\pi}{2}, 1\right) \qquad (\pi, 0) \qquad \left(\frac{3\pi}{2}, -1\right) \qquad \text{and} \qquad (2\pi, 0)$$

For $y = \cos x$, the key points are

$$(0, 1) \qquad \left(\frac{\pi}{2}, 0\right) \qquad (\pi, -1) \qquad \left(\frac{3\pi}{2}, 0\right) \qquad \text{and} \qquad (2\pi, 1)$$

Note how the *x*-coordinates of these points (which we will call *key numbers*) divide the period (2π) of sin *x* and cos *x* into *four* equal parts. Table 5.5 summarizes a scheme for identifying the key points on the graphs of $y = \sin x$ and $y = \cos x$.

TABLE 5.5
Key Points on Trigonometric Graphs

Function	Key Points	Pattern
$y = \sin x$ Period: 2π		i-M-i-m-i At increments of $\frac{1}{4}$ (period)
$y = \cos x$ Period: 2π		M-i-m-i-M At increments of $\frac{1}{4}$ (period)

EXAMPLE 1
Using Key Points to Sketch a Trigonometric Graph

Sketch a graph of $y = 2 \cos x$ on the interval $[-\pi, 4\pi]$.

Solution:
Note in this case that

$$y = 2 \cos x = 2(\cos x)$$

indicates that the *y*-values for the key points will have twice the magnitude of the graph of $y = \cos x$ (Figure 5.41). Thus, the key points for $y = 2 \cos x$ are

$$M \qquad i \qquad m \qquad i \qquad M$$

$$(0,\ 2) \qquad \left(\frac{\pi}{2},\ 0\right) \qquad (\pi,\ -2) \qquad \left(\frac{3\pi}{2},\ 0\right) \qquad (2\pi,\ 2)$$

By connecting these key points with a smooth curve and extending it in both directions over the interval $[-\pi,\ 4\pi]$, we obtain the graph shown in Figure 5.43.

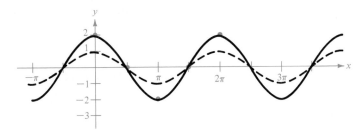

Graph of $y = 2 \cos x$ on $[-\pi,\ 4\pi]$

FIGURE 5.43

Remark: When using Table 5.5 to sketch graphs, you should confirm their shapes by using a calculator to find y-values at a few points between the key numbers.

For the rest of this section we look at variations in the graphs of the basic functions $y = \sin x$ and $y = \cos x$. In particular, we want to investigate the graphic effect of each of the constants a, b, and c in equations of the forms

$$y = a \sin (bx + c) \qquad \text{and} \qquad y = a \cos (bx + c)$$

In Figure 5.43, the constant factor 2 in $y = 2 \cos x$ acts as a vertical *stretch* factor, increasing the magnitude of the y-values. Similarly, the factor $\frac{1}{2}$ in $y = \frac{1}{2} \sin x$ *shrinks* the y-values so that y ranges between $-\frac{1}{2}$ and $\frac{1}{2}$ instead of between -1 and 1. Such factors are referred to as the **amplitudes** of the functions.

AMPLITUDE OF SINE AND COSINE

The **amplitude** of $y = a \sin x$ and $y = a \cos x$ is $|a|$.

EXAMPLE 2
**Using Amplitude
to Sketch Graphs**

On the same coordinate axes, sketch graphs of

$$y = \cos x, \qquad y = \frac{1}{2} \cos x, \qquad \text{and} \qquad y = 3 \cos x$$

Solution:

For one cycle of $y = \cos x$, we have

$$\text{amp} = 1 \implies \text{max} = 1, \text{min} = -1$$

$$\text{per} = 2\pi \implies \text{M-i-m-i-M numbers} = 0, \frac{\pi}{2}, \pi, \frac{3\pi}{2}, 2\pi$$

For one cycle of $y = \frac{1}{2} \cos x$, we have

$$\text{amp} = \frac{1}{2} \implies \text{max} = \frac{1}{2}, \text{min} = -\frac{1}{2}$$

$$\text{per} = 2\pi \implies \text{M-i-m-i-M numbers} = 0, \frac{\pi}{2}, \pi, \frac{3\pi}{2}, 2\pi$$

For one cycle of $y = 3 \cos x$, we have

$$\text{amp} = 3 \implies \text{max} = 3, \text{min} = -3$$

$$\text{per} = 2\pi \implies \text{M-i-m-i-M numbers} = 0, \frac{\pi}{2}, \pi, \frac{3\pi}{2}, 2\pi$$

The graphs of these three functions are shown in Figure 5.44.

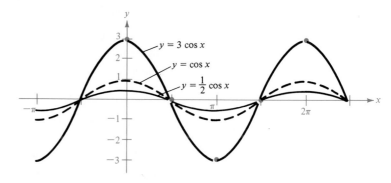

Amplitude Determines Vertical Stretch or Shrink

FIGURE 5.44

If $a < 0$ in the equation $y = a \cos x$, the resulting graph is a **reflection** through the x-axis of the case when $a > 0$. Compare the graphs of $y = 3 \cos x$ and $y = -3 \cos x$ in Figure 5.45.

EXAMPLE 3
Reflections in the x-Axis

On the same coordinate axes, sketch graphs of

$$y = 3 \cos x \qquad \text{and} \qquad y = -3 \cos x$$

Solution:

Refer to Figure 5.44 for a graph of $y = 3 \cos x$. For one cycle of $y = -3 \cos x$, we have

$$\text{amp} = |-3| = 3 \implies \text{max} = 3, \text{min} = -3$$

$$\text{per} = 2\pi \implies \text{m-i-M-i-m numbers} = 0, \frac{\pi}{2}, \pi, \frac{3\pi}{2}, 2\pi$$

Note that since

$$y = -3 \cos x = (-1)(3 \cos x)$$

the signs of the y-coordinates are the opposite of those for $y = 3 \cos x$. Thus, maximum points become minimum points, and vice versa. Using the scheme m-i-M-i-m in place of M-i-m-i-M, we obtain the graph shown in Figure 5.45.

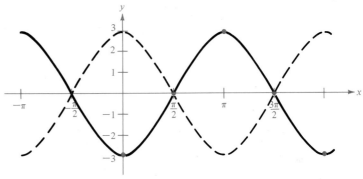

FIGURE 5.45   Reflection in the x–Axis

Next, we consider the effect of the constant b on the graph of

$$y = a \sin bx$$

Since $y = a \sin x$ completes one cycle from $x = 0$ to $x = 2\pi$ (per $= 2\pi$), it follows that $y = a \sin bx$ completes one cycle from $bx = 0$ to $bx = 2\pi$; that is

$$bx = 0 \quad \Rightarrow \quad x = 0$$

$$bx = 2\pi \quad \Rightarrow \quad x = \frac{2\pi}{b}$$

Consequently, $y = a \sin bx$ completes one cycle from $x = 0$ to $x = 2\pi/b$, and hence its period is $2\pi/b$. In general, we have:

PERIOD OF SINE AND COSINE

> The **period** of $y = a \sin bx$ and $y = a \cos bx$ is $2\pi/b$.

Remark: If $0 < b < 1$, then $(2\pi/b) > 2\pi$ and the b factor produces a *horizontal stretching* of each cycle. If $b > 1$, then $(2\pi/b) < 2\pi$ and this factor produces a *horizontal shrinking* of each cycle.

EXAMPLE 4
Functions with Period $2\pi/b$

Sketch graphs of

$$y = \sin \frac{x}{2} \quad \text{and} \quad y = \sin 3x$$

Solution:

For one cycle of $y = \sin \dfrac{x}{2} = \sin\left(\dfrac{1}{2}x\right)$, we have

amp $= 1$ $\Rightarrow$ max $= 1$, min $= -1$

per $= \dfrac{2\pi}{\frac{1}{2}}$

$= 4\pi$ $\Rightarrow$ i-M-i-m-i numbers $= 0, \dfrac{1}{4}(4\pi), \dfrac{2}{4}(4\pi), \dfrac{3}{4}(4\pi), 4\pi$

$= 0, \pi, 2\pi, 3\pi, 4\pi$

For one cycle of $y = \sin 3x$, we have

amp $= 1$ $\Rightarrow$ max $= 1$, min $= -1$

per $= \dfrac{2\pi}{3}$ $\Rightarrow$ i-M-i-m-i numbers $= 0, \dfrac{1}{4}\left(\dfrac{2\pi}{3}\right), \dfrac{2}{4}\left(\dfrac{2\pi}{3}\right), \dfrac{3}{4}\left(\dfrac{2\pi}{3}\right), \dfrac{2\pi}{3}$

$= 0, \dfrac{\pi}{6}, \dfrac{\pi}{3}, \dfrac{\pi}{2}, \dfrac{2\pi}{3}$

The graphs are shown in Figures 5.46 and 5.47, respectively.

Graph of $y = \sin \dfrac{x}{2}$
Period $= 4\pi$

FIGURE 5.46

Graph of $y = \sin 3x$
Period $= \dfrac{2\pi}{3}$

FIGURE 5.47

Phase Shift

We now consider the effect of the constant c in the general cases

$$y = a \sin(bx + c) \qquad \text{and} \qquad y = a \cos(bx + c)$$

Comparing $y = a \sin bx$ with $y = a \sin(bx + c)$, we find that $y = a \sin bx$ completes one cycle from

$$bx = 0 \quad \text{to} \quad bx = 2\pi$$

or

$$x = 0 \quad \text{to} \quad x = \frac{2\pi}{b}$$

whereas $y = a \sin(bx + c)$ completes one cycle from

$$bx + c = 0 \quad \text{to} \quad bx + c = 2\pi$$

or

$$x = \frac{-c}{b} \quad \text{to} \quad x = \frac{2\pi}{b} - \frac{c}{b}$$

We note two things about the function $y = a \sin(bx + c)$: Its period is $2\pi/b$, and its graph is shifted by an amount $-c/b$. We call the number $-c/b$ the **phase shift.** We suggest that you memorize the following analysis of the graphs of $y = a \sin(bx + c)$ and $y = a \cos(bx + c)$.

ANALYSIS OF GRAPHS OF $y = a \sin(bx + c)$
AND $y = a \cos(bx + c)$

The graphs of

$$y = a \sin(bx + c) \qquad \text{and} \qquad y = a \cos(bx + c)$$

have the following characteristics:

> **amplitude** $= |a|$.
>> **period** $= 2\pi/|b|$

The **phase shift** and resulting interval for one cycle are solutions to the equations:

$$bx + c = 0 \qquad \text{and} \qquad bx + c = 2\pi$$

The **key number increments** are $\frac{1}{4}$(period).

EXAMPLE 5
Using Amplitude, Period,
and Shift to Sketch Graphs

Sketch a graph of

$$y = \frac{1}{2} \sin\left(x - \frac{\pi}{3}\right)$$

Solution:

An analysis of the equation yields

$$\text{amp} = \frac{1}{2} \implies \text{max} = \frac{1}{2}, \text{min} = -\frac{1}{2}$$

$$\text{per} = \frac{2\pi}{b} = 2\pi \implies \text{increment} = \frac{1}{4}(2\pi) = \frac{\pi}{2}$$

$$\text{interval: } x - \frac{\pi}{3} = 0 \quad \text{to} \quad x - \frac{\pi}{3} = 2\pi$$

$$x = \frac{\pi}{3} \quad \text{to} \quad x = 2\pi + \frac{\pi}{3}$$

$$\text{shift} = \frac{\pi}{3}$$

$$\text{i-M-i-m-i numbers} = \frac{\pi}{3}, \frac{\pi}{3} + \frac{\pi}{2}, \frac{\pi}{3} + 2\left(\frac{\pi}{2}\right), \frac{\pi}{3} + 3\left(\frac{\pi}{2}\right), \frac{\pi}{3} + 2\pi$$

$$= \frac{\pi}{3}, \frac{5\pi}{6}, \frac{4\pi}{3}, \frac{11\pi}{6}, \frac{7\pi}{3}$$

Using this information, we obtain the graph shown in Figure 5.48. Its accuracy can be confirmed by computing a few points with your calculator.

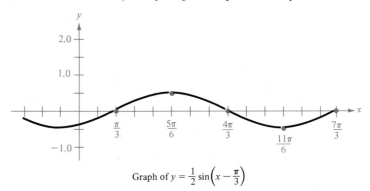

Graph of $y = \frac{1}{2}\sin\left(x - \frac{\pi}{3}\right)$

FIGURE 5.48

EXAMPLE 6
Using Amplitude, Period, and Shift to Sketch Graphs

Sketch a graph of

$$y = 2\cos(2x + 3)$$

Solution:
In this case, we have

$$\text{amp} = 2 \implies \text{max} = 2, \text{min} = -2$$

$$\text{per} = \frac{2\pi}{2} = \pi \implies \text{increment} = \frac{\pi}{4} \approx 0.7854$$

interval: $2x + 3 = 0$ to $2x + 3 = 2\pi$

$$x = \frac{-3}{2} \quad \text{to} \quad x = \pi - \frac{3}{2} \approx 1.6416$$

shift $= -\dfrac{3}{2}$

M-i-m-i-M numbers $= -\dfrac{3}{2}, \ -\dfrac{3}{2} + \dfrac{\pi}{4}, \ -\dfrac{3}{2} + \dfrac{2\pi}{4}, \ -\dfrac{3}{2} + \dfrac{3\pi}{4}, \ -\dfrac{3}{2} + \pi$

$= -1.5, \ -0.7146, \ 0.0708, \ 0.8562, \ 1.6416$

The resulting sketch is shown in Figure 5.49.

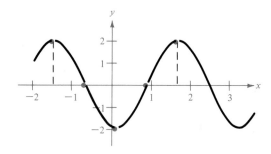

FIGURE 5.49   Graph of $y = 2 \cos(2x + 3)$

For our last example in this section, we reverse the situation and show how to determine the equation for a given graph.

EXAMPLE 7
Finding the Equation
for a Given Graph

Find the amplitude, period, and phase shift for the graph in Figure 5.50. Write an equation for this graph.

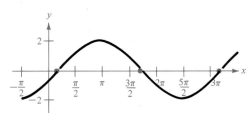

FIGURE 5.50   A Sine Curve

Solution:
For this sine curve, we have

$$\text{amp} = 2 \implies a = 2$$

$$\text{per} = \frac{19\pi}{6} - \frac{\pi}{6} = 3\pi = \frac{2\pi}{b} \implies b = \frac{2}{3}$$

$$\text{shift} = \frac{\pi}{6} = \frac{-c}{b} = \frac{-c}{\frac{2}{3}} \implies c = \frac{-\pi}{9}$$

Therefore, the equation for the graph in Figure 5.50 is

$$y = a \sin(bx + c) = 2 \sin\left(\frac{2}{3}x - \frac{\pi}{9}\right)$$

Finally, it is of interest to note that the graph of $y = \cos x$ corresponds (see Figure 5.51) to a left shift of $\pi/2$ units of the graph of $y = \sin x$, and conversely. This is consistent with the cofunction identities discussed in Section 5.2. That is,

$$\sin\left(\frac{\pi}{2} - x\right) = \cos x \quad \text{and} \quad \cos\left(\frac{\pi}{2} - x\right) = \sin x$$

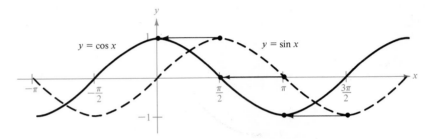

Shift $y = \sin x$ to Left $\frac{\pi}{2}$ Units to Obtain Graph of $y = \cos x$

FIGURE 5.51

Section Exercises 5.4

In Exercises 1–14, determine the period and amplitude of the given function.

1. $y = 2 \sin 2x$

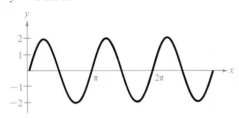

2. $y = 3 \cos 3x$

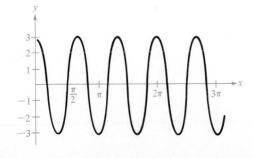

3. $y = \frac{3}{2} \cos \frac{x}{2}$

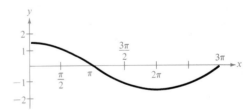

4. $y = -2 \sin \frac{x}{3}$

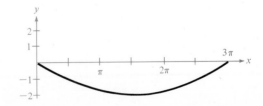

5. $y = \frac{1}{2} \sin \pi x$

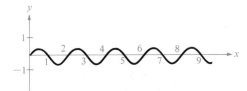

6. $y = \frac{5}{2} \cos \frac{\pi x}{2}$

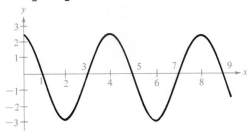

7. $y = 2 \sin x$

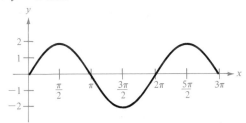

8. $y = -\cos \frac{2x}{3}$

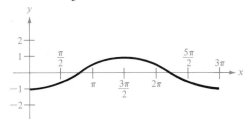

9. $y = -2 \sin 10x$ 10. $y = \frac{1}{3} \sin 8x$

11. $y = \frac{1}{2} \cos \frac{2x}{3}$ 12. $y = \frac{5}{2} \cos \frac{x}{4}$

13. $y = 3 \sin 4\pi x$ 14. $y = \frac{2}{3} \cos \frac{\pi x}{10}$

In Exercises 15–40, sketch the graph of the given function. (Include two full periods.)

15. $y = \sin \frac{x}{2}$ 16. $y = 4 \sin \frac{x}{3}$

17. $y = 2 \cos 2x$ 18. $y = \frac{3}{2} \cos \frac{2x}{3}$

19. $y = -2 \sin 6x$ 20. $y = -3 \cos 4x$

21. $y = \cos 2\pi x$ 22. $y = \frac{3}{2} \sin \frac{\pi x}{4}$

23. $y = -\sin \frac{2\pi x}{3}$ 24. $y = 10 \cos \frac{\pi x}{6}$

25. $y = \sin\left(x - \frac{\pi}{4}\right)$ 26. $y = \frac{1}{2} \sin(x - \pi)$

27. $y = 3 \cos(x + \pi)$ 28. $y = 4 \cos\left(x + \frac{\pi}{4}\right)$

29. $y = \frac{2}{3} \cos\left(\frac{x}{2} - \frac{\pi}{4}\right)$ 30. $y = -3 \cos(6x + \pi)$

31. $y = -2 \sin(4x + \pi)$ 32. $y = -4 \sin\left(\frac{2}{3}x - \frac{\pi}{3}\right)$

33. $y = \cos\left(2\pi x - \frac{\pi}{2}\right)$ 34. $y = 3 \cos\left(\frac{\pi x}{2} + \frac{\pi}{2}\right)$

35. $y = -0.1 \sin\left(\frac{\pi x}{10} + \pi\right)$ 36. $y = 5 \sin(\pi - 2x)$

37. $y = 5 \cos(\pi - 2x)$ 38. $y = \frac{1}{100} \sin(120\pi t)$

39. $y = \frac{1}{10} \cos(60\pi t)$ 40. $y = 5 \cos\left(\frac{\pi t}{12}\right)$

In Exercises 41–45, find a, b, and c so that the graph of the function matches the graph in the accompanying figure.

41. $y = a \sin(bx + c)$

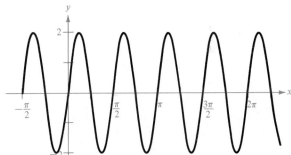

42. $y = a \sin(bx + c)$

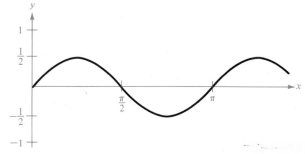

43. $y = a \cos(bx + c)$

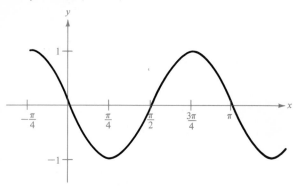

44. $y = a \cos(bx + c)$

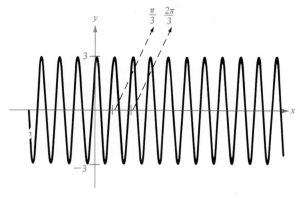

45. $y = a \sin(bx + c)$

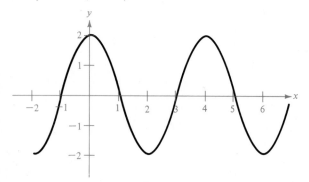

46. Show that $\sin x = -\cos\left(x + \dfrac{\pi}{2}\right)$ by graphing $y_1 = \sin x$ and $y_2 = -\cos\left(x + \dfrac{\pi}{2}\right)$.

47. For a person at rest, the velocity v (in liters per second) of air flow during a respiratory cycle is

$$v = 0.85 \sin \frac{\pi t}{3}$$

where t is the time in seconds. (Inhalation occurs when $v > 0$ and exhalation occurs when $v < 0$.)

(a) Find the time for one full respiratory cycle.
(b) Find the number of cycles per minute.
(c) Sketch the graph of the velocity function.

48. After exercising for a few minutes, a person has a respiratory cycle for which the velocity of air flow is approximated by

$$v = 1.75 \sin \frac{\pi t}{2}$$

Use this model to repeat Exercise 47.

49. When tuning a piano, a technician strikes a tuning fork for the A above middle C and sets up wave motion that can be approximated by

$$y = 0.001 \sin 880\pi t$$

where t is the time in seconds.

(a) What is the period p of this function?
(b) The frequency f is given by $f = 1/p$. What is the frequency of this note?
(c) Sketch the graph of this function.

Graphs of the Other Trigonometric Functions   5.5

In this section we continue our discussion of the graphs of the trigonometric functions, starting with the graph of $y = \tan x$. In Table 5.2 in Section 5.3, the value of $\tan \pi/2$ was listed as *undefined*. Let us examine this situation more carefully by calculating values for tan x as x gets closer and closer to $\pi/2$ (≈ 1.5708), as shown in Table 5.6.

TABLE 5.6
Values of tan x

x	1.5000	1.5700	1.5707	1.57079	$\rightarrow$	$\dfrac{\pi}{2}$	$\dfrac{\pi}{2}$	$\leftarrow$	1.57081	1.5710	1.5800
tan x	14.1	1256.8	10,385.3	158,714.0	$\rightarrow$	∞	$-\infty$	$\leftarrow$	$-79{,}988.3$	-4909.2	-108.6

We conclude from Table 5.6 that tan x *increases* without bound as x approaches $\pi/2$ from the left, and that tan x *decreases* without bound as x approaches $\pi/2$ from the right. In short, we write

$$\tan x \rightarrow \infty \qquad \text{as} \qquad x \rightarrow \frac{\pi}{2^-} \qquad\qquad \textit{(From the left)}$$

$$\tan x \rightarrow -\infty \qquad \text{as} \qquad x \rightarrow \frac{\pi}{2^+} \qquad\qquad \textit{(From the right)}$$

This suggests that $x = \pi/2$ is a *vertical asymptote* for the graph of $y = \tan x$. Using this fact and the values in Table 5.6, we obtain the graph shown in Figure 5.52.

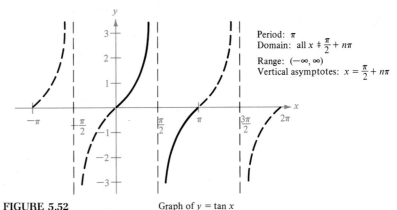

Period: π
Domain: all $x \neq \dfrac{\pi}{2} + n\pi$

Range: $(-\infty, \infty)$
Vertical asymptotes: $x = \dfrac{\pi}{2} + n\pi$

FIGURE 5.52   Graph of $y = \tan x$

From Figure 5.52 and the corresponding table of values, we can see that the tangent function begins to repeat itself for $x \geq \pi$, indicating a period of π for $y = \tan x$. For convenience, we consider *one cycle* of this tangent curve to be the portion on the interval $-\pi/2 < x < \pi/2$, where values of tan x range through the set of real numbers.

EXAMPLE 1
Graphing the Tangent Function

Sketch a graph of

$$y = \tan \frac{x}{2}$$

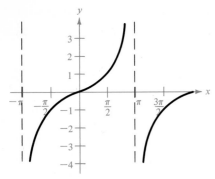

Graph of $y = \tan \frac{x}{2}$

FIGURE 5.53

EXAMPLE 2
Graphing the Tangent Function

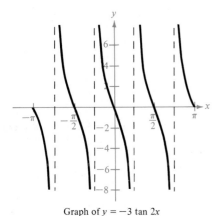

Graph of $y = -3 \tan 2x$

FIGURE 5.54

Solution:
As with sine and cosine, the period of

$$y = \tan \frac{x}{2} = \tan\left(\frac{1}{2}x\right)$$

is affected by the factor $\frac{1}{2}$. We have

$$\text{per} = \frac{\pi}{b} = \frac{\pi}{\frac{1}{2}} = 2\pi$$

Consequently, one complete cycle occurs from

$$x = -\pi \quad \text{to} \quad x = \pi$$

The graph is shown in Figure 5.53.

Sketch a graph of

$$y = -3 \tan 2x$$

Solution:
In this case, the factor, -3, produces a *reflection* in the x-axis. Furthermore,

$$\text{per} = \frac{\pi}{2}$$

$$\text{one cycle: } x = \frac{-\pi}{4} \quad \text{to} \quad x = \frac{\pi}{4}$$

The resulting graph is shown in Figure 5.54.

Graphs of the Reciprocal Functions

The graphs of the three remaining trigonometric functions can be readily obtained from the graphs of $y = \sin x$, $y = \cos x$, and $y = \tan x$, using the identities

$$\csc x = \frac{1}{\sin x}, \quad \sec x = \frac{1}{\cos x}, \quad \text{and} \quad \cot x = \frac{1}{\tan x}$$

For instance, at a given value for x, the y-coordinate for $\sec x$ is the reciprocal of the y-coordinate for $\cos x$ at that same x-value. Of course, when $\cos x = 0$, the reciprocal does not exist. At such values for x, the behavior of the secant function is similar to that of the tangent function. In other words,

$$\tan x = \frac{\sin x}{\cos x} \quad \text{and} \quad \sec x = \frac{1}{\cos x}$$

have vertical asymptotes at $x = (\pi/2) + n\pi$, because $\cos[(\pi/2) + n\pi] = 0$ for any integer n. Similarly,

$$\cot x = \frac{\cos x}{\sin x} \quad \text{and} \quad \csc x = \frac{1}{\sin x}$$

have vertical asymptotes where $\sin x = 0$; that is, at $x = n\pi$.

To sketch the graph of a reciprocal function such as $y = \csc x$, it is convenient to first sketch the graph of $y = \sin x$ with dashed lines, then take reciprocals of the y-coordinates to obtain points on the graph of $y = \csc x$. We use this procedure to obtain the graphs shown in Figure 5.55.

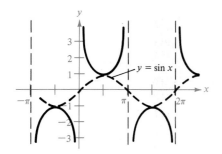

Period: 2π
Domain: all $x \neq n\pi$
Range: all y not in $(-1, 1)$

(a) Graph of $y = \csc x$

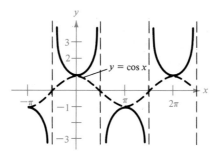

Period: 2π
Domain: all $x \neq \frac{\pi}{2} + n\pi$
Range: all y not in $(-1, 1)$

(b) Graph of $y = \sec x$

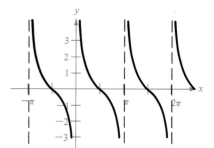

Period: π
Domain: all $x \neq n\pi$
Range: $(-\infty, \infty)$

(c) Graph of $y = \cot x$

FIGURE 5.55

In comparing the graphs of

$$y = \sin x \qquad \text{and} \qquad y = \cos x$$

with the graphs of their reciprocal functions

$$y = \csc x \qquad \text{and} \qquad y = \sec x$$

note the following changes that occur in the character of the key points:

Graphs of sin x and cos x		Graphs of csc x and sec x
intercepts	⇨	vertical asymptotes
maximum points	⇨	minimum points
minimum points	⇨	maximum points

EXAMPLE 3
Graphing the Cosecant Function

Sketch a graph of

$$y = 2 \csc\left(x + \frac{\pi}{4}\right)$$

Solution:
Using an analysis similar to that used with the sine function, we have

$$\text{per} = \frac{2\pi}{b} = 2\pi \quad \Rightarrow \quad \text{increment} = \frac{1}{4}(2\pi) = \frac{\pi}{2}$$

$$a = 2 \quad \Rightarrow \quad \text{range: } y \leq -2 \text{ and } y \geq 2$$

one cycle: $x + \dfrac{\pi}{4} = 0$ to $x + \dfrac{\pi}{4} = 2\pi$

$$x = -\dfrac{\pi}{4} \quad \text{to} \quad x = -\dfrac{\pi}{4} + 2\pi$$

shift $= -\dfrac{\pi}{4}$

vertical asymptotes: $x = -\dfrac{\pi}{4}, \ -\dfrac{\pi}{4} + \pi, \ -\dfrac{\pi}{4} + 2\pi$

The sketch is shown in Figure 5.56.

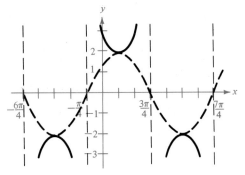

Graph of $y = 2 \csc\left(x + \dfrac{\pi}{4}\right)$

FIGURE 5.56

EXAMPLE 4
Graphing the Secant Function

Sketch a graph of

$$y = \sec\left(2x - \dfrac{\pi}{3}\right)$$

Solution:
Compared with the cosine function, we have

$$\text{per} = \dfrac{2\pi}{b} = \dfrac{2\pi}{2} = \pi \ \Rightarrow \ \text{increment} = \dfrac{1}{4}(\pi) = \dfrac{\pi}{4}$$

$\text{amp} = 1 \ \Rightarrow \ \text{range: } y \le -1 \text{ and } y \ge 1$

one cycle: $2x - \dfrac{\pi}{3} = 0$ to $2x - \dfrac{\pi}{3} = 2\pi$

$$x = \dfrac{\pi}{6} \quad \text{to} \quad x = \dfrac{\pi}{6} + \pi$$

shift $= \dfrac{\pi}{6}$

vertical asymptotes: $x = \dfrac{\pi}{6} + \dfrac{\pi}{4}, \dfrac{\pi}{6} + \dfrac{3\pi}{4}$

$$x = \dfrac{5\pi}{12}, \dfrac{11\pi}{12}$$

The sketch is shown in Figure 5.57.

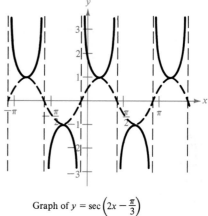

Graph of $y = \sec\left(2x - \dfrac{\pi}{3}\right)$

FIGURE 5.57

In Table 5.7, we summarize the significant graphical properties of the six basic trigonometric functions.

TABLE 5.7
Graphical Properties of the Trigonometric Functions

Function	Figure Number	Domain	Range	Positive Quadrant	Period	Negative Angles
$\sin x$	5.40	all reals	$[-1, 1]$	I, II	2π	$\sin(-x) = -\sin x$
$\cos x$	5.41	all reals	$[-1, 1]$	I, IV	2π	$\cos(-x) = \cos x$
$\tan x$	5.52	$x \neq \dfrac{\pi}{2} + n\pi$	$(-\infty, \infty)$	I, III	π	$\tan(-x) = -\tan x$
$\csc x$	5.55(a)	$x \neq n\pi$	$(-\infty, -1]$ and $[1, \infty)$	I, II	2π	$\csc(-x) = -\csc x$
$\sec x$	5.55(b)	$x \neq \dfrac{\pi}{2} + n\pi$	$(-\infty, -1]$ and $[1, \infty)$	I, IV	2π	$\sec(-x) = \sec x$
$\cot x$	5.55(c)	$x \neq n\pi$	$(-\infty, \infty)$	I, III	π	$\cot(-x) = -\cot x$

Section Exercises 5.5

In Exercises 1–10, match the trigonometric function with the correct graph and give the period of the function.

1. $y = \sec 2x$

2. $y = \tan 3x$

3. $y = \tan \dfrac{x}{2}$

4. $y = 2 \csc \dfrac{x}{2}$

5. $y = \dfrac{1}{2} \csc 2x$

6. $y = \dfrac{1}{2} \sec \pi x$

7. $y = \cot \pi x$

8. $y = 3 \cot \dfrac{\pi x}{2}$

9. $y = -\sec x$

10. $y = -2 \csc 2\pi x$

(a)

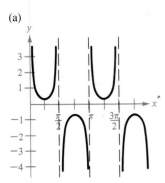

(b)

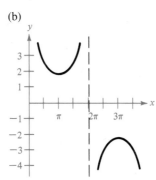

(c)

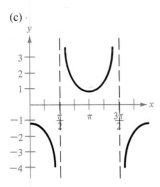

(d)

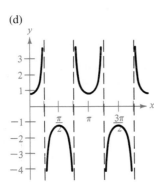

(e)

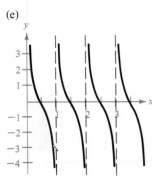

(f)

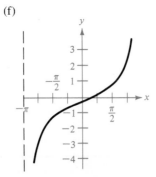

(g)

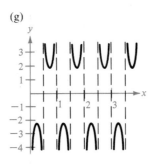

(h)

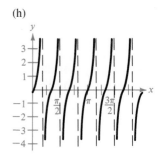

(i)

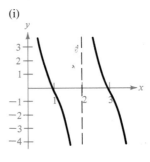

(j)

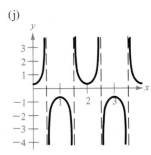

In Exercises 11–30, sketch the graph of the function through two periods.

11. $y = \tan 2x$

12. $y = 3 \tan \pi x$

13. $y = \sec \pi x$

14. $y = 2 \sec 4x$

15. $y = \csc \dfrac{x}{2}$

16. $y = \csc \dfrac{x}{3}$

17. $y = 2 \sec 2x$

18. $y = -\dfrac{1}{2} \tan x$

19. $y = \tan\left(x - \dfrac{\pi}{4}\right)$

20. $y = \sec(x + \pi)$

21. $y = \dfrac{1}{4} \csc\left(x + \dfrac{\pi}{4}\right)$

22. $y = -\csc(4x - \pi)$

23. $y = \dfrac{1}{4} \cot\left(x - \dfrac{\pi}{2}\right)$

24. $y = 2 \cot\left(x + \dfrac{\pi}{2}\right)$

25. $y = 2 \sec(2x - \pi)$

26. $y = \dfrac{1}{3} \sec\left(\dfrac{\pi x}{2} + \dfrac{\pi}{2}\right)$

27. $y = \tan\left(\dfrac{\pi}{4}x\right)$

28. $y = 0.1 \tan\left(\dfrac{\pi}{4}x + \dfrac{\pi}{4}\right)$

29. $y = \csc(\pi - x)$

30. $y = \sec(\pi - x)$

Additional Graphing Techniques

5.6

Addition of Ordinates

The behavior of some physical phenomena can be represented by more than one trigonometric function, or a combination of algebraic and trigonometric functions. To graph functions like

$$y = \sin x - \cos 2x \qquad \text{or} \qquad y = x + \cos x$$

we can make use of a technique known as **addition of ordinates.** For example, the graph of $y = x + \cos x$ can be obtained by first making dotted sketches of $y = x$ and $y = \cos x$ on the same set of axes and then geometrically adding the ordinates (*y*-values) for each x in a domain common to both functions. This addition of ordinates is aided by the use of a compass or ruler to measure the vertical displacements that are to be combined (see Figure 5.58).

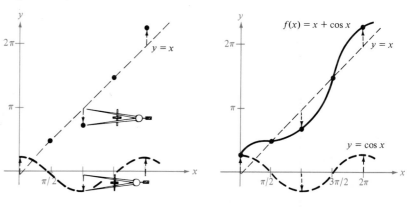

FIGURE 5.58   **FIGURE 5.59**   Addition of Ordinates

As with previous trigonometric graphs, the *key points* to plot are those for which one or both functions have an intercept, maximum point, or minimum point. Points of intersection are also useful points to plot with this method.

EXAMPLE 1
Graphing by Addition of Ordinates

Use addition of ordinates to sketch a graph of

$$y = x + \cos x$$

Solution:
First, we make dashed sketches of both

$$y = x \qquad \text{and} \qquad y = \cos x$$

on the same set of axes. For $y = \cos x$, the key numbers are $x = 0$, $\pi/2$, π, $3\pi/2$, and 2π. By geometrically adding the ordinates of the two functions at each of these key numbers and connecting the resulting points by a smooth curve, we obtain the graph shown in Figure 5.59.

Note that the function in Figure 5.59 is not periodic as we defined the concept. However, the next example involves two trigonometric functions whose resulting difference is a periodic function.

EXAMPLE 2
Graphing by Addition of Ordinates

Use addition of ordinates to sketch a graph of

$$y = \sin x - \cos 2x$$

Solution:
In this case, we make dashed sketches of

$$y = \sin x \quad \text{and} \quad y = -\cos 2x$$

noting that their respective periods are 2π and π. From the shorter period, we choose the *key numbers*

$$x = 0, \frac{\pi}{4}, \frac{\pi}{2}, \frac{3\pi}{4}, \pi$$

Adding ordinates at these numbers and at points of intersection, we obtain the graph shown in Figure 5.60. Note that this function has a period of 2π.

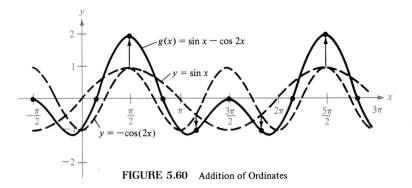

FIGURE 5.60   Addition of Ordinates

Vertical Translations

Addition of ordinates can also be used to graph functions like

$$y = 3 + \sin 2x \quad \text{and} \quad y = 2 + \cos\left(x - \frac{\pi}{4}\right)$$

Such equations have graphs with vertical translations like those discussed in Chapter 2. Adding a constant does not change the shape (or period) of a trigonometric graph—it only changes its vertical location. For example, the graph of $y = \sin 3x$ is given in Figure 5.47 in Section 5.4. We can readily graph

$$y = 2 + \sin 3x$$

by adding 2 to each ordinate of the given graph to obtain the graph shown in Figure 5.61.

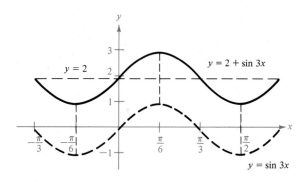

FIGURE 5.61

Products of Functions

A *product* of two functions can be graphed using properties of the individual functions involved. For instance, consider the function

$$f(x) = x \sin x, \qquad x \geq 0$$

as the product of the functions $y = x$ and $y = \sin x$. Using properties of absolute value and the fact that $|\sin x| \leq 1$, we have

$$|f(x)| = |x| \, |\sin x| \leq |x| = x, \qquad x \geq 0$$

Consequently,

$$-x \leq f(x) \leq x$$

which means that the graph of $f(x) = x \sin x$ lies between the lines $y = -x$

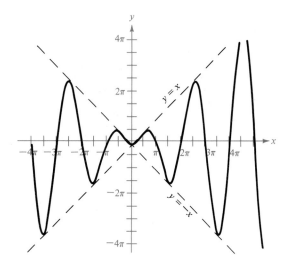

Graph of $y = x \sin x$

FIGURE 5.62

and $y = x$. Furthermore, since

$$f(x) = x \sin x = \pm x \quad \text{at} \quad x = \frac{\pi}{2} + n\pi$$

$$f(x) = x \sin x = 0 \quad \text{at} \quad x = n\pi$$

the graph of f touches the line $y = -x$ or the line $y = x$ at $x = (\pi/2) + n\pi$ and has x-intercepts at $x = n\pi$. A sketch of f is shown in Figure 5.62.
In the product function

$$f(x) = x \sin x$$

the factor x is sometimes referred to as the **damping factor.** In general, we can obtain compressed or expanded sine (or cosine) waves using damping factors.

We show another instance of a damping factor in the next example.

EXAMPLE 3
A Damped Cosine Wave

Sketch a graph of

$$f(x) = 2^{-x/2} \cos x$$

Solution:
Consider $f(x)$ as the product of the two functions

$$y = 2^{-x/2} \quad \text{and} \quad y = \cos x$$

each of which has the set of real numbers as its domain. We know that for any real number $2^{-x/2} \geq 0$ and $|\cos x| \leq 1$, and thus

$$|f(x)| = |2^{-x/2}| \, |\cos x| \leq 2^{-x/2}(1) = 2^{-x/2}$$

Consequently,

$$-2^{-x/2} \leq f(x) \leq 2^{-x/2}$$

Furthermore, since

$$f(x) = 2^{-x/2} \cos x = \pm 2^{-x/2} \quad \text{at} \quad x = n\pi$$

$$f(x) = 2^{-x/2} \cos/x = 0 \quad \text{at} \quad x = \frac{\pi}{2} + n\pi$$

The graph of f touches the curves $y = -2^{-x/2}$ and $y = 2^{-x/2}$ at $x = n\pi$ and has intercepts at $x = (\pi/2) + n\pi$. A sketch is shown in Figure 5.63.

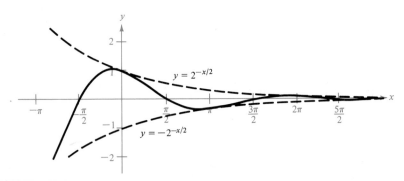

FIGURE 5.63   Graph of $y = 2^{-x/2} \cos x$

Section Exercises 5.6

In Exercises 1–20, use the method of addition of ordinates to sketch the graph of the function.

1. $y = x + \sin x$
2. $y = x + \cos x$
3. $y = \frac{1}{2}x - 2 \cos x$
4. $y = 2x - \sin x$
5. $y = \sin x + \cos x$
6. $y = \cos x + \cos 2x$
7. $y = 2 \sin x + \sin 2x$
8. $y = 2 \sin x + \cos 2x$
9. $y = \cos x - \cos \frac{x}{2}$
10. $y = \sin x - \frac{1}{2} \sin \frac{x}{2}$
11. $y = 2 - 2 \sin \frac{x}{2}$
12. $y = -3 + \cos\left(x + \frac{\pi}{4}\right)$

13. $y = -3 + \cos x + 2 \sin 2x$

14. $y = \sin \pi x + \sin \frac{\pi x}{2}$

15. $y = |x| + \frac{1}{2} \cos \pi x$

16. $y = 1 + \sin x + \cos\left(x + \frac{\pi}{2}\right)$

17. $y = 2 + \tan(\pi x)$
18. $y = -1 + \cot x$
19. $y = 1 + \csc x$
20. $y = 1 - \sec x$

In Exercises 21–26, sketch the graph of the function.

21. $y = x \cos x$
22. $y = |x| \sin x$
23. $y = |x| \cos x$
24. $y = 2^{-x/2} \cos x$
25. $y = e^{-x^2/2} \sin x$
26. $y = e^{-t} \cos t$
27. The monthly sales S in thousands of units of a seasonal product is approximated by

$$S = 74.50 + 43.75 \sin \frac{\pi t}{6}$$

where t is the time in months with $t = 1$ corresponding to January. Sketch the graph of this sales function over one year.

28. The function

$$P = 100 - 20 \cos \frac{5\pi t}{3}$$

approximates the blood pressure P (in millimeters of mercury) for a person at rest. (The time t is measured in seconds.) Sketch the graph of this function over a 10-second interval of time.

29. Suppose that the population of a certain predator at time t (in months) in a given region is estimated to be

$$P = 10,000 + 3000 \sin \frac{2\pi t}{24}$$

and the population of its primary food source (its prey) is estimated to be

$$p = 15,000 + 5000 \cos \frac{2\pi t}{24}$$

Sketch both of these functions on the same graph, and explain in the oscillations in the size of each population.

30. Use a calculator to evaluate the function

$$f(x) = \frac{\sin x}{x}$$

at several points in the interval $[-1, 1]$, and then use these points to sketch the function's graph. This function is undefined when $x = 0$. From your graph, estimate the value that $f(x)$ is approaching as x approaches 0.

x	-1.0	-0.9	-0.8	-0.7	-0.6	-0.5	-0.4	-0.3	-0.2	-0.1
$\dfrac{\sin x}{x}$										

x	0.1	0.2	0.3	0.4	0.5	0.6	0.7	0.8	0.9	1.0
$\dfrac{\sin x}{x}$										

Inverse Trigonometric Functions

5.7

Until now we have mainly been evaluating trigonometric functions at specified numbers or angles. However, in Section 5.3 we posed the *inverse* problem:

> *Given the value of sin x, find x.*

We used the calculator key $\boxed{\text{INV}}$ then, and promised to explain the functions involved later. We now investigate these functions, known as the **inverse trigonometric functions.**

Recall from Section 2.5 that in order for a function to have an inverse, it must be one-to-one. From Figure 5.64 it is obvious that $y = \sin x$ is not one-to-one. However, if we restrict the domain to the interval $-\pi/2 \le x \le \pi/2$ (corresponding to the solid portion of the graph in Figure 5.64), the following properties hold:

$y = \sin x$ is always increasing
y takes on its full range of values, $-1 \le \sin x \le 1$
$y = \sin x$ is a one-to-one function

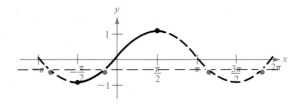

FIGURE 5.64 $\sin x = -\dfrac{1}{2}$ at $-\dfrac{5\pi}{6}, -\dfrac{\pi}{6}, \dfrac{7\pi}{6}, \dfrac{11\pi}{6}, \ldots$

Hence, on the restricted domain $-\pi/2 \le x \le \pi/2$, a unique inverse function exists called the **inverse sine function.** It is denoted by

$$y = \arcsin x \qquad \text{or} \qquad y = \sin^{-1} x$$

DEFINITION OF INVERSE SINE FUNCTION

> The **inverse sine function** is defined by
>
> $\quad y = \arcsin x \qquad \text{if and only if} \qquad \sin y = x$
>
> where $-1 \le x \le 1$ and $-\pi/2 \le y \le \pi/2$.

Remark: From an angle perspective, we read $y = \arcsin x$ as y *is an angle whose sine is x.* That is, arcsin x is an *angle* (in radian measure) such that

$$-\frac{\pi}{2} \le \arcsin x \le \frac{\pi}{2}$$

Both notations, arcsin and $\sin^{-1}$, are commonly used in mathematics, so remember that $\sin^{-1}$ means *inverse* rather than reciprocal.

EXAMPLE 1
Evaluating the Inverse Sine Function

Find the values of

(a) $\arcsin\left(-\dfrac{1}{2}\right)$

(b) $\sin^{-1}\left(\dfrac{\sqrt{3}}{2}\right)$

Solution:

(a) By definition, $y = \arcsin(-\frac{1}{2})$ implies that

$$\sin y = -\frac{1}{2}, \qquad \text{for } -\frac{\pi}{2} \le y \le \frac{\pi}{2}$$

From the special angle information in Table 5.1, we know that $\sin(\pi/6) = \frac{1}{2}$, and thus $\sin(-\pi/6) = -\frac{1}{2}$. Therefore,

$$\arcsin\left(-\frac{1}{2}\right) = -\frac{\pi}{6}$$

(b) By definition, $y = \sin^{-1}(\sqrt{3}/2)$ implies that

$$\sin y = \frac{\sqrt{3}}{2}, \qquad \text{for } -\frac{\pi}{2} \le y \le \frac{\pi}{2}$$

But $\sin(\pi/3) = \sqrt{3}/2$, and consequently

$$\sin^{-1}\left(\frac{\sqrt{3}}{2}\right) = \frac{\pi}{3}$$

From past experience we know that graphs of inverse functions are reflections of each other in the line $y = x$. In Example 2 we verify this for the solid portion of the graph shown in Figure 5.64.

EXAMPLE 2
Graphing the Arcsine Function

Sketch a graph of $y = \arcsin x$.

Solution:
By definition, the equations

$$y = \arcsin x \qquad \text{and} \qquad \sin y = x$$

are equivalent. Hence, their graphs are the same. By assigning values to y in the latter equation and adhering to the restrictions

$$-\frac{\pi}{2} \le y \le \frac{\pi}{2} \qquad \text{and} \qquad -1 \le x \le 1$$

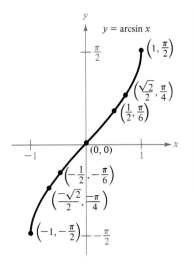

FIGURE 5.65

we can make the following table of values:

y	$-\dfrac{\pi}{2}$	$-\dfrac{\pi}{4}$	$-\dfrac{\pi}{6}$	0	$\dfrac{\pi}{6}$	$\dfrac{\pi}{4}$	$\dfrac{\pi}{2}$
x	-1	$-\dfrac{\sqrt{2}}{2}$	$-\dfrac{1}{2}$	0	$\dfrac{1}{2}$	$\dfrac{\sqrt{2}}{2}$	1

The resulting graph for $y = \arcsin x$ is shown in Figure 5.65. Note how it is the reflection (in line $y = x$) of the solid part of Figure 5.64.

On the interval $0 \leq x \leq \pi$, the function $y = \cos x$ is always decreasing (see Figure 5.66), taking on all values between -1 and 1. Consequently, on this interval a unique inverse function exists called the **inverse cosine function,** denoted by

$$y = \arccos x \qquad \text{or} \qquad y = \cos^{-1} x$$

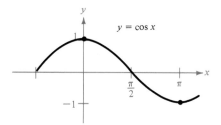

FIGURE 5.66

Similarly, for each of the other trigonometric functions, we can define an inverse function by a suitable restriction of the domain. The six inverse trigonometric functions are defined as follows.

DEFINITION OF THE INVERSE TRIGONOMETRIC FUNCTIONS

Function	Domain	Range
$y = $ **arcsin** x iff $\sin y = x$	$-1 \leq x \leq 1$	$-\pi/2 \leq y \leq \pi/2$
$y = $ **arccos** x iff $\cos y = x$	$-1 \leq x \leq 1$	$0 \leq y \leq \pi$
$y = $ **arctan** x iff $\tan y = x$	$-\infty < x < \infty$	$-\pi/2 < y < \pi/2$
$y = $ **arccot** x iff $\cot y = x$	$-\infty < x < \infty$	$0 < y < \pi$
$y = $ **arcsec** x iff $\sec y = x$	$x \leq -1, \quad 1 \leq x$	$0 \leq y \leq \pi, \quad y \neq \pi/2$
$y = $ **arccsc** x iff $\csc y = x$	$x \leq -1, \quad 1 \leq x$	$-\pi/2 \leq y \leq \pi/2, \quad y \neq 0$

See Figure 5.67 for their graphs.

Remark: We use "iff" to mean "if and only if."

Domain: $[-1, 1]$
Range: $[-\pi/2, \pi/2]$

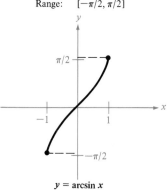

$y = \arcsin x$

Domain: $(-\infty, -1]$ and $[1, \infty)$
Range: $[-\pi/2, 0)$ and $(0, \pi/2]$

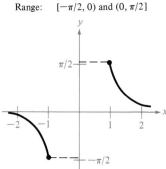

$y = \text{arccsc } x$

Domain: $(-\infty, \infty)$
Range: $(-\pi/2, \pi/2)$

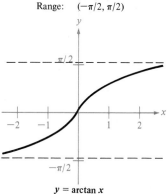

$y = \arctan x$

Domain: $[-1, 1]$
Range: $[0, \pi]$

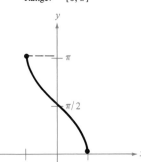

$y = \arccos x$

Domain: $(-\infty, -1]$ and $[1, \infty)$
Range: $[0, \pi/2)$ and $(\pi/2, \pi]$

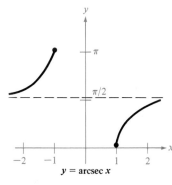

$y = \text{arcsec } x$

Domain: $(-\infty, \infty)$
Range: $(0, \pi)$

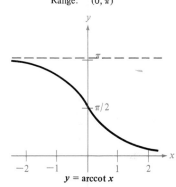

$y = \text{arccot } x$

FIGURE 5.67   *Inverse Trigonometric Functions*

Remark: Occasionally you will encounter variations in the ranges shown in Figure 5.67. In particular, the range of the arcsec function might be defined as

$$0 \le y < \frac{\pi}{2} \quad \text{and} \quad \pi \le y < \frac{3\pi}{2}$$

Recall from Section 2.5 that inverse functions (including inverse trigonometric functions) possess the properties

$$f(f^{-1}(x)) = x \quad \text{and} \quad f^{-1}(f(x)) = x$$

under suitable restrictions on x.

INVERSE PROPERTIES

If $-1 \le x \le 1$ and $-\pi/2 \le y \le \pi/2$, then

$\quad \sin(\arcsin x) = x \qquad$ and $\qquad \arcsin(\sin y) = y$

If $-1 \le x \le 1$ and $0 \le y \le \pi$, then

$\quad \cos(\arccos x) = x \qquad$ and $\qquad \arccos(\cos y) = y$

If $-\pi/2 < y < \pi/2$, then

$\quad \tan(\arctan x) = x \qquad$ and $\qquad \arctan(\tan y) = y$

EXAMPLE 3
Evaluating Inverse
Trigonometric Functions

Find the value of each of the following:

(a) $\arccos \dfrac{\sqrt{2}}{2}$ (b) $\arctan(-8.45)$ (c) $\arcsin(0.2447)$

Solution:

(a) If $y = \arccos(\sqrt{2}/2)$, then $\cos y = \sqrt{2}/2$, $0 \le y \le \pi$. From this we conclude that y is the special angle $\pi/4$. Thus, we have

$$\cos \frac{\pi}{4} = \frac{\sqrt{2}}{2} \qquad \text{or} \qquad \arccos \frac{\sqrt{2}}{2} = \frac{\pi}{4}$$

(b) We need tables or a calculator here. With a calculator

In Radian Mode	**In Degree Mode**
8.45 $\boxed{+/-}$ $\boxed{\text{INV}}$ $\boxed{\tan}$	8.45 $\boxed{+/-}$ $\boxed{\text{INV}}$ $\boxed{\tan}$
$\arctan(-8.45) = -1.453$ (radians)	$\arctan(-8.45) = -83.2508°$

(c) Let's use Appendix D this time. Since $y = \arcsin(0.2447)$ is equivalent to $\sin y = 0.2447$, we see that we are to find the *angle* value for y corresponding to the *function* value (0.2447) located in the interior of the table. Checking down the sine column, we locate 0.2447 and find to the left an angle value of $14° \, 10'$. Thus,

$\quad \arcsin(0.2447) = 14° \, 10' \approx 0.2472$ (radian)

Remark: In part (b) of Example 3, we used a calculator to obtain the value of the arctan *in degrees*. This convention is peculiar to calculators. By definition, the values of inverse trigonometric functions are always *in radians*.

EXAMPLE 4
Using Inverse Properties

Find the value of each of the following:

(a) $\tan(\arctan -5)$ (b) $\arcsin\left(\sin \dfrac{5\pi}{3}\right)$

Solution:

(a) Since -5 lies in the domain of the arctan x, the inverse property applies, and we have

$$\tan(\arctan -5) = -5$$

(b) In this case, $5\pi/3$ does not lie within the range of the arcsine function. However, $5\pi/3$ is coterminal with

$$\frac{5\pi}{3} - 2\pi = -\frac{\pi}{3}$$

which does lie in the range of the arcsine function, and we have

$$\arcsin\left(\sin\frac{5\pi}{3}\right) = \arcsin\left(\sin -\frac{\pi}{3}\right) = -\frac{\pi}{3}$$

Benefits without a Calculator

Although inverse trigonometric functions are easily evaluated with a calculator, often you can gain more understanding of the functions by not using a calculator.

EXAMPLE 5
Evaluating Functions of Inverse Trigonometric Functions

Evaluate each of the following without tables or a calculator:

(a) $\tan\left(\text{arcsec }\frac{3}{2}\right)$ (b) $\cos\left(\arcsin\frac{-3}{5}\right)$

Solution:

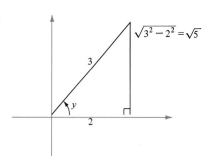

$\sqrt{3^2 - 2^2} = \sqrt{5}$

3

2

FIGURE 5.68

(a) If we let $y = \text{arcsec}(3/2)$ then sec $y = 3/2$. From this, we know that y is a *first* quadrant angle. We can sketch and label y as shown in Figure 5.68. Consequently,

$$\tan\left(\text{arcsec }\frac{3}{2}\right) = \tan y = \frac{\text{opp.}}{\text{adj.}} = \frac{\sqrt{5}}{2}$$

(b) Let $y = \arcsin(-3/5)$. Then sin $y = -3/5$. From this we know that y is a *fourth* quadrant angle. We can sketch and label y as shown in Figure 5.69. Consequently,

$$\cos\left(\arcsin\frac{-3}{5}\right) = \cos y = \frac{\text{adj.}}{\text{hyp.}} = \frac{4}{5}$$

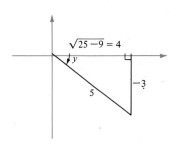

$\sqrt{25 - 9} = 4$

5

-3

FIGURE 5.69

Remark: In both parts of Example 5, we could also have used the Pythagorean Identities to find the solutions. For instance, since $\tan^2 y = \sec^2 y - 1$, we have (for y in Quadrant I)

$$\tan y = \sqrt{\sec^2 y - 1} = \sqrt{\sec^2(\text{arcsec }\tfrac{3}{2}) - 1}$$

$$= \sqrt{(\tfrac{3}{2})^2 - 1} = \frac{\sqrt{5}}{2}$$

Converting to Algebraic Expressions

In calculus it is sometimes beneficial to convert trigonometric expressions into algebraic expressions. The triangle technique shown in Example 5 is an effective procedure for such conversions. We demonstrate this in Example 6.

EXAMPLE 6
Some Problems from Calculus

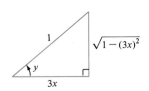

FIGURE 5.70

Write each of the following as an algebraic expression in x:

(a) $\sin(\arccos 3x)$

(b) $\cot(\arccos 3x)$

Solution:

Let $y = \arccos 3x$. Then we have $\cos y = 3x$. Since

$$\cos y = \frac{3x}{1} = \frac{\text{adj.}}{\text{hyp.}}$$

we can sketch a right triangle with acute angle y, as shown in Figure 5.70. From this triangle, we can readily convert each expression to algebraic form.

(a) $\sin(\arccos 3x) = \sin y = \dfrac{\text{opp.}}{\text{hyp.}} = \sqrt{1 - 9x^2}$

(b) $\cot(\arccos 3x) = \cot y = \dfrac{\text{adj.}}{\text{opp.}} = \dfrac{3x}{\sqrt{1 - 9x^2}}$

Most calculators do not have keys for cosecant, secant, and cotangent. Consequently, in evaluating these inverse functions the following reciprocal identities are useful.

RECIPROCAL IDENTITIES

If $x \leq -1$ or $x \geq 1$, then

$$\operatorname{arcsec} x = \arccos \frac{1}{x} \qquad \text{and} \qquad \operatorname{arccsc} x = \arcsin \frac{1}{x}$$

If $x > 0$, then

$$\operatorname{arccot} x = \arctan \frac{1}{x}$$

Proof:

To prove that

$$\operatorname{arcsec} x = \arccos \frac{1}{x}$$

let

$$y = \operatorname{arcsec} x$$

Then, for $0 \le y < \pi/2$ and $\pi/2 < y \le \pi$, we have sec $y = x$ or $1/\cos y = x$. Thus

$$\cos y = \frac{1}{x} \quad \text{or} \quad y = \arccos \frac{1}{x}$$

And finally,

$$\operatorname{arcsec} x = \arccos \frac{1}{x}, \quad |x| \ge 1$$

A similar argument can be used to prove the other two reciprocal identities.

EXAMPLE 7
Using Reciprocal Identities

Evaluate the following functions:

(a) $y = \operatorname{arcsec}(4.3)$ (b) $y = \operatorname{arccsc}(-2.1)$

Solution:

(a) To evaluate y, we use the identity $\operatorname{arcsec} x = \arccos(1/x)$, for $|x| \ge 1$. It follows then that

$$y = \operatorname{arcsec}(4.3) = \arccos\left(\frac{1}{4.3}\right)$$

which can be evaluated by putting a calculator in *radian mode* and pressing the following sequence:

Keystrokes	**Display**
4.3 $\boxed{1/x}$ $\boxed{\text{INV}}$ $\boxed{\cos}$	1.3361

(b) To evaluate y, we use the identity $\operatorname{arccsc} x = \arcsin(1/x)$, for $|x| \ge 1$. It follows then that

$$y = \operatorname{arccsc}(-2.1) = \arcsin\left(-\frac{1}{2.1}\right)$$

which can be evaluated by putting a calculator in *radian mode* and pressing the following sequence:

Keystrokes	**Display**
-2.1 $\boxed{+/-}$ $\boxed{1/x}$ $\boxed{\text{INV}}$ $\boxed{\sin}$	-0.49631

Section Exercises 5.7

In Exercises 1–20, evaluate the given expression without the use of a calculator or tables.

1. $\arcsin \dfrac{1}{2}$

2. $\arcsin 0$

3. $\arccos \dfrac{1}{2}$

4. $\arccos 0$

5. $\arctan \dfrac{\sqrt{3}}{3}$

6. $\operatorname{arccot}(-1)$

7. $\arccos\left(-\dfrac{\sqrt{3}}{2}\right)$

8. $\arcsin\left(-\dfrac{\sqrt{2}}{2}\right)$

9. $\text{arccot}(-\sqrt{3})$

10. $\arctan(-\sqrt{3})$

11. $\text{arcsec } 2$

12. $\text{arccsc } \sqrt{2}$

13. $\text{arccsc } \dfrac{2\sqrt{3}}{3}$

14. $\text{arcsec } 1$

15. $\text{arccot } 0$

16. $\arctan\left(-\dfrac{\sqrt{3}}{3}\right)$

17. $\sin[\arcsin(0.3)]$

18. $\tan(\arctan 25)$

19. $\sec[\text{arcsec}(-10)]$

20. $\csc[\text{arccsc}(-5)]$

In Exercises 21–32, evaluate the given inverse trigonometric function using a calculator or tables.

21. $\arccos(0.28)$ 22. $\arcsin(0.45)$ 23. $\arcsin(-0.75)$

24. $\arccos(-0.8)$ 25. $\arctan(-2)$ 26. $\arctan 15$

27. $\text{arccsc}(3.1)$ 28. $\text{arcsec}(2.6)$ 29. $\text{arcsec}(-4.1)$

30. $\text{arccsc}(-8)$ 31. $\text{arccot}(0.92)$ 32. $\text{arccot}(2.8)$

In Exercises 33–40, evaluate the given expression without the use of a calculator. (Hint: Make a sketch of a right triangle, as illustrated in Example 5.)

33. $\sin\left(\arctan \dfrac{3}{4}\right)$

34. $\sec\left(\arcsin \dfrac{4}{5}\right)$

35. $\tan(\text{arccot } 2)$

36. $\cos(\text{arcsec } \sqrt{5})$

37. $\cos\left(\arcsin \dfrac{5}{13}\right)$

38. $\csc\left[\arctan\left(-\dfrac{5}{12}\right)\right]$

39. $\sec\left[\arctan\left(-\dfrac{3}{5}\right)\right]$

40. $\tan\left[\arcsin\left(-\dfrac{5}{6}\right)\right]$

In Exercises 41–50, write an algebraic expression for the given expression.

41. $\tan(\arctan x)$

42. $\sin(\arccos x)$

43. $\cos(\arcsin 2x)$

44. $\sec(\arctan 3x)$

45. $\sin(\text{arcsec } x)$

46. $\cot(\text{arccot } x)$

47. $\tan\left(\text{arcsec } \dfrac{x}{3}\right)$

48. $\sec[\arcsin(x-1)]$

49. $\csc\left(\arctan \dfrac{x}{\sqrt{2}}\right)$

50. $\cos\left[\arcsin\left(\dfrac{x-h}{r}\right)\right]$

In Exercises 51–53, fill in the blanks.

51. $\arctan \dfrac{9}{x} = \arcsin(\underline{\hspace{1cm}}) = \text{arcsec}(\underline{\hspace{1cm}})$

52. $\arcsin \dfrac{\sqrt{36-x^2}}{6} = \arccos(\underline{\hspace{1cm}}) = \text{arccot}(\underline{\hspace{1cm}})$,
$|x| \le 6$

53. $\text{arcsec } \dfrac{\sqrt{x^2-2x+10}}{3} = \arcsin(\underline{\hspace{1cm}}) = \arctan(\underline{\hspace{1cm}})$

54. Verify the following:

(a) $\text{arccsc } x = \arcsin \dfrac{1}{x}, \quad |x| \ge 1$

(b) $\text{arccot } x = \arctan \dfrac{1}{x}, \quad x > 0$

55. Verify the following:
(a) $\arcsin(-u) = -\arcsin u, \quad |u| \le 1$
(b) $\arccos(-u) = \pi - \arccos u, \quad |u| \le 1$
(c) $\arctan(-u) = -\arctan u$

56. Is the following equation valid? Give reasons for your answer.

$$\arctan u = \frac{\arcsin u}{\arccos u}$$

In Exercises 57–60, sketch the graph of the function.

57. $f(x) = \arcsin(x-1)$

58. $f(x) = \arctan x + \dfrac{\pi}{2}$

59. $f(x) = \text{arcsec } 2x$

60. $f(x) = \arccos \dfrac{x}{4}$

Applications   5.8

In keeping with our twofold perspective on trigonometry, this section includes both triangle applications and applications that emphasize the periodic nature of the trigonometric functions. In a triangle we denote the three angles by the letters A, B, and C, and the lengths of the sides opposite these angles by the letters a, b, and c. For right triangles, C always denotes the right angle and c the length of the hypotenuse. An appropriate picture is essential to the solutions of triangle applications.

Solving a triangle means finding the measure of its three sides and three angles, given some of these quantities. In this section, we concentrate on right triangles and continue to use a calculator to obtain most results.

EXAMPLE 1

Solving a Right Triangle, Given
One Acute Angle and One Side

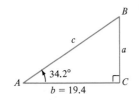

FIGURE 5.71

Solve the right triangle having $A = 34.2°$ and $b = 19.4$.

Solution:

First, make a sketch and label it, as shown in Figure 5.71. Since $C = 90°$,
it follows that

$$A + B = 90° \quad \text{and} \quad B = 90° - 34.2° = 55.8°$$

Now we choose two functions that involve the *known* side b.

$$\tan A = \frac{\text{opp.}}{\text{adj.}} = \frac{a}{b} \quad \text{and} \quad \cos A = \frac{\text{adj.}}{\text{hyp.}} = \frac{b}{c}$$

Solving for a and c, we have

$$a = b \tan A = 19.4(\tan 34.2°)$$
$$\approx 19.4(0.67959) \approx 13.184$$
$$c = \frac{b}{\cos A} = \frac{19.4}{\cos 34.2°}$$
$$\approx \frac{19.4}{0.82908} \approx 23.456$$

Note that in choosing functions that involve the known side b, it is convenient
to use sine, cosine, or tangent, since these three functions occur most commonly on calculators.

Remark: In Example 1 we followed what will be our general procedure in
this section of giving answers with five significant digits. If the example had
specifically stated that the given data were accurate to three significant digits,
we would then round off the answers to $a \approx 13.2$ and $c \approx 23.5$.

EXAMPLE 2

Finding the Side of a Triangle

A local fire safety regulation states that the maximum angle of elevation for
a rescue ladder is 72°. If the fire department's longest ladder is 110 feet, (to
three significant digits), what is the maximum safe rescue height under this
regulation?

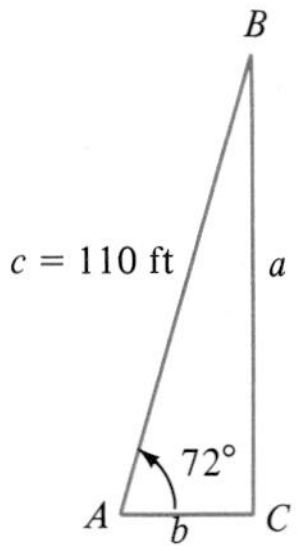

Solution:

A sketch is shown in Figure 5.72. We are only required to find a, so we
choose the function

$$\sin A = \frac{a}{c}$$

from which it follows that

$$a = c \sin A = 110(\sin 72°) \approx 110(0.95105)$$
$$a \approx 105 \text{ feet} \quad\quad\quad\quad\quad \textit{(Three significant digits)}$$

FIGURE 5.72

In problems involving the line of sight from an observer to an object, a
special name is given to the angle between the horizontal and the line of
sight. If the line of sight is downward from horizontal, the resulting angle is

called the **angle of depression.** If the line of sight is upward from horizontal, the angle is called the **angle of elevation.** (See Figure 5.73.)

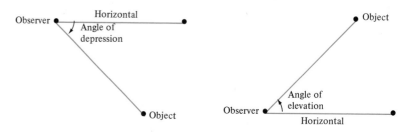

Angles of Depression and Elevation

FIGURE 5.73

EXAMPLE 3
Finding the Side of a Triangle

A smokestack stands on top of a building. From a point 200 feet from the base of the building, the angle of elevation to the *bottom* of the stack is 35°, while the angle of elevation to the *top* is 53°. Find the height of the stack alone.

Solution:
A sketch is shown in Figure 5.74. Let $A = 35°$ and $A' = 53°$. Then, by definition,

$$\tan A = \frac{a}{b} \quad \text{and} \quad \tan A' = \frac{a + s}{b}$$

To solve for s (the height of the stack), we proceed as follows:

$$a = b(\tan A)$$

Substituting this value into the second equation, we get

$$a + s = b \tan A'$$
$$s = b \tan A' - a = b \tan A' - b \tan A = b(\tan A' - \tan A)$$

or

$$s = 200(\tan 53° - \tan 35°) \approx 200(0.62683) \approx 125.37 \text{ feet}$$

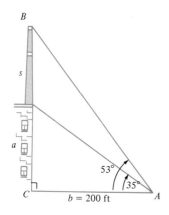

FIGURE 5.74

In Examples 1 through 3 we found the sides of a triangle, given at least one acute angle value. We can also find angles given only the lengths of sides. In the next two examples you will see how inverse trigonometric functions can be used in such situations.

EXAMPLE 4
Finding an Acute Angle of a Triangle

A swimming pool is 20 meters long and 12 meters wide. The bottom of the pool is slanted so that the water depth is 1.3 meters at the shallow end and 4 meters at the deep end. Find the angle of depression (in degrees) of the bottom of the pool.

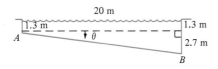

FIGURE 5.75

Solution:

A side view sketch of the pool is shown in Figure 5.75. In this case, we choose a function that involves the two given sides, $a = 2.7$ and $b = 20$. We choose

$$\tan A = \frac{a}{b} = \frac{2.7}{20} = 0.135$$

By definition of the inverse tangent, we have

$$A = \arctan 0.135$$
$$A \approx 0.13419 \text{ (radian)} \approx 7.6885°$$

Now, assuming a practical accuracy of two significant digits, the angle of depression of the pool is

$$A \approx 7.7°$$

In surveying and navigation, directions are generally given in terms of **bearings.** A bearing measures the acute angle a path or line of sight makes with the north-south line. For instance, in Figure 5.76 part (a), the bearing is S 35° E, meaning *35 degrees east of south*. Similarly, the bearings in parts (b) and (c) are, respectively, N 80° W and N 45° E.

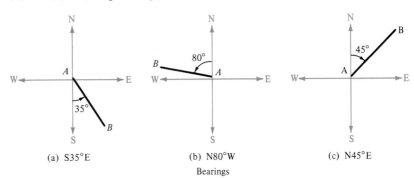

(a) S35°E   (b) N80°W   (c) N45°E

Bearings

FIGURE 5.76

EXAMPLE 5
Finding Directions
in Terms of Bearings

A ship leaves port at noon and heads due west at 20 knots. At 2 P.M., to avoid a storm, it changes course to N 54° W. Find the distance and bearing of the ship from the port of departure at 3 P.M.

Solution:
A sketch is shown in Figure 5.77.
In triangle BCD, we have $B = 90° - 54° = 36°$. To find angle A, we can find the lengths of sides b and d. We obtain them from the equations

$$\sin B = \frac{b}{20} \qquad\qquad \cos B = \frac{d}{20}$$
$$b = 20 \sin 36° \approx 11.756 \qquad\qquad d = 20 \cos 36° \approx 16.180$$

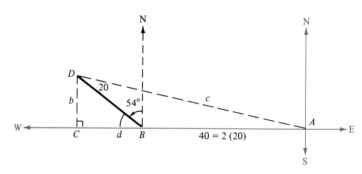

FIGURE 5.77

Now we know two sides of triangle *ACD*, from which we can determine angle *A* as follows:

$$\tan A = \frac{b}{d + 40} \approx \frac{11.756}{16.180 + 40} \approx 0.20925$$

$$A \approx \arctan 0.20925 \approx 11.819° \text{ or } 12°$$

The angle with the north-south line is $90° - 12° = 78°$, and therefore the bearing of the ship is N 78° W.

Again, from triangle *ACD*, we have

$$\sin A = \frac{b}{c}$$

$$c = \frac{b}{\sin A} = \frac{20 \sin 36°}{\sin(\arctan 0.20925)}$$

$$c \approx 57.397 \text{ nautical miles away from the port}$$

Harmonic Motion

The periodic nature of the trigonometric functions allows them to play a key role in describing the motion of a point on an object that vibrates, oscillates, rotates, or is moved by wave motion. For example, consider a ball that is bobbing up and down on the end of a spring, as shown in Figure 5.78.

Suppose that 10 centimeters is the maximum distance the ball moves vertically upward or downward from its equilibrium (at rest) position. Suppose further that the time it takes for the ball to move from its maximum displacement above to its maximum displacement below zero and back again is $t = 4$ seconds. Assuming the ideal conditions of perfect elasticity and no friction or air resistance, this ball would continue to move up and down in a uniform and regular manner.

From the given information about the spring we could conclude that the period (time for one complete cycle) of the motion is $t = 4$ and that its amplitude (maximum displacement from equilibrium) is 10. Motion of this nature can be described by a sine function, and it is referred to as **simple harmonic motion**.

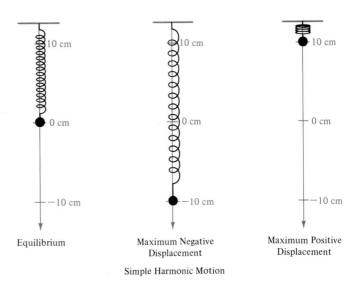

| Equilibrium | Maximum Negative Displacement | Maximum Positive Displacement |

Simple Harmonic Motion

FIGURE 5.78

DEFINITION OF SIMPLE HARMONIC MOTION

A point that moves on a coordinate line is said to be in **simple harmonic motion** if its distance d from the origin at time t is given by either

$$d = a \sin \omega t \qquad \text{or} \qquad d = a \cos \omega t$$

where a and ω are real numbers. The motion has **amplitude,** $|a|$; **period,** $2\pi/\omega$; and **frequency,** $\omega/2\pi$.

Remark: ω is the lowercase Greek letter omega.

EXAMPLE 6
Simple Harmonic Motion

Write the equation for the simple harmonic motion of the ball described above.

Solution:
Since the spring is at equilibrium ($d = 0$) when $t = 0$, we use the equation

$$d = a \sin \omega t$$

Since the maximum displacement from zero is 10 and the period is 4, we have

$$\text{amp} = |a| = 10$$

$$\text{per} = \frac{2\pi}{\omega} = 4 \qquad \Longrightarrow \qquad \omega = \frac{\pi}{2}$$

Consequently, the equation of motion is

$$d = 10 \sin \frac{\pi}{2} t$$

Note that the choice of $a = 10$ or $a = -10$ depends on whether the ball initially moves up or down.

A nice picture of the relation between sine waves and harmonic motion is seen in the wave motion resulting from dropping a stone into a calm pool of water. The waves move outward in roughly the shape of sine (or cosine) waves, as shown in Figure 5.79. Suppose you are fishing and your fishing bob is attached so that it doesn't move horizontally. As the waves move outward from the dropped stone, your fishing bob will move up and down in simple harmonic motion, as shown in Figure 5.80.

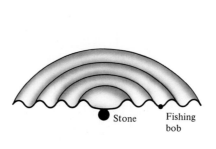

Waves Created in Calm Pool

FIGURE 5.79

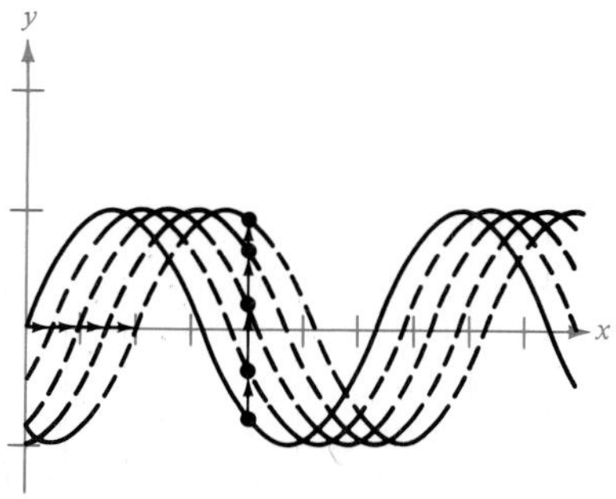

Fishing bob moves in a vertical direction as waves move to the right.

FIGURE 5.80

EXAMPLE 7
Simple Harmonic Motion

Given the equation for simple harmonic motion

$$d = 6 \cos \frac{3\pi}{4} t$$

find
(a) the maximum displacement
(b) the frequency
(c) the value of d when $t = 4$
(d) the least positive value of t for which $d = 0$

Solution:
The given equation has the form

$$d = a \cos \omega t$$

with $a = 6$ and $\omega = 3\pi/4$. Therefore, by definition, we have
(a) maximum displacement = amp = 6

(b) frequency $= \dfrac{\omega}{2\pi} = \dfrac{3\pi/4}{2\pi} = \dfrac{3}{8}$ cycles per unit of time

(c) When $t = 4$, we have

$$d = 6 \cos\left[\frac{3\pi}{4}(4)\right] = 6 \cos 3\pi = 6(-1) = -6$$

(d) To find the least positive value of t for which $d = 0$, we solve the following equation:

$$d = 6 \cos \frac{3\pi}{4}t = 0 \quad \Longrightarrow \quad \frac{3\pi}{4}t = \frac{\pi}{2}, \frac{3\pi}{2}, \frac{5\pi}{2}, \cdots$$

$$t = \frac{2}{3}, 2, \frac{10}{3}, \cdots$$

Thus, for the least positive value of t, we choose $t = \frac{2}{3}$.

Many other physical phenomena can be characterized by wave motion. These include electromagnetic waves like radio waves, television waves, and microwaves. The *carrier* waves sent out by a radio tower have special modifications so that they can carry sound waves to your radio. For an AM station, the *amplitude* of the wave is modified to carry sound; AM stands for **amplitude modulation** (see Figure 5.81). An FM radio signal has its *frequency* modified in order to carry sound; hence the term **frequency modulation** (FM).

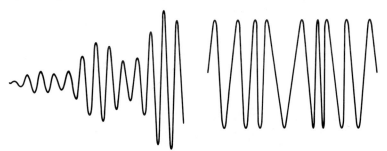

(a) AM: Amplitude Modulation   (b) FM: Frequency Modulation

Radio Waves

FIGURE 5.81

Section Exercises 5.8

In Exercises 1–10, solve the right triangle shown in Figure 5.82.

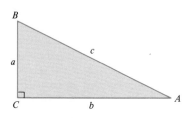

FIGURE 5.82

1. $A = 20°, b = 10$
2. $B = 54°, c = 15$
3. $B = 71°, b = 24$
4. $A = 8.4°, a = 40.5$
5. $A = 12° \ 15', c = 430.5$
6. $B = 65° \ 12', a = 14.2$
7. $a = 6, b = 10$
8. $a = 25, c = 35$
9. $b = 16, c = 52$
10. $b = 1.32, c = 9.45$
11. A ladder of length 16 feet leans against the side of a house. Find the height of the top of the ladder if the angle of elevation of the ladder is 74°.
12. The length of the shadow of a tree is approximately 125 feet when the angle of elevation of the sun is 33°. Approximate the height of the tree.

13. An amateur radio operator erects a 75-foot tower for his antenna. Find the angle of elevation to the top of the tower at a point 50 feet from the base.

14. Find the angle of depression from the top of a lighthouse 250 feet above water level to a ship 2 miles offshore.

15. Find the angle of depression from a spacecraft to the horizon if the vehicle is in a circular orbit 100 miles above the surface of the earth. Assume that the radius of the earth is 4000 miles.

16. A train travels 2.5 miles on a straight track with a grade of 1° 10'. What is the vertical rise of the train in that distance?

17. From a point 50 feet in front of a church, the angles of elevation to the base of the steeple and the top of the steeple are 35° and 47° 40', respectively. Find the height of the steeple.

18. From a point 100 feet in front of a public library, the angles of elevation to the base of a flagpole and the top of the pole are 28° and 39° 45', respectively. If the flagpole is mounted on the front of the library's roof, find the height of the pole.

19. An airplane flying at 550 miles per hour has a bearing of N 52° E. After flying 1.5 hours, how far north and how far east has the plane traveled from its point of departure?

20. A ship leaves port at noon and has a bearing of S 27° W. If the ship is sailing at 20 knots, how many nautical miles south and how many nautical miles west has the ship traveled by 6 P.M.?

21. A ship is 45 miles east and 30 miles south of port. If the captain wants to travel directly to port, what bearing should he take?

22. A plane is 120 miles north and 85 miles east of an airport. If the pilot wants to fly directly to the airport, what bearing should she take?

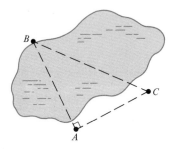

FIGURE 5.83

23. A surveyor wishes to find the distance from A to B across the swamp pictured in Figure 5.83. She begins by determining that the bearing from A to B is N 32° W. Then from A she walks 50 yards in a direction perpendicular to the line segment AB. At this point C the bearing to point B is N 68° W.
(a) Find the bearing that the surveyor walks from A to C.
(b) Find the distance from A to B.

24. Two fire towers are 20 miles apart, tower A being due north of tower B. A fire is spotted from the towers, and its bearings from A and B are S 14° E and N 34° E, respectively. Find the distance, d, of the fire from the line segment AB. [Hint: Show that $d = 20/(\cot 34° + \cot 14°)$.]

25. An observer in a lighthouse 300 feet above sea level spots two ships directly offshore. The angles of depression to the ships are 4° and 6.5°, respectively. How far apart are the ships?

26. A passenger in an airplane flying at 30,000 feet sees two towns directly to the left of the plane. The angles of depression to the towns are 28° and 55°, respectively. How far apart are the towns?

27. A plane is observed approaching your home, and you assume that it is traveling at approximately 550 miles per hour. If the angle of elevation of the plane is 16° at one time and one minute later the angle is 57°, approximate the altitude.

28. In traveling across relatively flat land, you notice a mountain directly in front of you. Its angle of elevation (to the peak) is 3.5°. After you drive 13 miles closer to the mountain, the angle of elevation is 9°. Approximate the height of the mountain.

29. A regular pentagon is inscribed in a circle of radius 25 inches. Find the length of the sides of the pentagon.

30. Repeat Exercise 29 using a hexagon.

In Exercises 31–34, for the simple harmonic motion described by the given trigonometric function, find (a) the maximum displacement, (b) the frequency, and (c) the least positive value of t for which $d = 0$.

31. $d = 4 \cos 8\pi t$
32. $d = \frac{1}{2} \cos 20\pi t$
33. $d = \frac{1}{16} \sin 120\pi t$
34. $d = \frac{1}{64} \sin 792\pi t$

35. A point on the end of a tuning fork moves in simple harmonic motion described by $d = a \sin \omega t$. If the tuning fork for middle C has a frequency of 264 vibrations per second, find ω.

36. A buoy oscillates in simple harmonic motion as waves move past. At a given time it is noted that the buoy moves a total of 3.5 feet from its low point to its high point, and that it returns to the high point every 10 seconds. Write an equation describing the motion of the buoy if at $t = 0$ it is at its high point.

Review Exercises / Chapter 5

1. Sketch the following angles in standard position, and give one positive and one negative coterminal angle:

    (a) $\dfrac{11\pi}{4}$
    (b) $-405°$

2. Convert the following angle measurements to decimal form:
    (a) $135° \ 16' \ 45''$
    (b) $-234° \ 50''$

3. Convert the following from radian to degree measure. List your answers to three decimal places.

    (a) $\dfrac{5\pi}{7}$
    (b) -3.5

4. Convert the following from degree to radian measure. List your answers to three decimal places.
    (a) $480°$
    (b) $-16.5°$

5. Find the reference angles for the following:

    (a) $-\dfrac{6\pi}{5}$
    (b) $640°$

6. Find the six trigonometric functions of the angle θ if it is in standard position and the terminal side passes through the given point.
    (a) $(-7, 2)$
    (b) $(4, -8)$

7. Find the remaining five trigonometric functions from the given information.
    (a) $\sec \theta = \frac{6}{5}, \quad \tan \theta < 0$
    (b) $\tan \theta = -\frac{12}{5}, \quad \sin \theta > 0$

8. Use a calculator to evaluate the given trigonometric functions.
    (a) $\tan 33°$
    (b) $\csc 105°$
    (c) $\sec \dfrac{12\pi}{5}$
    (d) $\sin\left(-\dfrac{\pi}{9}\right)$

9. Evaluate the given trigonometric functions without the use of a calculator or tables.

    (a) $\sin \dfrac{5\pi}{3}$
    (b) $\cot\left(-\dfrac{5\pi}{6}\right)$

    (c) $\cos 495°$
    (d) $\csc \dfrac{3\pi}{2}$

10. Find two values of θ in degrees ($0° \le \theta \le 360°$) and in radians ($0 \le \theta \le 2\pi$) without using a calculator or tables.

    (a) $\cos \theta = -\dfrac{\sqrt{2}}{2}$
    (b) $\sec \theta$ is undefined

    (c) $\csc \theta = -2$
    (d) $\tan \theta = \dfrac{\sqrt{3}}{3}$

11. Find two values of θ in degrees ($0° \le \theta \le 360°$) and in radians ($0 \le \theta \le 2\pi$) by using the inverse key on a calculator.
    (a) $\sin \theta = 0.8387$
    (b) $\cot \theta = -1.5399$
    (c) $\sec \theta = -1.0353$
    (d) $\csc \theta = 11.4737$

In Exercises 12–30, sketch the graph of the given function.

12. $f(x) = 5 \sin \dfrac{2x}{5}$

13. $f(x) = 8 \cos\left(-\dfrac{x}{4}\right)$

14. $f(x) = -\dfrac{1}{4} \cos\left(\dfrac{\pi x}{4}\right)$

15. $f(x) = -\tan\left(\dfrac{\pi x}{4}\right)$

16. $g(t) = \frac{5}{2} \sin (\pi - \pi t)$

17. $g(t) = \sec\left(t - \dfrac{\pi}{4}\right)$

18. $h(t) = \csc\left(3t - \dfrac{\pi}{2}\right)$

19. $h(t) = 3 \csc\left(2t + \dfrac{\pi}{4}\right)$

20. $f(\theta) = \cot\left(\dfrac{\pi \theta}{8}\right)$

21. $E(t) = 110 \cos\left(120\pi t - \dfrac{\pi}{3}\right)$

22. $f(x) = \dfrac{x}{4} - \sin x$

23. $g(x) = 3\left[\sin\left(\dfrac{\pi x}{3}\right) + 1\right]$

24. $h(\theta) = \theta \sin(\pi \theta)$

25. $f(t) = 2.5e^{-t/4} \sin 2\pi t$

26. $f(x) = \sqrt{x} \sin\left(\dfrac{\pi x}{2}\right)$

27. $f(x) = \arcsin\left(\dfrac{x}{2}\right)$

28. $f(x) = \operatorname{arcsec}(x - 1)$

29. $f(x) = \pi - \arctan x$

30. $2 \arccos x$

31. Write an algebraic expression for $\sec[\arcsin(x - 1)]$.

32. Write an algebraic expression for $\tan\left[\arccos\left(\dfrac{x}{2}\right)\right]$.

33. An observer 2.5 miles from the launch pad of a space shuttle measures the angle of elevation to the base of the vehicle to be $28°$ soon after liftoff. How high is the shuttle at that instant if you assume that the shuttle is still moving vertically?

34. In a flight from city A to city B, a plane flies 650 miles in a bearing that is N 48° E. From city B the plane flies 810 miles at a bearing of S 65° E to city C. Find the distance from A to C and the bearing from A to C.

CHAPTER

6 | ANALYTICAL TRIGONOMETRY

Fundamental Identities

6.1

In Chapter 5, we studied the basic definitions, properties, graphs, and applications of the individual trigonometric functions. In this chapter, we will study the uses of algebraic combinations of these functions.

We will show how to manipulate trigonometric expressions in order to prove new identities, solve trigonometric equations, and rewrite trigonometric expressions in equivalent forms. We begin with a review of the fundamental identities.

FUNDAMENTAL IDENTITIES

	Identity	**Alternative Forms**
Reciprocal:	$\csc x = \dfrac{1}{\sin x}$	$\sin x = \dfrac{1}{\csc x}$
	$\sec x = \dfrac{1}{\cos x}$	$\cos x = \dfrac{1}{\sec x}$
	$\cot x = \dfrac{1}{\tan x}$	$\tan x = \dfrac{1}{\cot x}$
	$\tan x = \dfrac{\sin x}{\cos x}$	$\cot x = \dfrac{\cos x}{\sin x}$
Pythagorean:	$\sin^2 x + \cos^2 x = 1$	$\sin^2 x = 1 - \cos^2 x, \ \cos^2 x = 1 - \sin^2 x$
	$\tan^2 x + 1 = \sec^2 x$	$\tan^2 x = \sec^2 x - 1$
	$\cot^2 x + 1 = \csc^2 x$	$\cot^2 x = \csc^2 x - 1$

Cofunction:	$\sin\left(\dfrac{\pi}{2} - x\right) = \cos x$	$\csc\left(\dfrac{\pi}{2} - x\right) = \sec x$
	$\cos\left(\dfrac{\pi}{2} - x\right) = \sin x$	$\sec\left(\dfrac{\pi}{2} - x\right) = \csc x$
	$\tan\left(\dfrac{\pi}{2} - x\right) = \cot x$	$\cot\left(\dfrac{\pi}{2} - x\right) = \tan x$
Negative Angle:	$\sin(-x) = -\sin x$	$\csc(-x) = -\csc x$
	$\cos(-x) = \cos x$	$\sec(-x) = \sec x$
	$\tan(-x) = -\tan x$	$\cot(-x) = -\cot x$

Remark: Pythagorean identities are sometimes used in radical form, such as

$$\sin x = \pm\sqrt{1 - \cos^2 x} \qquad \text{or} \qquad \tan x = \pm\sqrt{\sec^2 x - 1}$$

where the sign depends on the particular value of x. Note this use in the next example.

EXAMPLE 1
Using Fundamental Identities to Find Function Values

Use the fundamental identities to find $\csc x$, if

$$\tan x = -6 \qquad \text{and} \qquad \sin x > 0$$

Solution:
Since $\tan x < 0$ and $\sin x > 0$, it follows that x is in Quadrant II. Then,

$$\cot x = \frac{1}{\tan x} = \frac{1}{-6}$$

Consequently, from the radical form of a Pythagorean identity,

$$\csc x = \pm\sqrt{1 + \cot^2 x} = \pm\sqrt{1 + \left(\frac{-1}{6}\right)^2} = \pm\sqrt{\frac{37}{36}}$$

$$\csc x = \frac{\sqrt{37}}{6}, \qquad x \text{ in Quadrant II}$$

EXAMPLE 2
Using Fundamental Identities to Find Function Values

Use the fundamental identities to find the values of the other trigonometric functions, given that

$$\sec x = -\frac{3}{2} \qquad \text{and} \qquad \tan x > 0$$

Solution:
If $\sec x < 0$ and $\tan x > 0$, then x lies in Quadrant III. Then,

$$\cos x = \frac{1}{\sec x} = \frac{1}{-\frac{3}{2}} = -\frac{2}{3}$$

Thus, by a Pythagorean identity, we have

$$\sin^2 x = 1 - \cos^2 x = 1 - \left(-\frac{2}{3}\right)^2 = 1 - \frac{4}{9} = \frac{5}{9}$$

$$\sin x = -\frac{\sqrt{5}}{3}, \qquad x \text{ in Quadrant III}$$

Furthermore,

$$\tan x = \frac{\sin x}{\cos x} = \frac{-\sqrt{\frac{5}{9}}}{-\frac{2}{3}} = \frac{\sqrt{5}}{2}$$

$$\cot x = \frac{1}{\tan x} = \frac{2}{\sqrt{5}}$$

$$\csc x = \frac{1}{\sin x} = -\frac{3}{\sqrt{5}}$$

EXAMPLE 3
Evaluating Trigonometric
Expressions

If $\cot \theta = -2$, find all possible values for

$$\frac{\sin \theta + \cos \theta}{\tan \theta}$$

Solution:
Since $\cot \theta = -2$, θ lies in Quadrant II or IV, and we have

$$\tan \theta = \frac{1}{\cot \theta} = \frac{1}{-2}$$

$$\csc \theta = \pm\sqrt{1 + \cot^2 \theta} = \pm\sqrt{1 + 4} = \pm\sqrt{5}$$

Consequently,

$$\sin \theta = \frac{1}{\csc \theta} = \frac{\pm 1}{\sqrt{5}}$$

$$\cos \theta = \pm\sqrt{1 - \sin^2 \theta} = \pm\sqrt{1 - \frac{1}{5}} = \pm\sqrt{\frac{4}{5}} = \pm\frac{2}{\sqrt{5}}$$

Now, if θ lies in Quadrant II, then

$$\frac{\sin \theta + \cos \theta}{\tan \theta} = \frac{1/\sqrt{5} - 2/\sqrt{5}}{-1/2} = \frac{2}{\sqrt{5}}$$

If θ lies in Quadrant IV, then

$$\frac{\sin \theta + \cos \theta}{\tan \theta} = \frac{-1/\sqrt{5} + 2/\sqrt{5}}{-1/2} = -\frac{2}{\sqrt{5}}$$

In the next two examples, we show how to use fundamental identities to simplify trigonometric expressions or convert them to equivalent forms.

More details about such algebraic maneuvers will be given in Section 6.2. Note how algebraic factoring comes into play in the next example.

EXAMPLE 4
Converting to Equivalent Forms

Use fundamental identities to transform

$$\sec t + \tan t \quad \text{into} \quad \frac{\cos t}{1 - \sin t}$$

Solution:
First, converting all terms to sines and cosines, we have

$$\sec t + \tan t = \frac{1}{\cos t} + \frac{\sin t}{\cos t} = \frac{1 + \sin t}{\cos t}$$

Since we want a $\cos t$ factor in the numerator rather than the denominator, we multiply by $\cos t$, as follows:

$$\frac{1 + \sin t}{\cos t} = \frac{(1 + \sin t)(\cos t)}{\cos^2 t}$$

But $\cos^2 t = 1 - \sin^2 t$. Hence,

$$\frac{1 + \sin t}{\cos t} = \frac{(1 + \sin t)(\cos t)}{1 - \sin^2 t} \qquad \textit{Fund. identity}$$

$$= \frac{(1 + \sin t)(\cos t)}{(1 + \sin t)(1 - \sin t)} \qquad \textit{Factor}$$

$$= \frac{\cos t}{1 - \sin t} \qquad \textit{Reduce}$$

Therefore, we have shown that

$$\sec t + \tan t = \frac{\cos t}{1 - \sin t}$$

EXAMPLE 5
Simplifying a Trigonometric Expression

Use fundamental identities to simplify

$$\sin x + \cot x \cos x$$

Solution:
A common procedure for simplifying trigonometric expressions is to convert all terms to sines and cosines first, then simplify. In this case, we have

$$\sin x + (\cot x)\cos x = \sin x + \left(\frac{\cos x}{\sin x}\right)\cos x$$

$$= \frac{\sin^2 x + \cos^2 x}{\sin x} = \frac{1}{\sin x}$$

Thus,

$$\sin x + \cot x \cos x = \csc x$$

Section Exercises 6.1

In Exercises 1–14, use the fundamental identities to find the value of the other trigonometric functions.

1. $\sin x = \dfrac{1}{2}$, $\cos x = \dfrac{\sqrt{3}}{2}$

2. $\tan x = \dfrac{\sqrt{3}}{3}$, $\cos x = -\dfrac{\sqrt{3}}{2}$

3. $\sec \theta = \sqrt{2}$, $\sin \theta = -\dfrac{\sqrt{2}}{2}$

4. $\csc \theta = \dfrac{5}{3}$, $\tan \theta = \dfrac{3}{4}$

5. $\tan x = \dfrac{5}{12}$, $\sec x = -\dfrac{13}{12}$

6. $\cot \phi = -3$, $\sin \phi = \dfrac{\sqrt{10}}{10}$

7. $\sec \phi = -1$, $\sin \phi = 0$

8. $\cos\left(\dfrac{\pi}{2} - x\right) = \dfrac{3}{5}$, $\cos x = \dfrac{4}{5}$

9. $\sin(-x) = -\dfrac{2}{3}$, $\tan x = -\dfrac{2\sqrt{5}}{5}$

10. $\csc x = 5$, $\cos x > 0$ 11. $\tan \theta = 2$, $\sin \theta < 0$

12. $\sec \theta = -3$, $\tan \theta < 0$ 13. $\sin \theta = -1$, $\cot \theta = 0$

14. $\tan \theta$ is undefined, $\sin \theta > 0$

In Exercises 15–20, match the trigonometric expression with one of the following:

(a) -1 (b) $\cos x$ (c) $\cot x$

(d) 1 (e) $-\tan x$ (f) $\sin x$

15. $\sec x \cos x$

16. $\dfrac{\sin(-x)}{\cos(-x)}$

17. $\tan^2 x - \sec^2 x$

18. $\dfrac{1 - \cos^2 x}{\sin x}$

19. $\cot x \sin x$

20. $\dfrac{\sin[(\pi/2) - x]}{\cos[(\pi/2) - x]}$

In Exercises 21–26, match the trigonometric expression with one of the following:

(a) $\csc x$ (b) $\tan x$ (c) $\sin^2 x$

(d) $\sin x(\tan x)$ (e) $\sec^2 x$ (f) $\sec^2 x + \tan^2 x$

21. $\sin x \sec x$ 22. $\cos^2 x(\sec^2 x - 1)$

23. $\dfrac{\sec^2 x - 1}{\sin^2 x}$ 24. $\cot x \sec x$

25. $\sec^4 x - \tan^4 x$ 26. $\dfrac{\cos^2[(\pi/2) - x]}{\cos x}$

In Exercises 27–32, use the fundamental identities to simplify the trigonometric expression.

27. $\tan \phi \csc \phi$ 28. $\sin \phi(\csc \phi - \sin \phi)$

29. $\sec^2 x(1 - \sin^2 x)$ 30. $\dfrac{1}{\tan^2 x + 1}$

31. $\dfrac{\cos^2 y}{1 - \sin y}$ 32. $\cos t + \cos t \tan^2 t$

33. Express each of the other trigonometric functions of θ in terms of $\sin \theta$.

34. Rewrite the following expression in terms of $\sin \theta$ and $\cos \theta$:

$$\dfrac{\sec \theta(1 + \tan \theta)}{\sec \theta + \csc \theta}$$

35. Rewrite the following expression as a single trigonometric function of θ:

$$-\sqrt{\sec^2 \theta - 1}$$

In Exercises 36–38, for what values of θ, $0 \le \theta \le 2\pi$, are the equations true?

36. $\sec \theta = \sqrt{1 + \tan^2 \theta}$

37. $\cos \theta = -\sqrt{1 - \sin^2 \theta}$

38. $\cot \theta = -\sqrt{\csc^2 \theta - 1}$

In Exercises 39 and 40, rewrite the given expression as a single logarithm and simplify.

39. $\ln|\cos \theta| - \ln|\sin \theta|$

40. $\ln|\cot t| + \ln(1 + \tan^2 t)$

41. Determine if the following are true or false, and give a reason for your answer.

(a) $\dfrac{\sin k\theta}{\cos k\theta} = \tan \theta$, k is a constant

(b) $5 \sec \theta = \dfrac{1}{5 \cos \theta}$

(c) $\sin \theta \csc \phi = 1$

42. Use a calculator to demonstrate that $\csc^2 \theta - \cot^2 \theta = 1$ for

(a) $\theta = 132°$ (b) $\theta = \dfrac{2\pi}{7}$

Verifying Trigonometric Identities

6.2

A suitable alternative title for this section might be ''An Introduction to the Algebra of Trigonometry.'' A critical part of proving identities involves making extensive use of the rules of algebra in rewriting trigonometric expressions. By a **trigonometric expression,** we mean any algebraic combination of trigonometric functions alone, or in combination with other functions. Some examples are

$$2x - \cos x, \qquad \frac{1 - \sin x}{\tan^2 x}, \qquad \frac{\sqrt{\sin x} + x}{\sec x}$$

The skills developed in this section are needed to prove special identities (useful later in trigonometry and higher-level mathematics) and to solve trigonometric equations. The key to proving identities and solving equations lies in the ability to use the fundamental identities and the rules of algebra to rewrite trigonometric expressions in alternative (more useful) forms.

Let us review some distinctions among expressions, equations, and identities. An **expression** has no equal sign. It is merely a combination of functions. When simplifying expressions, we use an equal sign *only* to indicate the equivalence of the original expression and the new form.

Generally, the term **equation** means a statement containing an equal sign that is true for a specific set of values. In this sense, it is really a **conditional equation.** For example, the equation

$$\sin x = -1$$

is true only for $x = (3\pi/2) \pm 2n\pi$. Hence, it is a conditional equation. On the other hand, an equation that is true for all real values in the domain of the variable is called an **identity.** For example, the familiar equation

$$\sin^2 x = 1 - \cos^2 x$$

is true for all real x; hence, it is an identity.

EXAMPLE 1
Simplifying a
Trigonometric Expression

Perform the indicated addition and simplify the result:

$$\frac{\sin \theta}{1 + \cos \theta} + \frac{\cos \theta}{\sin \theta}$$

Solution:
Using the algebra definition of addition

$$\frac{a}{b} + \frac{c}{d} = \frac{ad + bc}{bd}$$

we can write

$$\frac{\sin\theta}{1+\cos\theta} + \frac{\cos\theta}{\sin\theta} = \frac{\sin\theta(\sin\theta) + \cos\theta(1+\cos\theta)}{(1+\cos\theta)\sin\theta} \qquad \textit{Add fractions}$$

$$= \frac{\sin^2\theta + \cos^2\theta + \cos\theta}{(1+\cos\theta)\sin\theta} \qquad \textit{Simplify}$$

$$= \frac{1+\cos\theta}{(1+\cos\theta)\sin\theta} \qquad \textit{Fund. identity}$$

$$= \frac{1}{\sin\theta} \qquad \textit{Reduce}$$

Therefore, in simplest form, we have

$$\frac{\sin\theta}{1+\cos\theta} + \frac{\cos\theta}{\sin\theta} = \csc\theta$$

Remark: To improve your skill in simplifying trigonometric expressions and proving identities, you may want to go back and briefly review some algebra, particularly operations with fractions, finding the LCD, special products, factoring quadratic expressions, reducing to lowest terms, and rationalizing denominators.

EXAMPLE 2
Factoring Trigonometric Expressions

Factor each of the following:

(a) $2\sin x\cos x - \sin x$　　(b) $\sec^2 x - 1$　　(c) $4\tan^2\theta + \tan\theta - 3$

Solution:

(a) In this case, we factor out a common monomial factor to get

$$2\sin x\cos x - \sin x = \sin x(2\cos x - 1)$$

(b) Here we have the difference of two squares, which factors as

$$\sec^2 x - 1 = (\sec x - 1)(\sec x + 1)$$

(c) This expression has the polynomial form $ax^2 + bx + c$, and it factors as

$$4\tan^2\theta + \tan\theta - 3 = (4\tan\theta - 3)(\tan\theta + 1)$$

In the remaining examples in this section, we will demonstrate some common techniques used to prove (*show* or *verify*) trigonometric identities. Verifying identities is not the same as solving equations. There is no well-defined set of rules to follow in verifying trigonometric identities, and the process is best learned by much practice. However, the following guidelines are helpful in this verification process.

GUIDELINES FOR VERIFYING TRIGONOMETRIC IDENTITIES

1. Work with one side of the equation at a time. It is often better to work with the more complicated side.
2. Look for opportunities to add fractions, square a binomial, or factor an expression.
3. Look for opportunities to use the fundamental identities or their alternative forms. Note which functions are in the final expression you want. Sines and cosines pair up well, as do secants and tangents, and cosecants and cotangents.
4. If the preceding guidelines don't help, try converting all terms on one side into sines and cosines and then combining them.
5. Don't just sit and stare at the problem. Try something! Even paths that lead to dead ends can give you insights.

EXAMPLE 3
Verifying a Trigonometric Identity

Verify the identity

$$\frac{\sec^2 \theta - 1}{\sec^2 \theta} = \sin^2 \theta$$

Solution:
Since the left side is more complicated, we will work with it. In the numerator, we see half of the identity $\tan^2 \theta = \sec^2 \theta - 1$. Thus,

$$\frac{\sec^2 \theta - 1}{\sec^2 \theta} = \frac{\tan^2 \theta}{\sec^2 \theta}$$

$$= \tan^2 \theta (\cos^2 \theta) \qquad \textit{Fund. identity}$$

$$= \frac{\sin^2 \theta}{\cos^2 \theta}(\cos^2 \theta) \qquad \textit{Fund. identity}$$

$$= \sin^2 \theta \qquad \textit{Simplify}$$

Alternative Solution:
Sometimes it is helpful to separate a fraction into two parts. In this case, we have

$$\frac{\sec^2 \theta - 1}{\sec^2 \theta} = \frac{\sec^2 \theta}{\sec^2 \theta} - \frac{1}{\sec^2 \theta} \qquad \textit{Separate fractions}$$

$$= 1 - \cos^2 \theta \qquad \textit{Fund. identity}$$

$$= \sin^2 \theta \qquad \textit{Fund. identity}$$

Remark: As you can see from Example 3, there can be more than one way to verify an identity. So don't be disturbed if your method is different from that used by your professor or fellow students. Here is a good chance to be creative and establish your own style, but try to be as efficient as possible.

EXAMPLE 4
Combining Fractions
Before Using Identities

Verify the identity

$$\frac{1}{1 - \sin \alpha} + \frac{1}{1 + \sin \alpha} = 2 \sec^2 \alpha$$

Solution:
Let us add the two fractions and see where we can go from there.

$$\frac{1}{1 - \sin \alpha} + \frac{1}{1 + \sin \alpha} = \frac{1 + \sin \alpha + 1 - \sin \alpha}{(1 - \sin \alpha)(1 + \sin \alpha)} \quad \text{\textit{Add fractions}}$$

$$= \frac{2}{1 - \sin^2 \alpha} \quad \text{\textit{Simplify}}$$

$$= \frac{2}{\cos^2 \alpha} \quad \text{\textit{Fund. identity}}$$

$$= 2 \sec^2 \alpha$$

EXAMPLE 5
Multiplying Before
Using Identities

Verify the identity

$$(\tan^2 x + 1)(\cos^2 x - 1) = 1 - \sec^2 x$$

Solution:
By multiplying first, we get

$$(\tan^2 x + 1)(\cos^2 x - 1) = \tan^2 x(\cos^2 x) - \tan^2 x + \cos^2 x - 1$$

$$= \frac{\sin^2 x}{\cos^2 x}(\cos^2 x) + \cos^2 x - (\tan^2 x + 1)$$

$$= (\sin^2 x + \cos^2 x) - (\tan^2 x + 1)$$

$$= 1 - \sec^2 x$$

EXAMPLE 6
Converting to Sines and Cosines

Verify the identity

$$\tan x + \cot x = \sec x \csc x$$

Solution:
In this case there appear to be no fractions to add, no products to find, and no opportunity to use one of the Pythagorean identities. Hence, we try converting the left side into sines and cosines to see what happens.

$$\tan x + \cot x = \frac{\sin x}{\cos x} + \frac{\cos x}{\sin x} = \frac{\sin^2 x + \cos^2 x}{\cos x \sin x}$$

$$= \frac{1}{\cos x \sin x} = \frac{1}{\cos x} \cdot \frac{1}{\sin x} = \sec x \csc x$$

Recall from algebra that *rationalizing the denominator* can be a powerful simplification technique. A form of this technique works for simplifying trigonometric expressions as well.

EXAMPLE 7
Rationalizing Denominators

Verify the identity

$$\sec y + \tan y = \frac{\cos y}{1 - \sin y}$$

Solution:

Let us work with the *right* side. In this case, rationalizing the denominator means multiplying by a factor that produces a perfect square in the denominator.

$$\frac{\cos y}{1 - \sin y} = \frac{\cos y}{1 - \sin y}\left(\frac{1 + \sin y}{1 + \sin y}\right)$$

$$= \frac{\cos y + \cos y \sin y}{1 - \sin^2 y}$$

$$= \frac{\cos y + \cos y \sin y}{\cos^2 y}$$

$$= \frac{\cos y}{\cos^2 y} + \frac{\cos y \sin y}{\cos^2 y}$$

$$= \frac{1}{\cos y} + \frac{\sin y}{\cos y}$$

$$= \sec y + \tan y$$

So far in this section, we have worked strictly with one side of an identity and converted it into a form identical to the opposite side. On occasion it is practical to work with each side *separately,* to obtain one common form equivalent to both sides. The work with each side must consist of a reversible sequence of steps. That is, from the common form, we can independently obtain each side as an equivalent form.

EXAMPLE 8
Working with Each
Side Separately

Verify the identity

$$\frac{\cot^2 \theta}{1 + \csc \theta} = \frac{1 - \sin \theta}{\sin \theta}$$

Solution:

(a) Working with the left side, we have

$$\frac{\cot^2 \theta}{1 + \csc \theta} = \frac{\csc^2 \theta - 1}{1 + \csc \theta}$$

$$= \frac{(\csc \theta - 1)(\csc \theta + 1)}{1 + \csc \theta}$$

$$= \csc \theta - 1$$

(b) Now, simplifying the right side, we have

$$\frac{1 - \sin \theta}{\sin \theta} = \frac{1}{\sin \theta} - \frac{\sin \theta}{\sin \theta}$$

$$= \csc \theta - 1$$

Consequently, the identity is verified, because

$$\frac{\cot^2 \theta}{1 + \csc \theta} = \csc \theta - 1 = \frac{1 - \sin \theta}{\sin \theta}$$

Have fun with the exercises for this section. Try to be both creative and efficient.

Section Exercises 6.2

In Exercises 1–8, factor each expression and simplify.

1. $\tan^2 x - \tan^2 x \sin^2 x$

2. $\sec^2 x \tan^2 x + \sec^2 x$

3. $\sin^2 x \sec^2 x - \sin^2 x$

4. $\dfrac{\sec^2 x - 1}{\sec x - 1}$

5. $\tan^4 x + 2 \tan^2 x + 1$

6. $1 - 2 \cos^2 x + \cos^4 x$

7. $\sin^4 x - \cos^4 x$

8. $\csc^3 x - \csc^2 x - \csc x + 1$

In Exercises 9–12, perform the multiplication and simplify.

9. $(\sin x + \cos x)^2$

10. $(\cot x + \csc x)(\cot x - \csc x)$

11. $(\sec x + 1)(\sec x - 1)$

12. $(3 - 2 \sin x)(2 + 3 \sin x)$

In Exercises 13–16, perform the addition and simplify.

13. $\dfrac{1}{1 + \cos x} + \dfrac{1}{1 - \cos x}$

14. $\dfrac{1}{\sec x + 1} - \dfrac{1}{\sec x - 1}$

15. $\dfrac{\cos x}{1 + \sin x} + \dfrac{1 + \sin x}{\cos x}$

16. $\tan x - \dfrac{\sec^2 x}{\tan x}$

In Exercises 17–20, rationalize the denominator and simplify.

17. $\dfrac{\sin^2 y}{1 - \cos y}$

18. $\dfrac{5}{\tan x + \sec x}$

19. $\dfrac{3}{\sec x - \tan x}$

20. $\dfrac{\tan^2 x}{\csc x + 1}$

In Exercises 21–66, verify the given identity.

21. $\dfrac{\sec^2 x}{\tan x} = \sec x \csc x$

22. $\dfrac{\cot^3 t}{\csc t} = \dfrac{\cos t}{\sin^2 t} - \cos t$

23. $\dfrac{\cot^2 t}{\csc t} = \csc t - \sin t$

24. $\dfrac{1}{\sin x} - \sin x = \dfrac{\cos^2 x}{\sin x}$

25. $\sin^{1/2} x \cos x - \sin^{5/2} x \cos x = \cos^3 x \sqrt{\sin x}$

26. $\dfrac{1}{\sec x \tan x} = \csc x - \sin x$

27. $\sec^6 x(\sec x \tan x) - \sec^4 x(\sec x \tan x) = \sec^5 x \tan^3 x$

28. $\dfrac{\sec \theta - 1}{1 - \cos \theta} = \sec \theta$

29. $\cos x + \sin x \tan x = \sec x$

30. $\sec x - \cos x = \sin x \tan x$

31. $\csc x - \sin x = \cos x \cot x$

32. $\dfrac{\sec x + \tan x}{\sec x - \tan x} = (\sec x + \tan x)^2$

33. $\dfrac{1}{\tan x} + \dfrac{1}{\cot x} = \tan x + \cot x$

34. $\dfrac{1}{\sin x} - \dfrac{1}{\csc x} = \csc x - \sin x$

35. $\dfrac{\cos \theta \cot \theta}{1 - \sin \theta} - 1 = \csc \theta$

36. $\dfrac{1 + \sin \theta}{\cos \theta} + \dfrac{\cos \theta}{1 + \sin \theta} = 2 \sec \theta$

37. $\dfrac{1}{\cot x + 1} + \dfrac{1}{\tan x + 1} = 1$

38. $\cos x - \dfrac{\cos x}{1 - \tan x} = \dfrac{\sin x \cos x}{\sin x - \cos x}$

39. $2 \sec^2 x - 2 \sec^2 x \sin^2 x - \sin^2 x - \cos^2 x = 1$

40. $\csc x(\csc x - \sin x) + \dfrac{\sin x - \cos x}{\sin x} + \cot x = \csc^2 x$

41. $2 + \cos^2 x - 3 \cos^4 x = \sin^2 x(2 + 3 \cos^2 x)$

42. $4 \tan^4 x + \tan^2 x - 3 = \sec^2 x(4 \tan^2 x - 3)$

43. $\csc^4 x - 2 \csc^2 x + 1 = \cot^4 x$

44. $\sin x(1 - 2 \cos^2 x + \cos^4 x) = \sin^5 x$

45. $\sec^4 \theta - \tan^4 \theta = 1 + 2 \tan^2 \theta$

46. $\csc^6 \theta - \cot^6 \theta = 3 \cot^4 \theta + 3 \cot^2 \theta + 1$

47. $\dfrac{\sin \beta}{1 - \cos \beta} = \dfrac{1 + \cos \beta}{\sin \beta}$

48. $\dfrac{\cot \alpha}{\csc \alpha - 1} = \dfrac{\csc \alpha + 1}{\cot \alpha}$

49. $\dfrac{\tan^3 \alpha - 1}{\tan \alpha - 1} = \tan^2 \alpha + \tan \alpha + 1$

50. $\dfrac{\sin^3 \beta + \cos^3 \beta}{\sin \beta + \cos \beta} = 1 - \sin \beta \cos \beta$

51. $\cos\left(\dfrac{\pi}{2} - x\right)\csc x = 1$

52. $\dfrac{\cos[(\pi/2) - x]}{\sin[(\pi/2) - x]} = \tan x$

53. $\dfrac{\csc(-x)}{\sec(-x)} = -\cot x$

54. $(1 + \sin y)[1 + \sin(-y)] = \cos^2 y$

55. $\dfrac{\cos(-\theta)}{1 + \sin(-\theta)} = \sec \theta + \tan \theta$

56. $\dfrac{1 + \sec(-\theta)}{\sin(-\theta) + \tan(-\theta)} = -\csc \theta$

57. $\dfrac{\sin x \cos y + \cos x \sin y}{\cos x \cos y - \sin x \sin y} = \dfrac{\tan x + \tan y}{1 - \tan x \tan y}$

58. $\dfrac{\tan x + \tan y}{1 - \tan x \tan y} = \dfrac{\cot x + \cot y}{\cot x \cot y - 1}$

59. $\dfrac{\tan x + \cot y}{\tan x \cot y} = \tan y + \cot x$

60. $\dfrac{\cos x - \cos y}{\sin x + \sin y} + \dfrac{\sin x - \sin y}{\cos x + \cos y} = 0$

61. $\sqrt{\dfrac{1 + \sin \theta}{1 - \sin \theta}} = \dfrac{1 + \sin \theta}{|\cos \theta|}$

62. $\sqrt{\dfrac{1 - \cos \theta}{1 + \cos \theta}} = \dfrac{1 - \cos \theta}{|\sin \theta|}$

63. $\ln|\tan \theta| = \ln|\sin \theta| - \ln|\cos \theta|$

64. $\ln|\sec \theta| = -\ln|\cos \theta|$

65. $-\ln(1 + \cos \theta) = \ln(1 - \cos \theta) - 2 \ln|\sin \theta|$

66. $-\ln|\sec \theta + \tan \theta| = \ln|\sec \theta - \tan \theta|$

67. Explain why the following are *not* identities:

   (a) $\sin \theta = \sqrt{1 - \cos^2 \theta}$

   (b) $\tan \theta = \sqrt{\sec^2 \theta - 1}$

Sum and Difference Formulas   6.3

In this and the following two sections, we show the derivations and uses of a number of specialized trigonometric identities that are important for rewriting trigonometric expressions in calculus and the physical sciences. These identities are often referred to as **trigonometric formulas.** You should memorize them.

In this section, we develop sum and difference formulas that express trigonometric functions of $(u \pm v)$ as functions of u and v alone.

SUM AND DIFFERENCE FORMULAS

Cosine:	$\cos(u + v) = \cos u \cos v - \sin u \sin v$
	$\cos(u - v) = \cos u \cos v + \sin u \sin v$
Sine:	$\sin(u + v) = \sin u \cos v + \cos u \sin v$
	$\sin(u - v) = \sin u \cos v - \cos u \sin v$
Tangent:	$\tan(u + v) = \dfrac{\tan u + \tan v}{1 - \tan u \tan v}$
	$\tan(u - v) = \dfrac{\tan u - \tan v}{1 + \tan u \tan v}$

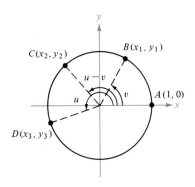

FIGURE 6.1

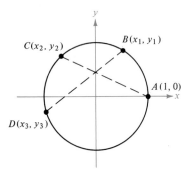

FIGURE 6.2

Proof:

We will begin by proving the formulas for $\cos(u \pm v)$, and postpone the proof of the remaining formulas until later in this section. In Figure 6.1, we let A be the point $(1, 0)$, and then use u and v to locate the points $B(x_1, y_1)$, $C(x_2, y_2)$, and $D(x_3, y_3)$ on the unit circle as indicated. For convenience, we assume that $0 < v < u < 2\pi$, though the results we get are valid for all real values of u and v.

From Figure 6.2, we note that arcs $\overset{\frown}{AC}$ and $\overset{\frown}{BD}$ have the same lengths. Hence, *line segments* $\overline{AC}$ and $\overline{BD}$ are also equal in length. Thus, by the distance formula, we have

$$\overline{AC} = \overline{BD}$$
$$\sqrt{(x_2 - 1)^2 + (y_2 - 0)^2} = \sqrt{(x_3 - x_1)^2 + (y_3 - y_1)^2}$$

By squaring both sides and removing parentheses, we obtain

$$x_2^2 - 2x_2 + 1 + y_2^2 = x_3^2 - 2x_1x_3 + x_1^2 + y_3^2 - 2y_1y_3 + y_1^2$$
$$(x_2^2 + y_2^2) + 1 - 2x_2 = (x_3^2 + y_3^2) + (x_1^2 + y_1^2) - 2x_1x_3 - 2y_1y_3$$

Since B, C, and D lie on the unit circle, it follows that

$$x_1^2 + y_1^2 = 1, \qquad x_2^2 + y_2^2 = 1, \qquad \text{and} \qquad x_3^2 + y_3^2 = 1$$

Making these substitutions, we obtain

$$1 + 1 - 2x_2 = 1 + 1 - 2x_1x_3 - 2y_1y_3$$
$$x_2 = x_3x_1 + y_3y_1$$

Finally, by backsubstituting the values

$$x_3 = \cos u, \qquad x_2 = \cos(u - v), \qquad x_1 = \cos v$$
$$y_3 = \sin u, \qquad y_1 = \sin v$$

we obtain the desired result

$$\cos(u - v) = \cos u \cos v + \sin u \sin v$$

The formula for $\cos(u + v)$ can now be easily established by considering $u + v = u - (-v)$ and substituting into the formula just derived. We obtain

$$\cos(u + v) = \cos[u - (-v)]$$
$$= \cos u \cos(-v) + \sin u \sin(-v)$$
$$= \cos u \cos v - \sin u \sin v$$

Recall from our list of fundamental identities that

$$\cos(-v) = \cos v \qquad \text{and} \qquad \sin(-v) = -\sin v$$

EXAMPLE 1
Using the Formulas
for cos(u + v)

Find the exact value for each of the following:

(a) $\cos 75°$ (b) $\cos \dfrac{\pi}{12}$

Solution:

(a) To find the *exact* value, we cannot use a calculator (since it would give us decimal approximations to the functional values). However, we note that

$$75° = 30° + 45°$$

Consequently, we can use the special angle information from Section 5.3 and the formula

$$\cos(u + v) = \cos u \cos v - \sin u \sin v$$

to get

$$\cos 75° = \cos(30° + 45°) = \cos 30° \cos 45° - \sin 30° \sin 45°$$
$$= \frac{\sqrt{3}}{2}\left(\frac{\sqrt{2}}{2}\right) - \frac{1}{2}\left(\frac{\sqrt{2}}{2}\right)$$
$$= \frac{\sqrt{6} - \sqrt{2}}{4}$$

(b) In a similar manner, we consider

$$\frac{\pi}{12} = \frac{\pi}{3} - \frac{\pi}{4}$$

and apply the difference formula

$$\cos(u - v) = \cos u \cos v + \sin u \sin v$$

to get

$$\cos \frac{\pi}{12} = \cos\left(\frac{\pi}{3} - \frac{\pi}{4}\right) = \cos \frac{\pi}{3} \cos \frac{\pi}{4} + \sin \frac{\pi}{3} \sin \frac{\pi}{4}$$
$$= \frac{1}{2}\left(\frac{\sqrt{2}}{2}\right) + \frac{\sqrt{3}}{2}\left(\frac{\sqrt{2}}{2}\right)$$
$$= \frac{\sqrt{2} + \sqrt{6}}{4}$$

EXAMPLE 2
Evaluating cos(u − v)

Find the value of $\cos(u - v)$, given the following conditions:

u lies in Quadrant III   **v lies in Quadrant I**

$$\cos u = -\frac{15}{17} \qquad\qquad \sin v = \frac{4}{5}$$

Solution:
Using the values $\cos u = -\frac{15}{17}$ and $\sin v = \frac{4}{5}$, we can sketch angles u and v, as shown in Figure 6.3. We obtain

$$\cos u = -\frac{15}{17}, \qquad \cos v = \frac{3}{5}, \qquad \sin u = -\frac{8}{17}, \qquad \sin v = \frac{4}{5}$$

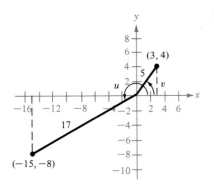

FIGURE 6.3

Therefore,

$$\cos(u - v) = \cos u \cos v + \sin u \sin v$$

$$= \left(-\frac{15}{17}\right)\left(\frac{3}{5}\right) + \left(-\frac{8}{17}\right)\left(\frac{4}{5}\right) = -\frac{77}{85}$$

The formulas for $\cos(u \pm v)$ can be used to prove many other identities, such as the following cofunction identities.

COFUNCTION IDENTITIES

$$\cos\left(\frac{\pi}{2} - u\right) = \sin u$$

$$\sin\left(\frac{\pi}{2} - u\right) = \cos u$$

$$\tan\left(\frac{\pi}{2} - u\right) = \cot u$$

Proof:
For the first identity, we use the difference formula to obtain

$$\cos\left(\frac{\pi}{2} - u\right) = \cos\frac{\pi}{2}\cos u + \sin\frac{\pi}{2}\sin u$$

$$= (0)\cos u + (1)\sin u$$

$$= \sin u$$

Now, we can use this result to prove the second identity. By replacing u with $(\pi/2) - u$ in this result, we get

$$\cos\left[\frac{\pi}{2} - \left(\frac{\pi}{2} - u\right)\right] = \sin\left(\frac{\pi}{2} - u\right)$$

$$\cos u = \sin\left(\frac{\pi}{2} - u\right)$$

Finally, for the third identity, we have

$$\tan\left(\frac{\pi}{2} - u\right) = \frac{\sin[(\pi/2) - u]}{\cos[(\pi/2) - u]} = \frac{\cos u}{\sin u} = \cot u$$

Remark: To memorize these cofunction identities, try remembering the phrase *cofunctions of complementary angles are equal.*

EXAMPLE 3
Using Cofunction Identities

Find an angle θ that makes each of the following true:

(a) $\sin \theta = \cos(3\theta - 10°)$ (b) $\tan \theta = \cot\left(\frac{\theta}{2} + \frac{\pi}{12}\right)$

Solution:

(a) Since we want *cofunction* values to be equal, the two angles must be *complementary*. That is,

$$\theta + (3\theta - 10°) = 90°$$
$$4\theta - 10° = 90°$$
$$\theta = 25°$$

(b) Again, the two angles must be complementary. Hence,

$$\theta + \left(\frac{\theta}{2} + \frac{\pi}{12}\right) = \frac{\pi}{2}$$

$$\frac{3\theta}{2} = \frac{\pi}{2} - \frac{\pi}{12} = \frac{5\pi}{12}$$

$$\theta = \frac{10\pi}{36} = \frac{5\pi}{18}$$

Sine Formulas

We are now ready to complete our derivations of the sum and difference formulas, using the cofunction identities.

$$\sin(u + v) = \cos\left[\frac{\pi}{2} - (u + v)\right] \qquad \textit{Cofunction identity}$$

$$= \cos\left[\left(\frac{\pi}{2} - u\right) - v\right]$$

$$= \cos\left(\frac{\pi}{2} - u\right)\cos v + \sin\left(\frac{\pi}{2} - u\right)\sin v$$

$$= \sin u \cos v + \cos u \sin v$$

Now, since $u - v = u + (-v)$, it follows that

$$\sin(u - v) = \sin[u + (-v)]$$

$$= \sin u \cos(-v) + \cos u \sin(-v)$$

$$= \sin u \cos v - \cos u \sin v$$

Tangent Formulas

We can verify the formula for $\tan(u + v)$ in the following manner:

$$\tan(u + v) = \frac{\sin(u + v)}{\cos(u + v)}$$

$$= \frac{\sin u \cos v + \cos u \sin v}{\cos u \cos v - \sin u \sin v}$$

Now, dividing each term by cos u cos v, we obtain

$$\tan(u + v) = \frac{\dfrac{\sin u}{\cos u}\left(\dfrac{\cos v}{\cos v}\right) + \dfrac{\cos u}{\cos u}\left(\dfrac{\sin v}{\cos v}\right)}{\dfrac{\cos u \cos v}{\cos u \cos v} - \dfrac{\sin u}{\cos u}\left(\dfrac{\sin v}{\cos v}\right)}$$

$$= \frac{\tan u + \tan v}{1 - \tan u \tan v}$$

The formula for $\tan(u - v)$ can be derived in the same manner as that for $\sin(u - v)$.

EXAMPLE 4
Using the Sum and Difference Formulas

Find the exact value for each of the following:

(a) $\sin 12° \cos 42° - \cos 12° \sin 42°$

(b) $\dfrac{\tan 80° + \tan 55°}{1 - \tan 80° \tan 55°}$

Solution:

(a) This combination of functions fits the difference formula

$$\sin(u - v) = \sin u \cos v - \cos u \sin v$$

Thus, we have

$$\sin 12° \cos 42° - \cos 12° \sin 42° = \sin(12° - 42°)$$
$$= \sin(-30°)$$
$$= -\sin 30° = -\frac{1}{2}$$

(b) In this case, the appropriate formula is

$$\tan(u + v) = \frac{\tan u + \tan v}{1 - \tan u \tan v}$$

Consequently,

$$\frac{\tan 80° + \tan 55°}{1 - \tan 80° \tan 55°} = \tan(80° + 55°)$$
$$= \tan 135°$$
$$= -\tan 45° = -1$$

EXAMPLE 5
Using the Sum and Difference Formulas

Find $\sin(u - v)$, $\cos(u - v)$, and $\tan(u - v)$, and determine the quadrant in which $(u - v)$ lies, given the following information:

$$\sec u = 3 \quad \text{and} \quad \sec v = \frac{3}{2}$$

$$\sin u < 0 \quad \text{and} \quad \sin v > 0$$

Solution:

From the signs of the given functions, we can conclude that u lies in Quadrant IV and v in Quadrant I. Since $\sec u = 3$, we have

$$\cos u = \frac{1}{3}$$

$$\sin u = -\sqrt{1 - \cos^2 u} = -\sqrt{1 - \frac{1}{9}} = -\frac{2\sqrt{2}}{3}$$

$$\tan u = \frac{\sin u}{\cos u} = -2\sqrt{2}$$

Since $\sec v = \frac{3}{2}$, we have

$$\cos v = \frac{2}{3}$$

$$\sin v = \sqrt{1 - \cos^2 v} = \sqrt{1 - \frac{4}{9}} = \frac{\sqrt{5}}{3}$$

$$\tan v = \frac{\sin v}{\cos v} = \frac{\sqrt{5}}{2}$$

Therefore,

$$\sin(u - v) = \sin u \cos v - \cos u \sin v$$

$$= \frac{-2\sqrt{2}}{3}\left(\frac{2}{3}\right) - \frac{1}{3}\left(\frac{\sqrt{5}}{3}\right) = \frac{-\sqrt{5} - 4\sqrt{2}}{9}$$

$$\cos(u - v) = \cos u \cos v + \sin u \sin v$$

$$= \frac{1}{3}\left(\frac{2}{3}\right) + \frac{-2\sqrt{2}}{3}\left(\frac{\sqrt{5}}{3}\right) = \frac{2 - 2\sqrt{10}}{9}$$

$$\tan(u - v) = \frac{\tan u - \tan v}{1 + \tan u \tan v}$$

$$= \frac{-2\sqrt{2} - (\sqrt{5}/2)}{1 + (-2\sqrt{2})(\sqrt{5}/2)} = \frac{-4\sqrt{2} - \sqrt{5}}{2 - 2\sqrt{10}}$$

Since

$$\cos(u - v) < 0 \qquad \text{and} \qquad \tan(u - v) > 0$$

we can conclude that $(u - v)$ lies in Quadrant III.

Our next example comes from calculus. It shows how a sum formula allows us to rewrite a trigonometric expression in a form that is useful in some calculus problems.

EXAMPLE 6
An Application from Calculus

Prove that

$$\frac{\sin(x+h) - \sin x}{h} = \cos x\left(\frac{\sin h}{h}\right) - \sin x\left(\frac{1 - \cos h}{h}\right)$$

where $h \neq 0$.

Solution:
Using the formula for $\sin(u + v)$, we have

$$\frac{\sin(x+h) - \sin x}{h} = \frac{\sin x \cos h + \cos x \sin h - \sin x}{h}$$

$$= \frac{\cos x \sin h - \sin x(1 - \cos h)}{h}$$

$$= \cos x\left(\frac{\sin h}{h}\right) - \sin x\left(\frac{1 - \cos h}{h}\right)$$

The sum and difference formulas can be used to rewrite expressions like

$$\sin\left(\theta + \frac{n\pi}{2}\right) \qquad \text{and} \qquad \cos\left(\theta + \frac{n\pi}{2}\right), \qquad n \text{ is any integer}$$

in terms of $\sin \theta$ or $\cos \theta$. The derived forms are referred to as **reduction formulas.** We show two such instances in our next example.

EXAMPLE 7
Deriving Reduction Formulas

Express

$$\cos\left(\theta - \frac{3\pi}{2}\right) \qquad \text{and} \qquad \sin(\theta + 3\pi)$$

in terms of a function of θ alone.

Solution:
By the difference formula for $\cos(u - v)$, we have

$$\cos\left(\theta - \frac{3\pi}{2}\right) = \cos \theta \cos \frac{3\pi}{2} + \sin \theta \sin \frac{3\pi}{2}$$

$$= (\cos \theta)(0) + (\sin \theta)(-1)$$

$$= -\sin \theta$$

Similarly, using the formula for $\sin(u + v)$, we have

$$\sin(\theta + 3\pi) = \sin \theta \cos 3\pi + \cos \theta \sin 3\pi$$

$$= (\sin \theta)(-1) + (\cos \theta)(0)$$

$$= -\sin \theta$$

Section Exercises 6.3

In Exercises 1–10, determine the exact values of the sine, cosine, and tangent of the given angle by using the sum or difference formulas.

1. $75° = 30° + 45°$
2. $15° = 45° - 30°$
3. $105° = 60° + 45°$
4. $165° = 135° + 30°$
5. $195° = 225° - 30°$
6. $255° = 300° - 45°$
7. $\dfrac{11\pi}{12} = \dfrac{3\pi}{4} + \dfrac{\pi}{6}$
8. $\dfrac{7\pi}{12} = \dfrac{\pi}{3} + \dfrac{\pi}{4}$
9. $\dfrac{17\pi}{12} = \dfrac{9\pi}{4} - \dfrac{5\pi}{6}$
10. $-\dfrac{\pi}{12} = \dfrac{\pi}{6} - \dfrac{\pi}{4}$

In Exercises 11–20, use the sum and difference formulas to simplify the given expression to a single term.

11. $\cos 25° \cos 15° - \sin 25° \sin 15°$
12. $\sin 140° \cos 50° + \cos 140° \sin 50°$
13. $\sin 230° \cos 30° - \cos 230° \sin 30°$
14. $\cos 20° \cos 30° + \sin 20° \sin 30°$
15. $\dfrac{\tan 325° - \tan 86°}{1 + \tan 325° \tan 86°}$
16. $\dfrac{\tan 140° - \tan 60°}{1 + \tan 140° \tan 60°}$
17. $\sin 3 \cos 1.2 - \cos 3 \sin 1.2$
18. $\cos \dfrac{\pi}{7} \cos \dfrac{\pi}{5} - \sin \dfrac{\pi}{7} \sin \dfrac{\pi}{5}$
19. $\dfrac{\tan 2x + \tan x}{1 - \tan 2x \tan x}$
20. $\cos 3x \cos 2y + \sin 3x \sin 2y$

In Exercises 21–40, verify the given identity.

21. $\sin\left(\dfrac{\pi}{2} + x\right) = \cos x$
22. $\sin(3\pi - x) = \sin x$
23. $\cos\left(\dfrac{3\pi}{2} - x\right) = -\sin x$
24. $\cos(\pi + x) = -\cos x$
25. $\sin\left(\dfrac{\pi}{6} + x\right) = \dfrac{1}{2}(\cos x + \sqrt{3} \sin x)$
26. $\cos\left(\dfrac{5\pi}{4} - x\right) = -\dfrac{\sqrt{2}}{2}(\cos x + \sin x)$
27. $\cos(\pi - \theta) + \sin\left(\dfrac{\pi}{2} + \theta\right) = 0$
28. $\sin\left(\dfrac{3\pi}{2} + \theta\right) + \sin(\pi - \theta) = \sin \theta - \cos \theta$
29. $\tan(\pi + \theta) = \tan \theta$

30. $\tan\left(\dfrac{\pi}{4} - \theta\right) = \dfrac{1 - \tan \theta}{1 + \tan \theta}$
31. $\cos(x + y) \cos(x - y) = \cos^2 x - \sin^2 y$
32. $\sin(x + y) \sin(x - y) = \sin^2 x - \sin^2 y$
33. $\sin(x + y) + \sin(x - y) = 2 \sin x \cos y$
34. $\cos(x + y) + \cos(x - y) = 2 \cos x \cos y$
35. $\sin(x + y + z) = \sin x \cos y \cos z + \sin y \cos x \cos z$
    $\qquad + \sin z \cos x \cos y - \sin x \sin y \sin z$
36. $\cos(x + y + z) = \cos x \cos y \cos z - \cos x \sin y \sin z$
    $\qquad - \cos y \sin x \sin z - \cos z \sin x \sin y$
37. $\cos(n\pi + \theta) = (-1)^n \cos \theta$, n is an integer
38. $\sin(n\pi + \theta) = (-1)^n \sin \theta$, n is an integer
39. $a \sin B\theta + b \cos B\theta = \sqrt{a^2 + b^2} \sin(B\theta + C)$, where
    $$C = \arctan \dfrac{b}{a}$$
40. $a \sin B\theta + b \cos B\theta = \sqrt{a^2 + b^2} \cos(B\theta - C)$, where
    $$C = \arctan \dfrac{a}{b}$$

In Exercises 41–46, use the reduction formulas given in Exercises 39 and 40.

41. Express $\sin \theta + \cos \theta$ in the form
    $$\sqrt{a^2 + b^2} \sin(B\theta + C)$$
42. Express $3 \sin 2\theta + 4 \cos 2\theta$ in the form
    $$\sqrt{a^2 + b^2} \sin(B\theta + C)$$
43. Express $12 \sin 3\theta + 5 \cos 3\theta$ in the form
    $$\sqrt{a^2 + b^2} \cos(B\theta - C)$$
44. Express $\sin 2\theta - \cos 2\theta$ in the form
    $$\sqrt{a^2 + b^2} \cos(B\theta - C)$$
45. Express $2 \sin\left(\theta + \dfrac{\pi}{4}\right)$ in the form $a \sin B\theta + b \cos B\theta$.
46. Express $5 \cos\left(\theta + \dfrac{3\pi}{4}\right)$ in the form $a \sin B\theta + b \cos B\theta$.
47. Write $\sin(\arcsin x + \arccos x)$ as an algebraic expression in x.
48. Write $\cos(\arctan x + \text{arccot } x)$ as an algebraic expression in x.

Multiple-Angle Formulas

6.4

In this section we focus on trigonometric formulas for functions of *multiple angles*, like sin ku and cos ku, where k is an integer or a fraction. In particular, for $k = 2$, we derive the following **double-angle** formulas, which you should memorize.

DOUBLE-ANGLE FORMULAS

$$\sin 2u = 2 \sin u \cos u \qquad \cos 2u = \cos^2 u - \sin^2 u$$
$$= 2 \cos^2 u - 1$$
$$= 1 - 2 \sin^2 u$$

$$\tan 2u = \frac{2 \tan u}{1 - \tan^2 u}$$

Proof:

To prove these formulas, we let $v = u$ in the formula for $\sin(u + v)$, and obtain

$$\sin 2u = \sin(u + u)$$
$$= \sin u \cos u + \cos u \sin u$$
$$= 2 \sin u \cos u$$

Similarly, for cos $2u$, we get

$$\cos 2u = \cos(u + u)$$
$$= \cos u \cos u - \sin u \sin u$$
$$= \cos^2 u - \sin^2 u$$

The alternative forms for cos $2u$ derive from the Pythagorean identity $\sin^2 u + \cos^2 u = 1$.

$$\cos 2u = \cos^2 u - \sin^2 u$$
$$= \cos^2 u - (1 - \cos^2 u)$$
$$= 2 \cos^2 u - 1$$

and

$$\cos 2u = (1 - \sin^2 u) - \sin^2 u$$
$$= 1 - 2 \sin^2 u$$

Finally, letting $v = u$ in the formula for $\tan(u + v)$, we get

$$\tan 2u = \tan(u + u)$$
$$= \frac{\tan u + \tan u}{1 - \tan u \tan u}$$
$$= \frac{2 \tan u}{1 - \tan^2 u}$$

You should consider using these and other trigonometric formulas in *both* directions. For example, in calculus there can be a distinct advantage to replacing

$$1 - 2 \sin^2 x \quad \text{by} \quad \cos 2x$$

whereas in factoring an expression like $2 \cos x + \sin 2x$, the advantage lies in replacing

$$\sin 2x \quad \text{by} \quad 2 \sin x \cos x$$

EXAMPLE 1
Evaluating Functions of 2θ

Find the values of $\sin 2\theta$, $\cos 2\theta$, and $\tan 2\theta$, given the following:

$$\cos \theta = \frac{5}{13} \quad \text{with} \quad \theta \text{ in Quadrant IV}$$

Solution:
From the triangle shown in Figure 6.4, we can see that

$$\sin \theta = -\frac{12}{13} \quad \text{and} \quad \tan \theta = -\frac{12}{5}$$

Consequently,

$$\sin 2\theta = 2 \sin \theta \cos \theta = 2\left(\frac{-12}{13}\right)\left(\frac{5}{13}\right) = \frac{-120}{169}$$

$$\cos 2\theta = 2 \cos^2 \theta - 1 = 2\left(\frac{25}{169}\right) - 1 = \frac{-119}{169}$$

$$\tan 2\theta = \frac{\sin 2\theta}{\cos 2\theta} = \frac{120}{119}$$

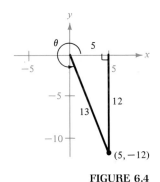

FIGURE 6.4

Remark: The double-angle formulas are not restricted to angles 2θ and θ. Other *double* combinations are also valid. For example, 4θ and 2θ or 6θ and 3θ can be used as a double combination, as shown in the following:

$$\sin 4\theta = 2 \sin 2\theta \cos 2\theta$$
$$\cos 6\theta = \cos^2 3\theta - \sin^2 3\theta$$

EXAMPLE 2
Using Sum and Double-Angle Formulas Together

Express $\sin 3x$ in terms of $\sin x$.

Solution:
Considering $3x = 2x + x$, we have

$$\sin 3x = \sin(2x + x)$$
$$= \sin 2x \cos x + \cos 2x \sin x$$
$$= 2 \sin x \cos x \cos x + (1 - 2 \sin^2 x)\sin x$$
$$= 2 \sin x \cos^2 x + \sin x - 2 \sin^3 x$$

$$= 2 \sin x (1 - \sin^2 x) + \sin x - 2 \sin^3 x$$
$$= 2 \sin x - 2 \sin^3 x + \sin x - 2 \sin^3 x$$
$$= 3 \sin x - 4 \sin^3 x$$

By rearranging the double-angle formulas, we obtain some identities that are useful for rewriting trigonometric functions raised to powers. These identities are sometimes referred to as **power-reducing identities.**

POWER-REDUCING IDENTITIES

$$\sin^2 u = \frac{1 - \cos 2u}{2} \qquad \cos^2 u = \frac{1 + \cos 2u}{2}$$

$$\tan^2 u = \frac{1 - \cos 2u}{1 + \cos 2u}$$

Proof:
The first two of these identities can be verified by solving for $\sin^2 u$ and $\cos^2 u$, respectively, in the equations

$$\cos 2u = 1 - 2 \sin^2 u \qquad \text{and} \qquad \cos 2u = 2 \cos^2 u - 1$$

The third identity arises from the fact that

$$\tan 2u = \frac{\sin 2u}{\cos 2u}$$

EXAMPLE 3
Reducing the Power of a
Trigonometric Function

Express $\sin^4 x$ in terms of the first power of the cosine.

Solution:
Note the repeated use of power-reducing identities in the following procedure:

$$\sin^4 x = (\sin^2 x)^2 = \left(\frac{1 - \cos 2x}{2} \right)^2$$

$$= \frac{1}{4}(1 - 2 \cos 2x + \cos^2 2x)$$

$$= \frac{1}{4}\left(1 - 2 \cos 2x + \frac{1 + \cos 4x}{2} \right)$$

$$= \frac{1}{4} - \frac{1}{2} \cos 2x + \frac{1}{8} + \frac{1}{8} \cos 4x$$

$$= \frac{3}{8} - \frac{1}{2} \cos 2x + \frac{1}{8} \cos 4x$$

A more common form of the power-reducing identities is obtained by replacing u with $u/2$. In this form, the identities are known as the **half-angle formulas,** listed next with some useful alternative forms.

HALF-ANGLE FORMULAS

Common Forms

$$\sin^2 \frac{u}{2} = \frac{1 - \cos u}{2}$$

$$\cos^2 \frac{u}{2} = \frac{1 + \cos u}{2}$$

$$\tan^2 \frac{u}{2} = \frac{1 - \cos u}{1 + \cos u}$$

Alternative Forms*

$$\sin \frac{u}{2} = \pm \sqrt{\frac{1 - \cos u}{2}}$$

$$\cos \frac{u}{2} = \pm \sqrt{\frac{1 + \cos u}{2}}$$

$$\tan \frac{u}{2} = \pm \sqrt{\frac{1 - \cos u}{1 + \cos u}}$$

$$= \frac{1 - \cos u}{\sin u} = \frac{\sin u}{1 + \cos u}$$

*Signs are selected on the basis of the quadrant in which $u/2$ lies.

Proofs of the last two alternative forms for $\tan u/2$ are given in Example 6; proofs of the other formulas follow from those for the power-reducing identities.

These half-angle formulas are valid for other half-angle combinations like $u/4$ and $u/2$, or $3u/2$ and $3u$. For instance, the following are valid:

$$\sin^2 \frac{3u}{2} = \frac{1 - \cos 3u}{2}$$

$$\cos \frac{u}{6} = \pm \sqrt{\frac{1 + \cos(u/3)}{2}}$$

$$\tan^2 3u = \frac{1 - \cos 6u}{1 + \cos 6u}$$

EXAMPLE 4
Using Half-Angle Formulas to Evaluate Trigonometric Functions

Find exact values for $\sin 105°$ and $\tan 105°$.

Solution:
We begin by noting that $105°$ is half of $210°$. Now, using the alternative half-angle formulas and the fact that $105°$ lies in Quadrant II, we have

$$\sin 105° = + \sqrt{\frac{1 - \cos 210°}{2}}$$

$$= \sqrt{\frac{1 - (-\cos 30°)}{2}}$$

$$= \sqrt{\frac{1 + (\sqrt{3}/2)}{2}} = \frac{\sqrt{2 + \sqrt{3}}}{2}$$

For $\tan 105°$, we have

$$\tan 105° = \frac{1 - \cos 210°}{\sin 210°} = \frac{1 + (\sqrt{3}/2)}{-1/2} = -2 - \sqrt{3}$$

EXAMPLE 5
**Using Half-Angle Formulas
to Verify an Identity**

Verify the identity

$$\sin^2 2t \cos^2 2t = \frac{1}{8}(1 - \cos 8t)$$

Solution:
Using half-angle formulas, we have

$$\sin^2 2t \cos^2 2t = \left(\frac{1 - \cos 4t}{2}\right)\left(\frac{1 + \cos 4t}{2}\right)$$

$$= \frac{1}{4}(1 - \cos^2 4t)$$

$$= \frac{1}{4}(\sin^2 4t)$$

$$= \frac{1}{4}\left(\frac{1 - \cos 8t}{2}\right)$$

$$= \frac{1}{8}(1 - \cos 8t)$$

EXAMPLE 6
**Proof of Alternative
Forms for tan(u/2)**

Verify the identities

$$\tan \frac{u}{2} = \frac{1 - \cos u}{\sin u} = \frac{\sin u}{1 + \cos u}$$

Solution:
By multiplying the numerator and denominator of the formula for $\tan^2(u/2)$ by $(1 - \cos u)$, we obtain

$$\tan^2 \frac{u}{2} = \frac{1 - \cos u}{1 + \cos u}\left(\frac{1 - \cos u}{1 - \cos u}\right)$$

$$= \frac{(1 - \cos u)^2}{1 - \cos^2 u} = \frac{(1 - \cos u)^2}{\sin^2 u}$$

Taking the square root of both sides, we have

$$\tan \frac{u}{2} = \pm\left(\frac{1 - \cos u}{\sin u}\right)$$

To determine the appropriate sign, we note that the numerator $(1 - \cos u)$ is never negative. Furthermore, $\tan(u/2)$ and $\sin u$ will always have the same sign, as the following argument indicates:

u	$sin\ u$	$u/2$	$tan(u/2)$
$0 < u < \pi$	$(+)$	$0 < u/2 < \pi/2$	$(+)$
$\pi < u < 2\pi$	$(-)$	$\pi/2 < u/2 < \pi$	$(-)$
$2\pi < u < 3\pi$	$(+)$	$\pi < u/2 < 3\pi/2$	$(+)$
$3\pi < u < 4\pi$	$(-)$	$3\pi/2 < u/2 < 2\pi$	$(-)$

Consequently, we can use the positive sign and write

$$\tan \frac{u}{2} = \frac{1 - \cos u}{\sin u}$$

Multiplying numerator and denominator by $(1 + \cos u)$ yields the final form

$$\tan \frac{u}{2} = \frac{1 - \cos u}{\sin u}\left(\frac{1 + \cos u}{1 + \cos u}\right)$$

$$= \frac{1 - \cos^2 u}{\sin u(1 + \cos u)}$$

$$= \frac{\sin^2 u}{\sin u(1 + \cos u)}$$

$$= \frac{\sin u}{1 + \cos u}$$

Section Exercises 6.4

In Exercises 1–6, use a double-angle formula to determine the exact values of the sine, cosine, and tangent of the given angle.

1. $90° = 2(45°)$
2. $180° = 2(90°)$
3. $60° = 2(30°)$
4. $120° = 2(60°)$
5. $\dfrac{2\pi}{3} = 2\left(\dfrac{\pi}{3}\right)$
6. $\dfrac{3\pi}{2} = 2\left(\dfrac{3\pi}{4}\right)$

In Exercises 7–12, find the exact value of $\sin 2u$, $\cos 2u$, and $\tan 2u$ by using the double-angle formulas and the given information.

7. $\sin u = \frac{3}{5}$, u lies in Quadrant I
8. $\cos u = -\frac{2}{3}$, u lies in Quadrant II
9. $\tan u = \frac{1}{2}$, u lies in Quadrant III
10. $\cot u = -4$, u lies in Quadrant IV
11. $\sec u = -\frac{5}{2}$, u lies in Quadrant II
12. $\csc u = 3$, u lies in Quadrant II

In Exercises 13–20, determine the exact values of the sine, cosine, and tangent of the given angle by using the half-angle formulas.

13. $105° = \dfrac{1}{2}(210°)$
14. $165° = \dfrac{1}{2}(330°)$
15. $112° \; 30' = \dfrac{1}{2}(225°)$
16. $67° \; 30' = \dfrac{1}{2}(135°)$
17. $\dfrac{\pi}{8} = \dfrac{1}{2}\left(\dfrac{\pi}{4}\right)$
18. $\dfrac{\pi}{12} = \dfrac{1}{2}\left(\dfrac{\pi}{6}\right)$
19. $52° \; 30' = \dfrac{1}{2}(105°)$ (Use the result of Exercise 13.)
20. $\dfrac{\pi}{24} = \dfrac{1}{2}\left(\dfrac{\pi}{12}\right)$ (Use the result of Exercise 18.)

In Exercises 21–26, find the exact values of $\sin(u/2)$, $\cos(u/2)$, and $\tan(u/2)$ by using the half-angle formulas and the given information.

21. $\sin u = \frac{5}{13}$, u lies in Quadrant II
22. $\cos u = \frac{3}{5}$, u lies in Quadrant I
23. $\tan u = -\frac{5}{8}$, u lies in Quadrant IV
24. $\cot u = 3$, u lies in Quadrant III
25. $\csc u = -\frac{5}{3}$, u lies in Quadrant III
26. $\sec u = -\frac{7}{2}$, u lies in Quadrant II

In Exercises 27–30, use the half-angle formulas to simplify the given expression.

27. $\sqrt{\dfrac{1 - \cos 6x}{2}}$
28. $\sqrt{\dfrac{1 + \cos 4x}{2}}$
29. $-\sqrt{\dfrac{1 - \cos 8x}{1 + \cos 8x}}$
30. $-\sqrt{\dfrac{1 - \cos(x - 1)}{2}}$

In Exercises 31–35, use the power-reducing identities to write each expression in terms of the first power of the cosine.

31. $\cos^4 x$
32. $\cos^6 x$
33. $\sin^6 x$
34. $\sin^2 x(\cos^2 x)$
35. $\sin^2 x(\cos^4 x)$

In Exercises 36–52, verify the given identity.

36. $\sec 2\theta = \dfrac{\sec^2 \theta}{2 - \sec^2 \theta}$
37. $\csc 2\theta = \dfrac{\csc \theta}{2 \cos \theta}$
38. $\sin\left(\dfrac{\alpha}{3}\right)\cos\left(\dfrac{\alpha}{3}\right) = \dfrac{1}{2}\sin\left(\dfrac{2\alpha}{3}\right)$

39. $\cos^2 2\alpha - \sin^2 2\alpha = \cos 4\alpha$
40. $\cos^4 x - \sin^4 x = \cos 2x$
41. $(\sin x + \cos x)^2 = 1 + \sin 2x$
42. $\sin 4\beta = 4 \sin \beta \cos \beta (1 - 2 \sin^2 \beta)$
43. $\cos 3\beta = \cos^3 \beta - 3 \sin^2 \beta \cos \beta$
44. $\dfrac{\cos 3\beta}{\cos \beta} = 1 - 4 \sin^2 \beta$
45. $1 + \cos 10y = 2 \cos^2 5y$
46. $2 \sin y \cos y \sec 2y = \tan 2y$

47. $\sec \dfrac{u}{2} = \pm \sqrt{\dfrac{2 \tan u}{\tan u + \sin u}}$

48. $\csc \dfrac{u}{2} = \pm \sqrt{\dfrac{2 \tan u}{\tan u - \sin u}}$

49. $\cot \dfrac{u}{2} = \pm \sqrt{\dfrac{\sec u + 1}{\sec u - 1}}$

50. $\tan \dfrac{u}{2} = \csc u - \cot u$

51. $\sin \dfrac{\theta}{2} + \cos \dfrac{\theta}{2} = \pm \sqrt{1 + \sin \theta}$

52. $\sin \dfrac{\theta}{2} - \cos \dfrac{\theta}{2} = \pm \sqrt{1 - \sin \theta}$

53. If ϕ and θ are complementary angles, show that
    (a) $\sin(\phi - \theta) = \cos 2\theta$ (b) $\cos(\phi - \theta) = \sin 2\theta$

54. Write each of the following as an algebraic expression in x:
    (a) $\sin(2 \arcsin x)$ (b) $\cos(2 \arccos x)$
    (c) $\cos(2 \arccos 2x)$

Product and Sum Formulas   6.5

The formulas we will derive in this section are not as widely applicable as those in the two preceding sections. However, they are used in higher-level math courses such as calculus to convert products of trigonometric functions into sums, thus often simplifying the operations of calculus. Rather than memorizing these **product-sum** formulas, you should be aware of them (by name) so that you can look them up when needed.

PRODUCT-SUM FORMULAS

$$\sin u \sin v = \tfrac{1}{2}[\cos(u - v) - \cos(u + v)]$$
$$\cos u \cos v = \tfrac{1}{2}[\cos(u - v) + \cos(u + v)]$$
$$\sin u \cos v = \tfrac{1}{2}[\sin(u + v) + \sin(u - v)]$$
$$\cos u \sin v = \tfrac{1}{2}[\sin(u + v) - \sin(u - v)]$$

Proof:
The first equation can be verified using the sum and difference formulas on the right side. We obtain

$$\frac{1}{2}[\cos(u - v) - \cos(u + v)]$$

$$= \frac{1}{2}[\cos u \cos v + \sin u \sin v - (\cos u \cos v - \sin u \sin v)]$$

$$= \frac{1}{2}[2 \sin u \sin v] = \sin u \sin v$$

The remaining three formulas can be easily verified in a similar manner. (See Exercises 36, 37, and 38.)

EXAMPLE 1
Writing Products as Sums

Express each of the following as a sum or difference:

(a) $\cos 5x \sin 4x$ (b) $\sin 15° \sin 45°$

Solution:
Using appropriate product-sum identities, we have

(a) $\cos 5x \sin 4x = \dfrac{1}{2}[\sin(5x + 4x) - \sin(5x - 4x)]$

$= \dfrac{1}{2} \sin 9x - \dfrac{1}{2} \sin x$

(b) $\sin 15° \sin 45° = \dfrac{1}{2}[\cos(15° - 45°) - \cos(15° + 45°)]$

$= \dfrac{1}{2}[\cos(-30°) - \cos(60°)]$

$= \dfrac{1}{2}\left(\dfrac{\sqrt{3}}{2} - \dfrac{1}{2}\right)$

$= \dfrac{\sqrt{3} - 1}{4}$

Occasionally, it is useful to write a sum of trigonometric functions as a product. Such identities can be obtained from the product-sum formulas, using the substitutions

$$x = u + v \quad \text{and} \quad y = u - v$$

Then,

$$u = \dfrac{x + y}{2} \quad \text{and} \quad v = \dfrac{x - y}{2}$$

Using these two values for u and v in the identity

$$\sin u \cos v = \tfrac{1}{2}[\sin(u + v) + \sin(u - v)]$$

we get

$$\sin\left(\dfrac{x + y}{2}\right) \cos\left(\dfrac{x - y}{2}\right) = \dfrac{1}{2}(\sin x + \sin y)$$

or, equivalently,

$$\sin x + \sin y = 2 \sin\left(\dfrac{x + y}{2}\right) \cos\left(\dfrac{x - y}{2}\right)$$

In a similar manner, each of the following **sum-product** formulas can be verified.

SUM-PRODUCT FORMULAS

$$\sin x + \sin y = 2 \sin\left(\frac{x + y}{2}\right) \cos\left(\frac{x - y}{2}\right)$$

$$\sin x - \sin y = 2 \cos\left(\frac{x + y}{2}\right) \sin\left(\frac{x - y}{2}\right)$$

$$\cos x + \cos y = 2 \cos\left(\frac{x + y}{2}\right) \cos\left(\frac{x - y}{2}\right)$$

$$\cos x - \cos y = -2 \sin\left(\frac{x + y}{2}\right) \sin\left(\frac{x - y}{2}\right)$$

EXAMPLE 2
Writing Sums as Products

Express each of the following as a product:

(a) $3 \sin 3x - 3 \sin 5x$ (b) $\cos 195° + \cos 105°$

Solution:
Using appropriate sum-product identities, we get

(a) $3 \sin 3x - 3 \sin 5x = 3(\sin 3x - \sin 5x)$

$$= 3(2) \cos\left(\frac{3x + 5x}{2}\right) \sin\left(\frac{3x - 5x}{2}\right)$$

$$= 6 \cos 4x[\sin(-x)]$$

$$= -6 \cos 4x \sin x$$

(b) $\cos 195° + \cos 105° = 2 \cos\left(\frac{195° + 105°}{2}\right) \cos\left(\frac{195° - 105°}{2}\right)$

$$= 2 \cos 150° \cos 45°$$

$$= 2\left(\frac{-\sqrt{3}}{2}\right)\left(\frac{\sqrt{2}}{2}\right)$$

$$= -\frac{\sqrt{6}}{2}$$

EXAMPLE 3
Using Sum-Product Identities

Verify the identity

$$\frac{\sin t + \sin 3t}{\cos t + \cos 3t} = \tan 2t$$

Solution:
Using appropriate sum-product identities, we have

$$\frac{\sin t + \sin 3t}{\cos t + \cos 3t} = \frac{2 \sin 2t \cos(-t)}{2 \cos 2t \cos(-t)} = \frac{\sin 2t}{\cos 2t} = \tan 2t$$

EXAMPLE 4
Using Multiple-Angle
and Sum-Product Identities

Verify the identity

$$\frac{\sin 6t + \sin 2t}{\sin 2t \cos^2 2t} = 4$$

Solution:

Using a sum-product identity, we have

$$\frac{\sin 6t + \sin 2t}{\sin 2t \cos^2 2t} = \frac{2 \sin 4t \cos 2t}{\sin 2t \cos^2 2t} \qquad \textit{Sum-product}$$

$$= \frac{2 \sin 4t}{\sin 2t \cos 2t} \qquad \textit{Reduce}$$

$$= \frac{2(2 \sin 2t \cos 2t)}{\sin 2t \cos 2t} \qquad \textit{Double-angle}$$

$$= 4 \qquad \textit{Reduce}$$

One final reminder: For your convenience, a complete list of the trigonometric identities and formulas is given inside the back cover of this book.

Section Exercises 6.5

In Exercises 1–10, rewrite the given product as a sum.

1. $6 \sin \dfrac{\pi}{4} \cos \dfrac{\pi}{4}$

2. $4 \sin \dfrac{\pi}{3} \cos \dfrac{5\pi}{6}$

3. $\sin 5\theta \cos 3\theta$

4. $3 \sin 2\alpha \sin 3\alpha$

5. $5 \cos(-5\beta) \cos 3\beta$

6. $\cos 2\theta \cos 4\theta$

7. $\sin(x + y) \sin(x - y)$

8. $\sin(x + y) \cos(x - y)$

9. $\sin(\theta + \pi) \cos(\theta - \pi)$

10. $10 \cos 75° \cos 15°$

In Exercises 11–20, express the given sum as a product.

11. $\sin 60° + \sin 30°$

12. $\cos 120° + \cos 30°$

13. $\cos \dfrac{3\pi}{4} - \cos \dfrac{\pi}{4}$

14. $\sin 5\theta - \sin 3\theta$

15. $\cos 6x + \cos 2x$

16. $\sin x + \sin 5x$

17. $\sin(\alpha + \beta) - \sin(\alpha - \beta)$

18. $\cos\left(\theta + \dfrac{\pi}{2}\right) - \cos\left(\theta - \dfrac{\pi}{2}\right)$

19. $\cos(\phi + 2\pi) + \cos \phi$

20. $\sin\left(x + \dfrac{\pi}{2}\right) + \sin\left(x - \dfrac{\pi}{2}\right)$

In Exercises 21–35, verify the given identity.

21. $\dfrac{\cos 4x + \cos 2x}{\sin 4x + \sin 2x} = \cot 3x$

22. $\dfrac{\cos 3x - \cos x}{\sin 3x - \sin x} = -\tan 2x$

23. $\dfrac{\cos 4x - \cos 2x}{2 \sin 3x} = -\sin x$

24. $\dfrac{\sin x \pm \sin y}{\cos x + \cos y} = \tan\left(\dfrac{x \pm y}{2}\right)$

25. $\dfrac{\sin x \pm \sin y}{\cos x - \cos y} = -\cot\left(\dfrac{x \mp y}{2}\right)$

26. $\dfrac{\sin x + \sin y}{\sin x - \sin y} = \dfrac{\tan[(x + y)/2]}{\tan[(x - y)/2]}$

27. $\sin 2x + \sin 4x - \sin 6x = 4 \sin x \sin 2x \sin 3x$

28. $1 + \cos 2x + \cos 4x + \cos 6x = 4 \cos x \cos 2x \cos 3x$

29. $1 - \cos 2x + \cos 4x - \cos 6x = 4 \sin x \cos 2x \sin 3x$

30. $\sin 2x + \sin 4x + \sin 6x = 4 \cos x \cos 2x \sin 3x$

31. $\dfrac{\cos t + \cos 3t}{\sin 3t - \sin t} = \cot t$

32. $\dfrac{\sin 6t - \sin 2t}{\cos 2t + \cos 6t} = \tan 2t$

33. $\sin^2 4x - \sin^2 2x = \sin 2x \sin 6x$

34. $\cos^2 5x - \cos^2 x = -\sin 4x \sin 6x$

35. $\sin\left(\dfrac{\pi}{6} + x\right) + \sin\left(\dfrac{\pi}{6} - x\right) = \cos x$

In Exercises 36–38, prove the given product-sum formulas.

36. $\cos u \cos v = \frac{1}{2}[\cos(u - v) + \cos(u + v)]$

37. $\sin u \cos v = \frac{1}{2}[\sin(u + v) + \sin(u - v)]$

38. $\cos u \sin v = \frac{1}{2}[\sin(u + v) - \sin(u - v)]$

Solving Trigonometric Equations

6.6

With this section, we switch from *verifying* trigonometric identities to *solving* trigonometric equations. As indicated at the start of Section 6.2, when we use the term *equation*, we usually mean a **conditional equation.** (A conditional equation is one that is true for a specific set of values, not for all values in the domain of the variable.)

Number of Solutions

The algebra needed to solve trigonometric equations includes the standard operations of collecting like terms, factoring, setting factors to zero, and solving for x. Recall that in algebra we normally have one, two, or just a few solutions, depending on the degree of the equation. However, with trigonometric equations there can be infinitely many solutions because of the periodic nature of trigonometric functions. For instance, the equation

$$\sin t = \frac{1}{2}$$

has two solutions, $t = \pi/6$ and $t = 5\pi/6$, in the interval $[0, 2\pi)$. However, since the sine has a period of 2π, we can add $2n\pi$ to either of these solutions to obtain infinitely many solutions of the form

$$t = \frac{\pi}{6} + 2n\pi \qquad \text{and} \qquad t = \frac{5\pi}{6} + 2n\pi$$

where n is an integer.

Remark: Instead of performing algebraic operations with just x as the variable, we will be making the same moves with *functions of x,* like $\sin x$, $\cos x$, $\tan x$, and so on. We start with a simple example of collecting like terms.

EXAMPLE 1
Collecting Like Terms

Find all solutions to the equation

$$\sin x + \sqrt{2} = -\sin x$$

Solution:
Collecting like terms and solving for $\sin x$, we obtain

$$\sin x + \sin x = -\sqrt{2}$$
$$2 \sin x = -\sqrt{2}$$
$$\sin x = \frac{-\sqrt{2}}{2}$$

Since $\sin x$ has a period of 2π, we first look for all solutions in the interval $[0, 2\pi)$:

$$x = \frac{5\pi}{4} \quad \text{and} \quad x = \frac{7\pi}{4}$$

Now we add $2n\pi$ to each of these solutions, and the complete solution set is

$$x = \frac{5\pi}{4} + 2n\pi \quad \text{and} \quad x = \frac{7\pi}{4} + 2n\pi$$

where n is any integer.

Remark: For convenience, when the solutions to trigonometric equations are to be in degrees, we will normally use variable θ. For real number solutions, or angles in radians, we will use variables x, y, or t.

EXAMPLE 2
Taking Square Roots

Solve the equation

$$3 \tan^2 x - 1 = 0$$

Solution:
Solving first for $\tan^2 x$, then taking square roots, we get

$$3 \tan^2 x = 1$$

$$\tan^2 x = \frac{1}{3}$$

$$\tan x = \pm \frac{1}{\sqrt{3}}$$

Since $\tan x$ has a period of π, we hunt for all solutions in the interval $[0, \pi)$:

$$x = \frac{\pi}{6} \quad \text{and} \quad x = \frac{5\pi}{6}$$

Now we add $n\pi$ to each, and the complete solution set is

$$x = \frac{\pi}{6} + n\pi \quad \text{and} \quad x = \frac{5\pi}{6} + n\pi$$

where n is any integer.

We added $2n\pi$ to form the general solution in Example 1 and $n\pi$ to form the general solution in Example 2, because the sine function (in Example 1) has a period of 2π and the tangent function (in Example 2) has a period of π. In general, we use the following procedure.

FORMING THE GENERAL SOLUTION

For equations of the form

$$\sin x = k \qquad \cos x = k \qquad \csc x = k \qquad \sec x = k$$

find all solutions in the interval $[0, 2\pi)$ and add $2n\pi$ to each to form the general solution.

For equations of the form

$$\tan x = k \qquad \cot x = k$$

find all solutions in the interval $[0, \pi)$ and add $n\pi$ to each to form the general solution.

In Examples 1 and 2, only one trigonometric function was involved. When two or more functions occur in the same problem, we collect all terms to one side and try to separate the functions by factoring. This may produce factors that yield no solutions, as illustrated in our next example.

EXAMPLE 3
Factoring

Solve the equation

$$\cot x \cos^2 x = 2 \cot x$$

Solution:

Collecting terms to one side and factoring, we get

$$\cot x \cos^2 x - 2 \cot x = 0$$
$$\cot x(\cos^2 x - 2) = 0$$

By setting each of these factors to zero, we obtain

$$\cot x = 0 \qquad \cos^2 x - 2 = 0$$
$$x = \frac{\pi}{2} \qquad \cos^2 x = 2$$
$$\cos x = \pm\sqrt{2}$$

No solution is obtained from $\cos x = \pm\sqrt{2}$, since $\pm\sqrt{2}$ is outside the range of the cosine function. Therefore, the complete solution set is obtained from the factor involving the cotangent. By adding multiples of π to $x = \pi/2$, we have

$$x = \frac{\pi}{2} \pm n\pi$$

where n is any integer.

Quadratic Forms

It is important to recognize trigonometric equations in quadratic form.

$$\tan^4 x - 2 \tan^2 x + 1 = 0 \quad \Longrightarrow \quad (\tan^2 x)^2 - 2(\tan^2 x) + 1 = 0$$
$$2 \sin^2 x - \sin x - 1 = 0 \quad \Longrightarrow \quad 2(\sin x)^2 - (\sin x) - 1 = 0$$
$$\sec^2 x - 3 \sec x - 2 = 0 \quad \Longrightarrow \quad (\sec x)^2 - 3(\sec x) - 2 = 0$$

While the first two equations are factorable (see Example 4), the third is not, so we make use of the *quadratic formula,* as shown in Example 7. Make sure that an equation in quadratic form involves *a single trigonometric function.* For instance, we need to rewrite the equation

$$2 \sin^2 x + 3 \cos x - 3 = 0$$

in terms of a single function to convert to quadratic form:

$$2(1 - \cos^2 x) + 3 \cos x - 3 = 0$$
$$2 \cos^2 x - 3 \cos x + 1 = 0$$

EXAMPLE 4
Factoring a Quadratic Form

Solve the equation

$$2 \sin^2 x - \sin x - 1 = 0$$

in the interval $[0, 2\pi)$.

Solution:
Factoring the given quadratic equation, we get

$$2 \sin^2 x - \sin x - 1 = 0$$
$$(2 \sin x + 1)(\sin x - 1) = 0$$

Setting each factor to zero, we obtain

$$2 \sin x + 1 = 0 \qquad\qquad \sin x - 1 = 0$$

$$\sin x = -\frac{1}{2} \qquad\qquad \sin x = 1$$

$$x = \frac{7\pi}{6}, \frac{11\pi}{6} \qquad\qquad x = \frac{\pi}{2}$$

Sometimes we must square both sides of an equation to obtain a quadratic, as demonstrated in the next example. Since this procedure can introduce extraneous solutions, you should check any solutions in the original equation to see if they are valid or extraneous.

EXAMPLE 5
Squaring and Converting to Quadratic Form

Find all solutions to

$$\sin x = \cos x + 1$$

in the interval $[0, 2\pi)$.

Solution:
Since the given equation is not factorable, we try squaring both sides to see if we can get a quadratic form. Squaring yields

$$\sin^2 x = \cos^2 x + 2 \cos x + 1 \qquad \textit{Square both sides}$$

$$1 - \cos^2 x = \cos^2 x + 2 \cos x + 1 \qquad \textit{Fund. identity}$$

$$0 = 2 \cos^2 x + 2 \cos x \qquad \textit{Collect terms}$$

$$0 = 2 \cos x(\cos x + 1) \qquad \textit{Factor}$$

Setting each factor to zero, we get

$$2 \cos x = 0 \qquad\qquad \cos x + 1 = 0$$

$$\cos x = 0 \qquad\qquad \cos x = -1$$

$$x = \frac{\pi}{2}, \frac{3\pi}{2} \qquad\qquad x = \pi$$

Since we squared the original equation, we check for extraneous solutions.

Check:
For $x = \pi/2$,

$$\sin \frac{\pi}{2} \stackrel{?}{=} \cos \frac{\pi}{2} + 1$$

$$1 = 0 + 1 \qquad\qquad \textit{Solution checks}$$

For $x = 3\pi/2$,

$$\sin \frac{3\pi}{2} \stackrel{?}{=} \cos \frac{3\pi}{2} + 1$$

$$-1 \neq 0 + 1 \qquad\qquad \textit{Extraneous solution}$$

For $x = \pi$,

$$\sin \pi \stackrel{?}{=} \cos \pi + 1$$

$$0 = -1 + 1 \qquad\qquad \textit{Solution checks}$$

Using Inverse Functions and a Calculator

Up to this point, we have carefully chosen our equations so that the solutions came out as special values (or angles), like 0, $\pm\pi/6$, $\pm\pi/4$, $\pm\pi/3$, $\pm\pi/2$, and so on. In the remaining examples, we use inverse functions and a calculator to solve more general trigonometric equations.

EXAMPLE 6
Using Inverse Functions

Find all solutions to

$$\sec^2 x - 2 \tan x = 4$$

in the interval $[0, \pi)$.

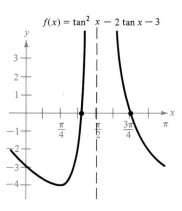

$f(x) = \tan^2 x - 2 \tan x - 3$

FIGURE 6.5

Solution:

Collecting terms and substituting $(1 + \tan^2 x)$ for $\sec^2 x$, we have

$$1 + \tan^2 x - 2 \tan x - 4 = 0$$

$$\tan^2 x - 2 \tan x - 3 = 0$$

$$(\tan x - 3)(\tan x + 1) = 0$$

Setting each factor equal to zero, we obtain

$$\tan x = 3 \qquad\qquad \tan x = -1$$

$$x = \arctan 3 \qquad\qquad x = \frac{3\pi}{4}$$

$$x \approx 1.2490 \qquad\qquad x \approx 2.3562$$

Thus, we have two solutions in the interval $[0, \pi)$, as shown in Figure 6.5.

Remark: If you have access to a computer with plotting capabilities, you might want to check your solutions graphically. For instance, the graph of the function

$$f(x) = \tan^2 x - 2 \tan x - 3$$

shown in Figure 6.5, has the two x-intercepts

$$x \approx 1.2490 \qquad \text{and} \qquad x \approx 2.3562$$

in the interval $[0, \pi)$.

When a calculator is used for $\arcsin x$, $\arccos x$, and $\arctan x$, the displayed solution may need to be adjusted to obtain solutions in the desired interval, as shown in the next example.

EXAMPLE 7
Using the Quadratic Formula

Solve the equation

$$\sec^2 t - 3 \sec t - 2 = 0$$

in the interval $[0, 2\pi)$.

Solution:

Since this equation is not factorable, we use the quadratic formula.

$$\sec t = \frac{-(-3) \pm \sqrt{(-3)^2 - 4(1)(-2)}}{2(1)} = \frac{3 \pm \sqrt{17}}{2}$$

$$\sec t \approx 3.5616 \text{ or } -0.56155$$

Since $|\sec t| \geq 1$, $\sec t = -0.56155$ has no solution. To solve the equation $\sec t = 3.5616$, we use a calculator and the inverse cosine function as follows:

$$t = \operatorname{arcsec}(3.5616) = \arccos\left(\frac{1}{3.5616}\right) \approx 1.2862$$

Finally, since the range of the inverse cosine is $[0, \pi)$, we must consider whether there are solutions between π and 2π. Using the formula

$$\cos t = \cos(2\pi - t)$$

we have

$$2\pi - t = 2\pi - 1.2862 \approx 4.9970$$

as the second solution.

When trigonometric equations involve functions of two different angles or functions of multiple angles, additional steps are needed. With two different angles, we try to convert to functions of one angle only (see Example 8). With multiple angles, extra care is needed to determine all possible solutions in a specified interval (see Example 9).

EXAMPLE 8
Functions of Different Angles

Find all solutions to

$$\sin 2\theta \cos \theta - \sin \theta = 0$$

in the interval $[0, 360°)$.

Solution:
Using a double-angle formula, we have

$$(2 \sin \theta \cos \theta)\cos \theta - \sin \theta = 0$$
$$\sin \theta(2 \cos^2 \theta - 1) = 0$$

Setting each factor to zero, we get

$$\sin \theta = 0 \qquad\qquad 2 \cos^2 \theta - 1 = 0$$
$$\theta = 0°, 180° \qquad\qquad \cos \theta = \pm\frac{1}{\sqrt{2}}$$
$$\theta = 45°, 135°, 225°, 315°$$

EXAMPLE 9
Functions of Multiple Angles

Find all values for t in $[0, 2\pi)$ that satisfy the equation

$$2 \cos 3t = 1$$

Solution:
Solving for $\cos 3t$, we have

$$2 \cos 3t = 1$$
$$\cos 3t = \frac{1}{2}$$

From this result we know that

$$3t = \frac{\pi}{3} + 2n\pi$$

$$t = \frac{\pi}{9} + \frac{2n\pi}{3}$$

or

$$3t = \frac{5\pi}{3} + 2n\pi$$

$$t = \frac{5\pi}{9} + \frac{2n\pi}{3}$$

Finally, to find all t in $[0, 2\pi)$, we begin with $n = 0$ and increment by 1 until we obtain t-values outside the desired interval. In this case, $n = 0, 1,$ and 2 work, and in the interval $[0, 2\pi)$ we have

$$t = \frac{\pi}{9}, \quad \frac{5\pi}{9}, \quad \frac{7\pi}{9}, \quad \frac{11\pi}{9}, \quad \frac{13\pi}{9}, \quad \text{or} \quad \frac{17\pi}{9}$$

Section Exercises 6.6

In Exercises 1–44, find all solutions in the interval $[0, 2\pi)$. Do not use a calculator or tables.

1. $2 \sin^2 x = 1$
2. $\tan^2 x = 3$
3. $3 \sec^2 x - 4 = 0$
4. $\csc^2 x - 2 = 0$
5. $\tan x(\tan x - 1) = 0$
6. $\sin^2 x = 3 \cos^2 x$
7. $2 \cos(x - \pi) = -\sqrt{3}$
8. $4 \sin^2 x - 3 = 0$
9. $\cos x(2 \cos x + 1) = 0$
10. $(3 \tan^2 x - 1)(\tan^2 x - 3) = 0$
11. $\sec x \csc x - 2 \csc x = 0$
12. $\sec^2 x - \sec x - 2 = 0$
13. $2 \sin^2 x + 3 \sin x + 1 = 0$
14. $3 \tan^3 x - \tan x = 0$
15. $\cos^3 x = \cos x$
16. $4 \sin^3 x + 2 \sin^2 x - 2 \sin x - 1 = 0$
17. $2 \sin^2 x = 2 + \cos x$
18. $\csc^2 x = (1 + \sqrt{3}) - (1 - \sqrt{3})\cot x$
19. $\sec^2 x = (1 + \sqrt{3}) - (1 - \sqrt{3})\tan x$
20. $2 \sec^2 x + \tan^2 x - 3 = 0$
21. $2 \sin x + \csc x = 0$
22. $\csc x + \cot x = 1$
23. $\sin 2x = -\dfrac{\sqrt{3}}{2}$
24. $\tan 3x = 1$
25. $\cos\left(\dfrac{x}{2}\right) = \dfrac{\sqrt{2}}{2}$
26. $\sec 4x = 2$
27. $4 \sin x \cos x = 1$
28. $\sin 2x \sin x = \cos x$
29. $\cos 2x = \cos x$
30. $\csc 2x + \cot 2x = 0$
31. $\csc 2x - \cot 2x = 1$
32. $\cos 3x - \cos x = 0$
33. $\sin 4x + 2 \sin 2x = 0$

34. $\tan\left(x + \dfrac{\pi}{4}\right) + \tan\left(x - \dfrac{\pi}{4}\right) = 2$
35. $\tan\left(x + \dfrac{\pi}{4}\right) - \tan\left(x - \dfrac{\pi}{4}\right) = 4$
36. $(\sin 2x + \cos 2x)^2 = 1$
37. $\cos 2x - \cos 6x = 0$
38. $\sin 6x - \sin 4x = 0$
39. $\sin 6x + \sin 2x = 0$
40. $\cos 7x + \cos 3x = 0$
41. $\dfrac{\cos 2x}{\sin 3x - \sin x} = 1$
42. $\sin^2 3x - \sin^2 x = 0$
43. $\sin 2x + \sin 4x + \sin 6x = 0$
44. $1 - \cos 2x + \cos 4x - \cos 6x = 0$

In Exercises 45–54, use a calculator or tables to find all solutions in the interval $[0, 2\pi)$.

45. $2 \tan^2 x + 7 \tan x - 15 = 0$
46. $3 \sec^2 x - 5 \sec x - 12 = 0$
47. $12 \sin^2 x - 13 \sin x + 3 = 0$
48. $4 \cot^2 x - 4 \cot x - 3 = 0$
49. $6 \cos^2 x - 13 \cos x + 6 = 0$
50. $\sin^2 x + \sin x - 20 = 0$
51. $\tan^2 x - 8 \tan x + 13 = 0$
52. $\sec^2 x - 6 \sec x - 2 = 0$
53. $\sin^2 x + 2 \sin x - 1 = 0$
54. $4 \cos^2 x - 4 \cos x - 1 = 0$
55. The function

$$f(x) = \sin x + \cos x$$

has maximum or minimum values when

$$\cos x - \sin x = 0$$

Find all solutions of this equation in the interval $[0, 2\pi)$, and sketch a graph of the function f.

56. The function

$$f(x) = 2 \sin x + \cos 2x$$

has maximum or minimum values when

$$2 \cos x - 2 \sin 2x = 0$$

Find all solutions of this equation in the interval $[0, 2\pi)$, and sketch a graph of the function f.

57. A 5-pound weight is oscillating on the end of a spring, and the position of the weight relative to the point of equilibrium is given by

$$h(t) = \tfrac{1}{4}(\cos 8t - 3 \sin 8t)$$

where t is the time in seconds. Find the times when the weight is at the point of equilibrium $[h(t) = 0]$ for $0 \le t \le 1$.

58. The monthly sales in thousands of units of a seasonal product is approximated by

$$S = 74.50 + 43.75 \sin \frac{\pi t}{6}$$

where t is the time in months with $t = 1$ corresponding to January. Determine the months when sales exceed 100,000 units.

59. A batted baseball leaves the bat at an angle of θ with the horizontal, with a velocity of $v_0 = 100$ feet per second, and is caught by an outfielder 300 feet from home plate. Find θ if the range r of a projectile is given by

$$r = \tfrac{1}{32}v_0{}^2 \sin 2\theta$$

60. A gun with a muzzle velocity of 1200 feet per second is pointed at a target 1000 yards away. Neglecting air resistance, what should be the minimum angle of elevation of the gun? (Use the formula for range given in Exercise 59.)

Review Exercises / Chapter 6

In Exercises 1–10, simplify the given expression.

1. $\dfrac{1}{\cot^2 x + 1}$

2. $\dfrac{\sin 2\alpha}{\cos^2 \alpha - \sin^2 \alpha}$

3. $\dfrac{\sin^2 \alpha - \cos^2 \alpha}{\sin^2 \alpha - \sin \alpha \cos \alpha}$

4. $\dfrac{\sin^3 \beta + \cos^3 \beta}{\sin \beta + \cos \beta}$

5. $\cos^2 \beta + \cos^2 \beta \tan^2 \beta$

6. $\dfrac{\sin \theta}{1 + \cos \theta} + \dfrac{1 + \cos \theta}{\sin \theta}$

7. $\tan^2 \theta(\csc^2 \theta - 1)$

8. $\dfrac{2 \tan(x + 1)}{1 - \tan^2(x + 1)}$

9. $1 - 4 \sin^2 x \cos^2 x$

10. $\sqrt{\dfrac{1 - \cos^2 x}{1 + \cos x}}$

In Exercises 11–30, verify the given identity.

11. $\tan x(1 - \sin^2 x) = \tfrac{1}{2} \sin 2x$
12. $\cos^3 x \sin^2 x = (\sin^2 x - \sin^4 x)\cos x$
13. $\sin^5 x \cos^2 x = (\cos^2 x - 2 \cos^4 x + \cos^6 x)\sin x$
14. $\sin^4 2x = \tfrac{1}{8}(\cos 8x - 4 \cos 4x + 3)$
15. $\sin^2 x \cos^4 x = \tfrac{1}{16}(1 - \cos 4x + 2 \sin^2 2x \cos 2x)$
16. $\sin 3x \cos 2x = \tfrac{1}{2}(\sin 5x + \sin x)$
17. $\sin 3\theta \sin \theta = \tfrac{1}{2}(\cos 2\theta - \cos 4\theta)$

18. $\sqrt{1 - \cos x} = \dfrac{|\sin x|}{\sqrt{1 + \cos x}}$

19. $\sqrt{\dfrac{1 - \sin \theta}{1 + \sin \theta}} = \dfrac{1 - \sin \theta}{|\cos \theta|}$

20. $\sin 4x = 8 \cos^3 x \sin x - 4 \cos x \sin x$
21. $\cos 3x = 4 \cos^3 x - 3 \cos x$
22. $\cos 4x = 8 \cos^4 x - 8 \cos^2 x + 1$

23. $\sin\left(x - \dfrac{3\pi}{2}\right) = \cos x$

24. $\cos\left(x + \dfrac{\pi}{2}\right) = -\sin x$

25. $\dfrac{\sec x - 1}{\tan x} = \tan \dfrac{x}{2}$

26. $\dfrac{2 \cos 3x}{\sin 4x - \sin 2x} = \csc x$

27. $\dfrac{\cos 3x - \cos x}{\sin 3x - \sin x} = -\tan 2x$

28. $1 - \cos 2x = 2 \sin^2 x$
29. $\sin(\pi - x) = \sin x$

30. $\cot\left(\dfrac{\pi}{2} - x\right) = \tan x$

In Exercises 31–40, find all solutions of the given equation in the interval $[0, 2\pi)$.

31. $\sin x - \tan x = 0$

32. $\csc x - 2 \cot x = 0$

33. $\dfrac{1 + \sin x}{\cos x} + \dfrac{\cos x}{1 + \sin x} = 4$

34. $\cos x = \cos \dfrac{x}{2}$

35. $\sin 2x + \sqrt{2} \sin x = 0$

36. $\cos 4x - 7 \cos 2x = 8$

37. $\cos^2 x + \sin x = 1$

38. $\sin 4x - \sin 2x = 0$

39. $\tan^3 x - \tan^2 x + 3 \tan x - 3 = 0$

40. $\sin x + \sin 3x + \sin 5x = 0$

41. Express each of the following as a sum or difference:

(a) $\sin 3\alpha(\sin 2\alpha)$

(b) $\cos x^2 \sin 2x^2$

(c) $\cos \dfrac{x}{2} \cos \dfrac{x}{4}$

42. Express each of the following as a product:

(a) $\cos 3\theta + \cos 2\theta$

(b) $\cos \theta - \cos 4\theta$

(c) $\sin\left(x + \dfrac{\pi}{4}\right) - \sin\left(x - \dfrac{\pi}{4}\right)$

CHAPTER

7 | ADDITIONAL APPLICATIONS OF TRIGONOMETRY

Law of Sines   7.1

In Chapter 5 we studied the trigonometry of right triangles. In this and the next section, we will solve triangles that have no right angle. Such triangles are called **oblique triangles.** As standard notation, we label the vertices of a triangle as A, B, and C, and their opposite sides as a, b, and c.

To solve a triangle using trigonometry, we need to know the measure of at least one side and any two other parts, either two sides, two angles, or one angle and one side. This breaks down into four possible cases.

SOLVING AN OBLIQUE TRIANGLE

We can find the remaining three parts of an oblique triangle, given

1. Two angles and any side (AAS or ASA)
2. Two sides and an angle opposite one of them (SSA)
3. Three sides (SSS)
4. Two sides and their included angle (SAS)

The first two cases can be readily solved using what is called the **Law of Sines,** while the last two cases require the **Law of Cosines.** Actually, all cases can be solved by appropriately subdividing an oblique triangle into right triangles, and then using right triangle procedures. However, this can be quite cumbersome, so we develop two new solution techniques. We begin with the Law of Sines.

LAW OF SINES

If *ABC* is any oblique triangle with sides *a*, *b*, and *c*, then

$$\frac{a}{\sin A} = \frac{b}{\sin B} = \frac{c}{\sin C}$$

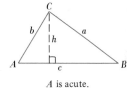 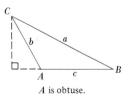

A is acute.   *A* is obtuse.

Oblique Triangles

FIGURE 7.1

Proof:

To prove the Law of Sines, it is convenient to construct the altitude *h* for each triangle shown in Figure 7.1. In either triangle, we have, by definition,

$$\sin A = \frac{h}{b} \quad \text{or} \quad h = b \sin A$$

$$\sin B = \frac{h}{a} \quad \text{or} \quad h = a \sin B$$

Consequently, by substitution for *h*, we have

$$a \sin B = b \sin A \quad \text{or} \quad \frac{a}{\sin A} = \frac{b}{\sin B}$$

Note that $\sin A \neq 0$ and $\sin B \neq 0$, since no angle of a triangle can have measure of 0° or 180°. By constructing an altitude from vertex *B* to extended side *AC*, we can establish by a similar argument that

$$\frac{a}{\sin A} = \frac{c}{\sin C}$$

Hence, by transitivity, the Law of Sines is established.

Rounding Rules

When using a calculator with the Law of Sines, remember to hold all intermediate calculations in the calculator without rounding. Then round the final result according to the following convention:

1. For sides, round to four significant digits.
2. For angles, round to the nearest tenth of a degree.

(We will assume all *given* sides and angles to have this same accuracy.)

EXAMPLE 1
Given Two Angles
and One Side—AAS

Given a triangle with $C = 72.3°$, $B = 28.7°$, and $b = 27.4$, find the remaining three parts.

Solution:
The third angle of the triangle is

$$A = 180° - B - C = 180° - 28.7° - 72.3° = 79.0°$$

To Find Side a

$$\frac{a}{\sin A} = \frac{b}{\sin B}$$

$$a = \frac{b}{\sin B}(\sin A)$$

$$= \frac{27.4}{\sin 28.7°}(\sin 79.0°)$$

$$a \approx 56.01$$

To Find Side c

$$\frac{c}{\sin C} = \frac{b}{\sin B}$$

$$c = \frac{b}{\sin B}(\sin C)$$

$$= \frac{27.4}{\sin 28.7°}(\sin 72.3°)$$

$$c \approx 54.36$$

Note that the ratio $b/\sin B$ occurs in both solutions. You can save time by storing this result for later multiplication.

An appropriate sketch is useful as a quick test for the feasibility of an answer. Remember that the longest side lies opposite the largest angle, and the shortest side opposite the smallest angle of a triangle. A sketch is especially important in applications like the following.

EXAMPLE 2
Given Two Angles
and One Side—ASA

A pole tilts *toward* the sun at an angle of 8° from vertical, and it casts a 22-foot shadow. The angle of elevation from the tip of the shadow to the top of the pole is 43°. How long is the pole?

Solution:
From Figure 7.2, we note that

$$c = 22', \qquad A = 43°, \qquad \text{and} \qquad B = 90° + 8° = 98°$$

Thus, the third angle is

$$C = 180° - A - B = 180° - 43° - 98° = 39°$$

To find side a (the length of the pole), we have

$$\frac{a}{\sin A} = \frac{c}{\sin C}$$

$$a = \frac{c}{\sin C}(\sin A) = \frac{22}{\sin 39°}(\sin 43°)$$

$$a \approx 23.84 \text{ feet}$$

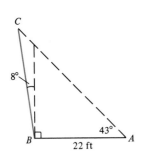

FIGURE 7.2

Remark: For practice, try reworking Example 2 for a pole that tilts *away from* the sun under the same conditions.

In Examples 1 and 2 we saw that any two angles (with sum $< 180°$) and any one side determine a unique triangle. However, if two sides and one opposite angle are given, three possible situations can arise: no such triangle exists, one such triangle exists, or two distinct triangles may satisfy these conditions. The possibilities in this *ambiguous* (SSA) case are summarized in Table 7.1.

TABLE 7.1

The Ambiguous Case (SSA) (Given: *a*, *b*, and *A*)

	A Is Acute				*A Is Obtuse*	
Sketch $(h = b \sin A)$	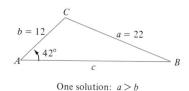					
Necessary Condition	$a < h$	$a = h$	$a > b$	$h < a < b$	$a \leq b$	$a > b$
Triangles Possible	None	One	One	Two	None	One

EXAMPLE 3
Determining the Number of Solutions in the Ambiguous Case

Determine how many triangles can satisfy each of the following conditions. Make a sketch in each case.

(a) $a = 22$, $b = 12$, $A = 42°$
(b) $a = 15$, $b = 25$, $A = 85°$
(c) $a = 15.2$, $b = 20$, $A = 110°$
(d) $a = 12$, $b = 31$, $A = 20.5°$

Solution:

(a) In this case, $a = 22$ is greater than $b = 12$; hence there is one solution, as indicated in Figure 7.3.

(b) In this case, $a = 15$ is less than $b = 25$; hence we must compare a with $h = b \sin A$.

$$h = b \sin A = 25(\sin 85°) \approx 24.90$$

Since $a < h$, we conclude that there is *no solution*. This conclusion is verified in Figure 7.4. Note that the Law of Sines would yield

$$\frac{15}{\sin 85°} = \frac{25}{\sin B}$$

$$\sin B = 25\left(\frac{\sin 85°}{15}\right) \approx 1.660 > 1$$

Since $|\sin B| \leq 1$, this further confirms our conclusion that there is no solution.

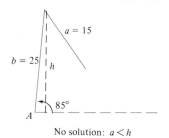

One solution: $a > b$

FIGURE 7.3

No solution: $a < h$

FIGURE 7.4

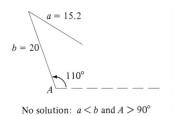

No solution: $a < b$ and $A > 90°$

FIGURE 7.5

EXAMPLE 4
Two-Solution Case

(c) Since A is obtuse, there are only two possibilities. And because $a = 15.2$ is less than $b = 20$, we conclude that there is *no* solution. (See Figure 7.5.)

(d) In this case,

$$a = 12$$
$$h = b \sin A = 31(\sin 20.5°) \approx 10.86$$

Now, since $a = 12$ lies between h and b, we have *two* possible solutions.

Find the two possible solutions for the conditions described in part (d) of Example 3.

Solution:
Here we have $a = 12$, $b = 31$, and $A = 20.5°$. By the Law of Sines, we obtain

$$\frac{a}{\sin A} = \frac{b}{\sin B}$$

$$\sin B = \frac{\sin A}{a}(b) = \frac{\sin 20.5°}{12}(31) \approx 0.9047$$

Now there are two angles between $0°$ and $180°$ whose sine is 0.9047:

$$B_1 \approx 64.8° \qquad \text{and} \qquad B_2 \approx 115.2°$$

The resulting triangles are shown in Figures 7.6 and 7.7.

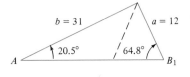

FIGURE 7.6

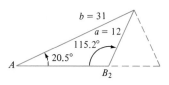

FIGURE 7.7

Case 1

$B_1 \approx 64.8°$

$C = 180° - 20.5° - 64.8°$
$\quad = 94.7°$

$c = \frac{a}{\sin A}(\sin C)$

$\quad = \frac{12}{\sin 20.5°}(\sin 94.7°)$

$\quad \approx 34.15$

Case 2

$B_2 \approx 115.2°$

$C = 180° - 20.5° - 115.2°$
$\quad = 44.3°$

$c = \frac{a}{\sin A}(\sin C)$

$\quad = \frac{12}{\sin 20.5°}(\sin 44.3°)$

$\quad \approx 23.93$

The procedure used to prove the Law of Sines also leads to a simple formula for the *area* of an oblique triangle. Referring to Figure 7.1, we note that each triangle has height

$$h = b \sin A$$

Consequently, the area of each triangle is given by

$$\text{Area} = \frac{1}{2}(\text{base})(\text{height})$$

$$= \frac{1}{2}(c)(b \sin A)$$

$$= \frac{1}{2}bc \sin A$$

By similar arguments, we can develop the formulas

$$\text{Area} = \frac{1}{2}ab \sin C = \frac{1}{2}ac \sin B$$

This leads to the following theorem.

AREA OF AN OBLIQUE TRIANGLE

> The area of any triangle is given by one-half the product of the lengths of two sides times the sine of their included angle.

EXAMPLE 5
Finding the Area of an Oblique Triangle

Find the area of a triangular lot having two sides of length 90 meters and 52 meters, with an included angle of 102°.

Solution:
Consider $a = 90$m, $b = 52$m, and angle $C = 102°$. Then, by the preceding theorem, the area of the triangle is

$$\text{Area} = \frac{1}{2}ab \sin C$$

$$= \frac{1}{2}(90)(52)(\sin 102°)$$

$$\approx 2289 \text{ square meters}$$

Section Exercises 7.1

In Exercises 1–16, find the remaining sides and angles of the triangle.

1.

2.
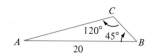

3.
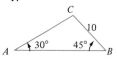

4.
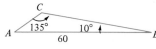

5.
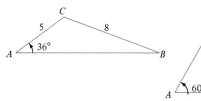

6.
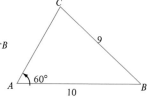

7. $A = 150°$, $C = 20°$, $a = 200$
8. $A = 24.3°$, $C = 54.6°$, $c = 2.68$
9. $A = 83° 20'$, $C = 54.6°$, $c = 18.1$
10. $A = 5° 40'$, $B = 8° 15'$, $b = 4.8$

11. $B = 15° 30'$, $a = 4.5$, $b = 6.8$
12. $C = 85° 20'$, $a = 35$, $c = 50$
13. $C = 145°$, $b = 4$, $c = 14$
14. $A = 100°$, $a = 125$, $c = 10$
15. $A = 110° 15'$, $a = 48$, $b = 16$
16. $B = 2° 45'$, $b = 6.2$, $c = 5.8$

In Exercises 17–22, determine the number of solutions to the triangle, and if solutions exist, find them.

17. $a = 4.5$, $b = 12.8$, $A = 58°$
18. $a = 11.4$, $b = 12.8$, $A = 58°$
19. $a = 4.5$, $b = 5$, $A = 58°$
20. $a = 42.4$, $b = 50$, $A = 58°$
21. $a = 125$, $b = 200$, $A = 110°$
22. $a = 125$, $b = 100$, $A = 110°$

23. Given a triangle with $A = 36°$ and $a = 5$, find a value of b so that the triangle has (a) one solution, (b) two solutions, and (c) no solution.
24. Given a triangle with $A = 60°$ and $a = 10$, find a value of b so that the triangle has (a) one solution, (b) two solutions, and (c) no solution.

In Exercises 25–30, find the area of the triangle.

25. $a = 4$, $b = 6$, $C = 120°$
26. $a = 105$, $c = 64$, $B = 72° 30'$
27. $b = 57$, $c = 85$, $A = 43° 45'$
28. $b = 4.5$, $c = 22$, $A = 5° 15'$
29. $a = 62$, $c = 20$, $B = 130°$
30. $a = 16$, $b = 20$, $C = 84° 30'$

31. Find the length d of the brace required to support the streetlight shown in Figure 7.8.

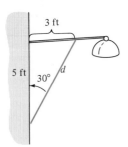

FIGURE 7.8

32. Because of prevailing winds, a tree grew so that it was leaning 6° from the vertical. Find the length of the tree if

the angle of elevation to the top of the tree is 22° 50' when one is a distance of 100 feet from the base of the tree, as shown in Figure 7.9.

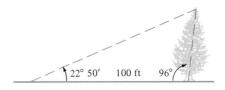

FIGURE 7.9

33. A bridge is to be built across a small lake from B to C, as shown in Figure 7.10. The bearing from B to C is S 41° W. From a point A, 100 yards from B, the bearings to B and C are S 74° E and S 28° E, respectively. Find the distance from B to C.

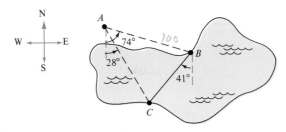

FIGURE 7.10

34. The angles of elevation to an airplane from two points A and B on level ground are 51° and 68°, respectively. If A and B are 6 miles apart and the airplane is between A and B in the same vertical plane, find the altitude of the airplane.
35. Two fire towers A and B are 18.5 miles apart. The bearing from A to B is N 65° E. A fire is spotted by the rangers in both towers, and its bearings from A and B are N 28° E and N 16.5° W, respectively. Find the distance of the fire from each tower, and then find the shortest distance from the fire to the line through A and B.
36. The angles of elevation to an airplane from points A and B on level ground are 51° and 68°, respectively. If A and B are 2.5 miles apart and the airplane is to the east of A and B in the same vertical plane, find the altitude of the airplane.
37. A boat is sailing due east parallel to the shoreline at a speed of 10 miles per hour. At a given time the bearing to a lighthouse is S 72° E, and 15 minutes later the bearing is S 66° E. Find the distance from the boat to the shoreline if the lighthouse is at the shoreline.
38. A family is traveling due west on a road that passes a famous landmark. At a given time the bearing to the landmark is N 62° W, and after the family travels 5 miles farther

the bearing is N 38° W. What is the closest the family will come to the landmark while on the road?

39. The following information about a triangular parcel of land is presented at a zoning hearing: "One side is 450 feet in length and another side is 120 feet in length. The angle opposite the shorter side is 30°." Could this information be correct?

40. A rescue vehicle is located near an apartment complex, and its emergency light is turning at the rate of 30 revolutions per minute. One-eighth second after illuminating the nearest point on the apartment complex, the lightbeam reaches a point 50 feet along the apartments. How far is the rescue truck from the apartment complex?

Law of Cosines   7.2

Two cases remain in our list of conditions needed to solve an oblique triangle—SSS and SAS. In Section 7.1, we said that the Law of Sines would not work in either of these cases. To see why, consider the three ratios given in the Law of Sines:

$$\frac{a}{\sin A} = \frac{b}{\sin B} = \frac{c}{\sin C}$$

To use this law you must be given a side, its opposite angle, and one other side or angle. Consequently, if you were given three sides (SSS), or two sides and their included angle (SAS), none of the above ratios would be complete. In such cases we rely on the **Law of Cosines.**

LAW OF COSINES

If *ABC* is any oblique triangle with sides *a*, *b*, and *c*, then the following equations are valid.

Standard Form   **Alternative Form**

$$a^2 = b^2 + c^2 - 2bc \cos A \qquad \cos A = \frac{b^2 + c^2 - a^2}{2bc}$$

$$b^2 = a^2 + c^2 - 2ac \cos B \qquad \cos B = \frac{a^2 + c^2 - b^2}{2ac}$$

$$c^2 = a^2 + b^2 - 2ab \cos C \qquad \cos C = \frac{a^2 + b^2 - c^2}{2ab}$$

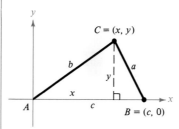 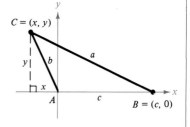

FIGURE 7.11

Proof:

In Figure 7.11, vertex B has coordinates $(c, 0)$. Furthermore, C has coordinates (x, y) where, by definition,

$$x = b \cos A$$

and

$$y = b \sin A$$

Now, since a is the distance from vertex C to vertex B, it follows that

$$a = \sqrt{(c - x)^2 + (0 - y)^2}$$
$$a^2 = (c - b \cos A)^2 + (-b \sin A)^2$$
$$= c^2 - 2bc \cos A + b^2 \cos^2 A + b^2 \sin^2 A$$
$$= b^2(\sin^2 A + \cos^2 A) + c^2 - 2bc \cos A$$

Using the identity $\sin^2 A + \cos^2 A = 1$, we then obtain

$$a^2 = b^2 + c^2 - 2bc \cos A$$

Similar arguments with angles B and C in standard position establish the other two equations.

EXAMPLE 1
Given Three Sides
of a Triangle—SSS

Solve the triangle ABC if $a = 8.65$, $b = 19.2$, and $c = 13.7$.

Solution:

In this case (SSS), it is a good idea first to find the angle opposite the longest side (side b in this case). Using the Law of Cosines, we find that

$$\cos B = \frac{a^2 + c^2 - b^2}{2ac}$$

$$= \frac{(8.65)^2 + (13.7)^2 - (19.2)^2}{2(8.65)(13.7)} \approx -0.4478$$

Since $\cos B < 0$, we know B is the *obtuse* angle

$$B \approx 116.6°$$

At this point we could use the Law of Cosines to find $\cos A$ and $\cos C$. However knowing that $B \approx 116.6°$, we can use the Law of Sines to obtain

$$\frac{b}{\sin B} = \frac{a}{\sin A}$$

$$\sin A = \frac{\sin B}{b}(a) \approx \frac{\sin 116.6°}{19.2}(8.65) \approx 0.40283$$

Since B is obtuse, we know that A must be acute, because a triangle can have at most one obtuse angle. Thus, $A \approx 23.8°$. Similarly,

$$\sin C \approx \frac{\sin 116.6°}{19.2}(13.7) \approx 0.63801$$

$$C \approx 39.6°$$

Remark: Do you see why it was wise to find the largest angle *first* in Example 1? Knowing the cosine of an angle, we can determine if the angle is acute or obtuse

$$\cos \theta < 0 \qquad \text{for } 90° < \theta < 180°$$

So in Example 1, once we found out that *B* was obtuse, we subsequently knew that angles *A* and *C* were both acute. Of course, if the largest angle is acute, the remaining two angles will be acute also.

EXAMPLE 2
Given Two Sides and
the Included Angle—SAS

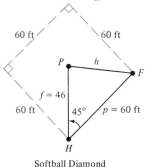

Softball Diamond

FIGURE 7.12

The pitcher's mound on a softball diamond is 46 feet from homeplate. If the distance between the bases is 60 feet, how far is the pitcher's mound from first base?

Solution:
Since a softball diamond has right angles at each base, we make the sketch shown in Figure 7.12. In triangle *HPF*, we have $H = 45°$ (line *HP* bisects the right angle at *H*), $f = 46$, and $p = 60$. Using the Law of Cosines on this SAS case, we have

$$h^2 = f^2 + p^2 - 2fp \cos H$$
$$= 46^2 + 60^2 - 2(46)(60) \cos 45° \approx 1812.8$$

Therefore, the approximate distance from the pitcher's mound to first base is

$$h \approx \sqrt{1812.8} \approx 42.58 \text{ feet}$$

EXAMPLE 3
Given Two Sides and
the Included Angle—SAS

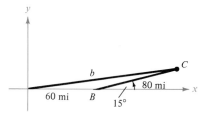

FIGURE 7.13

A ship travels 60 miles due east, then adjusts its course 15° northward. After traveling 80 miles in that direction, how far is the ship from its point of departure?

Solution:
Figure 7.13 shows a sketch of the situation. We have $c = 60$, $B = 180° - 15° = 165°$, and $a = 80$. Consequently, by the Law of Cosines, we find that

$$b^2 = a^2 + c^2 - 2ac \cos B$$
$$= 80^2 + 60^2 - 2(80)(60) \cos 165° \approx 19,273$$
$$b \approx \sqrt{19,273} \approx 138.8 \text{ miles}$$

The Law of Cosines can be used to establish the following formula for the area of any triangle, given its three sides.

HERON'S AREA FORMULA

> Given any triangle with sides of length a, b, and c, the area of the triangle is
>
> $$\text{Area} = \sqrt{s(s - a)(s - b)(s - c)}$$
>
> where $s = (a + b + c)/2$.

EXAMPLE 4
Using Heron's Area Formula

Find the area of the triangular region having sides of length $a = 47$ yards, $b = 58$ yards, and $c = 78.6$ yards.

Solution:
Since

$$s = \frac{1}{2}(a + b + c) = \frac{183.6}{2} = 91.8$$

Heron's Formula yields the area:

$$\begin{aligned}
\text{Area} &= \sqrt{s(s - a)(s - b)(s - c)} \\
&= \sqrt{91.8(44.8)(33.8)(13.2)} \\
&\approx 1{,}354.58 \text{ square yards}
\end{aligned}$$

Section Exercises 7.2

In Exercises 1–12, use the Law of Cosines to solve the given triangle.

1.
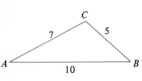

2.

3.
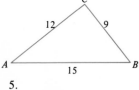

4.
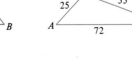

5.
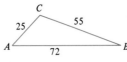

6.

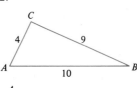

7. $a = 75.4$, $b = 52$, $c = 52$
8. $a = 1.42$, $b = 0.75$, $c = 1.25$
9. $B = 8°\ 45'$, $a = 25$, $c = 15$
10. $B = 75°\ 20'$, $a = 6.2$, $c = 9.5$
11. $C = 125°\ 40'$, $a = 32$, $b = 32$
12. $C = 15°$, $a = 6.25$, $b = 2.15$

In Exercises 13–18, use Heron's Formula to find the area of the triangle.

13. $a = 5$, $b = 7$, $c = 10$
14. $a = 2.5$, $b = 10.2$, $c = 9$
15. $a = 12$, $b = 15$, $c = 9$
16. $a = 75.4$, $b = 52$, $c = 52$
17. $a = 20$, $b = 20$, $c = 10$
18. $a = 4.25$, $b = 1.55$, $c = 3.00$

19. A boat race occurs along a triangular course marked by buoys *A*, *B*, and *C*. The race starts with the boats going 8000 feet in a northerly direction. The other two sides of the course lie to the east of the first side, and their lengths are 3500 feet and 6500 feet, as shown in Figure 7.14. Find the bearings for the last two legs of the race.

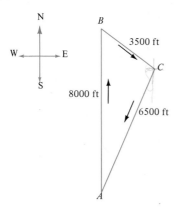

FIGURE 7.14

20. A plane flies 675 miles from *A* to *B* with a bearing of N 75° E. Then it flies 540 miles from *B* to *C* with a bearing of N 32° E, as shown in Figure 7.15. Find the straight line distance and bearing for the flight from *C* to *A*.

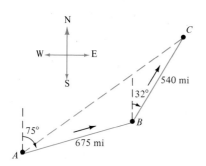

FIGURE 7.15

21. Two ships leave a port at 9 A.M. One travels at a bearing of N 53° W at 12 miles per hour, and the other at a bearing of S 67° W at 16 miles per hour. Approximate how far apart they are at noon of that day.

22. A triangular parcel of land has 375 feet of frontage, and the two other boundaries have lengths of 250 feet and 300 feet. What angle does the frontage make with the two other boundaries?

23. A vertical 100-foot tower is to be erected on the side of a hill that makes an 8° angle with the horizontal. Find the lengths of each of the two guy wires that will be anchored 75 feet directly above and directly below the base of the tower. (See Figure 7.16.)

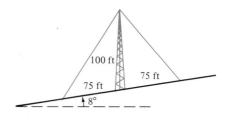

FIGURE 7.16

24. A baseball diamond is a square with sides 90 feet long and the pitcher's mound 60 feet from home plate.
    (a) How far is it from the pitcher's mound to third base?
    (b) When a runner is halfway from second to third, how far is he from the pitcher's mound?

25. On a certain map, Minneapolis is 6.5 inches due west of Albany, Phoenix is 8.5 inches from Minneapolis, and Phoenix is 14.5 inches from Albany. Find
    (a) the bearing of Minneapolis from Phoenix
    (b) the bearing of Albany from Phoenix

26. On a certain map, Orlando is 7 inches due south of Niagara Falls, Denver is 10.75 inches from Orlando, and Denver is 9.25 inches from Niagara Falls. Find
    (a) the bearing of Denver from Orlando
    (b) the bearing of Denver from Niagara Falls

27. Two trusses have lengths of 3 feet and 5 feet, respectively, and meet at an angle of 52°. Find the length of the span on the rafter supported by the trusses.

28. Given the triangle *ABC*, prove that if *R* is the radius of
    (a) the circumscribed circle, then

$$2R = \frac{a}{\sin A} = \frac{b}{\sin B} = \frac{c}{\sin C}$$

    (b) the inscribed circle, then

$$R = \sqrt{\frac{(s - a)(s - b)(s - c)}{s}}$$

where *s* is defined as in Heron's Formula.

In Exercises 29 and 30, use the results of Exercise 28.

29. Given the triangle with *a* = 25, *b* = 55, and *c* = 72, find the area of (a) the triangle, (b) the circumscribed circle, and (c) the inscribed circle.

30. Find the length of the largest circular track that can be built on a triangular piece of property whose sides are 200 feet, 250 feet, and 325 feet.

Vectors   7.3

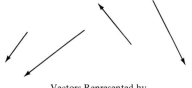

Vectors Represented by
Directed Line Segments
(Each specifies a magnitude and a direction.)

FIGURE 7.17

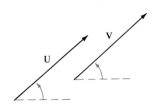

Equal vectors: U = V

FIGURE 7.18

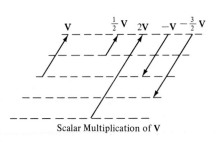

Scalar Multiplication of V

FIGURE 7.19

Many familiar quantities in geometry and physics can be characterized by a *single* real number scaled to an appropriate unit of measure. Some examples are circumference, area, volume, temperature, and time. Such quantities are called **scalar quantities,** and the real number associated with each is called a **scalar.**

Other physical and geometrical quantities cannot be completely characterized by a single real number, because they involve both magnitude and direction. For instance, to specify the movement of a ship, we use both *speed* (20 knots) and *direction* (W 30° N). Or, to characterize the force used to pull a sled by an attached rope, we need to specify the magnitude of the force (25 pounds) as well as the direction (40° above horizontal). Such quantities are called **vector quantities,** and the mathematical object used to describe each is called a **vector.**

Since the two fundamental characteristics of a vector are its *magnitude* and its *direction,* it is natural to represent it geometrically by a directed line segment (see Figure 7.17).

Two parallel line segments of equal length represent the same vector. Or, we can say that two vectors are **equal** if they have the same magnitude and direction (see Figure 7.18). In general, a vector can be represented by any one of a whole family of parallel line segments of equal length.

DEFINITION OF A VECTOR IN THE PLANE

> A **vector V** is the collection of all directed line segments in the plane having a given length and a given direction.

Remark: Though by definition a vector is an infinite collection of directed line segments, in practice we will work primarily with just a few *representatives* of the entire collection. To avoid the inconvenience of distinguishing between the entire family of directed line segments and just one of its representatives, we will use the term *vector* in both cases.

As a directed line segment, a vector has an **initial point** P and a **terminal point** Q. Thus, for the directed line segment $\overrightarrow{PQ}$ we use the vector notation $\overrightarrow{PQ}$. To denote general vectors for which no initial or terminal points are given, we use the boldface letters **U, V, W,** and **Z.**

Before we can make any meaningful applications of vectors, we need to define two basic vector operations. The first is **scalar multiplication,** in which the product of a scalar k and a vector **V** produce a new vector $k\mathbf{V}$. Some scalar multiples are shown in Figure 7.19. Note that if $k > 0$, then $k\mathbf{V}$ has the same direction as **V**, whereas if $k < 0$, then $k\mathbf{V}$ has the direction opposite that of **V**.

We can also generate new vectors by the operation of **vector addition.** To find the sum of two vectors **U** and **V**, we position them (without changing

To find **U** + **V** | move initial point of **V** to the terminal point of **U** | *or* | move initial point of **U** to the terminal point of **V**.

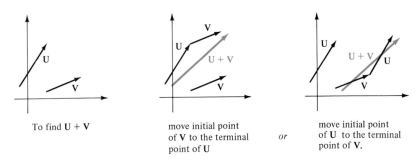

FIGURE 7.20

their magnitude or direction) so that the initial point of one coincides with the terminal point of the other. The resulting sum, **U** + **V** (or **V** + **U**), is the new vector formed by joining the initial point of the first vector with the terminal point of the second, as shown in Figure 7.20.

EXAMPLE 1
Adding Vectors Geometrically

FIGURE 7.21

Given the vectors **U** and **V**, as shown in Figure 7.21, make a sketch of the vectors **U** + **V** and **U** − **V**.

Solution:
Placing the initial point of **V** at the terminal point of **U**, we obtain **U** + **V**, as shown in Figure 7.22. Note in Figure 7.22 that we could just as easily have placed the initial point of **U** at the terminal point of **V**, to obtain the vector **V** + **U**.

Now, considering

$$\mathbf{U} - \mathbf{V} = \mathbf{U} + (-\mathbf{V})$$

we see that −**V** has the direction opposite that of **V**. Hence, **U** − **V** can be obtained by placing the initial point of −**V** at the terminal point of **U**, as shown in Figure 7.23.

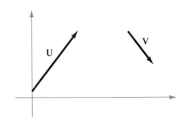

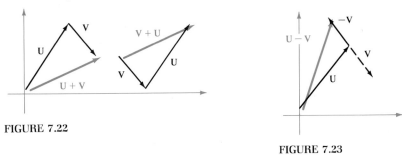

FIGURE 7.22

FIGURE 7.23

Unit Vectors **i** and **j**

FIGURE 7.24

So far our description of vectors has been primarily geometrical. To make meaningful applications of vectors, we also need an algebraic description. Such a description is based on the two **unit vectors i** and **j**, which have their initial points at the origin and their terminal points at (1, 0) and (0, 1), respectively (see Figure 7.24). By using a sum of scalar multiples of **i** and

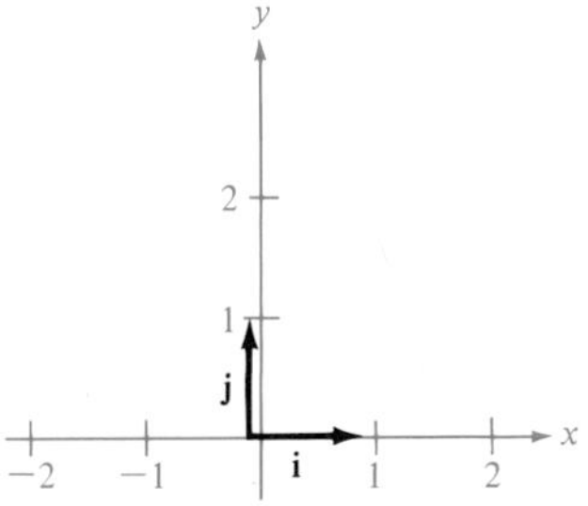

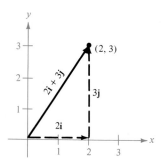

FIGURE 7.25

j, we can represent any vector in the plane. For instance, the vector $2\mathbf{i} + 3\mathbf{j}$ represents the vector with initial point $(0, 0)$ and terminal point $(2, 3)$, shown in Figure 7.25.

In general, a vector **V** can be written as a linear combination of vectors **i** and **j**, using the **component form**

$$\mathbf{V} = a\mathbf{i} + b\mathbf{j}$$

The procedure for obtaining this component form is incorporated in the following definition.

DEFINITION OF COMPONENT FORM OF A VECTOR

For any vector **V** represented by the directed line segment from point (x_1, y_1) to point (x_2, y_2), the **component form** of **V** is

$$\mathbf{V} = a\mathbf{i} + b\mathbf{j}$$

where

$$a = x_2 - x_1 \text{ is the \textbf{horizontal component} of } \mathbf{V}$$

and

$$b = y_2 - y_1 \text{ is the \textbf{vertical component} of } \mathbf{V}$$

Remark: Some texts represent vectors as ordered pairs, using the notation $\mathbf{V} = \langle a, b \rangle$ in place of $a\mathbf{i} + b\mathbf{j}$.

EXAMPLE 2
Finding Component
Forms of Vectors

Find the component form for each of the following vectors:

(a) Initial point $(0, 0)$; terminal point $(5, -3)$
(b) Initial point $(3, -7)$; terminal point $(-2, 5)$

Solution:

(a) By definition,

$$a = x_2 - x_1 = 5 - 0 = 5$$
$$b = y_2 - y_1 = -3 - 0 = -3$$

Hence, the component form is

$$a\mathbf{i} + b\mathbf{j} = 5\mathbf{i} - 3\mathbf{j}$$

(b) By definition,

$$a = x_2 - x_1 = -2 - 3 = -5$$
$$b = y_2 - y_1 = 5 - (-7) = 12$$

Therefore, the component form is

$$a\mathbf{i} + b\mathbf{j} = -5\mathbf{i} + 12\mathbf{j}$$

Note from the definition that if $(x_1, y_1) = (x_2, y_2)$ then the component form is

$$a\mathbf{i} + b\mathbf{j} = 0\mathbf{i} + 0\mathbf{j}$$

which is the **zero vector,** and we denote it simply as **0.**

To calculate the length of any vector in component form, we use the following rule.

DEFINITION OF MAGNITUDE OF A VECTOR

> The **magnitude (length)** of a vector $\mathbf{V} = a\mathbf{i} + b\mathbf{j}$ is denoted by $|\mathbf{V}|$ and is given by
>
> $$|\mathbf{V}| = \sqrt{a^2 + b^2}$$

A **unit vector** is a vector of magnitude 1. Clearly $\mathbf{i}$ and $\mathbf{j}$ are unit vectors, since

$$\mathbf{i} = 1\mathbf{i} + 0\mathbf{j} \quad \text{and} \quad |\mathbf{i}| = \sqrt{1^2 + 0^2} = 1$$
$$\mathbf{j} = 0\mathbf{i} + 1\mathbf{j} \quad \text{and} \quad |\mathbf{j}| = \sqrt{0^2 + 1^2} = 1$$

Furthermore,

$$\mathbf{V} = \frac{3}{5}\mathbf{i} - \frac{4}{5}\mathbf{j}$$

is a unit vector, since

$$|\mathbf{V}| = \sqrt{\frac{9}{25} + \frac{16}{25}} = \sqrt{\frac{25}{25}} = 1$$

EXAMPLE 3
Finding Component Form and Magnitude

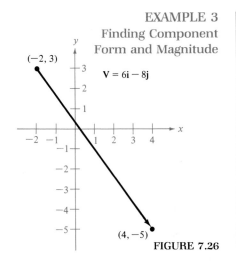

$\mathbf{V} = 6\mathbf{i} - 8\mathbf{j}$

$(-2, 3)$

$(4, -5)$

FIGURE 7.26

Write the component form of a vector $\mathbf{V}$ having initial point $(-2, 3)$ and terminal point $(4, -5)$. What is its length?

Solution:
By definition,

$$\mathbf{V} = (x_2 - x_1)\mathbf{i} + (y_2 - y_1)\mathbf{j}$$
$$= (4 + 2)\mathbf{i} + (-5 - 3)\mathbf{j} = 6\mathbf{i} - 8\mathbf{j}$$

Therefore, its length is

$$|\mathbf{V}| = \sqrt{6^2 + (-8)^2}$$
$$= \sqrt{36 + 64}$$
$$= \sqrt{100} = 10$$

(See Figure 7.26.)

The component form of a vector lends itself well to descriptions of the vector operations and to the verification of the following properties. Proofs are left as exercises.

VECTOR OPERATIONS AND PROPERTIES

Operations: Given $U = a_1\mathbf{i} + b_1\mathbf{j}$ and $V = a_2\mathbf{i} + b_2\mathbf{j}$,

1. $k\mathbf{U} = ka_1\mathbf{i} + kb_1\mathbf{j}$ *Scalar multiplication*
2. $\mathbf{U} + \mathbf{V} = (a_1 + a_2)\mathbf{i} + (b_1 + b_2)\mathbf{j}$ *Vector addition*
3. $\mathbf{U} - \mathbf{V} = (a_1 - a_2)\mathbf{i} + (b_1 - b_2)\mathbf{j}$ *Vector subtraction*

Properties: Given vectors **U**, **V**, and **W** and scalars a and b,

1. $\mathbf{U} + \mathbf{V} = \mathbf{V} + \mathbf{U}$
2. $\mathbf{U} + (\mathbf{V} + \mathbf{W}) = (\mathbf{U} + \mathbf{V}) + \mathbf{W}$
3. $(ab)\mathbf{V} = a(b\mathbf{V})$
4. $a(\mathbf{U} + \mathbf{V}) = a\mathbf{U} + a\mathbf{V}$
5. $\mathbf{V} + \mathbf{0} = \mathbf{V}$
6. $\mathbf{V} + (-\mathbf{V}) = \mathbf{0}$
7. $|a\mathbf{V}| = |a|\,|\mathbf{V}|$

EXAMPLE 4
Vector Operations

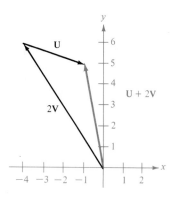

FIGURE 7.27

Given the vectors

$$\mathbf{U} = 3\mathbf{i} - \mathbf{j} \quad \text{and} \quad \mathbf{V} = -2\mathbf{i} + 3\mathbf{j}$$

find the component forms of $\mathbf{U} - \mathbf{V}$, $2\mathbf{V}$, and $\mathbf{U} + 2\mathbf{V}$. Make a sketch of $\mathbf{U} + 2\mathbf{V}$ and find its magnitude.

Solution:
By definition,

$$\mathbf{U} - \mathbf{V} = [3 - (-2)]\mathbf{i} + (-1 - 3)\mathbf{j} = 5\mathbf{i} - 4\mathbf{j}$$
$$2\mathbf{V} = 2(-2)\mathbf{i} + 2(3)\mathbf{j} = -4\mathbf{i} + 6\mathbf{j}$$
$$\mathbf{U} + 2\mathbf{V} = (3\mathbf{i} - \mathbf{j}) + (-4\mathbf{i} + 6\mathbf{j})$$
$$= (3 - 4)\mathbf{i} + (-1 + 6)\mathbf{j} = -\mathbf{i} + 5\mathbf{j}$$

A sketch of $\mathbf{U} + 2\mathbf{V}$ is shown in Figure 7.27. The magnitude of $\mathbf{U} + 2\mathbf{V}$ is

$$|\mathbf{U} + 2\mathbf{V}| = |-\mathbf{i} + 5\mathbf{j}| = \sqrt{(-1)^2 + (5)^2} = \sqrt{26}$$

So far we have been determining the component form of a vector, given its initial and terminal points. Frequently, vectors are described in terms of their magnitude r and a **direction angle** θ, which is the positive angle from the positive x-axis to the vector. With this information, the component form is determined by the following rule.

COMPONENT FORM, GIVEN MAGNITUDE AND DIRECTION

For any vector **V** with magnitude r and direction angle θ, the component form is determined by

$$\mathbf{V} = a\mathbf{i} + b\mathbf{j} = (r\cos\theta)\mathbf{i} + (r\sin\theta)\mathbf{j}$$

Moreover,

$$a^2 + b^2 = r^2 \quad \text{and} \quad \tan\theta = b/a$$

EXAMPLE 5
Finding Component Form,
Given Magnitude and Direction

Find the component form of the vector that represents the velocity of an airplane descending at a speed of 100 miles per hour at an angle 30° below horizontal.

Solution:
The velocity vector **V** has magnitude $r = 100$ and direction angle of 210° (see Figure 7.28). Hence, the component form of **V** is

$$\mathbf{V} = (r\cos\theta)\mathbf{i} + (r\sin\theta)\mathbf{j}$$
$$= (100\cos 210°)\mathbf{i} + (100\sin 210°)\mathbf{j}$$
$$= 100\left(\frac{-\sqrt{3}}{2}\right)\mathbf{i} + 100\left(\frac{-1}{2}\right)\mathbf{j}$$
$$= -50\sqrt{3}\,\mathbf{i} - 50\mathbf{j}$$

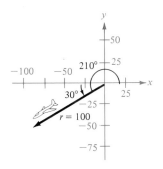

FIGURE 7.28

(You should check to see that $|\mathbf{V}| = 100$.)

EXAMPLE 6
Finding Magnitude and Direction

Find the magnitude r and direction angle θ for a vector **V** with initial point $(0, 0)$ and terminal point $(-4, 7)$.

Solution:
The component form is

$$\mathbf{V} = a\mathbf{i} + b\mathbf{j} = -4\mathbf{i} + 7\mathbf{j}$$

The magnitude is

$$r = |\mathbf{V}| = \sqrt{16 + 49} = \sqrt{65}$$

The direction angle θ is such that

$$\tan\theta = \frac{b}{a} = \frac{7}{-4}$$
$$\theta = \arctan\left(-\frac{7}{4}\right) \approx 119.7°$$

Note that θ must lie in the same quadrant as the point $(-4, 7)$.

Many applications of vectors involve the use of triangles and trigonometry in their solutions. Study the methods used in the next two examples.

EXAMPLE 7
Vector Application:
Similar Triangles

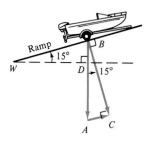

FIGURE 7.29

A force of 600 pounds is required to pull a boat and trailer up a ramp inclined at 15° from horizontal. Find the combined weight of the boat and trailer. (Assume no friction is involved.)

Solution:
Based on Figure 7.29, we make the following observations:

$$|\overrightarrow{BA}| = \text{force of gravity} = \text{weight of boat and trailer}$$

$$|\overrightarrow{BC}| = \text{force against ramp}$$

$$|\overrightarrow{AC}| = \text{force required to move boat up ramp} = 600 \text{ pounds}$$

By construction, triangles *WBD* and *ABC* are similar. Hence, angle *ABC* is 15°. Therefore, in triangle *ABC* we have

$$\sin 15° = \frac{|\overrightarrow{AC}|}{|\overrightarrow{BA}|} = \frac{600}{|\overrightarrow{BA}|}$$

$$|\overrightarrow{BA}| = \frac{600}{\sin 15°} \approx 2318$$

Consequently, the combined weight is approximately 2318 pounds.

EXAMPLE 8
Vector Application:
Law of Cosines

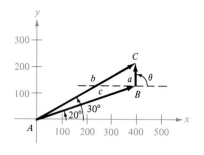

FIGURE 7.30

An airplane is flying in the direction N 70° E with an airspeed of 400 miles per hour. Because of wind force, its groundspeed and direction are 450 miles per hour and N 60° E, respectively. Find the direction and speed of the wind, and write the vector form of the velocity of the wind. (See Figure 7.30.)

Solution:
From Figure 7.30 we note the following about triangle *ABC*:

$$A = 10°$$
$$b = \text{groundspeed} = 450$$
$$c = \text{airspeed} = 400$$
$$B + \theta = 180° + 20° = 200°$$

Since $a = |\overrightarrow{BC}| = \text{speed of wind}$, we have, by the Law of Cosines,

$$a^2 = b^2 + c^2 - 2bc \cos A$$
$$= 450^2 + 400^2 - 2(450)(400) \cos 10°$$
$$a^2 \approx 7969.12$$
$$a \approx 89.27 = \text{wind speed in miles per hour}$$

Now, using the Law of Sines, we have

$$\frac{450}{\sin B} = \frac{89.27}{\sin 10°}$$

$$\sin B = \frac{\sin 10°}{89.27}(450) \approx 0.87533$$

$$B \approx 180° - 61.1° = 118.9°$$

Consequently, the direction angle θ is

$$\theta = 200° - B = 81.1°$$

Finally, the component form of the wind velocity is

$$\mathbf{W} = (r \cos \theta)\mathbf{i} + (r \sin \theta)\mathbf{j}$$
$$= 89.27(\cos 81.1°)\mathbf{i} + 89.27(\sin 81.1)\mathbf{j}$$
$$= 13.81\mathbf{i} + 88.20\mathbf{j}$$

Section Exercises 7.3

In Exercises 1–6, find the component form and the magnitude of the vector **V**.

1.

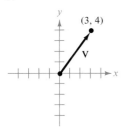

2.

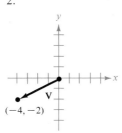

3.

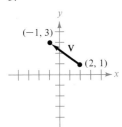

4.

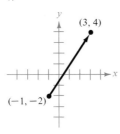

5.

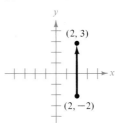

6.

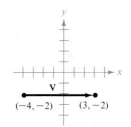

In Exercises 7–16, sketch **V** and find its component form. (Assume that angles are measured counterclockwise from the *x*-axis to the vector.)

7. **V** is a horizontal vector (pointing to the right) of length 3.
8. **V** is a unit vector making an angle of 45° with the positive *x*-axis.
9. **V** is a unit vector making an angle of 150° with the positive *x*-axis.
10. **V** is a vector of magnitude $\frac{5}{2}$ making an angle of 45° with the positive *x*-axis.
11. **V** is a vector of magnitude $3\sqrt{2}$ making an angle of 150° with the positive *x*-axis.

12. **V** is a unit vector in the direction of the vector **W** = **i** + 3**j**.

13. **V** is the sum of the vectors **V**$_1$ = 2**i** + **j** and **V**$_2$ = 3**i** + 5**j**.

14. **V** is −5**W**, where **W** = −**i** + 3**j**.

15. **V** is 3**V**$_1$ − 2**V**$_2$, where **V**$_1$ = 2**i** + **j** and **V**$_2$ = 3**i** + 5**j**.

16. **V** is a vertical vector (pointing upward) of magnitude 8.

In Exercises 17–22, find the component form of **V**, and illustrate the indicated vector operations geometrically, where **U** = 2**i** − **j** and **W** = **i** + 2**j**.

17. **V** = $\frac{3}{2}$**U**   18. **V** = **U** + **W**

19. **V** = **U** + 2**W**   20. **V** = −**U** + **W**

21. **V** = $\frac{1}{2}$(3**U** + **W**)   22. **V** = **U** − 2**W**

In Exercises 23–26, find a unit vector in the direction of the given vector.

23. **V** = 4**i** − 3**j**

24. **V** = **i** + **j**

25. **V** = 2**j**

26. **V** = **i** − 2**j**

In Exercises 27–34, use the Law of Cosines to find the angle θ between the given vectors. (Assume $\theta \leq 180°$.)

27. **V** = **i** + **j**, **W** = 2(**i** − **j**)

28. **V** = 3**i** + **j**, **W** = 2**i** − **j**

29. **V** = **i** + **j**, **W** = 3**i** − **j**

30. **V** = **i** + 2**j**, **W** = 2**i** − **j**

31. **V** = 2**i** − 3**j**, **W** = −9**i** − 6**j**

32. **V** = −**i** + 2**j**, **W** = 4**i** + 6**j**

33. **V** = 3**i** + **j**, **W** = −2**i** + 4**j**

34. **V** = **i** − 4**j**, **W** = 4**i** − 2**j**

35. Two forces, one of 35 pounds and the other of 50 pounds, act on the same object. If the angle between the forces is 30°, find the magnitude of the resultant (vector sum) of these forces.

36. Two forces, one of 100 pounds and the other of 150 pounds, act on the same object, at angles of 20° and 60°, respectively, with the positive *x*-axis. Find the direction and magnitude of the resultant (vector sum) of these forces.

37. Three forces of 75 pounds, 100 pounds, and 125 pounds act on the same object, at angles of 30°, 45°, and 120°, respectively, with the positive *x*-axis. Find the direction and magnitude of the resultant of these forces.

38. Three forces of 70 pounds, 40 pounds, and 60 pounds act on the same object, at angles of −30°, 45°, and 135°, respectively, with the positive *x*-axis. Find the direction and magnitude of the resultant of these forces.

39. A heavy implement is dragged 10 feet across the floor, using a force of 85 pounds. Find the work done if the direction of the force is 60° above the horizontal, as shown in Figure 7.31. (Use the formula for work, *W* = *FD*, where *F* is the horizontal component of force and *D* is the horizontal distance.)

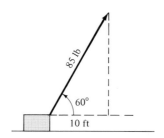

FIGURE 7.31

40. To carry a 100-pound cylindrical weight, two men lift on the ends of short ropes that are tied to an eyelet on the top center of the cylinder. If one of the ropes makes a 20° angle away from the vertical and the other a 30° angle, find the vertical component of each man's force. (See Figure 7.32.)

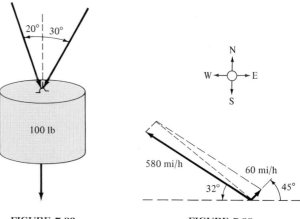

FIGURE 7.32   **FIGURE 7.33**

41. An airplane's velocity with respect to the air is 580 miles per hour, and it is headed W 32° N. The wind, at the altitude of the plane, is from the southwest and has a velocity of 60 miles per hour. What is the true direction of the plane, and what is its speed with respect to the ground? (See Figure 7.33.)

42. A ball is thrown into the air with an initial velocity of 80 feet per second, at an angle of 50° with the horizontal. Find the vertical and horizontal components of the velocity.

43. An airplane is flying in the direction S 32° E, with an airspeed of 540 miles per hour. Because of the wind, its groundspeed and direction are 500 miles per hour and S 40° E, respectively. Find the direction and speed of the wind.

Trigonometry and Complex Numbers 7.4

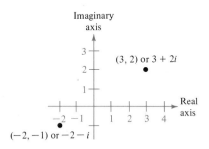

The Complex Plane

FIGURE 7.34

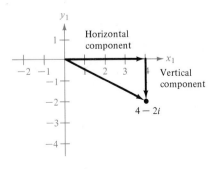

Vector Representation of a
Complex Number

FIGURE 7.35

In this section we develop the trigonometric form for representing complex numbers, but its usefulness will not be fully apparent until we introduce DeMoivre's Theorem in Section 7.5. The development of this form provides a good review and application of topics presented earlier in this text.

Since a complex number z can be represented in the form

$$z = a + bi$$

it is logical to consider such a number geometrically as an ordered pair (a, b), where a is the *real* part and b the *imaginary* part of the complex number. This concept suggests that we can graph a complex number $a + bi$ as an ordered pair (a, b) by slightly adapting the rectangular coordinate plane. In particular, we call the horizontal axis the **real axis** and the vertical axis the **imaginary axis.** When we use a rectangular coordinate plane in this manner to graph complex numbers, we call it the **complex plane.** (See Figure 7.34.)

This ordered pair representation of complex numbers leads naturally to a vector representation. The complex number $z = a + bi$ has a as its *horizontal* component and b as its *vertical* component. (See Figure 7.35.) The imaginary unit i should *not* be confused with the boldface unit vector **i.**

The vector representation of a complex number leads to the following definition of the **absolute value** of a complex number.

DEFINITION OF ABSOLUTE VALUE OF $a + bi$

> The **absolute value** of the complex number $z = a + bi$ is given by
>
> $$|a + bi| = \sqrt{a^2 + b^2}$$

EXAMPLE 1
Graphing Complex Numbers
and Finding Absolute Value

Given the complex numbers

$$z_1 = -3i \qquad \text{and} \qquad z_2 = -2 + 5i$$

(a) Sketch the graph of $z_1 + z_2$.
(b) Sketch the graph of $z_1 - z_2$.
(c) Find $|2z_1 + 3z_2|$.

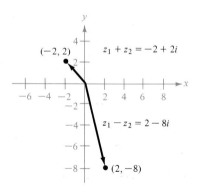

Graph of Complex Numbers

FIGURE 7.36

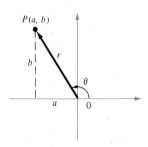

Complex Number: $a + bi$
Rectangular Form: (a, b)
Polar Form: (r, θ)

FIGURE 7.37

Solution:

(a) $z_1 + z_2 = -3i + (-2 + 5i) = -2 + 2i$
(b) $z_1 - z_2 = -3i - (-2 + 5i) = 2 - 8i$
   The graphs of parts (a) and (b) are shown in Figure 7.36.
(c) Since

$$2z_1 + 3z_2 = 2(-3i) + 3(-2 + 5i)$$
$$= -6i - 6 + 15i = -6 + 9i$$

it follows that

$$|2z_1 + 3z_2| = \sqrt{(-6)^2 + (9)^2} = \sqrt{117} = 3\sqrt{13}$$

To work effectively with *powers* and *roots* of complex numbers, we apply trigonometry to our vector representation. In Figure 7.37, consider the nonzero complex number $a + bi$. By letting the vector $\overrightarrow{OP}$ be the terminal side of angle θ in standard position, we have

$$a = r \cos \theta, \qquad b = r \sin \theta, \qquad \text{and} \qquad r = \sqrt{a^2 + b^2}$$

Consequently,

$$a + bi = (r \cos \theta) + (r \sin \theta)i$$

from which we obtain the following **trigonometric form of a complex number.**

TRIGONOMETRIC FORM OF A COMPLEX NUMBER

For any complex number $z = a + bi$, the trigonometric form is

$$z = r(\cos \theta + i \sin \theta)$$

where $a = r \cos \theta$, $b = r \sin \theta$, $r = \sqrt{a^2 + b^2}$, and $\tan \theta = b/a$. The number r is called the **modulus** (or **amplitude**) of z, and θ is called an **argument** of z.

Remark: The trigonometric form is often called the **polar form,** since any point in the complex plane can be located using distance r and angle θ.

Because there are infinitely many choices for θ, the trigonometric form of a complex number is not unique. Normally, we use θ values $0 \le \theta < 2\pi$, though on occasion we may use $\theta < 0$.

EXAMPLE 2
Converting Complex Numbers
to Alternative Forms

(a) Express $z = -2 - 2\sqrt{3}\, i$ in trigonometric form.
(b) Express $z = 6 + 2i$ in trigonometric form.
(c) Express the following complex number in standard form $a + bi$:

$$z = \sqrt{8}\left[\cos\left(\frac{-\pi}{3}\right) + i \sin\left(\frac{-\pi}{3}\right) \right]$$

Solution:

(a) In this case, we have

$$r = |-2 - 2\sqrt{3}\,i| = \sqrt{(-2)^2 + (-2\sqrt{3})^2} = \sqrt{16} = 4$$

$$\tan \theta = \frac{b}{a} = \frac{-2\sqrt{3}}{-2} = \sqrt{3} \qquad \textit{Reference angle: 60°}$$

Since $z = -2 - 2\sqrt{3}\,i$ lies in Quadrant III, we choose

$$\theta = 180° + 60° = 240°$$

Thus, we have

$$z = r(\cos \theta + i \sin \theta) = 4(\cos 240° + i \sin 240°)$$

(b) Here we have

$$r = |6 + 2i| = \sqrt{36 + 4} = \sqrt{40} = 2\sqrt{10}$$

$$\tan \theta = \frac{2}{6} = \frac{1}{3} \qquad \textit{θ in Quadrant I}$$

$$\theta = \arctan \frac{1}{3} \approx 18.4°$$

Therefore,

$$z = r(\cos \theta + i \sin \theta)$$

$$= 2\sqrt{10}\left[\cos\left(\arctan \frac{1}{3} \right) + i \sin\left(\arctan \frac{1}{3} \right) \right]$$

$$\approx 2\sqrt{10}(\cos 18.4° + i \sin 18.4°)$$

(c) Since

$$\cos\left(\frac{-\pi}{3} \right) = \frac{1}{2} \qquad \text{and} \qquad \sin\left(\frac{-\pi}{3} \right) = \frac{-\sqrt{3}}{2}$$

we obtain the standard form

$$z = \sqrt{8}\left[\cos\left(\frac{-\pi}{3} \right) + i \sin\left(\frac{-\pi}{3} \right) \right]$$

$$= \sqrt{8}\left[\frac{1}{2} - i\frac{\sqrt{3}}{2} \right] = 2\sqrt{2}\left[\frac{1}{2} - \frac{\sqrt{3}}{2}i \right]$$

$$= \sqrt{2} - \sqrt{6}\,i$$

The trigonometric form adapts nicely to multiplication and division of complex numbers. Suppose we are given two complex numbers

$$z_1 = r_1(\cos \theta_1 + i \sin \theta_1) \qquad \text{and} \qquad z_2 = r_2(\cos \theta_2 + i \sin \theta_2)$$

Their product is

$$z_1 z_2 = r_1 r_2 (\cos \theta_1 + i \sin \theta_1)(\cos \theta_2 + i \sin \theta_2)$$
$$= r_1 r_2 [(\cos \theta_1 \cos \theta_2 - \sin \theta_1 \sin \theta_2)$$
$$+ i(\sin \theta_1 \cos \theta_2 + \cos \theta_1 \sin \theta_2)]$$

Using the sum and difference formulas for cosine and sine, we can rewrite this equation as

$$z_1 z_2 = r_1 r_2 [\cos(\theta_1 + \theta_2) + i \sin(\theta_1 + \theta_2)]$$

This establishes the first part of the following theorem. The second part is left to you (see Exercise 47).

PRODUCT AND QUOTIENT OF TWO COMPLEX NUMBERS

Given two complex numbers in the form

$$z_1 = r_1(\cos \theta_1 + i \sin \theta_1) \qquad \text{and} \qquad z_2 = r_2(\cos \theta_2 + i \sin \theta_2)$$

their product and quotient are given by

$$z_1 z_2 = r_1 r_2 [\cos(\theta_1 + \theta_2) + i \sin(\theta_1 + \theta_2)] \qquad \qquad \textit{Product}$$

$$\frac{z_1}{z_2} = \frac{r_1}{r_2} [\cos(\theta_1 - \theta_2) + i \sin(\theta_1 - \theta_2)], \qquad z_2 \neq 0 \qquad \textit{Quotient}$$

Remark: This theorem says that to multiply two complex numbers we multiply moduli and add arguments, whereas to divide two complex numbers we divide moduli and subtract arguments.

EXAMPLE 3
Multiplying Complex Numbers
in Trigonometric Form

Use trigonometric forms to find $z_1 z_2$, given

$$z_1 = -1 + i\sqrt{3}$$

and

$$z_2 = 4\sqrt{3} - 4i$$

Solution:
For z_1, we have

$$r = \sqrt{(-1)^2 + (\sqrt{3})^2} = 2$$

$$\tan \theta = \frac{\sqrt{3}}{-1} = -\sqrt{3} \qquad \text{or} \qquad \theta = \frac{2\pi}{3} \qquad \textit{Quadrant II}$$

Thus,

$$z_1 = 2\left(\cos \frac{2\pi}{3} + i \sin \frac{2\pi}{3}\right)$$

For z_2, we have

$$r = \sqrt{(4\sqrt{3})^2 + (-4)^2} = \sqrt{64} = 8$$

$$\tan \theta = \frac{-4}{4\sqrt{3}} = \frac{-1}{\sqrt{3}} \quad \text{or} \quad \theta = \frac{11\pi}{6} \qquad \textit{Quadrant IV}$$

Thus,

$$z_2 = 8\left(\cos \frac{11\pi}{6} + i \sin \frac{11\pi}{6}\right)$$

Consequently, the product is

$$z_1 z_2 = 2\left(\cos \frac{2\pi}{3} + i \sin \frac{2\pi}{3}\right) \cdot 8\left(\cos \frac{11\pi}{6} + i \sin \frac{11\pi}{6}\right)$$

$$= 16\left[\cos\left(\frac{2\pi}{3} + \frac{11\pi}{6}\right) + i \sin\left(\frac{2\pi}{3} + \frac{11\pi}{6}\right)\right]$$

$$= 16\left[\cos\left(\frac{5\pi}{2}\right) + i \sin\left(\frac{5\pi}{2}\right)\right]$$

In standard form, we have

$$z_1 z_2 = 16[0 + i(1)] = 16i$$

Check this result using the multiplication rule given in Section 3.4.

EXAMPLE 4
Dividing Trigonometric Forms
of Complex Numbers

Find z_1/z_2 for the following two complex numbers:

$$z_1 = 24(\cos 300° + i \sin 300°)$$

$$z_2 = 8(\cos 75° + i \sin 75°)$$

Solution:
Direct application of the quotient theorem yields

$$\frac{z_1}{z_2} = \frac{24(\cos 300° + i \sin 300°)}{8(\cos 75° + i \sin 75°)}$$

$$= \frac{24}{8}[\cos(300° - 75°) + i \sin(300° - 75°)]$$

$$= 3[\cos 225° + i \sin 225°]$$

In standard form,

$$\frac{z_1}{z_2} = 3\left[\left(\frac{-1}{\sqrt{2}}\right) + i\left(\frac{-1}{\sqrt{2}}\right)\right] = \frac{-3}{\sqrt{2}}(1 + i)$$

$$= \frac{-3\sqrt{2}}{2} - \frac{3\sqrt{2}}{2}i$$

Section Exercises 7.4

In Exercises 1–4, express the complex number in trigonometric form.

1.

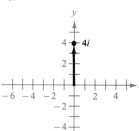

2.

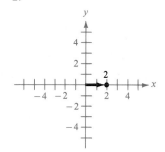

3.

4.

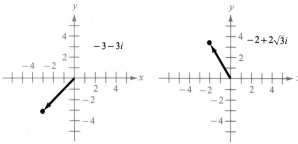

In Exercises 5–20, represent the complex numbers graphically, and give the trigonometric form of the number.

5. $3 - 3i$
6. $-2 - 2i$
7. $\sqrt{3} + i$
8. $-1 + \sqrt{3}\,i$
9. $-2(1 + \sqrt{3}\,i)$
10. $\frac{3}{2}(\sqrt{3} - i)$
11. $6i$
12. 4
13. $-7 + 4i$
14. $3 - i$
15. 7
16. $-2i$
17. $1 + 6i$
18. $2\sqrt{2} - i$
19. $-3 - i$
20. $1 + 3i$

In Exercises 21–30, represent the complex number graphically, and give the standard form of the number.

21. $2(\cos 150° + i \sin 150°)$
22. $5(\cos 135° + i \sin 135°)$
23. $\frac{3}{2}(\cos 300° + i \sin 300°)$
24. $\frac{3}{4}(\cos 315° + i \sin 315°)$
25. $3.75\left(\cos \dfrac{3\pi}{4} + i \sin \dfrac{3\pi}{4}\right)$
26. $8\left(\cos \dfrac{\pi}{12} + i \sin \dfrac{\pi}{12}\right)$
27. $4\left(\cos \dfrac{3\pi}{2} + i \sin \dfrac{3\pi}{2}\right)$
28. $6[\cos(230° \ 30') + i \sin(230° \ 30')]$
29. $3[\cos(18° \ 45') + i \sin(18° \ 45')]$
30. $7(\cos 0° + i \sin 0°)$

In Exercises 31–40, perform the indicated operation and leave the result in trigonometric form.

31. $[3(\cos 60° + i \sin 60°)][4(\cos 30° + i \sin 30°)]$
32. $[\frac{3}{2}(\cos 90° + i \sin 90°)][6(\cos 45° + i \sin 45°)]$
33. $[\frac{5}{3}(\cos 140° + i \sin 140°)][\frac{2}{3}(\cos 60° + i \sin 60°)]$
34. $[0.5(\cos 100° + i \sin 100°)][0.8(\cos 300° + i \sin 300°)]$
35. $[0.45(\cos 310° + i \sin 310°)][0.60(\cos 200° + i \sin 200°)]$
36. $\dfrac{2(\cos 120° + i \sin 120°)}{4(\cos 40° + i \sin 40°)}$
37. $\dfrac{\cos(5\pi/3) + i \sin(5\pi/3)}{\cos \pi + i \sin \pi}$
38. $\dfrac{5[\cos(4.3) + i \sin(4.3)]}{4[\cos(2.1) + i \sin(2.1)]}$
39. $\dfrac{12(\cos 52° + i \sin 52°)}{3(\cos 110° + i \sin 110°)}$
40. $\dfrac{9(\cos 20° + i \sin 20°)}{5(\cos 75° + i \sin 75°)}$

In Exercises 41–46, (a) give the trigonometric form of the complex number, (b) perform the indicated operation using the trigonometric form, and (c) perform the indicated operation using the standard form and check your result with the answer in part (b).

41. $(2 + 2i)(1 - i)$
42. $(\sqrt{3} + i)(1 + i)$
43. $-2i(1 + i)$
44. $\dfrac{3 + 4i}{1 - \sqrt{3}\,i}$
45. $\dfrac{5}{2 + 3i}$
46. $\dfrac{4i}{-4 + 2i}$

47. Given two complex numbers $z_1 = r_1(\cos \theta_1 + i \sin \theta_1)$ and $z_2 = r_2(\cos \theta_2 + i \sin \theta_2)$, $z_2 \neq 0$, prove that

$$\frac{z_1}{z_2} = \frac{r_1}{r_2}[\cos(\theta_1 - \theta_2) + i \sin(\theta_1 - \theta_2)]$$

48. Show that the complex conjugate of $z = r(\cos \theta + i \sin \theta)$ is $\bar{z} = r[\cos(-\theta) + i \sin(-\theta)]$.

49. Use the trigonometric form of z and $\bar{z}$ in Exercise 48 to find

(a) $z\bar{z}$ (b) $z/\bar{z}$, $\bar{z} \neq 0$

50. Show that the negative of $z = r(\cos \theta + i \sin \theta)$ is $-z = r[\cos(\theta + \pi) + i \sin(\theta + \pi)]$.

DeMoivre's Theorem   7.5

Our final look at complex numbers involves procedures for finding their powers and roots. Repeated use of the multiplication rule in the previous section yields

$$z = r(\cos \theta + i \sin \theta)$$
$$z^2 = r(\cos \theta + i \sin \theta)r(\cos \theta + i \sin \theta)$$
$$= r^2(\cos 2\theta + i \sin 2\theta)$$
$$z^3 = z^2(z) = r^2(\cos 2\theta + i \sin 2\theta)r(\cos \theta + i \sin \theta)$$
$$= r^3(\cos 3\theta + i \sin 3\theta)$$

Similarly,

$$z^4 = r^4(\cos 4\theta + i \sin 4\theta)$$
$$z^5 = r^5(\cos 5\theta + i \sin 5\theta)$$

$$\vdots$$

This pattern leads to the following important theorem, known as **DeMoivre's Theorem.**

DeMOIVRE'S THEOREM

> If $z = r(\cos \theta + i \sin \theta)$ is any complex number and n is any positive integer, then
>
> $$z^n = [r(\cos \theta + i \sin \theta)]^n = r^n(\cos n\theta + i \sin n\theta)$$

A complete proof of this theorem can be given using mathematical induction (see Section 10.4).

EXAMPLE 1
Finding Powers of a
Complex Number

Find $(-1 + \sqrt{3}\, i)^{12}$, and write the result in standard form.

Solution:
We first convert to trigonometric form. For $(-1 + \sqrt{3}\, i)$, we have

$$r = \sqrt{(-1)^2 + (\sqrt{3})^2} = 2$$
$$\tan \theta = \frac{\sqrt{3}}{-1} = -\sqrt{3} \quad \text{or} \quad \theta = \frac{2\pi}{3}$$

Therefore,

$$(-1 + \sqrt{3}\, i) = 2\left(\cos \frac{2\pi}{3} + i \sin \frac{2\pi}{3}\right)$$

By DeMoivre's Theorem, we have

$$(-1 + \sqrt{3}\, i)^{12} = \left[2\left(\cos \frac{2\pi}{3} + i \sin \frac{2\pi}{3} \right) \right]^{12}$$

$$= 2^{12} \left[\cos (12)\frac{2\pi}{3} + i \sin (12)\frac{2\pi}{3} \right]$$

$$= 4096(\cos 8\pi + i \sin 8\pi)$$

$$= 4096(\cos 0 + i \sin 0)$$

$$= 4096(1 + 0) = 4096$$

(Are you surprised to see a real number as the answer?)

Recall that a consequence of the Fundamental Theorem of Algebra (Section 3.5) is that a polynomial equation of degree n has n roots in the complex number system. Hence, an equation like $x^6 = 1$ has six roots, and in this case we can find the six roots by factoring and using the quadratic formula.

$$x^6 - 1 = (x^3 - 1)(x^3 + 1)$$

$$= (x - 1)(x^2 + x + 1)(x + 1)(x^2 - x + 1) = 0$$

Consequently, the roots are

$$x = \pm 1, \quad \frac{-1 \pm \sqrt{3}\, i}{2}, \quad \frac{1 \pm \sqrt{3}\, i}{2}$$

Each of these numbers is called a sixth root of 1. In general, we define the nth root of a complex number as follows.

DEFINITION OF nth ROOT OF A COMPLEX NUMBER

> The complex number $u = a + bi$ is an **nth root** of the complex number z if
>
> $$z = u^n = (a + bi)^n$$

To find a formula for an nth root of a complex number, we let u be an nth root of z. We can express u and z in trigonometric form, as follows:

$$u = s(\cos \beta + i \sin \beta) \quad \text{and} \quad z = r(\cos \theta + i \sin \theta)$$

Then, by DeMoivre's Theorem and the fact that $u^n = z$, we have

$$s^n(\cos n\beta + i \sin n\beta) = r(\cos \theta + i \sin \theta)$$

Now, taking the absolute value of both sides of this equation, it follows that

$$s^n = r$$

Substituting back into the previous equation and dividing by r, we get

$$\cos n\beta + i \sin n\beta = \cos \theta + i \sin \theta$$

Equating real parts and imaginary parts yields

$$\cos n\beta = \cos \theta \quad \text{and} \quad \sin n\beta = \sin \theta$$

Since both sine and cosine have a period of 2π, these last two equations have solutions if and only if the angles differ by a multiple of 2π. Consequently, for some integer k,

$$n\beta = \theta + 2\pi k \quad \text{or} \quad \beta = \frac{\theta + 2\pi k}{n}$$

Backsubstituting this value for β into the trigonometric form of u, we get the result stated in the following theorem.

nth ROOTS OF A COMPLEX NUMBER

For any positive integer n, the complex number

$$z = r(\cos \theta + i \sin \theta)$$

has exactly n distinct nth roots. These n roots are given by

$$\sqrt[n]{r}\left[\cos\left(\frac{\theta + 2\pi k}{n}\right) + i \sin\left(\frac{\theta + 2\pi k}{n}\right)\right]$$

where $k = 0, 1, 2, \ldots, n - 1$.

Remark: Note that when k exceeds $n - 1$, the roots begin to repeat. For instance, if $k = n$, the angle is

$$\frac{\theta + 2\pi n}{n} = \frac{\theta}{n} + 2\pi = \frac{\theta}{n}$$

which is also obtained when $k = 0$.

This formula for the nth roots of a complex number has a nice geometrical interpretation. Figure 7.38 shows the nth roots of z in the complex plane. Note that because these nth roots of z all have the same modulus (magnitude) $\sqrt[n]{r}$, they will all lie on a circle of radius $\sqrt[n]{r}$ with center at the origin. Furthermore, these roots are equally spaced along this circle, since successive nth roots have arguments that differ by $2\pi/n$.

We have already found the sixth roots of 1 by factoring and the quadratic formula. Now let's see how we can solve the same problem with the formula for nth roots.

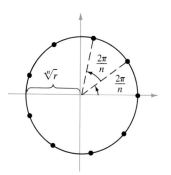

FIGURE 7.38

EXAMPLE 2
Finding nth Roots
of a Real Number

Find the six sixth roots of 1.

Solution:
First we write 1 in the trigonometric form

$$1 = 1(\cos 0° + i \sin 0°)$$

Then, by the *n*th root formula, with $n = 6$ and $r = 1$, the roots have the form

$$\sqrt[6]{1}\left[\cos\left(\frac{0° + 360°k}{6}\right) + i\sin\left(\frac{0° + 360°k}{6}\right)\right]$$

or simply

$$[\cos(60°k) + i\sin(60°k)]$$

Thus, for $k = 0, 1, 2, 3, 4, 5$, the respective sixth roots are

$$\cos 0° + i\sin 0° = 1$$

$$\cos 60° + i\sin 60° = \frac{1}{2} + \frac{\sqrt{3}}{2}i$$

$$\cos 120° + i\sin 120° = -\frac{1}{2} + \frac{\sqrt{3}}{2}i$$

$$\cos 180° + i\sin 180° = -1$$

$$\cos 240° + i\sin 240° = -\frac{1}{2} - \frac{\sqrt{3}}{2}i$$

$$\cos 300° + i\sin 300° = \frac{1}{2} - \frac{\sqrt{3}}{2}i$$

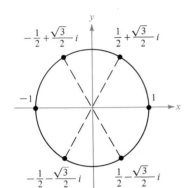

FIGURE 7.39

(See Figure 7.39.)

Remark: In Figure 7.39, notice that the roots obtained in Example 2 all have a magnitude of 1 and are equally spaced (60° apart) around this unit circle. Also, notice that the complex roots occur in conjugate pairs, as previously discussed in Section 3.5. We refer to the special case of the *n* distinct *n*th roots of 1 as the **nth roots of unity**.

EXAMPLE 3
Finding the *n*th Roots of a Complex Number

Find the three cube roots of

$$z = -2 + 2i$$

Solution:
In this case, we have

$$r = \sqrt{(-2)^2 + (2)^2} = \sqrt{8}$$

$$\theta = \arctan(-1) = 135° \qquad \text{*Quadrant II*}$$

Thus, the trigonometric form for z is

$$-2 + 2i = \sqrt{8}(\cos 135° + i\sin 135°)$$

By our formula for *n*th roots, the cube roots have the form

$$\sqrt[3]{8}\left[\cos\left(\frac{135° + 360°k}{3}\right) + i\sin\left(\frac{135° + 360°k}{3}\right)\right]$$

Finally, for $k = 0, 1, 2$, we get the roots

$$\sqrt{2}[\cos 45° + i \sin 45°] = 1 + i$$
$$\sqrt{2}[\cos 165° + i \sin 165°] \approx -1.3660 + 0.3660i$$
$$\sqrt{2}[\cos 285° + i \sin 285°] \approx 0.3660 - 1.3660i$$

The nth roots of a complex number can be quite useful in solving polynomial equations. Our last two examples are cases in point.

EXAMPLE 4
Finding the Roots of a Polynomial Equation

Find all solutions to the equation

$$x^4 + 16 = 0$$

Solution:
The given equation can be written as

$$x^4 = -16$$

and we can solve this equation by finding the four fourth roots of -16. First, we write -16 in trigonometric form

$$-16 = 16(\cos \pi + i \sin \pi)$$

Now, we see that the fourth roots of -16 have the form

$$\sqrt[4]{16}\left[\cos\left(\frac{\pi + 2\pi k}{4}\right) + i \sin\left(\frac{\pi + 2\pi k}{4}\right)\right]$$

Finally, using $k = 0, 1, 2, 3$, we obtain the roots

$$2\left[\cos \frac{\pi}{4} + i \sin \frac{\pi}{4}\right] = 2\left[\frac{\sqrt{2}}{2} + \frac{\sqrt{2}}{2}i\right] = \sqrt{2} + \sqrt{2}\, i$$

$$2\left[\cos \frac{3\pi}{4} + i \sin \frac{3\pi}{4}\right] = 2\left[\frac{-\sqrt{2}}{2} + \frac{\sqrt{2}}{2}i\right] = -\sqrt{2} + \sqrt{2}\, i$$

$$2\left[\cos \frac{5\pi}{4} + i \sin \frac{5\pi}{4}\right] = 2\left[\frac{-\sqrt{2}}{2} - \frac{\sqrt{2}}{2}i\right] = -\sqrt{2} - \sqrt{2}\, i$$

$$2\left[\cos \frac{7\pi}{4} + i \sin \frac{7\pi}{4}\right] = 2\left[\frac{\sqrt{2}}{2} - \frac{\sqrt{2}}{2}i\right] = \sqrt{2} - \sqrt{2}\, i$$

Section Exercises 7.5

In Exercises 1–12, use DeMoivre's Theorem to find the indicated powers of the given complex number. Express the result in standard form.

1. $(1 + i)^5$
2. $(2 + 2i)^6$
3. $(-1 + i)^{10}$
4. $(1 - i)^{12}$

5. $2(\sqrt{3} + i)^7$
6. $4(1 - \sqrt{3}\, i)^3$
7. $[5(\cos 20° + i \sin 20°)]^3$
8. $[3(\cos 150° + i \sin 150°)]^4$
9. $\left(\cos \frac{5\pi}{4} + i \sin \frac{5\pi}{4}\right)^{10}$
10. $\left[2\left(\cos \frac{\pi}{2} + i \sin \frac{\pi}{2}\right)\right]^8$

11. $[5(\cos 3.2 + i \sin 3.2)]^4$ 12. $(\cos 0 + i \sin 0)^{20}$

In Exercises 13–24, (a) use DeMoivre's Theorem to find the indicated roots, (b) represent each of the roots graphically, and (c) express each of the roots in standard form.

13. The square roots of $9(\cos 120° + i \sin 120°)$.
14. The square roots of $16(\cos 60° + i \sin 60°)$.
15. The fourth roots of $16\left(\cos \dfrac{4\pi}{3} + i \sin \dfrac{4\pi}{3}\right)$.
16. The fifth roots of $32\left(\cos \dfrac{5\pi}{6} + i \sin \dfrac{5\pi}{6}\right)$.
17. The square roots of $-25i$.
18. The fourth roots of $625i$.
19. The cube roots of $-\frac{125}{2}(1 + \sqrt{3}\, i)$.

20. The cube roots of $-4\sqrt{2}(1 - i)$.
21. The cube roots of 8.
22. The fourth roots of i.
23. The fifth roots of 1.
24. The cube roots of 1000.

In Exercises 25–30, find all the solutions to the given equation and represent your solutions graphically.

25. $x^4 - i = 0$
26. $x^3 + 1 = 0$
27. $x^5 + 243 = 0$
28. $x^4 - 81 = 0$
29. $x^3 + 64i = 0$
30. $x^6 - 64i = 0$

Review Exercises / **Chapter 7**

In Exercises 1–15, solve the triangle(s) (if possible) using the three given parts.

1. $a = 5, b = 8, c = 10$
2. $a = 6, b = 9, C = 45°$
3. $A = 12°, B = 58°, a = 5$
4. $B = 110°, C = 30°, c = 10.5$
5. $B = 110°, a = 4, c = 4$
6. $a = 80, b = 60, c = 100$
7. $A = 75°, a = 2.5, b = 16.5$
8. $A = 130°, a = 50, b = 30$
9. $B = 115°, a = 7, b = 14.5$
10. $C = 50°, a = 25, c = 22$
11. $A = 15°, a = 5, b = 10$
12. $B = 150°, a = 64, b = 10$
13. $B = 150°, a = 10, c = 20$
14. $a = 2.5, b = 15.0, c = 4.5$
15. $B = 25°, a = 6.2, b = 4$
16. Find the height of a tree that stands on a hillside of slope 32° (from the horizontal) if from a point 75 feet downhill the angle of elevation to the top of the tree is 48°. (See Figure 7.40.)

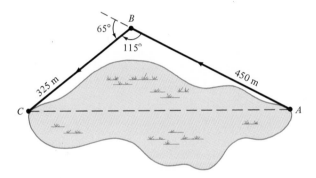

FIGURE 7.41

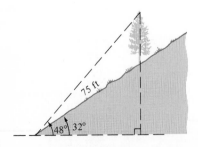

FIGURE 7.40

17. To approximate the length of a marsh, a surveyor walks 450 meters from point A to point B, then turns 65° and walks 325 meters to point C. Approximate the length $\overline{AC}$ of the marsh. (See Figure 7.41.)
18. Two planes leave an airport at approximately the same time. One is flying at 425 miles per hour at a bearing of N 5° W, and the other is flying at 530 miles per hour at a bearing of N 67° E. How far apart are the two planes after flying for 2 hours?
19. From a certain distance, the angle of elevation of the top of a building is 17°. At a point 50 meters closer to the building, the angle of elevation is 31°. Approximate the height of the building.
20. The lengths of the diagonals of a parallelogram are 10 feet and 16 feet. Find the lengths of the sides of the parallelogram if the diagonals intersect at an angle of 28°.

In Exercises 21–24, find the area of the triangle having the given parts.

21. $a = 4, b = 5, c = 7$ 22. $a = 15, b = 8, c = 10$
23. $A = 27°, b = 5, c = 8$ 24. $B = 80°, a = 4, c = 8$

In Exercises 25–28, let $\mathbf{U} = 6\mathbf{i} - 5\mathbf{j}$ and $\mathbf{V} = 10\mathbf{i} + 3\mathbf{j}$.

25. Find $|\mathbf{U}|$ and a unit vector in the direction of $\mathbf{U}$.
26. Find $3\mathbf{V}$.
27. Find $4\mathbf{U} - 5\mathbf{V}$.
28. Find a vector that has the direction opposite that of $\mathbf{V}$ and one-half the magnitude of $\mathbf{V}$.
29. Find the horizontal and vertical components of a vector of magnitude 650 that makes an angle of 210° (counterclockwise) with the positive x-axis.
30. Find the horizontal and vertical components of a force of 25 pounds, applied in a direction that makes an angle of 112° (counterclockwise) with the positive x-axis.
31. A 500-pound motorcycle is on a hill inclined at 12°. What force is required to keep the motorcycle from rolling back down the hill? (Assume that the force is parallel to the ground.)
32. An airplane has an airspeed of 450 miles per hour at a bearing of N 30° E. If the wind velocity is 20 miles per hour from the west, find the ground speed and the direction of the plane.
33. Two forces of 60 pounds and 100 pounds have a resultant force of magnitude 125 pounds. Find the angle between the directions of the two given forces.

In Exercises 34–36, (a) express the two given complex numbers in trigonometric form, and then (b) find $z_1 z_2$ and z_1/z_2 using the trigonometric form.

34. $z_1 = 2\sqrt{3} - 2i, z_2 = -10i$
35. $z_1 = -3(1 + i), z_2 = 2(\sqrt{3} + i)$
36. $z_1 = 5i, z_2 = 2 - 2i$

In Exercises 37–40, use DeMoivre's Theorem to find the indicated power of the given complex number. Express the result in standard form.

37. $\left[5\left(\cos \dfrac{\pi}{12} + i \sin \dfrac{\pi}{12}\right)\right]^4$

38. $\left[2\left(\cos \dfrac{4\pi}{15} + i \sin \dfrac{4\pi}{15}\right)\right]^5$

39. $(2 + 3i)^6$
40. $(1 - i)^8$

In Exercises 41–44, use DeMoivre's Theorem to find the roots of the given complex numbers.

41. The sixth roots of $-729i$.
42. The fourth roots of 256.
43. The cube roots of -1.
44. The fourth roots of $-1 + i$.

CHAPTER

8 SYSTEMS OF EQUATIONS AND INEQUALITIES

Systems of Equations   8.1

Recall from Chapter 1 that the graph of an equation in two variables consists of all points in the plane that satisfy the equation. We call such points **solution points** (or simply **solutions**) of the equation. For example, the equation

$$x + y = 4$$

is satisfied when $x = 3$ and $y = 1$. Thus, $(3, 1)$ is a solution point of the equation. Some additional solution points are $(-4, 8)$, $(0, 4)$, $(-2, 6)$, $(6, -2)$, and $(5, -1)$. On the other hand, $(2, 3)$ is not a solution point of the equation $x + y = 4$, since $x = 2$ and $y = 3$ yields

$$2 + 3 = 5 \neq 4$$

Keep in mind that in this chapter we are concerned only with the real number solutions to an equation.

Typically, a single equation in two variables will have infinitely many solutions. However, on occasion equations in two variables have only one solution or no solutions. For instance, the equation

$$x^2 + y^2 = 0$$

has $(0, 0)$ as its only solution, and the equation

$$x^2 + y^2 = -1$$

has no solutions.

In this section we will study techniques for finding solutions to systems of two or more equations in two or more variables. That is, we show how to find points that satisfy two or more equations simultaneously.

For example, consider the following problem:

Find two numbers whose sum is four and whose difference is two.

Using x and y to denote the numbers, we can translate the verbal statement into two equations, as follows:

Sum is four: $x + y = 4$ *Equation 1*

Difference is two: $x - y = 2$ *Equation 2*

We now have what is called a **system of two equations involving two variables, x and y.** By a **solution** of this system, we mean an ordered pair (x, y) that satisfies *both* equations. This particular system happens to have only one solution, and we can find it by the **substitution method,** shown in Example 1.

EXAMPLE 1
Solving a System of Two Equations in Two Variables

Find the solution(s) to the system of equations

$$x + y = 4 \quad \text{and} \quad x - y = 2$$

Solution:
We have

$$x + y = 4 \qquad \qquad \textit{Equation 1}$$
$$x - y = 2 \qquad \qquad \textit{Equation 2}$$

We can solve for y in Equation 1 to get

$$y = 4 - x$$

Substituting this value for y into Equation 2, we obtain the equation

$$x - (4 - x) = 2$$
$$x = 3$$

Finally, by backsubstituting $x = 3$ into the equation $y = 4 - x$, we get

$$y = 4 - x$$
$$y = 4 - 3 = 1$$

Thus, the solution to the given system is $x = 3$ and $y = 1$.

In the solution to Example 1, we can identify four basic steps in the **method of substitution,** shown schematically below.

The Method of Substitution

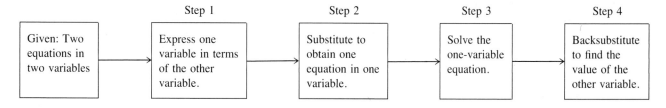

	Step 1	Step 2	Step 3	Step 4
Given: Two equations in two variables	Express one variable in terms of the other variable.	Substitute to obtain one equation in one variable.	Solve the one-variable equation.	Backsubstitute to find the value of the other variable.

EXAMPLE 2
Solving a System by Substitution:
One-Solution Case

Find the solution(s) to the following system:

$$x - y = 2 \qquad\qquad\qquad Equation\ 1$$

$$\frac{x}{2} + 3y = 1 \qquad\qquad\qquad Equation\ 2$$

Solution:

(1) Solve for x in Equation 1:

$$x = 2 + y$$

(2) Substitute into Equation 2:

$$\frac{2 + y}{2} + 3y = 1$$

(3) Solve for y:

$$2 + y + 6y = 2$$

$$7y = 0 \quad \Longrightarrow \quad y = 0$$

(4) Backsubstitute $y = 0$ to get

$$x = 2 + y = 2 + 0 = 2$$

The solution is

$$x = 2, \qquad y = 0$$

EXAMPLE 3
Solving a System by Substitution:
Two-Solution Case

Find the solution(s) to the system

$$x^2 - 2x + y^2 = 3 \qquad\qquad Equation\ 1$$

$$2x + 3y = -4 \qquad\qquad Equation\ 2$$

Solution:

(1) Solve for y in Equation 2:

$$3y = -4 - 2x$$

$$y = \frac{-4 - 2x}{3}$$

(2) Substitute into Equation 1:

$$x^2 - 2x + \left(\frac{-4 - 2x}{3}\right)^2 = 3$$

(3) Solve for x:

$$x^2 - 2x + \frac{16 + 16x + 4x^2}{9} = 3$$

$$9x^2 - 18x + 16 + 16x + 4x^2 = 27$$

$$13x^2 - 2x - 11 = 0$$

$$(13x + 11)(x - 1) = 0$$

$$x = 1 \text{ or } -\frac{11}{13}$$

(4) Backsubstitute $x = 1$:

$$y = \frac{-4 - 2(1)}{3} = \frac{-4 - 2}{3} = \frac{-6}{3} = -2$$

Backsubstitute $x = -\frac{11}{13}$:

$$y = \frac{-4 - 2(-11/13)}{3} = \frac{-52 + 22}{39} = \frac{-30}{39} = -\frac{10}{13}$$

The solutions are

$$x = 1, \qquad y = -2$$

$$x = -\frac{11}{13}, \qquad y = -\frac{10}{13}$$

EXAMPLE 4
Solving a System by Substitution:
No-Solution Case

Find the solution(s) to the system

$$x^2 + y^2 = 1 \qquad\qquad \textit{Equation 1}$$

$$x + y = 3 \qquad\qquad \textit{Equation 2}$$

Solution:

(1) Solve for y in Equation 2:

$$y = 3 - x$$

(2) Substitute into Equation 1:

$$x^2 + (3 - x)^2 = 1$$

(3) Solve for x:

$$x^2 + 9 - 6x + x^2 = 1$$

$$2x^2 - 6x + 8 = 0$$

$$x = \frac{6 \pm \sqrt{36 - 64}}{12}$$

Since the discriminant is negative, there are no real solutions for x, and hence this system has no (real) solutions.

Graphical Approach to Finding Solutions

From Examples 2, 3, and 4, we can see that a system of two equations in two unknowns can have exactly one solution, more than one solution, or no solutions. In practice, we can gain valuable insight about the number of solutions for a system of equations by graphing each of the equations on the same coordinate plane. We call the solutions of the system the **points of intersection** of the graphs. For instance, in Figure 8.1 we see that the two equations in Example 2 graph as two lines with a single point of intersection. Similarly, the two equations in Example 3 graph as a circle and a line, and it makes sense that there are two points of intersection.

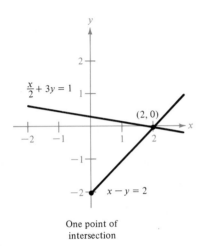

One point of
intersection

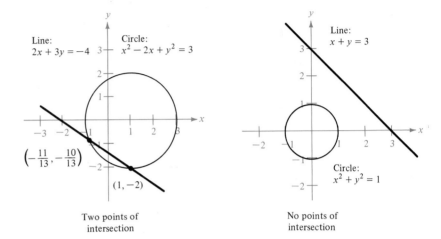

Two points of
intersection

No points of
intersection

FIGURE 8.1 *Points of Intersection*

When solving a system of equations in two unknowns, we can first use a graphical approach to get a good estimate for both the number and values of the solutions. Then to get exact solutions we use substitution.

FINDING POINTS OF INTERSECTION

1. Sketch the graph of each equation in the system.

2. Estimate the points of intersection of the graphs. Occasionally you can guess the exact coordinates from the sketch, but *be sure to test the points in each equation.*

3. Use substitution to find the exact solutions, and check to see that each of these solutions is represented on your sketch.

EXAMPLE 5
Finding Points of Intersection

Find the points of intersection of the graphs of

$$y = \ln x \quad \text{and} \quad x + y = 1$$

Solution:
The graph of each equation is shown in Figure 8.2. From this sketch it is clear that there is only one point of intersection. Also, it appears that (1, 0)

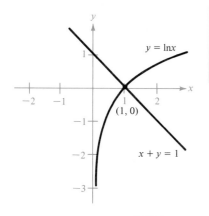

FIGURE 8.2

is the solution point, and we confirm this by checking these coordinates in *both* equations.

When $x = 1$ and $y = 0$, the equation $y = \ln x$ is true, since $0 = \ln(1)$. When $x = 1$ and $y = 0$, the equation $x + y = 1$ is true, since $1 + 0 = 1$. Hence $(1, 0)$ is the single point of intersection.

Remark: Example 5 shows us the value of a graphical approach to solving systems of equations in two variables. Notice what would have happened if we had tried only the substitution method in Example 5. By substituting $y = \ln x$ into $x + y = 1$, we obtain

$$x + \ln x = 1$$

You would be hard pressed to solve this equation for x using analytical techniques.

Economics Application

Many applications involve finding the point(s) of intersection of two graphs. A common one from business is called **break-even analysis.** The marketing of a new product typically requires a substantial investment to develop and produce the product. When enough units have been sold so that the total revenue has offset the total cost, we say that the sale of the product has reached the **break-even point.** We denote the **total cost** of producing x units of a product by C, and the **total revenue** received from selling x units of the product by R. Thus, we can find the break-even point by setting the cost C equal to the revenue R and solving for x. In other words, the break-even point corresponds to the point of intersection of the cost and revenue curves.

EXAMPLE 6
An Application: Break-Even Analysis

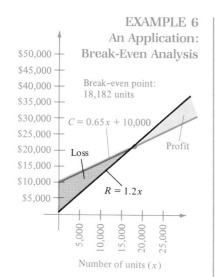

FIGURE 8.3

Roger Fisher is setting up a small business to manufacture and market an item he has developed. Roger has invested $10,000 in equipment and can produce each item for $0.65. If he can sell each item for $1.20, how many items must he sell before he breaks even?

Solution:
The total cost of producing x units is

$$C = 0.65x + 10,000$$

and the revenue obtained by selling x units is

$$R = 1.2x$$

Since the break-even point occurs when $R = C$, we have

$$1.2x = 0.65x + 10,000$$
$$0.55x = 10,000$$
$$x = \frac{10,000}{0.55} \approx 18,182 \text{ units}$$

Note in Figure 8.3 that sales less than the break-even point correspond to an overall loss, while sales greater than the break-even point correspond to a profit.

Solving Systems of Three Equations in Three Variables

We can extend the method of substitution to systems of equations involving more than two variables. The basic strategy is to reduce the system from three equations in three variables to a new system of two equations in two variables. The resulting new system is then solved with the usual substitution method.

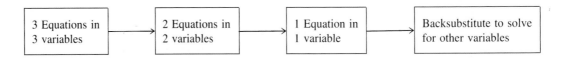

EXAMPLE 7
Solving a System Involving Three Variables

Find all solutions to the following system of equations:

$$x \ - 2y + z = -4 \qquad \text{\textit{Equation 1}}$$
$$2x + \ y - z = \ \ \ 5 \qquad \text{\textit{Equation 2}}$$
$$x^2 + \ y + z = \ \ \ 2 \qquad \text{\textit{Equation 3}}$$

Solution:

(1) Solve for z in Equation 1:

$$z = -x + 2y - 4$$

(2) Substitute this expression for z into the other two equations:

$$2x + y - z = 5 \qquad \text{\textit{Equation 2}}$$
$$2x + y - (-x + 2y - 4) = 5$$
$$3x - y = 1$$

$$x^2 + y + z = 2 \qquad \text{\textit{Equation 3}}$$
$$x^2 + y + (-x + 2y - 4) = 2$$
$$x^2 - x + 3y = 6$$

Now, we have succeeded in reducing the original system of three equations in three variables to a new system of two equations in two variables:

$$3x - \ y = 1$$
$$x^2 - \ x + 3y = 6$$

(3) By using substitution on this system, we have

$$3x - y = 1 \quad \Longrightarrow \quad y = 3x - 1$$

and substitution yields

$$x^2 - x + 3y = 6$$
$$x^2 - x + 3(3x - 1) = 6$$
$$x^2 + 8x - 9 = 0$$
$$(x - 1)(x + 9) = 0$$
$$x = 1 \text{ or } -9$$

(4) Backsubstitute to find y and z. If $x = 1$, then

$$y = 3x - 1 = 3(1) - 1 = 3 - 1 = 2$$
$$z = -x + 2y - 4 = -1 + 2(2) - 4 = -1 + 4 - 4 = -1$$

If $x = -9$, then

$$y = 3x - 1 = 3(-9) - 1 = -27 - 1 = -28$$
$$z = -x + 2y - 4$$
$$= -(-9) + 2(-28) - 4 = 9 - 56 - 4 = -51$$

Thus, we see that the original system has two solutions:

$$x = 1, \qquad y = 2, \qquad z = -1$$

and

$$x = -9, \qquad y = -28, \qquad z = -51$$

Section Exercises 8.1

In Exercises 1–10, find all solutions to the given system by the method of substitution.

1. $2x + y = 4$
   $-x + y = 1$

2. $x - y = -5$
   $x + 2y = 4$

3. $x - y = -3$
   $x^2 - y = -1$

4. $3x - y = -2$
   $x^3 - y = 0$

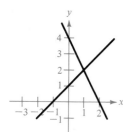

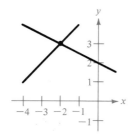

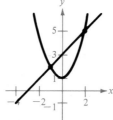

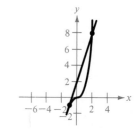

5. $x + 3y = 15$
   $x^2 + y^2 = 25$

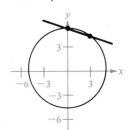

6. $x - y = 0$
   $x^3 - 5x + y = 0$

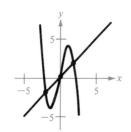

7. $x^2 \quad - y = 0$
   $x^2 - 4x + y = 0$

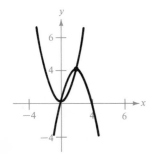

8. $x^2 + y = 1$
   $y = x^4 - 2x^2 + 1$

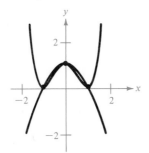

9. $x - 3y = -4$
   $x^2 - y^3 = 0$

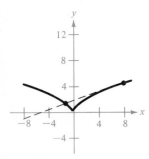

10. $y = x^3 - 3x^2 + 3$
    $2x + y = 3$

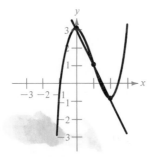

In Exercises 11–30, find all solutions to the system by the method of substitution.

11. $y = x$
    $5x - 3y = 10$

12. $x + 2y = 1$
    $5x - 4y = -23$

13. $2x - y + 2 = 0$
    $4x + y - 5 = 0$

14. $6x - 3y - 4 = 0$
    $x + 2y - 4 = 0$

15. $30x - 40y - 33 = 0$
    $10x + 20y - 21 = 0$

16. $1.5x + 0.8y = 2.3$
    $0.3x - 0.2y = 0.1$

17. $0.2x + 0.5y = 8$
    $x + \quad y = 20$

18. $\frac{1}{2}x + \frac{3}{4}y = 10$
    $\frac{3}{2}x - \quad y = 4$

19. $y = 2x$
    $y = x^2 + 1$

20. $x + y = 4$
    $x^2 - y = 2$

21. $3x - 7y + 6 = 0$
    $x^2 - y^2 = 4$

22. $x^2 + y^2 = 25$
    $2x + y = 10$

23. $x^2 + y^2 = 5$
    $x - y = 1$

24. $y = x^3 - 2x^2 + x - 1$
    $y = \quad - x^2 + 3x - 1$

25. $y = x^4 - 2x^2 + 1$
    $y = 1 - x^2$

26. $x^2 + y = 4$
    $2x - y = 1$

27. $x - y + z = 2$
    $2x + y - 3z = -5$
    $5x + y + z = 10$

28. $x - 2y - 2z = 9$
    $3x - y + z = 13$
    $2x + 3y + z = 8$

29. $xy - 1 = 0$
    $2x - 4y + 7 = 0$

30. $x - 2y = 1$
    $y = \sqrt{x - 1}$

In Exercises 31–38, find all points of intersection of the graphs of the given pair of equations.

31. $x + y = 4$
    $x^2 + y^2 - 4x = 0$

32. $3x - 2y = 0$
    $x^2 - y^2 = 4$

33. $2x - y + 3 = 0$
    $x^2 + y^2 - 4x = 0$

34. $x - y + 3 = 0$
    $x^2 - 4x + 7 = y$

35. $y = e^x$
    $x - y + 1 = 0$

36. $x^2 + y^2 = 8$
    $y = x^2$

37. $y = \cos x$
    $y = \sin 2x$

38. $x - y = 3$
    $x - y^2 = 1$

39. Andrea Jones is setting up a small business and has invested $16,000 to produce an item that will sell for $5.95. If each unit can be produced for $3.45, how many units must Andrea sell in order to break even?

40. Find two numbers whose sum is 20 and product is 96.

41. The sum of two numbers is 12 and the sum of their squares is 80. Find the numbers.

42. What are the dimensions of a rectangle if its perimeter is 40 miles and its area is 96 square miles?

In certain optimization problems in calculus, systems of equations (as found in Exercises 43–46) must be solved. The variable λ is called a Lagrange Multiplier. In the following exercises, find x, y, and λ satisfying the given system.

43. $y + \lambda = 0$
    $x + \lambda = 0$
    $x + y - 10 = 0$

44. $2x + \lambda = 0$
    $2y + \lambda = 0$
    $x + y - 4 = 0$

45. $2x - 2x\lambda = 0$
    $-2y + \lambda = 0$
    $y - x^2 = 0$

46. $2 + 2y + 2\lambda = 0$
    $2x + 1 + \lambda = 0$
    $2x + y - 100 = 0$

Use the following information in Exercises 47–50: The path of a projectile thrown at an angle of θ degrees with the horizontal and at an initial velocity of V_0 feet per second is given by

$$x = (V_0 \cos \theta)t \quad \text{and} \quad y = (V_0 \sin \theta)t - 16t^2$$

47. Find the initial velocity and travel time if $\theta = 45°$ and the projectile lands at the point (144, 0).
48. Find the initial velocity and travel time if $\theta = 30°$ and the projectile lands at the point (144, 0).

49. Find the initial velocity and the angle θ if the projectile lands at the point (100, 0) after traveling for 4 seconds.
50. Find the initial velocity and the angle θ if the projectile lands at the point (200, 0) after traveling for 4 seconds.

Systems of Linear Equations in Two Variables

8.2

In Section 2.3, we looked at equations of the form

$$ax + b = 0$$

where a and b are real numbers and $a \neq 0$. We call this type of equation a **linear equation in one variable.** We can generalize equations of this type to any number of variables:

$$ax + b = 0 \qquad \text{\textit{Linear equation in one variable}}$$

$$ax + by + c = 0 \qquad \text{\textit{Linear equation in two variables}}$$

$$ax + by + cz + d = 0 \qquad \text{\textit{Linear equation in three variables}}$$

$$a_n x_n + a_{n-1} x_{n-1} + \cdots + a_1 x_1 + a_0 = 0 \qquad \text{\textit{Linear equation in n variables}}$$

Note that we usually revert to a subscript notation for linear equations involving a large number of variables.

In Section 8.1 we used the method of substitution to solve various systems of equations. Now we restrict our study to linear systems and introduce a second method, which is called the method of **elimination.**

THE METHOD OF ELIMINATION

Given a linear system in two variables x and y:

1. Obtain coefficients for x (or y) that differ only in sign by multiplying all terms of one or both equations by suitably chosen constants.
2. Add the two equations to obtain a linear equation in (at most) one variable.
3. Solve the resulting linear equation in one variable.
4. Backsubstitute this solution into either of the original equations to solve for the other variable.
5. Check your solution in *both* of the original equations.

Remark: Note that the name *elimination* is derived from the fact that in step 2 one of the original variables is eliminated from the system.

EXAMPLE 1
Solving a Linear System
by Elimination

Solve the system

$$2x - 3y = -7 \qquad \text{\textit{Equation 1}}$$
$$3x + y = -5 \qquad \text{\textit{Equation 2}}$$

Solution:

(1) In Equation 2, if we change the coefficient of y to 3 by multiplying through by 3, then the coefficients of y in Equations 1 and 2 will differ only in sign. Multiply Equation 2 by 3:

$$3x + y = -5 \quad \Rightarrow \quad 9x + 3y = -15$$

(2) Add equations to eliminate y:

$$\begin{array}{r} 2x - 3y = -7 \\ 9x + 3y = -15 \\ \hline 11x = -22 \end{array}$$

(3) Solve for x:

$$11x = -22$$
$$x = -2$$

(4) Backsubstitute (into Equation 1) to solve for y:

$$2x - 3y = -7$$
$$2(-2) - 3y = -7$$
$$-3y = -3$$
$$y = 1$$

(5) Check the solution $x = -2$, $y = 1$.

$$2(-2) - 3(1) = -4 - 3 = -7 \quad \text{\textit{Equation 1}}$$
$$3(-2) + 1 = -6 + 1 = -5 \quad \text{\textit{Equation 2}}$$

EXAMPLE 2
Solving a Linear System
by Elimination

Solve the system

$$4x + 3y = 6 \qquad \text{\textit{Equation 1}}$$
$$5x + 2y = 11 \qquad \text{\textit{Equation 2}}$$

Solution:

(1) To eliminate y, we first change the coefficients of y to 6 and -6, as follows. Multiply Equation 1 by 2:

$$4x + 3y = 6 \quad \Rightarrow \quad 8x + 6y = 12$$

Multiply Equation 2 by -3:

$$5x + 2y = 11 \qquad \Longrightarrow \qquad -15x - 6y = -33$$

(2) Add equations to eliminate y:

$$
\begin{array}{r}
8x + 6y = 12 \\
-15x - 6y = -33 \\
\hline
-7x = -21
\end{array}
$$

(3) Solve for x:

$$-7x = -21$$
$$x = 3$$

(4) Backsubstitute (into Equation 1) to solve for y:

$$8x + 6y = 12$$
$$8(3) + 6y = 12$$
$$6y = -12$$
$$y = -2$$

(5) Check the solution $x = 3$, $y = -2$.

$$4(3) + 3(-2) = 12 - 6 = 6 \qquad \textit{Equation 1}$$
$$5(3) + 2(-2) = 15 - 4 = 11 \qquad \textit{Equation 2}$$

In Examples 1 and 2, both of the systems had exactly one solution. By sketching the graphs of these systems, we see that each one corresponds to two intersecting lines with a single point of intersection, as shown in Figure 8.4.

As we observed in Section 8.1, it is possible for a system of equations to have infinitely many solutions, no solution, or exactly one solution. In the case of two linear equations in two variables, we can interpret this result graphically as follows.

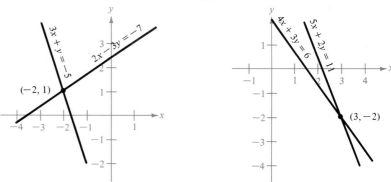

FIGURE 8.4   Linear Systems with a Single Point of Intersection

GRAPHIC INTERPRETATION OF SOLUTIONS

For a system of two linear equations in two variables, the number of solutions is given by one of the following:

Number of Solutions	**Graphical Interpretation**
Infinitely many solutions	The lines are identical.
No solution	The lines are parallel.
Exactly one solution	The lines intersect at one point.

Figure 8.5 illustrates the three cases described in this graphical interpretation of solutions.

FIGURE 8.5

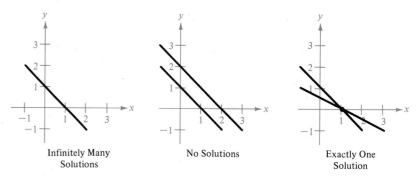

Infinitely Many Solutions   No Solutions   Exactly One Solution

We say that a system of two equations in two variables is **consistent** if it has at least one solution and is **inconsistent** if it has no solution. Moreover, a consistent system is said to be **dependent** if it has infinitely many solutions.

EXAMPLE 3
An Inconsistent System

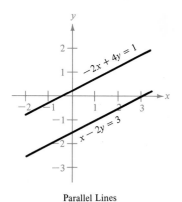

Parallel Lines

FIGURE 8.6

Find all solutions to the following system:

$$x - 2y = 3 \qquad\qquad \textit{Equation 1}$$
$$-2x + 4y = 1 \qquad\qquad \textit{Equation 2}$$

Solution:

(1) Multiply Equation 1 by 2:

$$x - 2y = 3 \quad \Longrightarrow \quad 2x - 4y = 6$$

(2) Add equations to eliminate x:

$$
\begin{array}{r}
2x - 4y = 6 \\
-2x + 4y = 1 \\
\hline
0 = 7
\end{array}
$$

From this absurd result, we conclude that there is no solution and the system is *inconsistent*. Figure 8.6 shows that these two equations correspond to two parallel lines that have no point of intersection.

EXAMPLE 4
A Dependent System

Find all solutions to the following system:

$$2x - y = 1 \qquad \qquad \textit{Equation 1}$$
$$-4x + 2y = -2 \qquad \textit{Equation 2}$$

Solution:

(1) Multiply Equation 2 by $\frac{1}{2}$:

$$-4x + 2y = -2 \qquad \Longrightarrow \qquad -2x + y = -1$$

(2) Add equations to eliminate x:

$$
\begin{array}{r}
2x - y = 1 \\
-2x + y = -1 \\
\hline
0 = 0
\end{array}
$$

Since *both* variables are eliminated *and* we obtained a valid equation, we conclude that the equations are dependent. Hence, there are infinitely many solutions to this system. Graphically, we interpret this result to mean that both equations graph as the same line, and thus *every* point on the line is a valid solution to the system. To represent this infinite solution set, we first solve for y:

$$2x - y = 1 \qquad \Longrightarrow \qquad y = 2x - 1$$

Now, let x be any real number a; then y is given by $y = 2a - 1$. In other words, every ordered pair of the form

$$(a, 2a - 1)$$

is a solution to the system. For instance, if $a = 1$, then we see that $(1, 1)$ is a solution. Similarly, if $a = 2$, we see that $(2, 3)$ is a solution.

It is important to properly interpret the results obtained with the elimination method.

1. If the elimination method leads to an equation of the form

$$0x + 0y = c, \qquad c \neq 0$$

the given system is *inconsistent* and has *no solution*.

2. If the elimination method leads to an equation of the form

$$0x + 0y = 0$$

the given system is *dependent* and has *infinitely many solutions*.

Systems of linear equations have a wide variety of applications. The following example shows how a system of linear equations may be used to solve a problem involving speed.

EXAMPLE 5
An Application Involving Rates

An airplane flying into a headwind travels the 2000-mile flying distance between two cities in 4 hours and 24 minutes. On the return flight, the same distance is traveled in 4 hours. Find the ground speed of the plane and the speed of the wind, assuming that both remain constant.

Solution:
The two unknown quantities in this problem are the speeds of the wind and the plane. We let

r_1 = rate of plane (in miles per hour)

r_2 = rate of wind (in miles per hour)

For the trip into the wind, the resulting speed is given by the difference of r_1 and r_2. For the trip with the wind, the resulting speed is given by the sum of the two speeds. (See Figure 8.7.) Using the formula

distance = (rate)(time)

we can write the following equations:

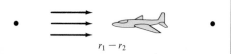

$r_1 - r_2$

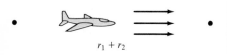

$r_1 + r_2$

FIGURE 8.7

$$2000 = (r_1 - r_2)\left(4 + \frac{24}{60}\right) \qquad \textit{Equation 1}$$

$$2000 = (r_1 + r_2)(4) \qquad \textit{Equation 2}$$

These two equations simplify to the following system:

$$5000 = 11r_1 - 11r_2 \qquad \textit{Equation 1}$$

$$500 = r_1 + r_2 \qquad \textit{Equation 2}$$

By elimination, the solution is

$$r_1 = \frac{5250}{11} \approx 477.27 \text{ miles per hour}$$

$$r_2 = \frac{250}{11} \approx 22.73 \text{ miles per hour}$$

Section Exercises 8.2 ———————————————

In Exercises 1–10, solve the linear system by elimination. Label each line on the graph with the appropriate equation.

1. $2x + y = 4$
   $x - y = 2$

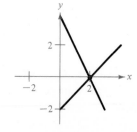

2. $x + 3y = 2$
   $-x + 2y = 2$

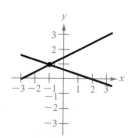

3. $x - y = 0$
   $3x - 2y = -1$

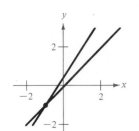

4. $2x - y = 2$
   $4x + 3y = 24$

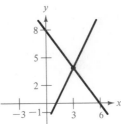

5. $x - y = 1$
   $-2x + 2y = 5$

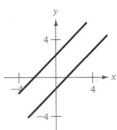

6. $3x + 2y = 2$
   $6x + 4y = 14$

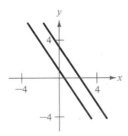

7. $\dfrac{x}{2} - \dfrac{y}{3} = 1$
   $-2x + \frac{4}{3}y = -4$

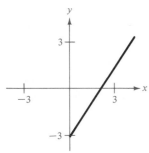

8. $x - 2y = 5$
   $6x + 2y = 7$

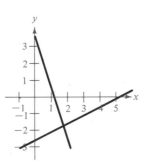

9. $9x - 3y = -1$
   $\frac{1}{5}x + \frac{2}{5}y = -\frac{1}{3}$

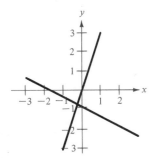

10. $5x + 3y = 18$
    $2x - 7y = -1$

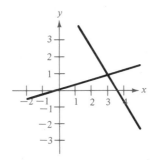

In Exercises 11–30, solve the system by elimination.

11. $x + 2y = 4$
    $x - 2y = 1$

12. $3x - 5y = 2$
    $2x + 5y = 13$

13. $2x + 3y = 18$
    $5x - y = 11$

14. $x + 7y = 12$
    $3x - 5y = 10$

15. $3x + 2y = 10$
    $2x + 5y = 3$

16. $8r + 16s = 20$
    $16r + 50s = 55$

17. $2u + v = 120$
    $u + 2v = 120$

18. $5u + 6v = 24$
    $3u + 5v = 18$

19. $6r - 5s = 3$
    $10s - 12r = 5$

20. $1.8x + 1.2y = 4$
    $9x + 6y = 3$

21. $2.5x - 3y = 1.5$
    $10x - 12y = 6$

22. $\frac{2}{3}x + \frac{1}{6}y = \frac{2}{3}$
    $4x + y = 4$

23. $\dfrac{x}{4} + \dfrac{y}{6} = 1$
    $x - y = 3$

24. $\dfrac{x-1}{2} + \dfrac{y+2}{3} = 4$
    $x - 2y = 5$

25. $\dfrac{x+3}{4} + \dfrac{y-1}{3} = 1$
    $2x - y = 12$

26. $0.02x - 0.05y = -0.19$
    $0.03x + 0.04y = 0.52$

27. $0.05x - 0.03y = 0.21$
    $0.07x + 0.02y = 0.16$

28. $3b + 3m = 7$
    $3b + 5m = 3$

29. $4b + 3m = 3$
    $3b + 11m = 13$

30. $\dfrac{12}{x} - \dfrac{12}{y} = 7$ (Hint: Let $X = 1/x$ and $Y = 1/y$.
    Then solve the linear system
    $\dfrac{3}{x} + \dfrac{4}{y} = 0$ in X and Y.)

In Exercises 31–38, solve the problem using a system of linear equations.

31. An airplane flying into a headwind travels the 1800-mile flying distance between two cities in 3 hours and 36 minutes. On the return flight, the distance is traveled in 3 hours. Find the ground speed of the plane and the speed of the wind, assuming that both remain constant.

32. Two planes start from the same airport and fly in opposite directions. The second plane starts one-half hour after the first plane, but its speed is 50 miles per hour faster. Find the ground speed of each plane if 2 hours after the first plane starts its flight the planes are 2000 miles apart.

33. Ten gallons of a 30% acid solution is obtained by mixing some 20% solution with some 50% solution. How much of each must be used?

34. Ten pounds of mixed nuts at a food co-op are to sell for $5.95 per pound. The mixture is obtained from two kinds of nuts, with one variety priced at $4.29 per pound and the other at $6.55 per pound. How much of each must be used?

35. Five hundred tickets were sold for a certain performance of a play. The tickets for adults and children sold for $2.50 and $2.00, respectively, and the receipts for the performance were $1187.50. How many of each kind of ticket were sold?

36. A woman invested an inheritance of $12,000 in two corporate bonds that pay 10.5% and 12% simple interest. If she receives $1380 in yearly interest, how much is invested in each bond?

37. The perimeter of a rectangle is 40 feet, and the length is 4 feet greater than the width. Find the rectangle's dimensions.

38. The sum of the digits of a given two-digit number is 12. If the digits are reversed, the number is increased by 36. Find the number.

The *least squares regression line*, $y = ax + b$, for the points

$$(x_1, y_1), (x_2, y_2), \ldots, (x_n, y_n)$$

is obtained by solving the linear system

$$nb + \left(\sum_{i=1}^{n} x_i\right) a = \sum_{i=1}^{n} y_i$$

$$\left(\sum_{i=1}^{n} x_i\right) b + \left(\sum_{i=1}^{n} x_i^2\right) a = \sum_{i=1}^{n} x_i y_i$$

for a and b, the slope and y-intercept. This gives the *best-fitting* line to the given points. For $n = 4$, recall that

$$\sum_{i=1}^{n} x_i = \sum_{i=1}^{4} x_i = x_1 + x_2 + x_3 + x_4$$

In Exercises 39–42, (a) find the least squares regression line, and (b) plot the given points and sketch the least squares regression line on the same axes.

39. $(-2, 0), (0, 1), (2, 3)$
40. $(-3, 0), (-1, 1), (1, 1), (3, 2)$
41. $(0, 4), (1, 3), (1, 1), (2, 0)$
42. $(1, 0), (2, 0), (3, 0), (3, 1), (4, 1), (4, 2), (5, 2), (6, 2)$

Systems of Linear Equations in More Than Two Variables

8.3

Triangular Form

The method of elimination can be applied to systems of equations in more than two variables. When we use this procedure, our goal is to transform the original system into an *equivalent* system in **triangular form.** (This may take several steps.) The following two systems are in triangular form:

$$
\begin{aligned}
x - 2y + 3z &= 9 \\
y + 2z &= 3 \\
z &= 2
\end{aligned}
\qquad
\begin{aligned}
2x - y + z - 2w &= 6 \\
2y + z - w &= 3 \\
z + w &= 1 \\
w &= -1
\end{aligned}
$$

Backsubstitution works well with a system in triangular form, since the last equation in the system gives us the value of the nth variable. By substituting this value back into the next to last equation, we find the value of an additional variable, and we continue backsubstituting until the value of each variable has been found.

Transforming a linear system into triangular form usually involves a *chain* of equivalent systems, each of which is obtained by using one or more of the following rules.

OPERATIONS THAT LEAD TO EQUIVALENT
SYSTEMS OF EQUATIONS

1. Interchange any two equations.
2. Multiply all terms of an equation by a nonzero constant.
3. Replace an equation by the sum of itself and a constant multiple of any other equation in the system.

Study Example 1 to see how we use backsubstitution to solve a system in triangular form.

EXAMPLE 1
Using Backsubstitution on a
System in Triangular Form

Solve the following system:

$$x - 2y + 3z = 9$$
$$y + 2z = 3$$
$$z = 2$$

Solution:

$$x - 2y + 3z = 9 \qquad \textit{Equation 1}$$
$$y + 2z = 3 \qquad \textit{Equation 2}$$
$$z = 2 \qquad \textit{Equation 3}$$

Substituting the value of z from the third equation into the second equation yields

$$y + 2z = 3$$
$$y + 2(2) = 3$$
$$y = -1$$

Next, substituting the values for both y and z into the first equation yields

$$x - 2y + 3z = 9$$
$$x - 2(-1) + 3(2) = 9$$
$$x = 1$$

Thus, the solution is $x = 1$, $y = -1$, and $z = 2$.

In transforming a system into triangular form, the goal is to reduce the size of the system by pairing equations and eliminating one of the variables. In the next example, note the operations used to obtain an equivalent triangular system.

EXAMPLE 2
Rewriting a System
in Triangular Form

Solve the following system:

$$x - 2y + 3z = 9 \qquad \textit{Equation 1}$$
$$-x + 3y - z = -6 \qquad \textit{Equation 2}$$
$$2x - 5y + 5z = 17 \qquad \textit{Equation 3}$$

Solution:

Though there are several ways to begin, we can eliminate x by adding the first two equations and by adding the first equation (multiplied by -2) to the third equation.

$$
\begin{array}{ll}
\begin{array}{r}
x - 2y + 3z = 9 \\
-x + 3y - z = -6 \\
\hline
y + 2z = 3
\end{array}
&
\begin{array}{r}
-2x + 4y - 6z = -18 \\
2x - 5y + 5z = 17 \\
\hline
-y - z = -1
\end{array}
\end{array}
$$

Now, we continue the process of elimination on these two new equations:

$$
\begin{array}{r}
y + 2z = 3 \\
-y - z = -1 \\
\hline
z = 2
\end{array}
$$

Finally, by choosing one equation from each system, we obtain the following triangular system:

$$
\begin{array}{r}
x - 2y + 3z = 9 \\
y + 2z = 3 \\
z = 2
\end{array}
$$

This is the same system we solved in Example 1, and we conclude that the solution is $x = 1$, $y = -1$, and $z = 2$.

From Example 2, we see a general procedure to follow.

THE METHOD OF ELIMINATION

> 1. Reduce the original system to triangular form by successively adding pairs of equations (or multiples of equations).
> 2. Backsubstitute to solve the triangular system of equations.
> 3. Check your solution in *each* of the original equations.

As with a system of linear equations in two variables, the solution(s) to a system of linear equations in more than two variables must fall into one and only one of the following categories:

1. There are infinitely many solutions (consistent system).
2. There is no solution (inconsistent system).
3. There is exactly one solution (consistent system).

When a system of equations has no solution, we simply state that it is inconsistent. If a system has exactly one solution, we list the value of each variable. However, for systems that have infinitely many solutions, we encounter a certain awkwardness in listing the solutions. For example, we might give the solutions to a system in three variables as

$$(a, a + 1, 2a), \qquad \text{where } a \text{ is any real number}$$

This means that for each value of a, we have a valid solution to the system. A few of the many possible solutions are $(-1, 0, -2)$, $(0, 1, 0)$, $(1, 2, 2)$, $(2, 3, 4)$, and so on.

Now consider the solutions represented by

$$(b - 1, b, 2b - 2), \qquad \text{where } b \text{ is any real number}$$

A few possible solutions are $(-1, 0, -2)$, $(0, 1, 0)$, $(1, 2, 2)$, $(2, 3, 4)$, and so on. Both descriptions result in the same collection of solutions. Thus, when comparing descriptions of infinite solution sets, bear in mind that there is more than one way to describe such sets.

EXAMPLE 3
A System with Infinitely Many Solutions

Solve the following system:

$$
\begin{aligned}
x + y - 3z &= -1 \qquad &\textit{Equation 1} \\
-x + 2y &= 1 \qquad &\textit{Equation 2} \\
y - z &= 0 \qquad &\textit{Equation 3}
\end{aligned}
$$

Solution:
We begin by adding the first two equations to eliminate x:

$$
\begin{aligned}
x + y - 3z &= -1 \\
-x + 2y &= 1 \\
\hline
3y - 3z &= 0
\end{aligned}
$$

Multiplying this new equation by $-\frac{1}{3}$ and adding it to Equation 3, we have

$$
\begin{aligned}
-y + z &= 0 \\
y - z &= 0 \\
\hline
0 &= 0
\end{aligned}
$$

This means that Equation 3 is *dependent* on Equations 1 and 2 in the sense that it gives us no additional information about the variables. Thus, the original system is equivalent to

$$
\begin{aligned}
x + y - 3z &= -1 \\
y - z &= 0
\end{aligned}
$$

We solve for y in terms of z:

$$y - z = 0 \qquad \Longrightarrow \qquad y = z$$

and backsubstitute to solve for x, also in terms of z:

$$x + y - 3z = -1$$
$$x + z - 3z = -1$$
$$x = 2z - 1$$

Finally, by letting z be any real number a, we have the following solution:

$$(2a - 1, a, a)$$

If we let $z = b/2$, the solution has the alternative form

$$\left(b - 1, \frac{b}{2}, \frac{b}{2}\right)$$

Nonsquare Systems

So far we have only considered **square** systems, for which the number of equations is equal to the number of variables. In a **nonsquare** system, the number of equations differs from the number of variables. In Chapter 9 we will prove that a system of linear equations cannot have a unique solution unless there are at least as many equations as there are variables in the system. Typically, systems with fewer equations than variables have either infinitely many solutions or no solutions.

EXAMPLE 4
A System with Fewer
Equations than Variables

Solve the following system:

$$x - 3y + z = 2 \qquad\qquad \textit{Equation 1}$$
$$2x + y - z = 1 \qquad\qquad \textit{Equation 2}$$

Solution:
To eliminate x, we multiply Equation 1 by (-2) and add the result to Equation 2:

$$
\begin{array}{r}
-2x + 6y - 2z = -4 \\
2x + y - z = 1 \\
\hline
7y - 3z = -3
\end{array}
$$

Thus, the following system is equivalent to the original system:

$$x - 3y + z = 2$$
$$7y - 3z = -3$$

Now we solve for both y and x in terms of z:

$$7y = 3z - 3$$
$$y = \frac{3z - 3}{7}$$

Backsubstitution into Equation 1 yields

$$x - 3\left(\frac{3z - 3}{7}\right) + z = 2$$

$$x = \frac{2z + 5}{7}$$

Finally, by letting z be any real number a, we have the solution

$$\left(\frac{2a + 5}{7}, \frac{3a - 3}{7}, a \right)$$

Homogeneous Systems

A system of linear equations in which the constant term in each equation is zero is called **homogeneous.** For example,

$$a_1 x + b_1 y + c_1 z = 0$$
$$a_2 x + b_2 y + c_2 z = 0$$
$$a_3 x + b_3 y + c_3 z = 0$$

The *trivial* (or obvious) solution to this homogeneous system is $(0, 0, 0)$. That means if all variables are given the value zero, then the equations must be satisfied. Often homogeneous systems will have nontrivial solutions also, and we can find these in the same way we find solutions for nonhomogeneous systems.

Applications

We conclude this section with two applications involving systems of linear equations in three variables.

EXAMPLE 5
An Application: Moving Object

The height at time t of an object that is moving in a (vertical) line with constant acceleration a is given by the **position equation**

$$s = \frac{1}{2} at^2 + v_0 t + s_0$$

The height s is measured in feet, t is measured in seconds, v_0 is the initial velocity (at time $t = 0$), and s_0 is the initial height. Find the values of a, v_0, and s_0 if at 1 second, $s = 52$ feet; at 2 seconds, $s = 52$ feet; and at 3 seconds, $s = 20$ feet.

Solution:
By substituting the three values of t and s into the equation for the height, we obtain three linear equations.

When $t = 1$: $\quad \frac{1}{2} a(1^2) + v_0(1) + s_0 = 52$

When $t = 2$: $\quad \frac{1}{2} a(2^2) + v_0(2) + s_0 = 52$

When $t = 3$: $\quad \frac{1}{2} a(3^2) + v_0(3) + s_0 = 20$

This system can be rewritten

$$a + 2v_0 + 2s_0 = 104 \qquad \textit{Equation 1}$$
$$2a + 2v_0 + s_0 = 52 \qquad \textit{Equation 2}$$
$$9a + 6v_0 + 2s_0 = 40 \qquad \textit{Equation 3}$$

By multiplying Equation 1 by (-2) and by (-9) and adding the results to the second and third equations respectively, we eliminate a.

$$
\begin{array}{ll}
-2a - 4v_0 - 4s_0 = -208 & -9a - 18v_0 - 18s_0 = -936 \\
\underline{2a + 2v_0 + s_0 = 52} & \underline{9a + 6v_0 + 2s_0 = 40} \\
 -2v_0 - 3s_0 = -156 & -12v_0 - 16s_0 = -896
\end{array}
$$

Now, by multiplying the first of these new equations by 3 and dividing the second by -2, we have

$$-6v_0 - 9s_0 = -468$$
$$\underline{6v_0 + 8s_0 = 448}$$
$$ -s_0 = -20$$
$$ s_0 = 20 \text{ feet}$$

By backsubstitution, we have

$$-2v_0 - 3s_0 = -156$$
$$-2v_0 - 3(20) = -156$$
$$v_0 = 48 \text{ feet per second}$$

and finally,

$$a + 2v_0 + 2s_0 = 104$$
$$a + 2(48) + 2(20) = 104$$
$$a = -32 \text{ feet per second squared}$$

This means that the position equation for this object is

$$s = -16t^2 + 48t + 20$$

In the next example we show how to fit a parabola through three given points in the plane. This procedure can be generalized to fit an nth degree polynomial function to $n + 1$ points in the plane. The only restriction to the procedure is that (since we are trying to fit a *function* to the points) every point must have a distinct x-coordinate.

EXAMPLE 6
An Application: Curve Fitting

Find an equation for the quadratic function

$$f(x) = ax^2 + bx + c$$

whose graph passes through the points $(-1, 3)$, $(1, 1)$, and $(2, 6)$.

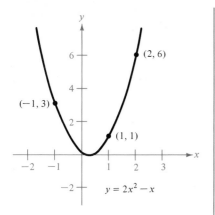

$y = 2x^2 - x$

FIGURE 8.8

Solution:
Since the graph of f passes through the points $(-1, 3)$, $(1, 1)$, and $(2, 6)$, we have

$$f(-1) = a(-1)^2 + b(-1) + c = 3$$
$$f(1) = a(1)^2 + b(1) + c = 1$$
$$f(2) = a(2)^2 + b(2) + c = 6$$

This produces the following system of linear equations in the variables a, b, and c:

$$a - b + c = 3 \qquad \text{Equation 1}$$
$$a + b + c = 1 \qquad \text{Equation 2}$$
$$4a + 2b + c = 6 \qquad \text{Equation 3}$$

The solution to this system turns out to be

$$a = 2, \qquad b = -1, \qquad \text{and} \qquad c = 0$$

Thus, the equation of the parabola passing through the three given points is

$$f(x) = 2x^2 - x$$

as shown in Figure 8.8.

Section Exercises 8.3

In Exercises 1–26, solve the systems of equations.

1. $\begin{aligned} x + y + z &= 6 \\ 2x - y + z &= 3 \\ 3x \quad\;\; - z &= 0 \end{aligned}$

2. $\begin{aligned} x + y + z &= 2 \\ -x + 3y + 2z &= 8 \\ 4x + y \quad\quad\;\; &= 4 \end{aligned}$

3. $\begin{aligned} 4x + y - 3z &= 11 \\ 2x - 3y + 2z &= 9 \\ x + y + z &= -3 \end{aligned}$

4. $\begin{aligned} 2x \quad\quad + 2z &= 2 \\ 5x + 3y \quad\quad &= 4 \\ 3y - 4z &= 4 \end{aligned}$

5. $\begin{aligned} 6y + 4z &= -12 \\ 3x + 3y \quad\quad &= 9 \\ 2x \quad\quad - 3z &= 10 \end{aligned}$

6. $\begin{aligned} 2x + 4y + z &= -4 \\ x - 2y + 3z &= \tfrac{13}{2} \\ 2x - y + \tfrac{1}{2}z &= 3 \end{aligned}$

7. $\begin{aligned} 3x - 2y + 4z &= 1 \\ x + y - 2z &= 3 \\ 2x - 3y + 6z &= 8 \end{aligned}$

8. $\begin{aligned} 5x - 3y + 2z &= 3 \\ 2x + 4y - z &= 7 \\ x - 11y + 4z &= 3 \end{aligned}$

9. $\begin{aligned} 3x + 3y + 5z &= 1 \\ 3x + 5y + 9z &= 0 \\ 5x + 9y + 17z &= 0 \end{aligned}$

10. $\begin{aligned} 2x + y + 3z &= 1 \\ 2x + 6y + 8z &= 3 \\ 6x + 8y + 18z &= 5 \end{aligned}$

11. $\begin{aligned} x + 2y - 7z &= -4 \\ 2x + y + z &= 13 \\ 3x + 9y - 36z &= -33 \end{aligned}$

12. $\begin{aligned} 2x + y - 3z &= 4 \\ 4x \quad\quad + 2z &= 10 \\ -2x + 3y - 13z &= -8 \end{aligned}$

13. $\begin{aligned} x \quad\quad + 4z &= 13 \\ 4x - 2y + z &= 7 \\ 2x - 2y - 7z &= -19 \end{aligned}$

14. $\begin{aligned} 4x - y + 5z &= 11 \\ x + 2y - z &= 5 \\ 5x - 8y + 13z &= 7 \end{aligned}$

15. $\begin{aligned} x - 2y + 5z &= 2 \\ 3x + 2y - z &= -2 \end{aligned}$

16. $\begin{aligned} x - 3y + 2z &= 18 \\ 5x - 13y + 12z &= 80 \end{aligned}$

17. $\begin{aligned} 2x - 3y + z &= -2 \\ -4x + 9y \quad\quad &= 7 \end{aligned}$

18. $\begin{aligned} 2x + 3y + 3z &= 7 \\ 4x + 18y + 15z &= 44 \end{aligned}$

19. $\begin{aligned} x \quad\quad + 4z &= 1 \\ x + y + 10z &= 10 \\ 2x - y + 2z &= -5 \end{aligned}$

20. $\begin{aligned} x + y + z + w &= 6 \\ 2x + 3y \quad\quad - w &= 0 \\ -3x + 4y + z + 2w &= 4 \\ x + 2y - z + w &= 0 \end{aligned}$

21. $\begin{aligned} x \quad\quad\quad + 3w &= 4 \\ 2y - z - w &= 0 \\ 3y \quad\quad - 2w &= 1 \\ 2x - y + 4z \quad\quad &= 5 \end{aligned}$

22. $\begin{aligned} 3x - 2y - 6z &= -4 \\ -3x + 2y + 6z &= 1 \\ x - y - 5z &= -3 \end{aligned}$

23. $\begin{aligned} 4x + 3y + 17z &= 0 \\ 5x + 4y + 22z &= 0 \\ 4x + 2y + 19z &= 0 \end{aligned}$ 24. $\begin{aligned} 2x + 3y &= 0 \\ 4x + 3y - z &= 0 \\ 8x + 3y + 3z &= 0 \end{aligned}$

25. $\begin{aligned} 5x + 5y - z &= 0 \\ 10x + 5y + 2z &= 0 \\ 5x + 15y - 9z &= 0 \end{aligned}$ 26. $\begin{aligned} 12x + 5y + z &= 0 \\ 12x + 4y - z &= 0 \end{aligned}$

In Exercises 27–30, find the equation of the parabola

$$y = ax^2 + bx + c$$

passing through the given points.

27. $(0, -4), (1, 1), (2, 10)$ 28. $(0, 5), (1, 6), (2, 5)$
29. $(1, 0), (3, 0), (2, -1)$ 30. $(1, 2), (2, 1), (3, -4)$

In Exercises 31–34, find an equation of the circle

$$x^2 + y^2 + Dx + Ey + F = 0$$

passing through the given points.

31. $(0, 0), (2, -2), (4, 0)$ 32. $(0, 0), (0, 6), (-3, 3)$
33. $(3, -1), (-2, 4), (6, 8)$ 34. $(0, 0), (0, 2), (3, 0)$

In Exercises 35–38, use the given information to find a, v_0, and s_0 in the position equation

$$s = \tfrac{1}{2}at^2 + v_0t + s_0$$

35. At $t = 1$ second, $s = 128$ feet; at $t = 2$ seconds, $s = 80$ feet; and at $t = 3$ seconds, $s = 0$.
36. At $t = 1$ second, $s = 48$ feet; at $t = 2$ seconds, $s = 64$ feet; and at $t = 3$ seconds, $s = 48$ feet.
37. At $t = 1$ second, $s = 452$ feet; at $t = 3$ seconds, $s = 260$ feet; and at $t = 4$ seconds, $s = 116$ feet.
38. At $t = 2$ seconds, $s = 132$ feet; at $t = 3$ seconds, $s = 100$ feet; and at $t = 4$ seconds, $s = 36$ feet.

In Exercises 39–42, use a system of linear equations to decompose each rational fraction into partial fractions. (See Section 3.7.)

39. $\dfrac{1}{x^3 - x} = \dfrac{A}{x} + \dfrac{B}{x - 1} + \dfrac{C}{x + 1}$

40. $\dfrac{x^2 + 4x - 1}{x^2 - x} = 1 + \dfrac{5x - 1}{x^2 - x} = 1 + \dfrac{A}{x} + \dfrac{B}{x - 1}$

41. $\dfrac{x^2 - 3x - 3}{x(x - 2)(x + 3)} = \dfrac{A}{x} + \dfrac{B}{x - 2} + \dfrac{C}{x + 3}$

42. $\dfrac{12}{x(x - 2)(x + 3)} = \dfrac{A}{x} + \dfrac{B}{x - 2} + \dfrac{C}{x + 3}$

43. A small company that manufactures products A and B has an order for 15 units of product A and 16 units of product B. The company has trucks of three different sizes that can haul the products, as shown in the following table:

		Product	
		A	B
Truck	Large	6	3
	Medium	4	4
	Small	0	3

How many trucks of each size are needed to deliver the order? (Give *two* possible solutions.)

44. A chemist needs 10 liters of a 25% acid solution. He has three solutions containing the acid, in which the concentration is 10%, 20%, and 50%, respectively. How many liters of each solution should the chemist use if he wants to use
(a) as little as possible of the 50% solution?
(b) as much as possible of the 50% solution?
(c) two liters of the 50% solution?

The Least Squares Regression Parabola, $y = ax^2 + bx + c$, for the points

$$(x_1, y_1), (x_2, y_2), \ldots, (x_n, y_n)$$

is obtained by solving the linear system

$$nc + \left(\sum_{i=1}^{n} x_i\right)b + \left(\sum_{i=1}^{n} x_i^2\right)a = \sum_{i=1}^{n} y_i$$

$$\left(\sum_{i=1}^{n} x_i\right)c + \left(\sum_{i=1}^{n} x_i^2\right)b + \left(\sum_{i=1}^{n} x_i^3\right)a = \sum_{i=1}^{n} x_iy_i$$

$$\left(\sum_{i=1}^{n} x_i^2\right)c + \left(\sum_{i=1}^{n} x_i^3\right)b + \left(\sum_{i=1}^{n} x_i^4\right)a = \sum_{i=1}^{n} x_i^2y_i$$

for a, b, and c. This gives the *best-fitting* parabola to the given points. In Exercises 45–48, (a) find the least squares regression parabola, and (b) plot the given points and sketch the least squares parabola on the same axes.

45. $(-2, 0), (-1, 0), (0, 1), (1, 2), (2, 5)$
46. $(-4, 5), (-2, 6), (2, 6), (4, 2)$
47. $(0, 0), (2, 2), (3, 6), (4, 12)$
48. $(0, 10), (1, 9), (2, 6), (3, 0)$

Systems of Inequalities and Linear Programming

8.4

The following statements are **inequalities in two variables:**

$$3x - 2y < 4 \qquad \text{and} \qquad 2x^2 + 3y^2 \geq 6$$

A **solution** to an inequality in two variables is an ordered pair (x, y) that satisfies the inequality. The **graph** of an inequality in two variables is the collection of all solutions. To sketch the graph of an inequality in two variables, begin by sketching the graph of the corresponding *equality* (or equation). This graph is made with a dashed line for the strict inequalities $<$ or $>$ and a solid line for the inequalities $\leq$ or $\geq$. The graph will normally divide the plane into two or more regions. In each such region, one of the following must be true:

1. All points in the region are solutions of the inequality, or
2. No points in the region are solutions of the inequality.

This means that we can determine whether or not the points in an entire region satisfy the inequality by simply testing *one* point in the region.

SKETCHING THE GRAPH OF AN INEQUALITY IN TWO VARIABLES

1. Replace the inequality sign by an equality sign, and sketch the graph of the resulting equation. (Use a dashed line for $<$ or $>$ and a solid line for $\leq$ or $\geq$.)
2. Test one point in each of the regions formed by the graph in step 1. If the point satisfies the inequality, then shade the entire region to denote that every point in the region satisfies the inequality.

EXAMPLE 1
Sketching the Graph of an Inequality

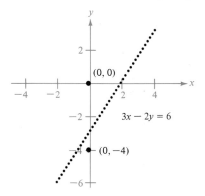

Graph of Equality   **FIGURE 8.9**

Sketch the graphs of the following inequalities:

(a) $3x - 2y < 6$ (b) $y \geq x^2 - 1$ (c) $xy \leq 1$

Solution:

(a) We begin by sketching the graph of the equation

$$3x - 2y = 6$$

as shown in Figure 8.9. Note that we use a dashed line to indicate that the points on the line *do not* satisfy the original inequality. Next, we choose two convenient test points, one from each of the regions determined in Figure 8.9. By testing these points, we have the following:

Test Point (0, 0): $3(0) - 2(0) = 0 < 6$

Test Point (0, -4): $3(0) - 2(-4) = 8 > 6$

Thus, we conclude that every point in the half-plane above the line satisfies the inequality and every point in the half-plane below (or on) the line does not satisfy the inequality, as shown in Figure 8.10.

(b) The graph of the equality is the parabola shown in Figure 8.11. By testing (0, 0) and (0, −2), we see that the points that satisfy the inequality are those lying above (or on) the parabola, as shown.

(c) The graph of the equality is the hyperbola shown in Figure 8.12. By testing (−2, −2), (0, 0), and (2, 2), we see that the points that satisfy the inequality are those lying between (or on) the two branches of the hyperbola, as shown.

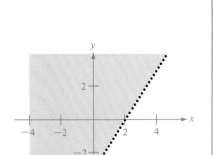

Graph of Inequality

FIGURE 8.10

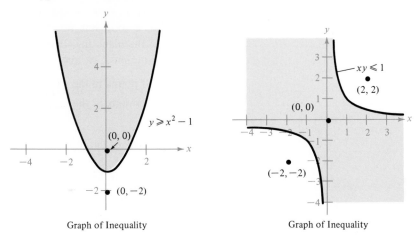

Graph of Inequality   Graph of Inequality

FIGURE 8.11   **FIGURE 8.12**

Remark: Remember that in general we can add or subtract variable quantities to both sides of an inequality, but we *cannot* multiply or divide both sides of an inequality by a variable quantity. For example,

$$xy \le 1 \quad \textit{is not} \text{ equivalent to} \quad y \le \frac{1}{x}$$

LINEAR INEQUALITIES IN TWO VARIABLES

A **linear inequality in two variables** has one of the following forms:

$$ax + by < c \qquad ax + by \le c$$
$$ax + by > c \qquad ax + by \ge c$$

The graph of a linear inequality is a half-plane.

Remark: Since the graph of a linear inequality must be a half-plane, it is actually unnecessary to test points in both half-planes. *However,* because this test is relatively simple, we suggest that you test a point in each half-plane as a check on your conclusion.

EXAMPLE 2
Graphing Linear Inequalities
in Two Variables

Sketch the following linear inequalities in the plane:

(a) $x - y \leq 2$ (b) $x > -2$ (c) $y \leq 3$

Solution:

(a) By replacing $\leq$ with $=$, we obtain the equation

$$x - y = 2 \quad \text{or} \quad y = x - 2$$

which is the line shown in Figure 8.13. Since the origin $(0, 0)$ satisfies the inequality, we see that the graph consists of the half-plane lying above this line, as shown. (Check a point in the other half-plane.)

(b) The equation $x = -2$ graphs as a vertical line. The points that satisfy the inequality $x > -2$ are the points lying to the right of this line, as shown in Figure 8.14.

(c) The equation $y = 3$ graphs as a horizontal line. The points that satisfy the inequality $y \leq 3$ are the points lying below (or on) this line, as shown in Figure 8.15.

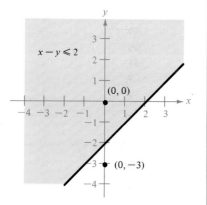

FIGURE 8.13

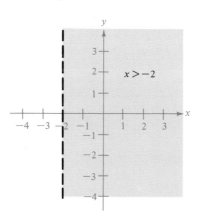

FIGURE 8.14

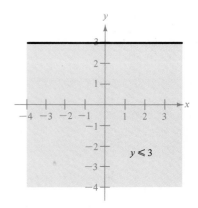

FIGURE 8.15

Systems of Inequalities

Graphing the solution of a *system* of inequalities in two variables is accomplished by graphing each inequality in the system and then hunting for the region in the plane that is common to every graph in the system.

EXAMPLE 3
Solving a System of Inequalities

Sketch the graph (and label the vertices) of the region containing all points that satisfy the following system:

$$x - y \leq 2$$
$$x > -2$$
$$y \leq 3$$

Solution:

We have already sketched the graph of each of these inequalities (in Example 2). To find the region common to all three graphs, we superimpose all three graphs on the same coordinate plane, as shown in Figure 8.16. To find the vertices of this triangular region, we solve the three systems of equations obtained by taking pairs of the equations that represent the borders of the region.

Inequality	**Equation of Border**
$x - y \leq 2$	$x - y = 2$
$x > -2$	$x = -2$
$y \leq 3$	$y = 3$

Vertex 1: $(-2, -4)$
Obtained by solving the system

$x - y = 2$
$x = -2$

Vertex 2: $(5, 3)$
Obtained by solving the system

$x - y = 2$
$y = 3$

Vertex 3: $(-2, 3)$
Obtained by solving the system

$x = -2$
$y = 3$

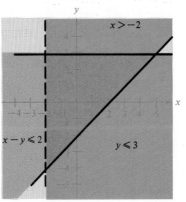
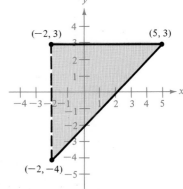

FIGURE 8.16

For the triangular region shown in Example 3, we found the vertices by intersecting each pair of border lines. With more complicated regions, two border lines can sometimes intersect at a point that is not a vertex of the region, as shown in Figure 8.17. In order to keep track of which points of intersection are actually vertices of the region, we suggest that you make a careful sketch of the region and refer to your sketch as you find each point of intersection.

In Example 3 (and in Figure 8.17), we dealt with systems of *linear* inequalities. Since the border equations for linear systems are straight lines, we can see that the enclosed region will always be a polygon. If one or more of the inequalities in the system are nonlinear, then the region will not (usually) be a polygon. Example 4 illustrates this more general type of region.

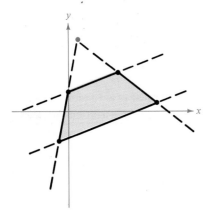

Border lines can intersect
at a point that is not a vertex.

FIGURE 8.17

EXAMPLE 4
Solving a System of Inequalities

Sketch the graph (and label the vertices) of the region containing all points that satisfy the following system:

$$x^2 + y^2 \geq 25$$
$$7x - y \leq 25$$

Solution:
As shown in Figure 8.18, the points that satisfy the inequality $x^2 + y^2 \geq 25$ are the points lying outside (or on) a circle of radius 5 centered at the origin. The points satisfying the inequality $7x - y \leq 25$ are the points lying above the line given by $7x - y = 25$.

To find the points of intersection, we can use substitution. We begin by solving for y in the linear *equation:*

$$7x - y = 25 \quad \Longrightarrow \quad y = 7x - 25$$

Then we substitute this value of y into the equation of the circle and solve for x:

$$x^2 + y^2 = 25$$
$$(7x - 25)^2 + x^2 = 25$$
$$49x^2 - 350x + 625 + x^2 - 25 = 0$$
$$50x^2 - 350x + 600 = 0$$
$$x^2 - 7x + 12 = 0$$
$$(x - 3)(x - 4) = 0$$
$$x = 3 \text{ or } 4$$

Finally, we backsubstitute to solve for y. If $x = 3$, then

$$y = 7(3) - 25 = -4$$

If $x = 4$, then

$$y = 7(4) - 25 = 3$$

Thus, the two points of intersection are $(3, -4)$ and $(4, 3)$, as shown in Figure 8.18.

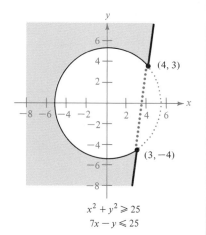

$$x^2 + y^2 \geq 25$$
$$7x - y \leq 25$$

FIGURE 8.18

Three types of unusual situations may arise in systems of inequalities:

1. Sometimes no points are common to a system. For example, the system

$$x + y > 3$$
$$x + y < -1$$

obviously has no solution points, since the quantity $(x + y)$ cannot be both less than -1 and greater than 3 for any values of x and y. (See Figure 8.19.)

2. A system of inequalities may describe a region that is unbounded in certain directions. For example, the system

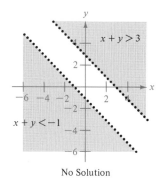

$x + y > 3$

$x + y < -1$

No Solution

FIGURE 8.19

$$x + y < 3$$
$$x + 2y > 3$$

has solutions that form an *infinite wedge,* as shown in Figure 8.20.

3. One or more of the inequalities in the system might be redundant. For example, the two systems

$$
\begin{array}{ll}
x + y < 2 & \quad x + y < 2 \\
x + y > 0 & \quad x > 0 \\
 x > 0 & \quad y > 0 \\
 y > 0 &
\end{array}
$$

have the same solution region. We can see in Figure 8.21 that the inequality $x + y > 0$ is redundant, since it adds no further restriction to that given by the other three inequalities.

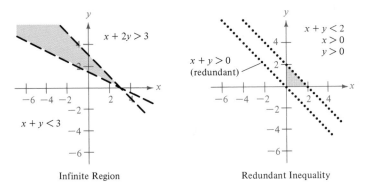

Infinite Region	Redundant Inequality
FIGURE 8.20	**FIGURE 8.21**

Linear Programming

In many applications in business or economics, we are concerned with a process called **optimization.** For example, an optimization problem can involve finding the minimum cost, the maximum profit, or the minimum use of resources. There are many different types of optimization problems, and many of these require calculus techniques. One type that does not require calculus is called **linear programming.** A linear programming problem consists of a linear **objective function** and a set of linear inequalities called **constraints.** The objective function gives the quantity that is to be minimized (or maximized), and the constraints determine the allowable solution points.

For example, consider the following linear programming problem, in which you are asked to maximize the value of C subject to the given constraints:

Objective Function: $C = 3x + 2y$

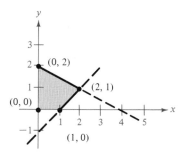

FIGURE 8.22

Constraints:

$$x \geq 0$$
$$y \geq 0$$
$$x + 2y \leq 4$$
$$x - y \leq 1$$

These constraints form the boundaries for the region shown in Figure 8.22. Every point in the region satisfies all four constraints, and since there are infinitely many points, it is not evident how we should go about finding the point that gives a maximum value of C. Fortunately, it can be shown that the *solution to a linear programming problem must occur at one of the vertices of the region bounded by the constraints*. This means that we can find the maximum value by testing C at each of the vertices. In this particular case we have the following:

At $(0, 0)$: $C = 3(0) + 2(0) = 0$
At $(1, 0)$: $C = 3(1) + 2(0) = 3$
At $(2, 1)$: $C = 3(2) + 2(1) = 8$ *(Maximum value of C)*
At $(0, 2)$: $C = 3(0) + 2(2) = 4$

Thus, the maximum value of C occurs when $x = 2$ and $y = 1$.

EXAMPLE 5
An Application of Linear Programming

A manufacturer wants to maximize the profit for two products. The first product yields a profit of $1.50 per unit, and the second product yields a profit of $2.00 per unit. Market tests and available resources have indicated the following constraints: (1) The combined production level should not exceed 1200 units per month, (2) the demand for product II is at most half of that for product I, and (3) the production level of product I can exceed three times the production level of product II by at most 600 units.

Solution:
If we let x be the number of units of product I and y be the number of units of product II, then the objective function (for the combined profit) is given by

$$P = 1.5x + 2y$$

The three given constraints can be interpreted as linear inequalities,

$$x + y \leq 1200 \tag{1}$$

$$y \leq \frac{x}{2} \qquad \text{or} \qquad x - 2y \geq 0 \tag{2}$$

$$x \leq 3y + 600 \qquad \text{or} \qquad x - 3y \leq 600 \tag{3}$$

Since neither x nor y can be negative, we also have the following two additional constraints:

$$x \geq 0 \qquad \text{and} \qquad y \geq 0$$

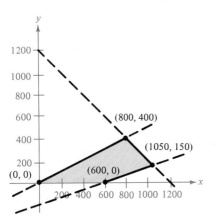

FIGURE 8.23

Figure 8.23 shows the region determined by these linear inequalities. To find the maximum profit, we test the value of *P* at the vertices of the region.

$$At\ (0,\ 0):\quad P = 1.5(0) + 2(0) = 0$$
$$At\ (600,\ 0):\quad P = 1.5(600) + 2(0) = \$900$$
$$At\ (1050,\ 150):\quad P = 1.5(1050) + 2(150) = \$1875$$
$$At\ (800,\ 400):\quad P = 1.5(800) + 2(400) = \$2000$$

Thus, the maximum profit is $2000 and occurs when the monthly production consists of 800 units of product I and 400 units of product II.

Section Exercises 8.4

In Exercises 1–10, match the inequality with the correct graph.

1. $x > 3$
2. $y \le 2$
3. $2x + 3y \le 6$
4. $2x - y \ge -2$
5. $x^2 + y^2 < 4$
6. $(x - 2)^2 + (y - 3)^2 > 4$
7. $xy > 2$
8. $y \le 4 - x^2$
9. $y^2 - x^2 \le 4$
10. $y \le e^{-x^2}$

(e)

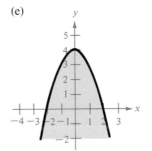

(f)

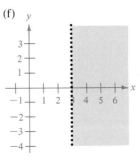

(a)

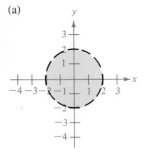

(b)

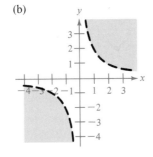

(g)

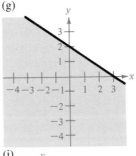

(h)

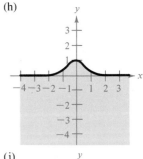

(c)

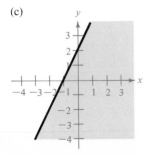

(d)

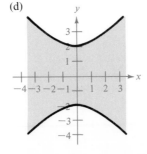

(i)

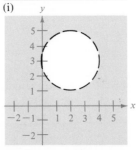

(j)

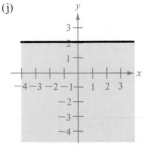

In Exercises 11–20, sketch the graph of the inequality.

11. $y \geq -1$

12. $x \leq 4$

13. $y < 2 - x$

14. $y > 2x - 4$

15. $2y - x \geq 4$

16. $5x + 3y \geq -15$

17. $(x + 1)^2 + (y - 2)^2 < 9$

18. $y^2 - x < 0$

19. $y \leq \dfrac{1}{1 + x^2}$

20. $y < \ln x$

In Exercises 21–40, sketch the graph of the solution of the system of inequalities.

21. $\begin{aligned} x + y &\leq 1 \\ -x + y &\leq 1 \\ y &\geq 0 \end{aligned}$

22. $\begin{aligned} 3x + 2y &< 6 \\ x &> 0 \\ y &> 0 \end{aligned}$

23. $\begin{aligned} x + y &\leq 5 \\ x &\geq 2 \\ y &\geq 0 \end{aligned}$

24. $\begin{aligned} 2x + y &\geq 2 \\ y &\leq 1 \\ x &\leq 2 \end{aligned}$

25. $\begin{aligned} -3x + 2y &< 6 \\ x + 4y &> -2 \\ 2x + y &< 3 \end{aligned}$

26. $\begin{aligned} x - 7y &> -36 \\ 5x + 2y &> 5 \\ 6x - 5y &> 6 \end{aligned}$

27. $\begin{aligned} 2x + y &> 2 \\ 6x + 3y &< 2 \end{aligned}$

28. $\begin{aligned} x - 2y &< -6 \\ 5x - 3y &> -9 \end{aligned}$

29. $\begin{aligned} x &\geq 1 \\ x - 2y &\leq 3 \\ 3x + 2y &\geq 9 \\ x + y &\leq 6 \end{aligned}$

30. $\begin{aligned} x - y^2 &> 0 \\ x - y &< 2 \end{aligned}$

31. $\begin{aligned} x^2 + y^2 &\leq 9 \\ x^2 + y^2 &\geq 1 \end{aligned}$

32. $\begin{aligned} x^2 + y^2 &\leq 25 \\ 4x - 3y &\leq 0 \end{aligned}$

33. $\begin{aligned} x &> y^2 \\ x &< y + 2 \end{aligned}$

34. $\begin{aligned} x &< 2y - y^2 \\ 0 &< x + y \end{aligned}$

35. $\begin{aligned} y &\leq \sqrt{3x} + 1 \\ y &\geq x + 1 \end{aligned}$

36. $\begin{aligned} y &< -x^2 + 2x + 3 \\ y &> x^2 - 4x + 3 \end{aligned}$

37. $\begin{aligned} y &< x^3 - 2x + 1 \\ y &> -2x \\ x &\leq 1 \end{aligned}$

38. $\begin{aligned} y &\geq x^4 - 2x^2 + 1 \\ y &\leq 1 - x^2 \end{aligned}$

39. $\begin{aligned} x^2 y &\geq 1 \\ 0 &< x \leq 4 \\ y &\leq 4 \end{aligned}$

40. $\begin{aligned} y &\leq e^{-x^2/2} \\ y &\geq 0 \\ -2 &\leq x \leq 2 \end{aligned}$

41. A furniture company can sell all the tables and chairs that it produces. Each table requires 1 hour in the assembly center and $1\frac{1}{3}$ hours in the finishing center. Each chair requires $1\frac{1}{2}$ hours in the assembly center and $1\frac{1}{2}$ hours in the finishing center. The company's assembly center is available 12 hours per day, and its finishing center is available 15 hours per day. If x is the number of tables produced per day and y is the number of chairs, find a system of inequalities describing all possible production levels. Sketch the graph of the system.

42. A store sells two models of a certain brand of computer.

Because of the demand, it is necessary to stock at least twice as many units of model A as units of model B. The cost to the store for the two models is $800 and $1200, respectively. The management does not want more than $20,000 in computer inventory at any time, and it wants at least 4 model A computers and 2 model B computers in inventory at all times. Devise a system of inequalities describing all possible inventory levels, and sketch the graph of the system.

In Exercises 43–48, derive a set of inequalities to describe the region.

43. Rectangle with vertices at $(2, 1)$, $(5, 1)$, $(5, 7)$, and $(2, 7)$.

44. Triangle with vertices at $(0, 0)$, $(5, 0)$, and $(2, 3)$.

45. Triangle with vertices at $(-1, 0)$, $(1, 0)$, and $(0, 1)$.

46. Segment of a circle

47. Segment of a circle

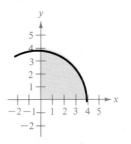

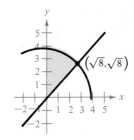

48. Parabolic region

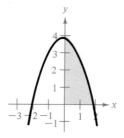

In Exercises 49–52, find the minimum and maximum values of *both* objective functions, subject to the given constraints.

49. Constraints:
$$\begin{aligned} x &\geq 0 \\ y &\geq 0 \\ x + 3y &\leq 15 \\ 4x + y &\leq 16 \end{aligned}$$

Objective function:

(a) $C = 3x + 2y$

(b) $C = 5x + \dfrac{y}{2}$

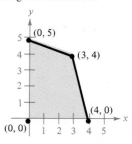

50. Constraints: $x \geq 0$
    $2x + 3y \geq 6$
    $3x - 2y \leq 9$
    $x + 5y \leq 20$

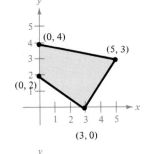

Objective function:
(a) $C = 4x + 3y$
(b) $C = x + 6y$

51. Constraints: $x \geq 0$
    $y \geq 0$
    $x \leq 60$
    $y \leq 45$
    $5x + 6y \leq 420$

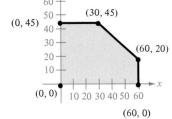

Objective function:
(a) $C = 10x + 7y$
(b) $C = 25x + 30y$

52. Constraints: $x \geq 0$
    $y \geq 0$
    $8x + 9y \leq 7200$
    $8x + 9y \geq 5400$

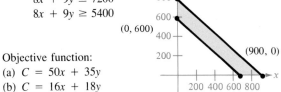

Objective function:
(a) $C = 50x + 35y$
(b) $C = 16x + 18y$

53. A merchant plans to sell two models of home computers that will cost $250 and $400, respectively, and he estimates

that the total monthly demand will not exceed 250 units. Find the number of units of each model that the merchant should stock in order to maximize profit if he does not want to invest more than $70,000 in computer inventory and if his profit on the two models is $45 (for the model costing $250) and $50 (for the model costing $400).

54. A fruit grower has 150 acres of land available to raise two crops, A and B. It takes 1 day to trim an acre of A and 2 days to trim an acre of B, and there are 240 days per year available for trimming. It takes $\frac{3}{10}$ day to pick an acre of A and $\frac{1}{10}$ day to pick an acre of B, and there are 30 days per year available for picking. Find the number of acres of each fruit that should be planted to maximize profit, if the profit for crops A and B is $140 per acre and $235 per acre, respectively.

55. Find the minimum value of the objective function $C = 4x + 5y$ subject to the constraints

$$x + y \geq 8$$
$$4x + 3y \geq 27$$
$$3x + 5y \geq 30$$
$$x \geq 0$$
$$y \geq 0$$

56. A farmer mixes two brands of cattle feed. Brand X costs $25 per bag and contains 2 units of nutritional element A, 2 units of element B, and 2 units of element C. Brand Y costs $20 per bag and contains 1 unit of nutritional element A, 9 units of element B, and 3 units of element C. Find the number of bags of each brand that should be mixed to produce a mixture having a minimum cost per bag. The minimum requirements of nutrients A, B, and C are 12 units, 36 units, and 24 units, respectively.

Review Exercises / Chapter 8

In Exercises 1–20, find all solutions to the system of equations.

1. $x + y = 2$
   $x - y = 0$

2. $2x = 3(y - 1)$
   $y = x$

3. $x^2 - y^2 = 9$
   $x - y = 1$

4. $x^2 + y^2 = 169$
   $3x + 2y = 39$

5. $y = 2x^2$
   $y = x^4 - 2x^2$

6. $x = y + 3$
   $x = y^2 + 1$

7. $y^2 - 2y + x = 0$
   $x + y = 0$

8. $y = 2x^2 - 4x + 1$
   $y = x^2 - 4x + 3$

9. $2x - y = 2$
   $6x + 8y = 39$

10. $40x + 30y = 24$
    $20x - 50y = -14$

11. $0.2x + 0.3y = 0.14$
    $0.4x + 0.5y = 0.20$

12. $12x + 42y = -17$
    $30x - 18y = 19$

13. $\dfrac{x}{2} - \dfrac{y}{3} = 0$
    $3x + 2(y + 5) = 10$

14. $\dfrac{x}{3} + \dfrac{4y}{7} = 3$
    $2x + 3y = 15$

15. $x + 2y + 6z = 4$
    $-3x + 2y - z = -4$
    $4x + 2z = 16$

16. $2x + 3y + 4z = 21$
    $5x + y - 2z = -17$
    $8x + 9y + z = -12$

17. $\begin{aligned} x - 2y + z &= -6 \\ 2x - 3y &= -7 \\ -x + 3y - 3z &= 11 \end{aligned}$

18. $\begin{aligned} 2x + 6z &= -9 \\ 3x - 2y + 11z &= -16 \\ 3x - y + 7z &= -11 \end{aligned}$

19. $\begin{aligned} 2x + 5y - 19z &= 34 \\ 3x + 8y - 31z &= 54 \end{aligned}$

20. $\begin{aligned} 2x + y + z + 2w &= -1 \\ 5x - 2y + z - 3w &= 0 \\ -x + 3y + 2z + 2w &= 1 \\ 3x + 2y + 3z - 5w &= 12 \end{aligned}$

In Exercises 21–28, sketch the graph of the solutions of the systems of inequalities.

21. $\begin{aligned} x + 2y &\le 160 \\ 3x + y &\le 180 \\ x &\ge 0 \\ y &\ge 0 \end{aligned}$

22. $\begin{aligned} 2x + 3y &\le 24 \\ 2x + y &\le 16 \\ x &\ge 0 \\ y &\ge 0 \end{aligned}$

23. $\begin{aligned} 3x + 2y &\ge 24 \\ x + 2y &\ge 12 \\ x &\ge 2 \\ x &\le 15 \\ y &\le 15 \end{aligned}$

24. $\begin{aligned} 2x + y &\ge 16 \\ x + 3y &\ge 18 \\ 0 \le x &\le 25 \\ 0 \le y &\le 25 \end{aligned}$

25. $\begin{aligned} y &< x + 1 \\ y &> x^2 - 1 \end{aligned}$

26. $\begin{aligned} y &\le 6 - 2x - x^2 \\ y &\ge x + 6 \end{aligned}$

27. $\begin{aligned} 2x - 3y &\ge 0 \\ 2x - y &\le 8 \\ y &\ge 0 \end{aligned}$

28. $\begin{aligned} x^2 + y^2 &\le 9 \\ (x - 3)^2 + y^2 &\le 9 \end{aligned}$

29. A mixture of 6 parts of chemical A, 8 parts of chemical B, and 13 parts of chemical C is required to kill a certain destructive crop insect. Commercial spray X contains 1, 2, and 2 parts, respectively, of these chemicals. Commercial spray Y contains only chemical C. Commercial spray Z contains chemicals A, B, and C in equal amounts. How much of each type of commercial spray is needed to get the desired mixture?

9

MATRICES AND DETERMINANTS

Matrix Solutions of Systems of Linear Equations

9.1

In mathematics we always look for a valid shortcut to solving problems. In this section we will look at a streamlined technique for solving systems of linear equations. Let's reconsider the system of equations in Example 2 of Section 8.3. (Note that the equations are arranged so that the variables line up vertically. This is important.)

$$x - 2y + 3z = 9$$
$$-x + 3y - z = -6$$
$$2x - 5y + 5z = 17$$

If you look back at the solution of Example 2, you will see that variables x, y, and z served mainly to keep track of the position of the coefficients. The actual decisions in the solution were based on the values of the coefficients. By writing only the coefficients, we obtain the following two-dimensional array, called a **matrix:**

$$\begin{bmatrix} 1 & -2 & 3 & 9 \\ -1 & 3 & -1 & -6 \\ 2 & -5 & 5 & 17 \end{bmatrix}$$

This particular matrix has three **rows** (numbers in horizontal lines) and four **columns** (numbers in vertical lines).

DEFINITION OF A MATRIX

If m and n are positive integers, then an $m \times n$ **matrix** is an array of the form

$$
\begin{bmatrix}
a_{11} & a_{12} & a_{13} & \cdots & a_{1n} \\
a_{21} & a_{22} & a_{23} & \cdots & a_{2n} \\
a_{31} & a_{32} & a_{33} & \cdots & a_{3n} \\
\cdot & \cdot & \cdot & & \cdot \\
\cdot & \cdot & \cdot & & \cdot \\
\cdot & \cdot & \cdot & & \cdot \\
a_{m1} & a_{m2} & a_{m3} & \cdots & a_{mn}
\end{bmatrix}
$$

where each **element a_{ij} of the matrix** is a real number.

An $m \times n$ matrix (read "m by n") has m rows and n columns and is said to be of *order $m \times n$*. If $m = n$, the matrix is **square** of order n. The element a_{ij} is located in the ith row and jth column. We call i the **row subscript** and j the **column subscript.** The elements $a_{11}, a_{22}, a_{33}, \ldots$ are called the **main diagonal elements.**

A matrix derived from a system of linear equations is called the **augmented matrix of the system.** Moreover, the matrix that is derived from the coefficients of the system (and does not include the constant terms) is called the **coefficient matrix of the system.** For example,

System

$$
\begin{aligned}
x - 4y + 3z &= 5 \\
-x + 3y - z &= -3 \\
2x \quad\quad - 4z &= 6
\end{aligned}
$$

Coefficient Matrix

$$
\begin{bmatrix}
1 & -4 & 3 \\
-1 & 3 & -1 \\
2 & 0 & -4
\end{bmatrix}
$$

Augmented Matrix

$$
\begin{bmatrix}
1 & -4 & 3 & 5 \\
-1 & 3 & -1 & -3 \\
2 & 0 & -4 & 6
\end{bmatrix}
$$

Remark: Note the use of 0 for the missing y-variable in both matrices, and note the fourth column of constant terms from the system in the augmented matrix.

Recall that two systems of linear equations are equivalent if they possess the same solutions. Moreover, we can construct an equivalent system in *triangular form* by adding equations, multiplying equations by constants, or interchanging the position of equations. In matrix terminology, these three types of changes correspond to what we call **elementary row operations.** Thus, an elementary row operation on an augmented matrix (corresponding to a given system of linear equations) produces a new augmented matrix corresponding to a new (but equivalent) system of linear equations.

ELEMENTARY ROW OPERATIONS

> If an augmented matrix (of a given system of linear equations) is changed by any of the following operations, it produces an augmented matrix of an *equivalent* system of linear equations:
>
> 1. Interchange any two rows of the matrix.
> 2. Multiply every element in a row by the same nonzero constant.
> 3. Replace a row by the sum of that row and a multiple of any other row.

Though each of the three elementary row operations is simple to perform, there is a considerable amount of arithmetic involved. It is easy to make a careless error, and we suggest that you use the following scheme to designate these operations:

Interchange two rows:

$$\begin{bmatrix} 0 & 1 & 3 & 4 \\ -1 & 2 & 0 & 3 \\ 2 & -3 & 4 & 1 \end{bmatrix} \quad \begin{matrix} R_2 \\ R_1 \end{matrix} \quad \begin{bmatrix} -1 & 2 & 0 & 3 \\ 0 & 1 & 3 & 4 \\ 2 & -3 & 4 & 1 \end{bmatrix}$$

Multiply a row by a (nonzero) constant ($\frac{1}{2}R_1$ is new R_1):

$$\begin{bmatrix} 2 & -4 & 6 & -2 \\ 1 & 3 & -3 & 0 \\ 5 & -2 & 1 & 2 \end{bmatrix} \quad \frac{1}{2}R_1 \quad \begin{bmatrix} 1 & -2 & 3 & -1 \\ 1 & 3 & -3 & 0 \\ 5 & -2 & 1 & 2 \end{bmatrix}$$

Add a row to another row ($R_1 + R_2$ is new R_2):

$$\begin{bmatrix} 1 & 2 & -4 & 3 \\ -1 & 3 & -2 & -1 \\ 2 & 1 & 5 & -2 \end{bmatrix} \quad R_1 + R_2 \quad \begin{bmatrix} 1 & 2 & -4 & 3 \\ 0 & 5 & -6 & 2 \\ 2 & 1 & 5 & -2 \end{bmatrix}$$

Add a multiple of a row to another row ($-2R_1 + R_3$ is new R_3):

$$\begin{bmatrix} 1 & 2 & -4 & 3 \\ -1 & 3 & -2 & -1 \\ 2 & 1 & 5 & -2 \end{bmatrix} \quad -2R_1 + R_3 \quad \begin{bmatrix} 1 & 2 & -4 & 3 \\ -1 & 3 & -2 & -1 \\ 0 & -3 & 13 & -8 \end{bmatrix}$$

Note that we write the elementary row operations beside the row that we are changing. To emphasize this, you may want to use the following notation (written to the left of the changed row):

$$\frac{1}{2}R_1 \rightarrow R_1, \qquad R_1 + R_2 \rightarrow R_2, \qquad -2R_1 + R_3 \rightarrow R_3$$

EXAMPLE 1
Using Elementary
Row Operations

Use matrices and elementary row operations to solve the following system:

$$x - 2y + 3z = 9$$
$$-x + 3y - z = -6$$
$$2x - 5y + 5z = 17$$

Solution:

We begin by forming the augmented matrix of this system.

$$\begin{bmatrix} 1 & -2 & 3 & 9 \\ -1 & 3 & -1 & -6 \\ 2 & -5 & 5 & 17 \end{bmatrix}$$

Our goal is to apply elementary row operations to obtain a matrix that corresponds to a system in *triangular form*. This means that we need to change the elements below the main diagonal to zeros (by means of elementary row operations). Study the following steps carefully to see how this is done.

$$\begin{bmatrix} 1 & -2 & 3 & 9 \\ -1 & 3 & -1 & -6 \\ 2 & -5 & 5 & 17 \end{bmatrix} \quad R_1 + R_2 \quad \begin{bmatrix} 1 & -2 & 3 & 9 \\ 0 & 1 & 2 & 3 \\ 2 & -5 & 5 & 17 \end{bmatrix}$$

$$-2R_1 + R_3 \quad \begin{bmatrix} 1 & -2 & 3 & 9 \\ 0 & 1 & 2 & 3 \\ 0 & -1 & -1 & -1 \end{bmatrix}$$

$$R_2 + R_3 \quad \begin{bmatrix} 1 & -2 & 3 & 9 \\ 0 & 1 & 2 & 3 \\ 0 & 0 & 1 & 2 \end{bmatrix}$$

Now, we write the system of linear equations corresponding to this final matrix and use backsubstitution to find the solution:

$$x - 2y + 3z = 9$$
$$y + 2z = 3$$
$$z = 2$$

This is the same system (in triangular form) that we solved in Example 1 in Section 8.3. The solution is

$$x = 1, \quad y = -1, \quad \text{and} \quad z = 2$$

The final matrix obtained in Example 1 is said to be in **echelon form.** The general procedure followed is

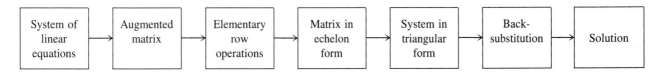

We can eliminate the need for backsubstitution by continuing to apply elementary row operations to the echelon (triangular) form so as to transform it into a matrix in **reduced echelon form.** To be in this form, a matrix must have the following properties.

$$\begin{bmatrix} 1 & 0 & 0 & 0 & a \\ 0 & 1 & 0 & 0 & b \\ 0 & 0 & 1 & 0 & c \\ 0 & 0 & 0 & 1 & d \end{bmatrix}$$

FIGURE 9.1
*Reduced Echelon Form for 4 × 5
Augmented Matrix*

REDUCED ECHELON FORM

1. All rows consisting entirely of zeros are placed at the bottom of the matrix.
2. The first nonzero entry in any row is 1 (called a leading 1).
3. Nonzero rows are arranged so that leading 1's occur farther to the right in each successive row.
4. Each column with a leading 1 has zeros as all other entries.

Once an augmented matrix is written in reduced echelon form, we can read the solution directly from the matrix. For example, in Figure 9.1, we can see that $x = a$, $y = b$, $z = c$, and $w = d$. This procedure of finding the solution to a linear system by writing the augmented matrix in reduced echelon form is called **Gauss-Jordan elimination,** summarized as follows.

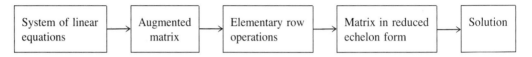

Had we applied Gauss-Jordan elimination to the system in Example 1, we would have obtained the reduced echelon form

$$\begin{bmatrix} 1 & 0 & 0 & 1 \\ 0 & 1 & 0 & -1 \\ 0 & 0 & 1 & 2 \end{bmatrix}$$

from which we obtain a system of equations that corresponds precisely to the solution $x = 1$, $y = -1$, and $z = 2$.

Remark: The order in which the elementary row operations are performed in the Gauss-Jordan elimination process is important. We move from *left to right by columns,* changing to zero all elements directly above or below the leading 1's.

The power of the Gauss-Jordan elimination method and the convenience of our notation become increasingly evident with linear systems involving a large number of variables. In the next example, note how several row operations are performed as single steps in the solution.

EXAMPLE 2
**The Gauss-Jordan
Elimination Method**

Use Gauss-Jordan elimination to solve the following system:

$$\begin{aligned} y + z - 2w &= -3 \\ x + 2y - z &= 2 \\ 2x + 4y + z - 3w &= -2 \\ x - 4y - 7z - w &= -19 \end{aligned}$$

Solution:

$$\begin{bmatrix} 0 & 1 & 1 & -2 & -3 \\ 1 & 2 & -1 & 0 & 2 \\ 2 & 4 & 1 & -3 & -2 \\ 1 & -4 & -7 & -1 & -19 \end{bmatrix}$$

$$\begin{matrix} \circlearrowright R_2 \\ \searrow R_1 \end{matrix} \quad \begin{bmatrix} 1 & 2 & -1 & 0 & 2 \\ 0 & 1 & 1 & -2 & -3 \\ 2 & 4 & 1 & -3 & -2 \\ 1 & -4 & -7 & -1 & -19 \end{bmatrix} \qquad \textit{Interchange rows}$$

$$\begin{matrix} \\ \\ -2R_1 + R_3 \\ -R_1 + R_4 \end{matrix} \quad \begin{bmatrix} 1 & 2 & -1 & 0 & 2 \\ 0 & 1 & 1 & -2 & -3 \\ 0 & 0 & 3 & -3 & -6 \\ 0 & -6 & -6 & -1 & -21 \end{bmatrix} \qquad \textit{Zeros in column 1}$$

$$\begin{matrix} -2R_2 + R_1 \\ \\ \\ 6R_2 + R_4 \end{matrix} \quad \begin{bmatrix} 1 & 0 & -3 & 4 & 8 \\ 0 & 1 & 1 & -2 & -3 \\ 0 & 0 & 3 & -3 & -6 \\ 0 & 0 & 0 & -13 & -39 \end{bmatrix} \qquad \textit{Zeros in column 2}$$

$$\begin{matrix} \\ \\ \frac{1}{3}R_3 \\ -\frac{1}{13}R_4 \end{matrix} \quad \begin{bmatrix} 1 & 0 & -3 & 4 & 8 \\ 0 & 1 & 1 & -2 & -3 \\ 0 & 0 & 1 & -1 & -2 \\ 0 & 0 & 0 & 1 & 3 \end{bmatrix} \qquad \textit{Ones on diagonal}$$

$$\begin{matrix} 3R_3 + R_1 \\ -R_3 + R_2 \\ \\ \end{matrix} \quad \begin{bmatrix} 1 & 0 & 0 & 1 & 2 \\ 0 & 1 & 0 & -1 & -1 \\ 0 & 0 & 1 & -1 & -2 \\ 0 & 0 & 0 & 1 & 3 \end{bmatrix} \qquad \textit{Zeros in column 3}$$

$$\begin{matrix} -R_4 + R_1 \\ R_4 + R_2 \\ R_4 + R_3 \\ \end{matrix} \quad \begin{bmatrix} 1 & 0 & 0 & 0 & -1 \\ 0 & 1 & 0 & 0 & 2 \\ 0 & 0 & 1 & 0 & 1 \\ 0 & 0 & 0 & 1 & 3 \end{bmatrix} \qquad \textit{Zeros in column 4}$$

Thus, the solution is

$$x = -1, \quad y = 2, \quad z = 1, \quad \text{and} \quad w = 3$$

Remark: We suggest that you repeat Example 2 using the echelon form and backsubstitution to see which method you prefer.

EXAMPLE 3
A System with Infinitely
Many Solutions

Solve the following system:

$$x - y + 2z = 4$$
$$2x + 4y + 4z = -1$$
$$x + y + 2z = 1$$

Solution:

$$\begin{bmatrix} 1 & -1 & 2 & 4 \\ 2 & 4 & 4 & -1 \\ 1 & 1 & 2 & 1 \end{bmatrix} \quad \begin{matrix} -2R_1 + R_2 \\ -R_1 + R_3 \end{matrix} \quad \begin{bmatrix} 1 & -1 & 2 & 4 \\ 0 & 6 & 0 & -9 \\ 0 & 2 & 0 & -3 \end{bmatrix}$$

$$\frac{1}{3}R_2 \quad \begin{bmatrix} 1 & -1 & 2 & 4 \\ 0 & 2 & 0 & -3 \\ 0 & 2 & 0 & -3 \end{bmatrix}$$

$$-R_2 + R_3 \quad \begin{bmatrix} 1 & -1 & 2 & 4 \\ 0 & 2 & 0 & -3 \\ 0 & 0 & 0 & 0 \end{bmatrix}$$

Now, since one of the rows has all zero elements, we can convert back to the corresponding system of linear equations and write the solution following the procedure given in Section 8.3.

$$x - y + 2z = 4$$
$$2y = -3$$

From this system, $y = -\frac{3}{2}$ and

$$x - \left(-\frac{3}{2}\right) + 2z = 4$$

$$x = \frac{-4z + 5}{2}$$

Letting z be any real number a, we have the following solution:

$$\left(\frac{-4a + 5}{2}, \frac{-3}{2}, a\right)$$

Remark: Note in Example 3 that we used the combined method of reducing to triangular form and then applying backsubstitution. As a general procedure, we have found this combined method to be more efficient than the Gauss-Jordan elimination method (involving the fully reduced echelon form).

EXAMPLE 4
A System with No Solutions

Solve the following system:

$$2x + 3y + z = 2$$
$$x - 3y + 2z = 4$$
$$x + z = 6$$

Solution:

$$\begin{bmatrix} 2 & 3 & 1 & 2 \\ 1 & -3 & 2 & 4 \\ 1 & 0 & 1 & 6 \end{bmatrix} \quad \begin{matrix} R_2 \\ R_1 \end{matrix} \quad \begin{bmatrix} 1 & -3 & 2 & 4 \\ 2 & 3 & 1 & 2 \\ 1 & 0 & 1 & 6 \end{bmatrix}$$

$$\begin{matrix} -2R_1 + R_2 \\ -R_1 + R_3 \end{matrix} \quad \begin{bmatrix} 1 & -3 & 2 & 4 \\ 0 & 9 & -3 & -6 \\ 0 & 3 & -1 & 2 \end{bmatrix}$$

$$\tfrac{1}{3}R_2 \quad \begin{bmatrix} 1 & -3 & 2 & 4 \\ 0 & 3 & -1 & -2 \\ 0 & 3 & -1 & 2 \end{bmatrix}$$

$$-R_2 + R_3 \quad \begin{bmatrix} 1 & -3 & 2 & 4 \\ 0 & 3 & -1 & -2 \\ 0 & 0 & 0 & 4 \end{bmatrix}$$

Now, by converting back to a system of linear equations, we have

$$\begin{aligned} x - 3y + 2z &= 4 \\ 3y - z &= -2 \\ 0 &= 4 \end{aligned}$$

Since the last of these equations is an absurdity, we conclude that the original system is *inconsistent* and has no solutions.

Section Exercises 9.1

1. Given the matrix

$$\begin{bmatrix} 1 & 2 & 3 \\ 2 & -1 & -4 \\ 3 & 1 & -1 \end{bmatrix}$$

perform the following *sequence* of elementary row operations to put the matrix in reduced echelon form:
   (a) Add (-2) times Row 1 to Row 2. (Only Row 2 should change.)
   (b) Add (-3) times Row 1 to Row 3. (Only Row 3 should change.)
   (c) Add (-1) times Row 2 to Row 3.
   (d) Multiply Row 2 by $(-\tfrac{1}{5})$.
   (e) Add (-2) times Row 2 to Row 1.

2. Given the matrix

$$\begin{bmatrix} 7 & 1 \\ 0 & 2 \\ -3 & 4 \\ 4 & 1 \end{bmatrix}$$

perform the following *sequence* of elementary row operations to put the matrix in reduced echelon form:
   (a) Add Row 3 to Row 4. (Only Row 4 should change.)
   (b) Interchange Rows 1 and 4. (Note that the first element in the matrix is now 1, and it was obtained without introducing fractions.)
   (c) Add (3) times Row 1 to Row 3.
   (d) Add (-7) times Row 1 to Row 4.
   (e) Multiply Row 2 by $\tfrac{1}{2}$.
   (f) Add the appropriate multiple of Row 2 to Rows 1, 3, and 4.

In Exercises 3–10, put the matrix in reduced echelon form.

3. $$\begin{bmatrix} 1 & 1 & 0 & 5 \\ -2 & -1 & 2 & -10 \\ 3 & 6 & 7 & 14 \end{bmatrix}$$

4. $$\begin{bmatrix} 1 & 2 & -1 & 3 \\ 3 & 7 & -5 & 14 \\ -2 & -1 & -3 & 8 \end{bmatrix}$$

5. $\begin{bmatrix} 1 & -1 & -1 & 1 \\ 5 & -4 & 1 & 8 \\ -6 & 8 & 18 & 0 \end{bmatrix}$

6. $\begin{bmatrix} 1 & -3 & 0 & -7 \\ -3 & 10 & 1 & 23 \\ 4 & -10 & 2 & -24 \end{bmatrix}$

7. $\begin{bmatrix} 3 & 3 & 3 \\ -1 & 0 & -4 \\ 2 & 4 & -2 \\ 5 & 6 & 12 \end{bmatrix}$

8. $\begin{bmatrix} 1 & 3 & 2 \\ 5 & 15 & 9 \\ 2 & 6 & 10 \end{bmatrix}$

9. $\begin{bmatrix} 1 & 2 & 3 & -5 \\ 1 & 2 & 4 & -9 \\ -2 & -4 & -4 & 3 \\ 4 & 8 & 11 & -14 \end{bmatrix}$

10. $\begin{bmatrix} 1 & -3 \\ -1 & 8 \\ 0 & 4 \\ -2 & 10 \end{bmatrix}$

In Exercises 11–30, use matrices to solve the system of equations.

11. $\begin{aligned} x + 2y &= 7 \\ 2x + y &= 8 \end{aligned}$

12. $\begin{aligned} 2x + 6y &= 16 \\ 2x + 3y &= 7 \end{aligned}$

13. $\begin{aligned} -3x + 5y &= -22 \\ 3x + 4y &= 4 \\ 4x - 8y &= 32 \end{aligned}$

14. $\begin{aligned} x + 2y &= 0 \\ x + y &= 6 \\ 3x - 2y &= 8 \end{aligned}$

15. $\begin{aligned} 8x - 4y &= 7 \\ 5x + 2y &= 1 \end{aligned}$

16. $\begin{aligned} 2x - y &= -0.1 \\ 3x + 2y &= 1.6 \end{aligned}$

17. $\begin{aligned} -x + 2y &= 1.5 \\ 2x - 4y &= 3 \end{aligned}$

18. $\begin{aligned} x - 3y &= 5 \\ -2x + 6y &= -10 \end{aligned}$

19. $\begin{aligned} x \quad\quad - 3z &= -2 \\ 3x + y - 2z &= 5 \\ 2x + 2y + z &= 4 \end{aligned}$

20. $\begin{aligned} 2x - y + 3z &= 24 \\ 2y - z &= 14 \\ 7x - 5y &= 6 \end{aligned}$

21. $\begin{aligned} x + y - 5z &= 3 \\ x \quad\quad - 2z &= 1 \\ 2x - y - z &= 0 \end{aligned}$

22. $\begin{aligned} 2x + \quad\quad 3z &= 3 \\ 4x - 3y + 7z &= 5 \\ 8x - 9y + 15z &= 9 \end{aligned}$

23. $\begin{aligned} x + 2y + z &= 8 \\ 3x + 7y + 6z &= 26 \end{aligned}$

24. $\begin{aligned} 4x + 12y - 7z - 20w &= 22 \\ 3x + 9y - 5z - 28w &= 30 \end{aligned}$

25. $\begin{aligned} 3x + 3y + 12z &= 6 \\ x + y + 4z &= 2 \\ 2x + 5y + 20z &= 10 \\ -x + 2y + 8z &= 4 \end{aligned}$

26. $\begin{aligned} 2x + 10y + 2z &= 6 \\ x + 5y + 2z &= 6 \\ x + 5y + z &= 3 \\ -3x - 15y - 3z &= -9 \end{aligned}$

27. $\begin{aligned} 2x + y - z + 2w &= -6 \\ 3x + 4y + w &= 1 \\ x + 5y + 2z + 6w &= -3 \\ 5x + 2y - z - w &= 3 \end{aligned}$

28. $\begin{aligned} x_1 + x_2 - 2x_3 + 3x_4 + 2x_5 &= 9 \\ 3x_1 + 3x_2 - x_3 + x_4 + x_5 &= 5 \\ 2x_1 + 2x_2 - x_3 + x_4 - 2x_5 &= 1 \\ 4x_1 + 4x_2 + x_3 - 3x_5 &= 4 \\ 8x_1 + 5x_2 - 2x_3 - x_4 + 2x_5 &= 3 \end{aligned}$

29. $\begin{aligned} x_1 - x_2 + 2x_3 + 2x_4 + 6x_5 &= 6 \\ 3x_1 - 2x_2 + 4x_3 + 4x_4 + 12x_5 &= 14 \\ x_2 - x_3 - x_4 - 3x_5 &= -3 \\ 2x_1 - 2x_2 + 4x_3 + 5x_4 + 15x_5 &= 10 \\ 2x_1 - 2x_2 + 4x_3 + 4x_4 + 13x_5 &= 13 \end{aligned}$

30. $2x_1 - 3x_2 + x_3 + 5x_4 = 1$

31. A small corporation borrowed $1,500,000 to expand its product line. Some of the money was borrowed at 8%, some at 9%, and some at 12%. How much was borrowed at each rate if the annual interest was $133,000 and the amount borrowed at 8% was 4 times the amount borrowed at 12%?

32. A grocer wishes to mix three kinds of nuts costing $3.50, $4.50, and $6.00 per pound, to obtain 50 pounds of a mixture costing $4.95 per pound. How many pounds of each variety should the grocer use if half of the mixture is composed of the two cheapest varieties?

33. Find a, b, and c for the quadratic function $f(x) = ax^2 + bx + c$, such that $f(1) = 1$, $f(-3) = 17$, and $f(2) = -\frac{1}{2}$.

34. Find a, b, c, and d for the cubic function $f(x) = ax^3 + bx^2 + cx + d$, such that $f(1) = 3$, $f(2) = 19$, $f(-1) = -11$, and $f(-2) = -33$.

35. Find D, E, and F for the circle $x^2 + y^2 + Dx + Ey + F = 0$ if the circle passes through the points $(1, 1)$, $(3, 3)$, and $(4, 2)$.

36. The sum of three positive numbers is 33. Find the three numbers if the second is 3 greater than the first and the third is 4 times the first.

The Algebra of Matrices

9.2

In Section 9.1 we looked at a matrix as if it were simply an array of real numbers. In reality, there is much more to matrices than that. In fact, there is a rich mathematical theory of matrices, and there are numerous practical applications of this theory. In this and the next section, we will study some of the basic portions of this theory. A more comprehensive study of the theory of matrices can be gained by taking a complete course in linear algebra.

In this section, we will look at the basic properties of **matrix addition and multiplication.** Throughout this chapter, when we refer to a matrix, we will mean a *real* matrix, whose elements are real numbers. Matrices can be represented in three convenient ways.

MATRIX NOTATION

1. A matrix can be denoted by an uppercase letter of the alphabet, such as A, B, C, ...
2. A matrix can be denoted by a representative element enclosed in brackets, such as $[a_{ij}]$, $[b_{ij}]$, $[c_{ij}]$, ...
3. A matrix can be denoted by an array of real numbers.

We say that two matrices are equal if and only if their corresponding elements are equal.

DEFINITION OF EQUALITY OF MATRICES

Two matrices $A = [a_{ij}]$ and $B = [b_{ij}]$ are equal if and only if

$$a_{ij} = b_{ij}$$

for every i and j. This definition implies that two matrices can be equal to each other only if they have the same order ($m \times n$).

We can add two matrices (of the same order) by adding their corresponding elements.

DEFINITION OF ADDITION OF MATRICES

If $A = [a_{ij}]$ and $B = [b_{ij}]$ are of the same order $m \times n$, then we define their sum to be the $m \times n$ matrix given by

$$A + B = [a_{ij} + b_{ij}]$$

We do not define the sum of two matrices of different orders.

Since addition of matrices is defined in terms of addition of real numbers, it is not surprising that many of the properties of addition of real numbers carry over to matrices.

PROPERTIES OF MATRIX ADDITION

If A, B, and C are matrices of order $m \times n$, then the following statements are true:

1. *Additive Identity:* The matrix whose elements are all zero is the additive identity, and we denote this matrix by 0.

$$A + 0 = A = 0 + A$$

2. *Commutative:* $A + B = B + A$

3. *Associative:* $A + (B + C) = (A + B) + C$

4. *Additive Inverse:* The additive inverse of $A = [a_{ij}]$ is given by $-A = [-a_{ij}]$ and has the property that

$$A + (-A) = 0 = -A + A$$

5. *Subtraction of Matrices:* The difference of two matrices is defined to be

$$A - B = A + (-B)$$

Proof: The proof of each of these five properties is quite straightforward and follows directly from the definition of addition of matrices together with the corresponding property of real numbers. For example, to prove the first property, we can use the fact that 0 is the additive identity in the real number system. Thus we have

$$A + 0 = \begin{bmatrix} a_{11} & a_{12} & \cdots & a_{1n} \\ a_{21} & a_{22} & \cdots & a_{2n} \\ \cdot & \cdot & & \cdot \\ \cdot & \cdot & & \cdot \\ \cdot & \cdot & & \cdot \\ a_{m1} & a_{m2} & \cdots & a_{mn} \end{bmatrix} + \begin{bmatrix} 0 & 0 & \cdots & 0 \\ 0 & 0 & \cdots & 0 \\ \cdot & \cdot & & \cdot \\ \cdot & \cdot & & \cdot \\ \cdot & \cdot & & \cdot \\ 0 & 0 & & 0 \end{bmatrix}$$

$$= \begin{bmatrix} a_{11} + 0 & a_{12} + 0 & \cdots & a_{1n} + 0 \\ a_{21} + 0 & a_{22} + 0 & \cdots & a_{2n} + 0 \\ \cdot & \cdot & & \cdot \\ \cdot & \cdot & & \cdot \\ \cdot & \cdot & & \cdot \\ a_{m1} + 0 & a_{m2} + 0 & \cdots & a_{mn} + 0 \end{bmatrix}$$

$$= \begin{bmatrix} a_{11} & a_{12} & \cdots & a_{1n} \\ a_{21} & a_{22} & \cdots & a_{2n} \\ \cdot & \cdot & & \cdot \\ \cdot & \cdot & & \cdot \\ \cdot & \cdot & & \cdot \\ a_{m1} & a_{m2} & \cdots & a_{mn} \end{bmatrix} = A$$

EXAMPLE 1
Addition of Matrices

Perform the following operations:

(a) $\begin{bmatrix} 0 & 1 & -2 \\ 1 & 2 & 3 \end{bmatrix} + \begin{bmatrix} 1 & -1 & 3 \\ 4 & -3 & 0 \end{bmatrix}$

(b) $\begin{bmatrix} 0 & 0 \\ 0 & 0 \end{bmatrix} + \begin{bmatrix} 1 & 3 \\ -1 & 2 \end{bmatrix}$

(c) $\begin{bmatrix} 3 & 1 & 4 \\ 2 & -1 & 2 \\ 0 & 5 & 3 \end{bmatrix} - \begin{bmatrix} 1 & 4 & -2 \\ 2 & 1 & 3 \\ 2 & -4 & 1 \end{bmatrix}$

(d) $\begin{bmatrix} 1 & -2 \\ 3 & 0 \end{bmatrix} + \begin{bmatrix} -1 & 2 \\ -3 & 0 \end{bmatrix}$

Solution:

(a) $\begin{bmatrix} 0 & 1 & -2 \\ 1 & 2 & 3 \end{bmatrix} + \begin{bmatrix} 1 & -1 & 3 \\ 4 & -3 & 0 \end{bmatrix}$

$$= \begin{bmatrix} 0+1 & 1-1 & -2+3 \\ 1+4 & 2-3 & 3+0 \end{bmatrix} = \begin{bmatrix} 1 & 0 & 1 \\ 5 & -1 & 3 \end{bmatrix}$$

(b) $\begin{bmatrix} 0 & 0 \\ 0 & 0 \end{bmatrix} + \begin{bmatrix} 1 & 3 \\ -1 & 2 \end{bmatrix} = \begin{bmatrix} 0+1 & 0+3 \\ 0-1 & 0+2 \end{bmatrix} = \begin{bmatrix} 1 & 3 \\ -1 & 2 \end{bmatrix}$

(c) $\begin{bmatrix} 3 & 1 & 4 \\ 2 & -1 & 2 \\ 0 & 5 & 3 \end{bmatrix} - \begin{bmatrix} 1 & 4 & -2 \\ 2 & 1 & 3 \\ 2 & -4 & 1 \end{bmatrix}$

$$= \begin{bmatrix} 3-1 & 1-4 & 4+2 \\ 2-2 & -1-1 & 2-3 \\ 0-2 & 5+4 & 3-1 \end{bmatrix} = \begin{bmatrix} 2 & -3 & 6 \\ 0 & -2 & -1 \\ -2 & 9 & 2 \end{bmatrix}$$

(d) $\begin{bmatrix} 1 & -2 \\ 3 & 0 \end{bmatrix} + \begin{bmatrix} -1 & 2 \\ -3 & 0 \end{bmatrix} = \begin{bmatrix} 1-1 & -2+2 \\ 3-3 & 0+0 \end{bmatrix} = \begin{bmatrix} 0 & 0 \\ 0 & 0 \end{bmatrix}$

We refer to the multiplication of a matrix by a real number as **scalar multiplication,** to distinguish it from the multiplication of two matrices.

DEFINITION OF SCALAR MULTIPLICATION

If $A = [a_{ij}]$ is an $m \times n$ matrix and c is a real number, then the product cA is defined to be the $m \times n$ matrix given by

$$cA = [ca_{ij}] = \begin{bmatrix} ca_{11} & ca_{12} & ca_{13} & \cdots & ca_{1n} \\ ca_{21} & ca_{22} & ca_{23} & \cdots & ca_{2n} \\ ca_{31} & ca_{32} & ca_{33} & \cdots & ca_{3n} \\ \cdot & \cdot & \cdot & & \cdot \\ \cdot & \cdot & \cdot & & \cdot \\ \cdot & \cdot & \cdot & & \cdot \\ ca_{m1} & ca_{m2} & ca_{m3} & \cdots & ca_{mn} \end{bmatrix}$$

PROPERTIES OF SCALAR MULTIPLICATION

If $A = [a_{ij}]$ and $B = [b_{ij}]$ are $m \times n$ matrices and c and d are real numbers, then the following statements are true:

1. Left Distributive: $c(A + B) = cA + cB$
2. Right Distributive: $(c + d)A = cA + dA$
3. Associative: $(cd)A = c(dA)$

EXAMPLE 2
Scalar Multiplication

Given

$$A = \begin{bmatrix} 1 & 2 & 4 \\ -3 & 0 & -1 \\ 2 & 1 & 2 \end{bmatrix} \qquad B = \begin{bmatrix} 2 & 0 & 0 \\ 1 & -4 & 3 \\ -1 & 3 & 2 \end{bmatrix}$$

perform the following operations:

(a) $3A$
(b) $A - 2B$

Solution:

(a) $3A = 3\begin{bmatrix} 1 & 2 & 4 \\ -3 & 0 & -1 \\ 2 & 1 & 2 \end{bmatrix} = \begin{bmatrix} 3(1) & 3(2) & 3(4) \\ 3(-3) & 3(0) & 3(-1) \\ 3(2) & 3(1) & 3(2) \end{bmatrix}$

$$= \begin{bmatrix} 3 & 6 & 12 \\ -9 & 0 & -3 \\ 6 & 3 & 6 \end{bmatrix}$$

(b) $A - 2B = \begin{bmatrix} 1 & 2 & 4 \\ -3 & 0 & -1 \\ 2 & 1 & 2 \end{bmatrix} - 2\begin{bmatrix} 2 & 0 & 0 \\ 1 & -4 & 3 \\ -1 & 3 & 2 \end{bmatrix}$

$$= \begin{bmatrix} 1 & 2 & 4 \\ -3 & 0 & -1 \\ 2 & 1 & 2 \end{bmatrix} - \begin{bmatrix} 4 & 0 & 0 \\ 2 & -8 & 6 \\ -2 & 6 & 4 \end{bmatrix}$$

$$= \begin{bmatrix} 1-4 & 2-0 & 4-0 \\ -3-2 & 0+8 & -1-6 \\ 2+2 & 1-6 & 2-4 \end{bmatrix}$$

$$= \begin{bmatrix} -3 & 2 & 4 \\ -5 & 8 & -7 \\ 4 & -5 & -2 \end{bmatrix}$$

EXAMPLE 3
Solving a Matrix Equation

Solve the following equation for X:

$$3X + A = B$$

where $A = \begin{bmatrix} 1 & -2 \\ 0 & 3 \end{bmatrix}$ and $B = \begin{bmatrix} -3 & 4 \\ 2 & 1 \end{bmatrix}$.

Solution: We have

$$3X + A = B$$

$$X = \frac{1}{3}(B - A)$$

Now, using the given values of A and B, we have

$$X = \frac{1}{3}\left(\begin{bmatrix} -3 & 4 \\ 2 & 1 \end{bmatrix} - \begin{bmatrix} 1 & -2 \\ 0 & 3 \end{bmatrix} \right)$$

$$= \frac{1}{3}\begin{bmatrix} -4 & 6 \\ 2 & -2 \end{bmatrix} = \begin{bmatrix} -\frac{4}{3} & 2 \\ \frac{2}{3} & -\frac{2}{3} \end{bmatrix}$$

The third matrix operation we will introduce in this section is matrix multiplication. Although more complicated than addition or scalar multiplication, this operation is very important and has numerous practical applications.

DEFINITION OF MATRIX MULTIPLICATION

If $A = [a_{ij}]$ is an $m \times n$ matrix and $B = [b_{ij}]$ is an $n \times p$ matrix, then the product AB is an $m \times p$ matrix

$$AB = [c_{ij}]$$

where $c_{ij} = a_{i1}b_{1j} + a_{i2}b_{2j} + a_{i3}b_{3j} + \cdots + a_{in}b_{nj}$.

Remark: This definition says that the element in the ith row and jth column of the product is obtained by multiplying the elements in the ith row of A by the corresponding elements in the *j*th column of *B* and adding the results. The following diagram should make this clearer.

To obtain the element in *i*th row and *j*th column of the product, use the *i*th row of *A* and the *j*th row of *B*.

$$
\begin{bmatrix}
a_{11} & a_{12} & a_{13} & \cdots & a_{1n} \\
a_{21} & a_{22} & a_{23} & \cdots & a_{2n} \\
\cdot & \cdot & \cdot & & \cdot \\
\cdot & \cdot & \cdot & & \cdot \\
a_{i1} & a_{i2} & a_{i3} & \cdots & a_{in} \\
\cdot & \cdot & \cdot & & \cdot \\
\cdot & \cdot & \cdot & & \cdot \\
a_{m1} & a_{m2} & a_{m3} & \cdots & a_{mn}
\end{bmatrix}
\begin{bmatrix}
b_{11} & b_{12} & \cdots & b_{1j} & \cdots & b_{1p} \\
b_{21} & b_{22} & \cdots & b_{2j} & \cdots & b_{2p} \\
b_{31} & b_{32} & \cdots & b_{3j} & \cdots & b_{3p} \\
\cdot & \cdot & & \cdot & & \cdot \\
\cdot & \cdot & & \cdot & & \cdot \\
b_{n1} & b_{n2} & \cdots & b_{nj} & \cdots & b_{np}
\end{bmatrix}
=
\begin{bmatrix}
c_{11} & c_{12} & \cdots & c_{1j} & \cdots & c_{1p} \\
c_{21} & c_{22} & \cdots & c_{2j} & \cdots & c_{2p} \\
\cdot & \cdot & & \cdot & & \cdot \\
\cdot & \cdot & & \cdot & & \cdot \\
c_{i1} & c_{i2} & \cdots & c_{ij} & \cdots & c_{ip} \\
\cdot & \cdot & & \cdot & & \cdot \\
\cdot & \cdot & & \cdot & & \cdot \\
c_{m1} & c_{m2} & \cdots & c_{mj} & \cdots & c_{mp}
\end{bmatrix}
$$

$$a_{i1}b_{1j} + a_{i2}b_{2j} + a_{i3}b_{3j} + \cdots + a_{in}b_{nj} = c_{ij}$$

Remark: In order for the product of two matrices to be defined, the number of columns of the first matrix must equal the number of rows of the second matrix.

$$
\begin{array}{ccccc}
A & B & = & C \\
m \times n & n \times p & & m \times p
\end{array}
$$

equal

order of *AB*

EXAMPLE 4
Finding the Product
of Two Matrices

Find the following products:

(a) $\begin{bmatrix} 1 & 2 \\ 0 & -1 \end{bmatrix} \begin{bmatrix} 3 & -1 \\ 2 & 1 \end{bmatrix}$ (b) $\begin{bmatrix} 1 & 0 & 3 \\ 2 & -1 & -2 \end{bmatrix} \begin{bmatrix} -2 & 4 & 2 \\ 1 & 0 & 0 \\ -1 & 1 & -1 \end{bmatrix}$

(c) $\begin{bmatrix} 3 & 4 \\ -2 & 5 \end{bmatrix} \begin{bmatrix} 1 & 0 \\ 0 & 1 \end{bmatrix}$ (d) $\begin{bmatrix} 1 & 2 \\ 1 & 1 \end{bmatrix} \begin{bmatrix} -1 & 2 \\ 1 & -1 \end{bmatrix}$

Solution:

(a) $\begin{bmatrix} 1 & 2 \\ 0 & -1 \end{bmatrix} \begin{bmatrix} 3 & -1 \\ 2 & 1 \end{bmatrix} = \begin{bmatrix} (1)(3) + (2)(2) & (1)(-1) + (2)(1) \\ (0)(3) + (-1)(2) & (0)(-1) + (-1)(1) \end{bmatrix}$

$= \begin{bmatrix} 7 & 1 \\ -2 & -1 \end{bmatrix}$

(b) $(1)(-2) + (0)(1) + (3)(-1) = -2 + 0 - 3$

$$\begin{bmatrix} 1 & 0 & 3 \\ 2 & -1 & -2 \end{bmatrix} \begin{bmatrix} -2 & 4 & 2 \\ 1 & 0 & 0 \\ -1 & 1 & -1 \end{bmatrix}$$

$$= \begin{bmatrix} -2 + 0 - 3 & 4 + 0 + 3 & 2 + 0 - 3 \\ -4 - 1 + 2 & 8 + 0 - 2 & 4 + 0 + 2 \end{bmatrix}$$

$$= \begin{bmatrix} -5 & 7 & -1 \\ -3 & 6 & 6 \end{bmatrix}$$

(c) $\begin{bmatrix} 3 & 4 \\ -2 & 5 \end{bmatrix} \begin{bmatrix} 1 & 0 \\ 0 & 1 \end{bmatrix} = \begin{bmatrix} 3 + 0 & 0 + 4 \\ -2 + 0 & 0 + 5 \end{bmatrix} = \begin{bmatrix} 3 & 4 \\ -2 & 5 \end{bmatrix}$

(d) $\begin{bmatrix} 1 & 2 \\ 1 & 1 \end{bmatrix} \begin{bmatrix} -1 & 2 \\ 1 & -1 \end{bmatrix} = \begin{bmatrix} -1 + 2 & 2 - 2 \\ -1 + 1 & 2 - 1 \end{bmatrix} = \begin{bmatrix} 1 & 0 \\ 0 & 1 \end{bmatrix}$

In part (c) of Example 4, notice that the matrix with 1's on its main diagonal and zeros elsewhere acts as an identity with respect to matrix multiplication. This and several other properties of matrix multiplication are summarized as follows.

THEOREM: PROPERTIES OF MATRIX MULTIPLICATION

For each of the following properties, assume that the matrices A, B, and C are of the appropriate orders.

1. Associative: $A(BC) = (AB)C$

2. Distributive: $A(B + C) = AB + AC$

    $(A + B)C = AC + BC$

3. Identity Matrix of Order n:

$$I_n = \begin{bmatrix} 1 & 0 & 0 & \cdots & 0 \\ 0 & 1 & 0 & \cdots & 0 \\ 0 & 0 & 1 & \cdots & 0 \\ \cdot & \cdot & \cdot & & \cdot \\ \cdot & \cdot & \cdot & & \cdot \\ \cdot & \cdot & \cdot & & \cdot \\ 0 & 0 & 0 & & 1 \end{bmatrix}$$

If A is of order $m \times n$ and B is of order $n \times p$, then we have

$$AI_n = A \quad \text{and} \quad I_n B = B$$

Remark: Note that we did *not* list a commutative property for matrix multiplication, because the product of two matrices is rarely commutative.

EXAMPLE 5
Noncommutativity
of Matrix Multiplication

Given the matrices

$$A = \begin{bmatrix} 1 & 3 \\ 2 & -1 \end{bmatrix}, \quad B = \begin{bmatrix} 2 & -1 \\ 0 & 2 \end{bmatrix},$$

$$C = \begin{bmatrix} 1 & 0 & 2 \\ 3 & -2 & 1 \end{bmatrix}, \quad D = \begin{bmatrix} -1 & 0 \\ 3 & 1 \\ 2 & 4 \end{bmatrix}$$

find the following products (if they exist):

(a) *AB* and *BA*   (b) *AC* and *CA*   (c) *CD* and *DC*

Solution:

(a) $AB = \begin{bmatrix} 1 & 3 \\ 2 & -1 \end{bmatrix} \begin{bmatrix} 2 & -1 \\ 0 & 2 \end{bmatrix} = \begin{bmatrix} 2+0 & -1+6 \\ 4+0 & -2-2 \end{bmatrix} = \begin{bmatrix} 2 & 5 \\ 4 & -4 \end{bmatrix}$

$BA = \begin{bmatrix} 2 & -1 \\ 0 & 2 \end{bmatrix} \begin{bmatrix} 1 & 3 \\ 2 & -1 \end{bmatrix} = \begin{bmatrix} 2-2 & 6+1 \\ 0+4 & 0-2 \end{bmatrix} = \begin{bmatrix} 0 & 7 \\ 4 & -2 \end{bmatrix}$

Note that $AB \neq BA$.

(b) $AC = \begin{bmatrix} 1 & 3 \\ 2 & -1 \end{bmatrix} \begin{bmatrix} 1 & 0 & 2 \\ 3 & -2 & 1 \end{bmatrix} = \begin{bmatrix} 1+9 & 0-6 & 2+3 \\ 2-3 & 0+2 & 4-1 \end{bmatrix}$

$$= \begin{bmatrix} 10 & -6 & 5 \\ -1 & 2 & 3 \end{bmatrix}$$

The product *CA* is *not defined*, since *C* is of order 2 × 3 and *A* is of order 2 × 2.

(c) $CD = \begin{bmatrix} 1 & 0 & 2 \\ 3 & -2 & 1 \end{bmatrix} \begin{bmatrix} -1 & 0 \\ 3 & 1 \\ 2 & 4 \end{bmatrix} = \begin{bmatrix} -1+0+4 & 0+0+8 \\ -3-6+2 & 0-2+4 \end{bmatrix}$

$$= \begin{bmatrix} 3 & 8 \\ -7 & 2 \end{bmatrix}$$

$DC = \begin{bmatrix} -1 & 0 \\ 3 & 1 \\ 2 & 4 \end{bmatrix} \begin{bmatrix} 1 & 0 & 2 \\ 3 & -2 & 1 \end{bmatrix} = \begin{bmatrix} -1+0 & 0+0 & -2+0 \\ 3+3 & 0-2 & 6+1 \\ 2+12 & 0-8 & 4+4 \end{bmatrix}$

$$= \begin{bmatrix} -1 & 0 & -2 \\ 6 & -2 & 7 \\ 14 & -8 & 8 \end{bmatrix}$$

Remark: You should not conclude from Example 5 that the product of two matrices is never commutative. Occasionally, it can happen that the products *AB* and *BA* are the same. (For example, try reversing the order of the product in part (d) of Example 4.) What you should conclude from Example 5 is that we cannot presume commutativity of matrix multiplication and we must compute both orders to see if they are equal or not.

Section Exercises 9.2

In Exercises 1–6, find (a) $A + B$, (b) $A - B$, (c) $3A$, and (d) $3A - 2B$.

1. $A = \begin{bmatrix} 1 & -1 \\ 2 & -1 \end{bmatrix}$, $B = \begin{bmatrix} 2 & -1 \\ -1 & 8 \end{bmatrix}$

2. $A = \begin{bmatrix} 1 & 2 \\ 2 & 1 \end{bmatrix}$, $B = \begin{bmatrix} -3 & -2 \\ 4 & 2 \end{bmatrix}$

3. $A = \begin{bmatrix} 6 & -1 \\ 2 & 4 \\ -3 & 5 \end{bmatrix}$, $B = \begin{bmatrix} 1 & 4 \\ -1 & 5 \\ 1 & 10 \end{bmatrix}$

4. $A = \begin{bmatrix} 2 & 1 & 1 \\ -1 & -1 & 4 \end{bmatrix}$, $B = \begin{bmatrix} 2 & -3 & 4 \\ -3 & 1 & -2 \end{bmatrix}$

5. $A = \begin{bmatrix} 2 & 2 & -1 & 0 & 1 \\ 1 & 1 & -2 & 0 & -1 \end{bmatrix}$,

$B = \begin{bmatrix} 1 & 1 & -1 & 1 & 0 \\ -3 & 4 & 9 & -6 & -7 \end{bmatrix}$

6. $A = \begin{bmatrix} 3 \\ 2 \\ -1 \end{bmatrix}$, $B = \begin{bmatrix} -4 \\ 6 \\ 2 \end{bmatrix}$

In Exercises 7–12, find (a) AB and (b) BA.

7. $A = \begin{bmatrix} 1 & 2 \\ 4 & 2 \end{bmatrix}$, $B = \begin{bmatrix} 2 & -1 \\ -1 & 8 \end{bmatrix}$

8. $A = \begin{bmatrix} 2 & -1 \\ 1 & 4 \end{bmatrix}$, $B = \begin{bmatrix} 0 & 0 \\ 3 & -3 \end{bmatrix}$

9. $A = \begin{bmatrix} 3 & -1 \\ 1 & 3 \end{bmatrix}$, $B = \begin{bmatrix} 1 & -3 \\ 3 & 1 \end{bmatrix}$

10. $A = \begin{bmatrix} 1 & -1 \\ 1 & 1 \end{bmatrix}$, $B = \begin{bmatrix} 1 & 3 \\ -3 & 1 \end{bmatrix}$

11. $A = \begin{bmatrix} 1 & -1 & 7 \\ 2 & -1 & 8 \\ 3 & 1 & -1 \end{bmatrix}$, $B = \begin{bmatrix} 1 & 1 & 2 \\ 2 & 1 & 1 \\ 1 & -3 & 2 \end{bmatrix}$

12. $A = \begin{bmatrix} 3 & 2 & 1 \end{bmatrix}$, $B = \begin{bmatrix} 2 \\ 3 \\ 0 \end{bmatrix}$

In Exercises 13–20, find AB, if possible.

13. $A = \begin{bmatrix} 2 & 1 \\ -3 & 4 \\ 1 & 6 \end{bmatrix}$, $B = \begin{bmatrix} 0 & -1 & 0 \\ 4 & 0 & 2 \\ 8 & -1 & 7 \end{bmatrix}$

14. $A = \begin{bmatrix} 0 & -1 & 0 \\ 4 & 0 & 2 \\ 8 & -1 & 7 \end{bmatrix}$, $B = \begin{bmatrix} 2 & 1 \\ -3 & 4 \\ 1 & 6 \end{bmatrix}$

15. $A = \begin{bmatrix} -1 & 3 \\ 4 & -5 \\ 0 & 2 \end{bmatrix}$, $B = \begin{bmatrix} 1 & 2 \\ 0 & 7 \end{bmatrix}$

16. $A = \begin{bmatrix} 1 & 0 & 0 \\ 0 & 4 & 0 \\ 0 & 0 & -2 \end{bmatrix}$, $B = \begin{bmatrix} 3 & 0 & 0 \\ 0 & -1 & 0 \\ 0 & 0 & 5 \end{bmatrix}$

17. $A = \begin{bmatrix} 5 & 0 & 0 \\ 0 & -8 & 0 \\ 0 & 0 & 7 \end{bmatrix}$, $B = \begin{bmatrix} \frac{1}{5} & 0 & 0 \\ 0 & -\frac{1}{8} & 0 \\ 0 & 0 & \frac{1}{2} \end{bmatrix}$

18. $A = \begin{bmatrix} 0 & 0 & 5 \\ 0 & 0 & -3 \\ 0 & 0 & 4 \end{bmatrix}$, $B = \begin{bmatrix} 6 & -11 & 4 \\ 8 & 16 & 4 \\ 0 & 0 & 0 \end{bmatrix}$

19. $A = \begin{bmatrix} 6 \\ -2 \\ 1 \\ 6 \end{bmatrix}$, $B = \begin{bmatrix} 10 & 12 \end{bmatrix}$

20. $A = \begin{bmatrix} 1 & 0 & 3 & -2 & 4 \\ 6 & 13 & 8 & -17 & 20 \end{bmatrix}$, $B = \begin{bmatrix} 1 & 6 \\ 4 & 2 \end{bmatrix}$

If

$$f(x) = a_0 + a_1 x + a_2 x^2 + \cdots + a_n x^n$$

then for an $n \times n$ matrix A, we define $f(A)$ to be

$$f(A) = a_0 I_n + a_1 A + a_2 A^2 + \cdots + a_n A^n$$

In Exercises 21–24, find $f(A)$.

21. $f(x) = x^2 - 5x + 2$, $A = \begin{bmatrix} 2 & 0 \\ 4 & 5 \end{bmatrix}$

22. $f(x) = x^2 - 7x + 6$, $A = \begin{bmatrix} 5 & 4 \\ 1 & 2 \end{bmatrix}$

23. $f(x) = x^3 - 10x^2 + 31x - 30$, $A = \begin{bmatrix} 3 & 1 & 4 \\ 0 & 2 & 6 \\ 0 & 0 & 5 \end{bmatrix}$

24. $f(x) = x^2 - 10x + 24$, $A = \begin{bmatrix} 8 & -4 \\ 2 & 2 \end{bmatrix}$

25. In matrix multiplication, if $AC = BC$, then A is *not* necessarily equal to B. Illustrate this using the matrices

$$A = \begin{bmatrix} 1 & 2 & 3 \\ 0 & 5 & 4 \\ 3 & -2 & 1 \end{bmatrix}, \quad B = \begin{bmatrix} 4 & -6 & 3 \\ 5 & 4 & 4 \\ -1 & 0 & 1 \end{bmatrix}$$

and

$$C = \begin{bmatrix} 0 & 0 & 0 \\ 0 & 0 & 0 \\ 4 & -2 & 3 \end{bmatrix}$$

26. In matrix multiplication, if $AB = 0$, then it is *not* necessarily true that $A = 0$ or $B = 0$. Illustrate this using the matrices

$$A = \begin{bmatrix} 3 & 3 \\ 4 & 4 \end{bmatrix} \quad \text{and} \quad B = \begin{bmatrix} 1 & -1 \\ -1 & 1 \end{bmatrix}$$

27. Explain why the following nonequalities are valid for matrices:

$$(A + B)(A - B) \neq A^2 - B^2$$

and

$$(A + B)(A + B) \neq A^2 + 2AB + B^2$$

28. A certain corporation has four factories, each of which produces two products. The number of units of product i produced at factory j in one day is represented by a_{ij} in the matrix

$$A = \begin{bmatrix} 100 & 90 & 70 & 30 \\ 40 & 20 & 60 & 60 \end{bmatrix}$$

Use scalar multiplication (multiply by 1.10) to give the production levels if production is increased by 10%.

29. A fruit grower raises two crops, which he ships to three outlets. The number of units of product i that are shipped to outlet j is represented by a_{ij} in the matrix

$$A = \begin{bmatrix} 100 & 75 & 75 \\ 125 & 150 & 100 \end{bmatrix}$$

The profit on one unit of product i is represented by b_{1i} in the matrix

$$B = [\$3.75 \quad \$7.00]$$

Find the product AB and state what each entry of the product represents.

30. The matrix

$$P = \begin{matrix} & \text{To } R & \text{To } D & \text{To } I \\ \text{From } R & \\ \text{From } D & \\ \text{From } I & \end{matrix}\begin{bmatrix} 0.75 & 0.15 & 0.10 \\ 0.20 & 0.60 & 0.20 \\ 0.30 & 0.40 & 0.30 \end{bmatrix}$$

is called a stochastic matrix. Each p_{ij} $(i \neq j)$ represents the proportion of the voting population that changes from party i to party j, and p_{ii} represents the proportion that remains loyal to the party from one election to the next. Find P^2. (This matrix gives the transition probabilities from the first election to the third.)

The Inverse of a Matrix   9.3

There are many similarities between the algebra of real numbers and the algebra of matrices. For example, compare the following solutions:

Real Numbers	**$m \times n$ Matrices**
(Solve for x)	**(Solve for X)**
$x + a = b$	$X + A = B$
$x + a + (-a) = b + (-a)$	$X + A + (-A) = B + (-A)$
$x + 0 = b - a$	$X + 0 = B - A$
$x = b - a$	$X = B - A$

The solution of equations involving multiplication is a bit more complicated. For example, we can solve the real number equation $ax = b$ for x and obtain $x = b/a = a^{-1}b$ *only if $a \neq 0$*. Here, a^{-1} is the *multiplicative inverse of a*, with the property that $(a^{-1})a = 1$. We use a similar definition for the multiplicative inverse of a matrix.

DEFINITION OF THE INVERSE OF A MATRIX

> If A is a square matrix of order n, then A^{-1} is called the **inverse of A** if it has the property that
>
> $$A(A^{-1}) = (A^{-1})A = I_n$$
>
> where I_n is the identity matrix of order n.

Remark: The symbol A^{-1} is read "*A* inverse." We do not use the reciprocal notation $1/A$ because matrix division is not defined.

EXAMPLE 1
The Inverse of a Matrix

Show that B is the inverse of A.

(a) $A = \begin{bmatrix} -1 & 2 \\ -1 & 1 \end{bmatrix}, B = \begin{bmatrix} 1 & -2 \\ 1 & -1 \end{bmatrix}$

(b) $A = \begin{bmatrix} 2 & 3 & 1 \\ 3 & 3 & 1 \\ 2 & 4 & 1 \end{bmatrix}, B = \begin{bmatrix} -1 & 1 & 0 \\ -1 & 0 & 1 \\ 6 & -2 & -3 \end{bmatrix}$

Solution:

(a) $AB = \begin{bmatrix} -1 & 2 \\ -1 & 1 \end{bmatrix}\begin{bmatrix} 1 & -2 \\ 1 & -1 \end{bmatrix} = \begin{bmatrix} -1+2 & 2-2 \\ -1+1 & 2-1 \end{bmatrix} = \begin{bmatrix} 1 & 0 \\ 0 & 1 \end{bmatrix}$

$BA = \begin{bmatrix} 1 & -2 \\ 1 & -1 \end{bmatrix}\begin{bmatrix} -1 & 2 \\ -1 & 1 \end{bmatrix} = \begin{bmatrix} -1+2 & 2-2 \\ -1+1 & 2-1 \end{bmatrix} = \begin{bmatrix} 1 & 0 \\ 0 & 1 \end{bmatrix}$

(b) $AB = \begin{bmatrix} 2 & 3 & 1 \\ 3 & 3 & 1 \\ 2 & 4 & 1 \end{bmatrix}\begin{bmatrix} -1 & 1 & 0 \\ -1 & 0 & 1 \\ 6 & -2 & -3 \end{bmatrix} = \begin{bmatrix} 1 & 0 & 0 \\ 0 & 1 & 0 \\ 0 & 0 & 1 \end{bmatrix}$

$BA = \begin{bmatrix} -1 & 1 & 0 \\ -1 & 0 & 1 \\ 6 & -2 & -3 \end{bmatrix}\begin{bmatrix} 2 & 3 & 1 \\ 3 & 3 & 1 \\ 2 & 4 & 1 \end{bmatrix} = \begin{bmatrix} 1 & 0 & 0 \\ 0 & 1 & 0 \\ 0 & 0 & 1 \end{bmatrix}$

Both matrices in Example 1 are square. If a matrix is not square, then it does not possess an inverse. But not all square matrices possess inverses. If a matrix does have an inverse, then we say that it is **invertible** (or **nonsingular**). A square matrix that has no inverse is said to be **singular**. We can use a system of equations to determine whether a matrix has an inverse.

EXAMPLE 2
A Singular Matrix

Show that the following matrix has no inverse:

$$A = \begin{bmatrix} 1 & -2 \\ -2 & 4 \end{bmatrix}$$

Solution:

To find an inverse for A, we try to solve the matrix equation

$$AB = I$$

$$\begin{bmatrix} 1 & -2 \\ -2 & 4 \end{bmatrix} \begin{bmatrix} a & b \\ c & d \end{bmatrix} = \begin{bmatrix} 1 & 0 \\ 0 & 1 \end{bmatrix}$$

Through matrix multiplication, this equation results in two systems of linear equations:

$$a - 2c = 1 \qquad b - 2d = 0$$
$$-2a + 4c = 0 \qquad -2b + 4d = 1$$

Neither of these systems has a solution. (Check this to convince yourself that this is true.) Thus, A has no inverse.

Example 2 shows us one way to find the inverse of a matrix if it exists. However, this procedure is somewhat cumbersome, and we recommend the following alternative procedure.

A PROCEDURE FOR FINDING THE INVERSE OF A MATRIX

If A is a square matrix of order n, we can find its inverse as follows:

1. Write the $n \times 2n$ matrix that consists of the given matrix A on the left augmented with the identity matrix I_n.

$$[A{:}I_n] = \begin{bmatrix} a_{11} & a_{12} & a_{13} & \cdots & a_{1n} & 1 & 0 & 0 & \cdots & 0 \\ a_{21} & a_{22} & a_{23} & \cdots & a_{2n} & 0 & 1 & 0 & \cdots & 0 \\ a_{31} & a_{32} & a_{33} & \cdots & a_{3n} & 0 & 0 & 1 & \cdots & 0 \\ \cdot & \cdot & \cdot & & \cdot & \cdot & \cdot & \cdot & & \cdot \\ \cdot & \cdot & \cdot & & \cdot & \cdot & \cdot & \cdot & & \cdot \\ \cdot & \cdot & \cdot & & \cdot & \cdot & \cdot & \cdot & & \cdot \\ a_{n1} & a_{n2} & a_{n3} & \cdots & a_{nn} & 0 & 0 & 0 & \cdots & 1 \end{bmatrix}$$

2. Transform A to I_n using elementary row operations *on the entire matrix* $[A : I_n]$. The result will be a matrix of the form

$$[I_n : B]$$

The matrix B is the inverse of A. That is, $B = A^{-1}$.

3. Check your work by multiplying to see that

$$A(A^{-1}) = I_n = (A^{-1})A$$

EXAMPLE 3
Finding the Inverse of a Matrix

Find the inverse of the following matrix:

$$A = \begin{bmatrix} 2 & 3 & 1 \\ 3 & 3 & 1 \\ 2 & 4 & 1 \end{bmatrix}$$

Solution:

We begin by adjoining the identity matrix, I_3, as follows:

$$\begin{bmatrix} 2 & 3 & 1 & 1 & 0 & 0 \\ 3 & 3 & 1 & 0 & 1 & 0 \\ 2 & 4 & 1 & 0 & 0 & 1 \end{bmatrix}$$

Now, using elementary row operations, we attempt to rewrite this matrix in the form

$$\begin{bmatrix} 1 & 0 & 0 & b_{11} & b_{12} & b_{13} \\ 0 & 1 & 0 & b_{21} & b_{22} & b_{23} \\ 0 & 0 & 1 & b_{31} & b_{32} & b_{33} \end{bmatrix}$$

We have

$$\begin{bmatrix} 2 & 3 & 1 & 1 & 0 & 0 \\ 3 & 3 & 1 & 0 & 1 & 0 \\ 2 & 4 & 1 & 0 & 0 & 1 \end{bmatrix} \begin{matrix} \uparrow R_2 \\ \downarrow R_1 \end{matrix} \begin{bmatrix} 3 & 3 & 1 & 0 & 1 & 0 \\ 2 & 3 & 1 & 1 & 0 & 0 \\ 2 & 4 & 1 & 0 & 0 & 1 \end{bmatrix}$$

$$-R_2 + R_1 \quad \begin{bmatrix} 1 & 0 & 0 & -1 & 1 & 0 \\ 2 & 3 & 1 & 1 & 0 & 0 \\ 0 & 1 & 0 & -1 & 0 & 1 \end{bmatrix}$$
$$-R_2 + R_3$$

$$\begin{matrix} \uparrow R_3 \\ \downarrow R_2 \end{matrix} \begin{bmatrix} 1 & 0 & 0 & -1 & 1 & 0 \\ 0 & 1 & 0 & -1 & 0 & 1 \\ 2 & 3 & 1 & 1 & 0 & 0 \end{bmatrix}$$

$$-2R_1 + R_3 \quad \begin{bmatrix} 1 & 0 & 0 & -1 & 1 & 0 \\ 0 & 1 & 0 & -1 & 0 & 1 \\ 0 & 3 & 1 & 3 & -2 & 0 \end{bmatrix}$$

$$-3R_2 + R_3 \quad \begin{bmatrix} 1 & 0 & 0 & -1 & 1 & 0 \\ 0 & 1 & 0 & -1 & 0 & 1 \\ 0 & 0 & 1 & 6 & -2 & -3 \end{bmatrix}$$

Thus, the inverse of A is given by

$$A^{-1} = B = \begin{bmatrix} -1 & 1 & 0 \\ -1 & 0 & 1 \\ 6 & -2 & -3 \end{bmatrix}$$

In part (b) of Example 1, we already verified that this is the correct inverse of A.

The process shown in Example 3 is general in the sense that it can be used both to determine if a matrix has an inverse and to find the inverse if it exists. For example, if we applied this process to the singular matrix in Example 2, we would obtain a row (in that part of the matrix occupied by A) with all zero elements:

$$[A : I] = \begin{bmatrix} 1 & -2 & 1 & 0 \\ -2 & 4 & 0 & 1 \end{bmatrix} \quad 2R_2 + R_1 \quad \begin{bmatrix} 1 & -2 & 1 & 0 \\ 0 & 0 & 2 & 1 \end{bmatrix}$$

This means that the matrix has no inverse. In general, when reducing the matrix $[A : I_n]$, we know A is not invertible if we arrive at a point at which all possible candidates for the next main diagonal element are zero.

One of the many practical applications involving the inverse of a matrix is solving a system of linear equations. Note how the system

$$2x + 3y + z = -1$$
$$3x + 3y + z = 1$$
$$2x + 4y + z = -2$$

can be written as a matrix equation using matrix multiplication.

$$\begin{bmatrix} 2 & 3 & 1 \\ 3 & 3 & 1 \\ 2 & 4 & 1 \end{bmatrix} \begin{bmatrix} x \\ y \\ z \end{bmatrix} = \begin{bmatrix} -1 \\ 1 \\ -2 \end{bmatrix}$$
$$\quad\quad A \quad\quad\quad X \;=\; B$$

To solve this system, we need A^{-1} to obtain

$$X = (A^{-1})B$$

For a *single* system of equations, this procedure would not be worth the effort. However, we often encounter several systems in which each system has the same coefficients but the right-hand constants differ. In this situation, it is convenient to find the inverse of the coefficient matrix and use this inverse to solve each system. This procedure is demonstrated in the next example.

EXAMPLE 4
Solving a System of Equations Using an Inverse

Use an inverse matrix to solve the following systems:

(a) $2x + 3y + z = -1$ (b) $2x + 3y + z = 4$
$\quad\;\; 3x + 3y + z = 1$ $\quad\;\;\; 3x + 3y + z = 8$
$\quad\;\; 2x + 4y + z = -2$ $\quad\;\;\; 2x + 4y + z = 5$
(c) $2x + 3y + z = 2$
$\quad\;\; 3x + 3y + z = -1$
$\quad\;\; 2x + 4y + z = 4$

Solution:
The coefficient matrix for each system is the same:

$$A = \begin{bmatrix} 2 & 3 & 1 \\ 3 & 3 & 1 \\ 2 & 4 & 1 \end{bmatrix}$$

From Example 3, we know that the inverse of this matrix is

$$A^{-1} = \begin{bmatrix} -1 & 1 & 0 \\ -1 & 0 & 1 \\ 6 & -2 & -3 \end{bmatrix}$$

(a) We let B be the following column matrix:

$$B = \begin{bmatrix} -1 \\ 1 \\ -2 \end{bmatrix}$$

Now, the solution to the matrix equation $AX = B$ is

$$X = (A^{-1})B = \begin{bmatrix} -1 & 1 & 0 \\ -1 & 0 & 1 \\ 6 & -2 & -3 \end{bmatrix} \begin{bmatrix} -1 \\ 1 \\ -2 \end{bmatrix}$$

$$= \begin{bmatrix} 1 + 1 + 0 \\ 1 + 0 - 2 \\ -6 - 2 + 6 \end{bmatrix} = \begin{bmatrix} 2 \\ -1 \\ -2 \end{bmatrix}$$

Thus, the solution to the system of linear equations is

$$x = 2, \qquad y = -1, \qquad \text{and} \qquad z = -2$$

(b) In this case,

$$B = \begin{bmatrix} 4 \\ 8 \\ 5 \end{bmatrix}$$

Therefore, the solution to the matrix equation $AX = B$ is

$$X = (A^{-1})B = \begin{bmatrix} -1 & 1 & 0 \\ -1 & 0 & 1 \\ 6 & -2 & -3 \end{bmatrix} \begin{bmatrix} 4 \\ 8 \\ 5 \end{bmatrix}$$

$$= \begin{bmatrix} -4 + 8 + 0 \\ -4 + 0 + 5 \\ 24 - 16 - 15 \end{bmatrix} = \begin{bmatrix} 4 \\ 1 \\ -7 \end{bmatrix}$$

Thus, the solution to the system of linear equations is

$$x = 4, \qquad y = 1, \qquad \text{and} \qquad z = -7$$

(c) Here we have

$$B = \begin{bmatrix} 2 \\ -1 \\ 4 \end{bmatrix}$$

Therefore, the solution to the matrix equation $AX = B$ is

$$X = (A^{-1})B = \begin{bmatrix} -1 & 1 & 0 \\ -1 & 0 & 1 \\ 6 & -2 & -3 \end{bmatrix} \begin{bmatrix} 2 \\ -1 \\ 4 \end{bmatrix}$$

$$X = (A^{-1})B = \begin{bmatrix} -2 - 1 + 0 \\ -2 + 0 + 4 \\ 12 + 2 - 12 \end{bmatrix} = \begin{bmatrix} -3 \\ 2 \\ 2 \end{bmatrix}$$

Thus, the solution to the system of linear equations is

$$x = -3, \qquad y = 2, \qquad \text{and} \qquad z = 2$$

Section Exercises 9.3

In Exercises 1–5, show that B is the inverse of A.

1. $A = \begin{bmatrix} 1 & 2 \\ 3 & 4 \end{bmatrix}$, $B = \begin{bmatrix} -2 & 1 \\ \frac{3}{2} & -\frac{1}{2} \end{bmatrix}$

2. $A = \begin{bmatrix} 1 & -1 \\ 2 & 3 \end{bmatrix}$, $B = \begin{bmatrix} \frac{3}{5} & \frac{1}{5} \\ -\frac{2}{5} & \frac{1}{5} \end{bmatrix}$

3. $A = \begin{bmatrix} -2 & 2 & 3 \\ 1 & -1 & 0 \\ 0 & 1 & 4 \end{bmatrix}$, $B = \frac{1}{3}\begin{bmatrix} -4 & -5 & 3 \\ -4 & -8 & 3 \\ 1 & 2 & 0 \end{bmatrix}$

4. $A = \begin{bmatrix} 2 & -17 & 11 \\ -1 & 11 & -7 \\ 0 & 3 & -2 \end{bmatrix}$, $B = \begin{bmatrix} 1 & 1 & 2 \\ 2 & 4 & -3 \\ 3 & 6 & -5 \end{bmatrix}$

5. $A = \begin{bmatrix} 1 & 2 & 0 & 0 \\ 2 & 1 & 2 & 0 \\ 0 & 2 & 1 & 2 \\ 0 & 0 & 2 & 1 \end{bmatrix}$,

$B = \frac{1}{5}\begin{bmatrix} -7 & 6 & 4 & -8 \\ 6 & -3 & -2 & 4 \\ 4 & -2 & -3 & 6 \\ -8 & 4 & 6 & -7 \end{bmatrix}$

In Exercises 6–30, find the inverse of the matrix (if it exists).

6. $\begin{bmatrix} 1 & 2 \\ 3 & 7 \end{bmatrix}$

7. $\begin{bmatrix} 1 & -2 \\ 2 & -3 \end{bmatrix}$

8. $\begin{bmatrix} -7 & 33 \\ 4 & -19 \end{bmatrix}$

9. $\begin{bmatrix} -1 & 1 \\ -2 & 1 \end{bmatrix}$

10. $\begin{bmatrix} 11 & 1 \\ -1 & 0 \end{bmatrix}$

11. $\begin{bmatrix} 2 & 4 \\ 4 & 8 \end{bmatrix}$

12. $\begin{bmatrix} 2 & 3 \\ 1 & 4 \end{bmatrix}$

13. $\begin{bmatrix} 2 & 7 \\ -3 & -9 \end{bmatrix}$

14. $\begin{bmatrix} - & 5 \\ 6 & -15 \end{bmatrix}$

15. $\begin{bmatrix} 1 & 1 & 1 \\ 3 & 5 & 4 \\ 3 & 6 & 5 \end{bmatrix}$

16. $\begin{bmatrix} 1 & 2 & 2 \\ 3 & 7 & 9 \\ -1 & -4 & -7 \end{bmatrix}$

17. $\begin{bmatrix} 1 & 2 & -1 \\ 3 & 7 & -10 \\ -5 & -7 & -15 \end{bmatrix}$

18. $\begin{bmatrix} 10 & 5 & -7 \\ -5 & 1 & 4 \\ 3 & 2 & -2 \end{bmatrix}$

19. $\begin{bmatrix} 1 & -2 & -1 & -2 \\ 3 & -5 & -2 & -3 \\ 2 & -5 & -2 & -5 \\ -1 & 4 & 4 & 11 \end{bmatrix}$

20. $\begin{bmatrix} 4 & 8 & -7 & 14 \\ 2 & 5 & -4 & 6 \\ 0 & 2 & 1 & -7 \\ 3 & 6 & -5 & 10 \end{bmatrix}$

21. $\begin{bmatrix} 1 & 1 & 2 \\ 3 & 1 & 0 \\ -2 & 0 & 3 \end{bmatrix}$

22. $\begin{bmatrix} 3 & 2 & 2 \\ 2 & 2 & 2 \\ -4 & 4 & 3 \end{bmatrix}$

23. $\begin{bmatrix} 0.1 & 0.2 & 0.3 \\ -0.3 & 0.2 & 0.2 \\ 0.5 & 0.4 & 0.4 \end{bmatrix}$

24. $\begin{bmatrix} 2 & 0 & 0 \\ 0 & 3 & 0 \\ 0 & 0 & 5 \end{bmatrix}$

25. $\begin{bmatrix} -8 & 0 & 0 & 0 \\ 0 & 1 & 0 & 0 \\ 0 & 0 & 4 & 0 \\ 0 & 0 & 0 & -5 \end{bmatrix}$

26. $\begin{bmatrix} 1 & 3 & -2 & 0 \\ 0 & 2 & 4 & 6 \\ 0 & 0 & -2 & 1 \\ 0 & 0 & 0 & 5 \end{bmatrix}$

27. $\begin{bmatrix} 1 & 0 & 0 \\ 3 & 4 & 0 \\ 2 & 5 & 5 \end{bmatrix}$

28. $\begin{bmatrix} 1 & 0 & 0 \\ 3 & 0 & 0 \\ 2 & 5 & 5 \end{bmatrix}$

29. $\begin{bmatrix} 1 & 0 & 3 & 0 \\ 0 & 2 & 0 & 4 \\ 1 & 0 & 3 & 0 \\ 0 & 2 & 0 & 4 \end{bmatrix}$

30. $\begin{bmatrix} 1/a & 0 \\ 0 & a \end{bmatrix}$

31. Use an inverse matrix to solve the following systems (see Exercise 9):
   (a) $\begin{aligned} -x + y &= 4 \\ -2x + y &= 0 \end{aligned}$ (b) $\begin{aligned} -x + y &= -3 \\ -2x + y &= 5 \end{aligned}$
   (c) $\begin{aligned} -x + y &= 0 \\ -2x + y &= 7 \end{aligned}$

32. Use an inverse matrix to solve the following systems (see Exercise 12):
   (a) $\begin{aligned} 2x + 3y &= 5 \\ x + 4y &= 10 \end{aligned}$ (b) $\begin{aligned} 2x + 3y &= 0 \\ x + 4y &= 3 \end{aligned}$
   (c) $\begin{aligned} 2x + 3y &= 1 \\ x + 4y &= -2 \end{aligned}$

33. Use an inverse matrix to solve the following systems (see Exercise 22):
   (a) $\begin{aligned} 3x + 2y + 2z &= 0 \\ 2x + 2y + 2z &= 5 \\ -4x + 4y + 3z &= 2 \end{aligned}$
   (b) $\begin{aligned} 3x + 2y + 2z &= -1 \\ 2x + 2y + 2z &= 2 \\ -4x + 4y + 3z &= 0 \end{aligned}$

(c) $3x + 2y + 2z = 0$
       $2x + 2y + 2z = 0$
       $-4x + 4y + 3z = 1$

34. Use an inverse matrix to solve the following systems (see Exercise 19):

(a) $x_1 - 2x_2 - x_3 - 2x_4 = 0$
       $3x_1 - 5x_2 - 2x_3 - 3x_4 = 1$

$2x_1 - 5x_2 - 2x_3 - 5x_4 = -1$
$-x_1 + 4x_2 + 4x_3 + 11x_4 = 2$

(b) $x_1 - 2x_2 - x_3 - 2x_4 = 1$
       $3x_1 - 5x_2 - 2x_3 - 3x_4 = -2$
       $2x_1 - 5x_2 - 2x_3 - 5x_4 = 0$
       $-x_1 + 4x_2 + 4x_3 + 11x_4 = -3$

Determinants   9.4

With every *square* matrix, we can associate a real number called its **determinant.** Determinants have many uses, and we will encounter a few of these in the next two sections. In this section we concentrate on procedures for computing (or evaluating) the determinant of a matrix. From this point on, we assume that all matrices under discussion are square matrices.

THE DETERMINANT OF A MATRIX OF ORDER 2

The **determinant** of the matrix

$$A = \begin{bmatrix} a_{11} & a_{12} \\ a_{21} & a_{22} \end{bmatrix}$$

is given by

$$|A| = \begin{vmatrix} a_{11} & a_{12} \\ a_{21} & a_{22} \end{vmatrix} = a_{11}a_{22} - a_{12}a_{21}$$

Remark: A convenient method for remembering this formula is shown in the following diagram. Note that the determinant is given by the difference of the products of the two diagonals of the matrix.

$$|A| = \begin{vmatrix} a_{11} & a_{12} \\ a_{21} & a_{22} \end{vmatrix} = a_{11}a_{22} - a_{12}a_{21}$$

EXAMPLE 1
The Determinant
of a Matrix of Order 2

Find the determinants of the following matrices:

(a) $A = \begin{bmatrix} 2 & -3 \\ 1 & 2 \end{bmatrix}$ (b) $B = \begin{bmatrix} 2 & 1 \\ 4 & 2 \end{bmatrix}$ (c) $C = \begin{bmatrix} 0 & 3 \\ 2 & 4 \end{bmatrix}$

Solution:

(a) $|A| = \begin{vmatrix} 2 & -3 \\ 1 & 2 \end{vmatrix} = 2(2) - 1(-3) = 4 + 3 = 7$

(b) $|B| = \begin{vmatrix} 2 & 1 \\ 4 & 2 \end{vmatrix} = 2(2) - 4(1) = 4 - 4 = 0$

(c) $|C| = \begin{vmatrix} 0 & 3 \\ 2 & 4 \end{vmatrix} = 0(4) - 2(3) = 0 - 6 = -6$

The determinant of a 3×3 matrix can be found as follows.

THE DETERMINANT OF A MATRIX OF ORDER 3

The determinant of the matrix

$$A = \begin{bmatrix} a_{11} & a_{12} & a_{13} \\ a_{21} & a_{22} & a_{23} \\ a_{31} & a_{32} & a_{33} \end{bmatrix}$$

is given by

$$|A| = \begin{vmatrix} a_{11} & a_{12} & a_{13} \\ a_{21} & a_{22} & a_{23} \\ a_{31} & a_{32} & a_{33} \end{vmatrix} = a_{11}a_{22}a_{33} + a_{12}a_{23}a_{31} + a_{13}a_{21}a_{32} \\ - a_{31}a_{22}a_{13} - a_{32}a_{23}a_{11} - a_{33}a_{21}a_{12}$$

The formula for the determinant of a 3×3 matrix would be difficult to memorize in terms of subscripts. A much simpler way to remember this formula is to recopy the first and second columns of the matrix (forming columns 4 and 5) and use products of diagonals, as shown in the following diagram.

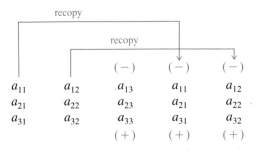

EXAMPLE 2
The Determinant
of a Matrix of Order 3

Find the determinant of the following matrix:

$$A = \begin{bmatrix} 0 & 2 & 1 \\ 3 & -1 & 2 \\ 4 & 0 & 1 \end{bmatrix}$$

Solution:
We begin by recopying the first two columns and computing the six "diagonal products," as follows:

$$-4(-1)(1) - 0(2)(0) - 1(3)(2) \quad \text{Negative diagonals}$$

$$\begin{vmatrix} 0 & 2 & 1 & 0 & 2 \\ 3 & -1 & 2 & 3 & -1 \\ 4 & 0 & 1 & 4 & 0 \end{vmatrix} = 0 + 16 + 0 - (-4) - 0 - 6$$

$$= 16 + 4 - 6 = 14$$

$$+0(-1)(1) + 2(2)(4) + 1(3)(0) \quad \text{Positive diagonals}$$

The diagonal product diagram used in Examples 1 and 2 is quite useful for finding the determinant for matrices of order 2 or 3. However, this scheme does not generalize to matrices of higher order. A more general method for finding determinants of matrices involves the concepts of **minors** and **cofactors.**

DEFINITION OF THE MINORS AND COFACTORS OF A MATRIX

$$M_{21} = \begin{bmatrix} a_{11} & a_{12} & a_{13} \\ a_{21} & a_{22} & a_{23} \\ a_{31} & a_{32} & a_{33} \end{bmatrix} = \begin{bmatrix} a_{12} & a_{13} \\ a_{32} & a_{33} \end{bmatrix}$$

FIGURE 9.2   *Minor of* a_{21}

If A is a square matrix (of order 3 or greater), then the **minor** M_{ij} of the element a_{ij} is the determinant of the matrix obtained by deleting the ith row and jth column of A. (See Figure 9.2.)

The **cofactor** C_{ij} is given by

$$C_{ij} = (-1)^{i+j} M_{ij}$$

Remark: Note that the minors and cofactors of a matrix differ at most in sign. To obtain the cofactors of a matrix, first find the minors and then apply the following "checkerboard" pattern of $+$'s and $-$'s.

Sign Pattern for Cofactors

3 × 3 Matrix

$$\begin{bmatrix} + & - & + \\ - & + & - \\ + & - & + \end{bmatrix}$$

4 × 4 Matrix

$$\begin{bmatrix} + & - & + & - \\ - & + & - & + \\ + & - & + & - \\ - & + & - & + \end{bmatrix}$$

Note that *odd* positions of a_{ij} (where $i + j$ is odd) have negative signs, and *even* positions of a_{ij} (where $i + j$ is even) have positive signs.

EXAMPLE 3
Finding the Minors
and Cofactors of a Matrix

Find all the minors and cofactors of the following matrix:

$$A = \begin{bmatrix} 0 & 2 & 1 \\ 3 & -1 & 2 \\ 4 & 0 & 1 \end{bmatrix}$$

Solution:
To find the minor M_{ij}, we delete the ith row and jth column and evaluate the

resulting determinant. To find M_{11}:

$$M_{11} = \begin{vmatrix} 0 & 2 & 1 \\ 3 & -1 & 2 \\ 4 & 0 & 1 \end{vmatrix} = \begin{vmatrix} -1 & 2 \\ 0 & 1 \end{vmatrix} = -1(1) - 0(2) = -1$$

To find M_{12}:

$$M_{12} = \begin{vmatrix} 0 & 2 & 1 \\ 3 & -1 & 2 \\ 4 & 0 & 1 \end{vmatrix} = \begin{vmatrix} 3 & 2 \\ 4 & 1 \end{vmatrix} = 3(1) - 4(2) = -5$$

By continuing this pattern, we can obtain the following *matrix of minors:*

$$\begin{bmatrix} M_{11} & M_{12} & M_{13} \\ M_{21} & M_{22} & M_{23} \\ M_{31} & M_{32} & M_{33} \end{bmatrix} = \begin{bmatrix} \begin{vmatrix} -1 & 2 \\ 0 & 1 \end{vmatrix} & \begin{vmatrix} 3 & 2 \\ 4 & 1 \end{vmatrix} & \begin{vmatrix} 3 & -1 \\ 4 & 0 \end{vmatrix} \\ \begin{vmatrix} 2 & 1 \\ 0 & 1 \end{vmatrix} & \begin{vmatrix} 0 & 1 \\ 4 & 1 \end{vmatrix} & \begin{vmatrix} 0 & 2 \\ 4 & 0 \end{vmatrix} \\ \begin{vmatrix} 2 & 1 \\ -1 & 2 \end{vmatrix} & \begin{vmatrix} 0 & 1 \\ 3 & 2 \end{vmatrix} & \begin{vmatrix} 0 & 2 \\ 3 & -1 \end{vmatrix} \end{bmatrix}$$

$$= \begin{bmatrix} -1 & -5 & 4 \\ 2 & -4 & -8 \\ 5 & -3 & -6 \end{bmatrix}$$

Now to find the cofactors, we combine our checkerboard pattern with these minors to obtain the *matrix of cofactors.*

$$\begin{bmatrix} C_{11} & C_{12} & C_{13} \\ C_{21} & C_{22} & C_{23} \\ C_{31} & C_{32} & C_{33} \end{bmatrix} = \begin{bmatrix} M_{11} & -M_{12} & M_{13} \\ -M_{21} & M_{22} & -M_{23} \\ M_{31} & -M_{32} & M_{33} \end{bmatrix}$$

$$= \begin{bmatrix} -1 & -(-5) & 4 \\ -(2) & -4 & -(-8) \\ 5 & -(-3) & -6 \end{bmatrix}$$

$$= \begin{bmatrix} -1 & 5 & 4 \\ -2 & -4 & 8 \\ 5 & 3 & -6 \end{bmatrix}$$

Remark: The matrix of cofactors for a given matrix A is useful for finding the inverse of A, as we will see in Section 9.5.

Now that we can find the minors and cofactors of a matrix, here is a general method for finding the determinant of a matrix of order 3 or greater. This method is referred to as **expansion by cofactors.**

FINDING DETERMINANTS BY EXPANSION BY COFACTORS

> If A is a matrix of order 3 or greater, then the determinant of A is found by adding the products of the elements in *any* row (or column) of A with its corresponding cofactors.

This method gives us many options to choose from when evaluating the determinant of a given matrix. As a case in point, let's return to the matrix shown in Examples 2 and 3 and find its determinant with this new method.

EXAMPLE 4
Expansion by Cofactors

Use expansion by cofactors to find the determinant of

$$A = \begin{bmatrix} 0 & 2 & 1 \\ 3 & -1 & 2 \\ 4 & 0 & 1 \end{bmatrix}$$

Solution:
From Example 3, the matrix of cofactors is

$$\begin{bmatrix} -1 & 5 & 4 \\ -2 & -4 & 8 \\ 5 & 3 & -6 \end{bmatrix}$$

(1) Using the *first row,* we have

$$|A| = 0(-1) + 2(5) + 1(4) = 10 + 4 = 14$$

(2) Using the *second row,* we have

$$|A| = 3(-2) + (-1)(-4) + 2(8) = -6 + 4 + 16 = 14$$

(3) Using the *first column,* we have

$$|A| = 0(-1) + 3(-2) + 4(5) = -6 + 20 = 14$$

You should try some other possibilities to see that the determinant of A can be evaluated by expanding by *any* row or column of A.

Remark: Note that in expansion by cofactors we are not using matrix multiplication, but are multiplying corresponding elements. For instance, in the expansion

$$|A| = a_{11}C_{11} + a_{21}C_{21} + \cdots + a_{n1}C_{n1}$$

along Row 1 of an $n \times n$ matrix A, we multiply each row entry by its corresponding cofactor (entry) in the matrix of cofactors of A.

In practice, the row (or column) containing the most zeros is the best choice for expansion by cofactors. Watch how this works in our next example.

EXAMPLE 5
Finding the Determinant
of a Matrix of Order 4

Use the method of expansion by cofactors to find the determinant of the following matrix:

$$A = \begin{bmatrix} 1 & -2 & 3 & 0 \\ -1 & 1 & 0 & 2 \\ 0 & 2 & 0 & 3 \\ 3 & 4 & 1 & -2 \end{bmatrix}$$

Solution:

After inspecting this matrix, we see that two of the elements in the third column are zero. Thus, we can eliminate some of the work in the expansion by using the third column.

$$|A| = a_{13}C_{13} + a_{23}C_{23} + a_{33}C_{33} + a_{43}C_{43}$$
$$= 3(C_{13}) + 0(C_{23}) + 0(C_{33}) + 1(C_{43})$$

Since two of the coefficients in this expansion are zero, we need only find the two cofactors C_{13} and C_{43}. For C_{13}, we have

$$C_{13} = (-1)^{1+3} \begin{vmatrix} 1 & -2 & 3 & 0 \\ -1 & 1 & 0 & 2 \\ 0 & 2 & 0 & 3 \\ 3 & 4 & 1 & -2 \end{vmatrix} = (1) \begin{vmatrix} -1 & 1 & 2 \\ 0 & 2 & 3 \\ 3 & 4 & -2 \end{vmatrix}$$

The diagonal procedure then yields

$$C_{13} = 4 + 0 + 9 - 12 + 12 + 0 = 13$$

For C_{43}, we have

$$C_{43} = (-1)^{4+3} \begin{vmatrix} 1 & -2 & 3 & 0 \\ -1 & 1 & 0 & 2 \\ 0 & 2 & 0 & 3 \\ 3 & 4 & 1 & -2 \end{vmatrix} = (-1) \begin{vmatrix} 1 & -2 & 0 \\ -1 & 1 & 2 \\ 0 & 2 & 3 \end{vmatrix}$$

Expanding the 3×3 determinant by minors along the first column yields

$$C_{43} = (-1)\left((1)(-1)^2 \begin{vmatrix} 1 & 2 \\ 2 & 3 \end{vmatrix} + (-1)(-1)^3 \begin{vmatrix} -2 & 0 \\ 2 & 3 \end{vmatrix} \right)$$
$$= -(3 - 4) - (-6 - 0) = 7$$

Thus, we have

$$|A| = 3(C_{13}) + 1(C_{43}) = 3(13) + 1(7) = 39 + 7 = 46$$

Section Exercises 9.4

In Exercises 1–20, find the determinant of the matrix.

1. $\begin{bmatrix} 2 & 1 \\ 3 & 4 \end{bmatrix}$ 2. $\begin{bmatrix} -3 & 1 \\ 5 & 2 \end{bmatrix}$ 3. $\begin{bmatrix} 5 & 2 \\ -6 & 3 \end{bmatrix}$ 4. $\begin{bmatrix} 2 & -2 \\ 4 & 3 \end{bmatrix}$ 5. $\begin{bmatrix} -7 & 6 \\ \frac{1}{2} & 3 \end{bmatrix}$ 6. $\begin{bmatrix} 4 & -3 \\ 0 & 0 \end{bmatrix}$

7. $\begin{bmatrix} 2 & 6 \\ 0 & 3 \end{bmatrix}$

8. $\begin{bmatrix} 2 & -3 \\ -6 & 9 \end{bmatrix}$

9. $\begin{bmatrix} 2 & -1 & 0 \\ 4 & 2 & 1 \\ 4 & 2 & 1 \end{bmatrix}$

10. $\begin{bmatrix} -2 & 2 & 3 \\ 1 & -1 & 0 \\ 0 & 1 & 4 \end{bmatrix}$

11. $\begin{bmatrix} 3 & 2 & 2 \\ 2 & 2 & 2 \\ -4 & 4 & 3 \end{bmatrix}$

12. $\begin{bmatrix} 0.1 & 0.2 & 0.3 \\ -0.3 & 0.2 & 0.2 \\ 0.5 & 0.4 & 0.4 \end{bmatrix}$

13. $\begin{bmatrix} 1 & 4 & -2 \\ 3 & 6 & -6 \\ -2 & 1 & 4 \end{bmatrix}$

14. $\begin{bmatrix} 2 & 3 & 1 \\ 0 & 5 & -2 \\ 0 & 0 & -2 \end{bmatrix}$

15. $\begin{bmatrix} 6 & 3 & -7 \\ 0 & 0 & 0 \\ 4 & -6 & 3 \end{bmatrix}$

16. $\begin{bmatrix} 1 & 1 & 2 \\ 3 & 1 & 0 \\ -2 & 0 & 3 \end{bmatrix}$

17. $\begin{bmatrix} -0.4 & 0.4 & 0.3 \\ 0.2 & 0.2 & 0.2 \\ 0.3 & 0.2 & 0.2 \end{bmatrix}$

18. $\begin{bmatrix} x & y & 1 \\ 2 & 3 & 1 \\ 0 & -1 & 1 \end{bmatrix}$

19. $\begin{bmatrix} x & y & 1 \\ -2 & -2 & 1 \\ 1 & 5 & 1 \end{bmatrix}$

20. $\begin{bmatrix} 3-\lambda & 2 \\ 4 & 1-\lambda \end{bmatrix}$

In Exercises 21–24, find (a) the matrix of minors, and (b) the matrix of cofactors of the given matrix (see Example 3).

21. $\begin{bmatrix} -3 & 2 & 1 \\ 4 & 5 & 6 \\ 2 & -3 & 1 \end{bmatrix}$

22. $\begin{bmatrix} 10 & 8 & 3 & -7 \\ 4 & 0 & 5 & -6 \\ 0 & 3 & 2 & 7 \\ 1 & 0 & -3 & 2 \end{bmatrix}$

23. $\begin{bmatrix} 6 & 0 & -3 & 5 \\ 4 & 13 & 6 & -8 \\ -1 & 0 & 7 & 4 \\ 8 & 6 & 0 & 2 \end{bmatrix}$

24. $\begin{bmatrix} -3 & 4 & 2 \\ 6 & 3 & 1 \\ 4 & -7 & -8 \end{bmatrix}$

25. Find the determinant of the matrix in Exercise 21 by the method of expansion by cofactors using (a) the first row and (b) the second column.

26. Find the determinant of the matrix in Exercise 22 by the method of expansion by cofactors using (a) the second column and (b) the first row.

27. Find the determinant of the matrix in Exercise 23 by the method of expansion by cofactors using (a) the second row and (b) the second column.

28. Find the determinant of the matrix in Exercise 24 by the method of expansion by cofactors using (a) the third row and (b) the first column.

In Exercises 29–40, find the determinant of the matrix by the method of expansion by cofactors. Choose the row or column that may simplify the calculations.

29. $\begin{bmatrix} 1 & 4 & -2 \\ 3 & 2 & 0 \\ -1 & 4 & 3 \end{bmatrix}$

30. $\begin{bmatrix} 2 & -1 & 3 \\ 1 & 4 & 4 \\ 1 & 0 & 2 \end{bmatrix}$

31. $\begin{bmatrix} 2 & 4 & 6 \\ 0 & 3 & 1 \\ 0 & 0 & -5 \end{bmatrix}$

32. $\begin{bmatrix} -3 & 0 & 0 \\ 7 & 11 & 0 \\ 1 & 2 & 2 \end{bmatrix}$

33. $\begin{bmatrix} 3 & 6 & -5 & 4 \\ -2 & 0 & 6 & 0 \\ 1 & 1 & 2 & 2 \\ 0 & 3 & -1 & -1 \end{bmatrix}$

34. $\begin{bmatrix} 2 & 6 & 6 & 2 \\ 2 & 7 & 3 & 6 \\ 1 & 5 & 0 & 1 \\ 3 & 7 & 0 & 7 \end{bmatrix}$

35. $\begin{bmatrix} 5 & 3 & 0 & 6 \\ 4 & 6 & 4 & 12 \\ 0 & 2 & -3 & 4 \\ 0 & 1 & -2 & 2 \end{bmatrix}$

36. $\begin{bmatrix} 1 & 4 & 3 & 2 \\ -5 & 6 & 2 & 1 \\ 0 & 0 & 0 & 0 \\ 3 & -2 & 1 & 5 \end{bmatrix}$

37. $\begin{bmatrix} 3 & 2 & 4 & -1 & 5 \\ -2 & 0 & 1 & 3 & 2 \\ 1 & 0 & 0 & 4 & 0 \\ 6 & 0 & 2 & -1 & 0 \\ 3 & 0 & 5 & 1 & 0 \end{bmatrix}$

38. $\begin{bmatrix} 5 & 2 & 0 & 0 & -2 \\ 0 & 1 & 4 & 3 & 2 \\ 0 & 0 & 2 & 6 & 3 \\ 0 & 0 & 3 & 4 & 1 \\ 0 & 0 & 0 & 0 & 2 \end{bmatrix}$

39. $\begin{bmatrix} 4 & 3 & -2 & 1 & 2 \\ 0 & 0 & 0 & 0 & 0 \\ 1 & 2 & -7 & 13 & 12 \\ 6 & -2 & 5 & 6 & 7 \\ 1 & 4 & 2 & 0 & 9 \end{bmatrix}$

40. $\begin{bmatrix} 3 & 0 & 7 & 0 \\ 2 & 6 & 11 & 12 \\ 4 & 1 & -1 & 2 \\ 1 & 5 & 2 & 10 \end{bmatrix}$

The equation of a line through the points (x_1, y_1) and (x_2, y_2) can be written

$$\begin{vmatrix} x & y & 1 \\ x_1 & y_1 & 1 \\ x_2 & y_2 & 1 \end{vmatrix} = 0$$

In Exercises 41–45, use this method to find an equation of the line through the given points (x_1, y_1) and (x_2, y_2).

41. $(0, 0), (5, 3)$

42. $(0, 0), (-2, 2)$

43. $(-4, 3), (2, 1)$

44. $(10, 7), (-2, -7)$

45. $(-\frac{1}{2}, 3), (\frac{5}{2}, 1)$

The area of a triangle with vertices (x_1, y_1), (x_2, y_2), and (x_3, y_3) is the absolute value of

$$\frac{1}{2}\begin{vmatrix} x_1 & y_1 & 1 \\ x_2 & y_2 & 1 \\ x_3 & y_3 & 1 \end{vmatrix}$$

In Exercises 46–50, use this method to find the area of the triangle with the given vertices.

46. $(0, 0)$, $(3, 1)$, $(1, 5)$ 47. $(0, 0)$, $(5, -2)$, $(4, 5)$
48. $(-2, -3)$, $(2, -3)$, $(0, 4)$
49. $(-2, 1)$, $(3, -1)$, $(1, 6)$
50. $(0, \frac{1}{2})$, $(\frac{5}{2}, 0)$, $(4, 3)$

Properties of Determinants   9.5

Which of the following determinants is easier to evaluate?

$$|A| = \begin{vmatrix} 4 & 1 & 3 & 2 \\ -2 & 2 & 1 & -3 \\ 1 & -2 & 3 & 1 \\ 3 & 1 & 3 & 2 \end{vmatrix} \qquad |B| = \begin{vmatrix} 0 & 9 & -9 & -2 \\ 0 & -2 & 7 & -1 \\ 1 & -2 & 3 & 1 \\ 0 & 7 & -6 & -1 \end{vmatrix}$$

From what we know about expansion by cofactors, it is clear that the second determinant is much simpler to evaluate, since three of the elements in the first column are zero. If we expand these two determinants by cofactors (using the first column), we get

$$|A| = (4)\begin{vmatrix} 2 & 1 & -3 \\ -2 & 3 & 1 \\ 1 & 3 & 2 \end{vmatrix} - (-2)\begin{vmatrix} 1 & 3 & 2 \\ -2 & 3 & 1 \\ 1 & 3 & 2 \end{vmatrix}$$

$$+ (1)\begin{vmatrix} 1 & 3 & 2 \\ 2 & 1 & -3 \\ 1 & 3 & 2 \end{vmatrix} - (3)\begin{vmatrix} 1 & 3 & 2 \\ 2 & 1 & -3 \\ -2 & 3 & 1 \end{vmatrix}$$

$$= 4(38) + 2(0) + 1(0) - 3(38) = 38$$

$$|B| = (0)C_{11} + (0)C_{12} + (1)\begin{vmatrix} 9 & -9 & -2 \\ -2 & 7 & -1 \\ 7 & -6 & -1 \end{vmatrix} + (0)C_{14}$$

$$= 1(38) = 38$$

It is not mere coincidence that these two determinants have the same value. In fact, we obtained the second determinant from the first by performing *elementary row operations* on the matrix *A*.

In this section, we identify the effect of elementary row operations on the value of a determinant. In addition, we discuss some special properties of determinants and their relationships to the existence and calculation of the inverse of a matrix.

ELEMENTARY ROW (COLUMN) OPERATIONS AND DETERMINANTS

If A is a square matrix of order 2 or greater, then the following elementary row (or column) operations produce the indicated change in $|A|$:

1. The interchange of two rows results in a change of signs of the determinant.

2. Adding a multiple of the elements of one row to the elements of another row leaves the determinant unchanged.

3. Multiplying the elements of one row by k produces a determinant that is k times the original.

Note that each of these three statements is still valid if the word "row" is changed to the word "column."

EXAMPLE 1
Applying Elementary Row Operations

Use elementary row (or column) operations to evaluate the following determinant:

$$|A| = \begin{vmatrix} 4 & 1 & 3 & 2 \\ -2 & 2 & 1 & -3 \\ 1 & -2 & 3 & 1 \\ 3 & 1 & 3 & 2 \end{vmatrix}$$

Solution:
There are many ways to approach this problem, but we begin by adding multiples of the third row to rows 1, 2, and 4, to create three zeros in the first column.

$$|A| = \begin{vmatrix} 4 & 1 & 3 & 2 \\ -2 & 2 & 1 & -3 \\ 1 & -2 & 3 & 1 \\ 3 & 1 & 3 & 2 \end{vmatrix} \begin{matrix} -4R_3 + R_1 \\ 2R_3 + R_2 \\ \\ -3R_3 + R_4 \end{matrix} \begin{vmatrix} 0 & 9 & -9 & -2 \\ 0 & -2 & 7 & -1 \\ 1 & -2 & 3 & 1 \\ 0 & 7 & -6 & -1 \end{vmatrix}$$

Now, we can expand by cofactors (as shown in the beginning of this section) to obtain

$$|A| = 38$$

EXAMPLE 2
Applying Elementary Row Operations

Use elementary row (or column) operations to evaluate the following determinants:

(a) $|A| = \begin{vmatrix} 1 & 4 & -2 \\ 2 & 5 & -4 \\ -3 & 0 & 6 \end{vmatrix}$ (b) $|B| = \begin{vmatrix} 2 & 0 & 1 & 3 & -2 \\ -2 & 1 & 0 & 1 & -1 \\ 1 & 0 & -1 & 2 & 3 \\ 3 & -1 & 2 & 4 & -3 \\ 1 & 1 & -1 & 2 & 0 \end{vmatrix}$

Solution:

(a) Notice that the third column is a multiple of the first column. This implies that the determinant is zero. To see this, note how we can use an elementary *column* operation to create *all* zeros in the third column, which then yields zero when expanded by cofactors.

$$|A| = \begin{vmatrix} 1 & 4 & -2 \\ 2 & 5 & -4 \\ -3 & 0 & 6 \end{vmatrix} = \begin{vmatrix} 1 & 4 & 0 \\ 2 & 5 & 0 \\ -3 & 0 & 0 \end{vmatrix} = 0$$

$2C_1 + C_3$

(b) Since the second column of this matrix already has two zeros, we choose it for the expansion.

$$|B| = \begin{vmatrix} 2 & 0 & 1 & 3 & -2 \\ -2 & 1 & 0 & 1 & -1 \\ 1 & 0 & -1 & 2 & 3 \\ 3 & -1 & 2 & 4 & -3 \\ 1 & 1 & -1 & 2 & 0 \end{vmatrix}$$

$$\begin{matrix} \\ \\ \\ 2R_2 + R_4 \\ -R_2 + R_5 \end{matrix} \begin{vmatrix} 2 & 0 & 1 & 3 & -2 \\ -2 & 1 & 0 & 1 & -1 \\ 1 & 0 & -1 & 2 & 3 \\ 1 & 0 & 2 & 5 & -4 \\ 3 & 0 & -1 & 1 & 1 \end{vmatrix}$$

$$= (-1)^4 \begin{vmatrix} 2 & 1 & 3 & -2 \\ 1 & -1 & 2 & 3 \\ 1 & 2 & 5 & -4 \\ 3 & -1 & 1 & 1 \end{vmatrix}$$

Now, suppose we choose to expand by the fourth row. We can create three zeros in the fourth row, as follows:

$$|B| = \begin{vmatrix} 2 & 1 & 3 & -2 \\ 1 & -1 & 2 & 3 \\ 1 & 2 & 5 & -4 \\ 3 & -1 & 1 & 1 \end{vmatrix} = \begin{vmatrix} 5 & 1 & 4 & -1 \\ -2 & -1 & 1 & 2 \\ 7 & 2 & 7 & -2 \\ 0 & -1 & 0 & 0 \end{vmatrix}$$

$3C_2 + C_1$
$C_2 + C_3$
$C_2 + C_4$

Consequently,

$$|B| = (-1)(-1)^{4+2} \begin{vmatrix} 5 & 4 & -1 \\ -2 & 1 & 2 \\ 7 & 7 & -2 \end{vmatrix}$$

$$= -[-10 + 14 + 56 - (-7) - 70 - 16] = 19$$

Note in Example 2, part (a), that if one column of the matrix is a multiple of another column, we can immediately conclude that the determinant is zero. This is one of three conditions, listed next, that yield a determinant of zero. Each condition is easily verified, using elementary row operations and expansion by cofactors.

CONDITIONS THAT YIELD A DETERMINANT OF ZERO

If A is a matrix of order n and any one of the following conditions is true, then $|A| = 0$.

1. An entire row (or column) is zero.

2. Two rows (or columns) have the same elements.

3. One row (or column) is a multiple of another row (or column).

When you are evaluating determinants, you can save considerable work by recognizing these conditions. Of course, these are not the only conditions that yield zero determinants. This fact is illustrated in the following example.

EXAMPLE 3
Evaluating Determinants

Show that each of the following determinants is zero:

(a) $|A| = \begin{vmatrix} 0 & 0 & 0 \\ 2 & 4 & -5 \\ 3 & -5 & 2 \end{vmatrix}$

(b) $|B| = \begin{vmatrix} 1 & -2 & 4 \\ 0 & 1 & 2 \\ 1 & -2 & 4 \end{vmatrix}$

(c) $|C| = \begin{vmatrix} 1 & 2 & -3 \\ 2 & -1 & -6 \\ -2 & 0 & 6 \end{vmatrix}$

(d) $|D| = \begin{vmatrix} 1 & 4 & 1 \\ 2 & -1 & 0 \\ 0 & -9 & -2 \end{vmatrix}$

Solution:

(a) $|A| = \begin{vmatrix} 0 & 0 & 0 \\ 2 & 4 & -5 \\ 3 & -5 & 2 \end{vmatrix} = 0$ *(Since the first row has all zero elements.)*

(b) $|B| = \begin{vmatrix} 1 & -2 & 4 \\ 0 & 1 & 2 \\ 1 & -2 & 4 \end{vmatrix} = 0$ *(Since the first and third rows have the same elements.)*

(c) $|C| = \begin{vmatrix} 1 & 2 & -3 \\ 2 & -1 & -6 \\ -2 & 0 & 6 \end{vmatrix} = 0$ *(Since the third column is a multiple of the first column.)*

(d) $|D| = \begin{vmatrix} 1 & 4 & 1 \\ 2 & -1 & 0 \\ 0 & -9 & -2 \end{vmatrix}$

It is not immediately evident that this determinant is zero. However, by adding (-2) times the first row to the second row, we have the following:

$$|D| = \begin{vmatrix} 1 & 4 & 1 \\ 2 & -1 & 0 \\ 0 & -9 & -2 \end{vmatrix}$$

$$= \begin{vmatrix} 1 & 4 & 1 \\ 0 & -9 & -2 \\ 0 & -9 & -2 \end{vmatrix}$$

Now, since the second and third rows have the same elements, the determinant is zero.

In Section 9.3 we saw that some square matrices have (multiplicative) inverses and some do not. One very useful application of determinants is that the matrices with zero determinants do not have inverses.

CLASSIFYING SINGULAR AND NONSINGULAR MATRICES

> If A is a square matrix of order n, then it has an inverse if and only if its determinant is nonzero.

From this result we can see that none of the matrices in Example 3 have inverses, since they all have zero determinants.

Another application of determinants is that they provide us with a convenient alternative procedure for finding the inverse of a nonsingular matrix, especially those of order 2 or 3.

COFACTOR FORM FOR INVERSE OF A MATRIX

> If A is a nonsingular matrix of order n, then its inverse is given by
>
> $$A^{-1} = \frac{1}{|A|} \begin{bmatrix} C_{11} & C_{21} & C_{31} & \cdots & C_{n1} \\ C_{12} & C_{22} & C_{32} & \cdots & C_{n2} \\ C_{13} & C_{23} & C_{33} & \cdots & C_{n3} \\ \cdot & \cdot & \cdot & & \cdot \\ \cdot & \cdot & \cdot & & \cdot \\ \cdot & \cdot & \cdot & & \cdot \\ C_{1n} & C_{2n} & C_{3n} & \cdots & C_{nn} \end{bmatrix}$$

Remark: Be sure you notice that the cofactor of a_{ij} appears in the *j*th row and *i*th column (rather than the *i*th row and *j*th column). That is, the rows and columns have been interchanged. This is called *transposing* a matrix. For instance, if

$$A = \begin{bmatrix} 1 & 2 & 3 \\ 4 & 3 & 2 \\ 1 & 1 & 3 \end{bmatrix} \quad \text{then} \quad A^T = \begin{bmatrix} 1 & 4 & 1 \\ 2 & 3 & 1 \\ 3 & 2 & 3 \end{bmatrix}$$

If $A = \begin{bmatrix} a & b \\ c & d \end{bmatrix}$, then the inverse of A is given by

$$A^{-1} = \frac{1}{|A|}\begin{bmatrix} C_{11} & C_{21} \\ C_{12} & C_{22} \end{bmatrix} = \frac{1}{|A|}\begin{bmatrix} d & -b \\ -c & a \end{bmatrix}$$

EXAMPLE 4
Finding the Inverse
of a Matrix by Cofactors

Use cofactors to find the inverses of the following matrices:

(a) $A = \begin{bmatrix} -1 & 2 \\ 3 & -8 \end{bmatrix}$

(b) $B = \begin{bmatrix} -1 & 3 & 2 \\ 0 & 1 & -2 \\ 1 & 4 & 1 \end{bmatrix}$

Solution:

(a) Since $|A| = 8 - 6 = 2$, the inverse of A is given by

$$A^{-1} = \frac{1}{2}\begin{bmatrix} -8 & -2 \\ -3 & -1 \end{bmatrix} = \begin{bmatrix} -4 & -1 \\ -\frac{3}{2} & -\frac{1}{2} \end{bmatrix}$$

(b) Since $|B| = -1 - 6 + 0 - 2 - 8 - 0 = -17$, the inverse of B is given by

$$B^{-1} = \frac{1}{-17}\begin{bmatrix} \begin{vmatrix} 1 & -2 \\ 4 & 1 \end{vmatrix} & -\begin{vmatrix} 3 & 2 \\ 4 & 1 \end{vmatrix} & \begin{vmatrix} 3 & 2 \\ 1 & -2 \end{vmatrix} \\ -\begin{vmatrix} 0 & -2 \\ 1 & 1 \end{vmatrix} & \begin{vmatrix} -1 & 2 \\ 1 & 1 \end{vmatrix} & -\begin{vmatrix} -1 & 2 \\ 0 & -2 \end{vmatrix} \\ \begin{vmatrix} 0 & 1 \\ 1 & 4 \end{vmatrix} & -\begin{vmatrix} -1 & 3 \\ 1 & 4 \end{vmatrix} & \begin{vmatrix} -1 & 3 \\ 0 & 1 \end{vmatrix} \end{bmatrix}$$

$$B^{-1} = \frac{1}{-17}\begin{bmatrix} 9 & 5 & -8 \\ -2 & -3 & -2 \\ -1 & 7 & -1 \end{bmatrix}$$

Section Exercises 9.5

In Exercises 1–15, state the property of determinants that allows you to determine that the equation is true without expanding the determinants.

1. $\begin{vmatrix} 2 & -6 \\ 1 & -3 \end{vmatrix} = 0$

2. $\begin{vmatrix} -4 & 5 \\ 12 & -15 \end{vmatrix} = 0$

3. $\begin{vmatrix} 1 & 4 & 2 \\ 0 & 0 & 0 \\ 5 & 6 & -7 \end{vmatrix} = 0$

4. $\begin{vmatrix} -4 & 3 & 2 \\ 8 & 0 & 0 \\ -4 & 3 & 2 \end{vmatrix} = 0$

5. $\begin{vmatrix} 1 & 3 & 4 \\ -7 & 2 & -5 \\ 6 & 1 & 2 \end{vmatrix} = -\begin{vmatrix} 1 & 4 & 3 \\ -7 & -5 & 2 \\ 6 & 2 & 1 \end{vmatrix}$

6. $\begin{vmatrix} 1 & 3 & 4 \\ -2 & 2 & 0 \\ 1 & 6 & 2 \end{vmatrix} = -\begin{vmatrix} 1 & 6 & 2 \\ -2 & 2 & 0 \\ 1 & 3 & 4 \end{vmatrix}$

7. $\begin{vmatrix} 5 & 10 \\ 2 & -7 \end{vmatrix} = 5\begin{vmatrix} 1 & 2 \\ 2 & -7 \end{vmatrix}$

8. $\begin{vmatrix} 1 & 8 & -3 \\ 3 & -12 & 6 \\ 7 & 4 & 9 \end{vmatrix} = 12\begin{vmatrix} 1 & 2 & -1 \\ 3 & -3 & 2 \\ 7 & 1 & 3 \end{vmatrix}$

9. $\begin{vmatrix} 5 & 0 & 10 \\ 25 & -30 & 40 \\ -15 & 5 & 20 \end{vmatrix} = 5^3\begin{vmatrix} 1 & 0 & 2 \\ 5 & -6 & 8 \\ -3 & 1 & 4 \end{vmatrix}$

10. $\begin{vmatrix} 6 & 0 & 0 & 0 \\ 0 & 6 & 0 & 0 \\ 0 & 0 & 6 & 0 \\ 0 & 0 & 0 & 6 \end{vmatrix} = 6^4 \begin{vmatrix} 1 & 0 & 0 & 0 \\ 0 & 1 & 0 & 0 \\ 0 & 0 & 1 & 0 \\ 0 & 0 & 0 & 1 \end{vmatrix}$

11. $\begin{vmatrix} 2 & -3 \\ 8 & 7 \end{vmatrix} = \begin{vmatrix} 2 & -3 \\ 0 & 19 \end{vmatrix}$

12. $\begin{vmatrix} 1 & -3 & 2 \\ 5 & 2 & -1 \\ -1 & 0 & 6 \end{vmatrix} = \begin{vmatrix} 1 & -3 & 2 \\ 0 & 17 & -11 \\ 0 & -3 & 8 \end{vmatrix}$

13. $\begin{vmatrix} 3 & 2 & 4 \\ -2 & 1 & 5 \\ 5 & -7 & -20 \end{vmatrix} = \begin{vmatrix} 7 & 2 & -6 \\ 0 & 1 & 0 \\ -9 & -7 & 15 \end{vmatrix}$

14. $\begin{vmatrix} 5 & 4 & 2 \\ 2 & -3 & 4 \\ 7 & 6 & 3 \end{vmatrix} = \begin{vmatrix} 1 & 10 & -6 \\ 2 & -3 & 4 \\ 7 & 6 & 3 \end{vmatrix}$

15. $\begin{vmatrix} 5 & 4 & 2 \\ 2 & -3 & 4 \\ 7 & 6 & 3 \end{vmatrix} = \begin{vmatrix} 2 & 4 & 5 \\ 3 & 6 & 7 \\ 4 & -3 & 2 \end{vmatrix}$

In Exercises 16–30, use elementary row (or column) operations as aids to evaluating the determinant.

16. $\begin{vmatrix} 1 & 7 & -3 \\ 1 & 3 & 1 \\ 4 & 8 & 1 \end{vmatrix}$

17. $\begin{vmatrix} 1 & 1 & 1 \\ 2 & -1 & -2 \\ 1 & -2 & -1 \end{vmatrix}$

18. $\begin{vmatrix} 2 & -1 & -1 \\ 1 & 3 & 2 \\ 1 & 1 & 3 \end{vmatrix}$

19. $\begin{vmatrix} 3 & -1 & -3 \\ -1 & -4 & -2 \\ 3 & -1 & -1 \end{vmatrix}$

20. $\begin{vmatrix} 4 & 3 & -2 \\ 5 & 4 & 1 \\ -2 & 3 & 4 \end{vmatrix}$

21. $\begin{vmatrix} 3 & 8 & -7 \\ 0 & -5 & 4 \\ 8 & 1 & 6 \end{vmatrix}$

22. $\begin{vmatrix} 5 & -8 & 0 \\ 9 & 7 & 4 \\ -8 & 7 & 1 \end{vmatrix}$

23. $\begin{vmatrix} 4 & -8 & 5 \\ 8 & -5 & 3 \\ 8 & 5 & 2 \end{vmatrix}$

24. $\begin{vmatrix} 4 & -7 & 9 & 1 \\ 6 & 2 & 7 & 0 \\ 3 & 6 & -3 & 3 \\ 0 & 7 & 4 & -1 \end{vmatrix}$

25. $\begin{vmatrix} 9 & -4 & 2 & 5 \\ 2 & 7 & 6 & -5 \\ 4 & 1 & -2 & 0 \\ 7 & 3 & 4 & 10 \end{vmatrix}$

26. $\begin{vmatrix} 1 & -2 & 7 & 9 \\ 3 & -4 & 5 & 5 \\ 3 & 6 & 1 & -1 \\ 4 & 5 & 3 & 2 \end{vmatrix}$

27. $\begin{vmatrix} 0 & -3 & 8 & 2 \\ 8 & 1 & -1 & 6 \\ -4 & 6 & 0 & 9 \\ -7 & 0 & 0 & 14 \end{vmatrix}$

28. $\begin{vmatrix} 1 & -1 & 8 & 4 & 2 \\ 2 & 6 & 0 & -4 & 3 \\ 2 & 0 & 2 & 6 & 2 \\ 0 & 2 & 8 & 0 & 0 \\ 0 & 1 & 1 & 2 & 2 \end{vmatrix}$

29. $\begin{vmatrix} 3 & -2 & 4 & 3 & 1 \\ -1 & 0 & 2 & 1 & 0 \\ 5 & -1 & 0 & 3 & 2 \\ 4 & 7 & -8 & 0 & 0 \\ 1 & 2 & 3 & 0 & 2 \end{vmatrix}$

30. $\begin{vmatrix} 4 & 2 & -1 & 0 & 3 \\ 0 & 1 & 1 & 2 & -3 \\ 0 & 0 & -2 & 8 & 12 \\ 0 & 0 & 0 & 5 & 13 \\ 0 & 0 & 0 & 0 & 3 \end{vmatrix}$

In Exercises 31–40, use cofactors to find the inverse (if it exists) of the matrix.

31. $\begin{bmatrix} 2 & 4 \\ -1 & 2 \end{bmatrix}$

32. $\begin{bmatrix} -3 & 4 \\ 2 & -4 \end{bmatrix}$

33. $\begin{bmatrix} 3 & 6 \\ 2 & 4 \end{bmatrix}$

34. $\begin{bmatrix} 2 & 7 \\ -3 & -9 \end{bmatrix}$

35. $\begin{bmatrix} 1 & 1 & 2 \\ 3 & 1 & 0 \\ -2 & 0 & 3 \end{bmatrix}$

36. $\begin{bmatrix} 3 & 2 & 2 \\ 2 & 2 & 2 \\ -4 & 4 & 3 \end{bmatrix}$

37. $\begin{bmatrix} 0.1 & 0.2 & 0.3 \\ -0.3 & 0.2 & 0.2 \\ 0.5 & 0.4 & 0.4 \end{bmatrix}$

38. $\begin{bmatrix} 1 & 0 & 0 \\ 3 & 4 & 0 \\ 2 & 5 & 5 \end{bmatrix}$

39. $\begin{bmatrix} 1 & 0 & -2 & 2 \\ 0 & 1 & 1 & -1 \\ -2 & 3 & 4 & 0 \\ 3 & 1 & -2 & -1 \end{bmatrix}$

40. $\begin{bmatrix} 1 & 0 & 3 & 0 \\ 0 & 2 & 0 & 4 \\ 1 & 0 & 3 & 0 \\ 0 & 2 & 0 & 4 \end{bmatrix}$

41. Show that $\begin{vmatrix} 1 & x & x^2 \\ 1 & y & y^2 \\ 1 & z & z^2 \end{vmatrix} = (y - x)(z - x)(z - y)$.

42. Show that $\begin{vmatrix} \cos \theta & \sin \theta \\ -\sin \theta & \cos \theta \end{vmatrix} = 1$.

43. Show that $\begin{vmatrix} a_{11} & 0 & 0 & \cdots & 0 \\ a_{21} & a_{22} & 0 & \cdots & 0 \\ \cdot & \cdot & & \cdot & \cdot \\ \cdot & \cdot & & & 0 \\ \cdot & \cdot & & & \cdot \\ a_{n1} & a_{n2} & a_{n3} & \cdots & a_{nn} \end{vmatrix} = a_{11}a_{22}a_{23} \cdots a_{nn}$.

Cramer's Rule   9.6

In Chapter 8 we looked at two general algebraic methods (substitution and elimination) for solving systems of equations. Then, in the beginning of this chapter, we studied some matrix methods for solving systems of linear equations. In this section we study Cramer's Rule, a method to use if the system of linear equations has a unique solution *and* the same number of equations as variables. This method can be used to find the *complete* solution or to single out a particular variable and solve for that variable alone.

To see how Cramer's Rule arises, let's look at the solution of a general system involving two linear equations in two unknowns.

$$a_{11}x + a_{12}y = c_1$$
$$a_{21}x + a_{22}y = c_2$$

If we solve this system by elimination, we obtain the following. Multiply the first equation by $-a_{21}$:

$$-a_{21}a_{11}x - a_{21}a_{12}y = -a_{21}c_1$$

Multiply the second equation by a_{11}:

$$a_{21}a_{11}x + a_{11}a_{22}y = a_{11}c_2$$

Add equations to eliminate x:

$$(a_{11}a_{22} - a_{21}a_{12})y = a_{11}c_2 - a_{21}c_1$$

Solve for y (provided $a_{11}a_{22} - a_{21}a_{12} \neq 0$):

$$y = \frac{a_{11}c_2 - a_{21}c_1}{a_{11}a_{22} - a_{21}a_{12}}$$

In a similar way, we can solve for x, to obtain

$$x = \frac{a_{22}c_1 - a_{12}c_2}{a_{11}a_{22} - a_{21}a_{12}}$$

Finally, by recognizing that the numerator and denominator for both x and y can be represented as determinants, we have

$$x = \frac{\begin{vmatrix} c_1 & a_{12} \\ c_2 & a_{22} \end{vmatrix}}{\begin{vmatrix} a_{11} & a_{12} \\ a_{21} & a_{22} \end{vmatrix}}, \quad y = \frac{\begin{vmatrix} a_{11} & c_1 \\ a_{21} & c_2 \end{vmatrix}}{\begin{vmatrix} a_{11} & a_{12} \\ a_{21} & a_{22} \end{vmatrix}}, \quad a_{11}a_{22} - a_{21}a_{12} \neq 0$$

The denominator for both x and y is simply the determinant of the coefficient matrix of the system, and we denote this determinant by

$$D = \begin{vmatrix} a_{11} & a_{12} \\ a_{21} & a_{22} \end{vmatrix}$$

The determinants that form the numerators for x and y can be obtained from D by replacing the first or second column by a column representing the constants of the system. We denote these two determinants by D_x and D_y, as follows:

$$D_x = \begin{vmatrix} c_1 & a_{12} \\ c_2 & a_{22} \end{vmatrix}, \qquad D_y = \begin{vmatrix} a_{11} & c_1 \\ a_{21} & c_2 \end{vmatrix}$$

This determinant form of the solution for x and y is called **Cramer's Rule.**

CRAMER'S RULE FOR TWO EQUATIONS IN TWO VARIABLES

The solution of a system of two linear equations in two variables is given by

$$x = \frac{|D_x|}{|D|} \quad \text{and} \quad y = \frac{|D_y|}{|D|}$$

provided $|D| \neq 0$.

Recall from our earlier discussion of systems of linear equations that a system of two equations in two variables can have infinitely many solutions, no solution, or precisely one solution. Our development of Cramer's Rule gives us the following quick test for determining if a system of n linear equations in n variables has a unique solution.

TEST FOR UNIQUE SOLUTION

A system of n linear equations in n variables has a unique solution if and only if the coefficient matrix has a nonzero determinant

$$|D| \neq 0$$

EXAMPLE 1
Solving a Linear System
with Cramer's Rule

Use Cramer's Rule to solve the following system:

$$4x - 2y = 10$$
$$3x - 5y = 11$$

Solution:
We have

$$|D| = \begin{vmatrix} 4 & -2 \\ 3 & -5 \end{vmatrix} = -20 + 6 = -14$$

Since $|D| \neq 0$, we know that the system has a unique solution. Using Cramer's Rule, we have

$$x = \frac{|D_x|}{|D|} = \frac{\begin{vmatrix} 10 & -2 \\ 11 & -5 \end{vmatrix}}{-14} = \frac{-50 + 22}{-14} = \frac{-28}{-14} = 2$$

$$y = \frac{|D_y|}{|D|} = \frac{\begin{vmatrix} 4 & 10 \\ 3 & 11 \end{vmatrix}}{-14} = \frac{44 - 30}{-14} = \frac{14}{-14} = -1$$

Cramer's Rule generalizes easily to systems of n linear equations in n variables. The denominator of the quotient is the determinant of the coefficient matrix, and the numerator is the determinant formed by replacing the column corresponding to a given variable (the one being solved for) with the column representing the constants in the system.

CRAMER'S RULE FOR n EQUATIONS IN n VARIABLES

The solution for a system of n linear equations in the n variables x_1, $x_2, \ldots, x_n$ is given by

$$x_1 = \frac{|D_{x_1}|}{|D|}, \qquad x_2 = \frac{|D_{x_2}|}{|D|}, \qquad \ldots, \qquad x_n = \frac{|D_{x_n}|}{|D|}$$

provided $|D| \neq 0$.

Remark: Before applying Cramer's Rule to any system of linear equations, you must write the system in the standard form:

$$a_{11}x_1 + a_{12}x_2 + a_{13}x_3 + \cdots + a_{1n}x_n = c_1$$
$$a_{21}x_1 + a_{22}x_2 + a_{23}x_3 + \cdots + a_{2n}x_n = c_2$$
$$a_{31}x_1 + a_{32}x_2 + a_{33}x_3 + \cdots + a_{3n}x_n = c_3$$
$$\vdots \qquad \vdots \qquad \vdots \qquad \qquad \vdots \qquad \vdots$$
$$a_{n1}x_1 + a_{n2}x_2 + a_{n3}x_3 + \cdots + a_{nn}x_n = c_n$$

EXAMPLE 2
Applying Cramer's Rule to a System of Three Equations

Use Cramer's Rule to solve the following system:

$$-x + 2y - 3z = 1$$
$$2x \qquad + z = 0$$
$$3x - 4y + 4z = 2$$

Solution:
We begin by evaluating the determinant of the coefficient matrix:

$$|D| = \begin{vmatrix} -1 & 2 & -3 \\ 2 & 0 & 1 \\ 3 & -4 & 4 \end{vmatrix} = 0 + 6 + 24 - 0 - 4 - 16 = 10$$

Since $|D| \neq 0$, we know that the solution is unique, and we obtain

$$x = \frac{\begin{vmatrix} 1 & 2 & -3 \\ 0 & 0 & 1 \\ 2 & -4 & 4 \end{vmatrix}}{10} = \frac{(-1)^5 \begin{vmatrix} 1 & 2 \\ 2 & -4 \end{vmatrix}}{10}$$

$$= \frac{(-1)(-4 - 4)}{10} = \frac{4}{5}$$

$$y = \frac{\begin{vmatrix} -1 & 1 & -3 \\ 2 & 0 & 1 \\ 3 & 2 & 4 \end{vmatrix}}{10} = \frac{0 + 3 - 12 - 0 + 2 - 8}{10}$$

$$= -\frac{15}{10} = -\frac{3}{2}$$

$$z = \frac{\begin{vmatrix} -1 & 2 & 1 \\ 2 & 0 & 0 \\ 3 & -4 & 2 \end{vmatrix}}{10} = \frac{2(-1)^3 \begin{vmatrix} 2 & 1 \\ -4 & 2 \end{vmatrix}}{10}$$

$$= \frac{-2(4 + 4)}{10} = -\frac{8}{5}$$

You should check this solution by substituting these values into *all* three equations.

EXAMPLE 3
Applying Cramer's Rule to a System of Four Equations

Solve for x in the following system:

$$1.3x - 2.7y + \quad\quad + 0.3w = \quad 2.5$$
$$-4.7x + 3.2y - 0.8z + 2.5w = -3.8$$
$$5.8x \quad\quad + 3.5z - 1.7w = \quad 5.8$$
$$2.4x - 1.2y \quad\quad + 2.3w = -3.3$$

Solution:
Expanding by cofactors along Column 3 yields

$$|D| = \begin{vmatrix} 1.3 & -2.7 & 0 & 0.3 \\ -4.7 & 3.2 & -0.8 & 2.5 \\ 5.8 & 0 & 3.5 & -1.7 \\ 2.4 & -1.2 & 0 & 2.3 \end{vmatrix} = -80.0233$$

(You should check this using your calculator.) Now, to solve for x, we again expand by cofactors, using Column 3:

$$|D_x| = \begin{vmatrix} 2.5 & -2.7 & 0 & 0.3 \\ -3.8 & 3.2 & -0.8 & 2.5 \\ 5.8 & 0 & 3.5 & -1.7 \\ -3.3 & -1.2 & 0 & 2.3 \end{vmatrix} = 112.8419$$

Finally, we have

$$x = \frac{112.8419}{-80.0233} \approx -1.410$$

We have now completed our discussion of *five* methods for solving systems of linear equations. To be efficient in solving systems of equations, you should keep each method fresh in your memory so that for any given system of equations you can make a wise choice of solution method. For your convenience, we list a summary of the methods studied, along with their unique advantages and/or disadvantages.

METHODS OF SOLVING SYSTEMS OF LINEAR EQUATIONS

Method	Advantage of Method
Substitution (Sections 8.1 and 8.2)	General method (can be applied to linear and nonlinear systems)
Elimination (Sections 8.2 and 8.3)	General linear method (can be used with any number of equations and variables)
Matrix solution with elementary row operations (Section 9.1)	Generalization of elimination method (easily adapted to computer techniques)
Solution by inverse matrix (Section 9.3)	Efficient method for several systems with *same* coefficient matrix (only applies to systems having n equations, n variables, and a unique solution)
Cramer's Rule (Section 9.6)	Efficient method for finding the value of a single variable in a solution (only applies to systems having n equations, n variables, and a unique solution)

Section Exercises 9.6

In Exercises 1–20, use Cramer's Rule to solve (if possible) the system of equations.

1. $x + 2y = 5$
   $-x + y = 1$

2. $2x - y = -10$
   $3x + 2y = -1$

3. $3x + 4y = -2$
   $5x + 3y = 4$

4. $18x + 12y = 13$
   $30x + 24y = 23$

5. $20x + 8y = 11$
   $12x - 24y = 21$

6. $13x - 6y = 17$
   $26x - 12y = 8$

7. $-0.4x + 0.8y = 1.6$
   $2x - 4y = 5$

8. $-0.4x + 0.8y = 1.6$
   $0.2x + 0.3y = 2.2$

9. $3x + 6y = 5$
   $6x + 14y = 11$

10. $3x + 2y = 1$
    $2x + 10y = 6$

11. $4x - y + z = -5$
    $2x + 2y + 3z = 10$
    $5x - 2y + 6z = 1$

12. $4x - 2y + 3z = -2$
    $2x + 2y + 5z = 16$
    $8x - 5y - 2z = 4$

13. $3x + 4y + 4z = 11$
    $4x - 4y + 6z = 11$
    $6x - 6y = 3$

14. $14x - 21y - 7z = 10$
    $-4x + 2y - 2z = 4$
    $56x - 21y + 7z = 5$

15. $3x + 3y + 5z = 1$
    $3x + 5y + 9z = 2$
    $5x + 9y + 17z = 4$

16. $2x + 3y + 5z = 4$
    $3x + 5y + 9z = 7$
    $5x + 9y + 17z = 13$

17. $7x - 3y + 2w = 41$
    $-2x + y - w = -13$
    $4x + -2w = 12$
    $-x + y - w = -8$

18. $2x + 5y + w = 11$
    $x + 4y + 2z - 2w = -7$

$2x - 2y + 5z + w = 3$
$x - 3w = 1$

19. $5x - 3y + 2z = 2$
    $2x + 2y - 3z = 3$
    $x - 7y + 8z = -4$

20. $3x + 2y + 5z = 4$
    $4x - 3y - 4z = 1$
    $-8x + 2y + 3z = 0$

In Exercises 21–24, find the least squares regression line for the given points. Use Cramer's Rule and the system of equations as given for Exercises 39–42 in Section 8.2.

21. $(1, 1), (2, 1), (2, 2), (4, 5)$
22. $(0, -3), (1, -2), (1, 1), (2, 4)$
23. $(0, 5), (0, 3), (1, 4), (2, 2), (3, 0)$
24. $(-2, 3), (-1, 3), (0, 1), (2, 0), (3, -1)$

In Exercises 25 and 26, find the least squares regression parabola for the given points. Use Cramer's Rule and the system of equations given for Exercises 45–48 in Section 8.3.

25. $(0, 1), (1, 0), (2, 3), (3, 5)$
26. $(-1, 1), (0, 4), (2, 2), (3, -1)$

Review Exercises / Chapter 9

In Exercises 1–12, use matrices and elementary row operations to solve the systems of equations.

1. $5x + 4y = 2$
   $-x + y = -22$

2. $2x - 5y = 2$
   $3x - 7y = 1$

3. $0.2x - 0.1y = 0.07$
   $0.4x - 0.5y = -0.01$

4. $2x + y = 0.3$
   $3x - y = -1.3$

5. $-x + y + 2z = 1$
   $2x + 3y + z = -2$
   $5x + 4y + 2z = 4$

6. $2x + 3y + z = 10$
   $2x - 3y - 3z = 22$
   $4x - 2y + 3z = -2$

7. $2x + 3y + 3z = 3$
   $6x + 6y + 12z = 13$
   $12x + 9y - z = 2$

8. $4x + 4y + 4z = 5$
   $4x - 2y - 8z = 1$
   $5x + 3y + 8z = 6$

9. $2x + y + 2z = 4$
   $2x + 2y = 5$
   $2x - y + 6z = 2$

10. $3x + 21y - 29z = -1$
    $2x + 15y - 21z = 0$

11. $x + 2y + 6z = 1$
    $2x + 5y + 15z = 4$
    $3x + y + 3z = -6$

12. $x_1 + 5x_2 + 3x_3 = 14$
    $4x_2 + 2x_3 + 5x_4 = 3$
    $3x_3 + 8x_4 + 6x_5 = 16$
    $2x_1 + 4x_2 - 2x_5 = 0$
    $2x_1 - x_3 = 0$

In Exercises 13–20, perform (if possible) the indicated matrix operations.

13. $\begin{bmatrix} 2 & 1 & 0 \\ 0 & 5 & -4 \end{bmatrix} - 3\begin{bmatrix} 5 & 3 & -6 \\ 0 & -2 & 5 \end{bmatrix}$

14. $-2\begin{bmatrix} 1 & 2 \\ 5 & -4 \\ 6 & 0 \end{bmatrix} + 8\begin{bmatrix} 7 & 1 \\ 1 & 2 \\ 1 & 4 \end{bmatrix}$

15. $\begin{bmatrix} 1 & 2 \\ 5 & -4 \\ 6 & 0 \end{bmatrix}\begin{bmatrix} 6 & -2 & 8 \\ 4 & 0 & 0 \end{bmatrix}$

16. $\begin{bmatrix} 1 & 5 & 6 \\ 2 & -4 & 0 \end{bmatrix}\begin{bmatrix} 6 & -2 & 8 \\ 4 & 0 & 0 \end{bmatrix}$

17. $\begin{bmatrix} 1 & 5 & 6 \\ 2 & -4 & 0 \end{bmatrix}\begin{bmatrix} 6 & 4 \\ -2 & 0 \\ 8 & 0 \end{bmatrix}$

18. $\begin{bmatrix} 4 \\ 6 \end{bmatrix} [6 \quad -2]$

19. $\begin{bmatrix} 1 & 3 & 2 \\ 0 & 2 & -4 \\ 0 & 0 & 3 \end{bmatrix}\begin{bmatrix} 4 & -3 & 2 \\ 0 & 3 & -1 \\ 0 & 0 & 2 \end{bmatrix}$

20. $\begin{bmatrix} 2 & 1 \\ 6 & 0 \end{bmatrix}\left\{ \begin{bmatrix} 4 & 2 \\ -3 & 1 \end{bmatrix} + \begin{bmatrix} -2 & 4 \\ 0 & 4 \end{bmatrix} \right\}$

21. Write out the system of linear equations represented by

$$\begin{bmatrix} 5 & 4 \\ -1 & 1 \end{bmatrix} \begin{bmatrix} x \\ y \end{bmatrix} = \begin{bmatrix} 2 \\ -22 \end{bmatrix}$$

22. Express the following system of linear equations in matrix form:

$$2x + 3y + z = 10$$
$$2x - 3y - 3z = 22$$
$$4x - 2y + 3z = -2$$

In Exercises 23–28, solve the system of linear equations using (a) the inverse of the coefficient matrix and (b) Cramer's Rule.

23. The system of equations in Exercise 1.
24. The system of equations in Exercise 2.
25. The system of equations in Exercise 5.
26. The system of equations in Exercise 6.
27. The system of equations in Exercise 7.
28. The system of equations in Exercise 8.
29. If A is a 3×3 matrix and $|A| = 2$, then what is the value of $|4A|$? Give the reason for your answer.

30. Prove that

$$\begin{vmatrix} a_{11} & a_{12} & a_{13} \\ a_{21} & a_{22} & a_{23} \\ a_{31} + x & a_{32} + y & a_{33} + z \end{vmatrix}$$

$$= \begin{vmatrix} a_{11} & a_{12} & a_{13} \\ a_{21} & a_{22} & a_{23} \\ a_{31} & a_{32} & a_{33} \end{vmatrix} + \begin{vmatrix} a_{11} & a_{12} & a_{13} \\ a_{21} & a_{22} & a_{23} \\ x & y & z \end{vmatrix}$$

The *characteristic equation* of a square matrix A is the equation

$$|A - \lambda I| = 0$$

Given the matrix

$$A = \begin{bmatrix} 3 & 2 \\ -1 & 4 \end{bmatrix}$$

the characteristic equation is

$$|A - \lambda I| = \begin{vmatrix} 3 - \lambda & 2 \\ -1 & 4 - \lambda \end{vmatrix} = \lambda^2 - 2\lambda + 14 = 0$$

It can be shown that a square matrix always satisfies its characteristic equation, so that in this case, $A^2 - 2A + 14I = 0$. In Exercises 31–34, find the characteristic equation and demonstrate that the matrix satisfies the equation.

31. $A = \begin{bmatrix} 2 & 1 \\ -3 & -1 \end{bmatrix}$

32. $A = \begin{bmatrix} 10 & 4 \\ 2 & 1 \end{bmatrix}$

33. $A = \begin{bmatrix} 2 & 1 & -1 \\ 0 & 1 & 5 \\ -1 & 5 & 2 \end{bmatrix}$

34. $A = \begin{bmatrix} 6 & 0 & 4 \\ -2 & 1 & 3 \\ -9 & 0 & 4 \end{bmatrix}$

10 SEQUENCES AND SERIES

Sequences and Summation Notation

10.1

In this chapter, we look at several special topics in algebra that are extensions or variations of concepts studied earlier, in the basic algebra portion of the text. We hope that you will find these special topics (many of which are used in calculus) to be both interesting and challenging. We begin with the notion of a **sequence.**

Infinite Sequences

In mathematics, the word "sequence" is used in much the same way as it is in ordinary English. When we say that a collection of objects is listed "in sequence," we usually mean that the collection is ordered in that it has an identified first member, second member, third member, and so on. Mathematically, we define a sequence as a *function* whose domain is the set of positive integers. For instance, the equation

$$a(n) = \frac{1}{2^n}$$

defines the sequence

$$\left\{ \frac{1}{2}, \frac{1}{4}, \frac{1}{8}, \frac{1}{16}, \ldots, \frac{1}{2^n}, \ldots \right\}$$

the terms of which correspond respectively to

$$\{a(1), a(2), a(3), a(4), \ldots, a(n), \ldots\}$$

We will generally write a sequence in the *subscript form*,

$$\{a_1, a_2, a_3, a_4, \ldots, a_n, \ldots\}$$

or we will denote it by $\{a_n\}$, where a_n is the **nth term** of the sequence.

DEFINITION OF A SEQUENCE

> A **sequence** $\{a_n\}$ is a function whose domain is the set of positive integers. The functional values $\{a_1, a_2, a_3, \ldots, a_n, \ldots\}$ are called the **terms** of the sequence.

Remark: Because the terms of a sequence comprise an infinite set, it is common to refer to a sequence as an **infinite sequence.** On occasion it is convenient to begin subscripting a sequence with zero instead of one.

EXAMPLE 1
Finding Terms in a Sequence

List the first four terms of the sequences with the following *n*th terms (in each case, assume *n* begins with 1):

(a) $a_n = 3 + (-1)^n$

(b) $b_n = \dfrac{2n}{1 + n}$

Solution:

(a) $\{a_n\} = \{3 + (-1)^n\}$

$$= \{3 + (-1)^1, 3 + (-1)^2, 3 + (-1)^3, 3 + (-1)^4, \ldots\}$$

$$= \{3 - 1, 3 + 1, 3 - 1, 3 + 1, \ldots\}$$

$$= \{2, 4, 2, 4, \ldots\}$$

(b) $\{b_n\} = \left\{\dfrac{2n}{1 + n}\right\}$

$$= \left\{\frac{2(1)}{1 + 1}, \frac{2(2)}{1 + 2}, \frac{2(3)}{1 + 3}, \frac{2(4)}{1 + 4}, \ldots\right\}$$

$$= \left\{\frac{2}{2}, \frac{4}{3}, \frac{6}{4}, \frac{8}{5}, \ldots\right\}$$

Some very important sequences in mathematics involve terms that are defined with special types of products, called **factorials.**

DEFINITION OF FACTORIAL

> If *n* is a positive integer, then *n* **factorial** is defined by
>
> $$n! = 1 \cdot 2 \cdot 3 \cdot 4 \cdots (n - 1) \cdot n$$
>
> If $n = 0$, then 0! is defined to be
>
> $$0! = 1$$

It is helpful to know the value of $n!$ for the following values of n:

$$0! = 1 \qquad\qquad 3! = 1 \cdot 2 \cdot 3 = 6$$
$$1! = 1 \qquad\qquad 4! = 1 \cdot 2 \cdot 3 \cdot 4 = 24$$
$$2! = 1 \cdot 2 = 2 \qquad 5! = 1 \cdot 2 \cdot 3 \cdot 4 \cdot 5 = 120$$

Factorials follow the same conventions for order of operation as do exponents.

$$2n! = 2(n!) = 2(1 \cdot 2 \cdot 3 \cdot 4 \cdots n)$$
$$(2n)! = 1 \cdot 2 \cdot 3 \cdot 4 \cdots n \cdot (n + 1) \cdots (2n)$$

EXAMPLE 2
Finding Terms of a Sequence

List the first four terms of the following sequences:

(a) $\{a_n\} = \left\{\dfrac{2^n}{n!}\right\}$ \qquad (Begin at $n = 0$)

(b) $\{b_n\} = \left\{\dfrac{(2n)!}{2^n n!}\right\}$ \qquad (Begin at $n = 1$)

Solution:

(a) $\{a_n\} = \left\{\dfrac{2^n}{n!}\right\} = \left\{\dfrac{2^0}{0!}, \dfrac{2^1}{1!}, \dfrac{2^2}{2!}, \dfrac{2^3}{3!}, \cdots\right\} = \left\{\dfrac{1}{1}, \dfrac{2}{1}, \dfrac{4}{2}, \dfrac{8}{6}, \cdots\right\}$

(b) $\{b_n\} = \left\{\dfrac{(2n)!}{2^n n!}\right\}$

$$b_1 = \frac{2!}{2^1(1!)} = \frac{1 \cdot 2}{2(1)} = \frac{1 \cdot \cancel{2}}{\cancel{2}} = 1$$

$$b_2 = \frac{4!}{2^2(2!)} = \frac{1 \cdot 2 \cdot 3 \cdot 4}{2 \cdot 2(1 \cdot 2)}$$
$$= \frac{1 \cdot \cancel{2} \cdot 3 \cdot \cancel{4}}{\cancel{2} \cdot \cancel{4}} = 1 \cdot 3$$

$$b_3 = \frac{6!}{2^3(3!)} = \frac{1 \cdot 2 \cdot 3 \cdot 4 \cdot 5 \cdot 6}{2 \cdot 2 \cdot 2(1 \cdot 2 \cdot 3)}$$
$$= \frac{1 \cdot \cancel{2} \cdot 3 \cdot \cancel{4} \cdot 5 \cdot \cancel{6}}{\cancel{2} \cdot \cancel{4} \cdot \cancel{6}} = 1 \cdot 3 \cdot 5$$

$$b_4 = \frac{8!}{2^4(4!)} = \frac{1 \cdot 2 \cdot 3 \cdot 4 \cdot 5 \cdot 6 \cdot 7 \cdot 8}{2 \cdot 2 \cdot 2 \cdot 2(1 \cdot 2 \cdot 3 \cdot 4)}$$
$$= \frac{1 \cdot \cancel{2} \cdot 3 \cdot \cancel{4} \cdot 5 \cdot \cancel{6} \cdot 7 \cdot \cancel{8}}{\cancel{2} \cdot \cancel{4} \cdot \cancel{6} \cdot \cancel{8}} = 1 \cdot 3 \cdot 5 \cdot 7$$

$$\{b_n\} = \{\,1, (1 \cdot 3), (1 \cdot 3 \cdot 5), (1 \cdot 3 \cdot 5 \cdot 7), \ldots\}$$

Remark: Notice the cancellation that took place in part (b) of Example 2. It often happens in work with factorials that appropriate cancellation will greatly simplify an expression. For instance, the following types of cancellations are common:

$$\frac{n!}{(n-1)!} = \frac{1 \cdot 2 \cdot 3 \cdots (n-1) \cdot n}{1 \cdot 2 \cdot 3 \cdots (n-1)} = n$$

$$\frac{(2n-1)!}{(2n+1)!} = \frac{1 \cdot 2 \cdot 3 \cdots (2n-2)(2n-1)}{1 \cdot 2 \cdot 3 \cdots (2n-2)(2n-1)(2n)(2n+1)}$$

$$= \frac{1}{2n(2n+1)}$$

It is important to realize that simply listing the first few terms is not sufficient to define a sequence—the nth term *must* be given. To see this, consider the following sequences, both of which have the same first three terms:

$$\{a_n\} = \left\{ \frac{1}{2}, \frac{1}{4}, \frac{1}{8}, \cdots, \frac{1}{2^n}, \cdots \right\}$$

$$\{b_n\} = \left\{ \frac{1}{2}, \frac{1}{4}, \frac{1}{8}, \cdots, \frac{6}{(n+1)(n^2-n+6)}, \cdots \right\}$$

Alternating Signs

The terms of a sequence need not all be positive. One rule used to *alternate* the signs of terms of a sequence is the following:

$$\{(-1)^{n+1}a_n\} = \{a_1, -a_2, a_3, -a_4, \ldots\}, \quad \text{where } n = 1, 2, 3, \ldots$$

Using $n + 1$ as the power of (-1) causes the sequence to begin with a positive sign. If you want to begin a sequence with a negative sign, you can use n as the power of (-1).

EXAMPLE 3
Sequence with Alternating Signs

Find the first four terms and the thirty-seventh term of the sequence

$$\{a_n\} = \left\{ \frac{(-1)^n}{2n-1} \right\}$$

Solution:
We have

$$a_1 = \frac{(-1)^1}{2(1) - 1} = \frac{-1}{1} = -1$$

$$a_2 = \frac{(-1)^2}{2(2) - 1} = \frac{1}{3}$$

$$a_3 = \frac{(-1)^3}{2(3) - 1} = \frac{-1}{5}$$

$$a_4 = \frac{(-1)^4}{2(4) - 1} = \frac{1}{7}$$

.
.
.

$$a_{37} = \frac{(-1)^{37}}{2(37) - 1} = \frac{-1}{73}$$

One of the most important types of sequences used in mathematics is the sequence obtained by taking successive sums of terms of a given sequence. Consider the sequence

$$\{a_n\} = \{a_1, a_2, a_3, a_4, \ldots, a_n, \ldots\}$$

By finding the successive sums

$$S_1 = a_1$$

$$S_2 = a_1 + a_2$$

$$S_3 = a_1 + a_2 + a_3$$

.
.
.

$$S_n = a_1 + a_2 + a_3 + a_4 + \cdots + a_n$$

we can form a new sequence $\{S_n\}$, which is called the **sequence of partial sums**. In particular, S_n is called the **nth partial sum** of the sequence $\{a_n\}$.

A convenient shorthand notation for representing sums is called **sigma notation**. The name for this notation comes from the use of the uppercase Greek letter sigma, written as Σ.

DEFINITION OF SIGMA NOTATION

If $\{a_n\}$ is a sequence, then the sum of the first n terms of the sequence is represented by

$$\sum_{i=1}^{n} a_i = a_1 + a_2 + a_3 + a_4 + \cdots + a_n$$

where i is called the **index of summation**, n is the **upper limit of summation**, and 1 is the **lower limit of summation**.

In calculus, we refer to the summation

$$\sum_{i=1}^{n} a_i = a_1 + a_2 + a_3 + a_4 + \cdots + a_n$$

as a **series.** And if the summation has infinitely many terms, we call it an **infinite series.** Consider the sequence of partial sums

$$\{S_n\} = \{S_1, S_2, S_3, S_4, \ldots, S_n, \ldots\}$$

If there is a number S such that the value of S_n approaches S as n increases without bound (symbolically we write $S_n \rightarrow S$ as $n \rightarrow \infty$), then we call S the **sum** of the infinite series and write

$$S = a_1 + a_2 + a_3 + a_4 + \cdots + a_n + \cdots = \sum_{i=1}^{\infty} a_i$$

Remark: This infinite summation process involves the concept of *convergence*, a topic that can be treated more precisely in calculus.

EXAMPLE 4
Finding a Sequence of Partial Sums

For the sequence

$$\{a_n\} = \left\{ \frac{3}{10^n} \right\}$$

find the first five partial sums, and use these results to try to evaluate the infinite summation

$$\sum_{i=1}^{\infty} \frac{3}{10^n}$$

Solution:
Since the terms of the given sequence are

$$\{a_n\} = \left\{ \frac{3}{10^1}, \frac{3}{10^2}, \frac{3}{10^3}, \frac{3}{10^4}, \frac{3}{10^5}, \cdots \right\}$$
$$= \{0.3, 0.03, 0.003, 0.0003, 0.00003, \ldots\}$$

it follows that

$$S_1 = 0.3$$
$$S_2 = 0.3 + 0.03 = 0.33$$
$$S_3 = 0.3 + 0.03 + 0.003 = 0.333$$
$$S_4 = 0.3 + 0.03 + 0.003 + 0.0003 = 0.3333$$
$$S_5 = 0.3 + 0.03 + 0.003 + 0.0003 + 0.00003 = 0.33333$$

From these results it appears that

$$S_n \rightarrow 0.3\overline{3} \quad \text{as} \quad n \rightarrow \infty \qquad (Recall\ that\ 0.3\overline{3} = 0.3333\ldots)$$

which suggests that

$$\sum_{i=1}^{\infty} \frac{3}{10^n} = 0.3\overline{3} = \frac{1}{3}$$

EXAMPLE 5
Sigma Notation for a Sum

Write out the terms and evaluate each of the following sums:

(a) $\displaystyle\sum_{i=1}^{5} i$ (b) $\displaystyle\sum_{k=3}^{6} (1 + k^2)$ (c) $\displaystyle\sum_{i=0}^{8} \frac{1}{i!}$

Solution:

(a) $\displaystyle\sum_{i=1}^{5} i = 1 + 2 + 3 + 4 + 5 = 15$

(b) $\displaystyle\sum_{k=3}^{6} (1 + k^2) = (1 + 3^2) + (1 + 4^2) + (1 + 5^2) + (1 + 6^2)$

$$= 10 + 17 + 26 + 37 = 90$$

(c) $\displaystyle\sum_{i=0}^{8} \frac{1}{i!} = \frac{1}{0!} + \frac{1}{1!} + \frac{1}{2!} + \frac{1}{3!} + \frac{1}{4!} + \frac{1}{5!} + \frac{1}{6!} + \frac{1}{7!} + \frac{1}{8!}$

$$= 1 + 1 + \frac{1}{2} + \frac{1}{6} + \frac{1}{24} + \frac{1}{120} + \frac{1}{720} + \frac{1}{5040} + \frac{1}{40320}$$

$$\approx 2.71828$$

Remark: Note that the lower index of a summation does not have to be 1. Also note that we can use any variable as the index. For instance, in part (b), the letter k is used.

The following properties of sigma notation are useful.

PROPERTIES OF SIGMA NOTATION

1. $\displaystyle\sum_{i=1}^{n} ca_i = c \sum_{i=1}^{n} a_i,$ c is any constant

2. $\displaystyle\sum_{i=1}^{n} (a_i + b_i) = \sum_{i=1}^{n} a_i + \sum_{i=1}^{n} b_i$

3. $\displaystyle\sum_{i=1}^{n} (a_i - b_i) = \sum_{i=1}^{n} a_i - \sum_{i=1}^{n} b_i$

Proof:
Each of these properties follows directly from the Associative, Commutative, and Distributive Properties of arithmetic. For example, note the use of the Distributive Property in the proof of Property 1.

$$\sum_{i=1}^{n} ca_i = ca_1 + ca_2 + ca_3 + \cdots + ca_n$$

$$= c(a_1 + a_2 + a_3 + \cdots + a_n)$$

$$= c \sum_{i=1}^{n} a_i$$

Be aware that variations in the upper and lower limits of summation can produce quite different-looking sigma notations for the same sum. For example,

$$\sum_{i=1}^{5} 3(2^i) = 3 \sum_{i=1}^{5} 2^i = 3(2^1 + 2^2 + 2^3 + 2^4 + 2^5)$$

$$\sum_{i=0}^{4} 3(2^{i+1}) = 3 \sum_{i=0}^{4} 2^{i+1} = 3(2^1 + 2^2 + 2^3 + 2^4 + 2^5)$$

Section Exercises 10.1

In Exercises 1–20, write out the first five terms of the specified sequence. (In each case, assume n starts with 1.)

1. $a_n = 2^n$

2. $a_n = \dfrac{n}{n+1}$

3. $a_n = (-\frac{1}{2})^n$

4. $a_n = \dfrac{n^2 - 1}{n^2 + 2}$

5. $a_n = \dfrac{3^n}{n!}$

6. $a_n = 5 - \dfrac{1}{n} + \dfrac{1}{n^2}$

7. $a_n = \dfrac{(-1)^n}{n^2}$

8. $a_n = \dfrac{3n!}{(n-1)!}$ ✓

9. $a_n = \dfrac{n+1}{n}$

10. $a_n = \dfrac{1}{n^{3/2}}$

11. $a_n = \dfrac{n^2 - 1}{n+1}$

12. $a_n = 1 + (-1)^n$

13. $a_n = \dfrac{1 + (-1)^n}{n}$

14. $a_n = \dfrac{n!}{n}$

15. $a_n = 3 - \dfrac{1}{2^n}$

16. $a_n = \dfrac{3^n}{4^n}$

17. $a_n = (-1)^n \left(\dfrac{n}{n+1}\right)$

18. $a_n = \dfrac{3n^2 - n + 4}{2n^2 + 1}$

19. $a_1 = 3$ and $a_{k+1} = 2(a_k - 1)$

20. $a_1 = 4$ and $a_{k+1} = a_k \left(\dfrac{k+1}{2}\right)$

In Exercises 21–36, write an expression for the nth term of the sequence.

21. $\{1, 4, 7, 10, 13, \ldots\}$

22. $\{3, 7, 11, 15, 19, \ldots\}$

23. $\{-1, 2, 7, 14, 23, \ldots\}$

24. $\left\{1, \dfrac{1}{4}, \dfrac{1}{9}, \dfrac{1}{16}, \dfrac{1}{25}, \ldots\right\}$

25. $\left\{\dfrac{2}{3}, \dfrac{3}{4}, \dfrac{4}{5}, \dfrac{5}{6}, \dfrac{6}{7}, \ldots\right\}$

26. $\left\{\dfrac{2}{1}, \dfrac{3}{3}, \dfrac{4}{5}, \dfrac{5}{7}, \dfrac{6}{9}, \ldots\right\}$

27. $\left\{\dfrac{-1}{1}, \dfrac{1}{2}, \dfrac{-1}{4}, \dfrac{1}{8}, \dfrac{-1}{16}, \ldots\right\}$

28. $\left\{\dfrac{1}{2}, \dfrac{1}{3}, \dfrac{2}{9}, \dfrac{4}{27}, \dfrac{8}{81}, \ldots\right\}$

29. $\left\{2, \left(1 + \dfrac{1}{2}\right), \left(1 + \dfrac{1}{3}\right), \left(1 + \dfrac{1}{4}\right), \left(1 + \dfrac{1}{5}\right), \ldots\right\}$

30. $\left\{\left(1 + \dfrac{1}{2}\right), \left(1 + \dfrac{3}{4}\right), \left(1 + \dfrac{7}{8}\right), \left(1 + \dfrac{15}{16}\right), \right.$
$\left. \left(1 + \dfrac{31}{32}\right), \ldots\right\}$

31. $\left\{\dfrac{1}{2 \cdot 3}, \dfrac{2}{3 \cdot 4}, \dfrac{3}{4 \cdot 5}, \dfrac{4}{5 \cdot 6}, \dfrac{5}{6 \cdot 7}, \ldots\right\}$

32. $\left\{1, \dfrac{1}{2}, \dfrac{1}{6}, \dfrac{1}{24}, \dfrac{1}{120}, \ldots\right\}$

33. $\left\{1, \dfrac{1}{1 \cdot 3}, \dfrac{1}{1 \cdot 3 \cdot 5}, \dfrac{1}{1 \cdot 3 \cdot 5 \cdot 7}, \right.$
$\left. \dfrac{1}{1 \cdot 3 \cdot 5 \cdot 7 \cdot 9}, \ldots\right\}$

34. $\{2, -4, 6, -8, 10, \ldots\}$

35. $\{1, -1, 1, -1, 1, \ldots\}$

36. $\left\{1, x, \dfrac{x^2}{2}, \dfrac{x^3}{6}, \dfrac{x^4}{24}, \dfrac{x^5}{120}, \ldots\right\}$

In Exercises 37–50, find the given sum.

37. $\displaystyle\sum_{i=1}^{5} (2i + 1)$

38. $\displaystyle\sum_{i=1}^{6} 2i$

39. $\displaystyle\sum_{k=0}^{4} \dfrac{1}{k^2 + 1}$

40. $\displaystyle\sum_{j=3}^{5} \dfrac{1}{j}$

41. $\displaystyle\sum_{k=1}^{4} 10$

42. $\displaystyle\sum_{n=1}^{10} \dfrac{3}{n+1}$

43. $\displaystyle\sum_{k=1}^{5} C$, C is constant

44. $\displaystyle\sum_{n=1}^{5} \dfrac{C}{n+1}$, C is constant

45. $\displaystyle\sum_{i=1}^{4} [(i - 1)^2 + (i + 1)^3]$

46. $\displaystyle\sum_{k=2}^{5} (k + 1)(k - 3)$

47. $\displaystyle\sum_{i=1}^{4} (x^2 + 2i)$ 48. $\displaystyle\sum_{y=0}^{5} (x^2 + y^2)$

49. $\displaystyle\sum_{x=1}^{3} (x^2 + 4ix)$ 50. $\displaystyle\sum_{j=1}^{4} (-2)^{j-1}$

In Exercises 51–60, use sigma notation to write the given sum.

51. $\dfrac{1}{3(1)} + \dfrac{1}{3(2)} + \dfrac{1}{3(3)} + \cdots + \dfrac{1}{3(9)}$

52. $\dfrac{5}{1 + 1} + \dfrac{5}{1 + 2} + \dfrac{5}{1 + 3} + \cdots + \dfrac{5}{1 + 15}$

53. $\left[2\left(\dfrac{1}{8}\right) + 3\right] + \left[2\left(\dfrac{2}{8}\right) + 3\right]$
$+ \left[2\left(\dfrac{3}{8}\right) + 3\right] + \cdots + \left[2\left(\dfrac{8}{8}\right) + 3\right]$

54. $\left[1 - \left(\dfrac{1}{4}\right)^2\right] + \left[1 - \left(\dfrac{2}{4}\right)^2\right]$
$+ \left[1 - \left(\dfrac{3}{4}\right)^2\right] + \left[1 - \left(\dfrac{4}{4}\right)^2\right]$

55. $3 - 9 + 27 - 81 + 243 - 729$

56. $1 - \dfrac{1}{2} + \dfrac{1}{4} - \dfrac{1}{8} + \cdots - \dfrac{1}{128}$

57. $\dfrac{1}{1^2} - \dfrac{1}{2^2} + \dfrac{1}{3^2} - \dfrac{1}{4^2} + \cdots - \dfrac{1}{20^2}$

58. $\dfrac{1}{1 \cdot 3} + \dfrac{1}{2 \cdot 4} + \dfrac{1}{3 \cdot 5} + \cdots + \dfrac{1}{10 \cdot 12}$

59. $\dfrac{1}{4} + \dfrac{3}{8} + \dfrac{7}{16} + \dfrac{15}{32} + \dfrac{31}{64}$

60. $\dfrac{1}{2} + \dfrac{2}{4} + \dfrac{6}{8} + \dfrac{24}{16} + \dfrac{120}{32} + \dfrac{720}{64}$

In Exercises 61 and 62, use the following definition of the arithmetic mean $\bar{x}$ of a set of n measurements $\{x_1, x_2, x_3, \ldots, x_n\}$:

$$\bar{x} = \frac{1}{n} \sum_{i=1}^{n} x_i$$

61. Prove that $\displaystyle\sum_{i=1}^{n} (x_i - \bar{x}) = 0$.

62. Prove that $\displaystyle\sum_{i=1}^{n} (x_i - \bar{x})^2 = \sum_{i=1}^{n} x_i^2 - \frac{1}{n}\left(\sum_{i=1}^{n} x_i\right)^2$.

63. A deposit of $5000 is made in an account that earns 8% interest compounded quarterly. The balance in the account after n quarters is given by

$$A_n = 5000\left(1 + \frac{0.08}{4}\right)^n, \qquad n = 1, 2, 3, \ldots$$

(a) Compute the first eight terms of this sequence.
(b) Find the balance in this account after 10 years by computing the fortieth term of the sequence.

64. A deposit of $100 is made *each* month in an account that earns 12% interest compounded monthly. The balance in the account after n months is given by

$$A_n = 100(101)[(1.01)^n - 1], \qquad n = 1, 2, 3, \ldots$$

(a) Compute the first six terms of this sequence.
(b) Find the balance after 5 years by computing the sixtieth term of the sequence.
(c) Find the balance after 20 years by computing the two hundred fortieth term of the sequence.

65. Simplify the following ratios of factorials:
(a) $\dfrac{10!}{8!}$ (b) $\dfrac{25!}{23!}$ (c) $\dfrac{n!}{(n - 2)!}$

Arithmetic Sequences and Series   10.2

In this section we look at a special type of sequence called an **arithmetic sequence** (or **arithmetic progression**). In an arithmetic sequence, consecutive terms have a common difference. For instance, the sequence

$$\{a_n\} = \{2 + 3n\}$$
$$= \{(2 + 3), (2 + 6), (2 + 9), (2 + 12), \ldots\}$$
$$= \{5, 8, 11, 14, \ldots\}$$

$$8 - 5 = 3 \qquad 11 - 8 = 3$$

has a common difference of 3 between consecutive terms.

DEFINITION OF AN ARITHMETIC SEQUENCE

> The sequence $\{a_n\}$ is called an **arithmetic sequence** if each pair of consecutive terms differs by the same amount
>
> $$d = a_{i+1} - a_i$$
>
> The number d is called the **common difference** of the sequence.

EXAMPLE 1
Finding the Common Difference
of an Arithmetic Sequence

Which of the following sequences are arithmetic? If arithmetic, find the common difference.

(a) $\{a_n\} = \{3 - 5n\}$ (b) $\{b_n\} = \{-2 + 3^n\}$

Solution:

(a) For the sequence

$$\{a_n\} = \{3 - 5n\} = \{3, -2, -7, -12, -17, \ldots\}$$

the difference between two consecutive terms is

$$a_{i+1} - a_i = [3 - 5(i + 1)] - [3 - 5(i)]$$
$$= 3 - 5i - 5 - 3 + 5i = -5$$

Thus, the sequence *is* arithmetic, and the common difference is $d = -5$.

(b) For the sequence

$$\{b_n\} = \{-2 + 3^n\} = \{(-2 + 3), (-2 + 9),$$
$$(-2 + 27), (-2 + 81), \ldots\}$$
$$= \{1, 7, 25, 79, \ldots\}$$

we see that the difference between the first two terms is

$$b_2 - b_1 = 7 - 1 = 6$$

and the difference between the second and third terms is

$$b_3 - b_2 = 25 - 7 = 18$$

Since these two differences are not the same, we conclude that the sequence is *not* arithmetic.

Remark: In Example 1, part (a), be sure you see that the common difference of an arithmetic sequence can be negative.

There is a simple method for writing out the terms of an arithmetic sequence. If you know the first term a_1 and the common difference d, then you can form the other terms by repeatedly adding d. For instance,

$$a_2 = a_1 + d, \qquad a_3 = a_2 + d, \qquad a_4 = a_3 + d$$

and, in general, we have

$$a_{i+1} = a_i + d$$

This is called a **recursive formula,** since it defines a given term by reference to the preceding term.

Using this recursive idea, it is possible to obtain a formula for finding the nth term of an arithmetic sequence *without* having to identify all the $n - 1$ preceding terms. Consider an arithmetic sequence with a_1 as its first term and d as the common difference. Then,

$$
\begin{aligned}
a_1 &= a_1 & &= a_1 + (0)d \\
a_2 &= a_1 + d & &= a_1 + (1)d \\
a_3 &= a_1 + d + d & &= a_1 + (2)d \\
a_4 &= a_1 + d + d + d & &= a_1 + (3)d \\
a_5 &= a_1 + d + d + d + d &&= a_1 + (4)d
\end{aligned}
$$

This leads to the following rule.

THE nTH TERM OF AN ARITHMETIC SEQUENCE

> The nth term of an arithmetic sequence whose first term is a_1 and whose common difference is d is given by
>
> $$a_n = a_1 + (n - 1)d$$

EXAMPLE 2
Finding the nth Term of an Arithmetic Sequence

Find the indicated term of the following arithmetic sequences:

(a) The sixty-fourth term of the sequence whose first three terms are

$$\{a_n\} = \{-1, 11, 23, \ldots\}$$

(b) The thirty-eighth term of the sequence whose first term is 8 and whose ith term is given by the recursive formula

$$a_{i+1} = a_i - 7$$

Solution:

(a) For the sequence

$$\{a_n\} = \{-1, 11, 23, \ldots\}$$

we find that $a_1 = -1$ and $d = 12$. Therefore, the sixty-fourth term of the sequence is

$$a_{64} = a_1 + (n - 1)d = -1 + 63(12) = -1 + 756 = 755$$

(b) From the recursive formula $a_{i+1} = a_i - 7$, we can see that $d = -7$. Since we are given that $a_1 = 8$, we find that the thirty-eighth term of the sequence is

$$a_{38} = a_1 + (n - 1)d = 8 + 37(-7) = 8 - 259 = -251$$

EXAMPLE 3
Finding the *n*th Term of an
Arithmetic Sequence

The seventh term of a certain arithmetic sequence is 55, and the twenty-second term is 145. Find the eighteenth term.

Solution:
Using the *n*th term formula, we have

$$a_7 = a_1 + 6d \qquad \text{and} \qquad a_{22} = a_1 + 21d$$
$$55 = a_1 + 6d \qquad\qquad 145 = a_1 + 21d$$

Solving this system of linear equations, we have

$$a_1 + 21d = 145$$
$$\underline{-a_1 - 6d = -55}$$
$$15d = 90 \qquad \text{or} \qquad d = 6$$

Backsubstitution yields $a_1 = 19$, and thus the eighteenth term is

$$a_{18} = a_1 + (n - 1)d = 19 + 17(6) = 121$$

Arithmetic Mean

Recall that $(a + b)/2$ is the midpoint between the two numbers a and b on the real number line. As a result, the terms

$$a, \frac{a + b}{2}, b$$

have a common difference. We call $(a + b)/2$ the **arithmetic mean** of the numbers a and b. We can generalize this concept by finding k numbers m_1, $m_2, m_3, \ldots, m_k$ between a and b such that the terms

$$a, m_1, m_2, m_3, \ldots, m_k, b$$

have a common difference. This process is referred to as **inserting k arithmetic means** between a and b.

EXAMPLE 4
Inserting *k* Arithmetic Means
Between Two Numbers

Insert three arithmetic means between 4 and 15.

Solution:
We need to find three numbers m_1, m_2, and m_3 such that the terms

$$4, m_1, m_2, m_3, 15$$

have a common difference. In this case we have $a_1 = 4$, $n = 5$, and $a_5 = 15$. Therefore,

$$a_5 = 15 = a_1 + (n - 1)d = 4 + 4d$$

Hence, $d = \frac{11}{4}$, and the three arithmetic means are

$$m_1 = a_1 + d = 4 + \tfrac{11}{4} = \tfrac{27}{4}$$
$$m_2 = m_1 + d = \tfrac{27}{4} + \tfrac{11}{4} = \tfrac{38}{4}$$
$$m_3 = m_2 + d = \tfrac{38}{4} + \tfrac{11}{4} = \tfrac{49}{4}$$

Arithmetic Series

A series that is formed from an arithmetic sequence is called an **arithmetic series.** For example, the arithmetic sequence

$$\{a_n\} = \{1, 3, 5, 7, 9, \ldots, 2n - 1, \ldots\}$$

gives rise to the following sequence of partial sums:

$$S_1 = 1 = 1 = 1^2$$
$$S_2 = 1 + 3 = 4 = 2^2$$
$$S_3 = 1 + 3 + 5 = 9 = 3^2$$
$$S_4 = 1 + 3 + 5 + 7 = 16 = 4^2$$
$$S_5 = 1 + 3 + 5 + 7 + 9 = 25 = 5^2$$

Judging from the pattern formed by these first five partial sums, the nth partial sum appears to be

$$S_n = 1 + 3 + 5 + 7 + 9 + \cdots + 2n - 1 = \sum_{i=1}^{n} (2i - 1) = n^2$$

We can verify this observation using the following formula for finding the nth partial sum of an arithmetic sequence.

THE nTH PARTIAL SUM OF AN ARITHMETIC SEQUENCE

The nth partial sum of the arithmetic sequence $\{a_n\}$ with common difference d is given by either of the following formulas:

1. $S_n = \dfrac{n}{2}(a_1 + a_n)$

2. $S_n = \dfrac{n}{2}[2a_1 + (n - 1)d]$

Proof:

There are two different ways to generate the terms of an arithmetic sequence. One way is to repeatedly add d to the first term, and another way is to repeatedly subtract d from the nth term. Thus, we have

$$S_n = a_1 + a_2 + a_3 + \cdots + a_{n-2} + a_{n-1} + a_n$$
$$= a_1 + [a_1 + d] + [a_1 + 2d] + \cdots + [a_1 + (n - 3)d]$$
$$+ [a_1 + (n - 2)d] + [a_1 + (n - 1)d]$$

and by adding terms in the opposite order, we have

$$S_n = a_n + a_{n-1} + a_{n-2} + \cdots + a_3 + a_2 + a_1$$
$$= a_n + [a_n - d] + [a_n - 2d] + \cdots + [a_n - (n - 3)d]$$
$$+ [a_n - (n - 2)d] + [a_n - (n - 1)d]$$

Now, the multiples of d cancel when these two versions of S_n are added, and we have

$$2S_n = \underbrace{(a_1 + a_n) + (a_1 + a_n) + (a_1 + a_n) + \cdots + (a_1 + a_n)}_{n \text{ terms}}$$

$$= n(a_1 + a_n)$$

Thus, we have

$$S_n = \frac{n}{2}(a_1 + a_n)$$

But since $a_n = a_1 + (n - 1)d$,

$$S_n = \frac{n}{2}[2a_1 + (n - 1)d]$$

Using this formula, we verify our observation that the sum of the first n odd integers is n^2.

$$S_n = 1 + 3 + 5 + \cdots + 2n - 1 = \frac{n}{2}[a_1 + a_n]$$

$$= \frac{n}{2}[1 + (2n - 1)] = \frac{n}{2}(2n) = n^2$$

EXAMPLE 5
Finding the *n*th Partial Sum
of an Arithmetic Sequence

Find the following partial sums:
(a) The sum of the first 99 terms of the sequence given by

$$\{a_n\} = \left\{7 + \frac{n}{2}\right\}$$

(b) The sum of the first 150 terms of the sequence given by

$$\{5, 16, 27, 38, 49, \ldots\}$$

Solution:

(a) The sum of the first 99 terms of this sequence is

$$S_{99} = \frac{n}{2}(a_1 + a_n) = \frac{99}{2}\left(\left[7 + \frac{1}{2}\right] + \left[7 + \frac{99}{2}\right]\right)$$

$$= \frac{99}{2}\left(\frac{15}{2} + \frac{113}{2}\right) = \frac{99}{2}\left(\frac{128}{2}\right) = 3168$$

(b) For the sequence

$$\{5, 16, 27, 38, 49, \ldots\}$$

we have $a_1 = 5$ and $d = 11$. Therefore, the sum of the first 150 terms is

$$S_{150} = \frac{n}{2}[2a_1 + (n - 1)d] = \frac{150}{2}[2(5) + 149(11)]$$

$$= 75(10 + 1639) = 123,675$$

EXAMPLE 6
An Application to Business

A small business sells $10,000 worth of products during its first year. The owner of the business has set a goal of increasing annual sales by $7500 each year for 9 years. Assuming that this goal is met, find the total sales during the first 10 years this business is in operation.

Solution:
The annual sales form an arithmetic sequence (since they increase by the same amount each year), and we can find the total sales for the first 10 years as follows:

$$S_{10} = \frac{n}{2}[2a_1 + (n - 1)d] = \frac{10}{2}[20,000 + 9(7500)]$$

$$= 5(87,500) = \$437,500$$

Section Exercises 10.2

In Exercises 1–10, determine if the sequence is arithmetic. If it is, find the common difference.

1. $\{4, 7, 10, 13, 16, \ldots\}$ 2. $\{10, 8, 6, 4, 2, \ldots\}$
3. $\{1, 2, 4, 8, 16, \ldots\}$ 4. $\{3, \frac{5}{2}, 2, \frac{3}{2}, 1, \ldots\}$
5. $\{\frac{9}{4}, 2, \frac{7}{4}, \frac{3}{2}, \frac{5}{4}, \ldots\}$
6. $\{-12, -8, -4, 0, 4, \ldots\}$
7. $\{\frac{1}{3}, \frac{2}{3}, \frac{4}{3}, \frac{8}{3}, \frac{16}{3}, \ldots\}$
8. $\{\ln 1, \ln 2, \ln 3, \ln 4, \ln 5, \ldots\}$
9. $\{5.3, 5.7, 6.1, 6.5, 6.9, \ldots\}$
10. $\{1^2, 2^2, 3^2, 4^2, 5^2, \ldots\}$

In Exercises 11–18, write the first five terms of the arithmetic sequence.

11. $a_1 = 5, d = 6$ 12. $a_1 = 5, d = -\frac{3}{4}$
13. $a_1 = -2.6, d = -0.4$ 14. $a_1 = 6, a_{k+1} = a_k + 12$
15. $a_1 = \frac{3}{2}, a_{k+1} = a_k - \frac{1}{4}$ 16. $a_5 = 28, a_{10} = 53$
17. $a_8 = 26, a_{12} = 42$ 18. $a_3 = 1.9, a_{15} = -1.7$

In Exercises 19–26, find a_n for the given arithmetic sequence.

19. $a_1 = 1, d = 3, n = 10$
20. $a_1 = 15, d = 4, n = 25$
21. $a_1 = 100, d = -8, n = 8$
22. $a_1 = 0, d = -\frac{2}{3}, n = 12$
23. $a_1 = x, d = 2x, n = 50$
24. $a_1 = -y, d = 5y, n = 10$
25. $\{4, \frac{3}{2}, -1, -\frac{7}{2}, \ldots\}, n = 10$
26. $\{10, 5, 0, -5, -10, \ldots\}, n = 50$

In Exercises 27–34, find the nth partial sum of the arithmetic sequence.

27. $\{8, 20, 32, 44, \ldots\}, n = 10$
28. $\{2, 8, 14, 20, \ldots\}, n = 25$
29. $\{-6, -2, 2, 6, \ldots\}, n = 50$
30. $\{0.5, 0.9, 1.3, 1.7, \ldots\}, n = 10$
31. $\{40, 37, 34, 31, \ldots\}, n = 10$
32. $\{1.50, 1.45, 1.40, 1.35, \ldots\}, n = 20$
33. $a_1 = 100, a_{25} = 220, n = 25$
34. $a_1 = 15, a_{100} = 307, n = 100$
35. Find the sum of the first 50 positive integers.
36. Find the sum of the first 100 positive even integers.
37. Find the sum of the first 100 positive multiples of 5.
38. Find the sum of the integers from 50 to 100 (inclusive).
39. A person accepts a position with a company and will receive a salary of $27,500 for the first year. The person is guaranteed a raise of $1500 per year for the first five years.
    (a) What will the salary be during the sixth year of employment?
    (b) How much will the company have paid the person by the end of the six years?

40. A freely falling object will fall 16 feet during the first second, 48 feet during the second second, 80 feet during the third second, 112 feet during the fourth second, and so on. What is the total distance the object will fall in 10 seconds if this pattern continues?

41. Determine the seating capacity of an auditorium with 30 rows of seats if there are 20 seats in the first row, 24 seats in the second row, 28 seats in the third row, and so on.

42. As a farmer bales a field of hay, each trip around the field gets shorter, since he is getting closer to the center. Suppose on the first round there were 267 bales and on the second round there were 253 bales. If you assume that the decrease will be the same on each round and there are 11 more trips, how many bales of hay will he get from the field?

In Exercises 43–46, find the given sum.

43. $\sum_{n=1}^{20} (2n + 5)$

44. $\sum_{n=1}^{100} \left(\dfrac{n + 4}{2}\right)$

45. $\sum_{n=0}^{50} (1000 - 5n)$

46. $\sum_{n=0}^{100} \left(\dfrac{8 - 3n}{16}\right)$

In Exercises 47–50, insert k arithmetic means between the given pair of numbers.

47. 5, 17; $k = 2$

48. 24, 56; $k = 3$

49. 3, 6; $k = 3$

50. 2, 5; $k = 4$

Geometric Sequences and Series   10.3

In this section we will study another important type of sequence called a **geometric sequence** (or **geometric progression**), characterized by the fact that consecutive terms have a common *ratio*. For instance, the sequence

$$\{a_n\} = \{3(2^{n-1})\}$$
$$= \{3(2^0), 3(2^1), 3(2^2), 3(2^3), \ldots\}$$
$$= \{3, 6, 12, 24, \ldots\}$$

$$\frac{6}{3} = 2 \qquad \frac{12}{6} = 2$$

has a common ratio of 2 between consecutive terms.

DEFINITION OF A GEOMETRIC SEQUENCE

The sequence $\{a_n\}$ is called a **geometric sequence** if each pair of consecutive terms has the same (nonzero) ratio

$$r = \frac{a_{i+1}}{a_i}, \qquad r \neq 0$$

The number r is called the **common ratio** of the sequence.

EXAMPLE 1
Finding the Common Ratio
of a Geometric Sequence

Determine which of the following sequences are geometric. For those that are geometric, find the common ratio.

(a) $\{a_n\} = \left\{\dfrac{2}{3^n}\right\}$

(b) $\{b_n\} = \{1, -1, 1, -1, 1, \ldots\}$

(c) $\{c_n\} = \{n!\}$

Solution:

(a) For the sequence

$$\{a_n\} = \left\{\frac{2}{3^n}\right\} = \left\{\frac{2}{3}, \frac{2}{9}, \frac{2}{27}, \frac{2}{81}, \ldots\right\}$$

the ratio of two consecutive terms is

$$r = \frac{a_{i+1}}{a_i} = \frac{2/3^{i+1}}{2/3^i} = \frac{2}{3^{i+1}}\left(\frac{3^i}{2}\right) = \frac{1}{3}$$

Thus, the sequence *is* geometric, and the common ratio is $r = \frac{1}{3}$.

(b) For the sequence

$$\{b_n\} = \{1, -1, 1, -1, 1, \ldots\}$$

the *n*th term is given by $b_n = (-1)^{n+1}$, and therefore the ratio of consecutive terms is

$$r = \frac{b_{i+1}}{b_i} = \frac{(-1)^{i+2}}{(-1)^{i+1}} = -1$$

Thus, the sequence *is* geometric, and the common ratio is $r = -1$.

(c) For the sequence

$$\{c_n\} = \{n!\} = \{1, 2, 6, 24, 120, 720, \ldots\}$$

the ratio of the first two terms is $2/1 = 2$ and the ratio of the third term to the second is $6/2 = 3$. Since these two ratios differ, we conclude that the sequence is *not* geometric.

By rewriting the equation for the common ratio r in the form

$$a_{i+1} = ra_i$$

we see that each successive term in a geometric sequence is an *r*-multiple of the preceding term. Consequently, starting with a_i and multiplying by r repeatedly, we obtain the geometric sequence

$$\{a_1, \ a_1r, \ a_1r^2, \ a_1r^3, \ a_1r^4, \ \ldots\}$$
$$\downarrow \quad \downarrow \quad \downarrow \quad \downarrow \quad \downarrow$$
$$\{a_1, \ a_2, \ a_3, \ a_4, \ a_5, \ \ldots\}$$

From this, it appears that the *n*th term of a geometric sequence has the following form.

THE *n*TH TERM OF A GEOMETRIC SEQUENCE

The *n*th term of a geometric sequence whose first term is a_1 and whose common ratio is r is given by

$$a_n = a_1 r^{n-1}$$

We postpone the formal proof of this formula until the next section on mathematical induction.

EXAMPLE 2
Finding the *n*th Term
of a Geometric Sequence

Find the indicated term of the following geometric sequences:
(a) The fifteenth term of the sequence whose first term is 20 and whose common ratio is 1.05.
(b) The twelfth term of the sequence whose first three terms are

$$\{5, -15, 45, \ldots\}$$

Solution:

(a) Since $a_1 = 20$ and $r = 1.05$, the fifteenth term is

$$a_{15} = a_1 r^{n-1} = 20(1.05)^{14} \approx 39.599$$

(b) For the geometric sequence

$$\{a_n\} = \{5, -15, 45, \ldots\}$$

the common ratio is $r = -3$. Therefore, since the first term is 5, we determine the twelfth term to be

$$a_{12} = a_1 r^{n-1} = 5(-3)^{11} = 5(-177,147) = -885,735$$

EXAMPLE 3
Finding the *n*th Term
of a Geometric Sequence

The fourth term of a geometric sequence is 125, and the tenth term is $125/64$. Find the fourteenth term.

Solution:
Using the *n*th term formula for a geometric sequence, we have

$$a_4 = a_1 r^3 \quad \text{and} \quad a_{10} = a_1 r^9$$

$$125 = a_1 r^3 \qquad \frac{125}{64} = a_1 r^9$$

We use substitution to solve this system of (nonlinear) equations. Solving for a_1 in the first equation, we have

$$125 = a_1 r^3 \quad \Longrightarrow \quad a_1 = \frac{125}{r^3}$$

Then by substituting in the second equation, we obtain

$$\frac{125}{64} = a_1 r^9 \quad \Longrightarrow \quad \frac{125}{64} = \left(\frac{125}{r^3}\right) r^9$$

$$\frac{1}{64} = r^6 \quad \Longrightarrow \quad r = \sqrt[6]{\frac{1}{64}} = \frac{1}{2}$$

By backsubstituting we find a_1 to be

$$125 = a_1 \left(\frac{1}{2}\right)^3$$

$$125(2^3) = a_1$$

$$1000 = a_1$$

Finally, using the nth term formula, we find the fourteenth term to be

$$a_{14} = a_1 r^{n-1} = 1000\left(\frac{1}{2}\right)^{13} = \frac{1000}{8192} = \frac{125}{1024}$$

The following formula gives us a simple method of calculating the nth partial sum of a geometric sequence.

THE nTH PARTIAL SUM OF A GEOMETRIC SEQUENCE

The nth partial sum of the geometric sequence $\{a_n\}$ with common ratio r is given by the following formula:

$$S_n = \frac{a_1(1 - r^n)}{1 - r}, \qquad r \neq 1$$

Proof:
We begin by writing out the nth partial sum, as follows:

$$S_n = a_1 + a_1 r + a_1 r^2 + \cdots + a_1 r^{n-2} + a_1 r^{n-1}$$

Multiplication by r gives us

$$r S_n = a_1 r + a_1 r^2 + a_1 r^3 + \cdots + a_1 r^{n-1} + a_1 r^n$$

By subtracting the second equation from the first, we have

$$
\begin{array}{l}
S_n = a_1 + a_1 r + a_1 r^2 + \cdots + a_1 r^{n-2} + a_1 r^{n-1} \\
-r S_n = \quad\;\; - a_1 r - a_1 r^2 - a_1 r^3 - \cdots - a_1 r^{n-1} - a_1 r^n \\
\hline
S_n - r S_n = a_1 - a_1 r^n
\end{array}
$$

Therefore, $S_n(1 - r) = a_1(1 - r^n)$, and since $r \neq 1$, we have

$$S_n = \frac{a_1(1 - r^n)}{1 - r}$$

Remark: When using this formula for the nth partial sum of a geometric sequence, be careful to check that the index begins at $i = 1$. It is often convenient to have the index of a geometric sequence begin at $i = 0$, and in such cases you must adjust the formula for the nth partial sum, as shown in the following example.

EXAMPLE 4
Finding the nth Partial Sum of a Geometric Sequence

Find the following sums:

(a) $\displaystyle\sum_{i=1}^{12} 4(0.3)^{i-1}$

(b) $\displaystyle\sum_{i=0}^{10} 10\left(-\frac{1}{2}\right)^{i}$

Solution:

(a) By writing out a few of the terms, we have

$$\sum_{i=1}^{12} 4(0.3)^{i-1} = 4 + 4(0.3) + 4(0.3)^2$$
$$+ 4(0.3)^3 + \cdots + 4(0.3)^{11}$$

Now, since $a_1 = 4$ and $r = 0.3$,

$$\sum_{i=1}^{12} 4(0.3)^{i-1} = \frac{a_1(1 - r^n)}{1 - r} = \frac{4[1 - (0.3)^{12}]}{1 - 0.3} \approx 5.714$$

(b) By writing out a few of the terms, we have

$$\sum_{i=0}^{10} 10\left(-\frac{1}{2}\right)^i = 10 - 10\left(\frac{1}{2}\right) + 10\left(\frac{1}{2}\right)^2$$
$$- 10\left(\frac{1}{2}\right)^3 + \cdots + 10\left(\frac{1}{2}\right)^{10}$$

Now, we see that the *first term* is $a_1 = 10$ and $r = -\frac{1}{2}$. Moreover, by starting with $i = 0$ and ending with $i = 10$, we are summing $n = 11$ terms, which means that the partial sum is

$$\sum_{i=0}^{10} 10\left(-\frac{1}{2}\right)^i = \frac{a_1(1 - r^n)}{1 - r} = \frac{10[1 - (-\frac{1}{2})^{11}]}{1 + (\frac{1}{2})} \approx 6.670$$

EXAMPLE 5
An Application:
Compound Interest

A deposit of $50.00 is made the first day of each month in a savings account that pays 12% compounded monthly. What is the balance at the end of two years?

Solution:
The formula for compound interest is

$$A = P\left(1 + \frac{r}{n}\right)^{tn}$$

where A is amount (balance), P is initial deposit (principal), r is annual percentage rate (in decimal form), n is number of compoundings per year, and t is time (in years). To find the balance in the account after 24 months, it is helpful to consider each of the 24 deposits separately. For example, the first deposit will gain interest for a full 24 months, and its balance will be

$$A_{24} = 50\left(1 + \frac{0.12}{12}\right)^{24}$$

The second deposit will gain interest for 23 months, and its balance will be

$$A_{23} = 50\left(1 + \frac{0.12}{12}\right)^{23}$$

The last (24th) deposit will gain interest for only 1 month, and its balance will be

$$A_1 = 50\left(1 + \frac{0.12}{12}\right)^1 = 50(1.01)$$

Finally, the total balance in the account will be the sum of the balances of the 24 deposits:

$$S_{24} = A_1 + A_2 + A_3 + \cdots + A_{23} + A_{24}$$

Using the formula for the nth partial sum of a geometric sequence, with $A_1 = 50(1.01)$ and $r = 1.01$, we have

$$S_{24} = 50 \, \frac{1.01[1 - (1.01)^{24}]}{1 - 1.01} = \$1362.16$$

Infinite Series

We now look at a general method for finding the sum of an infinite geometric series. For the nth partial sum

$$S_n = \frac{a_1(1 - r^n)}{1 - r}$$

The following statements are true as n increases without bound ($n \rightarrow \infty$):

1. If $r > 1$, then r^n becomes large without bound. This in turn means that the absolute value of $(1 - r^n)/(1 - r)$ increases without bound. Hence, S_n approaches no limiting value.

2. If $r < -1$, then r^n alternates between positive and negative numbers that are unbounded in absolute value. Again, S_n approaches no limiting value.

3. If $|r| < 1$, then r^n becomes arbitrarily close to zero, which means that

$$S_n \rightarrow \frac{a_1(1 - 0)}{1 - r} = \frac{a_1}{1 - r} \qquad \text{as} \qquad n \rightarrow \infty$$

The number $a_1/(1 - r)$ is called the **sum of the infinite geometric series**

$$\sum_{i=1}^{\infty} a_1 r^{n-1} = a_1 + a_1 r + a_1 r^2 + a_1 r^3 + \cdots + a_1 r^{n-1} + \cdots$$

as summarized in the following statement.

SUM OF AN INFINITE GEOMETRIC SERIES

If $|r| < 1$, then the infinite geometric series

$$a_1 + a_1 r + a_1 r^2 + a_1 r^3 + \cdots + a_1 r^{n-1} + \cdots$$

has the sum

$$\frac{a_1}{1 - r}$$

EXAMPLE 6
Finding the Sum of an
Infinite Geometric Series

Find the sum of the following infinite geometric series:

$$\sum_{n=1}^{\infty} 4(-0.6)^{n-1}$$

Solution:
Since $a_1 = 4$ and $r = -0.6$, we have

$$\sum_{n=1}^{\infty} 4(-0.6)^{n-1} = \frac{a_1}{1-r} = \frac{4}{1-(-0.6)} = 2.5$$

When we introduced the real number system in Section 1.1, we noted that if a real number has a decimal representation that (indefinitely) repeats a block of digits, then the real number must be rational. For example, the repeating decimals $0.1111 \ldots = 0.\overline{1}$ and $5.2135135135 \ldots = 5.2\overline{135}$ represent rational numbers. Irrational numbers must have nonrepeating (non-terminating) decimal representations. For example,

$$\pi = 3.141592653589793 \ldots \quad \text{and} \quad e = 2.718281828459045 \ldots$$

In order to write a repeating decimal as the ratio of two integers, we use a geometric series, as shown in the following example.

EXAMPLE 7
Writing a Repeating Decimal
as a Ratio of Two Integers

Write the following repeating decimals as the ratio of two integers:

(a) $0.2\overline{2}$ (b) $2.1\overline{425}$

Solution:

(a) As an infinite geometric sequence, we have

$$0.222\ldots = 0.2 + \quad 0.02 \quad + \quad 0.002 \quad + \cdots$$
$$\downarrow \qquad\quad \downarrow \qquad\qquad \downarrow$$
$$= 0.2 + 0.2(0.1) + 0.2(0.1)^2 + \cdots$$
$$+ \ 0.2(0.1)^{n-1} + \cdots$$

Thus, $a_1 = 0.2$ and $r = 0.1$, and by the formula for the sum of an infinite geometric series, we have

$$0.2\overline{2} = \frac{a_1}{1-r} = \frac{0.2}{1-0.1} = \frac{0.2}{0.9} = \frac{2}{9}$$

(b) Since the first two digits are not part of the repeating pattern, we pull them out of the representation, as follows:

$$2.1425425425\ldots = 2.1 + 0.0425425425\ldots$$

Now for the repeating portion, we write

$$0.0\overline{425} = 0.0425 + 0.0425(0.001) + 0.0425(0.001)^2 + \cdots$$
$$+ \ 0.0425(0.001)^{n-1} + \cdots$$

$$= \frac{a_1}{1-r} = \frac{0.0425}{1-0.001} = \frac{0.0425}{0.999} = \frac{425}{9990}$$

Finally, by adding 2.1 to this result, we have

$$2.1\overline{425} = \frac{21}{10} + \frac{425}{9990} = \frac{20979}{9990} + \frac{425}{9990}$$

$$= \frac{21404}{9990} = \frac{10702}{4995}$$

To conclude this section, we look at a technique for determining the sum of a nongeometric infinite series.

EXAMPLE 8
Finding the Sum
of an Infinite Series

Use the nth partial sum to show that the infinite series

$$\frac{2}{1 \cdot 3} + \frac{2}{3 \cdot 5} + \frac{2}{5 \cdot 7} + \frac{2}{7 \cdot 9} + \cdots + \frac{2}{(2n - 1)(2n + 1)} + \cdots$$

has a sum.

Solution:
Notice that the series is *not* geometric. The nth term has the *partial fraction* representation

$$a_n = \frac{2}{(2n - 1)(2n + 1)} = \frac{1}{2n - 1} - \frac{1}{2n + 1}$$

Consequently, we may write the terms as

$$a_1 = 1 - \frac{1}{3}, \qquad a_2 = \frac{1}{3} - \frac{1}{5}, \qquad a_3 = \frac{1}{5} - \frac{1}{7}, \qquad \cdots$$

Therefore, the nth partial sum is

$$S_n = a_1 + a_2 + a_3 + \cdots + a_n$$

$$= 1 - \frac{1}{3} + \frac{1}{3} - \frac{1}{5} + \frac{1}{5} - \frac{1}{7} + \cdots + \frac{1}{2n - 1} - \frac{1}{2n + 1}$$

$$= 1 - \frac{1}{2n + 1}$$

Note that all but the first and last terms cancel. Now, since $1/(2n + 1) \to 0$ as $n \to \infty$, it follows that $S_n \to 1$. Thus, we conclude that the sum is

$$1 = \frac{2}{1 \cdot 3} + \frac{2}{3 \cdot 5} + \frac{2}{5 \cdot 7} + \cdots + \frac{2}{(2n - 1)(2n + 1)} + \cdots$$

Section Exercises 10.3

In Exercises 1–10, determine whether or not the sequence is geometric. If it is, find its common ratio.

1. $\{5, 15, 45, 135, \ldots\}$
2. $\{3, 12, 48, 192, \ldots\}$
3. $\{3, 12, 21, 30, \ldots\}$
4. $\{1, -2, 4, -8, \ldots\}$
5. $\{1, -\frac{1}{2}, \frac{1}{4}, -\frac{1}{8}, \ldots\}$
6. $\{5, 1, 0.2, 0.04, \ldots\}$
7. $\{\frac{1}{2}, \frac{2}{3}, \frac{3}{4}, \frac{4}{5}, \ldots\}$
8. $\{9, -6, 4, -\frac{8}{3}, \ldots\}$
9. $\{1, \frac{1}{2}, \frac{1}{3}, \frac{1}{4}, \ldots\}$
10. $\{\frac{1}{5}, \frac{2}{3}, \frac{3}{9}, \frac{4}{11}, \ldots\}$

In Exercises 11–18, write the first five terms of the geometric sequence.

11. $a_1 = 2, r = 3$
12. $a_1 = 6, r = 2$
13. $a_1 = 1, r = \frac{1}{2}$
14. $a_1 = 1, r = \frac{1}{3}$
15. $a_1 = 5, r = -\frac{1}{10}$
16. $a_1 = 1, r = -x$
17. $a_1 = 1, r = x/2$
18. $a_1 = 2, r = \sqrt{3}$

In Exercises 19–26, find the nth term of the geometric sequence.

19. $a_1 = 4, r = \frac{1}{2}, n = 10$
20. $a_1 = 5, r = \frac{3}{2}, n = 8$
21. $a_1 = 6, r = -\frac{1}{3}, n = 12$
22. $a_1 = 1, r = -x/3, n = 7$
23. $a_1 = 100, r = e^x, n = 9$
24. $a_1 = 8, r = \sqrt{5}, n = 9$
25. $a_1 = 500, r = 1.02, n = 40$
26. $a_1 = 1000, r = 1.005, n = 60$

27. If $1000 is invested at 10% interest, find the amount after 10 years if the interest is compounded
    (a) annually   (b) semiannually   (c) quarterly
    (d) monthly   (e) daily

28. If $2500 is invested at 12% interest, find the amount after 20 years if the interest is compounded
    (a) annually   (b) semiannually   (c) quarterly
    (d) monthly   (e) daily

29. A company buys a machine for $135,000, and it depreciates at the rate of 30% per year. (In other words, at the end of each year its depreciated value is 70% of what it was at the beginning of the year.) Find the depreciated value of the machine after it has been used five full years.

30. A city of 250,000 people is growing at a rate of 1.3% per year. Estimate the population of the city 30 years from now.

In Exercises 31–36, use the formula for the nth partial sum of a geometric series to evaluate the given sum.

31. $\displaystyle\sum_{n=0}^{20} 3\left(\frac{3}{2}\right)^n$

32. $\displaystyle\sum_{n=0}^{15} 2\left(\frac{4}{3}\right)^n$

33. $\displaystyle\sum_{i=1}^{10} 8\left(\frac{-1}{4}\right)^{i-1}$

34. $\displaystyle\sum_{i=1}^{10} 5\left(\frac{-1}{3}\right)^{i-1}$

35. $\displaystyle\sum_{n=0}^{8} 2^n$

36. $\displaystyle\sum_{n=0}^{8} (-2)^n$

37. A deposit of $100 is made each month for five years in an account that pays 10%, compounded monthly. What is the balance A in the account at the end of five years?

$$A = 100\left(1 + \frac{0.10}{12}\right) + 100\left(1 + \frac{0.10}{12}\right)^2 + \cdots$$
$$+ 100\left(1 + \frac{0.10}{12}\right)^{60}$$

38. A deposit of $50 is made each month for five years in an account that pays 12%, compounded monthly. What is the balance A in the account at the end of five years?

$$A = 50\left(1 + \frac{0.12}{12}\right) + 50\left(1 + \frac{0.12}{12}\right)^2 + \cdots$$
$$+ 50\left(1 + \frac{0.12}{12}\right)^{120}$$

39. A deposit of P dollars is made every month for T years in an account that pays R percent interest compounded monthly. Let $N = 12T$ be the total number of deposits. The balance A after T years is

$$A = P\left(1 + \frac{R}{12}\right) + P\left(1 + \frac{R}{12}\right)^2 + \cdots$$
$$+ P\left(1 + \frac{R}{12}\right)^N$$

Show that the balance is given by

$$A = P\left[\left(1 + \frac{R}{12}\right)^N - 1\right]\left(1 + \frac{12}{R}\right)$$

40. Use the formula in Exercise 39 to find the amount in an account earning 9% compounded monthly, after monthly deposits of $50 have been made for 40 years.

41. Use the formula in Exercise 39 to find the amount in an account earning 12% compounded monthly, after monthly deposits of $50 have been made for 40 years.

42. Suppose that you went to work at a company that pays $0.01 for the first day, $0.02 for the second day, $0.04 for the third day, and so on. If the daily wage keeps doubling, what would your total income be for working 30 days?

In Exercises 43–52, find the sum of the infinite geometric series.

43. $1 + \frac{1}{2} + \frac{1}{4} + \frac{1}{8} + \cdots$
44. $2 + \frac{4}{3} + \frac{8}{9} + \frac{16}{27} + \cdots$
45. $1 - \frac{1}{2} + \frac{1}{4} - \frac{1}{8} + \cdots$
46. $2 - \frac{4}{3} + \frac{8}{9} - \frac{16}{27} + \cdots$
47. $4 + 1 + \frac{1}{4} + \frac{1}{16} + \cdots$
48. $1 + 0.1 + 0.01 + 0.001 + \cdots$
49. $8 + 6 + \frac{9}{2} + \frac{27}{8} + \cdots$
50. $3 - 1 + \frac{1}{3} - \frac{1}{9} + \cdots$
51. $4 - 2 + 1 - \frac{1}{2} + \cdots$
52. $2 + \sqrt{2} + 1 + 1/\sqrt{2} + \cdots$

In Exercises 53 and 54, find the sum of the infinite nongeometric series.

53. $\displaystyle\sum_{n=0}^{\infty} \left(\frac{1}{2^n} - \frac{1}{3^n}\right)$

54. $\displaystyle\sum_{n=0}^{\infty} [(0.7)^n + (0.9)^n]$

55. A ball is dropped from a height of 16 feet. Each time it drops h feet, it rebounds $0.81h$ feet. Find the total distance traveled by the ball.

56. The ball in Exercise 55 takes the following times for each fall:

$$s_1 = -16t^2 + 16, \qquad s_1 = 0 \text{ if } t = 1$$
$$s_2 = -16t^2 + 16(0.81), \qquad s_2 = 0 \text{ if } t = 0.9$$
$$s_3 = -16t^2 + 16(0.81)^2, \qquad s_3 = 0 \text{ if } t = (0.9)^2$$
$$s_4 = -16t^2 + 16(0.81)^3, \qquad s_4 = 0 \text{ if } t = (0.9)^3$$

.   .   .
.   .   .
.   .   .

$$s_n = -16t^2 + 16(0.81)^{n-1}, \qquad s_n = 0 \text{ if } t = (0.9)^{n-1}$$

Beginning with s_2, the ball takes the same amount of time to bounce up as it takes to fall, and thus the total time elapsed before it comes to rest is

$$t = 1 + 2 \sum_{n=1}^{\infty} (0.9)^n$$

Find this total.

In Exercises 57–64, write the repeating decimal as a rational number by considering it to be the sum of an infinite geometric series.

57. 0.1111. . .
58. 0.4444. . .
59. 0.363636. . .
60. 0.212121. . .
61. 0.432432432. . .
62. 0.46666. . .
63. 1.363636. . .
64. 1.185185185. . .

Mathematical Induction   10.4

In this section we look at a very important form of mathematical proof, called the principle of **mathematical induction.** It is important that you clearly see the logical need for it, so let's take a closer look at a problem we discussed earlier.

In Section 10.2 we looked at the following pattern for the sum of the first n odd integers:

$$S_1 = 1 = 1 = 1^2$$
$$S_2 = 1 + 3 = 4 = 2^2$$
$$S_3 = 1 + 3 + 5 = 9 = 3^2$$
$$S_4 = 1 + 3 + 5 + 7 = 16 = 4^2$$
$$S_5 = 1 + 3 + 5 + 7 + 9 = 25 = 5^2$$

Judging from the pattern formed by these first five partial sums, we decided that the nth partial sum was of the form

$$S_n = 1 + 3 + 5 + 7 + 9 + \cdots + 2n - 1 = n^2$$

Recognizing a pattern and then simply *jumping to the conclusion* that the pattern must be true for all values of n is *not* a logically valid method of proof. There are many examples in which a pattern appears to be developing for small values of n and then at some point the pattern fails. One of the most famous cases of this was the conjecture by a well-known mathematician named Pierre de Fermat, who speculated that all numbers of the form

$$F_n = 2^{2^n} + 1$$

are prime. For $n = 0, 1, 2, 3$, and 4, the conjecture is true.

$$F_0 = 3,$$
$$F_1 = 5,$$
$$F_2 = 17,$$
$$F_3 = 257,$$
$$F_4 = 65537$$

The size of the next Fermat number ($F_5 = 4,294,967,297$) is so great that it was difficult for Fermat to determine whether it was prime or not. However, another well-known mathematician, Leonhard Euler, later found a factorization

$$F_5 = 4,294,967,297 = 641(6,700,417)$$

which proved Fermat's conjecture to be false.

Here is the point: Just because a rule, pattern, or formula seems to work for several values of n, we cannot simply decide that it is valid for all values of n without going through a *legitimate proof*. Let's see how we can prove such statements by the principle of **mathematical induction.**

THE PRINCIPLE OF MATHEMATICAL INDUCTION

Let P_n be a statement involving the positive integer n. If

1. P_1 is true and
2. the truth of P_k implies the truth of P_{k+1}

then the statement must be true for all positive integers n.

EXAMPLE 1
Using Mathematical Induction

Use mathematical induction to prove the following formula:

$$S_n = 1 + 3 + 5 + 7 + 9 + \cdots + 2n - 1 = n^2$$

Solution:
Mathematical induction consists of two distinct parts. First, we must show that the formula is true when $n = 1$.
(1) When $n = 1$, the formula is valid, since

$$S_1 = 1 = 1^2$$

The second part of mathematical induction has two steps. The first step is to assume that the formula is valid for *some* integer k. The second step is to use this assumption to prove that the formula is valid for the integer $k + 1$.
(2) Assume that

$$S_k = 1 + 3 + 5 + 7 + 9 + \cdots + 2k - 1 = k^2$$

Now, we verify the formula for $k + 1$, as follows:

$$
\begin{aligned}
S_{k+1} &= 1 + 3 + 5 + 7 + 9 + \cdots \\
&\quad + 2k - 1 + 2(k + 1) - 1 \\
&= [1 + 3 + 5 + 7 + 9 + \cdots + 2k - 1] \\
&\quad + 2(k + 1) - 1 \\
&= [S_k] + 2k + 2 - 1 \qquad \textit{Substitute } S_k \\
&= [k^2] + 2k + 1 \qquad\quad\; \textit{Substitute } k^2 \\
&= (k + 1)^2
\end{aligned}
$$

Combining the result of parts (1) and (2), we conclude by mathematical induction that the formula is valid for *all* positive integer values of *n*.

Remark: When using mathematical induction to prove a *summation* formula (like the one in Example 1), it is helpful to think of S_{k+1} as

$$
S_{k+1} = S_k + a_{k+1}
$$

where a_{k+1} is the $(k + 1)$ term of the original sum.

A well-known illustration used to explain why the principle of mathematical induction works is that of an unending line of dominos, as shown in Figure 10.1. If the line actually contains infinitely many dominos, then it is clear that we could not knock the entire line down by knocking down only *one domino* at a time. However, suppose it were true that each domino would knock down the next one as it fell. Then we could knock them all down simply by pushing the first one and starting a chain reaction. This is the same way mathematical induction works. If the truth of P_k implies the truth of P_{k+1} and if P_1 is true, then the chain reaction proceeds as follows:

P_1 implies P_2

P_2 implies P_3

P_3 implies P_4

and so on.

FIGURE 10.1   The first domino knocks over the second which knocks over
the third which knocks over the fourth and so on.

It might happen that a statement involving natural numbers is not true for the first $k - 1$ positive integers but is true for all values of $n \geq k$. In these instances, we use a slight variation of the principle of mathematical induction in which we verify $P(k)$ rather than $P(1)$. To see the validity of this, note from Figure 10.1 that all but the first $k - 1$ dominos can be knocked down by knocking over the kth domino. This suggests that we can prove a statement $P(n)$ to be true for $n \geq k$ by showing that $P(k)$ is true and that $P(k)$ implies $P(k + 1)$. In Exercises 21-23 you are asked to apply this variation of mathematical induction.

EXAMPLE 2
Using Mathematical Induction

Use mathematical induction to prove the following formula:

$$S_n = 1^2 + 2^2 + 3^2 + 4^2 + \cdots + n^2 = \frac{n(n + 1)(2n + 1)}{6}$$

Solution:

(1) When $n = 1$, the formula is valid, since

$$S_1 = 1^2 = \frac{1(2)(3)}{6}$$

(2) Assume that

$$S_k = 1^2 + 2^2 + 3^2 + 4^2 + \cdots + k^2 = \frac{k(k + 1)(2k + 1)}{6}$$

Now, we verify the formula for $k + 1$, as follows:

$$\begin{aligned}
S_{k+1} &= 1^2 + 2^2 + 3^2 + 4^2 + \cdots + k^2 + (k + 1)^2 \\
&= [S_k] + a_{k+1} \\
&= [1^2 + 2^2 + 3^2 + 4^2 + \cdots + k^2] + (k + 1)^2 \\
&= \left[\frac{k(k + 1)(2k + 1)}{6} \right] + (k + 1)^2 \qquad \textit{By assumption} \\
&= \frac{k(k + 1)(2k + 1) + 6(k + 1)^2}{6} \\
&= \frac{(k + 1)[k(2k + 1) + 6(k + 1)]}{6} \\
&= \frac{(k + 1)[2k^2 + 7k + 6]}{6} \\
&= \frac{(k + 1)(k + 2)(2k + 3)}{6} \\
&= \frac{(k + 1)[(k + 1) + 1][2(k + 1) + 1]}{6}
\end{aligned}$$

Combining the result of parts (1) and (2), we conclude by mathematical induction that the formula is valid for *all* $n \geq 1$.

The formula in Example 2 is one of a collection of formulas that prove to be quite useful in calculus. We summarize this and other formulas dealing with the sum of various powers of the first n positive integers as follows.

SUMS OF POWERS OF INTEGERS

1. $1 + 2 + 3 + 4 + \cdots + n = \dfrac{n(n + 1)}{2}$

2. $1^2 + 2^2 + 3^2 + 4^2 + \cdots + n^2 = \dfrac{n(n + 1)(2n + 1)}{6}$

3. $1^3 + 2^3 + 3^3 + 4^3 + \cdots + n^3 = \dfrac{n^2(n + 1)^2}{4}$

4. $1^4 + 2^4 + 3^4 + 4^4 + \cdots + n^4$

$$= \dfrac{n(n + 1)(2n + 1)(3n^2 + 3n - 1)}{30}$$

5. $1^5 + 2^5 + 3^5 + 4^5 + \cdots + n^5 = \dfrac{n^2(n + 1)^2(2n^2 + 2n - 1)}{12}$

Remark: Each of these formulas for sums can be proved by mathematical induction. (See Exercises 7–11.)

Although choosing a pattern on the basis of a few observations does *not* guarantee the validity of the pattern, recognition *is* important. Once you have a pattern or formula that you think works, you can try using mathematical induction to prove your formula. We outline these steps as follows.

FINDING A FORMULA FOR THE nTH TERM OF A SEQUENCE

1. Calculate the first several terms of the sequence. (It is often a good idea to leave the terms in factored form *without* simplifying.)

2. Try to find a recognizable pattern from these terms, and write down a formula for the nth term of the sequence. (This is your hypothesis. You might try computing one or two more terms in the sequence to test your hypothesis.)

3. Use mathematical induction to prove (or disprove) your hypothesis.

EXAMPLE 3
Finding a Formula for the
nth Term of a Sequence

Find a formula for the nth partial sum of the sequence

$$\frac{1}{1 \cdot 2}, \frac{1}{2 \cdot 3}, \frac{1}{3 \cdot 4}, \frac{1}{4 \cdot 5}, \cdots, \frac{1}{n(n + 1)}, \cdots$$

Solution:
We begin by writing out a few of the partial sums:

$$S_1 = \frac{1}{1 \cdot 2} = \frac{1}{2}$$

$$S_2 = \frac{1}{1 \cdot 2} + \frac{1}{2 \cdot 3} = \frac{4}{6} = \frac{2}{3}$$

$$S_3 = \frac{1}{1 \cdot 2} + \frac{1}{2 \cdot 3} + \frac{1}{3 \cdot 4} = \frac{9}{12} = \frac{3}{4}$$

$$S_4 = \frac{1}{1 \cdot 2} + \frac{1}{2 \cdot 3} + \frac{1}{3 \cdot 4} + \frac{1}{4 \cdot 5} = \frac{48}{60} = \frac{4}{5}$$

Now, from these first four partial sums, it appears that the formula for the kth partial sum is

$$S_k = \frac{1}{1 \cdot 2} + \frac{1}{2 \cdot 3} + \frac{1}{3 \cdot 4} + \frac{1}{4 \cdot 5} + \cdots + \frac{1}{k(k+1)} = \frac{k}{k+1}$$

To prove the validity of this formula, we use mathematical induction, as follows. Note that we have already verified the formula for $n = 1$, so we begin by assuming that the formula is valid for $n = k$ and try to show that it is valid for $n = k + 1$.

$$S_{k+1} = \left[\frac{1}{1 \cdot 2} + \frac{1}{2 \cdot 3} + \cdots + \frac{1}{k(k+1)} \right] + \frac{1}{(k+1)(k+2)}$$

$$= \left[\frac{k}{k+1} \right] + \frac{1}{(k+1)(k+2)} \qquad \textit{By assumption}$$

$$= \frac{k(k+2)+1}{(k+1)(k+2)} = \frac{k^2 + 2k + 1}{(k+1)(k+2)}$$

$$= \frac{(k+1)^2}{(k+1)(k+2)} = \frac{k+1}{k+2}$$

Thus, the formula is valid.

EXAMPLE 4
Proving an Inequality by
Mathematical Induction

Prove that $n < 2^n$ for all positive integers n.

Solution:

(1) For $n = 1$, the formula is true, since

$$1 < 2^1$$

(2) Assume that

$$k < 2^k$$

Now, for $n = k + 1$, we have

$$2^{k+1} = 2[2^k] > 2[k] = 2k \qquad \textit{By assumption}$$

Since $2k = k + k > k + 1$ for all $k > 1$, it follows that

$$2^{k+1} > 2k > k + 1$$

or

$$k + 1 < 2^{k+1}$$

Hence, $n < 2^n$ for all integers $n \geq 1$.

Section Exercises 10.4

In Exercises 1–14, use mathematical induction to prove the given formula for every positive integer n.

1. $2 + 4 + 6 + 8 + \cdots + 2n = n(n + 1)$
2. $3 + 7 + 11 + 15 + \cdots + (4n - 1) = n(2n + 1)$
3. $2 + 7 + 12 + 17 + \cdots + (5n - 3) = \dfrac{n}{2}(5n - 1)$
4. $1 + 4 + 7 + 10 + \cdots + (3n - 2) = \dfrac{n}{2}(3n - 1)$
5. $1 + 2 + 2^2 + 2^3 + \cdots + 2^{n-1} = 2^n - 1$
6. $2(1 + 3 + 3^2 + 3^3 + \cdots + 3^{n-1}) = 3^n - 1$
7. $1 + 2 + 3 + 4 + \cdots + n = \dfrac{n(n + 1)}{2}$
8. $1^2 + 2^2 + 3^2 + 4^2 + \cdots + n^2 = \dfrac{n(n + 1)(2n + 1)}{6}$
9. $1^3 + 2^3 + 3^3 + 4^3 + \cdots + n^3 = \dfrac{n^2(n + 1)^2}{4}$
10. $\displaystyle\sum_{i=1}^{n} i^4 = \dfrac{n(n + 1)(2n + 1)(3n^2 + 3n - 1)}{30}$
11. $\displaystyle\sum_{i=1}^{n} i^5 = \dfrac{n^2(n + 1)^2(2n^2 + 2n - 1)}{12}$
12. $\left(1 + \dfrac{1}{1}\right)\left(1 + \dfrac{1}{2}\right)\left(1 + \dfrac{1}{3}\right) \cdots \left(1 + \dfrac{1}{n}\right) = n + 1$
13. $\displaystyle\sum_{i=1}^{n} i(i + 1) = \dfrac{n(n + 1)(n + 2)}{3}$
14. $\displaystyle\sum_{i=1}^{n} \dfrac{1}{(2i - 1)(2i + 1)} = \dfrac{n}{2n + 1}$

In Exercises 15–20, find a formula for the nth partial sum of the sequence.

15. $3, 7, 11, 15, \ldots$
16. $25, 22, 19, 16, \ldots$
17. $1, \dfrac{9}{10}, \dfrac{81}{100}, \dfrac{729}{1000}, \ldots$
18. $3, -\dfrac{9}{2}, \dfrac{27}{4}, -\dfrac{81}{8}, \ldots$
19. $\dfrac{1}{4}, \dfrac{1}{12}, \dfrac{1}{24}, \dfrac{1}{40}, \ldots, \dfrac{1}{2n(n + 1)}, \ldots$
20. $\dfrac{1}{2 \cdot 3}, \dfrac{1}{3 \cdot 4}, \dfrac{1}{4 \cdot 5}, \dfrac{1}{5 \cdot 6}, \ldots, \dfrac{1}{(n + 1)(n + 2)}, \ldots$

In Exercises 21–24, use mathematical induction to prove the given inequality for the indicated integer values of n.

21. $(\tfrac{4}{3})^n > n$, $\quad n \geq 7$
22. $\dfrac{1}{\sqrt{1}} + \dfrac{1}{\sqrt{2}} + \dfrac{1}{\sqrt{3}} + \cdots + \dfrac{1}{\sqrt{n}} > \sqrt{n}$, $\quad n \geq 2$
23. $n! > 2^n$, $\quad n \geq 4$
24. If $0 < x < y$, then $\left(\dfrac{x}{y}\right)^{n+1} < \left(\dfrac{x}{y}\right)^n$, $\quad n \geq 1$

In Exercises 25–33, use mathematical induction to prove the given property for all positive integers n.

25. $(ab)^n = a^n b^n$
26. $\left(\dfrac{a}{b}\right)^n = \dfrac{a^n}{b^n}$
27. If $x_1 \neq 0, x_2 \neq 0, x_3 \neq 0, \ldots, x_n \neq 0$, then
$$(x_1 x_2 x_3 \cdots x_n)^{-1} = x_1^{-1} x_2^{-1} x_3^{-1} \cdots x_n^{-1}$$
28. If $x_1 > 0, x_2 > 0, x_3 > 0, \ldots, x_n > 0$, then
$$\log_b(x_1 x_2 x_3 \cdots x_n) = \log_b(x_1) + \log_b(x_2) + \log_b(x_3) + \cdots + \log_b(x_n)$$
29. Generalized Distributive Law:
$$x(y_1 + y_2 + y_3 + \cdots + y_n)$$
$$= xy_1 + xy_2 + xy_3 + \cdots + xy_n$$
30. $x^n - y^n =$
$$(x - y)(x^{n-1} + x^{n-2}y + \cdots + xy^{n-2} + y^{n-1})$$
[Hint: $x^{n+1} - y^{n+1} = x^n(x - y) + y(x^n - y^n)$.]
31. $\sin(\theta + n\pi) = (-1)^n \sin\theta$
32. $\cos(\theta + n\pi) = (-1)^n \cos\theta$
33. DeMoivre's Theorem:
$$[r(\cos\theta + i\sin\theta)]^n = r^n[\cos n\theta + i\sin n\theta]$$
34. Prove that $(a + bi)^n$ and $(a - bi)^n$ are complex conjugates for $n \geq 1$.
35. Prove that 3 is a factor of $(n^3 + 3n^2 + 2n)$ for all $n \geq 1$.
36. Prove that 5 is a factor of $(2^{2n-1} + 3^{2n-1})$ for all $n \geq 1$.

The Binomial Theorem   10.5

Recall that a **binomial** is an expression that has two terms. In this section we will look at a formula that gives us a quick method of raising a binomial to a power.

Let's look at the expansion of $(x + y)^n$ for a few values of n.

$$(x + y)^0 = 1$$
$$(x + y)^1 = x + y$$
$$(x + y)^2 = x^2 + 2xy + y^2$$
$$(x + y)^3 = x^3 + 3x^2y + 3xy^2 + y^3$$
$$(x + y)^4 = x^4 + 4x^3y + 6x^2y^2 + 4xy^3 + y^4$$
$$(x + y)^5 = x^5 + 5x^4y + 10x^3y^2 + 10x^2y^3 + 5xy^4 + y^5$$

There are several observations we can make about these expansions of $(x + y)^n$:

1. In each expansion, there are $n + 1$ terms.

2. In each expansion, x and y have symmetrical roles. The powers of x decrease by 1 in each term, whereas the powers of y increase by 1.

3. The sum of the powers of each term in a binomial expansion is n. For example,

$$\overset{4 + 1 = 5}{} \qquad \overset{3 + 2 = 5}{}$$
$$(x + y)^5 = x^5 + 5x^4y^1 + 10x^3y^2 + 10x^2y^3 + 5xy^4 + y^5$$

4. The first term is x^n. The last term is y^n.

5. The second term is $nx^{n-1}y$. The next to last term is nxy^{n-1}.

The coefficients of the interior terms of the binomial expansion of $(x + y)^n$ are given by a well-known theorem called the **Binomial Theorem.**

THE BINOMIAL THEOREM

$$(x + y)^n = x^n + nx^{n-1}y + \frac{n(n - 1)}{2!} x^{n-2}y^2 + \cdots$$
$$+ C_r^n x^{n-r}y^r + \cdots + y^n$$

where the coefficient of $x^{n-r}y^r$ is given by

$$C_r^n = \frac{n!}{(n - r)!r!} = \frac{n(n - 1)(n - 2) \cdots (n - r + 1)}{r!}$$

Proof:

The Binomial Theorem can be proved quite nicely using mathematical induction. The steps are straightforward but look a little complex, so we will only present an outline of the proof.

(1) If $n = 1$, then we have

$$(x + y)^1 = x^1 + y^1 = C_0^1 x + C_1^1 y$$

and the formula is valid.

(2) Assuming the formula is true for $n = k$, then the coefficient of $x^{k-r}y^r$ is given by

$$C_r^k = \frac{k!}{(k-r)!r!} = \frac{k(k-1)(k-2)\cdots(k-r+1)}{r!}$$

To show that the formula is true for $n = k + 1$, we look at the coefficient of $x^{k+1-r}y^r$ in the expansion of

$$(x + y)^{k+1} = (x + y)^k(x + y)$$

From the right-hand side, we can determine that the term involving $x^{k+1-r}y^r$ is the sum of two products, as follows:

$$(C_r^k x^{k-r}y^r)(x) + (C_{r-1}^k x^{k+1-r}y^{r-1})(y)$$

$$= \left[\frac{k!}{(k-r)!r!} + \frac{k!}{(k-r+1)!(r-1)!}\right]x^{k+1-r}y^r$$

$$= \left[\frac{(k+1-r)k!}{(k+1-r)!r!} + \frac{k!r}{(k+1-r)!r!}\right]x^{k+1-r}y^r$$

$$= \left[\frac{k!(k+1-r+r)}{(k+1-r)!r!}\right]x^{k+1-r}y^r$$

$$= \left[\frac{(k+1)!}{(k+1-r)!r!}\right]x^{k+1-r}y^r$$

$$= C_r^{k+1}x^{k+1-r}y^r$$

Thus, by mathematical induction, we can assume that the Binomial Theorem is valid for all positive integers n.

The number C_r^n is called a **binomial coefficient.** Be sure you see that the expansion of $(x + y)^n$ has $n + 1$ coefficients. They are

$$C_0^n, \quad C_1^n, \quad C_2^n, \quad \cdots, \qquad C_r^n, \qquad \cdots, \quad C_{n-1}^n, \quad C_n^n$$

$$1, n, \quad \frac{n(n-1)}{2}, \ldots, \quad \frac{n(n-1)(n-2)\cdots(n-r+1)}{r!}, \ldots, n, 1$$

Remark: A common alternative notation for binomial coefficients is

$$C_r^n = \binom{n}{r}$$

EXAMPLE 1
Finding Binomial Coefficients

Find the following binomial coefficients:

(a) C_3^7 (b) C_5^{12} (c) C_7^{12} (d) C_0^8

Solution:

(a) $C_3^7 = \dfrac{7 \cdot 6 \cdot 5}{3 \cdot 2 \cdot 1} = 7(5) = 35$

Note that the denominator of a binomial coefficient is $r!$ and the numerator has r factors.

(b) $C_5^{12} = \dfrac{12 \cdot 11 \cdot 10 \cdot 9 \cdot 8}{5 \cdot 4 \cdot 3 \cdot 2 \cdot 1} = 11(9)(8) = 792$

(c) $C_7^{12} = \dfrac{12 \cdot 11 \cdot 10 \cdot 9 \cdot 8 \cdot 7 \cdot 6}{7 \cdot 6 \cdot 5 \cdot 4 \cdot 3 \cdot 2 \cdot 1} = 11(9)(8) = 792$

(d) $C_0^8 = \dfrac{8!}{8!(0!)} = 1$

Remark: In Example 1, note that the coefficients in parts (b) and (c) are the same. This is not a coincidence, since it is true in general that

$$C_r^n = C_{n-r}^n$$

This is in keeping with our earlier observation that the coefficients of a binomial expansion occur in a symmetrical pattern.

EXAMPLE 2
Using the Binomial Theorem

Use the Binomial Theorem to expand the following:

(a) $(x + 2)^5$ (b) $(1 - a^2)^4$

Solution:

(a) $(x + 2)^5 = C_0^5 x^5 \quad + C_1^5 x^4(2) + C_2^5 x^3(2^2) + C_3^5 x^2(2^3)$
$$+ C_4^5 x(2^4) + C_5^5(2^5)$$
$$= x^5 + 5(2)x^4 + 10(4)x^3 + 10(8)x^2 + 5(16)x + 32$$
$$= x^5 + 10x^4 + 40x^3 + 80x^2 + 80x + 32$$

(b) Since this binomial is a difference, rather than a sum, we write

$$(1 - a^2)^4 = [1 + (-a^2)]^4$$

From this we see that the signs of the terms will alternate, as follows:

$$(1 - a^2)^4 = C_0^4(1^4) \; - \; C_1^4(1^3)(a^2) + C_2^4(1^2)(a^2)^2$$
$$- C_3^4(1)(a^2)^3 + C_4^4(a^2)^4$$
$$= 1 - 4a^2 + 6a^4 - 4a^6 + a^8$$

Pascal's Triangle

There is an interesting way to remember the patterns for binomial coefficients. By arranging the coefficients in a triangular pattern, we have the following array, which is called **Pascal's Triangle:**

$$
\begin{array}{ccccccccccccccc}
 & & & & & & & 1 & & & & & & & \\
 & & & & & & 1 & & 1 & & & & & & \\
 & & & & & 1 & & 2 & & 1 & & & & & \\
 & & & & 1 & & 3 & & 3 & & 1 & & & & \\
 & & & 1 & & 4 & & 6 & & 4 & & 1 & & & \\
 & & 1 & & 5 & & 10 & & 10 & & 5 & & 1 & & \\
 & 1 & & 6 & & 15 & & 20 & & 15 & & 6 & & 1 & \\
1 & & 7 & & 21 & & 35 & & 35 & & 21 & & 7 & & 1
\end{array}
$$

The first and last number in each row is 1, and every other number in the triangle is formed by adding the two numbers immediately above that number. For example, the two numbers above 35 are 15 and 20:

$$
\begin{array}{cc}
15 & 20 \\
 \searrow \swarrow & \\
35 &
\end{array}
\qquad\qquad 15 + 20 = 35
$$

Since the top row corresponds to the binomial expansion $(x + y)^0 = 1$, we call it the **zero row.** Similarly, the next row corresponds to the binomial expansion $(x + y)^1 = 1(x) + 1(y)$, and we call it the **first row.** In general, the ***n*th row** in Pascal's Triangle gives us the coefficients of $(x + y)^n$.

EXAMPLE 3
Using Pascal's Triangle in a Binomial Expansion

Use Pascal's Triangle to expand the following:

$$\left(x - \frac{1}{2}\right)^6$$

Solution:

$$
\left(x - \frac{1}{2}\right)^6 = x^6 - 6x^5\left(\frac{1}{2}\right) + 15x^4\left(\frac{1}{2}\right)^2 - 20x^3\left(\frac{1}{2}\right)^3
$$

$$
+ 15x^2\left(\frac{1}{2}\right)^4 - 6x\left(\frac{1}{2}\right)^5 + \left(\frac{1}{2}\right)^6
$$

$$
= x^6 - \frac{6x^5}{2} + \frac{15x^4}{4} - \frac{20x^3}{8} + \frac{15x^2}{16} - \frac{6x}{32} + \frac{1}{64}
$$

$$
= x^6 - 3x^5 + \frac{15x^4}{4} - \frac{5x^3}{2} + \frac{15x^2}{16} - \frac{3x}{16} + \frac{1}{64}
$$

Remark: Remember that when expanding a binomial *difference* you must alternate signs, as shown in Example 3.

EXAMPLE 4
Finding a Specified Term in a Binomial Expansion

Find the sixth term in the expansion of $(3a + 2b)^{12}$.

Solution:

Using the Binomial Theorem, we let $x = 3a$ and $y = 2b$ and note that in the *sixth* term the exponent of y is $r = 5$ and the exponent of x is $12 - 5 = 7$. Consequently, the sixth term of the expansion is

$$C_5^{12} x^7 y^5 = \frac{12 \cdot 11 \cdot 10 \cdot 9 \cdot 8}{5!} (3a)^7 (2b)^5$$

Section Exercises 10.5

In Exercises 1–10, evaluate C_r^n.

1. C_3^5
2. C_6^8
3. C_0^{12}
4. C_{20}^{20}
5. C_{15}^{20}
6. C_5^{12}
7. C_{98}^{100}
8. C_4^{10}
9. C_2^{100}
10. C_6^{10}

In Exercises 11–24, use the Binomial Theorem to expand the given binomial. Simplify your answer.

11. $(x + y)^5$
12. $(x + y)^6$
13. $(a + 2)^4$
14. $(s + 3)^5$
15. $(r + 3s)^6$
16. $(x + 2y)^4$
17. $(x - y)^5$
18. $(2x - y)^5$
19. $(1 - 2x)^3$
20. $\left(\dfrac{x}{2} - 3y\right)^3$
21. $(x^2 + 5)^4$
22. $(x^2 + y^2)^6$
23. $\left(\dfrac{1}{x} + y\right)^5$
24. $\left(\dfrac{1}{x} + 2y\right)^6$

In Exercises 25–30, use the Binomial Theorem to expand the given complex number. Simplify your answer by using the fact that $i^2 = -1$.

25. $(1 + i)^4$
26. $(2 - i)^5$
27. $(2 - 3i)^6$
28. $(5 + \sqrt{-9})^3$
29. $\left(\dfrac{-1}{2} + \dfrac{\sqrt{3}}{2} i\right)^3$
30. $(5 - \sqrt{3} i)^4$

31. Find the term involving x^5 in the expansion of $(x + 3)^{12}$.
32. Find the term involving x^8 in the expansion of $(x^2 + 3)^{12}$.

33. Find the term involving x^8 in the expansion of $(x - 2y)^{10}$.
34. What is the coefficient of $x^2 y^8$ in the expansion of $(4x - y)^{10}$?
35. What is the coefficient of $x^4 y^{11}$ in the expansion of $(3x - 2y)^{15}$?
36. What is the middle term in the expansion of $(x^2 - 5)^8$?
37. What is the middle term in the expansion of $(\sqrt{x} + \sqrt{y})^{12}$?
38. What is the middle term in the expansion of $\left(\dfrac{2}{x} - \dfrac{3}{y}\right)^{10}$?
39. Use Pascal's Triangle to expand $(2t - s)^5$.
40. Use Pascal's Triangle to expand $\left(\dfrac{x}{2} + 2y\right)^6$.

In the study of probability, it is sometimes necessary to use the expansion of $(p + q)^n$, where $p + q = 1$. In Exercises 41–46, use the Binomial Theorem to expand the given expression.

41. $(\frac{1}{2} + \frac{1}{2})^7$
42. $(\frac{1}{4} + \frac{3}{4})^{10}$
43. $(\frac{1}{3} + \frac{2}{3})^8$
44. $(0.3 + 0.7)^{12}$
45. $(0.6 + 0.4)^5$
46. $(0.35 + 0.65)^6$

In Exercises 47–50, prove the given property for all integers r and n, $1 \le r \le n$.

47. $C_r^n = C_{n-r}^n$
48. $C_0^n - C_1^n + C_2^n - \cdots \pm C_n^n = 0$
49. $C_r^{n+1} = C_r^n + C_{r-1}^n$
50. $C_n^{2n} = (C_0^n)^2 + (C_1^n)^2 + (C_2^n)^2 + \cdots + (C_n^n)^2$
51. Prove that the sum of the numbers in the nth row of Pascal's Triangle is 2^n. [Hint: Consider $2^n = (1 + 1)^n$.]

Review Exercises / Chapter 10

In Exercises 1–14, find the sum.

1. $\displaystyle\sum_{i=1}^{6} 5$
2. $\displaystyle\sum_{k=2}^{5} 4k$
5. $\displaystyle\sum_{i=0}^{6} 2^i$
6. $\displaystyle\sum_{i=0}^{4} 3^i$
3. $\displaystyle\sum_{j=3}^{10} (2j - 3)$
4. $\displaystyle\sum_{j=1}^{8} (20 - 3j)$
7. $\displaystyle\sum_{i=0}^{\infty} (\tfrac{7}{8})^i$
8. $\displaystyle\sum_{i=0}^{\infty} (\tfrac{1}{3})^i$

9. $\displaystyle\sum_{k=0}^{\infty} 4(\tfrac{2}{3})^k$

10. $\displaystyle\sum_{k=0}^{\infty} 1.3(\tfrac{1}{10})^k$

11. $\displaystyle\sum_{k=1}^{11} (\tfrac{2}{3}k + 4)$

12. $\displaystyle\sum_{k=1}^{25} \left(\frac{3k+1}{4}\right)$

13. $\displaystyle\sum_{n=0}^{10} (n^2 + 3)$

14. $\displaystyle\sum_{n=1}^{100} \left(\frac{1}{n} - \frac{1}{n+1}\right)$

In Exercises 15–18, use mathematical induction to prove the given formula for every positive integer n.

15. $1 + 4 + 7 + \cdots + (3n - 2) = \dfrac{n}{2}(3n - 1)$

16. $1 + \dfrac{3}{2} + 2 + \dfrac{5}{2} + \cdots + \dfrac{1}{2}(n + 1) = \dfrac{n}{4}(n + 3)$

17. $\displaystyle\sum_{j=0}^{n-1} ar^j = \frac{a(1 - r^n)}{1 - r}$

18. $\displaystyle\sum_{k=0}^{n-1} [a + kd] = \frac{n}{2}[2a + (n - 1)d]$

In Exercises 19–24, write the repeating decimal as a rational number by considering each to be the sum of an infinite geometric series.

19. 0.454545. . .

20. 0.151515. . .

21. 1.0666. . .

22. 0.0222. . .

23. 0.01333. . .

24. 2.7333. . .

In Exercises 25–30, use the Binomial Theorem to expand the binomial. Simplify your answer.

25. $\left(\dfrac{x}{2} + y\right)^4$

26. $(a - 3b)^5$

27. $\left(\dfrac{2}{x} + 3x\right)^6$

28. $(3x - y^2)^7$

29. $(5 + 2i)^4$

30. $(4 - 5i)^3$

CHAPTER
11 | LINES IN THE PLANE AND CONICS

Lines in the Plane

11.1

Parallel and Perpendicular Lines

The slope of a line (Section 2.3) is a convenient tool for determining when two lines are parallel or perpendicular. This is seen in the following two rules.

PARALLEL LINES

> Two distinct nonvertical lines are parallel if and only if their slopes are equal.

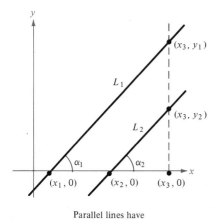

Parallel lines have
equal slopes.

FIGURE 11.1

Proof: Recall that the phrase "if and only if" is a way of stating two rules in one. One rule says, "If two nonvertical lines are parallel, then they must have equal slopes." The other rule is the converse, which says, "If two distinct lines have equal slopes, they must be parallel." We will prove the first of these two rules and leave the converse for you to prove.

Assume that we have two parallel lines L_1 and L_2 with slopes m_1 and m_2. If these lines are both horizontal, then $m_1 = m_2 = 0$, and the rule is established. If L_1 and L_2 are not horizontal, then they must intersect the x-axis at points $(x_1, 0)$ and $(x_2, 0)$, as shown in Figure 11.1. Since L_1 and L_2 are parallel, their intersection with the x-axis must produce equal angles α_1 and α_2. (α is the Greek lowercase letter alpha.) Therefore, the two right triangles with vertices

$$(x_1, 0), (x_3, 0), (x_3, y_1) \qquad \text{and} \qquad (x_2, 0) \ (x_3, 0), (x_3, y_2)$$

461

must be similar. From this we conclude that the ratios of their corresponding sides must be equal, and thus

$$m_1 = \frac{y_1 - 0}{x_3 - x_1} = \frac{y_2 - 0}{x_3 - x_2} = m_2$$

Hence the lines L_1 and L_2 must have equal slopes.

PERPENDICULAR LINES

> Two nonvertical lines are perpendicular if and only if their slopes are related by the equation
>
> $$m_1 = -\frac{1}{m_2}$$

Proof: As in the previous rule, we will prove only one direction of the rule and leave the other for you to prove. Let us assume that we are given two nonvertical perpendicular lines L_1 and L_2 with slopes m_1 and m_2. For simplicity's sake let these two lines intersect at the origin, as shown in Figure 11.2. The vertical line $x = 1$ will intersect L_1 and L_2 at the respective points $(1, m_1)$ and $(1, m_2)$. Since the triangle formed by these two points and the origin is a right triangle, we can apply the Pythagorean Theorem and conclude that

$$\left(\begin{matrix}\text{distance between}\\(0, 0) \text{ and } (1, m_1)\end{matrix}\right)^2 + \left(\begin{matrix}\text{distance between}\\(0, 0) \text{ and } (1, m_2)\end{matrix}\right)^2 = \left(\begin{matrix}\text{distance between}\\(1, m_1) \text{ and } (1, m_2)\end{matrix}\right)^2$$

Using the Distance Formula, we have

$$(\sqrt{1 + m_1^2})^2 + (\sqrt{1 + m_2^2})^2 = (\sqrt{0^2 + (m_1 - m_2)^2})^2$$
$$1 + m_1^2 + 1 + m_2^2 = (m_1 - m_2)^2$$
$$2 + m_1^2 + m_2^2 = m_1^2 - 2m_1m_2 + m_2^2$$
$$2 = -2m_1m_2$$
$$-1 = m_1m_2$$

or

$$-\frac{1}{m_2} = m_1$$

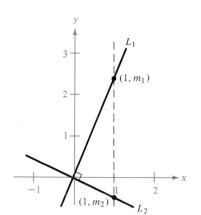

The slopes of perpendicular lines are negative reciprocals of each other.

FIGURE 11.2

EXAMPLE 1
Finding Parallel and Perpendicular Lines

Find the equation of the line that passes through the point $(2, -1)$ and is

(a) parallel to the line $2x - 3y = 5$
(b) perpendicular to the line $2x - 3y = 5$

Solution:
Writing the equation $2x - 3y = 5$ in slope-intercept form, we have

$$3y = 2x - 5$$

or

$$y = \frac{2}{3}x - \frac{5}{3}$$

Therefore, the given line has a slope of $m = \frac{2}{3}$.

(a) Any line parallel to the given line must also have a slope of $\frac{2}{3}$. Thus, the line through $(2, -1)$ that is parallel to the given line has an equation of the form

$$y - (-1) = \frac{2}{3}(x - 2)$$
$$3(y + 1) = 2(x - 2)$$
$$3y + 3 = 2x - 4$$
$$-2x + 3y = -7$$

or

$$2x - 3y = 7$$

(Note the similarity to the original equation $2x - 3y = 5$.)

(b) Any line perpendicular to the given line must have a slope of $-\frac{3}{2}$. Therefore, the line through $(2, -1)$ that is perpendicular to the given line has the equation

$$y - (-1) = -\frac{3}{2}(x - 2)$$
$$2(y + 1) = -3(x - 2)$$
$$3x + 2y = 4$$

See Figure 11.3.

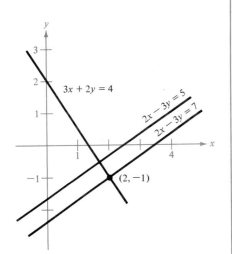

FIGURE 11.3

EXAMPLE 2
Verifying a Right Triangle

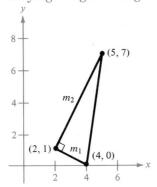

FIGURE 11.4

Show that the points $(2, 1)$, $(4, 0)$, and $(5, 7)$ are the vertices of a right triangle.

Solution:
From Figure 11.4 we find the slope of the line connecting $(4, 0)$ and $(2, 1)$ to be

$$m_1 = \frac{1 - 0}{2 - 4} = -\frac{1}{2}$$

Furthermore, the slope of the line connecting $(2, 1)$ and $(5, 7)$ is

$$m_2 = \frac{7 - 1}{5 - 2} = \frac{6}{3} = 2$$

Since m_1 and m_2 are negative reciprocals, we know that two sides of the triangle are perpendicular and hence the triangle is a right triangle.

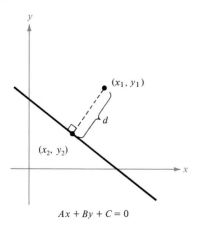

$Ax + By + C = 0$

FIGURE 11.5

Finding the distance between a line and a point not on the line is another application of perpendicular lines. We define this distance to be the length of the perpendicular line segment joining the point to the given line, as shown in Figure 11.5.

DISTANCE BETWEEN A POINT AND A LINE

The distance between the point (x_1, y_1) and the line given by $Ax + By + C = 0$ is

$$d = \frac{|Ax_1 + By_1 + C|}{\sqrt{A^2 + B^2}}$$

Proof:

For simplicity's sake, we assume that the given line is neither horizontal nor vertical. Writing the equation $Ax + By + C = 0$ in the slope-intercept form

$$y = -\frac{A}{B}x - \frac{C}{B}$$

we see that the line has a slope of

$$m = -\frac{A}{B}$$

Thus, the slope of the line passing through (x_1, y_1) and perpendicular to the given line is B/A, and its equation is

$$y - y_1 = \frac{B}{A}(x - x_1)$$

These two lines intersect at

$$x_2 = \frac{B(Bx_1 - Ay_1) - AC}{A^2 + B^2}, \qquad y_2 = \frac{A(-Bx_1 + Ay_1) - BC}{A^2 + B^2}$$

Finally, the distance between (x_1, y_1) and (x_2, y_2) is

$$d = \sqrt{(x_2 - x_1)^2 + (y_2 - y_1)^2}$$

$$= \sqrt{\left(\frac{B^2x_1 - ABy_1 - AC}{A^2 + B^2} - x_1\right)^2 + \left(\frac{-ABx_1 + A^2y_1 - BC}{A^2 + B^2} - y_1\right)^2}$$

$$= \sqrt{\frac{A^2(Ax_1 + By_1 + C)^2 + B^2(Ax_1 + By_1 + C)^2}{(A^2 + B^2)^2}}$$

$$= \frac{|Ax_1 + By_1 + C|}{\sqrt{A^2 + B^2}}$$

EXAMPLE 3
Finding the Distance
Between a Point and a Line

Find the distance between the point (4, 1) and the line given by $y = 2x + 1$.

Solution:
The general form of the given equation is

$$-2x + y - 1 = 0$$

Hence the required distance is

$$d = \frac{|-2(4) + 1(1) - 1|}{\sqrt{(-2)^2 + (1)^2}} = \frac{8}{\sqrt{5}} \approx 3.58$$

Every nonhorizontal line must intersect the *x*-axis. The angle formed by such an intersection determines the **inclination** of the line, as specified in the following definition.

DEFINITION OF INCLINATION

> The **inclination** of a nonhorizontal line is defined to be the angle θ between the line and the *x*-axis, where θ is measured counterclockwise from the *x*-axis to the line. (See Figure 11.6.)
>
> A horizontal line has an inclination of zero.

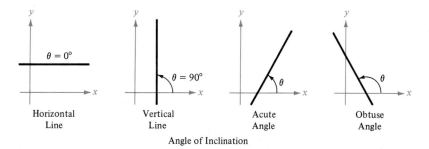

| Horizontal Line | Vertical Line | Acute Angle | Obtuse Angle |

Angle of Inclination

FIGURE 11.6

The inclination of a line is related to its slope in the following manner.

INCLINATION AND SLOPE

> If a line has inclination θ and slope m, then
>
> $$m = \tan \theta$$

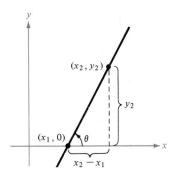

FIGURE 11.7

Proof:
If $m = 0$, then the line is *horizontal* and $\theta = 0$. Thus, the result is true since $0 = \tan 0$.

If the line has a *positive* slope, then it will intersect the *x*-axis. We label this point $(x_1, 0)$ as shown in Figure 11.7. If (x_2, y_2) is a second point on the line, then the slope is given by

$$m = \frac{y_2 - 0}{x_2 - x_1} = \frac{y_2}{x_2 - x_1} = \tan \theta$$

We leave the case in which the line has a *negative* slope for you to prove.

EXAMPLE 4
Finding the Inclination of a Line

Find the inclination of the line given by $2x + 3y = 6$.

Solution:
The slope of this line is $m = -\frac{2}{3}$. Thus, its inclination is determined from the equation

$$\tan \theta = -\frac{2}{3}$$

From Figure 11.8, we see that $90° < \theta < 180°$. This means that

$$\theta = 180° + \arctan\left(-\frac{2}{3}\right) \approx 180° - 33.69° \approx 146.31°$$

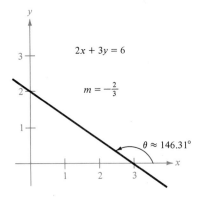

FIGURE 11.8

As a final topic in this section, we will look at the use of inclination to find the angle between two lines in the plane.

DEFINITION OF THE ANGLE BETWEEN TWO LINES

If two lines have inclinations θ_1 and θ_2, where $\theta_1 \leq \theta_2$, then the angle between the two lines is

$$\theta = \theta_2 - \theta_1$$

(See Figure 11.9.)

Remark: Note that this definition implies that the angle between two parallel lines is zero.

We can use the formula for the tangent of the difference of two angles to give us the following convenient formula for the angle between two lines.

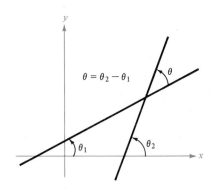

FIGURE 11.9

ANGLE BETWEEN TWO LINES

If two nonperpendicular lines have inclinations θ_1 and θ_2 and slopes m_1 and m_2, then the angle θ between the two lines is given by

$$\tan \theta = \frac{\tan \theta_2 - \tan \theta_1}{1 + \tan \theta_1 \tan \theta_2} = \frac{m_2 - m_1}{1 + m_1 m_2}$$

Proof:

This formula follows from the formula for the tangent of the difference of two angles.

$$\tan \theta = \tan (\theta_2 - \theta_1) = \frac{\tan \theta_2 - \tan \theta_1}{1 + \tan \theta_1 \tan \theta_2}$$

If $\theta_1 \le \theta_2$, then $\tan \theta_1 \le \tan \theta_2$ and the formula for θ matches the difference formula. To cover the case in which $\theta_2 \le \theta_1$, we add absolute value signs to obtain the general formula

$$\tan \theta = \frac{|\tan \theta_2 - \tan \theta_1|}{1 + \tan \theta_1 \tan \theta_2}$$

EXAMPLE 5
Finding the Angle Between Two Lines

Find the angle between the following two lines:

$$2x - y - 4 = 0 \qquad y = 2x + 4 \qquad Line \ 1$$
$$3x + 4y - 12 = 0 \qquad 4y = \frac{-3x + 12}{4} \qquad Line \ 2$$

Solution:

For Line 1, the slope is

$$m_1 = 2$$

For Line 2, the slope is

$$m_2 = -\tfrac{3}{4}$$

Thus, the angle between the two lines is given by

$$\tan \theta = \frac{m_2 - m_1}{1 + m_1 m_2} = \frac{(-\frac{3}{4}) - 2}{1 + (-\frac{3}{4})(2)}$$

$$= \frac{-\frac{11}{4}}{-\frac{2}{4}} = \frac{11}{2}$$

Finally, we conclude that the angle is

$$\theta = \arctan \left(\tfrac{11}{2}\right) \approx 79.70°$$

See Figure 11.10.

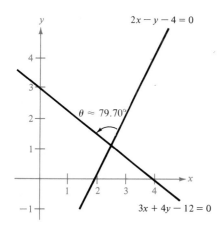

FIGURE 11.10

Section Exercises 11.1

In Exercises 1–6, write an equation of the line through the given point (a) parallel to the given line, (b) perpendicular to the given line.

1. $(2, 1)$, $4x - 2y = 3$
2. $(-3, 2)$, $x + y = 7$
3. $(\tfrac{7}{8}, \tfrac{3}{4})$, $5x + 3y = 0$
4. $(-6, 4)$, $3x + 4y = 7$
5. $(2, 5)$, $x = 4$
6. $(-1, 0)$, $y = -3$

In Exercises 7–12, determine if the three points are vertices of a right triangle.

7. $(4, 0)$, $(2, 1)$, $(-1, -5)$
8. $(-1, 1)$, $(5, -3)$, $(8, \tfrac{3}{2})$
9. $(1, -3)$, $(3, 2)$, $(-2, 4)$
10. $(1, 2)$, $(2, 1)$, $(3, 3)$
11. $(-2, -2)$, $(5, 1)$, $(3, \tfrac{17}{3})$
12. $(3, 7)$, $(4, 4)$, $(\tfrac{1}{3}, -\tfrac{2}{3})$

In Exercises 13–16, determine if the three given points are collinear (lie on the same straight line).

13. $(0, -4)$, $(2, 0)$, $(3, 2)$
14. $(0, 4)$, $(7, -6)$, $(-5, 11)$
15. $(-2, 1)$, $(-1, 0)$, $(2, -2)$
16. $(1, 1)$, $(2, \frac{7}{2})$, $(-1, -4)$

In Exercises 17–24, find the distance from the point to the line.

17. $(0, 0)$, $4x + 3y = 10$ 18. $(2, 3)$, $4x + 3y = 10$
19. $(-2, 1)$, $x - y - 2 = 0$ 20. $(6, 2)$, $x + 1 = 0$
21. $(\frac{1}{2}, \frac{2}{3})$, $y - 2 = 0$ 22. $(\frac{1}{2}, 4)$, $2x - 5y = -4$
23. $(\frac{3}{2}, \frac{1}{3})$, $8x + 9y = 15$ 24. $(-6, 4)$, $3x + 4y = 1$
25. Find the distance between the parallel lines $x + y = 1$ and $x + y = 5$.
26. Find the distance between the parallel lines $3x - 4y = 1$ and $3x - 4y = 10$.

In Exercises 27–34, find the inclination of the line.

27. $x - \sqrt{3}y = 0$ 28. $\sqrt{3}x + y = 3$
29. $x - y = 4$ 30. $x - 3y = 11$
31. $5x + 3y = 10$ 32. $-\frac{1}{3}x + \frac{5}{6}y = 1$

33. $0.02x + 0.15y = 0.25$ 34. $4x - 2y = 3$

In Exercises 35–40, find the angle between the lines.

35. $y = x + 2$ and $y = 3$
36. $2x - 3y = 1$ and $x + 5y = 2$
37. $4x + 3y + 2 = 0$ and $3x + 4y - 7 = 0$
38. $2x - y + 7 = 0$ and $x + y + 2 = 0$
39. $2x + 3y = 9$ and $\frac{4}{3}x + 2y = 4$
40. $5x - 6y + 12 = 0$ and $6x + 5y - 16 = 0$
41. Find an equation of the line through the point $(0, 6)$ with inclination 60°.
42. Find an equation of the line through the point $(-3, 4)$ with inclination 150°.
43. Write an equation of the line that bisects the angle between $2x - y + 7 = 0$ and $x + y + 2 = 0$.
44. Find the distance from the origin to the line $4x + 3y = 10$ by first finding the point of intersection of the given line and the line through the origin perpendicular to the given line. Then find the distance between the origin and the point of intersection. Compare the result with that of Exercise 17.

Circles and Introduction to Conics

11.2

Conic sections have a rich historical background, going back to early Greek mathematics. Initially, interest in conics centered around construction problems. With the advent of scientific discovery in the seventeenth century, the broad applicability of conics became apparent, and they played a prominent role in the early history of calculus.

The name **conic section,** or simply "conic," refers to the description of a conic as the intersection of a double-napped cone and a plane. Notice from Figure 11.11 that in the formation of the four basic conics, the intersecting plane does not pass through the vertex of the cone. When the plane does pass through the vertex, we call the resulting figure a **degenerate conic,** as shown in Figure 11.12.

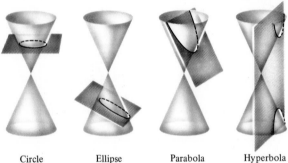

Circle   Ellipse   Parabola   Hyperbola

Conic Sections

FIGURE 11.11

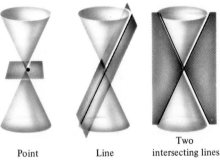

Point   Line   Two intersecting lines

Degenerate Conics

FIGURE 11.12

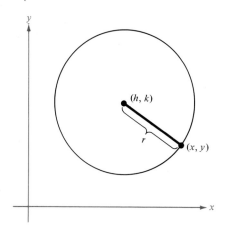

There are several ways to approach a study of the conics. We could begin by defining the conics in terms of the intersections of planes and cones, as the Greeks did. Or we could define them algebraically in terms of the general second-degree equation

$$Ax^2 + Bxy + Cy^2 + Dx + Ey + F = 0$$

a procedure to be discussed in Section 11.6. However, a third approach, in which each of the conics is defined as a "locus" (collection) of points satisfying a certain geometric property, suits our needs best. For example, in Section 1.6 we saw how the definition of a circle as "the collection of all points (x, y) that are equidistant from a fixed point (h, k)" led easily to the standard equation of a circle. (See Figure 11.13.)

FIGURE 11.13

STANDARD EQUATION OF A CIRCLE

The point (x, y) lies on the circle of radius r and center (h, k) if and only if

$$(x - h)^2 + (y - k)^2 = r^2$$

Remark: Recall that the equation of a circle with center at the origin is simply

$$x^2 + y^2 = r^2$$

EXAMPLE 1
Finding an Equation for a Circle

The point $(3, 4)$ lies on a circle whose center is at $(-1, 2)$ (Figure 11.14). Find an equation for the circle.

Solution:
The radius of the circle is the distance between $(-1, 2)$ and $(3, 4)$. Thus

$$r = \sqrt{[3 - (-1)]^2 + (4 - 2)^2} = \sqrt{16 + 4} = \sqrt{20}$$

Therefore, the standard equation for this circle is

$$[x - (-1)]^2 + (y - 2)^2 = (\sqrt{20})^2$$
$$(x + 1)^2 + (y - 2)^2 = 20$$

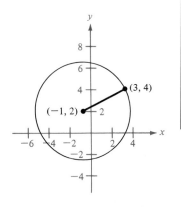

FIGURE 11.14

If we remove parentheses in the standard equation of Example 1, we obtain

$$(x + 1)^2 + (y - 2)^2 = 20$$
$$x^2 + 2x + 1 + y^2 - 4y + 4 = 20$$
$$x^2 + y^2 + 2x - 4y - 15 = 0$$

where the latter equation is in the **general form of the equation of a circle:**

$$Ax^2 + Ay^2 + Dx + Ey + F = 0, \qquad A \neq 0$$

The general form of the equation of a circle is less useful than the corresponding standard form. For instance, we know little about the circle with general equation

$$x^2 + y^2 - 6x + 10y + 24 = 0$$

However, from the corresponding standard form of this equation,

$$(x - 3)^2 + (y + 5)^2 = 10$$

we can readily see that the circle is centered at $(3, -5)$ and that its radius is $\sqrt{10}$. This observation suggests that to graph the equation of a circle, it is best to write the equation in standard form. This can be accomplished by using the algebraic process called **completing the square,** which we demonstrate in the following example.

EXAMPLE 2
Completing the Square

Sketch the graph of the circle whose general equation is

$$4x^2 + 4y^2 + 20x - 16y + 37 = 0$$

Solution:
To complete the square, we will first divide by 4 so that the coefficients of x^2 and y^2 are both 1. Thus we have

$$x^2 + y^2 + 5x - 4y + \tfrac{37}{4} = 0$$

Then we write

$$(x^2 + 5x + \quad) + (y^2 - 4y + \quad) = -\tfrac{37}{4}$$

reserving space to add the square of half the coefficient of x and the square of half the coefficient of y to both sides of the equation. Thus we obtain

$$(x^2 + 5x + \tfrac{25}{4}) + (y^2 - 4y + 4) = -\tfrac{37}{4} + \tfrac{25}{4} + 4$$
$$(x + \tfrac{5}{2})^2 + (y - 2)^2 = 1$$

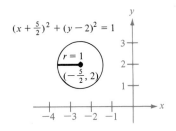

$(x + \tfrac{5}{2})^2 + (y - 2)^2 = 1$

$r = 1$

$(-\tfrac{5}{2}, 2)$

FIGURE 11.15

Therefore, the circle is centered at $(-\tfrac{5}{2}, 2)$ and its radius is 1 (Figure 11.15).

EXAMPLE 3
Writing an Equation in Standard Form

Discuss the graph of

$$3x^2 + 3y^2 + 24x - 6y + 51 = 0$$

Solution:
First, dividing by 3 we obtain

$$x^2 + y^2 + 8x - 2y + 17 = 0$$

Then we write

$$(x^2 + 8x + \quad) + (y^2 - 2y + \quad) = -17$$
$$(x^2 + 8x + 16) + (y^2 - 2y + 1) = -17 + 16 + 1$$
$$(x + 4)^2 + (y - 1)^2 = 0$$

The sum on the left side of the last equation can be zero only if both terms are zero. This is true only when $x = -4$ and $y = 1$. Therefore, the graph consists of the single point $(-4, 1)$.

Remark: Example 3 shows that the general equation $Ax^2 + Ay^2 + Dx + Ey + F = 0$ may not always represent a circle. In fact, such an equation may have no solution at all if the procedure of completing the square yields the impossible result

$$(x - h)^2 + (y - k)^2 = \text{(negative number)}$$

EXAMPLE 4
**Finding the Tangent
Line to a Circle**

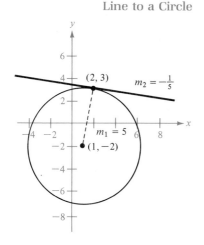

FIGURE 11.16

Find the equation of the tangent line to the circle

$$(x - 1)^2 + (y + 2)^2 = 26$$

at the point $(2, 3)$.

Solution:
From Figure 11.16, we see that the tangent line to the circle at $(2, 3)$ is perpendicular to the radial line passing through the center and the point $(2, 3)$. Since the slope of the radial line is

$$m_1 = \frac{3 - (-2)}{2 - 1} = 5$$

the slope of the tangent line is given by the negative reciprocal

$$m_2 = -\tfrac{1}{5}$$

Thus, the equation of the tangent line is

$$y - 3 = -\tfrac{1}{5}(x - 2)$$
$$x + 5y = 17$$

Section Exercises 11.2 ─────────────────────────

In Exercises 1–6, write the general equation of the specified circle.

1. Center $(0, 0)$; radius $= 3$
2. Center $(0, 0)$; radius $= 5$
3. Center $(2, -1)$; radius $= 4$
4. Center $(-4, 3)$; radius $= \tfrac{5}{8}$
5. Center $(-1, 2)$; point on circle $(0, 0)$
6. Center $(3, -2)$; point on circle $(-1, 1)$

In Exercises 7–14, write the given equation (of a circle) in standard form and sketch its graph.

7. $x^2 + y^2 - 2x + 6y + 6 = 0$

8. $x^2 + y^2 - 2x + 6y - 15 = 0$
9. $x^2 + y^2 - 2x + 6y + 10 = 0$
10. $3x^2 + 3y^2 - 6y - 1 = 0$
11. $2x^2 + 2y^2 - 2x - 2y - 3 = 0$
12. $4x^2 + 4y^2 - 4x + 2y - 1 = 0$
13. $16x^2 + 16y^2 + 16x + 40y - 7 = 0$
14. $x^2 + y^2 - 4x + 2y + 3 = 0$

In Exercises 15–20, the center of a circle and a tangent line to the circle are given. Write the general equation of the circle.

15. $(2, 3)$, $x = 5$
16. $(2, 3)$, $y = 6$
17. $(0, 0)$, $x - y = 10$
18. $(-1, -2)$, $x - y = 10$
19. $(4, -1)$, $x - 4y = 2$
20. $(-1, 5)$, $2x + 3y = 0$

In Exercises 21–26, find an equation of the tangent line to the circle at the given point.

21. $x^2 + y^2 + 24x - 10y = 0$, $(0, 0)$
22. $x^2 + y^2 - 8x - 2y + 12 = 0$, $(6, 2)$
23. $x^2 + y^2 - 7x + 5y - 14 = 0$, $(2, 3)$
24. $x^2 + y^2 - 4x - 2y - 3 = 0$, $(4, -1)$
25. $x^2 + y^2 + 2x + 6y - 40 = 0$, $(-2, 4)$
26. $3x^2 + 3y^2 - 14x - 67 = 0$, $(4, 5)$

In Exercises 27–30, sketch the set of all points for which the given inequality is valid.

27. $x^2 + y^2 - 4x + 2y + 1 \leq 0$
28. $x^2 + y^2 - 4x + 2y + 1 > 0$
29. $(x + 3)^2 + (y - 1)^2 > 9$
30. $(x - 1)^2 + (y - \frac{1}{2})^2 \leq 1$

31. Determine the x-intercepts of the circle $x^2 + y^2 + x + 2y - 6 = 0$.
32. Determine the y-intercepts of the circle $x^2 + y^2 - 4x + 2y - 8 = 0$.
33. Find the points of intersection of $x^2 + y^2 = 5$ and $x - y = 1$.
34. Find the points of intersection of $x^2 + y^2 = 25$ and $2x + y = 10$.
35. Find an equation of the curve such that for each point (x, y) on the curve the distance of the point from $(0, 4)$ is twice its distance from $(2, 0)$.
36. Find an equation of the curve such that for each point (x, y) on the curve the line segment joining (x, y) and the point $(-5, 0)$ is perpendicular to the line segment joining (x, y) and $(5, 0)$.

Parabolas 11.3

DEFINITION OF A PARABOLA

> A **parabola** is the set of all points (x, y) that are equidistant from a fixed line (**directrix**) and a fixed point (**focus**) not on the line. (See Figure 11.17.)

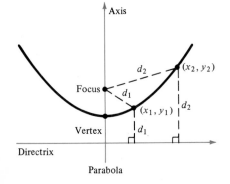

Axis

d_2 (x_2, y_2)

Focus d_1

(x_1, y_1) d_2

Vertex d_1

Directrix

Parabola

FIGURE 11.17

The midpoint between the focus and the directrix is called the **vertex**, and the line passing through the focus and the vertex is called the **axis** of the parabola. Note that a parabola is symmetric with respect to its axis.

Using this definition of a parabola, we derive the following result, which gives the standard form of the equation of a parabola whose directrix is parallel to the x-axis or to the y-axis.

STANDARD EQUATION OF A PARABOLA

> The standard form of the equation of a parabola with vertex at (h, k) and directrix $y = k - p$ is
>
> $$(x - h)^2 = 4p(y - k) \qquad \textit{(Vertical axis)}$$
>
> For directrix $x = h - p$ the equation is
>
> $$(y - k)^2 = 4p(x - h) \qquad \textit{(Horizontal axis)}$$
>
> The focus lies on the axis p units (*directed distance*) from the vertex.

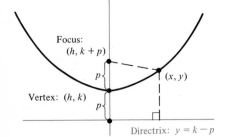

Axis

Focus: $(h, k + p)$

p (x, y)

Vertex: (h, k) p

Directrix: $y = k - p$

FIGURE 11.18

Proof:
Since the two cases are similar, we give a proof for the first case only. Suppose the directrix ($y = k - p$) is parallel to the x-axis. In Figure 11.18

we assume $p > 0$, and since p is the directed distance from the vertex to the focus, it follows that the focus lies *above* the vertex. Since by definition the point (x, y) is equidistant from $(h, k + p)$ and $y = k - p$, we have the equation

$$\sqrt{(x - h)^2 + [y - (k + p)]^2} = y - (k - p)$$

Squaring both sides of this equation yields

$$(x - h)^2 + [y - (k + p)]^2 = [y - (k - p)]^2$$
$$(x - h)^2 + y^2 - 2y(k + p) + (k + p)^2 = y^2 - 2y(k - p) + (k - p)^2$$
$$(x - h)^2 - 2py + 2pk = 2py - 2pk$$
$$(x - h)^2 = 4p(y - k)$$

EXAMPLE 1
Finding the Standard Equation of a Parabola

Find the standard form of the equation of the parabola with vertex $(2, 1)$ and focus $(2, 4)$. (See Figure 11.19.)

Solution:
Since the axis of the parabola is vertical, we consider the equation

$$(x - h)^2 = 4p(y - k)$$

where $h = 2$, $k = 1$, and $p = 3$. Thus the standard equation is

$$(x - 2)^2 = 12(y - 1)$$

By expanding this equation we come up with the more common quadratic form

$$y = (\tfrac{1}{12})(x^2 - 4x + 16)$$

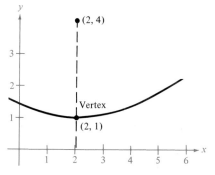

FIGURE 11.19

EXAMPLE 2
Finding the Focus of a Parabola

Find the focus of the parabola given by $y = \tfrac{1}{2}(1 - 2x - x^2)$.

Solution:
To find the focus, we convert the equation to standard form by completing the square. Thus

$$y = \tfrac{1}{2}(1 - 2x - x^2)$$
$$2y = 1 - (x^2 + 2x + \quad)$$
$$2y = 2 - (x^2 + 2x + 1)$$
$$(x + 1)^2 = -2(y - 1)$$

Comparing this equation to the standard form $(x - h)^2 = 4p(y - k)$, we conclude that

$$h = -1, \quad k = 1, \quad p = -\tfrac{1}{2}$$

Since p is negative, the parabola opens downward (see Figure 11.20), and thus the focus of the parabola is

$$(h, k + p) = (-1, \tfrac{1}{2})$$

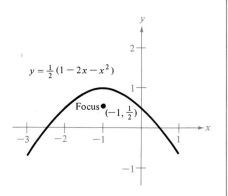

FIGURE 11.20

EXAMPLE 3
Finding the Standard
Equation of a Parabola

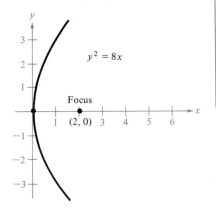

$y^2 = 8x$

Focus

$(2, 0)$ 3   4   5   6

FIGURE 11.21

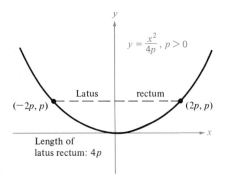

$y = \dfrac{x^2}{4p}$, $p > 0$

Latus   rectum

$(-2p, p)$ $(2p, p)$

Length of
latus rectum: $4p$

FIGURE 11.22

Write the standard equation of the parabola with its vertex at the origin and its focus at (2, 0).

Solution:
The axis of the parabola is horizontal, passing through (0, 0) and (2, 0). (See Figure 11.21.) Thus we consider the form

$$(y - k)^2 = 4p(x - h)$$

where $h = k = 0$ and $p = 2$. Therefore, our equation is

$$y^2 = 8x$$

The **latus rectum** of a parabola is the chord through the focus, perpendicular to the axis. For the parabola $y = x^2/4p$, the latus rectum is the line segment joining the points $(-2p, p)$ and $(2p, p)$. (See Figure 11.22.)

In Section 11.2 we were able to find the tangent line at a given point on a circle. Recall from geometry that a line is tangent to a circle if it intersects the circle only at the point of tangency. We can define the **tangent line** at a point on a parabola in a similar manner. That is, a line is tangent to a parabola at a point if the line does not cross the parabola and if it intersects the parabola only at the point of tangency. (See Figure 11.23.)

Tangent lines to parabolas have the following interesting property. If a line is drawn from the focus to a point (x_1, y_1) on the parabola, then the tangent line at (x_1, y_1) intersects this focal line and the axis of the parabola at equal angles. (See Figure 11.23.) This property of tangent lines is used in the construction of parabolic reflectors. In such a reflector, beams from a light source at the focus are reflected in a direction parallel to the axis of the parabola (see Figure 11.24). Conversely, the parabolic mirror of a reflecting telescope reflects through the focus all beams of light coming in parallel to the axis.

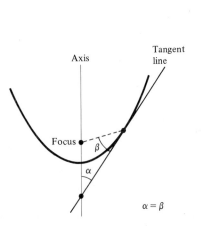

Axis   Tangent
line

Focus
β

α

$\alpha = \beta$

FIGURE 11.23

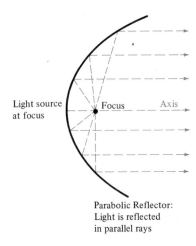

Light source
at focus   Focus   Axis

Parabolic Reflector:
Light is reflected
in parallel rays

FIGURE 11.24

EXAMPLE 4
Finding the Tangent Line
at a Point on a Parabola

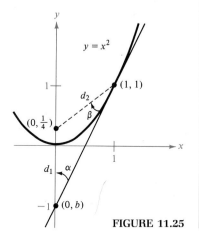

FIGURE 11.25

Find the tangent line to the parabola given by $y = x^2$ at the point $(1, 1)$.

Solution:
The focus of this parabola occurs at the point $(0, \frac{1}{4})$, as shown in Figure 11.25. We can find the y-intercept $(0, b)$ of the tangent line by equating the lengths of the two sides of the isosceles triangle shown in this figure.

$$d_1 = \tfrac{1}{4} - b \quad \text{and} \quad d_2 = \sqrt{(1 - 0)^2 + (1 - \tfrac{1}{4})^2}$$

$$\tfrac{1}{4} - b = \sqrt{1 + \tfrac{9}{16}} = \tfrac{5}{4}$$

$$b = -1$$

Thus, the slope of the tangent line is

$$m = \frac{1 - (-1)}{1 - 0} = 2$$

and the equation of the tangent line in the slope-intercept form is

$$y = 2x - 1$$

Section Exercises 11.3

In Exercises 1–20, find the vertex, focus, and directrix of the parabola.

1. $y = 4x^2$
2. $y = 2x^2$
3. $y^2 = -6x$
4. $y^2 = 3x$
5. $x^2 + 8y = 0$
6. $x + y^2 = 0$
7. $(x - 1)^2 + 8(y + 2) = 0$
8. $(x + 3) + (y - 2)^2 = 0$
9. $(y + \frac{1}{2})^2 = 2(x - 5)$
10. $(x + \frac{1}{2})^2 - 4(y - 3) = 0$
11. $y = \frac{1}{4}(x^2 - 2x + 5)$
12. $y = -\frac{1}{6}(x^2 + 4x - 2)$
13. $4x - y^2 - 2y - 33 = 0$
14. $y^2 + x + y = 0$
15. $y^2 + 6y + 8x + 25 = 0$
16. $x^2 - 2x + 8y + 9 = 0$
17. $y^2 - 4y - 4x = 0$
18. $y^2 - 4x - 4 = 0$
19. $x^2 + 4x + 4y - 4 = 0$
20. $y^2 + 4y + 8x - 12 = 0$

In Exercises 21–32, find an equation of the specified parabola.

21. Vertex: $(0, 0)$; focus: $(0, -\frac{3}{2})$
22. Vertex: $(0, 0)$; focus: $(2, 0)$
23. Vertex: $(3, 2)$; focus: $(1, 2)$
24. Vertex: $(-1, 2)$; focus: $(-1, 0)$
25. Vertex: $(0, 4)$; directrix: $y = 2$
26. Vertex: $(-2, 1)$; directrix: $x = 1$

27. Focus: $(0, 0)$; directrix: $y = 4$
28. Focus: $(2, 2)$; directrix: $x = -2$
29. Axis: parallel to y-axis; passes through points $(0, 3)$, $(3, 4)$, $(4, 11)$
30. Axis: parallel to x-axis; passes through points $(4, -2)$, $(0, 0)$, $(3, -3)$
31.

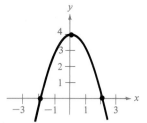

32.

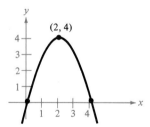

33. Find the equations of the parabolas with a common directrix $y = 1$ and a latus rectum of length 8.
34. Find an equation of the parabola with directrix $y = -2$ and latus rectum joining the points $(0, 2)$ and $(8, 2)$.
35. Each cable of a particular suspension bridge is suspended (in the shape of a parabola) between two towers that are 400 feet apart and 50 feet above the roadway. (See Figure 11.26.) If the cables touch the roadway midway between the towers, find an equation for the parabolic shape of each cable.

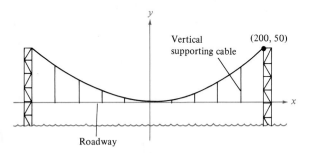

Vertical supporting cable   (200, 50)

Roadway

FIGURE 11.26

36. An earth satellite in a 100-mile-high circular orbit around the earth has a velocity of approximately 17,500 miles per hour. If this velocity is multiplied by $\sqrt{2}$, then the satellite will have the minimum velocity necessary to escape the earth's gravity and it will follow a parabolic path with the center of the earth as the focus. (See Figure 11.27.)
   (a) Find the escape velocity of the satellite.
   (b) Find an equation of its path (assume that the radius of the earth is 4000 miles).

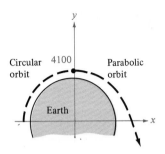

Circular orbit   4100   Parabolic orbit

Earth

FIGURE 11.27

The path of a projectile thrown horizontally with a velocity of v feet per second at a height of s feet is a parabola whose

equation is

$$y = -\frac{16}{v^2} x^2 + s$$

Use this result in Exercise 37–39.

37. Water is flowing from a horizontal pipe 48 feet above the ground at a rate of 10 feet per second. The falling stream of water has the shape of a parabola whose vertex, $(0, 48)$, is at the end of the pipe. (See Figure 11.28.)
   (a) Find the equation of the parabola.
   (b) Where does the water hit the ground?

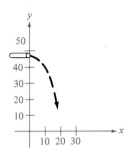

FIGURE 11.28

38. A bomber flying due east at 550 miles per hour at an altitude of 42,000 feet releases a bomb. Neglecting air resistance, determine how far the bomb travels horizontally before striking the ground.
39. A ball is thrown horizontally from the top of a 75-foot tower with a velocity of 32 feet per second.
   (a) Find the equation of the parabolic path.
   (b) How far does the ball travel horizontally before striking the ground?
40. Find the equation of the tangent line to the parabola $y = ax^2$ at $x = x_0$. Prove that the x-intercept of this tangent line is $(x_0/2, 0)$.

In Exercises 41–44, find the equation of the tangent line to the parabola at the given point.

41. $y = \frac{1}{2}x^2$, $(4, 8)$
42. $y = \frac{1}{2}x^2$, $(-3, \frac{9}{2})$
43. $y = -2x^2$, $(-1, -2)$
44. $y = -2x^2$, $(3, -18)$

Ellipses   11.4

DEFINITION OF AN ELLIPSE

An **ellipse** is the set of all points (x, y) the sum of whose distances from two distinct fixed points (**foci**) is constant. (See Figure 11.29.)

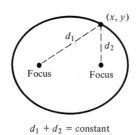
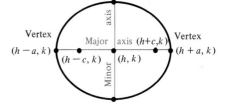

$d_1 + d_2$ = constant

FIGURE 11.29

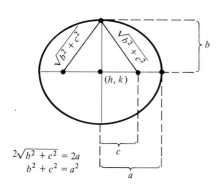

FIGURE 11.30

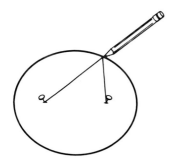

$2\sqrt{b^2 + c^2} = 2a$
$b^2 + c^2 = a^2$

FIGURE 11.31

One way to visualize this definition of an ellipse is to consider two thumbtacks placed at the foci (Figure 11.30). If we fasten the ends of a fixed length of string to the thumbtacks and draw the string taut with a pencil, then the traceable path of the pencil will be an ellipse.

Referring to Figure 11.29, the line through the foci intersects the ellipse at two points called the **vertices.** The chord joining the vertices is called the **major axis,** and its midpoint is called the **center** of the ellipse. The chord perpendicular to the major axis at the center is called the **minor axis** of the ellipse.

To derive the standard form of the equation of an ellipse, consider the ellipse in Figure 11.31 with the following points:

center (h, k);   vertices $(h \pm a, k)$;   foci $(h \pm c, k)$

From the definition of an ellipse, we know that the sum of the distances from any point on the ellipse to the two foci is constant. By considering the position of a taut string when the pencil is at one of the *vertices,* we can see that the length of the string would be

$$(a + c) + (a - c) = 2a$$

or simply the length of the major axis.

Now if we let (x, y) be any point on the ellipse, then the sum of the distances between this point and the two foci must also be $2a$. That is,

$$\sqrt{[x - (h - c)]^2 + (y - k)^2} + \sqrt{[x - (h + c)]^2 + (y - k)^2} = 2a$$

which reduces to

$$(a^2 - c^2)(x - h)^2 + a^2(y - k)^2 = a^2(a^2 - c^2)$$

However, from Figure 11.31 we can see that $b^2 = a^2 - c^2$, and therefore the equation of the ellipse is

$$b^2(x - h)^2 + a^2(y - k)^2 = a^2 b^2$$

$$\frac{(x - h)^2}{a^2} + \frac{(y - k)^2}{b^2} = 1$$

Had we chosen the major axis to be parallel to the y-axis, we would have obtained a similar equation.

STANDARD EQUATION OF AN ELLIPSE

The standard form of the equation of an ellipse, with center $(h,\ k)$ and major and minor axes of lengths $2a$ and $2b$, respectively, is

$$\frac{(x - h)^2}{a^2} + \frac{(y - k)^2}{b^2} = 1 \qquad \textit{(Major axis parallel to the x-axis)}$$

or

$$\frac{(x - h)^2}{b^2} + \frac{(y - k)^2}{a^2} = 1 \qquad \textit{(Major axis parallel to the y-axis)}$$

The foci lie on the major axis, c units from the center, with $c^2 = a^2 - b^2$.

EXAMPLE 1
Finding the Standard Equation of an Ellipse

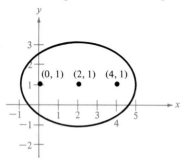

FIGURE 11.32

Find the standard form of the equation of the ellipse having foci at $(0,\ 1)$ and $(4,\ 1)$ and with a major axis of length 6. (See Figure 11.32.)

Solution:
Since the foci occur at $(0,\ 1)$ and $(4,\ 1)$, the center of the ellipse is $(2,\ 1)$, and $c = 2$ is the distance from the center to one of the foci. Furthermore, $2a = 6$ or $a = 3$, and since $c^2 = a^2 - b^2$, it follows that

$$b = \sqrt{a^2 - c^2} = \sqrt{9 - 4} = \sqrt{5}$$

Therefore, the standard equation of this ellipse, whose major axis is parallel to the x-axis, is

$$\frac{(x - 2)^2}{9} + \frac{(y - 1)^2}{5} = 1$$

EXAMPLE 2
Sketching an Ellipse

Sketch the graph of the ellipse whose equation is

$$x^2 + 4y^2 + 6x - 8y + 9 = 0$$

Solution:

$$x^2 + 4y^2 + 6x - 8y + 9 = 0 \qquad \textit{Given equation}$$

$$(x^2 + 6x +) + (4y^2 - 8y +) = -9 \qquad \textit{Group terms}$$

$$(x^2 + 6x +) + 4(y^2 - 2y +) = -9 \qquad \textit{Factor 4 out of y-terms}$$

$$(x^2 + 6x + 9) + 4(y^2 - 2y + 1) \qquad \textit{Add 9 and 4 to both sides}$$

$$= -9 + 9 + 4(1)$$

$$(x + 3)^2 + 4(y - 1)^2 = 4 \qquad \textit{Completed square form}$$

$$\frac{(x + 3)^2}{4} + \frac{(y - 1)^2}{1} = 1 \qquad \textit{Standard form}$$

Now we see that the center occurs at $(h, k) = (-3, 1)$. Since the denominator of the x-term is $4 = a^2 = 2^2$, we locate the endpoints of the major axis 2 units to the right and left of the center. Similarly, since the denominator of the y-term is $1 = b^2 = 1^2$, we locate the endpoints of the minor axis 1 unit up and down from the center. The graph of this ellipse is shown in Figure 11.33.

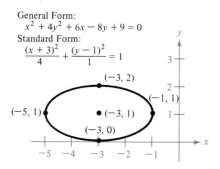

General Form:
$$x^2 + 4y^2 + 6x - 8y + 9 = 0$$
Standard Form:
$$\frac{(x + 3)^2}{4} + \frac{(y - 1)^2}{1} = 1$$

FIGURE 11.33

EXAMPLE 3
Finding the Center, Vertices, and Foci of an Ellipse

Find the center, vertices, and foci of the ellipse given by the equation

$$4x^2 - 8x + y^2 + 4y - 8 = 0$$

Solution:

By completing the square, we can write

$$4(x^2 - 2x + 1) + (y^2 + 4y + 4) = 8 + 4 + 4$$

$$4(x - 1)^2 + (y + 2)^2 = 16$$

$$\frac{(x - 1)^2}{4} + \frac{(y + 2)^2}{16} = 1$$

This is the equation of an ellipse with its major axis parallel to the y-axis, where $h = 1$, $k = -2$, $a = 4$, $b = 2$, and $c = \sqrt{16 - 4} = 2\sqrt{3}$. Therefore, we have

center $(1, -2)$;   vertices $(1, -6)$, $(1, 2)$

foci $(1, -2 - 2\sqrt{3})$, $(1, -2 + 2\sqrt{3})$

Various physical phenomena involve ellipses. Ellipses are used in acoustical design. Supporting arches and machine gears sometimes have elliptical shapes, and the orbits of satellites and planets are ellipses. In the next example we investigate the elliptical orbit of the moon about the earth.

EXAMPLE 4
An Application of Elliptical Orbits

The moon travels about the earth in an elliptical orbit with the earth at one focus. (See Figure 11.34.) If the major and minor axes of the orbit have lengths 774,000 kilometers and 773,000 kilometers, respectively, what are

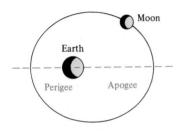

Earth

Moon

Perigee   Apogee

FIGURE 11.34

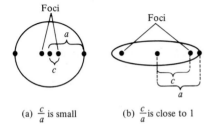

Foci

Foci

(a) $\frac{c}{a}$ is small   (b) $\frac{c}{a}$ is close to 1

FIGURE 11.35

the greatest and least distances (the apogee and perigee) from the earth's center to the moon's center?

Solution:
Since

$$2a = 774,000 \qquad \text{and} \qquad 2b = 773,000$$

we have

$$a = 387,000, \qquad b = 386,500, \qquad c = \sqrt{a^2 - b^2} \approx 19,500$$

Therefore, the greatest distance between the centers of the earth and the moon is

$$a + c \approx 406,500 \text{ kilometers}$$

and the least distance is

$$a - c \approx 367,000 \text{ kilometers}$$

The location of the foci of an ellipse relative to its center and vertices gives us a convenient way to classify ellipses. For instance, in Figure 11.35a the foci are relatively close to the center and the ellipse is almost circular. In Figure 11.35b, the foci are relatively close to the vertices and the ellipse is flat. Since c measures the distance between the center and the foci and a measures the distance between the center and the vertices, we can use the ratio of c to a to measure the flatness of an ellipse. We call this measure the **eccentricity** of an ellipse.

DEFINITION OF ECCENTRICITY

> The **eccentricity** of an ellipse is given by
>
> $$e = \frac{c}{a}$$

Remark: Note that $0 < e < 1$ for all ellipses. Furthermore, for ellipses that are nearly circular, the eccentricity is close to 0, and for ellipses that are very elongated, the eccentricity is close to 1.

EXAMPLE 5
Finding the Eccentricity of an Ellipse

Find the eccentricity of the following ellipses:

(a) $\dfrac{(x + 2)^2}{25} + \dfrac{(y - 1)^2}{24} = 1$ \qquad (b) $\dfrac{(x - 3)^2}{9} + \dfrac{(y + 2)^2}{25} = 1$

Solution:

(a) Since $a^2 = 25$ and $b^2 = 24$, we have

$$c^2 = a^2 - b^2 = 25 - 24 = 1$$

which implies that $c = 1$. Thus, the eccentricity is

$$e = \frac{c}{a} = \frac{1}{5}$$

From Figure 11.36 we see that this ellipse is almost circular. This corresponds with our notion of a small eccentricity.

(b) Since $a^2 = 25$ and $b^2 = 9$, we have

$$c^2 = a^2 - b^2 = 25 - 9 = 16$$

which implies that $c = 4$. Thus, the eccentricity is

$$e = \frac{c}{a} = \frac{4}{5}$$

From Figure 11.37 we see that this ellipse is quite elongated. This corresponds with our notion of eccentricities that are close to 1.

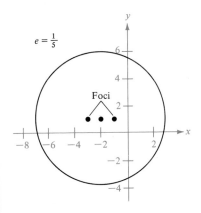

FIGURE 11.36

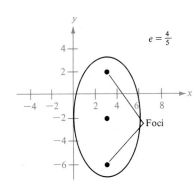

FIGURE 11.37

Section Exercises 11.4

In Exercises 1–20, find the center, foci, vertices, and eccentricity of each ellipse.

1. $\dfrac{x^2}{25} + \dfrac{y^2}{16} = 1$

2. $\dfrac{x^2}{144} + \dfrac{y^2}{169} = 1$

3. $\dfrac{x^2}{16} + \dfrac{y^2}{25} = 1$

4. $\dfrac{x^2}{169} + \dfrac{y^2}{144} = 1$

5. $\dfrac{x^2}{9} + \dfrac{y^2}{5} = 1$

6. $\dfrac{x^2}{28} + \dfrac{y^2}{64} = 1$

7. $x^2 + 4y^2 = 4$

8. $5x^2 + 3y^2 = 15$

9. $3x^2 + 2y^2 = 6$

10. $5x^2 + 7y^2 = 70$

11. $4x^2 + y^2 = 1$

12. $16x^2 + 25y^2 = 1$

13. $\dfrac{(x - 1)^2}{9} + \dfrac{(y - 5)^2}{25} = 1$

14. $(x + 2)^2 + 4(y + 4)^2 = 1$

15. $9x^2 + 4y^2 + 36x - 24y + 36 = 0$

16. $9x^2 + 4y^2 - 36x + 8y + 31 = 0$

17. $16x^2 + 25y^2 - 32x + 50y + 31 = 0$

18. $9x^2 + 25y^2 - 36x - 50y + 61 = 0$

19. $12x^2 + 20y^2 - 12x + 40y - 37 = 0$

20. $36x^2 + 9y^2 + 48x - 36y + 43 = 0$

In Exercises 21–30, find an equation for the specified ellipse.

21. Center (0, 0); focus (2, 0); vertex (3, 0)
22. Center (0, 0); vertex (2, 0); minor axis of length 3
23. Vertices (5, 0), (−5, 0); eccentricity $\frac{3}{5}$
24. Vertices (0, 8), (0, −8); eccentricity $\frac{1}{2}$
25. Vertices (0, 2), (4, 2); minor axis of length 2
26. Foci (−2, 0), (2, 0); major axis of length 8
27. Vertices (3, 1), (3, 9); minor axis of length 6
28. Center (0, 0); major axis horizontal; curve passes through points (3, 1) and (4, 0)
29. Foci (0, 5), (0, −5); sum of distances from the foci to any point on the ellipse is 14
30. Center (1, 2); major axis parallel to the *y*-axis; curve passes through points (1, 6) and (3, 2)
31. A fireplace arch is to be constructed in the shape of a semi-ellipse. The opening is to have a height of 2 feet at the center and a width of 5 feet along the base (see Figure 11.38). The contractor will first draw the form of the ellipse by the method shown in Figure 11.30. Where should the tacks be placed and how long should the piece of string be?
32. A line segment through a focus with endpoints on the ellipse and perpendicular to the major axis is called a **latus rectum** of the ellipse. Therefore, an ellipse has two latus recta. Knowing the length of the latus recta is helpful in sketching an ellipse because it yields another locus point on the curve.

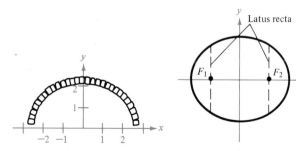

FIGURE 11.38

FIGURE 11.39

(See Figure 11.39.) Show that the length of each latus rectum is $2b^2/a$.

33. Sketch the graph of each ellipse, making use of the endpoints of the latus recta (see Exercise 32).

    (a) $\dfrac{x^2}{4} + \dfrac{y^2}{1} = 1$

    (b) $6x^2 + 4y^2 = 1$

    (c) $5x^2 + 3y^2 = 15$

34. Show that the eccentricity of the ellipse

$$\frac{x^2}{a^2} + \frac{y^2}{b^2} = 1$$

is identical to the eccentricity of

$$\frac{(tx)^2}{a^2} + \frac{(ty)^2}{b^2} = 1$$

for any real *t*. Give a geometrical explanation of this result.

35. The earth moves in an elliptical orbit with the sun at one of the foci. If the length of half the major axis is 93 million miles and the eccentricity is 0.017, find the least and greatest distances of the earth from the sun.

36. If the distances to the apogee and the perigee of an elliptical orbit of an earth satellite are measured from the center of the earth, show that the eccentricity of the orbit is given by $e = (A − P)/(A + P)$, where *A* and *P* are the apogee and perigee distances, respectively.

37. *Sputnik I*, orbited by the Russians in October 1957, had 583 miles and 132 miles above the earth's surface as the highest and lowest points of its elliptical orbit. What is the eccentricity of this orbit?

38. On November 26, 1963, the United States launched *Explorer 18*. Its low point over the surface of the earth was 119 miles and its high point was 122,000 miles from the surface of the earth.

    (a) Find the eccentricity of its elliptical orbit.

    (b) Find an equation that describes its orbit.

Hyperbolas   11.5

The last conic we will consider is the hyperbola. The definition of a hyperbola parallels that of an ellipse. The distinction is that for an ellipse the *sum* of the distances between the foci and a point on the ellipse is fixed, while for a hyperbola the *difference* of these distances is fixed.

DEFINITION OF A HYPERBOLA

A **hyperbola** is the set of all points (x, y) the difference of whose distances from two distinct fixed points (foci) is constant. (See Figure 11.40.)

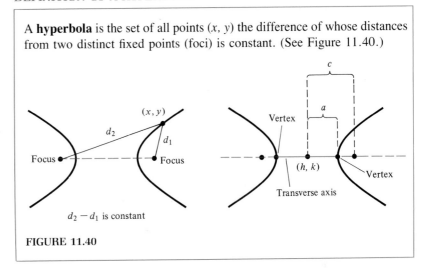

$d_2 - d_1$ is constant

FIGURE 11.40

One distinguishing feature of hyperbolas is that their graphs have two separate branches. Referring to Figure 11.40, the line through the two foci intersects the hyperbola at two points called the **vertices.** The line segment connecting the vertices is called the **transverse axis,** and the midpoint of the transverse axis is called the **center** of the hyperbola.

Remark: Note in Figure 11.40 that the fixed difference is $d_1 - d_2 = 2a$.

The development of the standard form of the equation of a hyperbola is similar to that for an ellipse, so we list the following result without proof.

STANDARD EQUATION OF A HYPERBOLA

The standard form of the equation of a hyperbola with center at (h, k) is

$$\frac{(x - h)^2}{a^2} - \frac{(y - k)^2}{b^2} = 1 \qquad \textit{(Transverse axis is horizontal)}$$

$$\frac{(y - k)^2}{a^2} - \frac{(x - h)^2}{b^2} = 1 \qquad \textit{(Transverse axis is vertical)}$$

where the vertices and foci are, respectively, a and c units from the center and $b^2 = c^2 - a^2$.

EXAMPLE 1
Finding the Standard
Equation of a Hyperbola

Find the standard form of the equation of the hyperbola with foci at $(-1, 2)$ and $(5, 2)$ and vertices at $(0, 2)$ and $(4, 2)$.

Solution:
By the Midpoint Rule (Section 1.5), the center of the hyperbola occurs at

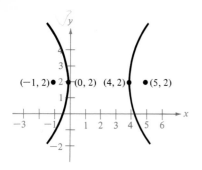

FIGURE 11.41

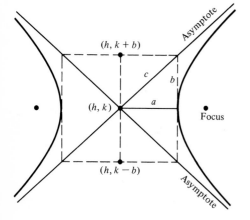

FIGURE 11.42

the point (2, 2). Furthermore, $c = 3$ and $a = 2$, and it follows that

$$b^2 = 3^2 - 2^2 = 9 - 4 = 5$$

Thus the equation of this hyperbola is

$$\frac{(x - 2)^2}{4} - \frac{(y - 2)^2}{5} = 1$$

(See Figure 11.41.)

Each hyperbola has two **asymptotes:** two straight lines that intersect at the center of the hyperbola. These asymptotes pass through the vertices of a rectangle of dimension $2a$ by $2b$ with its center at (h, k). The line segment of length $2b$ joining $(h, k + b)$ and $(h, k - b)$ is referred to as the **conjugate axis** of the hyperbola.

Remark: Note in Figure 11.42 that as you move out on the branches of a hyperbola, the branches approach the asymptotes of the hyperbola.

The following result gives the equations for the asymptotes of a hyperbola.

ASYMPTOTES OF A HYPERBOLA

If a hyperbola has a *horizontal* transverse axis, then its asymptotes are the lines

$$y = k + \frac{b}{a}(x - h) \qquad \text{and} \qquad y = k - \frac{b}{a}(x - h)$$

If a hyperbola has a *vertical* transverse axis, then its asymptotes are the lines

$$y = k + \frac{a}{b}(x - h) \qquad \text{and} \qquad y = k - \frac{a}{b}(x - h)$$

Note in Figure 11.42 that the asymptotes coincide with the diagonals of the rectangle centered at (h, k) with dimensions $2a$ and $2b$. This provides us with a quick means of sketching the asymptotes, which in turn aids us in sketching the hyperbola.

EXAMPLE 2
Sketching the Graph
of a Hyperbola

Sketch the graph of the hyperbola whose equation is

$$4x^2 - y^2 = 16$$

Solution:
Rewriting this equation in standard form, we have

$$\frac{4x^2}{16} - \frac{y^2}{16} = \frac{16}{16}$$

$$\frac{x^2}{2^2} - \frac{y^2}{4^2} = 1$$

From this, we conclude that the transverse axis is horizontal and the vertices occur at $(-2, 0)$ and $(2, 0)$. Moreover, the ends of the conjugate axis occur at $(0, -4)$ and $(0, 4)$, and we are able to sketch the rectangle shown in Figure 11.43. Finally, by drawing the asymptotes through the corners of this rectangle, we complete the sketch shown in Figure 11.44.

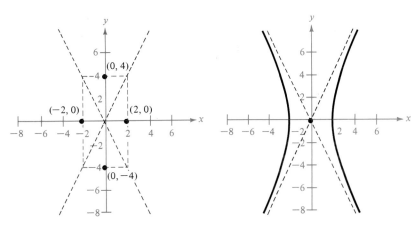

FIGURE 11.43   **FIGURE 11.44**

EXAMPLE 3
Finding the Asymptotes
of a Hyperbola

Find the asymptotes of the hyperbola given by

$$4x^2 + 8x - 3y^2 + 16 = 0$$

and then use the asymptotes as an aid to sketch a graph of the hyperbola.

Solution:
Rewriting this equation in standard form, we have

$$4x^2 + 8x - 3y^2 + 16 = 0$$
$$4(x^2 + 2x) - 3y^2 = -16$$
$$-4(x^2 + 2x + 1) + 3y^2 = 16 - 4$$
$$-4(x + 1)^2 + 3y^2 = 12$$
$$\frac{y^2}{4} - \frac{(x + 1)^2}{3} = 1$$

From this equation we conclude that the hyperbola is centered at $(-1, 0)$, has vertices at $(-1, 2)$ and $(-1, -2)$, and is asymptotic to the lines

$$y = \frac{2}{\sqrt{3}}(x + 1) \quad \text{and} \quad y = -\frac{2}{\sqrt{3}}(x + 1)$$

Once the rectangle, asymptotes, and vertices have been sketched, as shown in Figure 11.45, it is relatively easy to complete the sketch of the hyperbola.

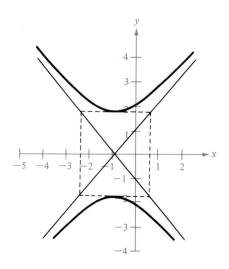

FIGURE 11.45

EXAMPLE 4
Finding the Standard
Equation of a Hyperbola

Find the standard form of the equation of the hyperbola having vertices at $(3, -5)$ and $(3, 1)$ and with asymptotes $y = 2x - 8$ and $y = -2x + 4$.

Solution:
By the Midpoint Rule, the center of the hyperbola is at $(3, -2)$. Furthermore, the hyperbola has a vertical transverse axis with $a = 3$. The asymptotes have equations whose slopes are, respectively,

$$m_1 = \frac{a}{b} = 2 \quad \text{and} \quad m_2 = -\frac{a}{b} = -2$$

Since $a = 3$, we conclude that $b = \frac{3}{2}$. Therefore, the standard equation of the hyperbola is

$$\frac{(y + 2)^2}{9} - \frac{(x - 3)^2}{\frac{9}{4}} = 1$$

or we write

$$\frac{(y + 2)^2}{9} - \frac{4(x - 3)^2}{9} = 1$$

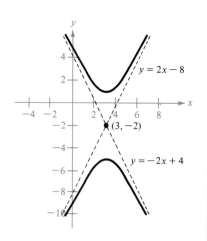

FIGURE 11.46

(See Figure 11.46.)

EXAMPLE 5

An Application Involving
the Travel of Sound

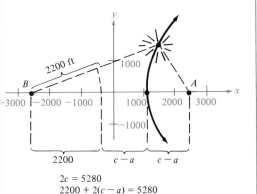

2c = 5280
2200 + 2(c − a) = 5280

FIGURE 11.47

Two microphones, 1 mile apart, record an explosion. Microphone A received the sound 2 seconds before microphone B. Where did the explosion come from?

Solution:

Since sound travels at 1100 feet per second, we know that the explosion took place 2200 feet further from B than from A. (See Figure 11.47.) The locus of all points that are 2200 feet closer to A than to B is, by definition, one branch of the hyperbola

$$\frac{x^2}{a^2} - \frac{y^2}{b^2} = 1$$

where

$$c = \frac{5280}{2} = 2640$$

$$a = \frac{2200}{2} = 1100$$

$$b^2 = c^2 - a^2 = 5{,}759{,}600$$

Thus the explosion occurred somewhere on the right branch of the hyperbola

$$\frac{x^2}{1{,}210{,}000} - \frac{y^2}{5{,}759{,}600} = 1$$

In Example 5 we were able to determine the hyperbola on which the explosion occurred but not the exact location of the explosion. If, however, we had received the sound from a third position C, two other hyperbolas would be determined. The exact location of the explosion would be the point where these three hyperbolas intersect.

Another interesting application of conic sections involves the orbits of comets in our solar system. Of the 566 comets identified prior to 1960, it was found that 211 have elliptical orbits, 290 have parabolic orbits, and 65 have hyperbolic orbits. The center of the sun is a focus point of each of these orbits, and each orbit has a vertex at the point where the comet is closest to the sun (Figure 11.48).

If p is the distance between the vertex and the focus, and v is the velocity of the comet at the vertex, then the orbit is

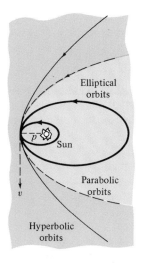

Elliptical
orbits

Sun

Parabolic
orbits

Hyperbolic
orbits

FIGURE 11.48

1. an ellipse if $v < \sqrt{\dfrac{2GM}{p}}$ 2. a parabola if $v = \sqrt{\dfrac{2GM}{p}}$

3. a hyperbola if $v > \sqrt{\dfrac{2GM}{p}}$

where M is the mass of the sun and $G \approx 6.67(10^{-8})$ cubic centimeters per gram second squared.

Section Exercises 11.5

In Exercises 1–20, find the center, vertices, and foci of the hyperbola and sketch its graph, using asymptotes as an aid.

1. $x^2 - y^2 = 1$

2. $\dfrac{x^2}{9} - \dfrac{y^2}{16} = 1$

3. $y^2 - \dfrac{x^2}{4} = 1$

4. $\dfrac{y^2}{9} - x^2 = 1$

5. $\dfrac{y^2}{25} - \dfrac{x^2}{144} = 1$

6. $\dfrac{x^2}{36} - \dfrac{y^2}{4} = 1$

7. $2x^2 - 3y^2 = 6$

8. $3y^2 = 5x^2 + 15$

9. $5y^2 = 4x^2 + 20$

10. $7x^2 - 3y^2 = 21$

11. $\dfrac{(x - 1)^2}{4} - \dfrac{(y + 2)^2}{1} = 1$

12. $\dfrac{(x + 1)^2}{144} - \dfrac{(y - 4)^2}{25} = 1$

13. $(y + 6)^2 - (x - 2)^2 = 1$

14. $\dfrac{(y - 1)^2}{\frac{1}{4}} - \dfrac{(x + 3)^2}{\frac{1}{9}} = 1$

15. $9x^2 - y^2 - 36x - 6y + 18 = 0$

16. $x^2 - 9y^2 + 36y - 72 = 0$

17. $9y^2 - x^2 + 2x + 54y + 62 = 0$

18. $16y^2 - x^2 + 2x + 64y + 63 = 0$

19. $x^2 - 9y^2 + 2x - 54y - 80 = 0$

20. $9x^2 - y^2 + 54x + 10y + 55 = 0$

In Exercises 21–30, find an equation for the specified hyperbola.

21. Center $(0, 0)$; one vertex $(0, 2)$; one focus $(0, 4)$

22. Center $(0, 0)$; one vertex $(3, 0)$; one focus $(5, 0)$

23. Vertices $(-1, 0)$, $(1, 0)$; asymptotes $y = \pm 3x$

24. Vertices $(0, -3)$, $(0, 3)$; asymptotes $y = \pm 3x$

25. Vertices $(0, 2)$, $(6, 2)$; asymptotes $y = \frac{2}{3}x$ and $y = 4 - \frac{2}{3}x$

26. Vertices $(2, 3)$, $(2, -3)$; foci $(2, 5)$, $(2, -5)$

27. Vertices $(2, 3)$, $(2, -3)$; passing through point $(0, 5)$

28. Asymptotes $y = \pm \frac{3}{4}x$; focus $(10, 0)$

29. For any point on the hyperbola, the difference of its distances from $(2, 2)$ and $(10, 2)$ is 6.

30. For any point on the hyperbola, the difference of its distances from $(-3, 0)$ and $(-3, 3)$ is 2.

31. Three listening stations located at $(4400, 0)$, $(4400, 1100)$, and $(-4400, 0)$ hear an explosion. If the latter two stations heard the sound 1 second and 5 seconds after the first, respectively, where did the explosion occur? Assume that the coordinate system is measured in feet and that sound travels at the rate of 1100 feet per second.

In Exercises 32–39, classify the graph of each equation as a circle, a parabola, an ellipse, or a hyperbola.

32. $x^2 + y^2 - 6x + 4y + 9 = 0$

33. $x^2 + 4y^2 - 6x + 16y + 21 = 0$

34. $4x^2 - y^2 - 4x - 3 = 0$

35. $y^2 - 4y - 4x = 0$

36. $4x^2 + 3y^2 + 8x - 24y + 51 = 0$

37. $4y^2 - 2x^2 - 4y - 8x - 15 = 0$

38. $25x^2 - 10x - 200y - 119 = 0$

39. $4x^2 + 4y^2 - 16y + 15 = 0$

Rotation and General Second-Degree Equations   11.6

In Sections 11.2 through 11.5 we have been studying conic sections whose axes are parallel to one of the coordinate axes. Circles, ellipses, and hyperbolas with centers at the origin are said to be in **standard position.** A parabola is said to be in **standard position** if its vertex is at the origin. We summarize the equations we have studied so far in the following table.

STANDARD FORMS OF EQUATIONS OF CONICS

	Standard Position	**With Vertex at (h, k)**	
Parabola			
Vertical axis	$x^2 = 4py$	$(x - h)^2 = 4p(y - k)$	

Horizontal axis	$y^2 = 4px$	$(y - k)^2 = 4p(x - h)$	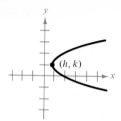

	Standard Position	**With Center at (h, k)**	
Circle	$x^2 + y^2 = r^2$	$(x - h)^2 + (y - k)^2 = r^2$	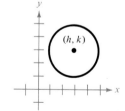

Ellipse

Horizontal major axis	$\dfrac{x^2}{a^2} + \dfrac{y^2}{b^2} = 1$	$\dfrac{(x - h)^2}{a^2} + \dfrac{(y - k)^2}{b^2} = 1$	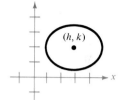
Vertical major axis	$\dfrac{x^2}{b^2} + \dfrac{y^2}{a^2} = 1$	$\dfrac{(x - h)^2}{b^2} + \dfrac{(y - k)^2}{a^2} = 1$	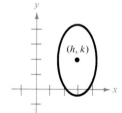

Hyperbola

Horizontal transverse axis	$\dfrac{x^2}{a^2} - \dfrac{y^2}{b^2} = 1$	$\dfrac{(x - h)^2}{a^2} - \dfrac{(y - k)^2}{b^2} = 1$	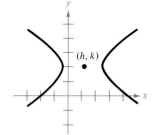
Vertical transverse axis	$\dfrac{y^2}{a^2} - \dfrac{x^2}{b^2} = 1$	$\dfrac{(y - k)^2}{a^2} - \dfrac{(x - h)^2}{b^2} = 1$	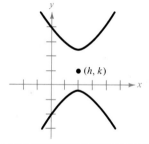

Each of the equations in this summary can be written in the *general* form

$$Ax^2 + Cy^2 + Dx + Ey + F = 0$$

In this section we investigate the equations of conics whose axes are not parallel to either the *x*-axis or the *y*-axis. Under these circumstances we will see that the general equation for such conics contains an *xy* term.

DEFINITION OF GENERAL SECOND-DEGREE EQUATION

The equation

$$Ax^2 + Bxy + Cy^2 + Dx + Ey + F = 0$$

where *A*, *B*, and *C* are not all zero, is called the **general second-degree equation.**

Every conic in the *xy*-plane possesses an equation that fits this general second-degree form. Furthermore, every second-degree equation that has solution points has a graph that is one of the four basic conics (circle, parabola, ellipse, or hyperbola) or one of the three degenerate forms (point, line, or two intersecting lines).

We already know how to sketch the graph of a second-degree equation that contains no *xy* term. We first write the equation in standard form by completing the square and then sketch the graph based on the information available from the standard form. However, for a general second-degree equation containing an *xy* term, it is not possible to obtain a standard form by completing the square. We can overcome this problem by a **rotation of axes,** which eliminates the *xy* term.

In this procedure we introduce a new coordinate system by rotating the *x*- and *y*-axes counterclockwise to a new position denoted by the *x'*- and *y'*-axes (see Figure 11.49). Therefore, if we are given a general equation of a conic,

$$Ax^2 + Bxy + Cy^2 + Dx + Ey + F = 0$$

our objective is to rotate the *x*- and *y*-axes until they are parallel to the axes of the conic. Then the equation of the conic in the new *x'y'* system will have the form

$$A'(x')^2 + C'(y')^2 + D'x' + E'y' + F' = 0$$

From this form, with no *x'y'* term, we can obtain a standard form by completing the square. A sketch of the conic can then be readily made in the *x'y'* system.

The following theorem identifies how much to rotate the axes to eliminate an *xy* term and also the equations for determining the new coefficients *A'*, *C'*, *D'*, *E'*, and *F'*.

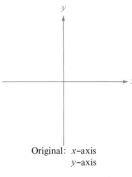

Original: *x*–axis
    y–axis

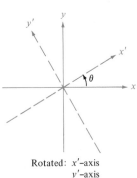

Rotated: *x'*–axis
    y'–axis

FIGURE 11.49

ROTATION OF AXES TO ELIMINATE AN xy TERM

The equation

$$Ax^2 + Bxy + Cy^2 + Dx + Ey + F = 0$$

can be rewritten as

$$A'(x')^2 + C'(y')^2 + D'x' + E'y' + F' = 0$$

by rotating the coordinate axes through an angle θ, where

$$\cot 2\theta = \frac{A - C}{B}$$

The coefficients of the new equation are given by

$$A' = A \cos^2 \theta + B \cos \theta \sin \theta + C \sin^2 \theta$$
$$C' = A \sin^2 \theta - B \cos \theta \sin \theta + C \cos^2 \theta$$
$$D' = D \cos \theta + E \sin \theta$$
$$E' = -D \sin \theta + E \cos \theta$$
$$F' = F$$

Proof:

Using Figure 11.50, we choose a point (x, y) in the original system and attempt to find its coordinates (x', y') in the rotated system. In either system the distance r between the point P and the origin is the same; thus the equations for x, y, x', and y' are those given in Figure 11.50.

Using the formulas for the sine and cosine of the difference of two angles, we have

$$x' = r \cos (\alpha - \theta) = r(\cos \alpha \cos \theta + \sin \alpha \sin \theta)$$
$$= r \cos \alpha \cos \theta + r \sin \alpha \sin \theta = x \cos \theta + y \sin \theta$$
$$y' = r \sin (\alpha - \theta) = r(\sin \alpha \cos \theta - \cos \alpha \sin \theta)$$
$$= r \sin \alpha \cos \theta - r \cos \alpha \sin \theta = y \cos \theta - x \sin \theta$$

Solving the system

$$x' = x \cos \theta + y \sin \theta$$
$$y' = -x \sin \theta + y \cos \theta$$

for x and y yields

$$x = x' \cos \theta - y' \sin \theta$$
$$y = x' \sin \theta + y' \cos \theta$$

Finally, by substituting these values for x and y into the equation

$$Ax^2 + Bxy + Cy^2 + Dx + Ey + F = 0$$

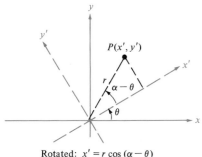

Rotated: $x' = r \cos (\alpha - \theta)$
$y' = r \sin (\alpha - \theta)$

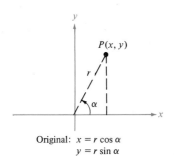

Original: $x = r \cos \alpha$
$y = r \sin \alpha$

FIGURE 11.50

and collecting terms, we obtain

$$[A \cos^2 \theta + B \sin \theta \cos \theta + C \sin^2 \theta](x')^2$$
$$+ [2(C - A) \sin \theta \cos \theta + B(\cos^2 \theta - \sin^2 \theta)](x'y')$$
$$+ [A \sin^2 \theta - B \sin \theta \cos \theta + C \cos^2 \theta](y')^2$$
$$+ [D \cos \theta + E \sin \theta](x')$$
$$+ [E \cos \theta - D \sin \theta](y') + F = 0$$

which is of the form

$$A'(x')^2 + B'x'y' + C'(y')^2 + D'x' + E'y' + F' = 0$$

Note that in order to eliminate the $x'y'$ term, we must select θ so that

$$B' = 2(C - A) \sin \theta \cos \theta + B(\cos^2 \theta - \sin^2 \theta) = 0$$

Since

$$B' = (C - A) \sin 2\theta + B \cos 2\theta = B(\sin 2\theta)\left(\frac{C - A}{B} + \cot 2\theta\right)$$

B' will be zero if we choose θ so that

$$\cot 2\theta = \frac{A - C}{B}, \quad B \neq 0$$

Thus we have established the desired results.

EXAMPLE 1
Writing an Equation
in Standard Form

Write the equation $xy - 1 = 0$ in standard form.

Solution:
Since $A = 0, B = 1, C = D = E = 0$, and $F = -1$, we have

$$\cot 2\theta = \frac{A - C}{B} = 0$$

Therefore, $2\theta = \pi/2$ and $\theta = \pi/4$. We then have

$$A' = A \cos^2 \theta + B \cos \theta \sin \theta + C \sin^2 \theta$$
$$= 0 + \left(\frac{\sqrt{2}}{2}\right)\left(\frac{\sqrt{2}}{2}\right) + 0 = \frac{1}{2}$$
$$C' = A \sin^2 \theta - B \cos \theta \sin \theta + C \cos^2 \theta$$
$$= 0 - \left(\frac{\sqrt{2}}{2}\right)\left(\frac{\sqrt{2}}{2}\right) + 0 = -\frac{1}{2}$$
$$D' = D \cos \theta + E \sin \theta = 0$$
$$E' = -D \sin \theta + E \cos \theta = 0$$
$$F' = F = -1$$

Hence the new equation is

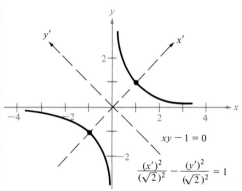

$xy - 1 = 0$

$$\frac{(x')^2}{(\sqrt{2})^2} - \frac{(y')^2}{(\sqrt{2})^2} = 1$$

Vertices: $(\sqrt{2}, 0), (-\sqrt{2}, 0)$ in $x'y'$–system
$(1, 1), (-1, -1)$ in xy–system

FIGURE 11.51

$$\frac{(x')^2}{2} - \frac{(y')^2}{2} - 1 = 0$$

whose standard form is

$$\frac{(x')^2}{(\sqrt{2})^2} - \frac{(y')^2}{(\sqrt{2})^2} = 1$$

This is the equation of a hyperbola centered at the origin with vertices at $(\pm\sqrt{2}, 0)$ in the $x'y'$ system (see Figure 11.51). From the equations

$$x = x' \cos \theta - y' \sin \theta$$
$$y = x' \sin \theta + y' \cos \theta$$

we can determine the vertices to be $(1, 1)$ and $(-1, -1)$ in the xy system.

Remark: In Example 1 note that the asymptotes of the hyperbola have equations $y' = \pm x'$, which correspond to the original x- and y-axes.

EXAMPLE 2
Sketching the Graph of a
Second-Degree Equation

Sketch the graph of the equation $7x^2 - 6\sqrt{3}xy + 13y^2 - 16 = 0$.

Solution:
Applying the Rotation of Axes Formula, we have

$$\cot 2\theta = \frac{A - C}{B} = \frac{7 - 13}{-6\sqrt{3}} = \frac{1}{\sqrt{3}}$$

Thus $2\theta = \pi/3$ and $\theta = \pi/6$. Solving for A', C', D', E', and F' yields

$$A' = A \cos^2 \theta + B \cos \theta \sin \theta + C \sin^2 \theta$$
$$= 7\left(\frac{\sqrt{3}}{2}\right)^2 - 6\sqrt{3}\left(\frac{\sqrt{3}}{2}\right)\left(\frac{1}{2}\right) + 13\left(\frac{1}{2}\right)^2$$
$$= \frac{21 - 18 + 13}{4} = 4$$

$$C' = A \sin^2 \theta - B \cos \theta \sin \theta + C \cos^2 \theta$$
$$= 7\left(\frac{1}{2}\right)^2 + 6\sqrt{3}\left(\frac{\sqrt{3}}{2}\right)\left(\frac{1}{2}\right) + 13\left(\frac{\sqrt{3}}{2}\right)^2$$
$$= \frac{7 + 18 + 39}{4} = 16$$

$$D' = D \cos \theta + E \sin \theta = 0$$
$$E' = -D \sin \theta + E \cos \theta = 0$$
$$F' = -16$$

Therefore, the new equation is

$$4(x')^2 + 16(y')^2 - 16 = 0$$

whose standard form is

$$\frac{(x')^2}{4} + \frac{(y')^2}{1} = 1$$

This is the equation of an ellipse centered at the origin with vertices at $(\pm 2, 0)$ in the $x'y'$ system (Figure 11.52).

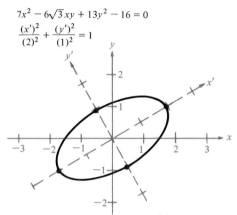

$$7x^2 - 6\sqrt{3}\,xy + 13y^2 - 16 = 0$$
$$\frac{(x')^2}{(2)^2} + \frac{(y')^2}{(1)^2} = 1$$

Vertices: $(\pm 2, 0), (0, \pm 1)$ in $x'y'$–system

$(\pm\sqrt{3}, \pm 1), (\pm\frac{1}{2}, \mp\frac{\sqrt{3}}{2})$ in xy–system

FIGURE 11.52

In Examples 1 and 2 we carefully chose the equations so that θ would turn out to be one of the common angles 30°, 45°, and so forth. In general, a second-degree equation will not yield such common solutions to the equation $\cot 2\theta = (A - C)/B$. Example 3 illustrates such a case.

EXAMPLE 3
Sketching the Graph of a
Second-Degree Equation

Sketch the graph of $x^2 - 4xy + 4y^2 + 5\sqrt{5}y + 1 = 0$.

Solution:
Since

$$\cot 2\theta = \frac{A - C}{B} \qquad \text{and} \qquad \cot 2\theta = \frac{\cot^2\theta - 1}{2\cot\theta}$$

we have

$$\cot 2\theta = \frac{1 - 4}{-4} = \frac{3}{4} = \frac{\cot^2\theta - 1}{2\cot\theta}$$

from which we obtain the equation

$$6\cot\theta = 4\cot^2\theta - 4$$
$$0 = 4\cot^2\theta - 6\cot\theta - 4$$
$$0 = (2\cot\theta - 4)(2\cot\theta + 1)$$

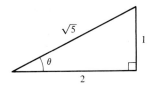

FIGURE 11.53

Considering $0 < \theta < \pi/2$, we have

$$2 \cot \theta = 4$$

$$\cot \theta = 2$$

$$\theta \approx 26.6°$$

From Figure 11.53 we obtain $\sin \theta = 1/\sqrt{5}$ and $\cos \theta = 2/\sqrt{5}$. Consequently,

$$A' = A \cos^2 \theta + B \cos \theta \sin \theta + C \sin^2 \theta$$

$$= \left(\frac{2}{\sqrt{5}}\right)^2 - 4\left(\frac{2}{\sqrt{5}}\right)\left(\frac{1}{\sqrt{5}}\right) + 4\left(\frac{1}{\sqrt{5}}\right)^2 = \frac{4 - 8 + 4}{5} = 0$$

$$C' = A \sin^2 \theta - B \cos \theta \sin \theta + C \cos^2 \theta$$

$$= \left(\frac{1}{\sqrt{5}}\right)^2 + 4\left(\frac{2}{\sqrt{5}}\right)\left(\frac{1}{\sqrt{5}}\right) + 4\left(\frac{2}{\sqrt{5}}\right)^2 = \frac{1 + 8 + 16}{5} = 5$$

$$D' = D \cos \theta + E \sin \theta = 0 + 5\sqrt{5}\left(\frac{1}{\sqrt{5}}\right) = 5$$

$$E' = -D \sin \theta + E \cos \theta = 0 + 5\sqrt{5}\left(\frac{2}{\sqrt{5}}\right) = 10$$

$$F' = F = 1$$

The new equation is

$$5(y')^2 + 5x' + 10y' + 1 = 0$$

By completing the square,

$$5(y' + 1)^2 = -5x' + 4$$

we obtain the standard form

$$(y' + 1)^2 = (-1)\left(x' - \frac{4}{5}\right)$$

We conclude that the graph of the equation is a parabola with its vertex at $(\frac{4}{5}, -1)$ and its axis parallel to the x'-axis in the $x'y'$ system (Figure 11.54).

$x^2 - 4xy + 4y^2 + 5\sqrt{5}y + 1 = 0$

$(y' + 1)^2 = 4(-\frac{1}{4})(x' - \frac{4}{5})$

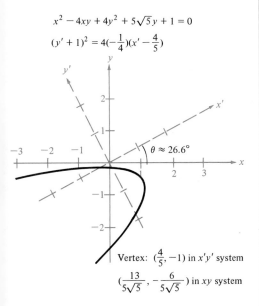

Vertex: $(\frac{4}{5}, -1)$ in $x'y'$ system

$(\frac{13}{5\sqrt{5}}, -\frac{6}{5\sqrt{5}})$ in xy system

FIGURE 11.54

Section Exercises 11.6

In Exercises 1–16, rotate the axes to eliminate the xy term. Sketch the graph of the resulting equation, showing both sets of axes.

1. $xy + 1 = 0$
2. $xy - 4 = 0$
3. $9x^2 + 24xy + 16y^2 + 90x - 130y = 0$
4. $9x^2 + 24xy + 16y^2 + 80x - 60y = 0$
5. $x^2 - 10xy + y^2 + 1 = 0$

6. $xy + x - 2y + 3 = 0$
7. $xy - 2y - 4x = 0$
8. $2x^2 - 3xy - 2y^2 + 10 = 0$
9. $5x^2 - 2xy + 5y^2 - 12 = 0$
10. $13x^2 + 6\sqrt{3}xy + 7y^2 - 16 = 0$
11. $3x^2 - 2\sqrt{3}xy + y^2 + 2x + 2\sqrt{3}y = 0$
12. $16x^2 - 24xy + 9y^2 - 60x - 80y + 100 = 0$
13. $17x^2 + 32xy - 7y^2 = 75$

14. $40x^2 + 36xy + 25y^2 = 52$

15. $32x^2 + 50xy + 7y^2 = 52$

16. $4x^2 - 12xy + 9y^2 + (4\sqrt{13} - 12)x - (6\sqrt{13} + 8)y = 91$

The discriminant of $Ax^2 + Bxy + Cy^2 + Dx + Ey + F = 0$ is given by $B^2 - 4AC$. The discriminant of a second-degree equation can be used as follows:

(a) If the equation graphs as a parabola, then $B^2 - 4AC = 0$.

(b) If the equation graphs as an ellipse, then $B^2 - 4AC < 0$.

(c) If the equation graphs as a hyperbola, then $B^2 - 4AC > 0$.

17. Each of the following three equations graphs is a nondegenerate conic. Use the discriminant of each equation to determine if the graph is a parabola, ellipse, or hyperbola.

(a) $16x^2 - 24xy + 9y^2 - 30x - 40y = 0$

(b) $x^2 - 4xy - 2y^2 - 6 = 0$

(c) $13x^2 - 8xy + 7y^2 - 45 = 0$

18. Show that the discriminant of the general second-degree equation is invariant (does not change) under rotation of axes.

19. Show that the equation $x^2 + y^2 = r^2$ is invariant under rotation of axes.

Review Exercises / **Chapter 11**

In Exercises 1–10, match the equation with the correct graph.

1. $\dfrac{x^2}{1} + \dfrac{y^2}{4} = 1$

2. $x^2 = 4y$

3. $\dfrac{x^2}{1} - \dfrac{y^2}{4} = 1$

4. $y^2 = -4x$

5. $\dfrac{x^2}{4} + \dfrac{y^2}{1} = 1$

6. $\dfrac{y^2}{4} - \dfrac{x^2}{1} = 1$

7. $x^2 = -6y$

8. $x^2 + 5y^2 = 10$

9. $x^2 - 5y^2 = -5$

10. $y^2 - 8x = 0$

(e)

(f)

(a)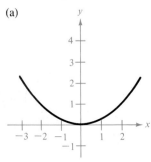

(b)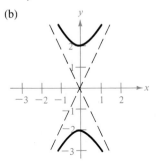

(g)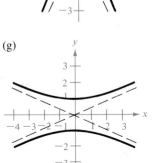

(h)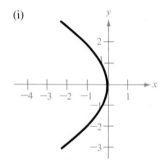

(c)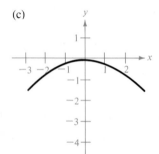

(d)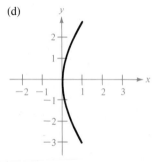

(i)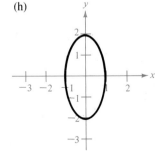

(j)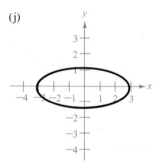

In Exercises 11–20, analyze each equation and sketch its graph.

11. $16x^2 + 16y^2 - 16x + 24y - 3 = 0$
12. $y^2 - 12y - 8x + 20 = 0$
13. $3x^2 - 2y^2 + 24x + 12y + 24 = 0$
14. $4x^2 + y^2 - 16x + 15 = 0$
15. $3x^2 + 2y^2 - 12x + 12y + 29 = 0$
16. $4x^2 - 4y^2 - 4x + 8y - 11 = 0$
17. $x^2 - 6x + 2y + 9 = 0$
18. $x^2 + y^2 - 2x - 4y + 5 = 0$
19. $x^2 + y^2 + 2xy + 2\sqrt{2}x - 2\sqrt{2}y + 2 = 0$
20. $9x^2 + 6y^2 + 4xy - 20 = 0$

In Exercises 21–30, find an equation of the specified conic.

21. Hyperbola; vertices $(0, \pm 1)$; foci $(0, \pm 3)$
22. Hyperbola; vertices $(2, 2)$, $(-2, 2)$; foci $(4, 2)$, $(-4, 2)$
23. The ellipse such that the sum of the distances from any of the points on its graph to the points $(0, 0)$ and $(4, 0)$ is 10
24. Parabola; vertex $(4, 2)$; focus $(4, 0)$
25. Parabola; vertex $(0, 0)$; focus $(1, 1)$
26. Ellipse; vertices $(2, 0)$, $(2, 4)$; foci $(2, 1)$, $(2, 3)$
27. Ellipse passing through points $(1, 2)$ and $(2, 0)$ with center at $(0, 0)$
28. Hyperbola; foci $(\pm 4, 0)$; asymptotes $y = \pm 2x$
29. The hyperbola such that the absolute value of difference of the distances from any point on its graph to the points $(4, 0)$ and $(-4, 0)$ is 4
30. Parabola; vertex $(0, 2)$; passes through point $(-1, 0)$; vertical axis
31. A large parabolic antenna is described as the surface formed by revolving the parabola $y = (1/200)x^2$ on the interval $[0, 100]$ about the y-axis. The receiving and transmitting equipment is positioned at the focus. Find the coordinates of the focus.

The equation of the tangent line to the conic

$$\frac{x^2}{a^2} \pm \frac{y^2}{b^2} = 1$$

at the point (x_0, y_0) is

$$\frac{x_0 x}{a^2} \pm \frac{y_0 y}{b^2} = 1$$

Use this result in Exercises 32–36.

32. Find an equation of the tangent line to the ellipse $\dfrac{x^2}{100} + \dfrac{y^2}{25} = 1$ at the point $(-8, 3)$.

33. Find an equation of the tangent line to the ellipse $x^2 + 7y^2 = 16$ at the point $(3, 1)$.
34. Find an equation of the tangent line to the hyperbola $\dfrac{x^2}{9} - y^2 = 1$ at the point $(6, \sqrt{3})$.
35. Find an equation of the tangent line to the hyperbola $\dfrac{x^2}{4} - \dfrac{y^2}{2} = 1$ at the point $(6, 4)$.
36. Prove that the ellipse

$$\frac{x^2}{a^2} + \frac{y^2}{b^2} = 1$$

and the hyperbola

$$\frac{x^2}{a^2 - 2b^2} - \frac{y^2}{b^2} = 1$$

intersect at right angles.

37. A circle with center $(4, \frac{3}{2})$ is tangent to the line $x + 2y = 0$. Write the general equation of the circle.
38. A circle with center $(2, -4)$ passes through the point $(5, 0)$.
   (a) Write a general equation of the circle.
   (b) Find an equation of the tangent line to the circle at the point $(5, 0)$.
39. Find the points of intersection (if any exist) of the conics $x^2 = 3y$ and $3x^2 + y^2 - 9y - 25 = 0$.
40. Find the points of intersection (if any exist) of the conics $\dfrac{x^2}{4} + \dfrac{y^2}{2} = 1$ and $x^2 - y^2 = 2$.
41. Find the distance from the point $(3, 4)$ to the line $5x - 12y + 20 = 0$.
42. Find the distance between the parallel lines $5x - 12y + 20 = 0$ and $5x - 12y - 2 = 0$.
43. Find the angle of inclination of the line $x + \sqrt{3}\,y = 4$.
44. Find the angle of inclination of the line $5x - 3y = 7$.
45. Find the angle of inclination of the line $2x - 5 = 0$.
46. Find the angle between the lines $4x + 2y = 1$ and $x - y = 3$.
47. Given a triangle with vertices $(1, 1)$, $(4, 2)$ and $(7, 10)$, find
   (a) the equations of the lines forming the sides of the triangle.
   (b) the interior angles of the triangle by finding the angle between the lines of part (a).
48. Repeat Exercise 47 if the vertices of the triangle are $(-5, -3)$, $(2, 6)$, and $(3, 3)$.

POLAR COORDINATES AND PARAMETRIC EQUATIONS

Polar Coordinates

12.1

We have been representing a point in the plane by means of an ordered pair (x, y) of rectangular coordinates, which measure the distance of the point from the y- and x-axes, respectively. An alternative way of representing points in the plane is by **polar coordinates.** In this system a point (r, θ) is identified by its distance r from the origin of the system and an angle θ. We make this more specific in the following definition.

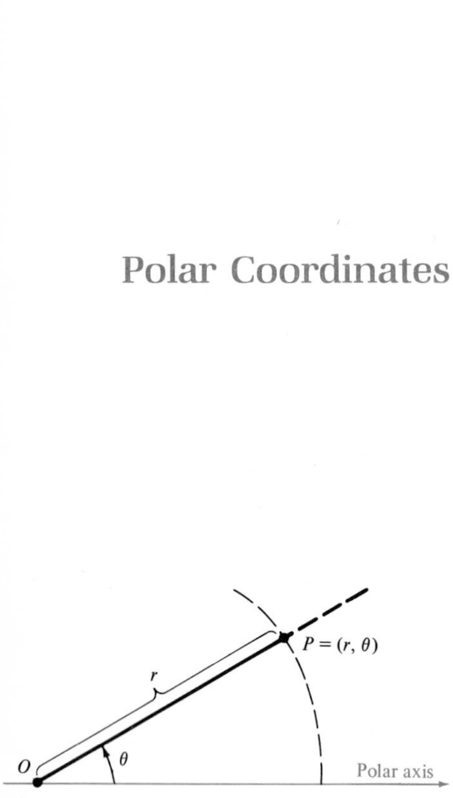

DEFINITION OF POLAR COORDINATE SYSTEM

> The **polar coordinate system** is a system of coordinates in which a point in the plane is identified by its distance r along a ray from a fixed point (the **pole**) and by the angle θ between a fixed line (the **polar axis**) and the ray. (See Figure 12.1.)

FIGURE 12.1

Every point P, other than the pole, can be represented by an ordered pair of real numbers (r, θ), where

r = directed distance from O to P

θ = directed angle from polar axis to OP

For an angle θ, the positive direction is counterclockwise and the negative direction is clockwise.

EXAMPLE 1
Plotting Points in the Polar
Coordinate System

Plot the points whose polar coordinates are $(2, \pi/4)$, $(3, -\pi/6)$, $(2, 9\pi/4)$, and $(-2, \pi/6)$.

Solution:
Even though there is only one axis in the polar coordinate system, we usually include a second reference axis corresponding to $\theta = \pi/2$, as shown in Figure 12.2.

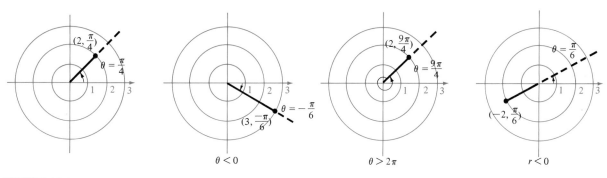

FIGURE 12.2

Remark: In Figure 12.2 note that $\theta = \pi/6$ is a first quadrant angle and yet the point $(-2, \pi/6)$ lies in the third quadrant. This occurs because r is negative. In general, a point (r, θ) lies in the same quadrant as θ if r is positive and in the quadrant on the opposite side of the pole if r is negative.

In the rectangular coordinate system, each point in the plane has a unique representation, but in the polar coordinate system, each point in the plane has many representations. For example, the polar coordinates $(2, \pi/4)$ and $(2, 9\pi/4)$ represent the same point (Figure 12.2). As a matter of fact, the polar coordinates

$$\left(2, \frac{\pi}{4} + 2n\pi\right) \qquad \text{and} \qquad \left(-2, \frac{5\pi}{4} + 2n\pi\right)$$

all represent the same point for any integer value of n. We summarize this situation as follows.

MULTIPLE REPRESENTATION OF POINTS IN POLAR COORDINATES

The point (r, θ) can also be represented by

$$(r, \theta + 2n\pi) \qquad \text{or} \qquad (-r, \theta + (2n + 1)\pi)$$

where n is an integer. Furthermore, the pole can be represented by $(0, \theta)$ for any angle θ.

EXAMPLE 2
Multiple Representations
of a Point

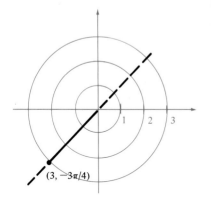

(3, −3π/4)

FIGURE 12.3

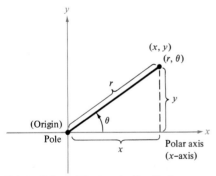

Relating Polar and Rectangular Coordinates

FIGURE 12.4

Plot the point $(3, -3\pi/4)$ and find three additional polar coordinate representations of this point, using $-2\pi < \theta < 2\pi$.

Solution:
The point is plotted in Figure 12.3. Three other representations for $(3, -3\pi/4)$ are

$$\left(3, \frac{-3\pi}{4} + 2\pi\right) = \left(3, \frac{5\pi}{4}\right)$$

$$\left(-3, \frac{-3\pi}{4} - \pi\right) = \left(-3, \frac{-7\pi}{4}\right)$$

$$\left(-3, \frac{-3\pi}{4} + \pi\right) = \left(-3, \frac{\pi}{4}\right)$$

To establish the relationship between polar and rectangular coordinates, we let the polar axis coincide with the x-axis and the pole with the origin. From Figure 12.4 we see that the following relationships hold:

$$\tan \theta = \frac{y}{x}$$

$$r^2 = x^2 + y^2$$

$$x = r \cos \theta$$

$$y = r \sin \theta$$

These relationships allow us to convert points or equations from one system to the other. Since the representation of a point in rectangular coordinates is unique, the conversion from polar to rectangular coordinates is straightforward and is indicated in the following rule.

POLAR-TO-RECTANGULAR CONVERSION

To convert a point (r, θ) to the rectangular form (x, y) use the equations

$$x = r \cos \theta \qquad \text{and} \qquad y = r \sin \theta$$

EXAMPLE 3
Conversion from Polar
to Rectangular Form

Find the rectangular coordinates of the points whose polar coordinates are

(a) $(-2, 5\pi/6)$ (b) $(3, 4\pi/3)$

Solution:

(a) For $(r, \theta) = (-2, 5\pi/6)$, we have

$$x = r \cos \theta = -2 \cos \frac{5\pi}{6} = -2\left(-\frac{\sqrt{3}}{2}\right) = \sqrt{3}$$

$$y = r \sin \theta = -2 \sin \frac{5\pi}{6} = -2\left(\frac{1}{2}\right) = -1$$

Therefore, the rectangular coordinates are $(\sqrt{3},\ -1)$ (see Figure 12.5).
(b) For $(r,\ \theta)\ =\ (3,\ 4\pi/3)$, we have

$$x = r \cos \theta = 3 \cos \frac{4\pi}{3} = 3\left(-\frac{1}{2}\right) = -\frac{3}{2}$$

$$y = r \sin \theta = 3 \sin \frac{4\pi}{3} = 3\left(-\frac{\sqrt{3}}{2}\right) = -\frac{3\sqrt{3}}{2}$$

and the rectangular coordinates are $(-\frac{3}{2},\ -\frac{3}{2}\sqrt{3})$ (see Figure 12.6).

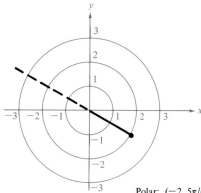

Polar: $(-2, 5\pi/6)$
Rectangular: $(\sqrt{3}, -1)$

FIGURE 12.5

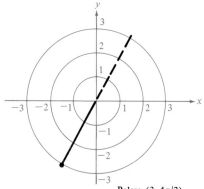

Polar: $(3, 4\pi/3)$
Rectangular: $(-3/2, -3\sqrt{3}/2)$

FIGURE 12.6

Conversion from rectangular to polar coordinates is complicated some-what by the fact that points in the polar coordinate system have multiple representations.

RECTANGULAR-TO-POLAR CONVERSION

To convert a point $(x,\ y)$ to the polar form $(r,\ \theta)$ use the equations

$$\tan \theta = \frac{y}{x} \quad \text{and} \quad r = \pm\sqrt{x^2 + y^2}$$

For values of θ in the same quadrant as $(x,\ y)$, choose r to be positive. For values of θ in the opposite quadrant from $(x,\ y)$, choose r to be negative.

EXAMPLE 4
Conversion from Rectangular to Polar Form

Convert from rectangular to polar form:

(a) $(1,\ -1)$

(b) $(-\frac{1}{2},\ \sqrt{3}/2)$

Solution:

(a) For $(x, y) = (1, -1)$, we have

$$r = \pm\sqrt{x^2 + y^2} = \pm\sqrt{(1)^2 + (-1)^2} = \pm\sqrt{2}$$

$$\tan\theta = \frac{y}{x} = \frac{-1}{1} = -1$$

and thus

$$\theta = \ldots, \frac{-\pi}{4}, \frac{3\pi}{4}, \frac{7\pi}{4}, \ldots$$

Two possible polar coordinate representations are

$$\left(\sqrt{2}, \frac{-\pi}{4}\right) \quad \text{and} \quad \left(-\sqrt{2}, \frac{3\pi}{4}\right)$$

(b) For $(x, y) = (-\frac{1}{2}, \sqrt{3}/2)$, we have

$$r = \pm\sqrt{x^2 + y^2} = \pm\sqrt{\frac{1}{4} + \frac{3}{4}} = \pm 1$$

$$\tan\theta = \frac{\sqrt{3}/2}{-1/2} = -\sqrt{3}$$

and hence

$$\theta = \ldots, \frac{2\pi}{3}, \frac{5\pi}{3}, \ldots$$

Two possible polar representations are

$$\left(1, \frac{2\pi}{3}\right) \quad \text{and} \quad \left(-1, \frac{5\pi}{3}\right)$$

In the remaining discussion of polar coordinates, we will assume, unless otherwise specified, that whenever the coordinates of a point are given, these numbers are *polar* coordinates of the point. Furthermore, we refer to an equation involving the variables r and θ as a **polar equation.**

The relationships between polar and rectangular coordinates also allow us to convert *equations* from one form to another. The next two examples illustrate the procedure.

EXAMPLE 5
Conversion from Polar
to Rectangular Form

Find a rectangular equation that has the same graph as the polar equation $r = 2\cos\theta$.

Solution:
With the conversion equations in mind, we note that it is convenient to multiply both sides of the equation by r. Then we have

$$r^2 = 2(r\cos\theta)$$

Since $x^2 + y^2 = r^2$ and $x = r \cos \theta$ we have

$$x^2 + y^2 = 2x$$

Remark: In Example 5 note that in the polar equation $r = 2 \cos \theta$, r is a *function* of θ, but in the rectangular equation $x^2 + y^2 = 2x$, neither variable is a function of the other.

EXAMPLE 6
Conversion from Rectangular to Polar Form

Find a polar equation that has the same graph as the rectangular equation $y = x^3$.

Solution:
Substituting $x = r \cos \theta$ and $y = r \sin \theta$, we have

$$y = x^3$$
$$r \sin \theta = (r \cos \theta)^3$$

Although this equation is acceptable as it stands, it can be simplified by rewriting it as

$$r(\sin \theta - r^2 \cos^3 \theta) = 0$$

This implies that

$$r = 0 \quad \text{or} \quad \sin \theta - r^2 \cos^3 \theta = 0$$

Since the pole is the only point satisfying the equation $r = 0$, and since the coordinates $(0, 0)$ also satisfy the second equation, we can drop $r = 0$, so we are left with the polar equation

$$\sin \theta - r^2 \cos^3 \theta = 0$$

At this point you might be asking, "Why another coordinate system?" One reason is that some curves in the plane have simpler polar equations than rectangular ones. For example, the circle $x^2 + y^2 = 4$ has the simple polar equation $r = 2$. In the next section we will show that conics with a focus at the pole have quite simple polar equations. In general, polar coordinates are most useful with graphs that have an *axis of symmetry* that passes through the pole. This will be seen more clearly in Section 12.3.

Section Exercises 12.1 ——————————————————————

In Exercises 1–8, the polar coordinates of a point are given. Plot the point and find the rectangular coordinates for the same point.

1. $(4, 3\pi/6)$
2. $(4, 3\pi/2)$
3. $(-1, 5\pi/4)$
4. $(0, -\pi)$
5. $(4, -\pi/3)$
6. $(-1, -3\pi/4)$
7. $(\sqrt{2}, 2.36)$
8. $(-3, -1.57)$

In Exercises 9–16, the rectangular coordinates of a point are given. In each case find two sets of polar coordinates for the point, using $0 \leq \theta < 2\pi$.

9. $(1, 1)$
10. $(0, -5)$
11. $(-3, 4)$
12. $(3, -1)$
13. $(-\sqrt{3}, -\sqrt{3})$
14. $(-2, 0)$
15. $(4, 6)$
16. $(5, 12)$

In Exercises 17–23, find a polar equation of the graph having the given rectangular equation.

17. $x^2 + y^2 = a^2$ 18. $x^2 + y^2 - 2ay = 0$

19. $(x^2 + y^2)^2 - 9(x^2 - y^2) = 0$

20. $4x + 7y - 2 = 0$ 21. $x^2 + y^2 - 2ax = 0$

22. $y^2 - 8x - 16 = 0$ 23. $x^2 - 4ay - 4a^2 = 0$

In Exercises 24–31, find a rectangular equation of the graph having the given polar equation.

24. $r = 4 \cos \theta$ 25. $r = 4 \sin \theta$

26. $r = \dfrac{1}{1 - \cos \theta}$ 27. $r = 1 - 2 \sin \theta$

28. $r^2 = \sin 2\theta$ 29. $r = 2 \sin 3\theta$

30. $r = \dfrac{6}{2 - 3 \sin \theta}$ 31. $r = \dfrac{6}{2 \cos \theta - 3 \sin \theta}$

32. Show that the distance between (r_1, θ_1) and (r_2, θ_2) is

$$\sqrt{r_1^2 + r_2^2 - 2r_1 r_2 \cos(\theta_1 - \theta_2)}$$

33. Show that (r_1, θ_1), (r_2, θ_2), and (r_3, θ_3) are collinear if and only if

$$r_2 r_3 \sin(\theta_3 - \theta_2) + r_3 r_1 \sin(\theta_1 - \theta_3)$$
$$+ \, r_1 r_2 \sin(\theta_2 - \theta_1) = 0$$

34. (a) Find a polar equation of the vertical line whose rectangular equation is $x = k$.

    (b) Find a polar equation of the horizontal line whose rectangular equation is $y = k$.

Polar Equations of Conics   12.2

In Chapter 11 we studied each of the conic sections and gave a separate definition to each one. The following *general* definition lends itself well to polar representation of ellipses, parabolas, and hyperbolas.

GENERAL DEFINITION OF CONIC SECTION

> The set of all points whose distances from a fixed point are in constant ratio to their distances from a fixed line is called a **conic section.** The fixed point is called a **focus** of the conic, the fixed line is called the **directrix,** and the constant ratio is called the **eccentricity,** usually denoted by e.

The value of the eccentricity e determines whether the conic is a parabola, an ellipse, or a hyperbola.

If $e < 1$, the conic is an *ellipse*.

If $e = 1$, the conic is a *parabola*.

If $e > 1$, the conic is a *hyperbola*.

Recall from Chapter 11 that the rectangular equation of a conic takes a simpler form for conics whose axes are parallel to the coordinate axes. A comparable situation exists in polar coordinates. That is, if we assume that the conic has the pole as a focus and has a directrix parallel to the *x*- or *y*-axis, then the polar equation of the conic takes on a simple form. For

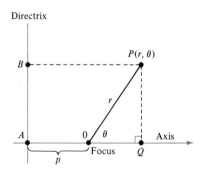

Directrix

$P(r, \theta)$

B

r

A

0 θ Axis

Focus Q

p

FIGURE 12.7

instance, in Figure 12.7 we let the directrix of the conic be vertical and lie to the left of the pole. To derive the polar equation for this conic, we need to find the set of points $P(r, \theta)$ so that the ratio $\overline{OP}/\overline{BP} = e$ is constant. We note that

$$\overline{BP} = p + \overline{OQ} = p + r \cos \theta$$

Since $r = \overline{OP} = e\overline{BP}$, it follows that

$$r = e(p + r \cos \theta) = ep + er \cos \theta$$
$$r(1 - e \cos \theta) = ep$$

$$r = \frac{ep}{1 - e \cos \theta}$$

If the directrix were to the *right* of the pole, a similar argument would yield the form

$$r = \frac{ep}{1 + e \cos \theta}$$

Finally, if the directrix were horizontal, then the polar equation would be of the form

$$r = \frac{ep}{1 \pm e \sin \theta}$$

We summarize these results as follows.

POLAR EQUATIONS OF CONICS

Polar equations of the form

$$r = \frac{ep}{1 \pm e \cos \theta}$$
Horizontal axis
Vertices $\theta = 0$ and/or $\theta = \pi$

$$r = \frac{ep}{1 \pm e \sin \theta}$$
Vertical axis
Vertices $\theta = \pi/2$ and/or $\theta = 3\pi/2$

represent conics with a *focus at the pole* and an axis on one of the coordinate axes. The eccentricity is e, and p is the distance between the focus and the directrix.

Remark: The number of directrices and axes of a conic corresponds to the number of foci. That is, a parabola has one directrix and one axis, while an ellipse (or a hyperbola) has two directrices and two axes.

EXAMPLE 1
Sketching the Graph of a
Conic in Polar Form

Describe and sketch the graph of the equation

$$r = \frac{4}{2 - 2 \sin \theta}$$

Solution:
Dividing each term by 2, we obtain

$$r = \frac{2}{1 - \sin \theta} = \frac{ep}{1 - e \sin \theta}$$

Thus, we conclude that $e = 1$ and the graph is a parabola with its focus at the pole and the y-axis as its axis. Since r is undefined at $\theta = \pi/2$, the vertex is at $(1, 3\pi/2)$. Plotting the vertex and the points $(2, 0)$ and $(2, \pi)$, we obtain the sketch shown in Figure 12.8.

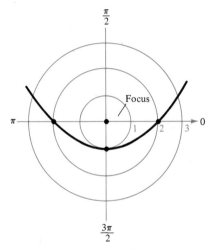

Parabola: $r = \dfrac{4}{2 - 2 \sin \theta}$

FIGURE 12.8

In our next example we use the fact that the eccentricity of an ellipse or hyperbola is $e = c/a$, where c and a are the distances from the center to a focus and the center to a vertex, respectively.

EXAMPLE 2
Sketching the Graph of a
Conic in Polar Form

Sketch the graph of the polar equation

$$r = \frac{32}{3 + 5 \cos \theta}$$

Solution:
Dividing each term by 3, we have

$$r = \frac{\frac{32}{3}}{1 + \left(\frac{5}{3}\right) \cos \theta}$$

Since $e = \frac{5}{3} > 1$, the graph of this equation is a hyperbola for which the pole is a focus. The transverse axis of the hyperbola lies on the x-axis, and we can determine the vertices by

$$\theta = 0, \qquad r = \frac{\frac{32}{3}}{1 + \frac{5}{3}} = \frac{\frac{32}{3}}{\frac{8}{3}} = 4$$

$$\theta = \pi, \qquad r = \frac{\frac{32}{3}}{1 - \frac{5}{3}} = \frac{\frac{32}{3}}{-\frac{2}{3}} = -16$$

From Figure 12.9 we see that the distance $2a$ between the vertices is 12, and thus $a = 6$. Furthermore, since $e = \frac{5}{3} = c/a$, we know that $c = 10$; and finally, $b = \sqrt{c^2 - a^2} = \sqrt{64} = 8$. Therefore, we have the sketch shown in Figure 12.9.

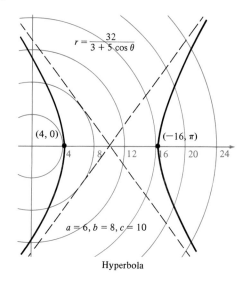

$$r = \frac{32}{3 + 5\cos\theta}$$

$(4, 0)$ $(-16, \pi)$

$a = 6, b = 8, c = 10$

Hyperbola

FIGURE 12.9

EXAMPLE 3
Finding a Polar Equation
for a Conic

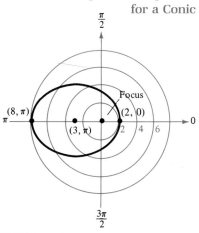

FIGURE 12.10   Ellipse

Find a polar equation for the ellipse having a focus at the pole and vertices at $(2, 0)$ and $(8, \pi)$, as shown in Figure 12.10.

Solution:
In this case, the major axis is horizontal and the equation is of the form

$$r = \frac{ep}{1 + e\cos\theta}$$

(We chose the $+$ sign because r increases as θ increases from 0 to π.) Since the ellipse is centered at $(3, \pi)$, we have $a = 5$ and $c = 3$, and thus, $e = \frac{3}{5}$. Therefore, our equation becomes

$$r = \frac{\frac{3}{5}p}{1 + \frac{3}{5}\cos 0}$$

Now since $(2, 0)$ lies on the ellipse, we obtain

$$2 = \frac{\frac{3}{5}p}{1 + \frac{3}{5}\cos\theta} = \frac{\frac{3}{5}p}{\frac{8}{5}} \qquad \text{or} \qquad p = \frac{16}{3}$$

Finally, the polar equation for the ellipse is

$$r = \frac{\frac{16}{5}}{1 + \frac{3}{5}\cos\theta} = \frac{16}{5 + 3\cos\theta}$$

Each of the polar equations in Examples 1, 2, and 3 could be converted to rectangular form to obtain the standard forms for conics studied in Chapter 11. We illustrate this procedure in Example 4.

EXAMPLE 4
Converting from Polar
to Rectangular Form

Convert the following polar equation of a conic to rectangular form:

$$r = \frac{4}{2 - 2\sin\theta}$$

Solution:
The object is to manipulate the given equation to form expressions involving r^2, $r\cos\theta$, and $r\sin\theta$.

$$r = \frac{2}{1 - \sin\theta}$$

$$r - r\sin\theta = 2$$

$$r = 2 + r\sin\theta$$

$$r^2 = (2 + r\sin\theta)^2$$

Now, using $r^2 = x^2 + y^2$ and $r\sin\theta = y$, we have

$$x^2 + y^2 = (2 + y)^2 = 4 + 2y + y^2$$

$$x^2 = 4 + 2y$$

In rectangular form, we recognize this as the equation of a parabola.

Section Exercises 12.2

In Exercises 1–6, match the given polar equation with one of the graphs given below and on page 509.

1. $r = \dfrac{6}{1 - \cos\theta}$

2. $r = \dfrac{2}{2 - \cos\theta}$

3. $r = \dfrac{3}{1 - 2\sin\theta}$

4. $r = \dfrac{2}{1 + \sin\theta}$

5. $r = \dfrac{6}{2 - \sin\theta}$

6. $r = \dfrac{2}{2 + 3\cos\theta}$

(a)

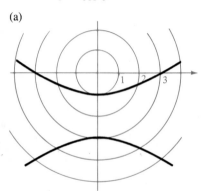

(b)

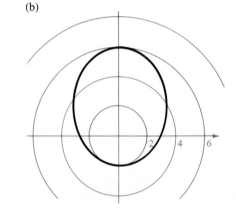

(c)

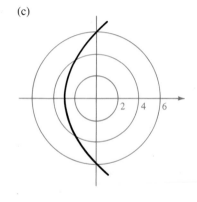

(d)

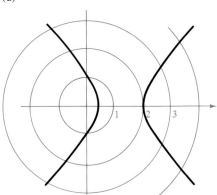

(e)

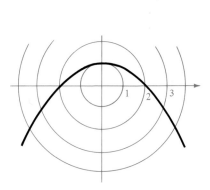

(f)

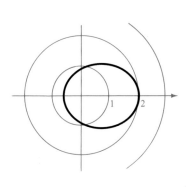

In Exercises 7–16, find a polar equation of the specified conic with focus at (0, 0).

7. Parabola; vertex $\left(1, -\dfrac{\pi}{2}\right)$

8. Ellipse; vertices $(2, 0)$, $(8, \pi)$

9. Ellipse; vertices $\left(2, \dfrac{\pi}{2}\right)$, $\left(4, \dfrac{3\pi}{2}\right)$

10. Parabola; vertex $(4, 0)$

11. Hyperbola; vertices $\left(1, \dfrac{3\pi}{2}\right)$, $\left(9, \dfrac{3\pi}{2}\right)$

12. Hyperbola; vertices $\left(4, \dfrac{\pi}{2}\right)$, $\left(-1, \dfrac{3\pi}{2}\right)$

13. Hyperbola; vertices $(2, 0)$, $(10, 0)$

14. Ellipse; vertices $(20, 0)$, $(4, \pi)$

15. Parabola; vertex $(5, \pi)$

16. Parabola; vertex $\left(10, \dfrac{\pi}{2}\right)$

In Exercises 17–26, classify the graph of each equation as a parabola, ellipse, or hyperbola. In each case write the equation of the conic in rectangular form.

17. $r = \dfrac{2}{1 - \cos \theta}$

18. $r(1 - 2 \cos \theta) = 4$

19. $r = \dfrac{2}{2 - \cos \theta}$

20. $r(3 + 2 \sin \theta) = 3$

21. $r = \dfrac{3}{2 - 6 \sin \theta}$

22. $r = \dfrac{4}{2 - 3 \cos \theta}$

23. $r(3 + 3 \sin \theta) = 2$

24. $r = \dfrac{10}{3 - 2 \sin \theta}$

25. $r(4 + 3 \cos \theta) = 12$

26. $r = \dfrac{4}{5 - 5 \sin \theta}$

27. Show that the corresponding polar equation of the ellipse $(x^2/a^2) + (y^2/b^2) = 1$ is

$$r^2 = \frac{b^2}{1 - e^2 \cos^2 \theta}$$

28. Show that the corresponding polar equation of the hyperbola $(x^2/a^2) - (y^2/b^2) = 1$ is

$$r^2 = \frac{-b^2}{1 - e^2 \cos^2 \theta}$$

In Exercises 29–34, use the results of Exercises 27 and 28 to write the polar form of the equation of the conic.

29. $\dfrac{x^2}{169} + \dfrac{y^2}{144} = 1$

30. $\dfrac{x^2}{25} + \dfrac{y^2}{16} = 1$

31. $\dfrac{x^2}{9} - \dfrac{y^2}{16} = 1$

32. $\dfrac{x^2}{36} - \dfrac{y^2}{4} = 1$

33. Hyperbola; one focus $\left(5, \dfrac{\pi}{2}\right)$;

vertices $\left(4, \dfrac{\pi}{2}\right)$, $\left(4, -\dfrac{\pi}{2}\right)$

34. Ellipse; one focus $(4, 0)$; vertices $(5, 0)$, $(5, \pi)$

35. On November 26, 1963, the United States launched *Explorer 18*. Its low point in its elliptical orbit over the surface of the earth was 119 miles, and its high point was 122,000 miles from the surface of the earth. (The center of the earth is the focus of the orbit.) Find the polar equation that describes its orbit. (See Exercises 36–38 of Section 11.4.)

36. Use the equation of Exercise 35 to find the distance from the earth's surface to the satellite when the angle between the line from the center of the earth to the satellite and the major axis of the ellipse is 60°.

37. Find the polar equation for the parabola described in Exercise 36 of Section 11.3.

Graphs of Polar Equations

12.3

In previous chapters of this text we devoted a considerable amount of time to curve sketching in rectangular coordinates. We began with the basic point-plotting method. We approach curve sketching in polar coordinates in a similar manner.

A useful aid in sketching the graphs of many common polar equations is their periodic nature. For example, if $r = f(\theta)$ has a period of 2π, then we can obtain its *entire* graph using only the interval $[0, 2\pi]$, since the graph merely retraces itself for θ outside this interval.

EXAMPLE 1
Sketching a Polar Curve

Use the point-plotting method to sketch the graph of the polar equation $r = 2 + \sin \theta$.

Solution:
Since $r = 2 + \sin \theta$ has a period of 2π, we can restrict θ in our table of coordinates to the interval $[0, 2\pi]$. After plotting and connecting these points, we obtain the graph shown in Figure 12.11.

θ	0	$\dfrac{\pi}{6}$	$\dfrac{\pi}{3}$	$\dfrac{\pi}{2}$	$\dfrac{2\pi}{3}$	$\dfrac{5\pi}{6}$	π	$\dfrac{7\pi}{6}$	$\dfrac{4\pi}{3}$	$\dfrac{3\pi}{2}$	$\dfrac{5\pi}{3}$	$\dfrac{11\pi}{6}$	2π
r	2	2.5	2.87	3	2.87	2.5	2	1.5	1.13	1	1.13	1.5	2

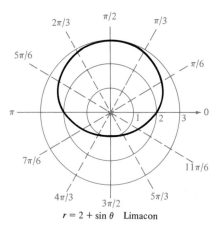

$r = 2 + \sin \theta$ Limacon

FIGURE 12.11

It appears that the graph in Figure 12.11 is symmetric with respect to the vertical axis, $\theta = \pi/2$. Had we known about this symmetry beforehand, our sketch could have been obtained from half as many points. As with rectangular coordinates, there are some simple tests for symmetry in polar coordinates. Figure 12.12 gives a graphic description of the three types of symmetry we will consider: symmetry with respect to the line $\theta = \pi/2$, the polar axis, and the pole.

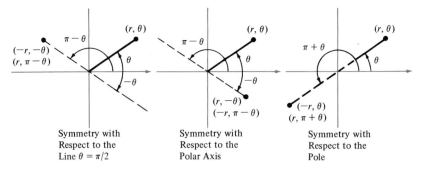

Symmetry with Respect to the Line $\theta = \pi/2$

Symmetry with Respect to the Polar Axis

Symmetry with Respect to the Pole

FIGURE 12.12

TESTS FOR SYMMETRY IN POLAR COORDINATES

> The graph of a polar equation is symmetric with respect to:
> 1. *the line* $\theta = \pi/2$ if replacing (r, θ) by either $(r, \pi - \theta)$ or $(-r, -\theta)$ yields an equivalent equation
> 2. *the polar axis* if replacing (r, θ) by either $(r, -\theta)$ or $(-r, \pi - \theta)$ yields an equivalent equation
> 3. *the pole* if replacing (r, θ) by either $(r, \pi + \theta)$ or $(-r, \theta)$ yields an equivalent equation

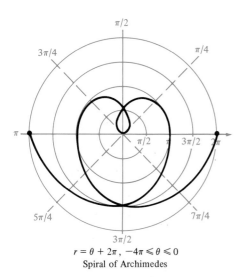

$r = \theta + 2\pi, \ -4\pi \leq \theta \leq 0$
Spiral of Archimedes

FIGURE 12.13

When applying this test it is important to remember that we are using only two of the many possible polar representations of the points under consideration. Consequently, these conditions are *sufficient* to guarantee the various types of symmetry, but they are not necessary. Thus, if a test indicates symmetry, then such symmetry will exist. On the other hand, if a test fails (suggesting no symmetry), the graph may still have that type of symmetry. For instance, Figure 12.13 shows the graph of $r = \theta + 2\pi$ to be symmetrical with respect to the line $\theta = \pi/2$. Yet the symmetry test fails to indicate symmetry. [Try to convince yourself of this by showing that the equation $r = \theta + 2\pi$ changes when (r, θ) is replaced by either $(-r, -\theta)$ or $(r, \pi - \theta)$.]

EXAMPLE 2
Testing for Symmetry

Use symmetry tests to show the following:

(a) The graph of $r = 2 + \sin \theta$ is symmetric with respect to the line $\theta = \pi/2$.
(b) The graph of $r = 1 - 2 \cos \theta$ is symmetric with respect to the polar axis.
(c) The graph of $r = 3 \sin 2\theta$ is symmetric with respect to the line $\theta = \pi/2$, the polar axis, and the pole.

Solution:

(a) If we replace (r, θ) by $(r, \pi - \theta)$, we obtain the equation

$$r = 2 + \sin (\pi - \theta) = 2 + \sin \theta$$

which is the same as the original equation, and therefore, the graph is symmetric with respect to the line $\theta = \pi/2$.

(b) If we replace (r, θ) by $(r, -\theta)$, we obtain the equation

$$r = 1 - 2 \cos (-\theta) = 1 - 2 \cos \theta$$

which is the same as the original equation, and therefore, the graph is symmetric with respect to the polar axis.

(c) If we replace (r, θ) by $(r, \pi - \theta)$, we obtain

$$r = 3 \sin (2\pi - 2\theta) = -3 \sin 2\theta$$

which is different from the original equation. Using the other substitution, we replace (r, θ) by $(-r, -\theta)$ and obtain

$$-r = 3 \sin(-2\theta) = -3 \sin 2\theta$$

which is equivalent to the original equation, and hence the graph is symmetric with respect to the line $\theta = \pi/2$. For polar axis symmetry, the second substitution yields an equivalent equation. Finally, since the graph is symmetric with respect to both the line $\theta = \pi/2$ and the polar axis, it must also be symmetric with respect to the pole.

As a final comment about symmetry in polar coordinates, we point out the following quick test for symmetry.

SYMMETRY FOR FUNCTIONS OF SINE AND COSINE

1. If r is a function of $\sin \theta$, the graph of r will be symmetric with respect to the line $\theta = \pi/2$.

2. If r is a function of $\cos \theta$, the graph of r will be symmetric with respect to the polar axis.

For example, the graphs of $r = 2 \sin \theta + \sin^2 \theta$, $r = 1/(3 + \sin \theta)$, and $r = \sqrt{\sin \theta + 1}$ all have vertical axis symmetry, and the graphs of $r = 2 \cos \theta + \cos^2 \theta$, $r = 1/(3 + \cos \theta)$, and $r = \sqrt{\cos \theta + 1}$ all have polar axis symmetry.

An additional aid to sketching polar graphs is knowing the θ values that yield the maximum value for $|r|$. These values can often be found by inspection. For instance, in Example 1 the maximum value of $r = 2 + \sin \theta$ occurs when $\theta = \pi/2$, as shown in Figure 12.11.

EXAMPLE 3
Sketching a Polar Curve

Sketch a graph of $r = 1 - 2 \cos \theta$.

Solution:
We have

Period: 2π

Symmetry: with respect to the polar axis

Maximum value of $|r|$: $r = 3$ when $\theta = \pi$

Making a table of points for several θ values in the interval $[0, \pi]$ and plotting them, we obtain the graph shown in Figure 12.14.

θ	0	$\dfrac{\pi}{6}$	$\dfrac{\pi}{3}$	$\dfrac{\pi}{2}$	$\dfrac{2\pi}{3}$	$\dfrac{5\pi}{6}$	π
r	-1	-0.73	0	1	2	2.73	3

Remark: Note how the negative *r* values determine the *inner loop* of the graph in Figure 12.14.

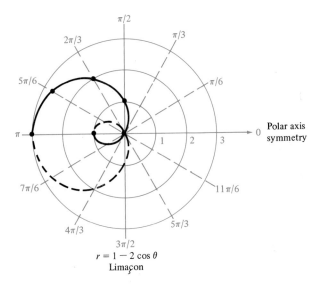

$$r = 1 - 2 \cos \theta$$
Limaçon

FIGURE 12.14

Some curves reach their maximum *r* values at more than one point. We show how to handle this situation in the next example.

EXAMPLE 4
Sketching a Polar Curve

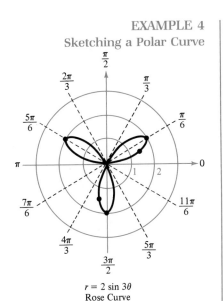

$$r = 2 \sin 3\theta$$
Rose Curve

FIGURE 12.15

Sketch a graph of $r = 2 \sin 3\theta$.

Solution:
We have

Period: $2\pi/3$

Symmetry: with respect to the line $\theta = \pi/2$

Maximum value of $|r|$: $r = 2$ when $3\theta = \pi/2, 3\pi/2, 5\pi/2$

or $\qquad \theta = \pi/6, \pi/2, 5\pi/6$

θ	0	$\dfrac{\pi}{12}$	$\dfrac{\pi}{6}$	$\dfrac{\pi}{3}$	$\dfrac{5\pi}{12}$	$\dfrac{\pi}{2}$	$\dfrac{5\pi}{6}$
r	0	1.41	2	0	-1.41	-2	2

Plotting these points and using the specified symmetry, we obtain the graph shown in Figure 12.15.

Each of the curves discussed in Examples 1 through 4 can be classified as either a limaçon or a rose curve. Some of the characteristics of these two types of curves are outlined in the following summary.

SPECIAL POLAR CURVES

1. The graphs of equations of the form

$$r = a \pm b \cos \theta \qquad \text{or} \qquad r = a \pm b \sin \theta$$

are called **limaçons.** Furthermore,

    i.   if $|a| < |b|$, the limaçon has two loops (Figure 12.14),
    ii.   if $|a| = |b|$, the limaçon has one heart-shaped loop and is called a **cardioid,**
    iii.   if $|a| > |b|$, the limaçon has one flattened loop (Figure 12.11).

2. The graphs of equations of the form

$$r = a \cos n\theta \qquad \text{or} \qquad r = a \sin n\theta \qquad (n \geq 2)$$

are called **rose curves.** Furthermore,

    i.   if n is odd, the curve has n petals (Figure 12.15),
    ii.   if n is even, the curve has $2n$ petals (Figure 12.16).

If r is replaced by r^2 in the equation for a rose curve, the number of petals will change to $2n$ if n is odd and to n if n is even. Two-petal curves resulting from this substitution are called **lemniscates.**

EXAMPLE 5
Sketching a Polar Curve

Sketch the graph of $r = 3 \cos 2\theta$.

Solution:
We have

    Type of curve: rose curve with $2n = 4$ petals
    Period: π
    Symmetry: with respect to the polar axis and the line $\theta = \pi/2$
    Maximum value of $|r|$: when $2\theta = 0, \pi, 2\pi, 3\pi$
$$\text{or} \qquad \theta = 0, \pi/2, \pi, 3\pi/2$$

θ	0	$\dfrac{\pi}{6}$	$\dfrac{\pi}{4}$	$\dfrac{\pi}{3}$	$\dfrac{\pi}{2}$
r	3	1.5	0	-1.5	-3

The graph is shown in Figure 12.16.

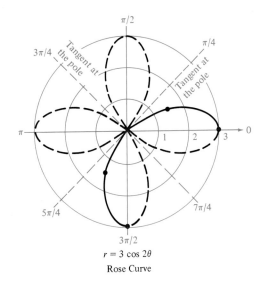

$r = 3 \cos 2\theta$

Rose Curve

FIGURE 12.16

Points of Intersection

Two polar equations can have the same polar graph even though they differ algebraically. Perhaps the simplest example of this phenomenon is seen in the two polar equations $r = 1$ and $r = -1$. These two equations are not equivalent, and yet the polar graph of each equation is the unit circle with its center at the pole.

This possibility of representing a graph by more than one polar equation leads to complications when we try to find the points of intersection of two polar graphs. As a case in point, let us consider the points of intersection of the graphs of

$$r = 1 - 2 \cos \theta \qquad \text{and} \qquad r = 1$$

as shown in Figure 12.17.

If, as with rectangular equations, we attempt to find the points of intersection by solving the two equations simultaneously, we have

$$1 = 1 - 2 \cos \theta$$
$$2 \cos \theta = 0$$
$$\cos \theta = 0$$
$$\theta = \frac{\pi}{2}, \frac{3\pi}{2} \qquad (0 \le \theta < 2\pi)$$

from which we obtain the two points

$$\left(1, \frac{\pi}{2}\right) \qquad \text{and} \qquad \left(1, \frac{3\pi}{2}\right)$$

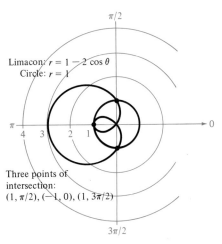

Limaçon: $r = 1 - 2 \cos \theta$
Circle: $r = 1$

Three points of intersection: $(1, \pi/2), (-1, 0), (1, 3\pi/2)$

FIGURE 12.17

To try to find the third point, we can choose an *alternative* polar equation for one of the graphs. For example, if we replace the polar equation $r = 1$ by $r = -1$ and again solve simultaneously, we have

$$-1 = 1 - 2 \cos \theta$$
$$2 \cos \theta = 2$$
$$\cos \theta = 1$$
$$\theta = 0 \qquad (0 \leq \theta < 2\pi)$$

which gives us the third point of intersection, $(-1, 0)$.

We can compare the problem of finding the points of intersection of two polar graphs with the situation of two satellites in orbit about the earth. (See Figure 12.18.) It is entirely possible for the paths of the two satellites to intersect without there being a collision as long as the satellites reach the points of intersection at different times. A collision will occur only if the satellites reach a particular point of intersection at the *same* time. Similarly, when we solve two polar equations simultaneously, we find only those points of intersection that have the same θ-values. With this in mind, we suggest the following procedure for finding the points of intersection of two polar graphs:

FIGURE 12.18

FINDING POINTS OF INTERSECTION IN POLAR COORDINATES

1. Solve the given polar equations simultaneously and determine the resulting points of intersection.
2. Replace r by $-r$ and θ by $\theta + \pi$ in *one* equation and solve simultaneously with the *other* equation to determine additional points of intersection.
3. Test to see if the pole is a point of intersection.
4. Make a rough sketch of both graphs to decide if any points of intersection have been missed.

EXAMPLE 6
Finding Points of Intersection

Find the points of intersection of the graphs of

$$r = 1 + 3 \cos \theta \qquad \text{and} \qquad r = \frac{2}{2 - \cos \theta}$$

Solution:
Solving simultaneously, we have

$$1 + 3 \cos \theta = \frac{2}{2 - \cos \theta}$$
$$-3 \cos^2 \theta + 5 \cos \theta + 2 = 2$$
$$\cos \theta(-3 \cos \theta + 5) = 0$$

The second factor is never zero, and the first factor is zero when $\theta = \pi/2$ or $\theta = 3\pi/2$. Hence we have the two points of intersection

$$\left(1, \frac{\pi}{2}\right) \qquad \text{and} \qquad \left(1, \frac{3\pi}{2}\right)$$

To test for other points of intersection, we replace r by $-r$ and θ by $\theta + \pi$ in the equation $r = 1 + 3\cos\theta$. Thus we have

$$-r = 1 + 3\cos(\theta + \pi) \qquad \text{or} \qquad r = -1 + 3\cos\theta$$

Now, solving the equations

$$r = -1 + 3\cos\theta \qquad \text{and} \qquad r = \frac{2}{2 - \cos\theta}$$

simultaneously, we have

$$-1 + 3\cos\theta = \frac{2}{2 - \cos\theta}$$

$$-3\cos^2\theta + 7\cos\theta - 2 = 2$$

$$-3\cos^2\theta + 7\cos\theta - 4 = 0$$

$$(\cos\theta - 1)(-3\cos\theta + 4) = 0$$

Again, the second factor is never zero, but this time the first factor is zero when $\theta = 0$. Hence $(2, 0)$ is another point of intersection. [Note that $(2, 0)$ coincides with $(-2, \pi)$.] Finally, from Figure 12.19 we can see that there are only three points of intersection: $(1, \pi/2)$, $(1, 3\pi/2)$, and $(2, 0)$.

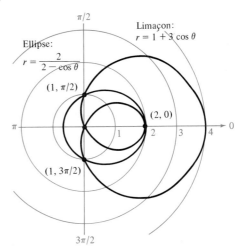

FIGURE 12.19

EXAMPLE 7
Finding Points of Intersection

Find the points of intersection of the graphs given by

$$r = \sin\theta \qquad \text{and} \qquad r = \cos\left(\theta + \frac{\pi}{6}\right)$$

Solution:

Solving simultaneously, we have

$$\sin \theta = \cos \left(\theta + \frac{\pi}{6} \right)$$

$$\sin \theta = \cos \theta \cos \frac{\pi}{6} - \sin \theta \sin \frac{\pi}{6}$$

$$\sin \theta = \frac{\sqrt{3}}{2} \cos \theta - \frac{1}{2} \sin \theta$$

$$\frac{3}{2} \sin \theta = \frac{\sqrt{3}}{2} \cos \theta$$

$$\tan \theta = \frac{\sqrt{3}}{3}$$

$$\theta = \frac{\pi}{6}, \frac{7\pi}{6} \qquad (0 \le \theta < 2\pi)$$

As it turns out, both of these values of θ yield the same point of intersection, $(\frac{1}{2}, \pi/6)$.

Now when we try to construct alternative equations of $r = \sin \theta$ or of $r = \cos [\theta + (\pi/6)]$, we find that the alternative equations are the same as the original ones. Nevertheless, from a sketch of the graphs of these two equations, we can see that the pole is a second point of intersection (Figure 12.20).

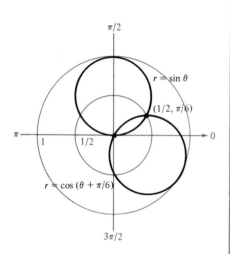

FIGURE 12.20

Section Exercises 12.3

In Exercises 1–24, sketch the graph of each equation and identify all limaçons, cardioids, rose curves, and conics.

1. $r = 3(1 - \cos \theta)$
2. $r = 2(1 - \sin \theta)$
3. $r = 2 + 3 \sin \theta$
4. $r = 4 + 5 \cos \theta$
5. $r = 3 - 2 \cos \theta$
6. $r = 5 - 4 \sin \theta$
7. $r = 5$
8. $r = -2$
9. $r = \theta$
10. $r = 3 \cos \theta$
11. $r = 3 \sin \theta$
12. $r = 4 \sin 2\theta$
13. $r = 2 \cos 3\theta$
14. $r = -\sin 5\theta$
15. $r = 3 \cos 2\theta$
16. $r^2 = 4 \sin \theta$
17. $r^2 = 4 \cos 2\theta$
18. $r^2 = \cos 3\theta$
19. $r^2 = 9 \sin 3\theta$
20. $r = \dfrac{6}{2 \sin \theta - 3 \cos \theta}$
21. $r = 2 \sec \theta$
22. $r = \dfrac{6}{1 - \cos \theta}$
23. $r = \dfrac{2}{2 - \cos \theta}$
24. $r = \dfrac{4}{6 + \sin \theta}$
25. Sketch the graph of $r = 2 \cos (3\theta/2)$.
26. Sketch the graph of $r = 3 \sin (5\theta/2)$.

In Exercises 27–38, find the points of intersection of the graphs of the given equations.

27. $r = 1 + \cos \theta; r = 1 - \cos \theta$
28. $r = 3(1 + \sin \theta); r = 3(1 - \sin \theta)$
29. $r = 1 + \cos \theta; r = 1 - \sin \theta$
30. $r = 2 - 3 \cos \theta; r = \cos \theta$
31. $r = 4 - 5 \sin \theta; r = 3 \sin \theta$
32. $r = 4 \sin 2\theta; r = 2$
33. $r = 4 \sin 2\theta; r = 2 \sin \theta$
34. $\theta = \dfrac{\pi}{4}; r = 2$
35. $r = \dfrac{\theta}{2}; r = 2$
36. $r = 3 + \sin \theta; r = 2 \csc \theta$
37. $r = 2 + 3 \cos \theta; r = \dfrac{\sec \theta}{2}$
38. $r = 3(1 - \cos \theta); r = \dfrac{6}{1 - \cos \theta}$

Plane Curves and Parametric Equations

12.4

Up to this point we have been representing all curves (graphs) in the plane by a single equation in two variables. In many applications it is convenient to think of a graph as the curve traced out by a moving point in the plane. In this context, we use three variables: the usual position coordinates x and y and a third variable called a **parameter.** Frequently, this parameter is the *time t* that a moving object is at a particular position. Such motion is referred to as **curvilinear motion** in the plane. In such a setting a curve in the plane is represented by two **parametric equations,** one expressing the x-coordinate and the other the y-coordinate in terms of the parameter.

Consider the example of a projectile shot into the air at an angle of 45°. It will travel along the parabolic path given by

$$y = -\frac{x^2}{72} + x$$

With time t as the parameter, this path can also be represented by the parametric equations

$$x = 24\sqrt{2}t, \qquad y = -16t^2 + 24\sqrt{2}t$$

With these two equations, we can easily determine the following positions of the projectile, as shown in Figure 12.21:

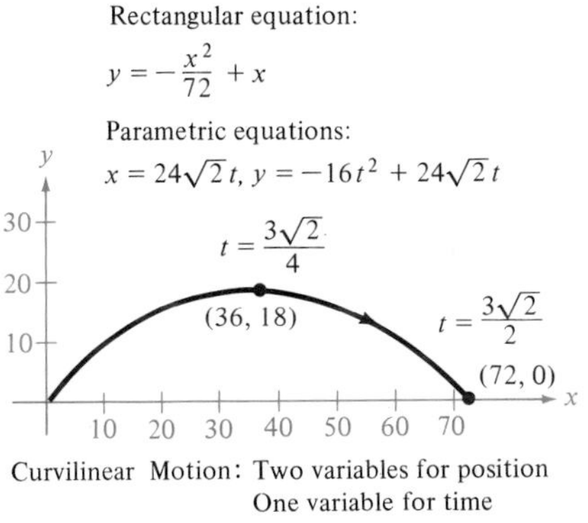

When $t = 0,$ $\qquad (x, y) = (0, 0).$

When $t = 3\sqrt{2}/4,$ $\qquad (x, y) = (36, 18).$

When $t = 3\sqrt{2}/2,$ $\qquad (x, y) = (72, 0).$

Although we have been informally using the terms *curve* and *graph* interchangeably, a curve is actually a special type of graph. Note that the following definition requires that f and g be *continuous* functions. A technical discussion of this term is covered in calculus. However, for our purposes it is sufficient to say that a function is continuous if its graph can be traced with a pencil without lifting the pencil from the paper.

Rectangular equation:

$$y = -\frac{x^2}{72} + x$$

Parametric equations:

$$x = 24\sqrt{2}\,t, \; y = -16t^2 + 24\sqrt{2}\,t$$

$t = \dfrac{3\sqrt{2}}{4}$

$(36, 18)$

$t = \dfrac{3\sqrt{2}}{2}$

$(72, 0)$

Curvilinear Motion: Two variables for position
One variable for time

FIGURE 12.21

DEFINITION OF PLANE CURVE

A **plane curve** C is the set of all points $(f(t), g(t))$ satisfying the **parametric equations**

$$x = f(t) \qquad \text{and} \qquad y = g(t)$$

where f and g are continuous functions of the **parameter** t on an interval I.

Remark: When sketching the curve represented by a pair of parametric equations, we still plot points (x, y) with each coordinate being determined from the value chosen for the parameter t. Once the points are plotted, we connect them by a smooth curve in the order of increasing (or decreasing) values of t.

EXAMPLE 1
Sketching a Plane Curve

Sketch the curve described by the parametric equations

$$x = \frac{t}{2} \quad \text{and} \quad y = t^2 - 4$$

for $-2 \le t \le 3$.

Solution:
From the table of values, we make the sketch shown in Figure 12.22. Note that we have labeled several points with their corresponding value for t. From this we see how a point P traces out the curve as we increase the value of t.

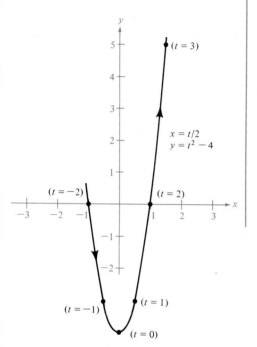

FIGURE 12.22

t	-2	-1	0	1	2	3
x	-1	$-\frac{1}{2}$	0	$\frac{1}{2}$	1	$\frac{3}{2}$
y	0	-3	-4	-3	0	5

From Figure 12.22 it appears that the curve represented by the parametric equations $x = t/2$ and $y = t^2 - 4$ is a parabola. To verify this, we can find the rectangular equation of the curve by a process called **eliminating the parameter.** In this particular instance we can eliminate the parameter by solving for t in the equation $x = t/2$ and substituting this value for t into the equation $y = t^2 - 4$. We suggest the following scheme for eliminating a parameter and obtaining the rectangular equation of a curve:

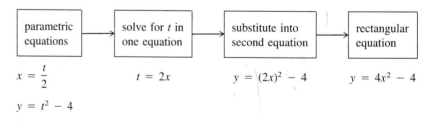

parametric equations	solve for t in one equation	substitute into second equation	rectangular equation
$x = \dfrac{t}{2}$ $y = t^2 - 4$	$t = 2x$	$y = (2x)^2 - 4$	$y = 4x^2 - 4$

Remark: Parametric equations for a curve C are not unique. For instance, the curve in Example 1 can also be represented parametrically by

$$x = 3t, \quad y = 36t^2 - 4 \quad \text{where} \quad -\frac{1}{3} \le t \le \frac{1}{2}$$

You should verify the equivalence of this representation by eliminating the parameter.

To eliminate the parameter in equations involving trigonometric functions, we suggest a slightly different approach, one that is based on familiar trigonometric identities. This approach is demonstrated in the next example.

EXAMPLE 2
Sketching a Plane Curve

Given the parametric equations

$$x = 3 \cos t \quad \text{and} \quad y = 4 \sin t$$

(a) Eliminate the parameter and determine the corresponding rectangular equation.
(b) Sketch a graph of the curve represented by the given parametric equations.

Solution:

(a) Since the parametric equations involve the sine and cosine, we first think of a trigonometric identity that involves these two functions. In this case we consider the identity $\sin^2 t + \cos^2 t = 1$. From the parametric equations we have

$$\frac{y}{4} = \sin t \quad \text{and} \quad \frac{x}{3} = \cos t$$

Therefore, using this familiar trigonometric identity, we obtain

$$\sin^2 t + \cos^2 t = \left(\frac{y}{4}\right)^2 + \left(\frac{x}{3}\right)^2 = 1$$

and the rectangular equation is

$$\frac{y^2}{16} + \frac{x^2}{9} = 1$$

(b) Now from this rectangular equation we recognize the curve to be an ellipse with a vertical major axis, where $a = 4$ and $b = 3$. (See Figure 12.23.)

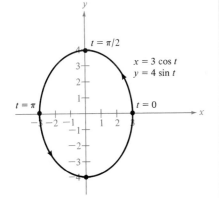

FIGURE 12.23

In addition to the methods illustrated in Examples 1 and 2, there are many other ingenious devices for eliminating the parameter from two parametric equations. For instance, we can sometimes eliminate the parameter by adding or subtracting multiples of the two equations. However, you should be aware that in some instances it may be impossible to eliminate the parameter. In such cases the curve must be plotted point by point directly from the parametric equations.

One word of caution is in order when converting equations from parametric to rectangular form: the range of x and y implied by the parametric equations may be altered by the change to rectangular form. In such instances it is necessary to alter the domain of the rectangular equation so that its graph matches the graph of the parametric equations. Such a situation is demonstrated in the next example.

EXAMPLE 3
**Comparing Graphs
of Rectangular and
Parametric Forms**

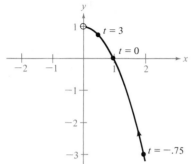

Parametric Equations:
$$x = \frac{1}{\sqrt{t + 1}}, \ y = \frac{t}{t + 1}$$

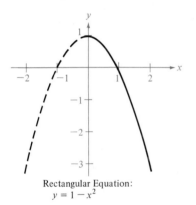

Rectangular Equation:
$y = 1 - x^2$

FIGURE 12.24

Given the parametric equations

$$x = \frac{1}{\sqrt{t + 1}} \quad \text{and} \quad y = \frac{t}{t + 1}$$

(a) Find the corresponding rectangular equation.
(b) Sketch the graph of the curve.

Solution:

(a) Solving for t in the equation

$$x = \frac{1}{\sqrt{t + 1}}, \quad -1 < t < \infty$$

yields

$$x^2 = \frac{1}{t + 1} \quad \text{or} \quad t = \frac{1 - x^2}{x^2}$$

By substituting this value for t into the given equation for y, we obtain

$$y = \frac{t}{t + 1} = \frac{(1 - x^2)/x^2}{[(1 - x^2)/x^2] + 1} = 1 - x^2$$

(b) The rectangular equation, $y = 1 - x^2$, is defined for *all* values of x, but from the parametric equations we see that x is defined only when $-1 < t < \infty$. Thus we restrict the domain of x to

$$x = \frac{1}{\sqrt{t + 1}} > 0$$

as shown in Figure 12.24.

In Examples 1 through 3 it is important to realize that eliminating the parameter is primarily an aid to curve sketching. In the case of moving objects, once we know the path of the object, we still need the parametric equations to tell us the position, direction, and speed at a given time. For instance, in Example 3 the parametric equations tell us that the object is moving up its parabolic path toward the limiting position of (0, 1) with an ever-decreasing speed. If this same path were traveled in a different manner, the parametric equations describing its position in time would be different.

Of course, it is not necessary that the parameter in a set of parametric equations represent time. The next example uses an angle θ as the parameter.

EXAMPLE 4
**Determining a Set of
Parametric Equations**

Determine the path traced out by a point P on the circumference of a circle of radius a as the circle rolls along a straight line in a plane. Such a path is called a **cycloid.**

Solution:
As our parameter, we let θ be the measure of the circle's rotation, and we let the point P begin at the origin. Thus when $\theta = 0$, P is at the origin;

when $\theta = \pi$, P is at a maximum point $(\pi a, 2a)$; and when $\theta = 2\pi$, P is back on the x-axis at $(2\pi a, 0)$.

In Figure 12.25, $\angle APC = 180° - \theta$. Hence

$$\sin \theta = \sin (180° - \theta) = \sin (\angle APC) = \frac{\overline{AC}}{a} = \frac{\overline{BD}}{a}$$

$$\cos \theta = -\cos (180° - \theta) = -\cos (\angle APC) = \frac{\overline{AP}}{-a}$$

which implies that

$$\overline{AP} = -a \cos \theta \qquad \text{and} \qquad \overline{BD} = a \sin \theta$$

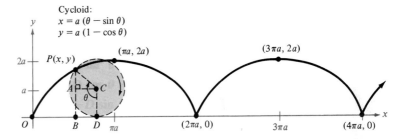

Cycloid:
$x = a (\theta - \sin \theta)$
$y = a (1 - \cos \theta)$

FIGURE 12.25

Now since the circle rolls along the x-axis, we know that $\overline{OD} = \overline{PD} = a\theta$. Furthermore, since $\overline{BA} = \overline{DC} = a$ we have

$$x = \overline{OD} - \overline{BD} = a\theta - a \sin \theta$$
$$y = \overline{BA} + \overline{AP} = a - a \cos \theta$$

Therefore, the parametric equations are

$$x = a(\theta - \sin \theta) \qquad \text{and} \qquad y = a(1 - \cos \theta)$$

We should point out that in Example 4 it is *convenient* but not *necessary* to choose as the parameter a measure of the rotation of the circle. Had we chosen a different parameter, we would have obtained a different set of parametric equations to represent the cycloid.

Section Exercises 12.4

In Exercises 1–20, sketch the curve represented by the parametric equations and write the corresponding rectangular equation by eliminating the parameter.

1. $x = 3t - 1$, $y = 2t + 1$
2. $x = 3 - 2t$, $y = 2 + 3t$

3. $x = t + 1$, $y = t^2$
4. $x = t + 1$, $y = t^3$
5. $x = t^3$, $y = \dfrac{t^2}{2}$

6. $x = 1 + \dfrac{1}{t}, \; y = t - 1$

7. $x = t - 1, \; y = \dfrac{t}{t - 1}$

8. $x = t^2 + t, \; y = t^2 - t$

9. $x = 3 \cos \theta, \; y = 3 \sin \theta$

10. $x = \cos \theta, \; y = 3 \sin \theta$

11. $x = 4 \sin 2\theta, \; y = 2 \cos 2\theta$

12. $x = \cos \theta, \; y = 2 \sin 2\theta$

13. $x = \cos \theta, \; y = 2 \sin^2 \theta$

14. $x = 4 \cos^2 \theta, \; y = 2 \sin \theta$

15. $x = 4 + 2 \cos \theta, \; y = -1 + \sin \theta$

16. $x = 4 + 2 \cos \theta, \; y = -1 + 2 \sin \theta$

17. $x = 4 + 2 \cos \theta, \; y = -1 + 4 \sin \theta$

18. $x = \sec \theta, \; y = \tan \theta$

19. $x = 4 \sec \theta, \; y = 3 \tan \theta$

20. $x = \cos^3 \theta, \; y = \sin^3 \theta$

21. Show that the parametric equations

$$x = x_1 + t(x_2 - x_1) \quad \text{and} \quad y = y_1 + t(y_2 - y_1)$$

represent the equation of the line passing through the points (x_1, y_1) and (x_2, y_2).

22. Find parametric equations for the line passing through the points $(-1, 1)$ and $(2, 3)$.

23. Find parametric equations for the line passing through the points $(0, 0)$ and $(5, -2)$.

24. Eliminate the parameter from the equations

$$x = h + a \cos \theta \quad \text{and} \quad y = k + b \sin \theta$$

to obtain the standard form of the equation of an ellipse.

25. Eliminate the parameter from the equations

$$x = h + a \sec \theta \quad \text{and} \quad y = k + b \tan \theta$$

to obtain the standard form of the equation of a hyperbola.

26. Eliminate the parameter from the equations

$$x = h + a\sqrt{t + 1} \quad \text{and} \quad y = k + b\sqrt{t}$$

and compare the result with that of Exercise 25.

27. Find two different pairs of parametric equations for the rectangular equation $y = x^3$.

In Exercises 28–32, sketch the curve represented by the given equations.

28. $x = \theta + \sin \theta, \; y = 1 - \cos \theta$ (cycloid)

29. $x = 2 \cot \theta, \; y = 2 \sin^2 \theta$ (witch of Agnesi)

30. $x = 2\theta - \sin \theta, \; y = 2 - \cos \theta$ (curtate cycloid)

31. $x = \theta - \frac{3}{2} \sin \theta, \; y = 1 - \frac{3}{2} \cos \theta$ (prolate cycloid)

32. $x = \dfrac{3t}{1 + t^3}, \; y = \dfrac{3t^2}{1 + t^3}$ (folium of Descartes)

33. A wheel of radius a rolls along a straight line without slipping. Find the parametric equation for the curve described by a point P that is b units from the center of the wheel. This curve is called a *curtate cycloid* when $b < a$. (See Figure 12.26.)

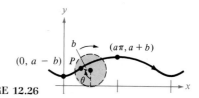

FIGURE 12.26

34. A circle of radius 1 rolls around the outside of a circle of radius 2 without slipping. Show that the parametric equations for the curve described by a point on the circumference of the rolling wheel are $x = 3 \cos \theta - \cos 3\theta$, $y = 3 \sin \theta - \sin 3\theta$. This curve is called an *epicycloid*. (See Figure 12.27.)

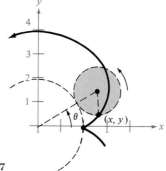

FIGURE 12.27

Review Exercises / Chapter 12

In Exercises 1–18, sketch the graph of the given equation.

1. $r = -2(1 + \cos \theta)$

2. $r = 3 - 4 \cos \theta$

3. $r = 4 - 3 \cos \theta$

4. $r = \cos 5\theta$

5. $r = -3 \cos 2\theta$

6. $r^2 = \cos 2\theta$

7. $r = 4$

8. $r = 2\theta$

9. $r = \dfrac{3}{\cos (\theta - \pi/4)}$

10. $r = \dfrac{4}{\cos (\theta + \pi/3)}$

11. $r = -\sec \theta$

12. $r = 3 \csc \theta$

13. $r^2 = 4 \sin^2 2\theta$

14. $r = \dfrac{4}{5 - 3 \cos \theta}$

15. $r = \dfrac{2}{1 - \sin \theta}$

16. $r = 2 \sin \theta \cos^2 \theta$

17. $r = 4(\sec \theta - \cos \theta)$

18. $r = 4 \cos 2\theta \sec \theta$

In Exercises 19–23, find a rectangular equation that has the same graph as the given polar equation.

19. $r = 3 \cos \theta$

20. $r = 4 \sec\left(\theta - \dfrac{\pi}{3}\right)$

21. $r = -2(1 + \cos \theta)$

22. $r = 1 + \tan \theta$

23. $r = 4 \cos 2\theta \sec \theta$

24. Find a polar equation of the curve whose rectangular equation is $(x^2 + y^2)^2 = ax^2y$.

25. Find a polar equation of the circle $x^2 + y^2 - 4x = 0$.

26. Find a polar equation of the parabola with focus at the pole and vertex at $(2, \pi)$.

27. Find a polar equation of the ellipse with a focus at the pole and vertices at $(5, 0)$ and $(1, \pi)$.

28. Find a polar equation of the hyperbola with a focus at the pole and vertices at $(1, 0)$ and $(7, 0)$.

29. Find a polar equation of the line with intercepts $(3, 0)$ and $(0, 4)$.

30. Verify that if the curve whose polar equation is $r = f(\theta)$ is rotated about the pole through an angle ϕ, then an equation for the rotated curve is $r = f(\theta - \phi)$.

31. If the polar form of an equation for a curve is $r = f(\sin \theta)$, then show that the form becomes the following:
   (a) $r = f(-\cos \theta)$ if the curve is rotated counterclockwise $\pi/2$ radians about the pole
   (b) $r = f(-\sin \theta)$ if the curve is rotated counterclockwise π radians about the pole
   (c) $r = f(\cos \theta)$ if the curve is rotated counterclockwise $3\pi/2$ radians about the pole

32. Use the results of Exercises 30 and 31 to write the equation of the limaçon $r = 2 - \sin \theta$ after it has been rotated counterclockwise through the following:
   (a) $\dfrac{\pi}{4}$ radians   (b) $\dfrac{\pi}{2}$ radians
   (c) π radians   (d) $\dfrac{3\pi}{2}$ radians

33. Use the result of Exercise 31 to write an equation of the parabola with focus at the pole and vertex at $(2, 7\pi/6)$.

In Exercises 34–43, sketch the curve represented by the parametric equations and, if possible, write the corresponding rectangular equation by eliminating the parameter.

34. $x = 1 + 4t, \ y = 2 - 3t$

35. $x = e^t, \ y = e^{-t}$

36. $x = 3 + 2 \cos \theta, \ y = 2 + 5 \sin \theta$

37. $x = t^2 - 3t + 2, \ y = t^3 - 3t^2 + 2$

38. $x = \dfrac{1}{t}, \ y = 2t + 3$

39. $x = 2t - 1, \ y = \dfrac{1}{t^2 - 2t}$

40. $x = \dfrac{1}{2t + 1}, \ y = \dfrac{2t(t + 1)}{2t + 1}$

41. $x = \cot \theta, \ y = \sin 2\theta$

42. $x = \cos^3 \theta, \ y = 4 \sin^3 \theta$

43. $x = 2\theta - \sin \theta, \ y = 2 - \cos \theta$

44. Show that the Cartesian equation of a cycloid is

$$x = a \arccos [(a - y)/a] \pm \sqrt{2ay - y^2}$$

45. Find a parametric representation of the ellipse with center at $(-3, 4)$, major axis horizontal and 8 units in length, and minor axis 6 units in length.

46. Find a parametric representation of the hyperbola with vertices at $(0, \pm 4)$ and foci at $(0, \pm 5)$.

47. The involute of a circle is described by the endpoint P of a string that is held taut as it is unwound from a spool (the spool does not rotate; see Figure 12.28). Show that a parametric representation of the involute of a circle is given by

$$x = r(\cos \theta + \theta \sin \theta)$$
$$y = r(\sin \theta - \theta \cos \theta)$$

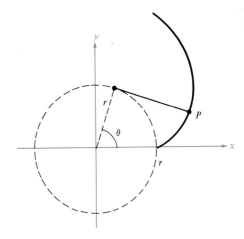

FIGURE 12.28

APPENDIX A

EXPONENTIAL TABLES

x	e^x	e^{-x}	x	e^x	e^{-x}
0.0	1.0000	1.0000	3.0	20.086	0.0498
0.1	1.1052	0.9048	3.1	22.198	0.0450
0.2	1.2214	0.8187	3.2	24.533	0.0408
0.3	1.3499	0.7408	3.3	27.113	0.0369
0.4	1.4918	0.6703	3.4	29.964	0.0334
0.5	1.6487	0.6065	3.5	33.115	0.0302
0.6	1.8221	0.5488	3.6	36.598	0.0273
0.7	2.0138	0.4966	3.7	40.447	0.0247
0.8	2.2255	0.4493	3.8	44.701	0.0224
0.9	2.4596	0.4066	3.9	49.402	0.0202
1.0	2.7183	0.3679	4.0	54.598	0.0183
1.1	3.0042	0.3329	4.1	60.340	0.0166
1.2	3.3201	0.3012	4.2	66.686	0.0150
1.3	3.6693	0.2725	4.3	73.700	0.0136
1.4	4.0552	0.2466	4.4	81.451	0.0123
1.5	4.4817	0.2231	4.5	90.017	0.0111
1.6	4.9530	0.2019	4.6	99.484	0.0101
1.7	5.4739	0.1827	4.7	109.95	0.0091
1.8	6.0496	0.1653	4.8	121.51	0.0082
1.9	6.6859	0.1496	4.9	134.29	0.0074
2.0	7.3891	0.1353	5.0	148.41	0.0067
2.1	8.1662	0.1225	5.1	164.02	0.0061
2.2	9.0250	0.1108	5.2	181.27	0.0055
2.3	9.9742	0.1003	5.3	200.34	0.0050
2.4	11.023	0.0907	5.4	221.41	0.0045
2.5	12.182	0.0821	5.5	244.69	0.0041
2.6	13.464	0.0743	5.6	270.43	0.0037
2.7	14.880	0.0672	5.7	298.87	0.0033
2.8	16.445	0.0608	5.8	330.30	0.0030
2.9	18.174	0.0550	5.9	365.04	0.0027

EXPONENTIAL TABLES (Continued)

x	e^x	e^{-x}	x	e^x	e^{-x}
6.0	403.43	0.0025	8.0	2980.96	0.0003
6.1	445.86	0.0022	8.1	3294.47	0.0003
6.2	492.75	0.0020	8.2	3640.95	0.0003
6.3	544.57	0.0018	8.3	4023.87	0.0002
6.4	601.85	0.0017	8.4	4447.07	0.0002
6.5	665.14	0.0015	8.5	4914.77	0.0002
6.6	735.10	0.0014	8.6	5431.66	0.0002
6.7	812.41	0.0012	8.7	6002.91	0.0002
6.8	897.85	0.0011	8.8	6634.24	0.0002
6.9	992.27	0.0010	8.9	7331.97	0.0001
7.0	1096.63	0.0009	9.0	8103.08	0.0001
7.1	1211.97	0.0008	9.1	8955.29	0.0001
7.2	1339.43	0.0007	9.2	9897.13	0.0001
7.3	1480.30	0.0007	9.3	10938.02	0.0001
7.4	1635.98	0.0006	9.4	12088.38	0.0001
7.5	1808.04	0.0006	9.5	13359.73	0.0001
7.6	1998.20	0.0005	9.6	14764.78	0.0001
7.7	2208.35	0.0005	9.7	16317.61	0.0001
7.8	2440.60	0.0004	9.8	18033.74	0.0001
7.9	2697.28	0.0004	9.9	19930.37	0.0001
			10.0	22026.47	0.0000

APPENDIX B

NATURAL LOGARITHMIC TABLES

	0.00	0.01	0.02	0.03	0.04	0.05	0.06	0.07	0.08	0.09
1.0	0.0000	0.0100	0.0198	0.0296	0.0392	0.0488	0.0583	0.0677	0.0770	0.0862
1.1	0.0953	0.1044	0.1133	0.1222	0.1310	0.1398	0.1484	0.1570	0.1655	0.1740
1.2	0.1823	0.1906	0.1989	0.2070	0.2151	0.2231	0.2311	0.2390	0.2469	0.2546
1.3	0.2624	0.2700	0.2776	0.2852	0.2927	0.3001	0.3075	0.3148	0.3221	0.3293
1.4	0.3365	0.3436	0.3507	0.3577	0.3646	0.3716	0.3784	0.3853	0.3920	0.3988
1.5	0.4055	0.4121	0.4187	0.4253	0.4318	0.4383	0.4447	0.4511	0.4574	0.4637
1.6	0.4700	0.4762	0.4824	0.4886	0.4947	0.5008	0.5068	0.5128	0.5188	0.5247
1.7	0.5306	0.5365	0.5423	0.5481	0.5539	0.5596	0.5653	0.5710	0.5766	0.5822
1.8	0.5878	0.5933	0.5988	0.6043	0.6098	0.6152	0.6206	0.6259	0.6313	0.6366
1.9	0.6419	0.6471	0.6523	0.6575	0.6627	0.6678	0.6729	0.6780	0.6831	0.6881
2.0	0.6931	0.6981	0.7031	0.7080	0.7129	0.7178	0.7227	0.7275	0.7324	0.7372
2.1	0.7419	0.7467	0.7514	0.7561	0.7608	0.7655	0.7701	0.7747	0.7793	0.7839
2.2	0.7885	0.7930	0.7975	0.8020	0.8065	0.8109	0.8154	0.8198	0.8242	0.8286
2.3	0.8329	0.8372	0.8416	0.8459	0.8502	0.8544	0.8587	0.8629	0.8671	0.8713
2.4	0.8755	0.8796	0.8838	0.8879	0.8920	0.8961	0.9002	0.9042	0.9083	0.9123
2.5	0.9163	0.9203	0.9243	0.9282	0.9322	0.9361	0.9400	0.9439	0.9478	0.9517
2.6	0.9555	0.9594	0.9632	0.9670	0.9708	0.9746	0.9783	0.9821	0.9858	0.9895
2.7	0.9933	0.9969	1.0006	1.0043	1.0080	1.0116	1.0152	1.0188	1.0225	1.0260
2.8	1.0296	1.0332	1.0367	1.0403	1.0438	1.0473	1.0508	1.0543	1.0578	1.0613
2.9	1.0647	1.0682	1.0716	1.0750	1.0784	1.0818	1.0852	1.0886	1.0919	1.0953
3.0	1.0986	1.1019	1.1053	1.1086	1.1119	1.1151	1.1184	1.1217	1.1249	1.1282
3.1	1.1314	1.1346	1.1378	1.1410	1.1442	1.1474	1.1506	1.1537	1.1569	1.1600
3.2	1.1632	1.1663	1.1694	1.1725	1.1756	1.1787	1.1817	1.1848	1.1878	1.1909
3.3	1.1939	1.1969	1.2000	1.2030	1.2060	1.2090	1.2119	1.2149	1.2179	1.2208
3.4	1.2238	1.2267	1.2296	1.2326	1.2355	1.2384	1.2413	1.2442	1.2470	1.2499
3.5	1.2528	1.2556	1.2585	1.2613	1.2641	1.2669	1.2698	1.2726	1.2754	1.2782
3.6	1.2809	1.2837	1.2865	1.2892	1.2920	1.2947	1.2975	1.3002	1.3029	1.3056
3.7	1.3083	1.3110	1.3137	1.3164	1.3191	1.3218	1.3244	1.3271	1.3297	1.3324
3.8	1.3350	1.3376	1.3403	1.3429	1.3455	1.3481	1.3507	1.3533	1.3558	1.3584
3.9	1.3610	1.3635	1.3661	1.3686	1.3712	1.3737	1.3762	1.3788	1.3813	1.3838

NATURAL LOGARITHMIC TABLES (Continued)

	0.00	0.01	0.02	0.03	0.04	0.05	0.06	0.07	0.08	0.09
4.0	1.3863	1.3888	1.3913	1.3938	1.3962	1.3987	1.4012	1.4036	1.4061	1.4085
4.1	1.4110	1.4134	1.4159	1.4183	1.4207	1.4231	1.4255	1.4279	1.4303	1.4327
4.2	1.4351	1.4375	1.4398	1.4422	1.4446	1.4469	1.4493	1.4516	1.4540	1.4563
4.3	1.4586	1.4609	1.4633	1.4656	1.4679	1.4702	1.4725	1.4748	1.4770	1.4793
4.4	1.4816	1.4839	1.4861	1.4884	1.4907	1.4929	1.4951	1.4974	1.4996	1.5019
4.5	1.5041	1.5063	1.5085	1.5107	1.5129	1.5151	1.5173	1.5195	1.5217	1.5239
4.6	1.5261	1.5282	1.5304	1.5326	1.5347	1.5369	1.5390	1.5412	1.5433	1.5454
4.7	1.5476	1.5497	1.5518	1.5539	1.5560	1.5581	1.5602	1.5623	1.5644	1.5665
4.8	1.5686	1.5707	1.5728	1.5748	1.5769	1.5790	1.5810	1.5831	1.5851	1.5872
4.9	1.5892	1.5913	1.5933	1.5953	1.5974	1.5994	1.6014	1.6034	1.6054	1.6074
5.0	1.6094	1.6114	1.6134	1.6154	1.6174	1.6194	1.6214	1.6233	1.6253	1.6273
5.1	1.6292	1.6312	1.6332	1.6351	1.6371	1.6390	1.6409	1.6429	1.6448	1.6467
5.2	1.6487	1.6506	1.6525	1.6544	1.6563	1.6582	1.6601	1.6620	1.6639	1.6658
5.3	1.6677	1.6696	1.6715	1.6734	1.6752	1.6771	1.6790	1.6808	1.6827	1.6845
5.4	1.6864	1.6882	1.6901	1.6919	1.6938	1.6956	1.6974	1.6993	1.7011	1.7029
5.5	1.7047	1.7066	1.7084	1.7102	1.7120	1.7138	1.7156	1.7174	1.7192	1.7210
5.6	1.7228	1.7246	1.7263	1.7281	1.7299	1.7317	1.7334	1.7352	1.7370	1.7387
5.7	1.7405	1.7422	1.7440	1.7457	1.7475	1.7492	1.7509	1.7527	1.7544	1.7561
5.8	1.7579	1.7596	1.7613	1.7630	1.7647	1.7664	1.7681	1.7699	1.7716	1.7733
5.9	1.7750	1.7766	1.7783	1.7800	1.7817	1.7834	1.7851	1.7867	1.7884	1.7901
6.0	1.7918	1.7934	1.7951	1.7967	1.7984	1.8001	1.8017	1.8034	1.8050	1.8066
6.1	1.8083	1.8099	1.8116	1.8132	1.8148	1.8165	1.8181	1.8197	1.8213	1.8229
6.2	1.8245	1.8262	1.8278	1.8294	1.8310	1.8326	1.8342	1.8358	1.8374	1.8390
6.3	1.8405	1.8421	1.8437	1.8453	1.8469	1.8485	1.8500	1.8516	1.8532	1.8547
6.4	1.8563	1.8579	1.8594	1.8610	1.8625	1.8641	1.8656	1.8672	1.8687	1.8703
6.5	1.8718	1.8733	1.8749	1.8764	1.8779	1.8795	1.8810	1.8825	1.8840	1.8856
6.6	1.8871	1.8886	1.8901	1.8916	1.8931	1.8946	1.8961	1.8976	1.8991	1.9006
6.7	1.9021	1.9036	1.9051	1.9066	1.9081	1.9095	1.9110	1.9125	1.9140	1.9155
6.8	1.9169	1.9184	1.9199	1.9213	1.9228	1.9242	1.9257	1.9272	1.9286	1.9301
6.9	1.9315	1.9330	1.9344	1.9359	1.9373	1.9387	1.9402	1.9416	1.9430	1.9445
7.0	1.9459	1.9473	1.9488	1.9502	1.9516	1.9530	1.9544	1.9559	1.9573	1.9587
7.1	1.9601	1.9615	1.9629	1.9643	1.9657	1.9671	1.9685	1.9699	1.9713	1.9727
7.2	1.9741	1.9755	1.9769	1.9782	1.9796	1.9810	1.9824	1.9838	1.9851	1.9865
7.3	1.9879	1.9892	1.9906	1.9920	1.9933	1.9947	1.9961	1.9974	1.9988	2.0001
7.4	2.0015	2.0028	2.0042	2.0055	2.0069	2.0082	2.0096	2.0109	2.0122	2.0136
7.5	2.0149	2.0162	2.0176	2.0189	2.0202	2.0215	2.0229	2.0242	2.0255	2.0268
7.6	2.0281	2.0295	2.0308	2.0321	2.0334	2.0347	2.0360	2.0373	2.0386	2.0399
7.7	2.0412	2.0425	2.0438	2.0451	2.0464	2.0477	2.0490	2.0503	2.0516	2.0528
7.8	2.0541	2.0554	2.0567	2.0580	2.0592	2.0605	2.0618	2.0631	2.0643	2.0656
7.9	2.0669	2.0681	2.0694	2.0707	2.0719	2.0732	2.0744	2.0757	2.0769	2.0782

NATURAL LOGARITHMIC TABLES (Continued)

	0.00	0.01	0.02	0.03	0.04	0.05	0.06	0.07	0.08	0.09
8.0	2.0794	2.0807	2.0819	2.0832	2.0844	2.0857	2.0869	2.0882	2.0894	2.0906
8.1	2.0919	2.0931	2.0943	2.0956	2.0968	2.0980	2.0992	2.1005	2.1017	2.1029
8.2	2.1041	2.1054	2.1066	2.1078	2.1090	2.1102	2.1114	2.1126	2.1138	2.1150
8.3	2.1163	2.1175	2.1187	2.1199	2.1211	2.1223	2.1235	2.1247	2.1258	2.1270
8.4	2.1282	2.1294	2.1306	2.1318	2.1330	2.1342	2.1353	2.1365	2.1377	2.1389
8.5	2.1401	2.1412	2.1424	2.1436	2.1448	2.1459	2.1471	2.1483	2.1494	2.1506
8.6	2.1518	2.1529	2.1541	2.1552	2.1564	2.1576	2.1587	2.1599	2.1610	2.1622
8.7	2.1633	2.1645	2.1656	2.1668	2.1679	2.1691	2.1702	2.1713	2.1725	2.1736
8.8	2.1748	2.1759	2.1770	2.1782	2.1793	2.1804	2.1815	2.1827	2.1838	2.1849
8.9	2.1861	2.1872	2.1883	2.1894	2.1905	2.1917	2.1928	2.1939	2.1950	2.1961
9.0	2.1972	2.1983	2.1994	2.2006	2.2017	2.2028	2.2039	2.2050	2.2061	2.2072
9.1	2.2083	2.2094	2.2105	2.2116	2.2127	2.2138	2.2148	2.2159	2.2170	2.2181
9.2	2.2192	2.2203	2.2214	2.2225	2.2235	2.2246	2.2257	2.2268	2.2279	2.2289
9.3	2.2300	2.2311	2.2322	2.2332	2.2343	2.2354	2.2364	2.2375	2.2386	2.2396
9.4	2.2407	2.2418	2.2428	2.2439	2.2450	2.2460	2.2471	2.2481	2.2492	2.2502
9.5	2.2513	2.2523	2.2534	2.2544	2.2555	2.2565	2.2576	2.2586	2.2597	2.2607
9.6	2.2618	2.2628	2.2638	2.2649	2.2659	2.2670	2.2680	2.2690	2.2701	2.2711
9.7	2.2721	2.2732	2.2742	2.2752	2.2762	2.2773	2.2783	2.2793	2.2803	2.2814
9.8	2.2824	2.2834	2.2844	2.2854	2.2865	2.2875	2.2885	2.2895	2.2905	2.2915
9.9	2.2925	2.2935	2.2946	2.2956	2.2966	2.2976	2.2986	2.2996	2.3006	2.3016

APPENDIX C

COMMON LOGARITHMIC TABLES

	0.00	0.01	0.02	0.03	0.04	0.05	0.06	0.07	0.08	0.09
1.0	0.0000	0.0043	0.0086	0.0128	0.0170	0.0212	0.0253	0.0294	0.0334	0.0374
1.1	0.0414	0.0453	0.0492	0.0531	0.0569	0.0607	0.0645	0.0682	0.0719	0.0755
1.2	0.0792	0.0828	0.0864	0.0899	0.0934	0.0969	0.1004	0.1038	0.1072	0.1106
1.3	0.1139	0.1173	0.1206	0.1239	0.1271	0.1303	0.1335	0.1367	0.1399	0.1430
1.4	0.1461	0.1492	0.1523	0.1553	0.1584	0.1614	0.1644	0.1673	0.1703	0.1732
1.5	0.1761	0.1790	0.1818	0.1847	0.1875	0.1903	0.1931	0.1959	0.1987	0.2014
1.6	0.2041	0.2068	0.2095	0.2122	0.2148	0.2175	0.2201	0.2227	0.2253	0.2279
1.7	0.2304	0.2330	0.2355	0.2380	0.2405	0.2430	0.2455	0.2480	0.2504	0.2529
1.8	0.2553	0.2577	0.2601	0.2625	0.2648	0.2672	0.2695	0.2718	0.2742	0.2765
1.9	0.2788	0.2810	0.2833	0.2856	0.2878	0.2900	0.2923	0.2945	0.2967	0.2989
2.0	0.3010	0.3032	0.3054	0.3075	0.3096	0.3118	0.3139	0.3160	0.3181	0.3201
2.1	0.3222	0.3243	0.3263	0.3284	0.3304	0.3324	0.3345	0.3365	0.3385	0.3404
2.2	0.3424	0.3444	0.3464	0.3483	0.3502	0.3522	0.3541	0.3560	0.3579	0.3598
2.3	0.3617	0.3636	0.3655	0.3674	0.3692	0.3711	0.3729	0.3747	0.3766	0.3784
2.4	0.3802	0.3820	0.3838	0.3856	0.3874	0.3892	0.3909	0.3927	0.3945	0.3962
2.5	0.3979	0.3997	0.4014	0.4031	0.4048	0.4065	0.4082	0.4099	0.4116	0.4133
2.6	0.4150	0.4166	0.4183	0.4200	0.4216	0.4232	0.4249	0.4265	0.4281	0.4298
2.7	0.4314	0.4330	0.4346	0.4362	0.4378	0.4393	0.4409	0.4425	0.4440	0.4456
2.8	0.4472	0.4487	0.4502	0.4518	0.4533	0.4548	0.4564	0.4579	0.4594	0.4609
2.9	0.4624	0.4639	0.4654	0.4669	0.4683	0.4698	0.4713	0.4728	0.4742	0.4757
3.0	0.4771	0.4786	0.4800	0.4814	0.4829	0.4843	0.4857	0.4871	0.4886	0.4900
3.1	0.4914	0.4928	0.4942	0.4955	0.4969	0.4983	0.4997	0.5011	0.5024	0.5038
3.2	0.5052	0.5065	0.5079	0.5092	0.5105	0.5119	0.5132	0.5145	0.5159	0.5172
3.3	0.5185	0.5198	0.5211	0.5224	0.5237	0.5250	0.5263	0.5276	0.5289	0.5302
3.4	0.5315	0.5328	0.5340	0.5353	0.5366	0.5378	0.5391	0.5403	0.5416	0.5428
3.5	0.5441	0.5453	0.5465	0.5478	0.5490	0.5502	0.5514	0.5527	0.5539	0.5551
3.6	0.5563	0.5575	0.5587	0.5599	0.5611	0.5623	0.5635	0.5647	0.5658	0.5670
3.7	0.5682	0.5694	0.5705	0.5717	0.5729	0.5740	0.5752	0.5763	0.5775	0.5786
3.8	0.5798	0.5809	0.5821	0.5832	0.5843	0.5855	0.5866	0.5877	0.5888	0.5899
3.9	0.5911	0.5922	0.5933	0.5944	0.5955	0.5966	0.5977	0.5988	0.5999	0.6010

COMMON LOGARITHMIC TABLES (Continued)

	0.00	0.01	0.02	0.03	0.04	0.05	0.06	0.07	0.08	0.09
4.0	0.6021	0.6031	0.6042	0.6053	0.6064	0.6075	0.6085	0.6096	0.6107	0.6117
4.1	0.6128	0.6138	0.6149	0.6160	0.6170	0.6180	0.6191	0.6201	0.6212	0.6222
4.2	0.6232	0.6243	0.6253	0.6263	0.6274	0.6284	0.6294	0.6304	0.6314	0.6325
4.3	0.6335	0.6345	0.6355	0.6365	0.6375	0.6385	0.6395	0.6405	0.6415	0.6425
4.4	0.6435	0.6444	0.6454	0.6464	0.6474	0.6484	0.6493	0.6503	0.6513	0.6522
4.5	0.6532	0.6542	0.6551	0.6561	0.6571	0.6580	0.6590	0.6599	0.6609	0.6618
4.6	0.6628	0.6637	0.6646	0.6656	0.6665	0.6675	0.6684	0.6693	0.6702	0.6712
4.7	0.6721	0.6730	0.6739	0.6749	0.6758	0.6767	0.6776	0.6785	0.6794	0.6803
4.8	0.6812	0.6821	0.6830	0.6839	0.6848	0.6857	0.6866	0.6875	0.6884	0.6893
4.9	0.6902	0.6911	0.6920	0.6928	0.6937	0.6946	0.6955	0.6964	0.6972	0.6981
5.0	0.6990	0.6998	0.7007	0.7016	0.7024	0.7033	0.7042	0.7050	0.7059	0.7067
5.1	0.7076	0.7084	0.7093	0.7101	0.7110	0.7118	0.7126	0.7135	0.7143	0.7152
5.2	0.7160	0.7168	0.7177	0.7185	0.7193	0.7202	0.7210	0.7218	0.7226	0.7235
5.3	0.7243	0.7251	0.7259	0.7267	0.7275	0.7284	0.7292	0.7300	0.7308	0.7316
5.4	0.7324	0.7332	0.7340	0.7348	0.7356	0.7364	0.7372	0.7380	0.7388	0.7396
5.5	0.7404	0.7412	0.7419	0.7427	0.7435	0.7443	0.7451	0.7459	0.7466	0.7474
5.6	0.7482	0.7490	0.7497	0.7505	0.7513	0.7520	0.7528	0.7536	0.7543	0.7551
5.7	0.7559	0.7566	0.7574	0.7582	0.7589	0.7597	0.7604	0.7612	0.7619	0.7627
5.8	0.7634	0.7642	0.7649	0.7657	0.7664	0.7672	0.7679	0.7686	0.7694	0.7701
5.9	0.7709	0.7716	0.7723	0.7731	0.7738	0.7745	0.7752	0.7760	0.7767	0.7774
6.0	0.7782	0.7789	0.7796	0.7803	0.7810	0.7818	0.7825	0.7832	0.7839	0.7846
6.1	0.7853	0.7860	0.7868	0.7875	0.7882	0.7889	0.7896	0.7903	0.7910	0.7917
6.2	0.7924	0.7931	0.7938	0.7945	0.7952	0.7959	0.7966	0.7973	0.7980	0.7987
6.3	0.7993	0.8000	0.8007	0.8014	0.8021	0.8028	0.8035	0.8041	0.8048	0.8055
6.4	0.8062	0.8069	0.8075	0.8082	0.8089	0.8096	0.8102	0.8109	0.8116	0.8122
6.5	0.8129	0.8136	0.8142	0.8149	0.8156	0.8162	0.8169	0.8176	0.8182	0.8189
6.6	0.8195	0.8202	0.8209	0.8215	0.8222	0.8228	0.8235	0.8241	0.8248	0.8254
6.7	0.8261	0.8267	0.8274	0.8280	0.8287	0.8293	0.8299	0.8306	0.8312	0.8319
6.8	0.8325	0.8331	0.8338	0.8344	0.8351	0.8357	0.8363	0.8370	0.8376	0.8382
6.9	0.8388	0.8395	0.8401	0.8407	0.8414	0.8420	0.8426	0.8432	0.8439	0.8445
7.0	0.8451	0.8457	0.8463	0.8470	0.8476	0.8482	0.8488	0.8494	0.8500	0.8506
7.1	0.8513	0.8519	0.8525	0.8531	0.8537	0.8543	0.8549	0.8555	0.8561	0.8567
7.2	0.8573	0.8579	0.8585	0.8591	0.8597	0.8603	0.8609	0.8615	0.8621	0.8627
7.3	0.8633	0.8639	0.8645	0.8651	0.8657	0.8663	0.8669	0.8675	0.8681	0.8686
7.4	0.8692	0.8698	0.8704	0.8710	0.8716	0.8722	0.8727	0.8733	0.8739	0.8745
7.5	0.8751	0.8756	0.8762	0.8768	0.8774	0.8779	0.8785	0.8791	0.8797	0.8802
7.6	0.8808	0.8814	0.8820	0.8825	0.8831	0.8837	0.8842	0.8848	0.8854	0.8859
7.7	0.8865	0.8871	0.8876	0.8882	0.8887	0.8893	0.8899	0.8904	0.8910	0.8915
7.8	0.8921	0.8927	0.8932	0.8938	0.8943	0.8949	0.8954	0.8960	0.8965	0.8971
7.9	0.8976	0.8982	0.8987	0.8993	0.8998	0.9004	0.9009	0.9015	0.9020	0.9025

COMMON LOGARITHMIC TABLES (Continued)

	0.00	0.01	0.02	0.03	0.04	0.05	0.06	0.07	0.08	0.09
8.0	0.9031	0.9036	0.9042	0.9047	0.9053	0.9058	0.9063	0.9069	0.9074	0.9079
8.1	0.9085	0.9090	0.9096	0.9101	0.9106	0.9112	0.9117	0.9122	0.9128	0.9133
8.2	0.9138	0.9143	0.9149	0.9154	0.9159	0.9165	0.9170	0.9175	0.9180	0.9186
8.3	0.9191	0.9196	0.9201	0.9206	0.9212	0.9217	0.9222	0.9227	0.9232	0.9238
8.4	0.9243	0.9248	0.9253	0.9258	0.9263	0.9269	0.9274	0.9279	0.9284	0.9289
8.5	0.9294	0.9299	0.9304	0.9309	0.9315	0.9320	0.9325	0.9330	0.9335	0.9340
8.6	0.9345	0.9350	0.9355	0.9360	0.9365	0.9370	0.9375	0.9380	0.9385	0.9390
8.7	0.9395	0.9400	0.9405	0.9410	0.9415	0.9420	0.9425	0.9430	0.9435	0.9440
8.8	0.9445	0.9450	0.9455	0.9460	0.9465	0.9469	0.9474	0.9479	0.9484	0.9489
8.9	0.9494	0.9499	0.9504	0.9509	0.9513	0.9518	0.9523	0.9528	0.9533	0.9538
9.0	0.9542	0.9547	0.9552	0.9557	0.9562	0.9566	0.9571	0.9576	0.9581	0.9586
9.1	0.9590	0.9595	0.9600	0.9605	0.9609	0.9614	0.9619	0.9624	0.9628	0.9633
9.2	0.9638	0.9643	0.9647	0.9652	0.9657	0.9661	0.9666	0.9671	0.9675	0.9680
9.3	0.9685	0.9689	0.9694	0.9699	0.9703	0.9708	0.9713	0.9717	0.9722	0.9727
9.4	0.9731	0.9736	0.9741	0.9745	0.9750	0.9754	0.9759	0.9764	0.9768	0.9773
9.5	0.9777	0.9782	0.9786	0.9791	0.9795	0.9800	0.9805	0.9809	0.9814	0.9818
9.6	0.9823	0.9827	0.9832	0.9836	0.9841	0.9845	0.9850	0.9854	0.9859	0.9863
9.7	0.9868	0.9872	0.9877	0.9881	0.9886	0.9890	0.9894	0.9899	0.9903	0.9908
9.8	0.9912	0.9917	0.9921	0.9926	0.9930	0.9934	0.9939	0.9943	0.9948	0.9952
9.9	0.9956	0.9961	0.9965	0.9969	0.9974	0.9978	0.9983	0.9987	0.9991	0.9996

APPENDIX D

TRIGONOMETRIC TABLES

1 degree ≈ 0.01745 radians
1 radian ≈ 57.29578 degrees

For $0 \leqslant \theta \leqslant 45$, read from upper left.
For $45 \leqslant \theta \leqslant 90$, read from lower right.
For $90 \leqslant \theta \leqslant 360$, use the identities:

θ	Quadrant II	Quadrant III	Quadrant IV
$\sin \theta$	$\sin(180-\theta)$	$-\sin(\theta-180)$	$-\sin(360-\theta)$
$\cos \theta$	$-\cos(180-\theta)$	$-\cos(\theta-180)$	$\cos(360-\theta)$
$\tan \theta$	$-\tan(180-\theta)$	$\tan(\theta-180)$	$-\tan(360-\theta)$
$\cot \theta$	$-\cot(180-\theta)$	$\cot(\theta-180)$	$-\cot(360-\theta)$

Degrees	Radians	sin	cos	tan	cot			Degrees	Radians	sin	cos	tan	cot		
0° 00′	.0000	.0000	1.0000	.0000	—	1.5708	90° 00′	4° 00′	.0698	.0698	.9976	.0699	14.301	1.5010	86° 00′
10	.0029	.0029	1.0000	.0029	343.774	1.5679	50	10	.0727	.0727	.9974	.0729	13.727	1.4981	50
20	.0058	.0058	1.0000	.0058	171.885	1.5650	40	20	.0756	.0756	.9971	.0758	13.197	1.4952	40
30	.0087	.0087	1.0000	.0087	114.589	1.5621	30	30	.0785	.0785	.9969	.0787	12.706	1.4923	30
40	.0116	.0116	.9999	.0116	85.940	1.5592	20	40	.0814	.0814	.9967	.0816	12.251	1.4893	20
50	.0145	.0145	.9999	.0145	68.750	1.5563	10	50	.0844	.0843	.9964	.0846	11.826	1.4864	10
1° 00′	.0175	.0175	.9998	.0175	57.290	1.5533	89° 00′	5° 00′	.0873	.0872	.9962	.0875	11.430	1.4835	85° 00′
10	.0204	.0204	.9998	.0204	49.104	1.5504	50	10	.0902	.0901	.9959	.0904	11.059	1.4806	50
20	.0233	.0233	.9997	.0233	42.964	1.5475	40	20	.0931	.0929	.9957	.0934	10.712	1.4777	40
30	.0262	.0262	.9997	.0262	38.188	1.5446	30	30	.0960	.0958	.9954	.0963	10.385	1.4748	30
40	.0291	.0291	.9996	.0291	34.368	1.5417	20	40	.0989	.0987	.9951	.0992	10.078	1.4719	20
50	.0320	.0320	.9995	.0320	31.242	1.5388	10	50	.1018	.1016	.9948	.1022	9.788	1.4690	10
2° 00′	.0349	.0349	.9994	.0349	28.636	1.5359	88° 00′	6° 00′	.1047	.1045	.9945	.1051	9.514	1.4661	84° 00′
10	.0378	.0378	.9993	.0378	26.432	1.5330	50	10	.1076	.1074	.9942	.1080	9.255	1.4632	50
20	.0407	.0407	.9992	.0407	24.542	1.5301	40	20	.1105	.1103	.9939	.1110	9.010	1.4603	40
30	.0436	.0436	.9990	.0437	22.904	1.5272	30	30	.1134	.1132	.9936	.1139	8.777	1.4573	30
40	.0465	.0465	.9989	.0466	21.470	1.5243	20	40	.1164	.1161	.9932	.1169	8.556	1.4544	20
50	.0495	.0494	.9988	.0495	20.206	1.5213	10	50	.1193	.1190	.9929	.1198	8.345	1.4515	10
3° 00′	.0524	.0523	.9986	.0524	19.081	1.5184	87° 00′	7° 00′	.1222	.1219	.9925	.1228	8.144	1.4486	83° 00′
10	.0553	.0552	.9985	.0553	18.075	1.5155	50	10	.1251	.1248	.9922	.1257	7.953	1.4457	50
20	.0582	.0581	.9983	.0582	17.169	1.5126	40	20	.1280	.1276	.9918	.1287	7.770	1.4428	40
30	.0611	.0610	.9981	.0612	16.350	1.5097	30	30	.1309	.1305	.9914	.1317	7.596	1.4399	30
40	.0640	.0640	.9980	.0641	15.605	1.5068	20	40	.1338	.1334	.9911	.1346	7.429	1.4370	20
50	.0669	.0669	.9978	.0670	14.924	1.5039	10	50	.1367	.1363	.9907	.1376	7.269	1.4341	10
		cos	sin	cot	tan	Radians	Degrees			cos	sin	cot	tan	Radians	Degrees

TRIGONOMETRIC TABLES (Continued)

Degrees	Radians	sin	cos	tan	cot		
8°00'	.1396	.1392	.9903	.1405	7.115	1.4312	82°00'
10	.1425	.1421	.9899	.1435	6.968	1.4283	50
20	.1454	.1449	.9894	.1465	6.827	1.4254	40
30	.1484	.1478	.9890	.1495	6.691	1.4224	30
40	.1513	.1507	.9886	.1524	6.561	1.4195	20
50	.1542	.1536	.9881	.1554	6.435	1.4166	10
9°00'	.1571	.1564	.9877	.1584	6.314	1.4137	81°00'
10	.1600	.1593	.9872	.1614	6.197	1.4108	50
20	.1629	.1622	.9868	.1644	6.084	1.4079	40
30	.1658	.1650	.9863	.1673	5.976	1.4050	30
40	.1687	.1679	.9858	.1703	5.871	1.4021	20
50	.1716	.1708	.9853	.1733	5.769	1.3992	10
10°00'	.1745	.1736	.9848	.1763	5.671	1.3963	80°00'
10	.1774	.1765	.9843	.1793	5.576	1.3934	50
20	.1804	.1794	.9838	.1823	5.485	1.3904	40
30	.1833	.1822	.9833	.1853	5.396	1.3875	30
40	.1862	.1851	.9827	.1883	5.309	1.3846	20
50	.1891	.1880	.9822	.1914	5.226	1.3817	10
11°00'	.1920	.1908	.9816	.1944	5.145	1.3788	79°00'
10	.1949	.1937	.9811	.1974	5.066	1.3759	50
20	.1978	.1965	.9805	.2004	4.989	1.3730	40
30	.2007	.1994	.9799	.2035	4.915	1.3701	30
40	.2036	.2022	.9793	.2065	4.843	1.3672	20
50	.2065	.2051	.9787	.2095	4.773	1.3643	10
12°00'	.2094	.2079	.9781	.2126	4.705	1.3614	78°00'
10	.2123	.2108	.9775	.2156	4.638	1.3584	50
20	.2153	.2136	.9769	.2186	4.574	1.3555	40
30	.2182	.2164	.9763	.2217	4.511	1.3526	30
40	.2211	.2193	.9757	.2247	4.449	1.3497	20
50	.2240	.2221	.9750	.2278	4.390	1.3468	10
13°00'	.2269	.2250	.9744	.2309	4.331	1.3439	77°00'
10	.2298	.2278	.9737	.2339	4.275	1.3410	50
20	.2327	.2306	.9730	.2370	4.219	1.3381	40
30	.2356	.2334	.9724	.2401	4.165	1.3352	30
40	.2385	.2363	.9717	.2432	4.113	1.3323	20
50	.2414	.2391	.9710	.2462	4.061	1.3294	10
14°00'	.2443	.2419	.9703	.2493	4.011	1.3265	76°00'
10	.2473	.2447	.9696	.2524	3.962	1.3235	50
20	.2502	.2476	.9689	.2555	3.914	1.3206	40
30	.2531	.2504	.9681	.2586	3.867	1.3177	30
40	.2560	.2532	.9674	.2617	3.821	1.3148	20
50	.2589	.2560	.9667	.2648	3.776	1.3119	10
15°00'	.2618	.2588	.9659	.2679	3.732	1.3090	75°00'
10	.2647	.2616	.9652	.2711	3.689	1.3061	50
20	.2676	.2644	.9644	.2742	3.647	1.3032	40
30	.2705	.2672	.9636	.2773	3.606	1.3003	30
40	.2734	.2700	.9628	.2805	3.566	1.2974	20
50	.2763	.2728	.9621	.2836	3.526	1.2945	10
16°00'	.2793	.2756	.9613	.2867	3.487	1.2915	74°00'
10	.2822	.2784	.9605	.2899	3.450	1.2886	50
20	.2851	.2812	.9596	.2931	3.412	1.2857	40
30	.2880	.2840	.9588	.2962	3.376	1.2828	30
40	.2909	.2868	.9580	.2994	3.340	1.2799	20
50	.2938	.2896	.9572	.3026	3.305	1.2770	10
17°00'	.2967	.2924	.9563	.3057	3.271	1.2741	73°00'
10	.2996	.2952	.9555	.3089	3.237	1.2712	50
20	.3025	.2979	.9546	.3121	3.204	1.2683	40
30	.3054	.3007	.9537	.3153	3.172	1.2654	30
40	.3083	.3035	.9528	.3185	3.140	1.2625	20
50	.3113	.3062	.9520	.3217	3.108	1.2595	10
		cos	sin	cot	tan	Radians	Degrees

Degrees	Radians	sin	cos	tan	cot		
18°00'	.3142	.3090	.9511	.3249	3.078	1.2566	72°00'
10	.3171	.3118	.9502	.3281	3.047	1.2537	50
20	.3200	.3145	.9492	.3314	3.018	1.2508	40
30	.3229	.3173	.9483	.3346	2.989	1.2479	30
40	.3258	.3201	.9474	.3378	2.960	1.2450	20
50	.3287	.3228	.9465	.3411	2.932	1.2421	10
19°00'	.3316	.3256	.9455	.3443	2.904	1.2392	71°00'
10	.3345	.3283	.9446	.3476	2.877	1.2363	50
20	.3374	.3311	.9436	.3508	2.850	1.2334	40
30	.3403	.3338	.9426	.3541	2.824	1.2305	30
40	.3432	.3365	.9417	.3574	2.798	1.2275	20
50	.3462	.3393	.9407	.3607	2.773	1.2246	10
20°00'	.3491	.3420	.9397	.3640	2.747	1.2217	70°00'
10	.3520	.3448	.9387	.3673	2.723	1.2188	50
20	.3549	.3475	.9377	.3706	2.699	1.2159	40
30	.3578	.3502	.9367	.3739	2.675	1.2130	30
40	.3607	.3529	.9356	.3772	2.651	1.2101	20
50	.3636	.3557	.9346	.3805	2.628	1.2072	10
21°00'	.3665	.3584	.9336	.3839	2.605	1.2043	69°00'
10	.3694	.3611	.9325	.3872	2.583	1.2014	50
20	.3723	.3638	.9315	.3906	2.560	1.1985	40
30	.3752	.3665	.9304	.3939	2.539	1.1956	30
40	.3782	.3692	.9293	.3973	2.517	1.1926	20
50	.3811	.3719	.9283	.4006	2.496	1.1897	10
22°00'	.3840	.3746	.9272	.4040	2.475	1.1868	68°00'
10	.3869	.3773	.9261	.4074	2.455	1.1839	50
20	.3898	.3800	.9250	.4108	2.434	1.1810	40
30	.3927	.3827	.9239	.4142	2.414	1.1781	30
40	.3956	.3854	.9228	.4176	2.394	1.1752	20
50	.3985	.3881	.9216	.4210	2.375	1.1723	10
23°00'	.4014	.3907	.9205	.4245	2.356	1.1694	67°00'
10	.4043	.3934	.9194	.4279	2.337	1.1665	50
20	.4072	.3961	.9182	.4314	2.318	1.1636	40
30	.4102	.3987	.9171	.4348	2.300	1.1606	30
40	.4131	.4014	.9159	.4383	2.282	1.1577	20
50	.4160	.4041	.9147	.4417	2.264	1.1548	10
24°00'	.4189	.4067	.9135	.4452	2.246	1.1519	66°00'
10	.4218	.4094	.9124	.4487	2.229	1.1490	50
20	.4247	.4120	.9112	.4522	2.211	1.1461	40
30	.4276	.4147	.9100	.4557	2.194	1.1432	30
40	.4305	.4173	.9088	.4592	2.177	1.1403	20
50	.4334	.4200	.9075	.4628	2.161	1.1374	10
25°00'	.4363	.4226	.9063	.4663	2.145	1.1345	65°00'
10	.4392	.4253	.9051	.4699	2.128	1.1316	50
20	.4422	.4279	.9038	.4734	2.112	1.1286	40
30	.4451	.4305	.9026	.4770	2.097	1.1257	30
40	.4480	.4331	.9013	.4806	2.081	1.1228	20
50	.4509	.4358	.9001	.4841	2.066	1.1199	10
26°00'	.4538	.4384	.8988	.4877	2.050	1.1170	64°00'
10	.4567	.4410	.8975	.4913	2.035	1.1141	50
20	.4596	.4436	.8962	.4950	2.020	1.1112	40
30	.4625	.4462	.8949	.4986	2.006	1.1083	30
40	.4654	.4488	.8936	.5022	1.991	1.1054	20
50	.4683	.4514	.8923	.5059	1.977	1.1025	10
27°00'	.4712	.4540	.8910	.5095	1.963	1.0996	63°00'
10	.4741	.4566	.8897	.5132	1.949	1.0966	50
20	.4771	.4592	.8884	.5169	1.935	1.0937	40
30	.4800	.4617	.8870	.5206	1.921	1.0908	30
40	.4829	.4643	.8857	.5243	1.907	1.0879	20
50	.4858	.4669	.8843	.5280	1.894	1.0850	10
		cos	sin	cot	tan	Radians	Degrees

TRIGONOMETRIC TABLES (Continued)

Degrees	Radians	sin	cos	tan	cot		Degrees
28°00′	.4887	.4695	.8829	.5317	1.881	1.0821	62°00′
10	.4916	.4720	.8816	.5354	1.868	1.0792	50
20	.4945	.4746	.8802	.5392	1.855	1.0763	40
30	.4974	.4772	.8788	.5430	1.842	1.0734	30
40	.5003	.4797	.8774	.5467	1.829	1.0705	20
50	.5032	.4823	.8760	.5505	1.816	1.0676	10
29°00′	.5061	.4848	.8746	.5543	1.804	1.0647	61°00′
10	.5091	.4874	.8732	.5581	1.792	1.0617	50
20	.5120	.4899	.8718	.5619	1.780	1.0588	40
30	.5149	.4924	.8704	.5658	1.767	1.0559	30
40	.5178	.4950	.8689	.5696	1.756	1.0530	20
50	.5207	.4975	.8675	.5735	1.744	1.0501	10
30°00′	.5236	.5000	.8660	.5774	1.732	1.0472	60°00′
10	.5265	.5025	.8646	.5812	1.720	1.0443	50
20	.5294	.5050	.8631	.5851	1.709	1.0414	40
30	.5323	.5075	.8616	.5890	1.698	1.0385	30
40	.5325	.5100	.8601	.5930	1.686	1.0356	20
50	.5381	.5125	.8587	.5969	1.675	1.0327	10
31°00′	.5411	.5150	.8572	.6009	1.664	1.0297	59°00′
10	.5440	.5175	.8557	.6048	1.653	1.0268	50
20	.5469	.5200	.8542	.6088	1.643	1.0239	40
30	.5498	.5225	.8526	.6128	1.632	1.0210	30
40	.5527	.5250	.8511	.6168	1.621	1.0181	20
50	.5556	.5275	.8496	.6208	1.611	1.0152	10
32°00′	.5585	.5299	.8480	.6249	1.600	1.0123	58°00′
10	.5614	.5324	.8465	.6289	1.590	1.0094	50
20	.5643	.5348	.8450	.6330	1.580	1.0065	40
30	.5672	.5373	.8434	.6371	1.570	1.0036	30
40	.5701	.5398	.8418	.6412	1.560	1.0007	20
50	.5730	.5422	.8403	.6453	1.550	.9977	10
33°00′	.5760	.5446	.8387	.6494	1.540	.9948	57°00′
10	.5789	.5471	.8371	.6536	1.530	.9919	50
20	.5818	.5495	.8355	.6577	1.520	.9890	40
30	.5847	.5519	.8339	.6619	1.511	.9861	30
40	.5876	.5544	.8323	.6661	1.501	.9832	20
50	.5905	.5568	.8307	.6703	1.492	.9803	10
34°00′	.5934	.5592	.8290	.6745	1.483	.9774	56°00′
10	.5963	.5616	.8274	.6787	1.473	.9745	50
20	.5992	.5640	.8258	.6830	1.464	.9716	40
30	.6021	.5664	.8241	.6873	1.455	.9687	30
40	.6050	.5688	.8225	.6916	1.446	.9657	20
50	.6080	.5712	.8208	.6959	1.437	.9628	10
35°00′	.6109	.5736	.8192	.7002	1.428	.9599	55°00′
10	.6138	.5760	.8175	.7046	1.419	.9570	50
20	.6167	.5783	.8158	.7089	1.411	.9541	40
30	.6196	.5807	.8141	.7133	1.402	.9512	30
40	.6225	.5831	.8124	.7177	1.393	.9483	20
50	.6254	.5854	.8107	.7221	1.385	.9454	10
36°00′	.6283	.5878	.8090	.7265	1.376	.9425	54°00′
10	.6312	.5901	.8073	.7310	1.368	.9396	50
20	.6341	.5925	.8056	.7355	1.360	.9367	40
30	.6370	.5948	.8039	.7400	1.351	.9338	30
40	.6400	.5972	.8021	.7445	1.343	.9308	20
50	.6429	.5995	.8004	.7490	1.335	.9279	10
		cos	sin	cot	tan	Radians	Degrees

Degrees	Radians	sin	cos	tan	cot		Degrees
37°00′	.6458	.6018	.7986	.7536	1.327	.9250	53°00′
10	.6487	.6041	.7969	.7581	1.319	.9221	50
20	.6516	.6065	.7951	.7627	1.311	.9192	40
30	.6545	.6088	.7934	.7673	1.303	.9163	30
40	.6574	.6111	.7916	.7720	1.295	.9134	20
50	.6603	.6134	.7898	.7766	1.288	.9105	10
38°00′	.6632	.6157	.7880	.7813	1.280	.9076	52°00′
10	.6661	.6180	.7862	.7860	1.272	.9047	50
20	.6690	.6202	.7844	.7907	1.265	.9018	40
30	.6720	.6225	.7826	.7954	1.257	.8988	30
40	.6749	.6248	.7808	.8002	1.250	.8959	20
50	.6778	.6271	.7790	.8050	1.242	.8930	10
39°00′	.6807	.6293	.7771	.8098	1.235	.8901	51°00′
10	.6836	.6316	.7753	.8146	1.228	.8872	50
20	.6865	.6338	.7735	.8195	1.220	.8843	40
30	.6894	.6361	.7716	.8243	1.213	.8814	30
40	.6923	.6383	.7698	.8292	1.206	.8785	20
50	.6952	.6406	.7679	.8342	1.199	.8756	10
40°00′	.6981	.6428	.7660	.8391	1.192	.8727	50°00′
10	.7010	.6450	.7642	.8441	1.185	.8698	50
20	.7039	.6472	.7623	.8491	1.178	.8668	40
30	.7069	.6494	.7604	.8541	1.171	.8639	30
40	.7098	.6517	.7585	.8591	1.164	.8610	20
50	.7127	.6539	.7566	.8642	1.157	.8581	10
41°00′	.7156	.6561	.7547	.8693	1.150	.8552	49°00′
10	.7185	.6583	.7528	.8744	1.144	.8523	50
20	.7214	.6604	.7509	.8796	1.137	.8494	40
30	.7243	.6626	.7490	.8847	1.130	.8465	30
40	.7272	.6648	.7470	.8899	1.124	.8436	20
50	.7301	.6670	.7451	.8952	1.117	.8407	10
42°00′	.7330	.6691	.7431	.9004	1.111	.8378	48°00′
10	.7359	.6713	.7412	.9057	1.104	.8348	50
20	.7389	.6734	.7392	.9110	1.098	.8319	40
30	.7418	.6756	.7373	.9163	1.091	.8290	30
40	.7447	.6777	.7353	.9217	1.085	.8261	20
50	.7476	.6799	.7333	.9271	1.079	.8232	10
43°00′	.7505	.6820	.7314	.9325	1.072	.8203	47°00′
10	.7534	.6841	.7294	.9380	1.066	.8174	50
20	.7563	.6862	.7274	.9435	1.060	.8145	40
30	.7592	.6884	.7254	.9490	1.054	.8116	30
40	.7621	.6905	.7234	.9545	1.048	.8087	20
50	.7650	.6926	.7214	.9601	1.042	.8058	10
44°00′	.7679	.6947	.7193	.9657	1.036	.8029	46°00′
10	.7709	.6967	.7173	.9713	1.030	.7999	50
10	.7738	.6988	.7153	.9770	1.024	.7970	40
30	.7767	.7009	.7133	.9827	1.018	.7941	30
40	.7796	.7030	.7112	.9884	1.012	.7912	20
50	.7825	.7050	.7092	.9942	1.006	.7883	10
45°00′	.7854	.7071	.7071	1.0000	1.000	.7854	45°00′
		cos	sin	cot	tan	Radians	Degrees

APPENDIX E

TABLE OF SQUARE ROOTS AND CUBE ROOTS

n	$\sqrt{n}$	$\sqrt[3]{n}$	n	$\sqrt{n}$	$\sqrt[3]{n}$	n	$\sqrt{n}$	$\sqrt[3]{n}$
1	1.00000	1.00000	31	5.56776	3.14138	61	7.81025	3.93650
2	1.41421	1.25992	32	5.65685	3.17480	62	7.87401	3.95789
3	1.73205	1.44225	33	5.74456	3.20753	63	7.93725	3.97906
4	2.00000	1.58740	34	5.83095	3.23961	64	8.00000	4.00000
5	2.23607	1.70998	35	5.91608	3.27107	65	8.06226	4.02073
6	2.44949	1.81712	36	6.00000	3.30193	66	8.12404	4.04124
7	2.64575	1.91293	37	6.08276	3.33222	67	8.18535	4.06155
8	2.82843	2.00000	38	6.16441	3.36198	68	8.24621	4.08166
9	3.00000	2.08008	39	6.24500	3.39121	69	8.30662	4.10157
10	3.16228	2.15443	40	6.32456	3.41995	70	8.36660	4.12129
11	3.31662	2.22398	41	6.40312	3.44822	71	8.42615	4.14082
12	3.46410	2.28943	42	6.48074	3.47603	72	8.48528	4.16017
13	3.60555	2.35133	43	6.55744	3.50340	73	8.54400	4.17934
14	3.74166	2.41014	44	6.63325	3.53035	74	8.60233	4.19834
15	3.87298	2.46621	45	6.70820	3.55689	75	8.66025	4.21716
16	4.00000	2.51984	46	6.78233	3.58305	76	8.71780	4.23582
17	4.12311	2.57128	47	6.85565	3.60883	77	8.77496	4.25432
18	4.24264	2.62074	48	6.92820	3.63424	78	8.83176	4.27266
19	4.35890	2.66840	49	7.00000	3.65931	79	8.88819	4.29084
20	4.47214	2.71442	50	7.07107	3.68403	80	8.94427	4.30887
21	4.58258	2.75892	51	7.14143	3.70843	81	9.00000	4.32675
22	4.69042	2.80204	52	7.21110	3.73251	82	9.05539	4.34448
23	4.79583	2.84387	53	7.28011	3.75629	83	9.11043	4.36207
24	4.89898	2.88450	54	7.34847	3.77976	84	9.16515	4.37952
25	5.00000	2.92402	55	7.41620	3.80295	85	9.21954	4.39683
26	5.09902	2.96250	56	7.48331	3.82586	86	9.27362	4.41400
27	5.19615	3.00000	57	7.54983	3.84850	87	9.32738	4.43105
28	5.29150	3.03659	58	7.61577	3.87088	88	9.38083	4.44796
29	5.38516	3.07232	59	7.68115	3.89300	89	9.43398	4.46475
30	5.47723	3.10723	60	7.74597	3.91487	90	9.48683	4.48140

TABLE OF SQUARE ROOTS AND CUBE ROOTS (Continued)

n	$\sqrt{n}$	$\sqrt[3]{n}$	n	$\sqrt{n}$	$\sqrt[3]{n}$	n	$\sqrt{n}$	$\sqrt[3]{n}$
91	9.53939	4.49794	128	11.3137	5.03968	165	12.8452	5.48481
92	9.59166	4.51436	129	11.3578	5.05277	166	12.8841	5.49586
93	9.64365	4.53065	130	11.4018	5.06580	167	12.9228	5.50688
94	9.69536	4.54684	131	11.4455	5.07875	168	12.9615	5.51785
95	9.74679	4.56290	132	11.4891	5.09164	169	13.0000	5.52877
96	9.79796	4.57886	133	11.5326	5.10447	170	13.0384	5.53966
97	9.84886	4.59470	134	11.5758	5.11723	171	13.0767	5.55050
98	9.89949	4.61044	135	11.6190	5.12993	172	13.1149	5.56130
99	9.94987	4.62606	136	11.6619	5.14256	173	13.1529	5.57205
100	10.0000	4.64159	137	11.7047	5.15514	174	13.1909	5.58277
101	10.0499	4.65701	138	11.7473	5.16765	175	13.2288	5.59344
102	10.0995	4.67233	139	11.7898	5.18010	176	13.2665	5.60408
103	10.1489	4.68755	140	11.8322	5.19249	177	13.3041	5.61467
104	10.1980	4.70267	141	11.8743	5.20483	178	13.3417	5.62523
105	10.2470	4.71769	142	11.9164	5.21710	179	13.3791	5.63574
106	10.2956	4.73262	143	11.9583	5.22932	180	13.4164	5.64622
107	10.3441	4.74746	144	12.0000	5.24148	181	13.4536	5.65665
108	10.3923	4.76220	145	12.0416	5.25359	182	13.4907	5.66705
109	10.4403	4.77686	146	12.0830	5.26564	183	13.5277	5.67741
110	10.4881	4.79142	147	12.1244	5.27763	184	13.5647	5.68773
111	10.5357	4.80590	148	12.1655	5.28957	185	13.6015	5.69802
112	10.5830	4.82028	149	12.2066	5.30146	186	13.6382	5.70827
113	10.6301	4.83459	150	12.2474	5.31329	187	13.6748	5.71848
114	10.6771	4.84881	151	12.2882	5.32507	188	13.7113	5.72865
115	10.7238	4.86294	152	12.3288	5.33680	189	13.7477	5.73879
116	10.7703	4.87700	153	12.3693	5.34848	190	13.7840	5.74890
117	10.8167	4.89097	154	12.4097	5.36011	191	13.8203	5.75897
118	10.8628	4.90487	155	12.4499	5.37169	192	13.8564	5.76900
119	10.9087	4.91868	156	12.4900	5.38321	193	13.8924	5.77900
120	10.9545	4.93242	157	12.5300	5.39469	194	13.9284	5.78896
121	11.0000	4.94609	158	12.5698	5.40612	195	13.9642	5.79889
122	11.0454	4.95968	159	12.6095	5.41750	196	14.0000	5.80879
123	11.0905	4.97319	160	12.6491	5.42884	197	14.0357	5.81865
124	11.1355	4.98663	161	12.6886	5.44012	198	14.0712	5.82848
125	11.1803	5.00000	162	12.7279	5.45136	199	14.1067	5.83827
126	11.2250	5.01330	163	12.7671	5.46256	200	14.1421	5.84804
127	11.2694	5.02653	164	12.8062	5.47370			

ANSWERS TO ODD-NUMBERED EXERCISES

CHAPTER 1

Section 1.1

1.

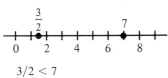

$3/2 < 7$

3.

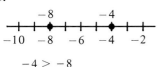

$\pi > -6$

5.

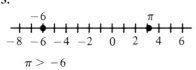

$-4 > -8$

7.

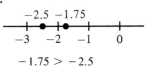

$5/6 > 2/3$

9.

$-1.75 > -2.5$

11. $[-2, 0)$ $-2 \leq x < 0$

13. $[3, 11/2]$ $3 \leq x \leq 11/2$

15. $[100, \infty)$ $100 \leq x$

17. $(\sqrt{2}, 8]$ $\sqrt{2} < x \leq 8$

19. $x < 0$ **21.** $A \geq 30$ **23.** $1/3\% \leq R \leq 1\%$

25. 4   **27.** 23/4   **29.** 51   **31.** 14.99

33. $|x - 5| \leq 3$ **35.** $|z - 3/2| > 1$ **37.** $|y| < |y - 8|$

39. (a) No **(b)** No **(c)** Solution **(d)** No

41. (a) Solution **(b)** Solution **(c)** No **(d)** No

43. (a) Solution **(b)** No **(c)** No **(d)** No

45. (a) No **(b)** No **(c)** Solution **(d)** Solution

47. (a) No **(b)** No **(c)** No **(d)** Solution

49. Conditional   **51.** Identity   **53.** Conditional

55. Conditional   **57.** Identity   **59.** $x = 3$

61. $x = -3/2$ **63.** $x = 9$

65. $x = -26$ **67.** $z = -6/5$ **69.** $x = 9$

71. $u = 10$ **73.** $x = 4$ **75.** $x = 3$ **77.** $x = 11/6$

79. $x = 5$ **81.** $x = 0$ **83.** Identity

85. No solution   **87.** $x = \dfrac{4b + 1}{a + 2}$ **89.** $-3/4, 1/4$

91. $1 \pm \sqrt{3}$ **93.** $-7 \pm \sqrt{5}$ **95.** $2(-2 \pm \sqrt{5})$

97. $\dfrac{2 \pm \sqrt{7}}{3}$ **99.** $\dfrac{-2 \pm \sqrt{11}}{6}$ **101. (a)** 127/90,

584/413, 7071/5000, $\sqrt{2}$, 47/33   **(b)** 7071/5000

103. 138.889   **105.** 62.372   **107.** $-0.643, 0.976$

109. 1.794, 0.310

Section 1.2

1. -24　**3.** $1/2$　**5.** 5　**7.** $5x^6$　**9.** $24y^{10}$
11. $10x^4$　**13.** $7x^5$　**15.** $(4/3)(x + y)^2$　**17.** $-2x^3$
19. $x^2/9z^4$　**21.** $a^6/64b^9$　**23.** 1　**25.** x^4/y^4
27. b^5/a^5　**29.** 1　**31.** (a) 9.3×10^7　(b) 9.0×10^8
(c) 4.35×10^{-6}　**33.** (a) $1,910,000$
(b) $234,500,000,000$　(c) 6.21　(d) 0.00852
(e) 0.00007021　(f) 0.00000003798

	$\sqrt[n]{b^m} = a$	$b^{m/n} = a$	$a^{n/m} = b$
35.	$\sqrt{9} = 3$	$9^{1/2} = 3$	$3^2 = 9$
37.	$\sqrt[5]{32} = 2$	$32^{1/5} = 2$	$2^5 = 32$
39.	$\sqrt{196} = 14$	$196^{1/2} = 14$	$14^2 = 196$
41.	$\sqrt[3]{-216} = -6$	$(-216)^{1/3} = -6$	$(-6)^3 = -216$
43.	$\sqrt[3]{27^2} = 9$	$27^{2/3} = 9$	$9^{3/2} = 27$
45.	$\sqrt[4]{81^3} = 27$	$81^{3/4} = 27$	$27^{4/3} = 81$

47. (a) $3/2$　(b) $3/2$　**49.** (a) 64　(b) 4　**51.** (a) 326
(b) 562　**53.** (a) $1/16$　(b) $2/3$　**55.** (a) $2\sqrt{2}$
(b) $3\sqrt{2}$　**57.** (a) $2x\sqrt[3]{2x^2}$　(b) $2xz\sqrt[4]{2z}$

59. (a) $\dfrac{5x}{y^2}\sqrt{3}$　(b) $(x - y)\sqrt{5(x - y)}$　**61.** (a) $\sqrt{x}$

(b) $\sqrt[3]{x}$　**63.** (a) $\sqrt[4]{3x^2y^3}$　(b) $xy^2\sqrt{6}$　**65.** (a) $\dfrac{\sqrt{3}}{3}$

(b) $\dfrac{\sqrt{21}}{7}$　**67.** (a) $4\sqrt[3]{4}$　(b) $\dfrac{\sqrt[3]{18}}{6}$

69. (a) $\dfrac{x(5 + \sqrt{3})}{11}$　(b) $\dfrac{8(6 - \sqrt{10})}{13}$

71. (a) $\dfrac{3(\sqrt{x} - \sqrt{y})}{x - y}$　(b) $-(4/5)(\sqrt{2} + 2\sqrt{3})$

73. (a) $\dfrac{13}{2\sqrt{13}}$　(b) $\dfrac{2}{3\sqrt{2}}$　**75.** (a) $\dfrac{1}{x(\sqrt{3} + \sqrt{2})}$

(b) $\dfrac{1}{2(\sqrt{15} - 3)}$　**77.** (a) $2\sqrt{x}$　(b) $13\sqrt{2}$

79. (a) $8\sqrt{xy}$　(b) $(13/2)\sqrt{2ab}$　**81.** (a) $|x|y\sqrt{15}$
(b) $3\sqrt{3}/|a|$　**83.** $2\sqrt[4]{2}$　(b) $a\sqrt[6]{10ab}$
85. (a) 7697.13　(b) 954.45　(c) 1.479　(d) 1.11×10^8
87. (a) 3.0981×10^6　(b) 3.0769×10^{10}
89. $8\frac{1}{3}$ min　**91.** (a) 7.5498　(b) 3.5477　(c) 12.6515
(d) 6.3096　(e) 0.0221　(f) 9609.4958

93.

t	5	10	20
A	$910.97	$1659.73	$5509.41

t	30	40	50
A	$18,288.29	$60,707.30	$201,515.58

95. (a) $R \approx 1.72$　(b) $R \approx 1.08$

Section 1.3

1. $-3x^2 + x - 8$
3. $8x^3 + 29x^2 + 11$
5. $9x^5 + 5x^4 + 19x^3 + 19x^2 + x - 9$
7. $25x^7 + 16x^3 - 26$　**9.** $x^3 - 12xy - 9y^2$
11. $x^4 - 5x^3 - 2x^2 + 11x - 5$
13. $x^4 - x^3 + 5x^2 - 9x - 36$　**15.** $x^3 + 27$
17. $x^4 - 1$　**19.** $x^3 + 4x^2 - 5x - 20$
21. $x^2 + 7x + 12$　**23.** $6x^2 - 7x - 5$
25. $x^2 + 12x + 36$　**27.** $4x^2 - 20xy + 25y^2$
29. $x^2 + 2xy + y^2 - 6x - 6y + 9$　**31.** $x^2 - 4y^2$
33. $4r^4 - 25$　**35.** $x^3 + 3x^2 + 3x + 1$
37. $8x^3 - 12x^2y + 6xy^2 - y^3$
39. $x^3 - 3x^2y + 3xy^2 - y^3 + 3x^2 - 6xy + 3y^2 + 3x -$
$3y + 1$　**41.** $x - y$　**43.** $16r^6 - 24r^3s^2 + 9s^4$
45. $30m^2 + 31mn - 12n^2$
47. $3(x + 2)$　**49.** $x(y - z)$　**51.** $3ab(3a - 4b^2)$
53. $(x + y)(z^2 - 1)$　**55.** $(x + 6)(x - 6)$
57. $(4y - 3)(4y + 3)$　**59.** $(x + 1)(x - 3)$
61. $(9 + y)(9 - y)$　**63.** $(x - 2)^2$
65. $(2x + 1)^2$　**67.** $(x - 2y)^2$　**69.** $(ab - c)^2$
71. $(x + 2)(x - 1)$　**73.** $(x - 2)(x - 3)$
75. $(3x - 2)(x - 1)$　**77.** $(2y - 3z)(y + 12z)$
79. $(x - 2)(x^2 + 2x + 4)$　**81.** $(y + 4)(y^2 - 4y + 16)$
83. $(x - 3y)(x^2 + 3xy + 9y^2)$
85. $(x^2 + 4)(x^4 - 4x^2 + 16)$　**87.** $(x - 1)(y + z)$
89. $(5x - 3)(y + 2)$　**91.** $(10x - 7)(y - 1)$
93. $2x(x + 1)(x - 2)$　**95.** $7r(3s + r)(3s - r)$
97. $xy(2x - y)^2$　**99.** $6x(x - 2y)(x^2 + 2xy + 4y^2)$
101. $(x + 2y)(x - 2y)(x + y)(x - y)$
103. $2x(2x - 1)(4x - 1)$　**105.** $2xz(x + 2y)(z + 1)$
107. $m + n$　**109.** (a) 0.300　(b) -6.666
(c) Undefined　(d) 1.125　**111.** $(3x + y)^3$

Section 1.4

1. (a) Yes　(b) No　(c) Yes　(d) No　**3.** (a) Yes
(b) No　(c) No　(d) Yes

5. $x \geq 12$

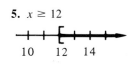

10   12   14

7. $x < -1/2$

$-3\ -2\ -1\quad 0\quad 1$

9. $x \geq 1/2$

$-1\quad 0\quad 1\quad 2\quad 3$

11. $x > 1/2$

$-1\quad 0\quad 1\quad 2\quad 3$

13. $-9/2 < x < 15/2$

$-6\ -4\ -2\quad 0\quad 2\quad 4\quad 6\quad 8$

15. $-3/4 < x < -1/4$

$-2\quad -1\quad 0\quad 1$

17. $-5 < x < 5$

$-6\ -4\ -2\quad 0\quad 2\quad 4\quad 6$

19. $x < -6, x > 6$

$-8\quad -4\quad 0\quad 4\quad 8$

21. $-7 < x < 3$

$-8\ -6\ -4\ -2\quad 0\quad 2\quad 4$

23. $x \leq -7, x \geq 13$

$-8\ -4\quad 0\quad 4\quad 8\quad 12\ 16$

25. $4 < x < 5$

$3\quad 4\quad 5\quad 6$

27. $-3 \leq x \leq 3$

$-4\ -2\quad 0\quad 2\quad 4$

29. $x < -2, x > 2$

$-4\quad -2\quad 0\quad 2\quad 4$

31. $-7 < x < 3$

$-8\ -6\ -4\ -2\quad 0\quad 2\quad 4$

33. $x \leq -7/2, x \geq -1/2$

$-4\quad -2\quad 0$

35. $-3 < x < 2$

$-4\quad -2\quad 0\quad 2$

37. $x < -1, x > 1$

$-2\quad 0\quad 2$

39. $x < 0, 0 < x < 3/2$

$-2\quad 0\quad 2$

41. $x \leq 3, x \geq 4$ **43.** $-2 \leq x \leq 2$ **45.** $-4 \leq x \leq 3$
47. $-\infty < x < \infty$ **49. (a)** $|x| \leq 2$ **(b)** $|x| > 2$
51. $x < -1, x > 4$ **53.** $5 < x < 15$
55. $-5 < x < -3/2, x > -1$
57. $-9/7 < x < 6/5, x \geq 36$ **59.** $x \geq -1/2$
61. (a) 10 sec **(b)** $4 < t < 6$ **63.** $r > 12.5\%$
65. $x > 36$ units   **67.** $65.8 \leq h \leq 71.2$

Section 1.5

1.

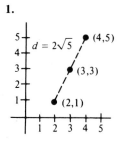

$d = 2\sqrt{5}$ (4,5)

(3,3)

(2,1)

$1\quad 2\quad 3\quad 4\quad 5$

3.

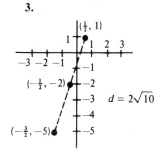

$(\frac{1}{2}, 1)$

$1\quad 2\quad 3$

$-3\ -2\ -1$

$(-\frac{1}{2}, -2)$

$d = 2\sqrt{10}$

$(-\frac{3}{2}, -5)$

5.

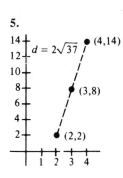

$d = 2\sqrt{37}$ (4,14)

(3,8)

(2,2)

$1\quad 2\quad 3\quad 4$

7

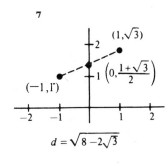

$(1, \sqrt{3})$

$(-1, 1)$

$\left(0, \dfrac{1+\sqrt{3}}{2}\right)$

$-2\quad -1\quad 1\quad 2$

$d = \sqrt{8 - 2\sqrt{3}}$

9.

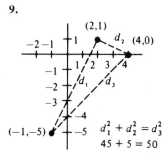

(2,1)

d_2 (4,0)

$-2\ -1$

$1\quad 2\quad 3\quad 4$

$d_1\quad d_3$

$(-1, -5)$

$d_1^2 + d_2^2 = d_3^2$

$45 + 5 = 50$

11.

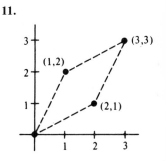

(3,3)

(1,2)

(2,1)

$1\quad 2\quad 3$

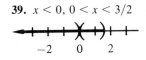

The length of each side
is $\sqrt{5}$

13.

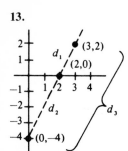

Collinear since
$$d_1 + d_2 = d_3$$
$$2\sqrt{5} + \sqrt{5} = 3\sqrt{5}$$

15.

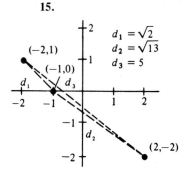

$d_1 = \sqrt{2}$
$d_2 = \sqrt{13}$
$d_3 = 5$

Not on a line since
$$d_1 + d_2 > d_3$$

31. $y = 1 - x^2$

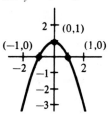

33. $y = x^3 + 2$

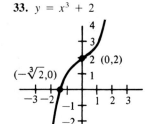

35. $y = (x + 2)^2$

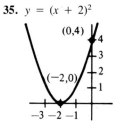

37. $y = 2x^4$

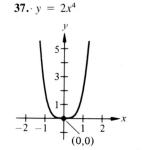

17. $x = \pm 3$ **19.** $2y = 3x - 1$

21. $\left(\dfrac{3x_1 + x_2}{4}, \dfrac{3y_1 + y_2}{4}\right), \left(\dfrac{x_1 + x_2}{2}, \dfrac{y_1 + y_2}{2}\right),$

$\left(\dfrac{x_1 + 3x_2}{4}, \dfrac{y_1 + 3y_2}{4}\right)$ **25. (a)** 15.5% **(b)** 14.8%

(c) 13.3% **(d)** 16.1%   **27. (a)** 175 **(b)** 700 **(c)** 280
(d) 750   **29.** 1972, 1973, 1976   **31. (a)** (1.25, 3.6)
(b) 10.534   **33. (a)** (6, −45) **(b)** 99.860
35. (a) (0.1880, −0.5995) **(b)** 1.471

39. $y = \sqrt{x - 3}$

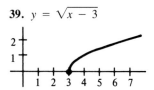

41. (1, 2) is not on the graph
(1, −1) is on the graph
(4, 5) is on the graph

43. (1, 1/5) is on the graph
(2, 1/2) is on the graph
(−1, −2) is not on the graph

Section 1.6

1. c   **3.** b   **5.** a   **7.** (0, −3), (3/2, 0)
9. (0, −2), (−2, 0), (1, 0)   **11.** (0, 0), (−3, 0), (3, 0)
13. (0, 1/2), (1, 0)   **15.** (0, 0)   **17.** Symmetric with
respect to the y-axis   **19.** Symmetric with respect to the
y-axis   **21.** Symmetric with respect to the x-axis
23. Symmetric with respect to the origin   **25.** Symmetric
with respect to the origin
27. $y = x$

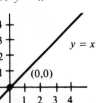

29. $y = -3x + 2$

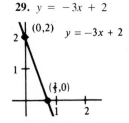

$y = -3x + 2$

45. (a)

Year	1970	1971	1972	1973	1974
t	0	1	2	3	4
CPI	114.4	120.8	128.3	136.9	146.6

Year	1975	1976	1977	1978	1979
t	5	6	7	8	9
CPI	157.4	169.3	182.3	196.4	211.6

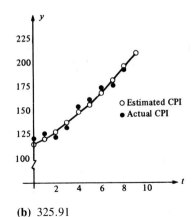

(b) 325.91

47. (a)

Year	1950	1955	1960	1965
t	0	5	10	15
Percentage	20.4	11.3	7.8	6.0

Year	1970	1975	1979
t	20	25	29
Percentage	4.8	4.1	3.6

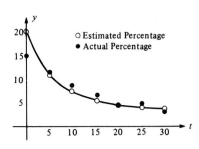

(b) 2.74

55. (b) $\left(\dfrac{\sqrt{2+\sqrt{2}}}{2}, \dfrac{\sqrt{2-\sqrt{2}}}{2}\right)$ **(c)** $\left(\pm\dfrac{\sqrt{2}}{2}, \pm\dfrac{\sqrt{2}}{2}\right)$,

$\left(\pm\dfrac{2+\sqrt{2}}{\sqrt{(2+\sqrt{2})^2+2}}, \pm\dfrac{\sqrt{2}}{\sqrt{(2+\sqrt{2})^2+2}}\right)$,

$\left(\pm\dfrac{\sqrt{2}}{\sqrt{(2+\sqrt{2})^2+2}}, \pm\dfrac{2+\sqrt{2}}{\sqrt{(2+\sqrt{2})^2+2}}\right)$

Section 1.7

1. $2x - 3y - 4$ **3.** $5z + 3x - 6$ **5.** $\dfrac{3-x}{x-1}$

7. ax/y **9.** $16x^2$ **11.** $\sqrt{x+9}$ **13.** $\dfrac{6x+y}{6x-y}$

15. $\dfrac{y}{xy+1}$ **17.** $\dfrac{4y+xy}{x}$ **19.** $\sqrt[3]{x^2(x+7)}$

21. $\dfrac{4x+3y}{xy}$ **23.** $1/2y$ **25.** $(1/3)(2x^2 + x + 15)$

27. $\sqrt{x}(1+x)$ **29.** $(1/2)x^{-1/2}(5x^2 + 1)$

31. $(1/3)(1 - 2x)^{-2/3}(3 - 8x)$ **33.** $\dfrac{-1}{x^2\sqrt{x^2+1}}$

35. $(1/15)(2x + 1)^{3/2}(3x - 1)$

37. $(3/28)(t + 1)^{4/3}(4t - 3)$

39. $(1/15)(2x - 1)^{1/2}(3x^2 + 2x - 13)$ **41.** $-1/4$

43. 2   **45.** $1/2$ **47.** $25/9, 49/16$ **49.** $1/(2x^2)$

51. $16x^{-1} - 5 - x$ **53.** $4x^{8/3} - 7x^{5/3} + x^{-1/3}$

55. $\dfrac{x^2}{x^4+1} + \dfrac{4x}{x^4+1} + \dfrac{8}{x^4+1}$

Chapter 1 Review Exercises

1. Do not add denominators. $7/16 + 3/16 = 10/16 = 5/8$
3. Multiplication is not distributive over multiplication.
$10(4 \cdot 7) = 40 \cdot 7 = 4 \cdot 70 = 280$ **5.** Only the
numerator is multiplied by 4. $4\left(\dfrac{3}{7}\right) = \dfrac{4 \cdot 3}{7} = \dfrac{12}{7}$
7. Invert the divisor before multiplying.
$15/16 \div 2/3 = 15/16 \cdot 3/2 = 45/32$
9. Multiply before adding. $12 + 8 \times 6 = 12 + 48 = 60$
11. Distribute the negative.
$2[5 - (3 - 2)] = 2[5 - 3 + 2] = 2[4] = 8$
13. Exponent applies to each factor in the base.
$(2x)^4 = 2^4x^4 = 16x^4$ **15.** Order of operations; perform
the operations within the symbol of grouping first.
$(5 + 8)^2 = 13^2 = 169$ **17.** Multiply the exponents.
$(3^4)^4 = 3^{16}$ **19.** Perform the operations on the radicand
first. $\sqrt{3^2 + 4^2} = \sqrt{25} = 5$ **21.** Radicals cannot be
multiplied if the indices are not the same. $(7x)^{1/2}(2)^{1/3} =$
$(7x)^{3/6}(2)^{2/6} = \sqrt[6]{(7x)^3(4)} = \sqrt[6]{1372x^3}$ **23.** Multiplication
by a negative number changes the direction of the inequality.
$-5 < -3,$ $-2(-5) > -2(-3),$ $10 > 6$
25. -20 **27.** -11 **29.** -11 **31.** 25
33. -144 **35.** $(5/3)^6$ **37.** 50   **39.** 9×10^8

41. $|x - 7| \geq 4$ **43.** $|y + 30| < 5$

45.

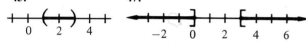

47.

49. (a) 6 (b) (3, 0)   **51.** (a) 5 (b) (0, 1/2)

53. (a) 13 (b) (8, 7/2)   **55.** (a) 8 (b) (5, 2)

57. (a) $\sqrt{53}$ (b) (5/2, 1)   **59.** (a) $\sqrt{65}/15$
(b) (11/30, 16/15)

61. $x^5 - 2x^4 + x^3 - x^2 + 2x - 1$

63. $x^4 + x^2 - x - \dfrac{1}{x}$ **65.** $y^6 + y^4 - y^3 - y$

67. $\dfrac{1}{x^2(x^2 + x + 1)}$ **69.** $\dfrac{x^2(x^2 + 4x + 2)}{(x + 2)(x - 2)(x^2 + 2)^2}$

71. $\dfrac{3}{(x - 1)(x + 2)}$ **73.** $\dfrac{x^3 - x + 3}{(x - 1)(x + 2)}$

75. $\dfrac{x^4}{(x - 1)^3}$ **77.** $\dfrac{x + 1}{x(x^2 + 1)}$ **79.** $\dfrac{2x^3}{(x^2 - 4)^2}$

81. $\dfrac{x(x + 1)}{x^2 + x + 1}$ **83.** $(1/5)x(5x - 6)$ **85.** $\dfrac{-1}{xy(x + y)}$

87. $\dfrac{3ax^2}{(a^2 - x)(a - x)}$ **89.** $\dfrac{1}{x(x - 4)(x - 5)}$

91. $(x - 1)(x^2 + x + 1)$ **93.** $(x + y)^2(x - y)^2$

95. $(1/12)(9x^2 - 10x + 48)$ **97.** $\dfrac{1}{2(x + 1)^{3/2}}(x + 2)$

99. $(z - 5)(z + 2)(z^2 - 2z + 4)$

101.

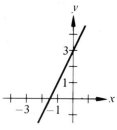

103.

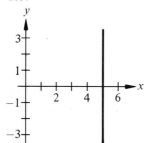

105.

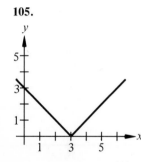

107.

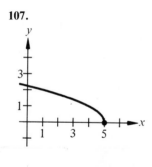

109.

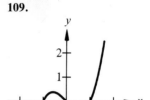

111.

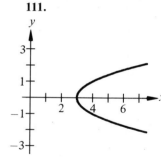

113.

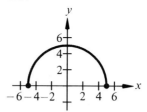

115. 50625   **117.** 0

119.

n	1	10	10^2
$5/n$	5	0.5	0.05

n	10^4	10^6	10^{10}
$5/n$	0.0005	5×10^{-6}	5×10^{-10}

CHAPTER 2

Section 2.1

1. (a) -1 (b) -3 (c) -9 (d) $2b - 3$ (e) $2x - 5$
(f) $-5/2$ **3.** (a) 0 (b) 1 (c) $\sqrt{3}$ (d) 3
(e) $\sqrt{x + \Delta x + 3}$ (f) $\sqrt{c + 3}$ **5.** (a) 1 (b) -1
(c) -1 (d) 1 (e) 1 (f) $|x - 1|/(x - 1)$ **7.** $3 + \Delta x$

9. $3x^2 + 3x\,\Delta x + (\Delta x)^2$ **11.** $\dfrac{-1}{\sqrt{x - 1}(1 + \sqrt{x - 1})}$

13. ± 3 **15.** 10/7

17. Domain: $[1, \infty)$ **19.** Domain: $(-\infty, \infty)$
     Range: $[0, \infty)$ Range: $[0, \infty)$

21. Domain: $[-3, 3]$
　　Range:　$[0, 3]$
23. Domain:　$(-\infty, 0)$ and $(0, \infty)$
　　Range:　　$(0, \infty)$
25. Domain:　$(-\infty, 0)$ and $(0, \infty)$
　　Range:　　-1 and 1
27. y is not a function of x　**29.** y is a function of x
31. y is a function of x　**33.** y is not a function of x
35. y is a function of x　**37.** $V = 1750a + 500,000, \ a > 0$
39. (a) $C = 12.30x + 98,000$ **(b)** $R = 17.98x$

(c) $P = 5.68x - 98,000$　**41. (a)** $x = \dfrac{100(14.75 - p)}{p}$

(b) $x = 47.5$　**43. (a)** $p = 91 - 0.01x, \ 100 < x < 1600$
(b) $P = 31x - 0.01x^2, \ 100 < x < 1600$　**(c)** \$21,000.00

Section 2.2

1. y is a function of x　**3.** y is not a function of x
5. y is a function of x　**7.** y is a function of x
9. y is not a function of x　**11. (a)** Increasing on $(-\infty, \infty)$
(b) Odd function　**13. (a)** Increasing on $(-\infty, 0)$ and
$(2, \infty)$, decreasing on $(0, 2)$　**(b)** Neither even nor odd
15. (a) Increasing on $(-1, 0)$ and $(1, \infty)$, decreasing on
$(-\infty, -1)$ and $(0, 1)$　**(b)** Even function
17. (a) Increasing on $(1, \infty)$, decreasing on $(-\infty, -1)$,
constant on $(-1, 1)$　**(b)** Even function
19. (a) Increasing on $(-2, \infty)$, decreasing on $(-3, -2)$
(b) Neither even nor odd

21.

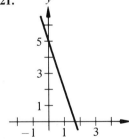

23.

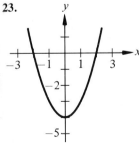

25.

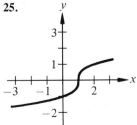

27.

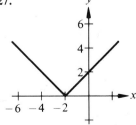

29.

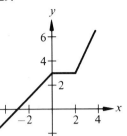

(b)

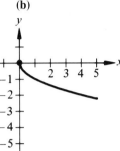

(d)

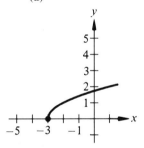

(f)

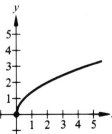

31. (a)

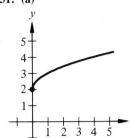

(c)

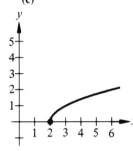

(e)

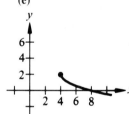

31. (g)

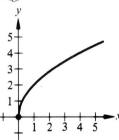

33. (a) $y = (x + 1)\sqrt{x + 4}$　**(b)** $y = x\sqrt{x + 3} + 2$
(c) $y = -x\sqrt{x + 3}$　**(d)** $y = (1 - x)\sqrt{x + 2}$

Section 2.3

1. 1   **3.** 0   **5.** −3

7. $m = 3$

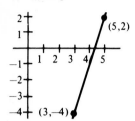

9. $m = 0$

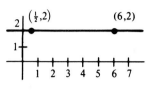

11. m is undefined

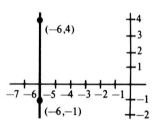

13. $m = 4/3$

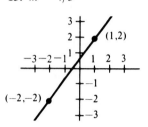

15. $2x - y - 3 = 0$

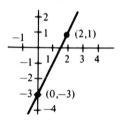

17. $3x + y = 0$

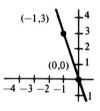

19. $x - 2 = 0$

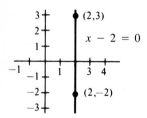

21. $y + 2 = 0$

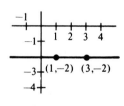

23. $3x - 4y + 12 = 0$

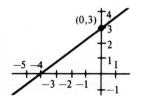

25. $2x - 3y = 0$

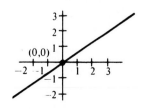

27. $2x + y - 5 = 0$

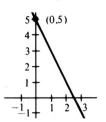

29. $4x - y + 2 = 0$

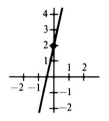

31. $9x - 12y + 8 = 0$

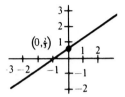

33. $x - 3 = 0$

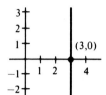

35. $3x + 2y = 6$ **37.** $12x + 3y = -2$

39. $x + y = 3$ **41.** $2x + 4y = 5$

43. $5F - 9C = 160$ **45.** $W = 4.50 + 0.75x$

47. $V = 825,000 - 30,000t$

49. (a) $y = (76/3)t + 96$

(b) 1976: 121.33, 1977: 146.67   **(c)** 1980: 222.67

(d) Amount spent for energy imports was increasing at the rate of 25.33 billion dollars per year.

51. (a) $C = 26,500 + 14.75t$ **(b)** $R = 25t$

(c) $P = 10.25t - 26,500$ **(d)** $t = 2585.4$ hr

Section 2.4

1. h   **3.** d   **5.** g   **7.** a   **9.** c

11. (a) $f(x) = (x - 2)^2$ **(b)** $f(x) = x^2 - 4x + 4$

13. (a) $f(x) = -(x + 2)^2 + 4$ **(b)** $f(x) = -x^2 - 4x$

15. (a) $f(x) = -(x + 3)^2 + 3$ **(b)** $f(x) = -x^2 - 6x - 6$

17.

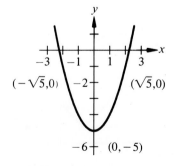

19.

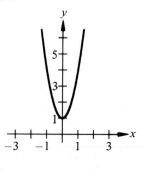

21.

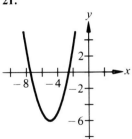

23.

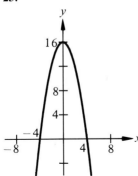

25.

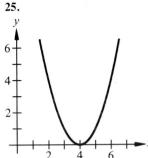

27.

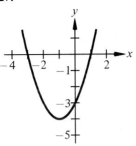

29.

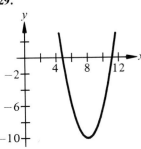

31.

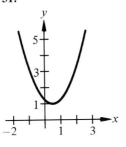

33.

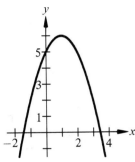

35.

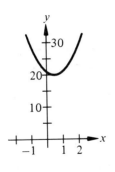

37. $x = 25$ **39.** $x = 20$ **41.** 200

43. $f(x) = (3/4)(x - 5)^2 + 12$

45. $f(x) = (1/2)(2x^2 + x - 10)$

$g(x) = (-1/2)(2x^2 + x - 10)$

Section 2.5

1. (a) $2x$ **(b)** 2 **(c)** $x^2 - 1$ **(d)** $\dfrac{x + 1}{x - 1}$ **(e)** x **(f)** x

3. (a) $x^2 - x + 1$ **(b)** $x^2 + x - 1$ **(c)** $x^2 - x^3$

(d) $\dfrac{x^2}{1 - x}$ **(e)** $(1 - x)^2$ **(f)** $1 - x^2$

5. (a) $x^2 + 5 + \sqrt{1 - x}$ **(b)** $x^2 + 5 - \sqrt{1 - x}$

(c) $(x^2 + 5)\sqrt{1 - x}$ **(d)** $\dfrac{x^2 + 5}{\sqrt{1 - x}}$ **(e)** $6 - x$

(f) Undefined   **7. (a)** $\dfrac{x + 1}{x^2}$ **(b)** $\dfrac{x - 1}{x^2}$ **(c)** $\dfrac{1}{x^3}$

(d) x **(e)** x^2 **(f)** x^2

9.

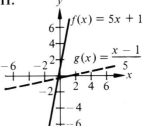

11.

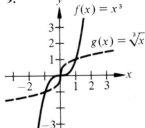

13.

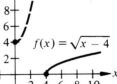

15.

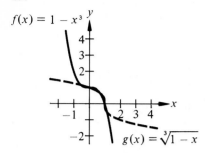

$f(x) = 1 - x^3$

$g(x) = \sqrt[3]{1 - x}$

17. $f^{-1}(x) = \dfrac{x + 3}{2}$

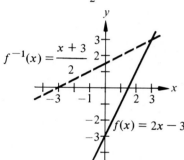

$f^{-1}(x) = \dfrac{x + 3}{2}$

$f(x) = 2x - 3$

19. $f^{-1}(x) = \sqrt[5]{x}$

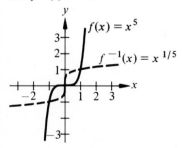

$f(x) = x^5$

$f^{-1}(x) = x^{1/5}$

21. $f^{-1}(x) = x^2, \ 0 \le x$

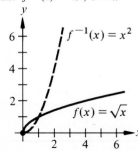

$f^{-1}(x) = x^2$

$f(x) = \sqrt{x}$

23. $f^{-1}(x) = \sqrt{4 - x^2}, \ 0 \le x \le 2$

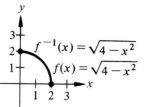

$f^{-1}(x) = \sqrt{4 - x^2}$

$f(x) = \sqrt{4 - x^2}$

25. $f^{-1}(x) = x^3 + 1$

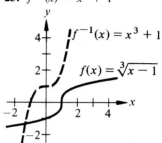

$f^{-1}(x) = x^3 + 1$

$f(x) = \sqrt[3]{x - 1}$

27. $f^{-1}(x) = x^{3/2}$

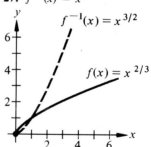

$f^{-1}(x) = x^{3/2}$

$f(x) = x^{2/3}$

29. Yes, as long as $a \ne 0$, $f^{-1}(x) = \dfrac{x - b}{a}$ **31.** No

33. Yes, $f^{-1}(x) = \dfrac{x\sqrt{5}}{\sqrt{1 - x^2}}$ **35.** No

37. Yes, $f^{-1}(x) = \dfrac{1}{x}$ **39.** Yes, $f^{-1}(x) = \dfrac{x^2 - 3}{2}$

Chapter 2 Review Exercises

1. $y = 0$ **3.** $3x - 4y + 2 = 0$

5. $5x - 12y + 2 = 0$ **7.** $x = 5$

9. $2x - 7y + 2 = 0$ **11.** $7/3$

15. Domain: $[-5, 5]$; range: $[0, 5]$ **17.** (a) 5 (b) 17

(c) $t^4 + 1$ (d) $-x^2 - 1$ (e) $x^2 + 2x\Delta x + (\Delta x)^2 + 1$

19. (a) $f[g(x)] = \dfrac{1}{x^2 + 1}$ (b) $g[f(x)] = \dfrac{x^2 + 1}{x^2}$

21.

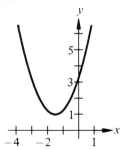

23.

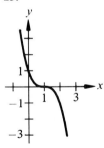

41. (a) $f^{-1}(x) = x^2 - 1$

(b)

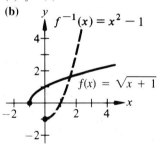

25.

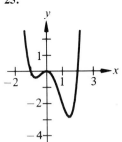

27.

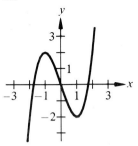

43. (a) $f^{-1}(x) = \sqrt{x + 5}$

(b)

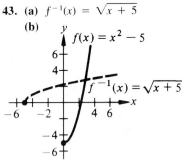

29.

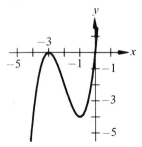

45. $x \geq 4$, $f^{-1}(x) = \dfrac{\sqrt{2x + 8}}{2}$

CHAPTER 3

Section 3.1

1. g   **3.** d   **5.** h   **7.** i   **9.** j   **11.** Up to the right and to the left   **13.** Down to the right and to the left   **15.** Up to the right and down to the left   **17.** Down to the right and up to the left   **19.** Down to the right and to the left   **21.** ± 5 **23.** 3   **25.** -2, 1   **27.** $2 \pm \sqrt{3}$ **29.** 0, 2   **31.** ± 1 **33.** $\pm \sqrt{5}$ **35.** No real roots   **37.** $f(x) = x^2 - 10x$ **39.** $f(x) = x^2 + 4x - 12$ **41.** $f(x) = x^3 + 5x^2 + 6x$ **43.** $f(x) = x^4 - 4x^3 - 9x^2 + 36x$ **45. (a)** y   **(b) (i)** y

31. $(1, -1)$ Minimum   **33.** $(3, 9)$ Maximum
35. $(-3/2, -41/4)$ Minimum   **37.** $x = 4500$ units
39. (a) $f^{-1}(x) = 2x + 6$

(b)

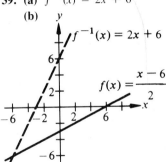

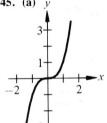

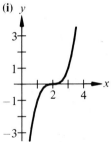

(ii)

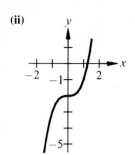

(iii)

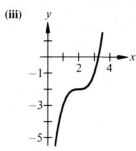

(iv)

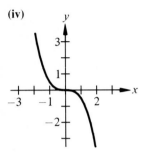

47.

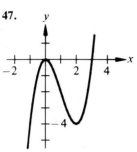

49.

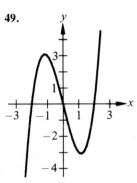

51.

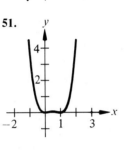

53.

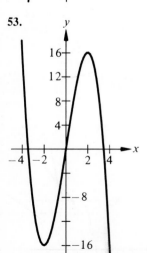

55.

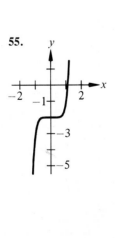

57. 0.7   **59.** 3.3

Section 3.2

1. $x^2 - 3x + 1$ **3.** $x^3 + 3x^2 - 1$ **5.** $3x^2 + 3x - 4$

7. $7 - \dfrac{11}{x + 2}$ **9.** $x + \dfrac{5x}{x^2 - 5}$ **11.** $x - \dfrac{x + 9}{x^2 + 1}$

13. $x^2 + \dfrac{x^2 + 7}{x^3 - 1}$ **15.** $2x^2 - 6x + 21 - \dfrac{144x - 68}{2x^2 + 6x - 3}$

17. $-x - 4 + \dfrac{20}{5 + 4x - x^2}$ **19.** $2x + \dfrac{x + 5}{x^2 - 2x - 8}$

21. $3x^2 - 2x + 5$ **23.** $4x^2 - 9$

25. $-x^2 + 10x - 25$ **27.** $5x^2 + 14x + 56 + \dfrac{232}{x - 4}$

29. $10x^3 + 10x^2 + 60x + 360 + \dfrac{1360}{x - 6}$

31. $x^4 + 2x^3 + 4x^2 + 8x + 16$ **33.** $x^2 - 8x + 64$

35. $-3x^3 - 6x^2 - 12x - 24 - \dfrac{48}{x - 2}$

37. $-x^2 + 3x - 6 + \dfrac{11}{x + 1}$ **39.** $4x^2 + 14x - 30$

41. (a) 1   **(b)** 4   **(c)** 4   **(d)** 1954

43. $(x - 2)(x + 3)(x - 1)$ **45.** $(2x - 1)(x - 5)(x - 2)$

47. $(x - 0.4)(x - 0.5)(x - 1)$
49. $[x - (1 + \sqrt{3})][x - (1 - \sqrt{3})](x - 1)$
51. $27, -274.625$ **53.** $11.705, -5646.972$

Section 3.3

1. 1, 2, 3   **3.** -1, 1, 4   **5.** -10, -1
7. 2   **9.** 1, 2   **11.** -1, 1/2   **13.** $-1/2$, 1
15. $-3/4$, 1/2, 1   **17.** $-3/4$ **19.** ± 1, $\pm \sqrt{2}$
21. -1, 2   **23.** ± 1, ± 2 **25.** -3, -1, 0, 4
27. -2, $-1/2$, 4   **29.** $\pm \sqrt{2}$, 0, 3, 4
31. (a) ± 1, $\pm 1/2$, $\pm 1/4$, $\pm 1/8$, $\pm 1/16$, $\pm 1/32$, ± 3, $\pm 3/2$, $\pm 3/4$, $\pm 3/8$, $\pm 3/16$, $\pm 3/32$

(b)   **(c)** $-1/8$, 3/4, 1

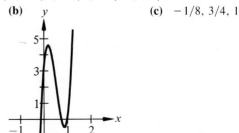

33. (a) ± 1, ± 2, ± 3, ± 6, ± 9, ± 18, $\pm 1/2$, $\pm 3/2$, $\pm 9/2$, $\pm 1/4$, $\pm 3/4$, $\pm 9/4$

(b)

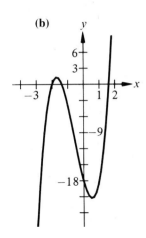

(c) $-2, \dfrac{1 \pm \sqrt{145}}{8}$

35. (a) $\pm 1, \pm 2, \pm 5, \pm 10, \pm 1/2, \pm 5/2, \pm 1/4, \pm 5/4$
(b) *y*
(c) $-2, -1/2, 5/2$

37. $-1, 1/4, 1$ **39.** $-2/3, 1/3, 1$
41. (a) Upper bound **(b)** Lower bound **(c)** Neither
43. (a) Neither **(b)** Lower bound **(c)** Upper bound

Section 3.4

1. $\overbrace{i, -1, -i, 1,}\ \overbrace{i, -1, -i, 1,}\ \overbrace{i, -1, -i, 1,}\ \overbrace{i, -1, -i, 1,}$
$\overbrace{i, -1, -i, 1}$ **3.** $4 + 3i, 4 - 3i$
5. $-7 - 8i, -7 + 8i$ **7.** $-7 + 9i, -7 - 9i$
9. $2 - 3\sqrt{3}\,i, 2 + 3\sqrt{3}\,i$ **11.** $5\sqrt{3}\,i, -5\sqrt{3}\,i$
13. $-1 - 6i, -1 + 6i$ **15.** $8 + 7\sqrt{7}\,i, 8 - 7\sqrt{7}\,i$
17. $8, 8$ **19.** $5 - 3\sqrt{3}, 5 - 3\sqrt{3}$ **21.** $11 - i$
23. 4 **25.** $-5 + 3i$ **27.** $(3 - 5\sqrt{2}) + 2\sqrt{2}\,i$
29. $1/6 + (7/6)i$ **31.** $5 + i$ **33.** 41
35. $12 + 30i$ **37.** -90 **39.** $-40 + 16i$
41. 24 **43.** $-9 + 40i$ **45.** $-236 + 115i$
47. $16/41 + (20/41)i$ **49.** $3/5 + (4/5)i$
51. $-7 - 6i$ **53.** $31/5 + (2/5)i$ **55.** $(1/8)i$

57. $-5/4 - (5/4)i$ **59.** $35/29 + (595/29)i$
61. $a = -10, b = 6$ **63.** $a = 6, b = 5$

Section 3.5

1. 1 **3.** 2 **5.** 0 **7.** 2 **9.** $\pm\sqrt{5}$
11. $-5 \pm \sqrt{6}$ **13.** 4 **15.** $-1 \pm 2i$
17. $\dfrac{1}{2} \pm i$ **19.** $20 \pm 2\sqrt{215}$
21. $f(x) = x^3 - x^2 + 25x - 25$
23. $f(x) = x^3 - 10x^2 + 33x - 34$
25. $f(x) = x^4 + 37x^2 + 36$
27. $f(x) = x^4 + 8x^3 + 9x^2 - 10x + 100$
29. $f(x) = 16x^4 + 36x^3 + 16x^2 + x - 30$
31. $-11, -7; (x + 7)(x + 11)$ **33.** $2 \pm \sqrt{3};$
$[x - (2 - \sqrt{3})][x - (2 + \sqrt{3})]$
35. $\pm 5i; (x - 5i)(x + 5i)$
37. $\pm 3, \pm 3i; (x + 3)(x - 3)(x + 3i)(x - 3i)$
39. $1 \pm i; [x - (1 + i)][x - (1 - i)]$
41. $2, 2 \pm i; (x - 2)[x - (2 + i)][x - (2 - i)]$
43. $-5, 4 \pm 3i; (x + 5)[x - (4 + 3i)][x - (4 - 3i)]$
45. $-10, -7 \pm 5i; (x + 10)[x - (-7 + 5i)]$
$[x - (-7 - 5i)]$
47. $-\dfrac{3}{4}, 1 \pm \dfrac{1}{2}i; (4x + 3)(2x - 2 + i)(2x - 2 - i)$
49. $-2, 1 \pm \sqrt{2}i; (x + 2)[x - (1 + \sqrt{2}i)]$
$[x - (1 - \sqrt{2}i)]$
51. $-1/5, 1 \pm \sqrt{5}i; (5x + 1)[x - (1 + \sqrt{5}i)]$
$[x - (1 - \sqrt{5}i)]$
53. $2, \pm 2i; (x - 2)^2(x + 2i)(x - 2i)$
55. $\pm i, \pm 3i; (x + i)(x - i)(x + 3i)(x - 3i)$
57. $-1, \pm 2, \pm 3; (x + 1)(x + 2)(x - 2)(x + 3)(x - 3)$
59. $-2, 1 \pm \sqrt{2}i; (x + 2)^2[x - (1 + \sqrt{2}i)]$
$[x - (1 - \sqrt{2}i)]$
61. (a) $(x^2 + 9)(x^2 - 3)$ **(b)** $(x^2 + 9)(x + \sqrt{3})(x - \sqrt{3})$
(c) $(x + 3i)(x - 3i)(x + \sqrt{3})(x - \sqrt{3})$
63. (a) $(x^2 - 2x - 2)(x^2 - 2x + 3)$
(b) $[x - (1 + \sqrt{3})][x - (1 - \sqrt{3})](x^2 - 2x + 3)$
(c) $[x - (1 + \sqrt{3})][x - (1 - \sqrt{3})][x - (1 + \sqrt{2}i)]$
$[x - (1 - \sqrt{2}i)]$

Section 3.6

1. f **3.** a **5.** j **7.** c **9.** h

11.

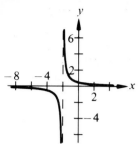

13.

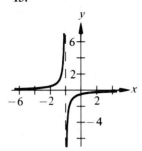

27.

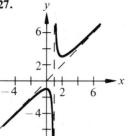

29.

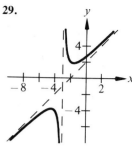

31. Horizontal asymptote: $y = 0$
      Vertical asymptote: $x = 0$
33. Horizontal asymptote: $y = -1$
      Vertical asymptote: $x = 2$
35. Vertical asymptotes: $x = \pm 1$
      Slant asymptote: $y = x$
37. Horizontal asymptote: $y = 3$
39. No asymptotes

15.

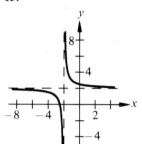

17.

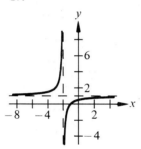

41.

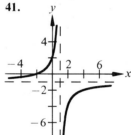

43.

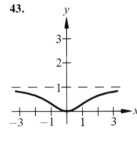

19.

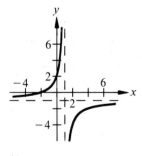

21.

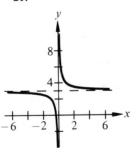

45.

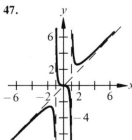

47.

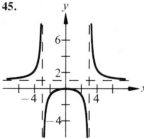

23.

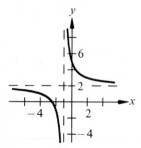

25.

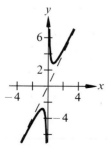

49.

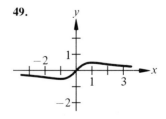

51.

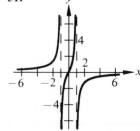

53.

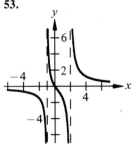

55.

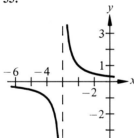

57.

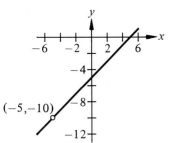

59.

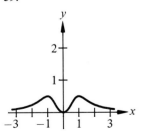

61. (c) $a = 4$

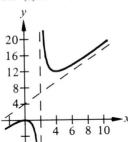

63. $8/3$

Section 3.7

1. $\dfrac{1}{2}\left[\dfrac{1}{x-1} - \dfrac{1}{x+1}\right]$ **3.** $\dfrac{1}{x-1} - \dfrac{1}{x+2}$

5. $\dfrac{3}{2x-1} - \dfrac{2}{x+1}$ **7.** $\dfrac{5}{x-2} - \dfrac{1}{x+2} - \dfrac{3}{x}$

9. $2x + \dfrac{3/2}{x-4} - \dfrac{1/2}{x+2}$ **11.** $\dfrac{3}{x} - \dfrac{1}{x^2} + \dfrac{1}{x+1}$

13. $x + 3 + \dfrac{6}{x-1} + \dfrac{4}{(x-1)^2} + \dfrac{1}{(x-1)^3}$

15. $3\left[\dfrac{1}{x-3} + \dfrac{3}{(x-3)^2}\right]$ **17.** $\dfrac{2x}{x^2+1} - \dfrac{1}{x}$

19. $\dfrac{1}{6}\left[\dfrac{1}{x-2} - \dfrac{1}{x+2} + \dfrac{2}{x^2+2}\right]$

21. $\dfrac{1}{8}\left[\dfrac{1}{2x-1} + \dfrac{1}{2x+1} - \dfrac{4x}{4x^2+1}\right]$

23. $\dfrac{1}{x^2+2} + \dfrac{x}{(x^2+2)^2}$ **25.** $\dfrac{1}{x+1} + \dfrac{2}{x^2-2x+3}$

27. $\dfrac{1}{x} + \dfrac{1-x}{x^2+1}$ **29.** $\dfrac{1}{L}\left(\dfrac{1}{y} + \dfrac{1}{L-y}\right)$

Chapter 3 Review Exercises

1.

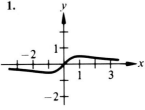

3.

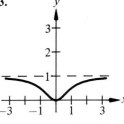

5.

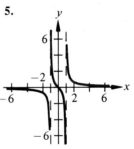

7.

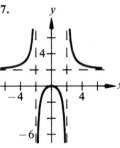

9.

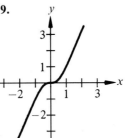

11.

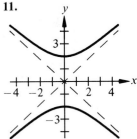

13.

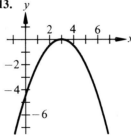

15.

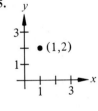

17. $1 - \dfrac{1}{8}\left[\dfrac{25}{x+5} - \dfrac{9}{x-3}\right]$

19. $\dfrac{1}{2}\left[\dfrac{3}{x-1} - \dfrac{x-3}{x^2+1}\right]$

21. $\dfrac{3x}{x^2+1} + \dfrac{x}{(x^2+1)^2}$ **23.** $3 + 7i$ **25.** $-\sqrt{2}\,i$

27. $40 + 65i$ **29.** $-4 - 46i$ **31.** $1 - 6i$ **33.** $\dfrac{4}{3}i$

35. $x^2 - 2$ **37.** $3x^3 + 6x^2 - 4x - 31 - \dfrac{46x - 129}{x^2 - 2x + 4}$

39. $\dfrac{3}{2} + \dfrac{x-2}{2(2x^3+x)}$

41. $\dfrac{1}{4}x^3 - \dfrac{7}{2}x^2 - 7x - 14 - \dfrac{28}{x-2}$ **43.** $6x^3 - 27x$

45. $2x^2 + (-3 + 4i)x + (1 - 2i)$ **47.** $1,\ 3/4$

49. $5/6,\ \pm 2i$ **51.** $-1, \dfrac{2}{3}, \dfrac{3}{2}, 3$

53. $6x^4 + 13x^3 + 7x^2 - x - 1$ **55.** $y_0 = 11.30$

CHAPTER 4

Section 4.1

1. i   **3.** c   **5.** a   **7.** f   **9.** e   **11.** $x = 4$
13. $x = -2$ **15.** $x = 2$ **17.** $x = 16$
19. $x = -2$

21.

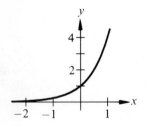

23.

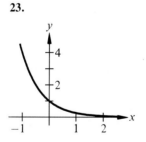

25.

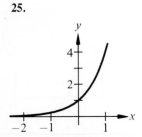

27.

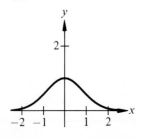

29.

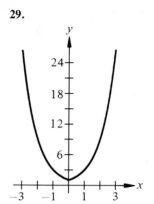

31.

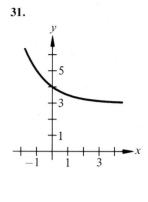

33.

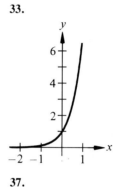

35.

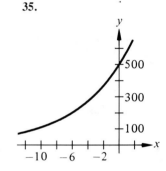

37.

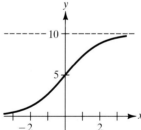

39. (a) \$24,115.73   (b) \$25,714.29   (c) \$26,602.23
(d) \$27,231.38   (e) \$27,547.07   (f) \$27,557.94
41. (a) \$472.70   (b) \$298.29   **43.** (a) 100   (b) 300
(c) 900   **45.** 96%   **47.** (a) 0.15   (b) 0.49   (c) 0.81
49. 229.2 units/mL

Section 4.2

1. $2^4 = 16$ **3.** $5^{-2} = 1/25$ **5.** $16^{1/2} = 4$
7. $7^0 = 1$ **9.** $e^2 = e^2$ **11.** $3 = \log_5 125$
13. $1/4 = \log_{81} 3$ **15.** $-2 = \log_6(1/36)$

17. $3 = \ln(20.0855 \ldots)$ **19.** $v = \log_u w$ **21.** $x = 3$
23. $x = -3$ **25.** $x = 1/3$ **27.** $b = 3$ **29.** $x = 3$
31. $x = -1, x = 2$ **33.** $x = 8$ **35.** 0.9208
37. 0.2084 **39.** 1.6542 **41.** 0.1781 **43.** 1.8957
45. -0.7124 **47.** 0.91355 **49.** 2.0367
51. $\log_2 x + \log_2 y + \log_2 z$ **53.** $(1/2)\ln(a - 1)$
55. $\ln z + 2\ln(z - 1)$ **57.** $2\log_b x - 2\log_b y - 3\log_b z$
59. $(1/3)[\ln x + (1/2)\ln y]$ **61.** $\log_3\left(\dfrac{x - 2}{x + 2}\right)$
63. $\ln \sqrt[3]{\dfrac{x(x + 3)^2}{x^2 - 1}}$ **65.** $\ln \dfrac{9}{\sqrt{x^2 + 1}}$ **67.** 0.7124
69. -0.4312 **71.** 1.6117 **73.** 3.8227
75. -1.4436 **77.** 23.68 yr   **79.** (a) 7.9 (b) 7.7
81. 1.6×10^{-6} **83.** 4.64 **85.** (a) 2.6201
(b) 1.6201 (c) -1.3799 **87.** (a) 3.6420 (b) 1.6420
(c) -2.3580 **89.** (a) 426 (b) 4.26 (c) 0.00426
91. 18.11 **93.** 2.079 **95.** 4.423 **97.** 0.4194
99. 0.2761

29. (a)

K	1	2	3	4
t	0	7.30	11.56	14.59

K	6	8	10	12
t	18.86	21.89	24.24	26.16

(b)

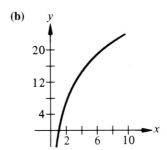

Section 4.3

1.

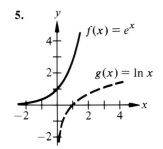

3.

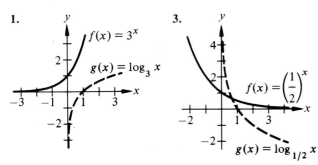

31.

x	y	$\dfrac{\ln x}{\ln y}$	$\ln \dfrac{x}{y}$	$\ln x - \ln y$
1	2	0	-0.6931	-0.6931
3	4	0.7925	-0.2877	-0.2877
10	5	1.4307	0.6931	0.6931
4	0.5	-2.0000	2.0794	2.0794

5.

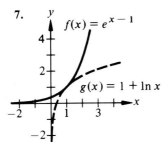

7.

9. d   **11.** j   **13.** g   **15.** a   **17.** h
19. $(1/2, \infty)$ **21.** $(-\infty, 0)$ **23.** $(-2, 2)$
25. $(-\infty, -2), (-2, \infty)$ **27.** (a) 20 (b) 70 (c) 95
(d) 120

Section 4.4

1. $x = -2$ **3.** $x = 5$ **5.** $x = \dfrac{\log 12}{\log 4} = 1.7925$
7. $-\dfrac{\ln 360}{\ln 8.5} = -2.7504$ **9.** $t = \dfrac{\ln 3}{0.09} = 12.2068$
11. $t = \dfrac{\ln 2}{12\ln(1 + 0.10/12)} = 6.9603$
13. $t = 5(\ln 19 - \ln 4) = 7.7907$
15. $N = -\dfrac{\ln 0.2247}{\ln 1.0775} = 20.0016$ **17.** $x = 3.7087$
19. $x = 3$ ($x = -1$ is not in the domain of $\log x$)
21. $x = -1, x = 3$ **23.** $x = 3$ **25.** $x = e^2, 1$
27. $x = (1/2)\ln 3$ **29.** $x = \pm 2$ **31.** (a) 6.64 yr
(b) 6.33 yr (c) 6.30 yr (d) 6.30 yr

33.

r	2%	4%	6%	8%	10%	12%
t	54.93	27.47	18.31	13.73	10.99	9.16

35. (a) 303 units   **(b)** 528 units   **37. (a)** 29.33 yr
(b) 39.79 yr   **39.** 7.79 months   **41.** 22.4°

Chapter 4 Review Exercises

1.

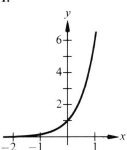

3.

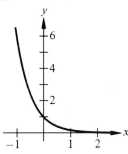

5.

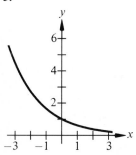

7.

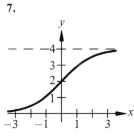

9.

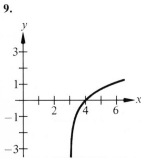

11.

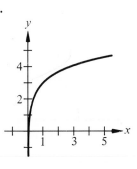

13.

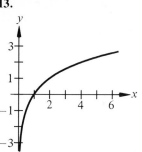

15.

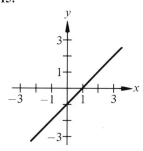

17. $\ln(x^2 + 1) - \ln|x|$ **19.** $\ln(x^2 + 1) + \ln|x - 1|$
21. $\ln \dfrac{\sqrt{2x - 1}}{(x + 1)^2}$ **23.** $\ln(x^2\sqrt[3]{y})$ **25.** $\ln \dfrac{3\sqrt[3]{4 - x^2}}{x}$
27. False   **29.** True   **31. (a)** $x = 1151.3$ units
(b) $x = 1324.6$ units   **33.** $y = 2e^{0.1014t}$

35. $y = 4e^{-0.4159t}$ **37. (a)** $r = \dfrac{\ln 2}{7.75} = 9\%$

(b) \$1834.37   **39. (a)** $N = 30(1 - e^{-0.0502t})$

(b) $t = \dfrac{\ln 6}{0.0502} = 36$ days

41. (a) $S = 30(1 - e^{-0.1823t})$ **(b)** 17,944 units

(c)

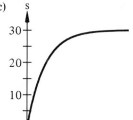

43. (a) 30
(b) 25
(c) a
(d) 0

CHAPTER 5

Section 5.1

1.

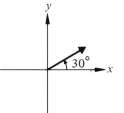

3.

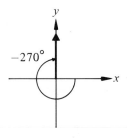

5.

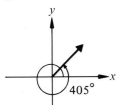

7.

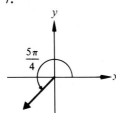

77.

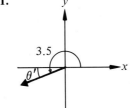

(c) $\theta' = \pi/3$

79.

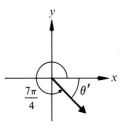

(c) $\theta' = \pi/4$

9.

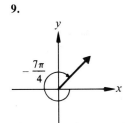

11.

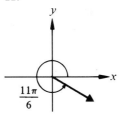

13. 396°, −324°   **15.** 240°, −480°   **17.** 660°, −60°
19. 300°, −60°   **21.** 19π/9, − 17π/9
23. 23π/6, −π/6   **25.** 7π/4, −π/4
27. 26π/9, −10π/9   **29.** π/6   **31.** 7π/4
33. −π/9   **35.** −3π/2   **37.** 270°   **39.** −105°
41. 420°   **43.** 330°   **45.** 2.007   **47.** −3.776
49. 9.285   **51.** −0.014   **53.** 6.021   **55.** 25.714°
57. 337.500°   **59.** −756°   **61.** −114.592°
63. 429.718°   **65. (a)** 54.75° **(b)** −128.5°
67. (a) 85.308° **(b)** 330.007°

81.

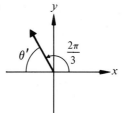

(c) $\theta' = 3.5 - \pi$

83.

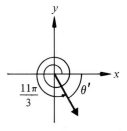

(c) $\theta' = \pi/3$

85.

r	8 ft	15 in	85 cm
s	12 ft	24 in	63.75π cm
θ	1.5	1.6	3π/4

r	24 in	12963/π mi
s	96 in	8642 mi
θ	4	2π/3

87. (a) $\dfrac{5\pi}{12}$ **(b)** 7.8125π in   **89.** 4.655°
91. (a) 560.225 rev/min **(b)** 3520 rad/min

69.

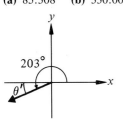

(c) $\theta' = 23°$

71.

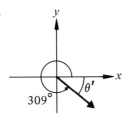

(c) $\theta' = 51°$

73.

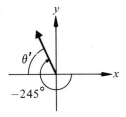

(c) $\theta' = 65°$

75.

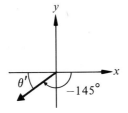

(c) $\theta' = 35°$

Section 5.2

1. (a) $\sin 30° = \dfrac{1}{2}$ $\csc 30° = 2$

$\cos 30° = \dfrac{\sqrt{3}}{2}$ $\sec 30° = \dfrac{2\sqrt{3}}{3}$

$\tan 30° = \dfrac{\sqrt{3}}{3}$ $\cot 30° = \sqrt{3}$

(b) $\sin 45° = \dfrac{\sqrt{2}}{2}$ $\csc 45° = \sqrt{2}$

$\cos 45° = \dfrac{\sqrt{2}}{2}$ $\sec 45° = \sqrt{2}$

$\tan 45° = 1$ $\cot 45° = 1$

3. (a) $\sin 60° = \dfrac{\sqrt{3}}{2}$ $\csc 60° = \dfrac{2\sqrt{3}}{3}$

$\cos 60° = \dfrac{1}{2}$ $\sec 60° = 2$

$\tan 60° = \sqrt{3}$ $\cot 60° = \dfrac{\sqrt{3}}{3}$

(b) $\sin 90° = 1$ $\csc 90° = 1$

$\cos 90° = 0$ $\sec 90°$ undefined

$\tan 90°$ undefined $\cot 90° = 0$

5. (a) $\sin \pi = 0$ $\csc \pi$ undefined

$\cos \pi = -1$ $\sec \pi = -1$

$\tan \pi = 0$ $\cot \pi$ undefined

(b) $\sin \dfrac{3\pi}{2} = -1$ $\csc \dfrac{3\pi}{2} = -1$

$\cos \dfrac{3\pi}{2} = 0$ $\sec \dfrac{3\pi}{2}$ undefined

$\tan \dfrac{3\pi}{2}$ undefined $\cot \dfrac{3\pi}{2} = 0$

7. (a) $\sin \dfrac{17\pi}{4} = \dfrac{\sqrt{2}}{2}$ $\csc \dfrac{17\pi}{4} = \sqrt{2}$

$\cos \dfrac{17\pi}{4} = \dfrac{\sqrt{2}}{2}$ $\sec \dfrac{17\pi}{4} = \sqrt{2}$

$\tan \dfrac{17\pi}{4} = 1$ $\cot \dfrac{17\pi}{4} = 1$

(b) $\sin \dfrac{7\pi}{3} = \dfrac{\sqrt{3}}{2}$ $\csc \dfrac{7\pi}{3} = \dfrac{2\sqrt{3}}{3}$

$\cos \dfrac{7\pi}{3} = \dfrac{1}{2}$ $\sec \dfrac{7\pi}{3} = 2$

$\tan \dfrac{7\pi}{3} = \sqrt{3}$ $\cot \dfrac{7\pi}{3} = \dfrac{\sqrt{3}}{3}$

9. III   **11.** I   **13.** II   **15.** IV   **17.** III

19. (a)

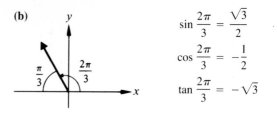

$\sin \dfrac{4\pi}{3} = -\dfrac{\sqrt{3}}{2}$

$\cos \dfrac{4\pi}{3} = -\dfrac{1}{2}$

$\tan \dfrac{4\pi}{3} = \sqrt{3}$

(b)

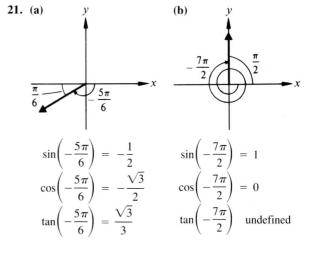

$\sin \dfrac{2\pi}{3} = \dfrac{\sqrt{3}}{2}$

$\cos \dfrac{2\pi}{3} = -\dfrac{1}{2}$

$\tan \dfrac{2\pi}{3} = -\sqrt{3}$

21. (a)   **(b)**

$\sin\left(-\dfrac{5\pi}{6}\right) = -\dfrac{1}{2}$ $\sin\left(-\dfrac{7\pi}{2}\right) = 1$

$\cos\left(-\dfrac{5\pi}{6}\right) = -\dfrac{\sqrt{3}}{2}$ $\cos\left(-\dfrac{7\pi}{2}\right) = 0$

$\tan\left(-\dfrac{5\pi}{6}\right) = \dfrac{\sqrt{3}}{3}$ $\tan\left(-\dfrac{7\pi}{2}\right)$ undefined

23. (a)   **(b)**

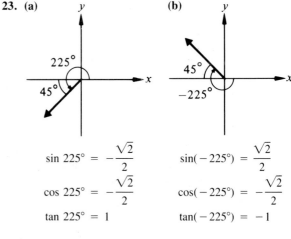

$\sin 225° = -\dfrac{\sqrt{2}}{2}$ $\sin(-225°) = \dfrac{\sqrt{2}}{2}$

$\cos 225° = -\dfrac{\sqrt{2}}{2}$ $\cos(-225°) = -\dfrac{\sqrt{2}}{2}$

$\tan 225° = 1$ $\tan(-225°) = -1$

25. (a)

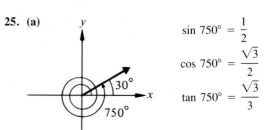

$\sin 750° = \dfrac{1}{2}$

$\cos 750° = \dfrac{\sqrt{3}}{2}$

$\tan 750° = \dfrac{\sqrt{3}}{3}$

(b)

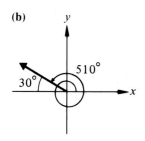

$$\sin 510° = \frac{1}{2}$$

$$\cos 510° = -\frac{\sqrt{3}}{2}$$

$$\tan 510° = -\frac{\sqrt{3}}{3}$$

27. $\sin \theta = \sqrt{3}/2$ $\csc \theta = 2\sqrt{3}/3$
     $\cos \theta = -1/2$ $\sec \theta = -2$
     $\tan \theta = -\sqrt{3}$ $\cot \theta = -\sqrt{3}/3$

29. $\sin \theta = 0$ $\csc \theta$ undefined
     $\cos \theta = -1$ $\sec \theta = -1$
     $\tan \theta = 0$ $\cot \theta$ undefined

31. $\sin \theta = 0$ $\csc \theta$ undefined
     $\cos \theta = 1$ $\sec \theta = 1$
     $\tan \theta = 0$ $\cot \theta$ undefined

	Degrees	*Radians*
33.	**(a)** 30°, 150°	$\pi/6, 5\pi/6$
	(b) 210°, 330°	$7\pi/6, 11\pi/6$
35.	**(a)** 60°, 120°	$\pi/3, 2\pi/3$
	(b) 135°, 315°	$3\pi/4, 7\pi/4$
37.	**(a)** 45°, 225°	$\pi/4, 5\pi/4$
	(b) 150°, 330°	$5\pi/6, 11\pi/6$

Section 5.3

1. 2   **3.** 4/3   **5.** 17/15

7.   **(a)** $c_1 = 5, b_2 = 12, c_2 = 15$
(b) $\sin \alpha_1 = \sin \alpha_2 = 3/5$ $\csc \alpha_1 = \csc \alpha_2 = 5/3$
     $\cos \alpha_1 = \cos \alpha_2 = 4/5$ $\sec \alpha_1 = \sec \alpha_2 = 5/4$
     $\tan \alpha_1 = \tan \alpha_2 = 3/4$ $\cot \alpha_1 = \cot \alpha_2 = 4/3$

9.   **(a)** $b_1 = \sqrt{3}, a_2 = \dfrac{5\sqrt{3}}{3}, c_2 = \dfrac{10\sqrt{3}}{3}$
(b) $\sin \alpha_1 = \sin \alpha_2 = 1/2$ $\csc \alpha_1 = \csc \alpha_2 = 2$
     $\cos \alpha_1 = \cos \alpha_2 = \sqrt{3}/2$ $\sec \alpha_1 = \sec \alpha_2 = 2\sqrt{3}/3$
     $\tan \alpha_1 = \tan \alpha_2 = \sqrt{3}/3$ $\cot \alpha_1 = \cot \alpha_2 = \sqrt{3}$

11.

$\sin \theta = 2/3$ $\csc \theta = 3/2$
$\cos \theta = \sqrt{5}/3$ $\sec \theta = 3\sqrt{5}/5$
$\tan \theta = 2\sqrt{5}/5$ $\cot \theta = \sqrt{5}/2$

13.

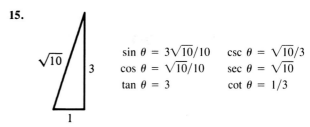

$\sin \theta = \sqrt{3}/2$ $\csc \theta = 2\sqrt{3}/3$
$\cos \theta = 1/2$ $\sec \theta = 2$
$\tan \theta = \sqrt{3}$ $\cot \theta = \sqrt{3}/3$

15.

$\sin \theta = 3\sqrt{10}/10$ $\csc \theta = \sqrt{10}/3$
$\cos \theta = \sqrt{10}/10$ $\sec \theta = \sqrt{10}$
$\tan \theta = 3$ $\cot \theta = 1/3$

17. (a) $1/2$ **(b)** $\sqrt{3}$ **(c)** $2\sqrt{3}/3$ **(d)** $\sqrt{3}/2$
19. (a) $2\sqrt{2}$ **(b)** $1/3$ **(c)** $\sqrt{2}/4$ **(d)** 3
21. $\sin \theta = \sqrt{2}/2$
     $\cos \theta = \sqrt{2}/2$
     $\tan \theta = 1$
     $\csc \theta = \sqrt{2}$
     $\sec \theta = \sqrt{2}$
     $\cot \theta = 1$
23. (a) 0.1736   **(b)** 0.1736   **25. (a)** 1.3499   **(b)** 1.3432
27. (a) 0.0755   **(b)** 0.0857   **29. (a)** 1.2803   **(b)** 0.7811
31. (a) 5.0273   **(b)** 0.1989   **33. (a)** 1.1884   **(b)** 1.1884
35. (a) $\theta = 30° = \pi/6$ **(b)** $\theta = 30° = \pi/6$
37. (a) $\theta = 60° = \pi/3$ **(b)** $\theta = 45° = \pi/4$
39. (a) $\theta = 60° = \pi/3$ **(b)** $\theta = 45° = \pi/4$
41. (a) $\theta = 10° = \pi/18$ **(b)** $\theta = 29° = 29\pi/180$
43. (a) $\theta = 43° = 43\pi/180$ **(b)** $\theta = 82° = 41\pi/90$
45. (a) $\theta = 89.3° = 1.56$ **(b)** $\theta = 75° = 5\pi/12$
47. $h = 15$ ft   **49.** $100\sqrt{3}/3$ **51.** $25\sqrt{3}/3$
53. 15.5572   **55.** 9.1925
57. $\sin \theta = 4/5$ $\csc \theta = 5/4$
     $\cos \theta = 3/5$ $\sec \theta = 5/3$
     $\tan \theta = 4/3$ $\cot \theta = 3/4$
59. $\sin \theta = -5/13$ $\csc \theta = -13/5$
     $\cos \theta = -12/13$ $\sec \theta = -13/12$
     $\tan \theta = 5/12$ $\cot \theta = 12/5$
61. $\sin \theta = 1/2$ $\csc \theta = 2$
     $\cos \theta = -\sqrt{3}/2$ $\sec \theta = -2\sqrt{3}/3$
     $\tan \theta = -\sqrt{3}/3$ $\cot \theta = -\sqrt{3}$
63. $\sin \theta = -2\sqrt{5}/5$ $\csc \theta = -\sqrt{5}/2$
     $\cos \theta = -\sqrt{5}/5$ $\sec \theta = -\sqrt{5}$
     $\tan \theta = 2$ $\cot \theta = 1/2$
65. $h = 19.3$ ft   **67.** 2145.1 ft   **69.** True
71. False   **73.** False

Section 5.4

1. Period: π
Amplitude: 2

3. Period: 4π
Amplitude: 3/2

5. Period: 2
Amplitude: 1/2

7. Period: 2π
Amplitude: 2

9. Period: $\pi/5$
Amplitude: 2

11. Period: 3π
Amplitude: 1/2

13. Period: 1/2
Amplitude: 3

15.

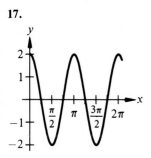

17.

19.

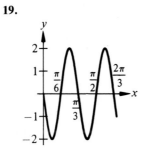

21.

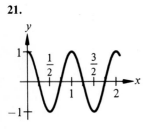

23.

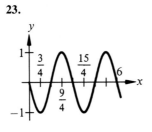

25.

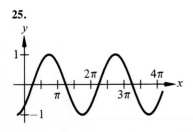

27.

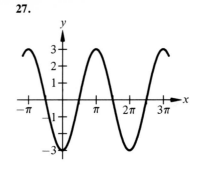

29.

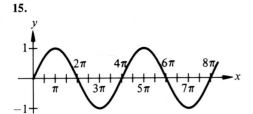

31.

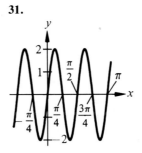

33.

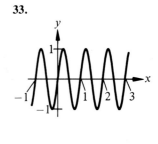

35.

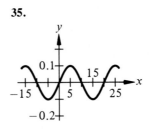

37.

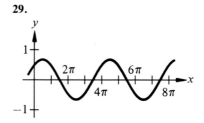

39.

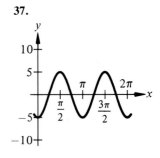

41. $y = 2 \sin 4x$

43. $y = \cos\left(2x + \dfrac{\pi}{2}\right)$

45. $y = 2 \sin\left(\dfrac{\pi x}{2} + \dfrac{\pi}{2}\right)$

47. (a) $t = 6$ sec (b) 10

(c)

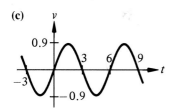

49. (a) $p = 1/440$ (b) $f = 440$

(c)

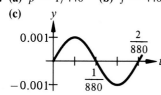

Section 5.5

1. d **3.** f **5.** a **7.** e **9.** c

11.

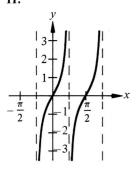

13.

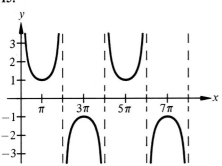

15.

17.

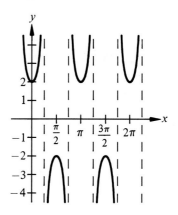

19.

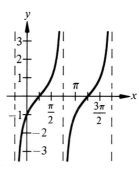

21.

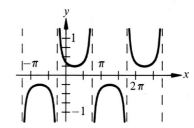

23.

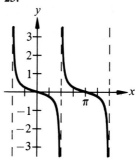

25.

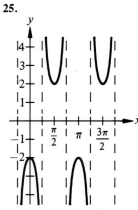

27.

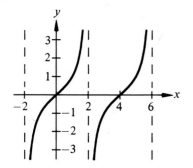

29.

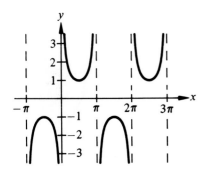

9.

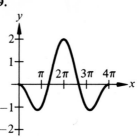

11.

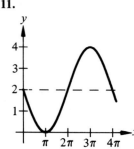

13.

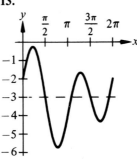

15.

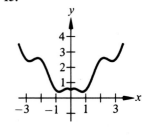

Section 5.6

1.

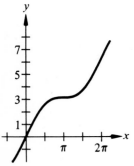

3.

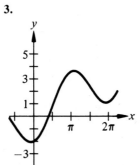

17.

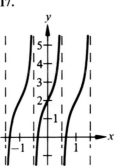

19.

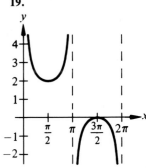

5.

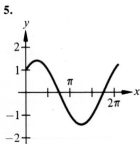

7.

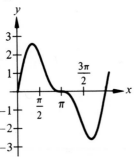

21.

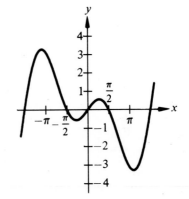

23.

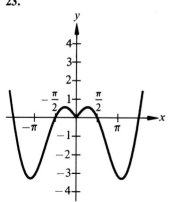

29.

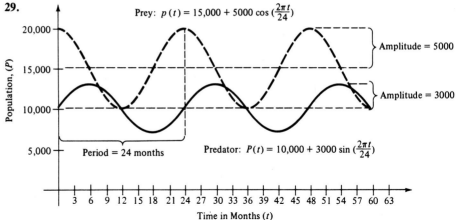

Prey: $p(t) = 15,000 + 5000 \cos\left(\frac{2\pi t}{24}\right)$

Amplitude = 5000

Amplitude = 3000

Period = 24 months

Predator: $P(t) = 10,000 + 3000 \sin\left(\frac{2\pi t}{24}\right)$

Time in Months (t)

Population, (P)

25.

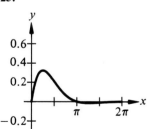

We can explain the cycles of this predator-prey population by noting the following cause and effect pattern:

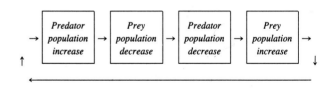

27.

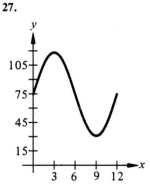

Section 5.7

1. $\pi/6$ **3.** $\pi/3$ **5.** $\pi/6$ **7.** $5\pi/6$ **9.** $5\pi/6$

11. $\pi/3$ **13.** $\pi/3$ **15.** $\pi/2$ **17.** 0.3 **19.** -10

21. 1.287 **23.** -0.848 **25.** -1.107 **27.** 0.328

29. 1.817 **31.** 0.827 **33.** $3/5$ **35.** $1/2$

37. $12/13$ **39.** $\sqrt{34}/5$ **41.** x **43.** $\sqrt{1 - 4x^2}$

45. $\dfrac{\sqrt{x^2 - 1}}{x}$ **47.** $\dfrac{\sqrt{x^2 - 9}}{3}$ **49.** $\dfrac{\sqrt{x^2 + 2}}{x}$

51. $\arctan\left(\dfrac{9}{x}\right) = \arcsin\left(\dfrac{9}{\sqrt{x^2 + 81}}\right) = \text{arcsec}\left(\dfrac{\sqrt{x^2 + 81}}{x}\right)$

53. $\text{arcsec}\left(\dfrac{\sqrt{x^2 - 2x + 10}}{3}\right) = \arcsin\left(\dfrac{x - 1}{\sqrt{x^2 - 2x + 10}}\right)$

$= \arctan\left(\dfrac{x - 1}{3}\right)$

57.

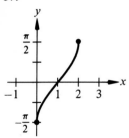

59.

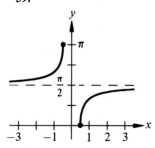

3. (a) 128.571° **(b)** −200.535°   **5. (a)** $\pi/5$ **(b)** 80°

7. (a) $\sin \theta = -\sqrt{11}/6$ $\csc \theta = -6\sqrt{11}/11$

       $\cos \theta = 5/6$ $\sec \theta = 6/5$

       $\tan \theta = -\sqrt{11}/5$ $\cot \theta = -5\sqrt{11}/11$

  (b) $\sin \theta = 12/13$ $\csc \theta = 13/12$

       $\cos \theta = -5/13$ $\sec \theta = -13/5$

       $\tan \theta = -12/5$ $\cot \theta = -5/12$

9. (a) $-\sqrt{3}/2$ **(b)** $\sqrt{3}$ **(c)** $-\sqrt{2}/2$ **(d)** -1

11.

Degrees	Radians
(a) 57°, 123°	$19\pi/60,\ 41\pi/60$
(b) 147°, 327°	$49\pi/60,\ 109\pi/60$
(c) 165°, 195°	$11\pi/12,\ 13\pi/12$
(d) 5°, 175°	$\pi/36,\ 35\pi/36$

Section 5.8

1. $B = 70°$ $a = 3.6397$ $c = 10.6418$

3. $A = 19°$ $a = 8.2639$ $c = 25.3829$

5. $B = 77° \ 45'$ $a = 91.3425$ $b = 420.6980$

7. $A = 30.96°$ $B = 59.04°$ $c = 2\sqrt{34}$

9. $A = 72.08°$ $B = 17.92°$ $a = 12\sqrt{17}$

11. $h = 16 \sin 74° = 15.38$ ft

13. $\theta = \arctan (3/2) = 56.31°$

15. $\theta = 90° - \arcsin (40/41) = 12.68°$

17. $h = 50(\tan 47° \ 40' - \tan 35°) = 19.87$ ft

19. $825 \cos 52° = 507.92$ mi north

     $825 \sin 52° = 650.11$ mi east

21. N 56° 18′ 36″ W   **23. (a)** N 58° E   **(b)** 68.82 yd

25. $300(\cot 4° - \cot 6.5°) = 1657.13$ ft

27. $\dfrac{48,400}{\cot 16° - \cot 57°} = 17,054$ ft   **29.** 29.39 in

31. (a) 4   **(b)** 4 cycles per unit of time   **(c)** $t = 1/16$

33. (a) 1/16   **(b)** 60 cycles per unit of time   **(c)** $t = 1/120$

35. $\omega = 528\pi$

13.

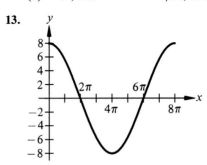

15.

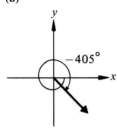

Chapter 5 Review Exercises

1. (a)

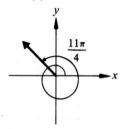

$3\pi/4,\ -5\pi/4$

(b)

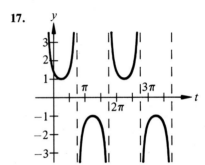

$315°,\ -45°$

17.

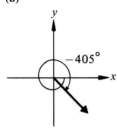

19.

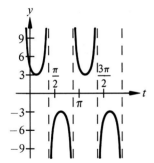

21.

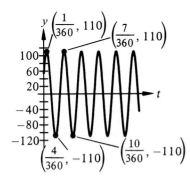

23.

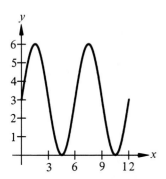

25.

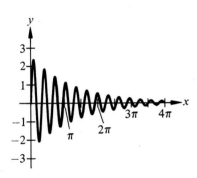

27.

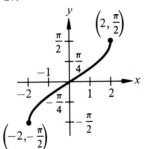

29.

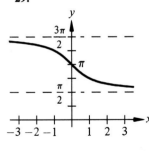

31. $\dfrac{1}{\sqrt{2x - x^2}}$ **33.** $h = 2.5 \tan 28° = 1.33$ mi

CHAPTER 6

Section 6.1

1. $\sin x = 1/2$ $\csc x = 2$
  $\cos x = \sqrt{3}/2$ $\sec x = 2\sqrt{3}/3$
  $\tan x = \sqrt{3}/3$ $\cot x = \sqrt{3}$

3. $\sin \theta = -\sqrt{2}/2$ $\csc \theta = -\sqrt{2}$
  $\cos \theta = \sqrt{2}/2$ $\sec \theta = \sqrt{2}$
  $\tan \theta = -1$ $\cot \theta = -1$

5. $\sin x = -5/13$ $\csc x = -13/5$
  $\cos x = -12/13$ $\sec x = -13/12$
  $\tan x = 5/12$ $\cot x = 12/5$

7. $\sin \phi = 0$ $\csc \phi$ undefined
  $\cos \phi = -1$ $\sec \phi = -1$
  $\tan \phi = 0$ $\cot \phi$ undefined

9. $\sin x = 2/3$ $\csc x = 3/2$
  $\cos x = -\sqrt{5}/3$ $\sec x = -3\sqrt{5}/5$
  $\tan x = -2\sqrt{5}/5$ $\cot x = -\sqrt{5}/2$

11. $\sin \theta = -2\sqrt{5}/5$ $\csc \theta = -\sqrt{5}/2$
  $\cos \theta = -\sqrt{5}/5$ $\sec \theta = -\sqrt{5}$
  $\tan \theta = 2$ $\cot \theta = 1/2$

13. $\sin \theta = -1$ $\csc \theta = -1$
  $\cos \theta = 0$ $\sec \theta$ undefined
  $\tan \theta$ undefined $\cot \theta = 0$

15. d   **17.** a   **19.** b   **21.** b   **23.** e   **25.** f
27. $\sec \phi$ **29.** 1   **31.** $1 + \sin y$

33. $\csc \theta = \dfrac{1}{\sin \theta}$ $\sec \theta = \dfrac{\pm 1}{\sqrt{1 - \sin^2\theta}}$

$\cos \theta = \pm\sqrt{1 - \sin^2\theta}$ $\cot \theta = \dfrac{\pm\sqrt{1 - \sin^2\theta}}{\sin \theta}$

$\tan \theta = \dfrac{\pm \sin \theta}{\sqrt{1 - \sin^2\theta}}$

35. $-\tan \theta$ **37.** $\pi/2 \le \theta \le 3\pi/2$ **39.** $\ln |\cot \theta|$

41. (a) False, $\sin k\theta/\cos k\theta = \tan k\theta$

(b) False, $5 \sec \theta = 5/\cos \theta$ **(c)** False, $\sin \theta \csc \theta = 1$

Section 6.2

1. $\sin^2 x$ **3.** $\sin^2 x \tan^2 x$ **5.** $\sec^4 x$

7. $\sin^2 x - \cos^2 x$ **9.** $1 + 2 \sin x \cos x$ **11.** $\tan^2 x$

13. $2 \csc^2 x$ **15.** $2 \sec x$ **17.** $1 + \cos y$

19. $3(\sec x + \tan x)$ **67. (a)** $\sin \theta = \pm\sqrt{1 - \cos^2 \theta}$

(b) $\tan \theta = \pm\sqrt{\sec^2 \theta - 1}$

Section 6.3

1. $\sin 75° = (\sqrt{2}/4)(1 + \sqrt{3})$
$\cos 75° = (\sqrt{2}/4)(\sqrt{3} - 1)$
$\tan 75° = (1/2)(\sqrt{3} + 1)^2$

3. $\sin 105° = (\sqrt{2}/4)(\sqrt{3} + 1)$
$\cos 105° = (\sqrt{2}/4)(1 - \sqrt{3})$
$\tan 105° = (-1/2)(\sqrt{3} + 1)^2$

5. $\sin 195° = (\sqrt{2}/4)(1 - \sqrt{3})$
$\cos 195° = (-\sqrt{2}/4)(1 + \sqrt{3})$
$\tan 195° = (1/2)(\sqrt{3} - 1)^2$

7. $\sin (11\pi/12) = (\sqrt{2}/4)(\sqrt{3} - 1)$
$\cos (11\pi/12) = (-\sqrt{2}/4)(\sqrt{3} + 1)$
$\tan (11\pi/12) = (-1/2)(1 - \sqrt{3})^2$

9. $\sin (17\pi/12) = (-\sqrt{2}/4)(\sqrt{3} + 1)$
$\cos (17\pi/12) = (\sqrt{2}/4)(1 - \sqrt{3})$
$\tan (17\pi/12) = (1/2)(\sqrt{3} + 1)^2$

11. $\cos 40°$ **13.** $\sin 200°$ **15.** $\tan 239°$ **17.** $\sin 1.8$

19. $\tan 3x$ **41.** $\sqrt{2} \sin\left(\theta + \dfrac{\pi}{4}\right)$

43. $13 \cos\left(3\theta - \arctan \dfrac{12}{5}\right)$ **45.** $\sqrt{2} \sin \theta + \sqrt{2} \cos \theta$

47. 1

5. $\sin (2\pi/3) = \sqrt{3}/2$
$\cos (2\pi/3) = -1/2$
$\tan (2\pi/3) = -\sqrt{3}$

7. $\sin 2u = 24/25$
$\cos 2u = 7/25$
$\tan 2u = 24/7$

9. $\sin 2u = 4/5$
$\cos 2u = 3/5$
$\tan 2u = 4/3$

11. $\sin 2u = -4\sqrt{21}/25$
$\cos 2u = -17/25$
$\tan 2u = 4\sqrt{21}/17$

13. $\sin 105° = \frac{1}{2}\sqrt{2 + \sqrt{3}}$
$\cos 105° = -\frac{1}{2}\sqrt{2 - \sqrt{3}}$
$\tan 105° = -2 - \sqrt{3}$

15. $\sin(112° \, 30') = \frac{1}{2}\sqrt{2 + \sqrt{2}}$
$\cos(112° \, 30') = -\frac{1}{2}\sqrt{2 - \sqrt{2}}$
$\tan(112° \, 30') = -\sqrt{2} - 1$

17. $\sin(\pi/8) = \dfrac{\sqrt{2 - \sqrt{2}}}{2}$

$\cos(\pi/8) = \dfrac{\sqrt{2 + \sqrt{2}}}{2}$

$\tan(\pi/8) = \sqrt{2} - 1$

19. $\sin(52° \, 30') = \frac{1}{2}\sqrt{2 + \sqrt{2 - \sqrt{3}}}$
$\cos(52° \, 30') = \frac{1}{2}\sqrt{2 - \sqrt{2 - \sqrt{3}}}$
$\tan(52° \, 30') = \dfrac{2 + \sqrt{2 - \sqrt{3}}}{\sqrt{2 + \sqrt{3}}}$

21. $\sin(u/2) = 5\sqrt{26}/26$
$\cos(u/2) = \sqrt{26}/26$
$\tan(u/2) = 5$

23. $\sin(u/2) = \sqrt{\dfrac{\sqrt{89} - 8}{2\sqrt{89}}}$

$\cos(u/2) = -\sqrt{\dfrac{\sqrt{89} + 8}{2\sqrt{89}}}$

$\tan(u/2) = (1/5)(8 - \sqrt{89})$

25. $\sin(u/2) = 3\sqrt{10}/10$
$\cos(u/2) = -\sqrt{10}/10$
$\tan(u/2) = -3$

27. $\sin 3x$ **29.** $\tan 4x$

31. $(1/8)(3 + 4 \cos 2x + \cos 4x)$

33. $(1/16)[2 - 6 \cos 2x + (3 - \cos 2x)(1 + \cos 4x)]$

35. $(1/16)(1 - \cos 2x)(3 + 4 \cos 2x + \cos 4x)$

Section 6.4

1. $\sin 90° = 1$
$\cos 90° = 0$
$\tan 90°$ undefined

3. $\sin 60° = \sqrt{3}/2$
$\cos 60° = 1/2$
$\tan 60° = \sqrt{3}$

Section 6.5

1. $3\left(\sin \dfrac{\pi}{2} + \sin 0\right)$ **3.** $(1/2)(\sin 8\theta + \sin 2\theta)$

5. $(5/2)[\cos(8\beta) + \cos(2\beta)]$

7. $(1/2)(\cos 2y - \cos 2x)$ **9.** $(1/2)(\sin 2\theta + \sin 2\pi)$

11. $2 \sin 45° \cos 15°$ **13.** $-2 \sin \dfrac{\pi}{2} \sin \dfrac{\pi}{4}$

15. $2 \cos 4x \cos 2x$ **17.** $2 \cos \alpha \sin \beta$

19. $2 \cos(\phi + \pi) \cos \pi$

Section 6.6

1. $\pi/4, 3\pi/4, 5\pi/4, 7\pi/4$ **3.** $\pi/6, 5\pi/6, 7\pi/6, 11\pi/6$

5. $0, \pi/4, \pi, 5\pi/4$ **7.** $\pi/6, 11\pi/6$

9. $\pi/2, 2\pi/3, 4\pi/3, 3\pi/2$ **11.** $\pi/3, 5\pi/3$

13. $7\pi/6, 3\pi/2, 11\pi/6$ **15.** $0, \pi/2, \pi, 3\pi/2$

17. $\pi/2, 2\pi/3, 3\pi/2, 4\pi/3$ **19.** $\pi/3, 3\pi/4, 4\pi/3, 7\pi/4$

21. No solution **23.** $2\pi/3, 5\pi/6, 5\pi/3, 11\pi/6$

25. $\pi/2$ **27.** $\pi/12, 5\pi/12, 13\pi/12, 17\pi/12$

29. $0, 2\pi/3, 4\pi/3$ **31.** $\pi/4, 5\pi/4$

33. $0, \pi/2, \pi, 3\pi/2$ **35.** $\pi/6, 5\pi/6, 7\pi/6, 11\pi/6$

37. $0, \pi/4, \pi/2, 3\pi/4, \pi, 5\pi/4, 3\pi/2, 7\pi/4$

39. $0, \pi/4, \pi/2, 3\pi/4, \pi, 5\pi/4, 3\pi/2, 7\pi/4$ **41.** $\pi/6,$ $5\pi/6$ **43.** $0, \pi/4, \pi/3, \pi/2, 2\pi/3, 3\pi/4, \pi, 5\pi/4,$ $4\pi/3, 3\pi/2, 5\pi/3, 7\pi/4$ **45.** $0.9828, 1.7682, 4.1244,$ 4.9098 **47.** $0.3398, 0.8481, 2.2935, 2.8018$

49. $0.8411, 5.4421$ **51.** $1.3981, 1.1555, 4.2971, 4.5397$

53. $0.4271, 2.7145$

55. $\pi/4, 5\pi/4$

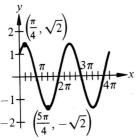

57. $0.04, 0.43, 0.83$

59. $36.9°$

Chapter 6 Review Exercises

1. $\sin^2 x$ **3.** $1 + \cot \alpha$ **5.** 1 **7.** 1 **9.** $\cos^2 2x$

31. $0, \pi$ **33.** $\pi/3, 5\pi/3$

35. $0, 3\pi/4, \pi, 5\pi/4$ **37.** $0, \pi/2, \pi$ **39.** $\pi/4, 5\pi/4$

41. (a) $(1/2)(\cos \alpha - \cos 5\alpha)$

  (b) $(1/2)(\sin 3x^2 + \sin x^2)$

  (c) $(1/2)\left(\cos \dfrac{3x}{4} + \cos \dfrac{x}{4}\right)$

CHAPTER 7

Section 7.1

1. $C = 105°, b = 14.1, c = 19.3$

3. $C = 110°, b = 22.4, c = 24.4$

5. $B = 21.6°, C = 122.4°, c = 11.5$

7. $B = 10°, b = 69.5, c = 136.8$

9. $B = 42° 4', a = 22.1, b = 14.9$

11. $A = 10° 11', C = 154° 19', c = 11.0$

13. $A = 25.6°, B = 9.4°, a = 10.5$

15. $B = 18° 13', C = 51° 32', c = 40.1$

17. No solution **19.** Two solutions $B = 70.4°,$ $C = 51.6°, c = 4.2; \quad B = 109.6°, C = 12.4°, c = 1.1$

21. No solution **23. (a)** $b = 5/\sin 36°$

(b) $b < 5/\sin 36°$ **(c)** $b > 5/\sin 36°$ **25.** 10.4

27. 1675.2 **29.** 474.9 **31.** 6.0 **33.** 77 yd

35. 26.1 mi, 15.9 mi, 15.7 mi **37.** 3 mi **39.** No

Section 7.2

1. $A = 27.7°, B = 40.5°, C = 111.8°$

3. $A = 36.9°, B = 53.1°, C = 90°$

5. $B = 23.8°, C = 126.2°, a = 12.4$

7. $A = 92.9°, B = 43.5°, C = 43.5°$

9. $A = 158° 36', C = 12° 38', b = 10.4$

11. $A = 20°10', B = 20°10', c = 56.9$ **13.** 16.25

15. 54 **17.** 96.82 **19.** S $52° 37'$ E, S $25° 20'$ W

21. 43.3 mi **23.** 116.35 ft, 133.09 ft

25. (a) N $60.1°$ E **(b)** N $73.0°$ E **27.** 3.94 ft

29. (a) 570.60 **(b)** 5910.08 **(c)** 177.09

Section 7.3

1. $\mathbf{V} = 3\mathbf{i} + 4\mathbf{j}, |\mathbf{V}| = 5$ **3.** $\mathbf{V} = -3\mathbf{i} + 2\mathbf{j}, |\mathbf{V}| = \sqrt{13}$

5. $\mathbf{V} = 5\mathbf{j}, |\mathbf{V}| = 5$

7. $\mathbf{V} = 3\mathbf{i}$ **9.** $\mathbf{V} = \dfrac{-\sqrt{3}\mathbf{i} + \mathbf{j}}{2}$

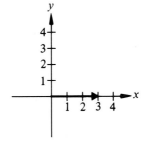
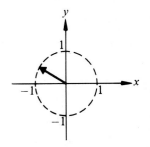

11. $\mathbf{V} = \dfrac{3\sqrt{2}}{2}(-\sqrt{3}\mathbf{i} \pm \mathbf{j})$

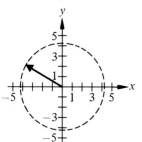

13. $\mathbf{V} = 5\mathbf{i} + 6\mathbf{j}$

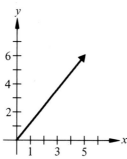

5. $3\sqrt{2}\left(\cos\dfrac{7\pi}{4} + i\sin\dfrac{7\pi}{4}\right)$

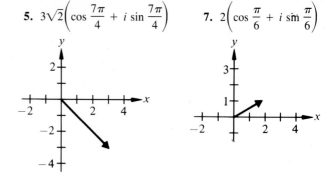

7. $2\left(\cos\dfrac{\pi}{6} + i\sin\dfrac{\pi}{6}\right)$

15. $\mathbf{V} = -7\mathbf{j}$

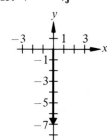

17. $\mathbf{V} = 3\mathbf{i} - (3/2)\mathbf{j}$

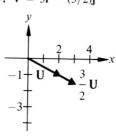

9. $4\left(\cos\dfrac{4\pi}{3} + i\sin\dfrac{4\pi}{3}\right)$

11. $6\left(\cos\dfrac{\pi}{2} + i\sin\dfrac{\pi}{2}\right)$

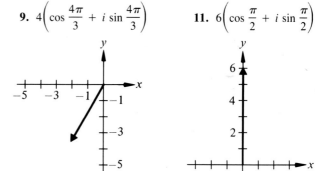

19. $\mathbf{V} = 4\mathbf{i} + 3\mathbf{j}$

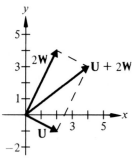

21. $\mathbf{V} = \dfrac{7\mathbf{i} - \mathbf{j}}{2}$

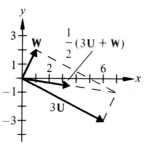

13. $\sqrt{65}[\cos(2.6224) + i\sin(2.6224)]$

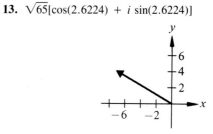

23. $(4\mathbf{i} - 3\mathbf{j})/5$ **25.** $\mathbf{j}$ **27.** $\pi/2$

29. $\arccos(1/\sqrt{5}) = 63.4°$ **31.** $\pi/2$

33. $\arccos(-\sqrt{2}/10) = 98.1°$ **35.** 82.20 lb

37. 228.5 lb, $71.3°$ **39.** 425 ft-lb

41. W $37.9°$ N, 569.5 mi/hr   **43.** N $25.2°$ E, 82.8 mi/hr

15. $7(\cos 0 + i\sin 0)$

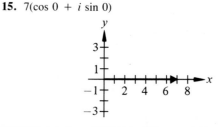

Section 7.4

1. $4\left(\cos\dfrac{\pi}{2} + i\sin\dfrac{\pi}{2}\right)$ **3.** $3\sqrt{2}\left(\cos\dfrac{5\pi}{4} + i\sin\dfrac{5\pi}{4}\right)$

17. $\sqrt{37}[\cos(1.4056) + i\sin(1.4056)]$

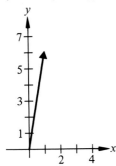

19. $\sqrt{10}[\cos(3.4633) + i\sin(3.4633)]$

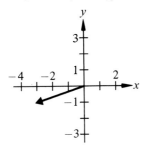

21. $-\sqrt{3} + i$

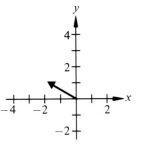

23. $\dfrac{3}{4} - \dfrac{3\sqrt{3}}{4}i$

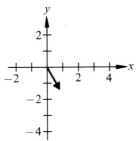

25. $\dfrac{-15\sqrt{2}}{8} + \dfrac{15\sqrt{2}}{8}i$

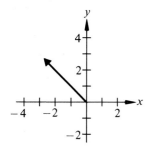

27. $-4i$

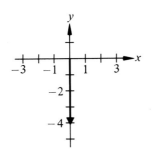

29. $2.8408 + 0.9643i$

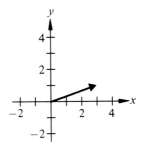

31. $12(\cos 90° + i\sin 90°)$

33. $(10/9)(\cos 200° + i\sin 200°)$

35. $0.27(\cos 150° + i\sin 150°)$ **37.** $\cos\dfrac{2\pi}{3} + i\sin\dfrac{2\pi}{3}$

39. $4[\cos(-58°) + i\sin(-58°)]$

41. $4(\cos 0° + i\sin 0°) = 4$

43. $2\sqrt{2}\left(\cos\dfrac{7\pi}{4} + i\sin\dfrac{7\pi}{4}\right) = 2 - 2i$

45. $(5\sqrt{13}/13)[\cos(-56.3°) + i\sin(-56.3°)] =$
$(5/13)(2 - 3i)$ **49. (a)** r^2 **(b)** $\cos 2\theta + i\sin 2\theta$

Section 7.5

1. $-4 - 4i$ **3.** $-32i$ **5.** $-128\sqrt{3} - 128i$

7. $\dfrac{125}{2} + \dfrac{125\sqrt{3}}{2}i$ **9.** i **11.** $608.02 + 144.69i$

13. $3\left(\cos\dfrac{\pi}{3} + i\sin\dfrac{\pi}{3}\right) = \dfrac{3}{2} + \dfrac{3\sqrt{3}}{2}i$

$3\left(\cos\dfrac{4\pi}{3} + i\sin\dfrac{4\pi}{3}\right) = -\dfrac{3}{2} - \dfrac{3\sqrt{3}}{2}i$

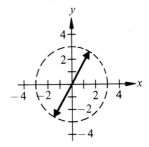

15. $2\left(\cos\dfrac{\pi}{3} + i\sin\dfrac{\pi}{3}\right) = 1 + \sqrt{3}\,i$

$2\left(\cos\dfrac{5\pi}{6} + i\sin\dfrac{5\pi}{6}\right) = -\sqrt{3} + i$

$2\left(\cos\dfrac{4\pi}{3} + i\sin\dfrac{4\pi}{3}\right) = -1 - \sqrt{3}\,i$

$2\left(\cos\dfrac{11\pi}{6} + i\sin\dfrac{11\pi}{6}\right) = \sqrt{3} - i$

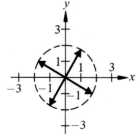

21. $2(\cos 0 + i\sin 0) = 2$

$2\left(\cos\dfrac{2\pi}{3} + i\sin\dfrac{2\pi}{3}\right) = -1 + \sqrt{3}\,i$

$2\left(\cos\dfrac{4\pi}{3} + i\sin\dfrac{2\pi}{3}\right) = -1 - \sqrt{3}\,i$

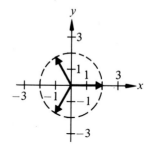

17. $5\left(\cos\dfrac{3\pi}{4} + i\sin\dfrac{3\pi}{4}\right) = -\dfrac{5\sqrt{2}}{2} + \dfrac{5\sqrt{2}}{2}\,i$

$5\left(\cos\dfrac{7\pi}{4} + i\sin\dfrac{7\pi}{4}\right) = \dfrac{5\sqrt{2}}{2} - \dfrac{5\sqrt{2}}{2}\,i$

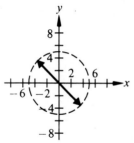

23. $\cos 0 + i\sin 0 = 1$

$\cos\dfrac{2\pi}{5} + i\sin\dfrac{2\pi}{5} = 0.3090 + 0.9510i$

$\cos\dfrac{4\pi}{5} + i\sin\dfrac{4\pi}{5} = -0.8090 + 0.5878i$

$\cos\dfrac{6\pi}{5} + i\sin\dfrac{6\pi}{5} = -0.8090 - 0.5878i$

$\cos\dfrac{8\pi}{5} + i\sin\dfrac{8\pi}{5} = 0.3090 - 0.9510i$

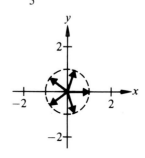

19. $5\left(\cos\dfrac{4\pi}{9} + i\sin\dfrac{4\pi}{9}\right) = 0.8682 + 4.9240i$

$5\left(\cos\dfrac{10\pi}{9} + i\sin\dfrac{10\pi}{9}\right) = -4.6985 - 1.7101i$

$5\left(\cos\dfrac{16\pi}{9} + i\sin\dfrac{16\pi}{9}\right) = 3.8302 - 3.2139i$

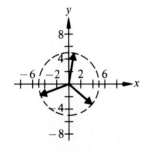

25. $\cos\dfrac{\pi}{8} + i\sin\dfrac{\pi}{8}$

$\cos\dfrac{5\pi}{8} + i\sin\dfrac{5\pi}{8}$

$\cos\dfrac{9\pi}{8} + i\sin\dfrac{9\pi}{8}$

$\cos\dfrac{13\pi}{8} + i\sin\dfrac{13\pi}{8}$

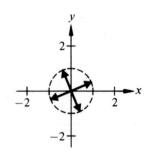

27. $3\left(\cos\dfrac{\pi}{5} + i\sin\dfrac{\pi}{5}\right)$

$3\left(\cos\dfrac{3\pi}{5} + i\sin\dfrac{3\pi}{5}\right)$

$3(\cos\pi + i\sin\pi)$

$3\left(\cos\dfrac{7\pi}{5} + i\sin\dfrac{7\pi}{5}\right)$

$3\left(\cos\dfrac{9\pi}{5} + i\sin\dfrac{9\pi}{5}\right)$

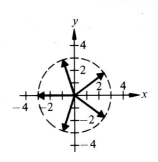

29. $4\left(\cos\dfrac{\pi}{2} + i\sin\dfrac{\pi}{2}\right)$

$4\left(\cos\dfrac{7\pi}{6} + i\sin\dfrac{7\pi}{6}\right)$

$4\left(\cos\dfrac{11\pi}{6} + i\sin\dfrac{11\pi}{6}\right)$

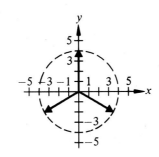

Chapter 7 Review Exercises

1. $A = 29.7°$, $B = 52.4°$, $C = 97.9°$
3. $C = 110°$, $b = 20.4$, $c = 22.6$
5. $A = 35°$, $C = 35°$, $b = 6.6$ **7.** No solution
9. $A = 25.9°$, $C = 39.1°$, $c = 10.1$
11. $B = 31.2°$, $C = 133.8°$, $c = 13.9$; $B = 148.8°$,
$C = 16.2°$, $c = 5.38$ **13.** $A = 9.9°$, $C = 20.1°$,
$b = 29.1$ **15.** $A = 40.9°$, $C = 114.1°$, $c = 8.6$;
$A = 139.1°$, $C = 15.9°$, $c = 2.6$ **17.** 657 m
19. 31.1 m   **21.** $4\sqrt{6}$ **23.** 9.08
25. $|\mathbf{U}| = \sqrt{61}$, $(1/\sqrt{61})(6\mathbf{i} - 5\mathbf{j})$ **27.** $-26\mathbf{i} - 35\mathbf{j}$
29. $-325(\sqrt{3}\,\mathbf{i} + \mathbf{j})$ **31.** 104 lb   **33.** 80.3°
35. (a) $z_1 = 3\sqrt{2}\left(\cos\dfrac{5\pi}{4} + i\sin\dfrac{5\pi}{4}\right)$

$z_2 = 4\left(\cos\dfrac{\pi}{6} + i\sin\dfrac{\pi}{6}\right)$

  (b) $z_1 z_2 = 12\sqrt{2}\left(\cos\dfrac{17\pi}{12} + i\sin\dfrac{17\pi}{12}\right)$

$\dfrac{z_1}{z_2} = \dfrac{3\sqrt{2}}{4}\left(\cos\dfrac{13\pi}{12} + i\sin\dfrac{13\pi}{12}\right)$

37. $\dfrac{5^4}{2} + \dfrac{5^4\sqrt{3}}{2}\,i$

39. $2035 - 828i$

41. $3\left(\cos\dfrac{\pi}{4} + i\sin\dfrac{\pi}{4}\right)$, $3\left(\cos\dfrac{7\pi}{12} + i\sin\dfrac{7\pi}{12}\right)$

$3\left(\cos\dfrac{11\pi}{12} + i\sin\dfrac{11\pi}{12}\right)$

$3\left(\cos\dfrac{5\pi}{4} + i\sin\dfrac{5\pi}{4}\right)$

$3\left(\cos\dfrac{19\pi}{12} + i\sin\dfrac{19\pi}{12}\right)$

$3\left(\cos\dfrac{23\pi}{12} + i\sin\dfrac{23\pi}{12}\right)$

43. $\cos\dfrac{\pi}{3} + i\sin\dfrac{\pi}{3} = \dfrac{1}{2} + \dfrac{\sqrt{3}}{2}\,i$

$\cos\pi + i\sin\pi = -1$

$\cos\dfrac{5\pi}{3} + i\sin\dfrac{5\pi}{3} = \dfrac{1}{2} - \dfrac{\sqrt{3}}{2}\,i$

CHAPTER 8

Section 8.1

1. $(1, 2)$ **3.** $(-1, 2)$, $(2, 5)$ **5.** $(0, 5)$, $(3, 4)$
7. $(0, 0)$, $(2, 4)$ **9.** $(-1, 1)$, $(8, 4)$ **11.** $(5, 5)$
13. $(1/2, 3)$ **15.** $(1.5, 0.3)$ **17.** $(20/3, 40/3)$
19. $(1, 2)$ **21.** $(29/10, 21/10)$, $(-2, 0)$
23. $(-1, -2)$, $(2, 1)$ **25.** $(-1, 0)$, $(0, 1)$, $(1, 0)$
27. $(1, 2, 3)$ **29.** $(1/2, 2)$, $(-4, -1/4)$
31. $(2, 2)$, $(4, 0)$ **33.** No points of intersection
35. $(0, 1)$ **37.** $(\pi/2, 0)$, $(3\pi/2, 0)$, $(\pi/6, \sqrt{3}/2)$,
$(5\pi/6, -\sqrt{3}/2)$ **39.** 6400 units   **41.** 4, 8
43. $x = 5$, $y = 5$, $\lambda = -5$
45. $x = \pm\sqrt{2}/2$, $y = 1/2$, $\lambda = 1$ **47.** $48\sqrt{2}$ ft/sec, $3s$
49. 68.71 ft/sec, arctan $(64/25) = 68.66°$

Section 8.2

1. $(2, 0)$ **3.** $(-1, -1)$ **5.** No solution
7. Dependent   **9.** $(-1/3, -2/3)$ **11.** $(5/2, 3/4)$
13. $(3, 4)$ **15.** $(4, -1)$ **17.** $(40, 40)$
19. No solution   **21.** Dependent   **23.** $(18/5, 3/5)$
25. $(5, -2)$ **27.** $(90/31, -67/31)$
29. $(-6/35, 43/35)$ **31.** 550 mi/hr, 50 mi/hr
33. 20/3 gal of 20% solution, 10/3 gal of 50% solution
35. 375 adults, 125 children   **37.** 12×8 ft

39. (a) $y = (3/4)x + (4/3)$

(b)

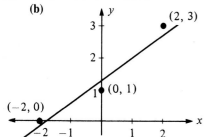

41. (a) $y = -2x + 4$

(b)

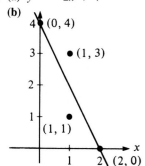

Section 8.3

1. $(1, 2, 3)$ **3.** $(2, -3, -2)$ **5.** $(5, -2, 0)$

7. Inconsistent   **9.** $(1, -3/2, 1/2)$

11. $(-3a + 10, 5a - 7, a)$

13. $\left(13 - 4a, \dfrac{45}{2} - \dfrac{15}{2}a, a\right)$

15. $(-a, 2a - 1, a)$ **17.** $\left(\dfrac{1}{2} - \dfrac{3}{2}a, 1 - \dfrac{2}{3}a, a\right)$

19. Inconsistent   **21.** $(1, 1, 1, 1)$

23. $(0, 0, 0)$ **25.** $(-\dfrac{3}{5}a, \dfrac{4}{5}a, a)$

27. $y = 2x^2 + 3x - 4$ **29.** $y = x^2 - 4x + 3$

31. $x^2 + y^2 - 4x = 0$ **33.** $x^2 + y^2 - 6x - 8y = 0$

35. $a = -32, v_0 = 0, s_0 = 144$

37. $a = -32, v_0 = -32, s_0 = 500$

39. $\dfrac{1}{2}\left(-\dfrac{2}{x} + \dfrac{1}{x - 1} + \dfrac{1}{x + 1}\right)$

41. $\dfrac{1}{2}\left(\dfrac{1}{x} - \dfrac{1}{x - 2} + \dfrac{2}{x + 3}\right)$

43. 4 medium *or* 2 large, 1 medium, 2 small

45. $y = \dfrac{3}{7}x^2 + \dfrac{6}{5}x + \dfrac{26}{35}$ **47.** $y = x^2 - x$

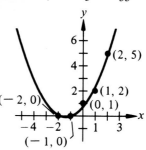

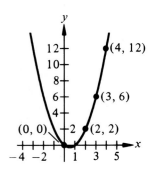

Section 8.4

1. f   **3.** g   **5.** a   **7.** b   **9.** d

11.

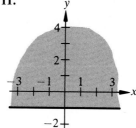

13.

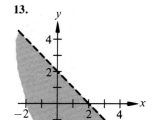

15.

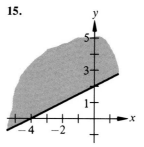

17.

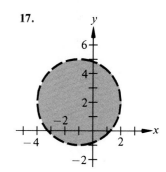

19.

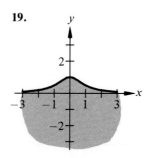

21.

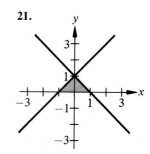

23.

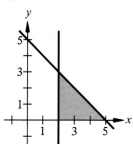

25.

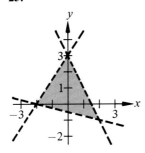

39.

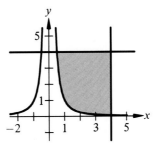

27.

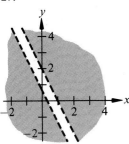

29.

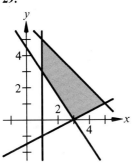

41. $x + (3/2)y \le 12$
   $(4/3)x + (3/2)y \le 15$
   $x \ge 0$
   $y \ge 0$

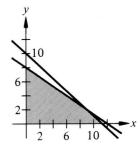

43. $x \ge 2, x \le 5, y \ge 1, y \le 7$
45. $y \le x + 1, y \le -x + 1, y \ge 0$
47. $x^2 + y^2 \le 16, y \ge x, x \ge 0$
49. **(a)** Minimum at (0, 0): 0
      Maximum at (3, 4): 17
   (b) Minimum at (0, 0): 0
      Maximum at (4, 0): 20
51. **(a)** Minimum at (0, 0): 0
      Maximum at (60, 20): 740
   (b) Minimum at (0, 0): 0
      Maximum at any point along line segment connecting
      (60, 20) and (30, 45): 2100
53. 200 units (at $250)
   50 units (at $400)
55. Minimum at (5, 3): 35

31.

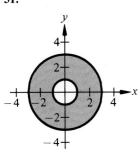

33.

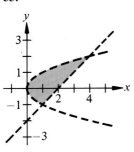

Chapter 8 Review Exercises

1. (1, 1)   **3.** (5, 4)
5. (0, 0), (2, 8), (−2, 8)   **7.** (0, 0), (−3, 3)
9. (5/2, 3)   **11.** (−0.5, 0.8)
13. (0, 0)   **15.** (4.8, 4.4, −1.6)
17. $(3a + 4, 2a + 5, a)$
19. $(-3a + 2, 5a + 6, a)$

35.

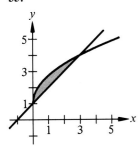

37.

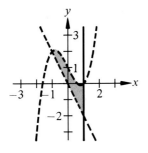

21.

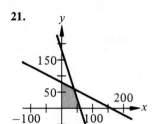

23.

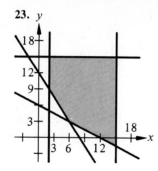

25.

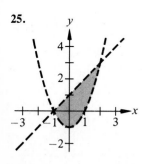

29. $(X, Y, Z) = (2, 5, 4)$

CHAPTER 9

Section 9.1

1. (a) $\begin{bmatrix} 1 & 2 & 3 \\ 0 & -5 & -10 \\ 3 & 1 & -1 \end{bmatrix}$ (b) $\begin{bmatrix} 1 & 2 & 3 \\ 0 & -5 & -10 \\ 0 & -5 & -10 \end{bmatrix}$

(c) $\begin{bmatrix} 1 & 2 & 3 \\ 0 & -5 & -10 \\ 0 & 0 & 0 \end{bmatrix}$ (d) $\begin{bmatrix} 1 & 2 & 3 \\ 0 & 1 & 2 \\ 0 & 0 & 0 \end{bmatrix}$

(e) $\begin{bmatrix} 1 & 0 & -1 \\ 0 & 1 & 2 \\ 0 & 0 & 0 \end{bmatrix}$ **3.** $\begin{bmatrix} 1 & 0 & 0 & 3 \\ 0 & 1 & 0 & 2 \\ 0 & 0 & 1 & -1 \end{bmatrix}$

5. $\begin{bmatrix} 1 & 0 & 5 & 4 \\ 0 & 1 & 6 & 3 \\ 0 & 0 & 0 & 0 \end{bmatrix}$ **7.** $\begin{bmatrix} 1 & 0 & 0 \\ 0 & 1 & 0 \\ 0 & 0 & 1 \\ 0 & 0 & 0 \end{bmatrix}$

9. $\begin{bmatrix} 1 & 2 & 0 & 0 \\ 0 & 0 & 1 & 0 \\ 0 & 0 & 0 & 1 \\ 0 & 0 & 0 & 0 \end{bmatrix}$ **11.** $(3, 2)$ **13.** $(4, -2)$

15. $(1/2, -3/4)$ **17.** Inconsistent   **19.** $(4, -3, 2)$
21. $(2a + 1, 3a + 2, a)$ **23.** $(5a + 4, -3a + 2, a)$
25. $(0, 2 - 4a, a)$ **27.** $(1, 0, 4, -2)$
29. $(2, -2, 3, -5, 1)$ **31.** \$800,000 at 8%, \$500,000 at 9%, \$200,000 at 12%   **33.** $a = 1/2, b = -3, c = 7/2$
35. $D = -5, E = -3, F = 6$

Section 9.2

1. (a) $\begin{bmatrix} 3 & -2 \\ 1 & 7 \end{bmatrix}$ (b) $\begin{bmatrix} -1 & 0 \\ 3 & -9 \end{bmatrix}$ (c) $\begin{bmatrix} 3 & -3 \\ 6 & -3 \end{bmatrix}$

(d) $\begin{bmatrix} -1 & -1 \\ 8 & -19 \end{bmatrix}$ **3.** (a) $\begin{bmatrix} 7 & 3 \\ 1 & 9 \\ -2 & 15 \end{bmatrix}$

(b) $\begin{bmatrix} 5 & -5 \\ 3 & -1 \\ -4 & -5 \end{bmatrix}$ (c) $\begin{bmatrix} 18 & -3 \\ 6 & 12 \\ -9 & 15 \end{bmatrix}$ (d) $\begin{bmatrix} 16 & -11 \\ 8 & 2 \\ -11 & -5 \end{bmatrix}$

5. (a) $\begin{bmatrix} 3 & 3 & -2 & 1 & 1 \\ -2 & 5 & 7 & -6 & -8 \end{bmatrix}$

(b) $\begin{bmatrix} 1 & 1 & 0 & -1 & 1 \\ 4 & -3 & -11 & 6 & 6 \end{bmatrix}$

(c) $\begin{bmatrix} 6 & 6 & -3 & 0 & 3 \\ 3 & 3 & -6 & 0 & -3 \end{bmatrix}$

(d) $\begin{bmatrix} 4 & 4 & -1 & -2 & 3 \\ 9 & -5 & -24 & 12 & 11 \end{bmatrix}$ **7.** (a) $\begin{bmatrix} 0 & 15 \\ 6 & 12 \end{bmatrix}$

(b) $\begin{bmatrix} -2 & 2 \\ 31 & 14 \end{bmatrix}$ **9.** (a) $\begin{bmatrix} 0 & -10 \\ 10 & 0 \end{bmatrix}$ (b) $\begin{bmatrix} 0 & -10 \\ 10 & 0 \end{bmatrix}$

11. (a) $\begin{bmatrix} 6 & -21 & 15 \\ 8 & -23 & 19 \\ 4 & 7 & 5 \end{bmatrix}$ (b) $\begin{bmatrix} 9 & 0 & 13 \\ 7 & -2 & 21 \\ 1 & 4 & -19 \end{bmatrix}$

13. Not possible   **15.** $\begin{bmatrix} -1 & 19 \\ 4 & -27 \\ 0 & 14 \end{bmatrix}$

17. $\begin{bmatrix} 1 & 0 & 0 \\ 0 & 1 & 0 \\ 0 & 0 & 7/2 \end{bmatrix}$ **19.** $\begin{bmatrix} 60 & 72 \\ -20 & -24 \\ 10 & 12 \\ 60 & 72 \end{bmatrix}$

21. $\begin{bmatrix} -4 & 0 \\ 8 & 2 \end{bmatrix}$ **23.** $\begin{bmatrix} 0 & 0 & 0 \\ 0 & 0 & 0 \\ 0 & 0 & 0 \end{bmatrix}$

25. $AC = BC = \begin{bmatrix} 12 & -6 & 9 \\ 16 & -8 & 12 \\ 4 & -2 & 3 \end{bmatrix}$ **27.** $AB \neq BA$

29. $AB = [\$1250 \quad \$1331.25 \quad \$981.25]$

Section 9.3

7. $\begin{bmatrix} -3 & 2 \\ -2 & 1 \end{bmatrix}$ **9.** $\begin{bmatrix} 1 & -1 \\ 2 & -1 \end{bmatrix}$ **11.** Does not exist

13. $\dfrac{1}{3}\begin{bmatrix} -9 & -7 \\ 3 & 2 \end{bmatrix}$ **15.** $\begin{bmatrix} 1 & 1 & -1 \\ -3 & 2 & -1 \\ 3 & -3 & 2 \end{bmatrix}$

17. $\begin{bmatrix} -175 & 37 & -13 \\ 95 & -20 & 7 \\ 14 & -3 & 1 \end{bmatrix}$

19. $\begin{bmatrix} -24 & 7 & 1 & -2 \\ -10 & 3 & 0 & -1 \\ -29 & 7 & 3 & -2 \\ 12 & -3 & -1 & 1 \end{bmatrix}$ **21.** $\dfrac{1}{2}\begin{bmatrix} -3 & 3 & 2 \\ 9 & -7 & -6 \\ -2 & 2 & 2 \end{bmatrix}$

23. $\dfrac{5}{11}\begin{bmatrix} 0 & -4 & 2 \\ -22 & 11 & 11 \\ 22 & -6 & -8 \end{bmatrix}$

25. $\begin{bmatrix} -1/8 & 0 & 0 & 0 \\ 0 & 1 & 0 & 0 \\ 0 & 0 & 1/4 & 0 \\ 0 & 0 & 0 & -1/5 \end{bmatrix}$

27. $\begin{bmatrix} 1 & 0 & 0 \\ -0.75 & 0.25 & 0 \\ 0.35 & -0.25 & 0.2 \end{bmatrix}$ **29.** Does not exist

31. (a) $(4, 8)$ (b) $(-8, -11)$ (c) $(-7, -7)$
33. (a) $(-5, -81/2, 48)$ (b) $(-3, -24, 28)$
(c) $(0, 1, -1)$

Section 9.4

1. 5   **3.** 27   **5.** -24 **7.** 6   **9.** 0   **11.** -2
13. 0   **15.** 0   **17.** 0.002   **19.** $-7x + 3y - 8$

21. (a) $\begin{bmatrix} 23 & -8 & -22 \\ 5 & -5 & 5 \\ 7 & -22 & -23 \end{bmatrix}$ (b) $\begin{bmatrix} 23 & 8 & -22 \\ -5 & -5 & -5 \\ 7 & 22 & -23 \end{bmatrix}$

23. (a) $\begin{bmatrix} 662 & 708 & 394 & 524 \\ -282 & -298 & -174 & -234 \\ 42 & 48 & 44 & 24 \\ 611 & 674 & 377 & 507 \end{bmatrix}$

(b) $\begin{bmatrix} 662 & -708 & 394 & -524 \\ 282 & -298 & 174 & -234 \\ 42 & -48 & 44 & -24 \\ -611 & 674 & -377 & 507 \end{bmatrix}$ **25.** -75

27. 170   **29.** -58 **31.** -30 **33.** -108 **35.** 0
37. 412   **39.** 0   **41.** $3x - 5y = 0$
43. $x + 3y - 5 = 0$ **45.** $2x + 3y - 8 = 0$
47. $33/2$ **49.** $31/2$

Section 9.5

1. Column 2 is a multiple of Column 1.   **3.** Row 2 has only zero elements.   **5.** The interchange of Columns 2 and 3 results in a change of sign of the determinant.
7. Multiplying the elements of Row 1 by $1/5$ produces a determinant that is $1/5$ times the original.   **9.** Multiplying the elements of all three rows by $1/5$ produces a determinant that is $(1/5)^3$ times the original.   **11.** Adding -4 times the elements of Row 1 to the elements of Row 2 leaves the determinant unchanged.   **13.** Adding multiples of Column 2 to Columns 1 and 3 leaves the determinant unchanged.
15. The interchange of two rows (or columns) results in a change of sign of the determinant. Rows 2 and 3 were interchanged, followed by an interchange of Columns 1 and 3.
17. -6 **19.** -26 **21.** -126 **23.** 236

25. -3740 **27.** 7441   **29.** 410   **31.** $\dfrac{1}{8}\begin{bmatrix} 2 & -4 \\ 1 & 2 \end{bmatrix}$

33. Inverse does not exist.   **35.** $\dfrac{1}{2}\begin{bmatrix} -3 & 3 & 2 \\ 9 & -7 & -6 \\ -2 & 2 & 2 \end{bmatrix}$

37. $\dfrac{1}{11}\begin{bmatrix} 0 & -20 & 10 \\ -110 & 55 & 55 \\ 110 & -30 & -40 \end{bmatrix}$

39. $\dfrac{1}{3}\begin{bmatrix} -9 & -26 & 6 & 8 \\ 6 & 16 & -3 & -4 \\ -9 & -25 & 6 & 7 \\ -3 & -12 & 3 & 3 \end{bmatrix}$

Section 9.6

1. $(1, 2)$ **3.** $(2, -2)$ **5.** $(3/4, -1/2)$ **7.** $|D| = 0$
9. $(2/3, 1/2)$ **11.** $(-1, 3, 2)$ **13.** $(1, 1/2, 3/2)$
15. $(0, -1/2, 1/2)$ **17.** $(5, 0, -2, 3)$ **19.** $|D| = 0$
21. $y = (27/19)x - (18/19)$
23. $y = -(22/17)x + (74/17)$
25. $y = (3/4)(x^2 - x + 1)$

Chapter 9 Review Exercises

1. $(10, -12)$ **3.** $(0.6, 0.5)$ **5.** $(2, -3, 3)$

7. $(1/2, -1/3, 1)$ **9.** $\left(-2a + \dfrac{3}{2}, 2a + 1, a\right)$

11. Inconsistent   **13.** $\begin{bmatrix} -13 & -8 & 18 \\ 0 & 11 & -19 \end{bmatrix}$

15. $\begin{bmatrix} 14 & -2 & 8 \\ 14 & -10 & 40 \\ 36 & -12 & 48 \end{bmatrix}$ **17.** $\begin{bmatrix} 44 & 4 \\ 20 & 8 \end{bmatrix}$

19. $\begin{bmatrix} 4 & 6 & 3 \\ 0 & 6 & -10 \\ 0 & 0 & 6 \end{bmatrix}$

21. $5x + 4y = 2, -x + y = -22$ **23.** $(10, -12)$
25. $(2, -3, 3)$ **27.** $(1/2, -1/3, 1)$ **29.** 128
31. $\lambda^2 - \lambda + 1 = 0$
33. $-\lambda^3 + 5\lambda^2 + 18\lambda - 52 = 0$

CHAPTER 10

Section 10.1

1. $2, 4, 8, 16, 32$ **3.** $-1/2, 1/4, -1/8, 1/16, -1/32$
5. $3, 9/2, 9/2, 27/8, 81/40$
7. $-1, 1/4, -1/9, 1/16, -1/25$
9. $2, 3/2, 4/3, 5/4, 6/5$ **11.** $0, 1, 2, 3, 4$
13. $0, 1, 0, 1/2, 0$ **15.** $5/2, 11/4, 23/8, 47/16, 95/32$
17. $-1/2, 2/3, -3/4, 4/5, -5/6$ **19.** $3, 4, 6, 10, 18$

21. $a_n = 3n - 2$ **23.** $a_n = n^2 - 2$ **25.** $a_n = \dfrac{n + 1}{n + 2}$

27. $a_n = \dfrac{(-1)^n}{2^{n-1}}$ **29.** $a_n = 1 + \dfrac{1}{n}$

31. $a_n = \dfrac{n}{(n + 1)(n + 2)}$ **33.** $a_n = \dfrac{2^n n!}{(2n)!}$

35. $a_n = (-1)^{n+1}$ **37.** 35 **39.** $158/85$ **41.** 40
43. $5C$ **45.** 238 **47.** $4x^2 + 20$ **49.** $14 + 24i$

51. $\displaystyle\sum_{i=1}^{9} \frac{1}{3i}$ **53.** $\displaystyle\sum_{i=1}^{8} \left[2\left(\frac{i}{8}\right) + 3\right]$

55. $\displaystyle\sum_{i=1}^{6} (-1)^{i+1} 3^i$ **57.** $\displaystyle\sum_{i=1}^{20} \frac{(-1)^{i+1}}{i^2}$

59. $\displaystyle\sum_{i=1}^{5} \frac{2^i - 1}{2^{i+1}}$

63. (a) $A_1 = \$5100.00, A_2 = \$5202.00,$
$A_3 = \$5306.04, A_4 = \$5412.16, A_5 = \$5520.40,$
$A_6 = \$5630.81, A_7 = \$5743.43, A_8 = \$5858.30$
(b) $\$11,040.20$ **65.** (a) $10 \cdot 9 = 90$
(b) $25 \cdot 24 = 600$ (c) $n(n - 1)$

Section 10.2

1. Arithmetic sequence, $d = 3$ **3.** Not an arithmetic
sequence   **5.** Arithmetic sequence, $d = -1/4$ **7.** Not
an arithmetic sequence   **9.** Arithmetic sequence, $d = 0.4$
11. $\{5, 11, 17, 23, 29, \ldots\}$
13. $\{-2.6, -3.0, -3.4, -3.8, -4.2, \ldots\}$
15. $\{3/2, 5/4, 1, 3/4, 1/2, \ldots\}$
17. $\{-2, 2, 6, 10, 14, \ldots\}$ **19.** 28 **21.** 44
23. $99x$ **25.** $-37/2$ **27.** 620 **29.** 4600
31. 265 **33.** 4000 **35.** 1275 **37.** $25,250$
39. (a) $\$35,000$ (b) $\$187,500$ **41.** 2340 **43.** 520
45. $44,625$ **47.** $9, 13$ **49.** $15/4, 9/2, 21/4$

Section 10.3

1. Geometric sequence, $r = 3$ **3.** Not a geometric
sequence   **5.** Geometric sequence, $r = -1/2$ **7.** Not
a geometric sequence   **9.** Not a geometric sequence
11. $\{2, 6, 18, 54, 162, \ldots\}$
13. $\{1, 1/2, 1/4, 1/8, 1/16, \ldots\}$
15. $\{5, -1/2, 1/20, -1/200, 1/2000, \ldots\}$
17. $\{1, x/2, x^2/4, x^3/8, x^4/16, \ldots\}$ **19.** $(1/2)^7$
21. $-2(-1/3)^{10}$ **23.** $100e^{8x}$ **25.** $500(1.02)^{39}$
27. (a) $\$2593.74$ (b) $\$2653.30$ (c) $\$2685.06$
(d) $\$2707.04$ (e) $\$2717.91$ **29.** $\$22,689.45$
31. $29,921.31$ **33.** 6.400 **35.** 511 **37.** $\$7808.24$
41. $\$594,121.01$ **43.** 2 **45.** $2/3$ **47.** $16/3$
49. 32 **51.** $8/3$ **53.** $1/2$ **55.** 152.42 ft
57. $1/9$ **59.** $4/11$ **61.** $16/37$ **63.** $15/11$

Section 10.4

15. $n(2n + 1)$ **17.** $10[1 - (0.9)^n]$ **19.** $\dfrac{n}{2(n + 1)}$

Section 10.5

1. 10   **3.** 1   **5.** 15,504   **7.** 4950   **9.** 4950

11. $x^5 + 5x^4y + 10x^3y^2 + 10x^2y^3 + 5xy^4 + y^5$

13. $a^4 + 8a^3 + 24a^2 + 32a + 16$

15. $r^6 + 18r^5s + 135r^4s^2 + 540r^3s^3 + 1215r^2s^4 + 1458rs^5 + 729s^6$

17. $x^5 - 5x^4y + 10x^3y^2 - 10x^2y^3 + 5xy^4 - y^5$

19. $1 - 6x + 12x^2 - 8x^3$

21. $x^8 + 20x^6 + 150x^4 + 500x^2 + 625$

23. $\dfrac{1}{x^5} + \dfrac{5y}{x^4} + \dfrac{10y^2}{x^3} + \dfrac{10y^3}{x^2} + \dfrac{5y^4}{x} + y^5$ **25.** -4

27. $2035 + 828i$ **29.** 1   **31.** $1{,}732{,}104x^5$

33. $180x^8y^2$ **35.** $-226{,}437{,}120$ **37.** $924x^3y^3$

39. $32t^5 - 80t^4s + 80t^3s^2 - 40t^2s^3 + 10ts^4 - s^5$

41. $\dfrac{1}{128} + \dfrac{7}{128} + \dfrac{21}{128} + \dfrac{35}{128} + \dfrac{35}{128} + \dfrac{21}{128} + \dfrac{7}{128} + \dfrac{1}{128}$

43. $\dfrac{1}{6561} + \dfrac{16}{6561} + \dfrac{112}{6561} + \dfrac{448}{6561} + \dfrac{1120}{6561} + \dfrac{1792}{6561} +$

$\dfrac{1792}{6561} + \dfrac{1024}{6561} + \dfrac{256}{6561}$ **45.** $0.07776 + 0.25920 +$

$0.34560 + 0.23040 + 0.07680 + 0.01024$

Chapter 10 Review Exercises

1. 30   **3.** 80   **5.** 127   **7.** 8   **9.** 12   **11.** 88

13. 418   **19.** 5/11   **21.** 16/15   **23.** 1/75

25. $\dfrac{x^4}{16} + \dfrac{x^3y}{2} + \dfrac{3x^2y^2}{2} + 2xy^3 + y^4$

27. $\dfrac{64}{x^6} + \dfrac{576}{x^4} + \dfrac{2160}{x^2} + 4320 + 4860x^2 + 2916x^4 + 729x^6$

29. $41 + 840i$

CHAPTER 11

Section 11.1

1. (a) $2x - y = 3$ **(b)** $x + 2y = 4$

3. (a) $40x + 24y = 53$ **(b)** $24x - 40y = -9$

5. (a) $x = 2$ **(b)** $y = 5$ **7.** Vertices of a right triangle

9. Vertices of a right triangle

11. Vertices of a right triangle   **13.** Collinear

15. Not collinear   **17.** 2   **19.** $\dfrac{5\sqrt{2}}{2}$

21. $\dfrac{4}{3}$ **23.** 0   **25.** $2\sqrt{2}$ **27.** $30°$ **29.** $45°$

31. $\arctan\left(-\dfrac{5}{3}\right) = 120.96°$

33. $\arctan\left(-\dfrac{2}{15}\right) = 172.41°$ **35.** $45°$

37. $\arctan\dfrac{7}{24} = 16.26°$ **39.** $0°$

41. $\sqrt{3}x - y + 6 = 0$

43. $y - 1 = \left[\tan\left(\arctan 2 + \dfrac{1}{2}\arctan 3\right)\right](x + 3)$

$\qquad y - 1 = -6.16(x + 3)$

Section 11.2

1. $x^2 + y^2 - 9 = 0$ **3.** $x^2 + y^2 - 4x + 2y - 11 = 0$

5. $x^2 + y^2 + 2x - 4y = 0$

7. $(x - 1)^2 + (y + 3)^2 = 4$

9. $(x - 1)^2 + (y + 3)^2 = 0$

11. $\left(x - \dfrac{1}{2}\right)^2 + \left(y - \dfrac{1}{2}\right)^2 = 2$

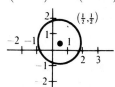

13. $\left(x + \dfrac{1}{2}\right)^2 + \left(y + \dfrac{5}{4}\right)^2 = \dfrac{9}{4}$

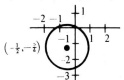

15. $x^2 + y^2 - 4x - 6y + 4 = 0$

17. $x^2 + y^2 - 50 = 0$

19. $17x^2 + 17y^2 - 136x + 34y + 253 = 0$

21. $12x - 5y = 0$ **23.** $3x - 11y + 27 = 0$

25. $x - 7y + 30 = 0$

27.   **29.**

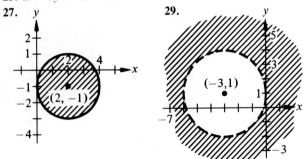

31. $(-3, 0), (2, 0)$ **33.** $(-1, -2), (2, 1)$

35. $3x^2 + 3y^2 - 16x + 8y = 0$

Section 11.3

1. Vertex: $(0, 0)$
  Focus: $(0, 1/16)$
  Directrix: $y = -1/16$

3. Vertex: $(0, 0)$
  Focus: $(-3/2, 0)$
  Directrix: $x = 3/2$

5. Vertex: $(0, 0)$
  Focus: $(0, -2)$
  Directrix: $y = 2$

7. Vertex: $(1, -2)$
  Focus: $(1, -4)$
  Directrix: $y = 0$

9. Vertex: $(5, -1/2)$
  Focus: $(11/2, -1/2)$
  Directrix: $x = 9/2$

11. Vertex: $(1, 1)$
  Focus: $(1, 2)$
  Directrix: $y = 0$

13. Vertex: $(8, -1)$
  Focus: $(9, -1)$
  Directrix: $x = 7$

15. Vertex: $(-2, -3)$
  Focus: $(-4, -3)$
  Directrix: $x = 0$

17. Vertex: $(-1, 2)$
  Focus: $(0, 2)$
  Directrix: $x = -2$

19. Vertex: $(-2, 2)$
  Focus: $(-2, 1)$
  Directrix: $y = 3$

21. $x^2 + 6y = 0$ **23.** $y^2 - 4y + 8x - 20 = 0$

25. $x^2 - 8y + 32 = 0$ **27.** $x^2 + 8y - 16 = 0$

29. $5x^2 - 14x - 3y + 9 = 0$ **31.** $x^2 + y - 4 = 0$

33. For vertex at $(h, 3)$: $(x - h)^2 = 8(y - 3)$
  For vertex at $(h, -1)$: $(x - h)^2 = -8(y + 1)$

35. $x^2 - 800y = 0$

37. **(a)** $4x^2 + 25y - 1200 = 0$
  (b) $10\sqrt{3} = 17.32$ feet from the point directly below the end of the pipe

39. **(a)** $16x^2 + 1024y - 76,800 = 0$ **(b)** $40\sqrt{3} = 69.28$ ft

41. $4x - y - 8 = 0$

43. $4x - y + 2 = 0$

Section 11.4

1. Center: $(0, 0)$
  Foci: $(\pm 3, 0)$
  Vertices: $(\pm 5, 0)$
  $e = 3/5$

3. Center: $(0, 0)$
  Foci: $(0, \pm 3)$
  Vertices: $(0, \pm 5)$
  $e = 3/5$

5. Center: $(0, 0)$
  Foci: $(\pm 2, 0)$
  Vertices: $(\pm 3, 0)$
  $e = 2/3$

7. Center: $(0, 0)$
  Foci: $(\pm \sqrt{3}, 0)$
  Vertices: $(\pm 2, 0)$
  $e = \dfrac{\sqrt{3}}{2}$

9. Center: $(0, 0)$
  Foci: $(0, \pm 1)$
  Vertices: $(0, \pm \sqrt{3})$
  $e = \dfrac{\sqrt{3}}{3}$

11. Center: $(0, 0)$
  Foci: $\left(0, \pm \dfrac{\sqrt{3}}{2}\right)$
  Vertices: $(0, \pm 1)$
  $e = \dfrac{\sqrt{3}}{2}$

13. Center: $(1, 5)$
  Foci: $(1, 9), (1, 1)$
  Vertices: $(1, 10), (1, 0)$
  $e = 4/5$

15. Center: $(-2, 3)$
  Foci: $(-2, 3 \pm \sqrt{5})$
  Vertices: $(-2, 6), (-2, 0)$
  $e = \dfrac{\sqrt{5}}{3}$

17. Center: $(1, -1)$
  Foci: $\left(1 \pm \dfrac{3\sqrt{10}}{20}, -1\right)$
  Vertices: $\left(1 \pm \dfrac{\sqrt{10}}{4}, -1\right)$
  $e = 3/5$

19. Center: $(1/2, -1)$
  Foci: $(1/2 \pm \sqrt{2}, -1)$
  Vertices: $(1/2 \pm \sqrt{5}, -1)$
  $e = \dfrac{\sqrt{10}}{5}$

21. $\dfrac{x^2}{9} + \dfrac{y^2}{5} = 1$ **23.** $\dfrac{x^2}{25} + \dfrac{y^2}{16} = 1$

25. $\dfrac{(x-2)^2}{4} + \dfrac{(y-2)^2}{1} = 1$

27. $\dfrac{(x-3)^2}{9} + \dfrac{(y-5)^2}{16} = 1$

29. $\dfrac{x^2}{24} + \dfrac{y^2}{49} = 1$ **31.** 3/2 ft from the center, 5 ft

33. (a)

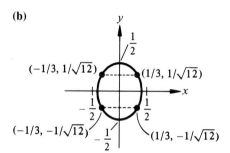

(b)

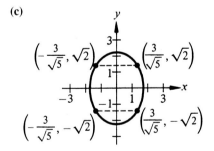

(c)

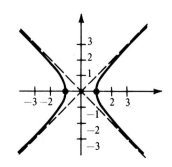

35. 91,419,000 mi, 94,581,000 mi **37.** $e = 0.052$

Section 11.5

1. Center: $(0, 0)$
Vertices: $(\pm 1, 0)$
Foci: $(\pm \sqrt{2}, 0)$

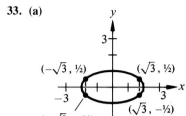

3. Center: $(0, 0)$
Vertices: $(0, \pm 1)$
Foci: $(0, \pm \sqrt{5})$

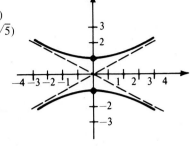

5. Center: $(0, 0)$
Vertices: $(0, \pm 5)$
Foci: $(0, \pm 13)$

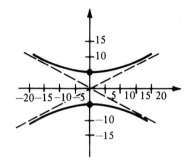

7. Center: $(0, 0)$
Vertices: $(\pm \sqrt{3}, 0)$
Foci: $(\pm \sqrt{5}, 0)$

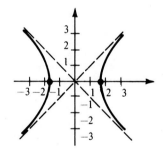

9. Center: $(0, 0)$
Vertices: $(0, \pm 2)$
Foci: $(0, \pm 3)$

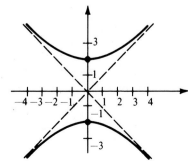

11. Center: $(1, -2)$
  Vertices: $(-1, -2), (3, -2)$
  Foci: $(1 \pm \sqrt{5}, -2)$

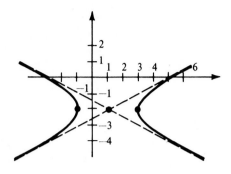

13. Center: $(2, -6)$
  Vertices: $(2, -5), (2, -7)$
  Foci: $(2, -6 \pm \sqrt{2})$

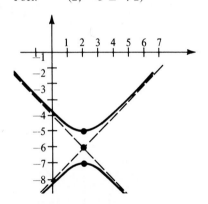

15. Center: $(2, -3)$
  Vertices: $(1, -3), (3, -3)$
  Foci: $(2 \pm \sqrt{10}, -3)$

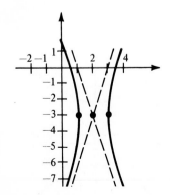

17. Center: $(1, -3)$
  Vertices: $(1, -3 \pm \sqrt{2})$
  Foci: $(1, -3 \pm 2\sqrt{5})$

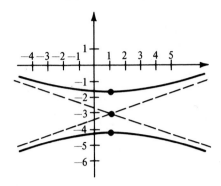

19. Degenerate hyperbola: graph is two intersecting lines with center at $(-1, -3)$

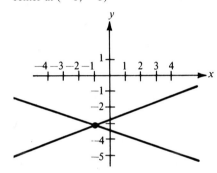

21. $\dfrac{y^2}{4} - \dfrac{x^2}{12} = 1$

23. $\dfrac{x^2}{1} - \dfrac{y^2}{9} = 1$

25. $\dfrac{(x - 3)^2}{9} - \dfrac{(y - 2)^2}{4} = 1$

27. $\dfrac{y^2}{9} - \dfrac{(x - 2)^2}{9/4} = 1$

29. $\dfrac{(x - 6)^2}{9} - \dfrac{(y - 2)^2}{7} = 1$

31. $(4400, -4290)$

33. Ellipse   **35.** Parabola   **37.** Hyperbola

39. Circle

Section 11.6

1. $\dfrac{(y')^2}{2} - \dfrac{(x')^2}{2} = 1$

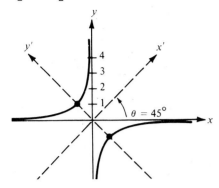

3. $y' = \dfrac{(x')^2}{6} - \dfrac{x'}{3}$

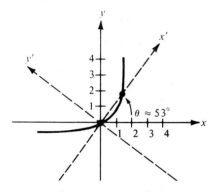

5. $\dfrac{(x')^2}{1/4} - \dfrac{(y')^2}{1/6} = 1$

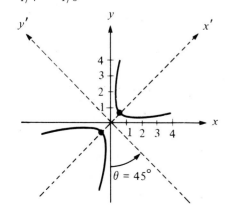

7. $\dfrac{(x' - 3\sqrt{2})^2}{16} - \dfrac{(y' - \sqrt{2})^2}{16} = 1$

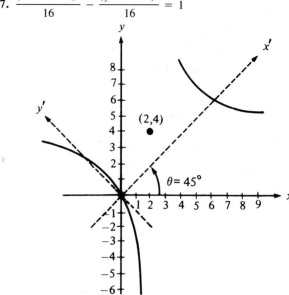

9. $\dfrac{(x')^2}{3} + \dfrac{(y')^2}{2} = 1$

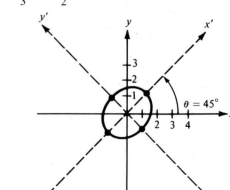

11. $x' = -(y')^2$

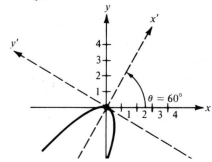

13. $\dfrac{(x')^2}{3} - \dfrac{(y')^2}{5} = 1$

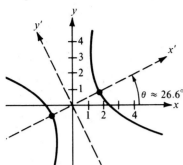

15. $\dfrac{(x')^2}{1.096} - \dfrac{(y')^2}{6.153} = 1$

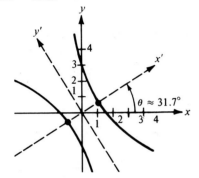

17. (a) Parabola   **(b)** Hyperbola   **(c)** Ellipse

Chapter 11 Review Exercises

1. h   **3.** e   **5.** f   **7.** c   **9.** g

11. Circle

Center: $(1/2, -3/4)$

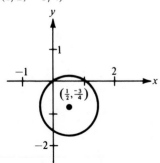

13. Hyperbola

Center: $(-4, 3)$

Vertices: $(-4 \pm \sqrt{2}, 3)$

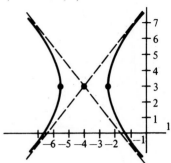

15. Ellipse

Center: $(2, -3)$

Vertices: $\left(2, -3 \pm \dfrac{\sqrt{2}}{2}\right)$

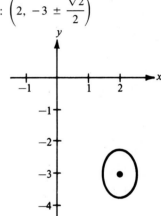

17. Parabola

Vertex: $(3, 0)$

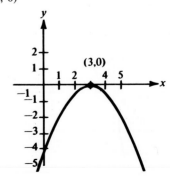

19. Parabola

Vertex: $\left(-\dfrac{1}{\sqrt{8}}, \dfrac{1}{\sqrt{8}}\right)$

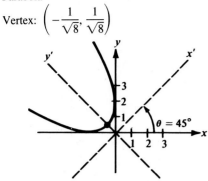

21. $\dfrac{y^2}{1} - \dfrac{x^2}{8} = 1$ **23.** $\dfrac{(x-2)^2}{25} + \dfrac{y^2}{21} = 1$

25. $x^2 - 2xy + y^2 - 8x - 8y = 0$

27. $\dfrac{x^2}{4} + \dfrac{y^2}{16/3} = 1$ **29.** $\dfrac{x^2}{4} - \dfrac{y^2}{12} = 1$

31. $(0, 50)$ **33.** $3x + 7y - 16 = 0$

35. $3x - 4y - 2 = 0$

37. $20x^2 + 20y^2 - 160x - 60y + 169 = 0$

39. $(\pm\sqrt{15}, 5)$ **41.** 1 **43.** $150°$ **45.** $90°$

47. (a) $x - 3y + 2 = 0$

$8x - 3y - 26 = 0$

$3x - 2y - 1 = 0$

(b) $\arctan \dfrac{7}{9} \approx 37.87°$

$180° - \arctan \dfrac{21}{17} \approx 128.99°$

$\arctan \dfrac{7}{30} \approx 13.13°$

CHAPTER 12

Section 12.1

1.

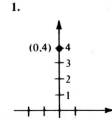

3.

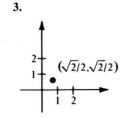

5.

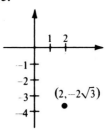

7.

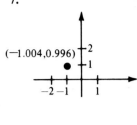

9. $\left(\sqrt{2}, \dfrac{\pi}{4}\right), \left(-\sqrt{2}, \dfrac{5\pi}{4}\right)$ **11.** $(5, 2.214), (-5, 5.356)$

13. $\left(\sqrt{6}, \dfrac{5\pi}{4}\right), \left(-\sqrt{6}, \dfrac{\pi}{4}\right)$

15. $(2\sqrt{13}, 0.983), (-2\sqrt{13}, 4.124)$

17. $r = a$ **19.** $r^2 = 9\cos(2\theta)$ **21.** $r = 2a\cos\theta$

23. $r = \dfrac{2a}{1 - \sin\theta}$ or $r = \dfrac{-2a}{1 + \sin\theta}$

25. $x^2 + y^2 - 4y = 0$ **27.** $(x^2 + y^2 + 2y)^2 = x^2 + y^2$

29. $(x^2 + y^2)^2 = 2y(3x^2 - y^2)$ **31.** $2x - 3y = 6$

Section 12.2

1. c   **3.** a   **5.** b

7. $r = \dfrac{2}{1 - \sin\theta}$ or $r = \dfrac{-2}{1 + \sin\theta}$

9. $r = \dfrac{8}{3 + \sin\theta}$ or $r = \dfrac{-8}{3 - \sin\theta}$

11. $r = \dfrac{-9}{4 + 5\sin\theta}$ or $r = \dfrac{9}{4 - 5\sin\theta}$

13. $r = \dfrac{10}{2 + 3\cos\theta}$ or $r = \dfrac{-10}{2 - 3\cos\theta}$

15. $r = \dfrac{10}{1 - \cos\theta}$ or $r = \dfrac{-10}{1 + \cos\theta}$

17. $y^2 - 4x - 4 = 0$ Parabola

19. $3x^2 + 4y^2 - 4x - 4 = 0$ Ellipse

21. $32y^2 - 4x^2 + 36y + 9 = 0$ Hyperbola

23. $9x^2 + 12y - 4 = 0$ Parabola

25. $7x^2 + 16y^2 + 72x - 144 = 0$ Ellipse

29. $r^2 = \dfrac{24{,}336}{169 - 25\cos^2\theta}$ **31.** $r^2 = \dfrac{144}{25\cos^2\theta - 9}$

33. $r^2 = \dfrac{144}{25\sin^2\theta - 16}$ **35.** $r = \dfrac{7975.8}{1 + 0.937\cos\theta}$

37. $r = \dfrac{8200}{1 - \sin\theta}$

Section 12.3

1. Cardioid

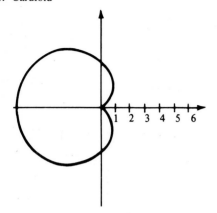

3. Limaçon

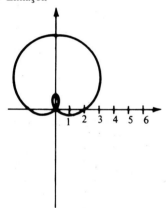

5. Limaçon

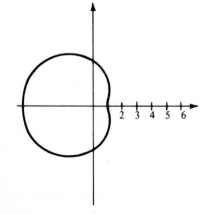

7. Circle

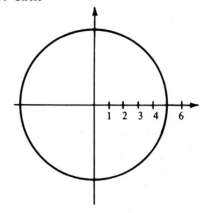

9. Spiral

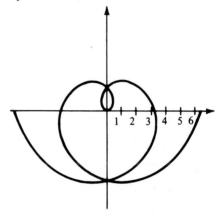

11. Circle

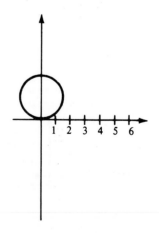

13. Rose curve

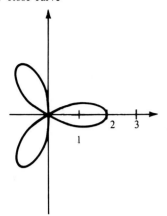

15. Rose curve

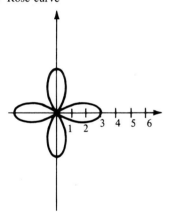

17. Lemniscate

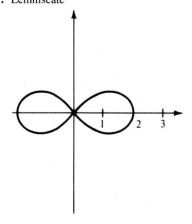

19. Rose curve

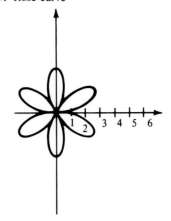

21. Line

23. Ellipse

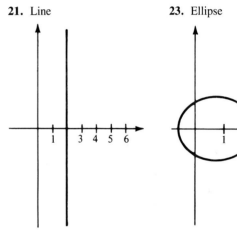

25.

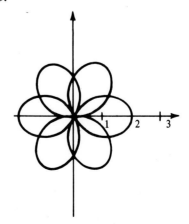

27. $\left(1, \dfrac{\pi}{2}\right), \left(1, \dfrac{3\pi}{2}\right), (0, 0)$

29. $\left(\dfrac{2 - \sqrt{2}}{2}, \dfrac{3\pi}{4}\right), \left(\dfrac{2 + \sqrt{2}}{2}, \dfrac{7\pi}{4}\right), (0, 0)$

31. $\left(\dfrac{3}{2}, \dfrac{\pi}{6}\right), \left(\dfrac{3}{2}, \dfrac{5\pi}{6}\right), (0, 0)$

33. $(1.936, 1.318), (1.936, 1.824), (0, 0)$

35. $(2, 4), (-2, -4)$

37. $(0.581, \pm 0.535), (2.581, \pm 1.376)$

9. $x^2 + y^2 = 9$

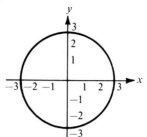

11. $\dfrac{x^2}{16} + \dfrac{y^2}{4} = 1$

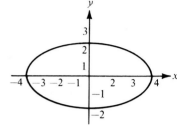

Section 12.4

1. $2x - 3y + 5 = 0$

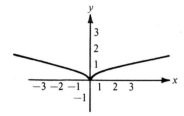

3. $y = (x - 1)^2$

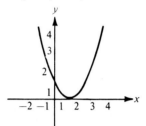

13. $y = 2 - 2x^2, \; -1 \le x \le 1$

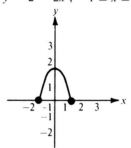

5. $y = \dfrac{1}{2}x^{2/3}$

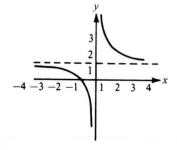

15. $\dfrac{(x - 4)^2}{4} + \dfrac{(y + 1)^2}{1} = 1$

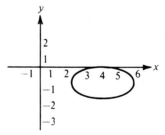

7. $y = \dfrac{x + 1}{x}$

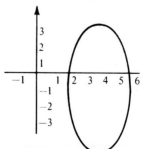

17. $\dfrac{(x - 4)^2}{4} + \dfrac{(y + 1)^2}{16} = 1$

19. $\dfrac{x^2}{16} - \dfrac{y^2}{9} = 1$

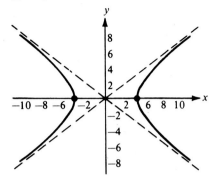

23. $x = 5t, \; y = -2t$ (Solution is not unique.)

25. $\dfrac{(x - h)^2}{a^2} - \dfrac{(y - k)^2}{b^2} = 1$

27. $x = t, \; y = t^3$
$x = \sqrt[3]{s}, \; y = s$ (Solution is not unique.)

29.

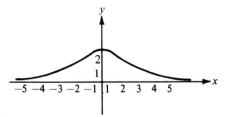

31.

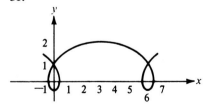

33. $x = a\theta - b \sin \theta$
$ y = a - b \cos \theta$

Chapter 12 Review Exercises

1. Cardioid

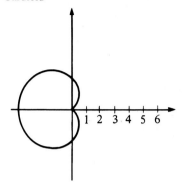

3. Limaçon

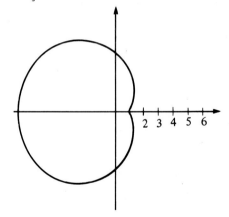

5. Rose curve

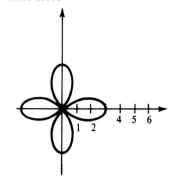

7. Circle

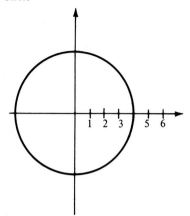

9. Line

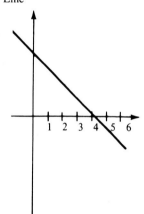

11. Line

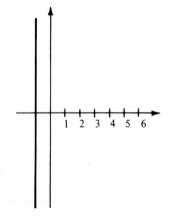

13. Rose curve

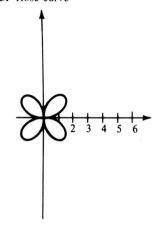

15. Parabola

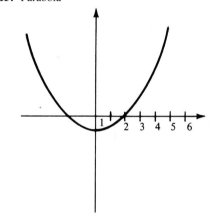

17. Semicubical parabola

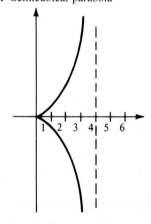

19. $x^2 + y^2 - 3x = 0$

21. $(x^2 + y^2 + 2x)^2 = 4(x^2 + y^2)$

23. $y^2 = x^2\left(\dfrac{4-x}{4+x}\right)$ **25.** $r = 4\cos\theta$

27. $r = \dfrac{5}{3 - 2\cos\theta}$ **29.** $r = \dfrac{12}{3\sin\theta + 4\cos\theta}$

33. $r = \dfrac{4}{1 - \cos[\theta - (\pi/6)]}$

35. $y = \dfrac{1}{x}$

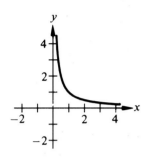

37. $(2y - 3x + 2)^2 = (1 + 4x)(x - 2)^2$

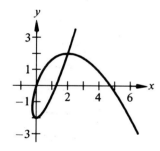

39. $y = \dfrac{4}{(x + 1)(x - 3)}$

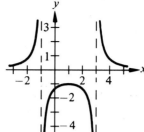

41. $y = \dfrac{2x}{1 + x^2}$

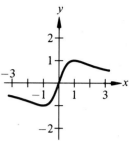

43. $2 - y = \cos(x - \sqrt{1 - (2 - y)^2})$

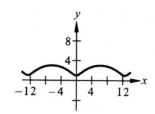

45. $x = -3 + 4\cos\theta$
     $y = 4 + 3\sin\theta$

INDEX

FORMULAS FROM GEOMETRY

Triangle:

$h = a \sin\theta$

$\text{Area} = \dfrac{1}{2} bh$

(Law of Cosines)

$c^2 = a^2 + b^2 - 2ab \cos\theta$

Right Triangle:

(Pythagorean Theorem)

$c^2 = a^2 + b^2$

Equilateral Triangle:

$h = \dfrac{\sqrt{3}\,s}{2}$

$\text{Area} = \dfrac{\sqrt{3}\,s^2}{4}$

Parallelogram:

$\text{Area} = bh$

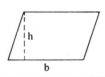

Trapezoid:

$\text{Area} = \dfrac{h}{2}(a+b)$

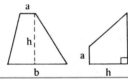

Circle:

$\text{Area} = \pi r^2$

$\text{Circumference} = 2\pi r$

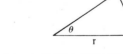

Sector of Circle:

(θ in radians)

$\text{Area} = \dfrac{\theta r^2}{2}$

$s = r\theta$

Circular Ring:

(p = average radius,
w = width of ring)

$\text{Area} = \pi(R^2 - r^2)$

$\quad\quad = 2\pi pw$

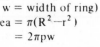

Sector of Circular Ring:

(p = average radius,
w = width of ring,
θ in radians)

$\text{Area} = \theta pw$

Ellipse:

$\text{Area} = \pi ab$

$\text{Circumference} \approx 2\pi \sqrt{\dfrac{a^2 + b^2}{2}}$

Cone:

(A = area of base)

$\text{Volume} = \dfrac{Ah}{3}$

Right Circular Cone:

$\text{Volume} = \dfrac{\pi r^2 h}{3}$

$\text{Lateral Surface Area} = \pi r\sqrt{r^2 + h^2}$

Frustum of Right Circular Cone:

$\text{Volume} = \dfrac{\pi(r^2 + rR + R^2)h}{3}$

$\text{Lateral Surface Area} = \pi s(R+r)$

Right Circular Cylinder:

$\text{Volume} = \pi r^2 h$

$\text{Lateral Surface Area} = 2\pi rh$

Sphere:

$\text{Volume} = \dfrac{4}{3}\pi r^3$

$\text{Surface Area} = 4\pi r^2$

Wedge:

(A = area of upper face,
B = area of base)

$A = B \sec\theta$

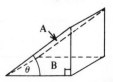